# AN INTRODUCTION T

## Concepts and Applications

# AN INTRODUCTION TO COMBUSTION

## Concepts and Applications

### THIRD EDITION

**Stephen R. Turns**

*Propulsion Engineering Research Center*
*and*
*Department of Mechanical and Nuclear Engineering*
*The Pennsylvania State University*

Mc Graw Hill

*Connect*
*Learn*
*Succeed*™

The **McGraw·Hill** Companies

Mc Graw Hill · **Connect Learn Succeed**™

AN INTRODUCTION TO COMBUSTION: CONCEPTS AND APPLICATIONS, THIRD EDITION
International Edition 2012

10  09  08  07  06  05  04  03  02  01
20  15  14  13  12  11
CTP   BJE

**When ordering this title, use ISBN 978-007-108687-5 or MHID 007-108687-0**

1006375403

Printed in Singapore

www.mhhe.com

# ABOUT THE AUTHOR

**Stephen R. Turns** received degrees in mechanical engineering from The Pennsylvania State University (B.S., 1970), Wayne State University (M.S., 1974), and the University of Wisconsin at Madison (Ph.D., 1979). He was a research engineer at General Motors Research Laboratories from 1970 to 1975. He joined the Penn State faculty in 1979 and is currently Professor of Mechanical Engineering. Dr. Turns teaches a wide variety of courses in the thermal sciences and has received several awards for teaching excellence at Penn State. In 2009, he received the American Society of Engineering Education's Ralph Coats Roe award. Dr. Turns had conducted research in several combustion-related areas. He is a member of The Combustion Institute, the American Institute of Aeronautics, the American Society of Engineering Education, and the Society of Automotive Engineers. Dr. Turns is a Fellow of the American Society of Mechanical Engineers.

*This book is dedicated to Joan Turns.*

By contrast, the first fires flickering at a cave mouth are our own discovery, our own triumph, our grasp upon invisible chemical power. Fire contained, in that place of brutal darkness and leaping shadows, the crucible and the chemical retort, steam and industry. It contained the entire human future.

*Loren Eiseley*
*The Unexpected Universe*

# Preface to the Third Edition

The third edition retains the same primary objectives as previous editions: first, to present basic combustion concepts using relatively simple and easy-to-understand analyses; and second, to introduce a wide variety of practical applications that motivate or relate to the various theoretical concepts. The overarching goal is to provide a textbook that is useful for both formal undergraduate and introductory graduate study in mechanical engineering and related fields, and informal study by practicing engineers.

The overarching theme of the revisions in this edition is the addition and updating of specific topics related to energy use; protection of the environment, including climate change; and fuels. The largest single change is the addition of a new chapter dedicated to fuels. Highlights of these changes and a brief discussion of the new chapter follow.

Chapter 1 includes more detailed information on energy sources and use and electricity generation and use. Chapter 4 contains new sections devoted to reduced mechanisms and to catalysis and heterogeneous reactions. As detailed chemical mechanisms for combustion and pollutant formation have grown in complexity, the need for robust reduced mechanisms has grown. Catalytic exhaust aftertreatment has become the standard approach to controlling emissions from spark-ignition engines and is making inroads for controlling diesel engine emissions. Catalytic combustion has also been of interest in some applications. These factors were the drivers for the new sections of Chapter 4. Changes in Chapter 5 reflect the progress that has been made in developing detailed mechanisms for realistic transportation fuels. Other changes include updating the detailed methane combustion kinetics (GRI Mech) to include detailed nitrogen chemistry and the addition of a major new section presenting a reduced mechanism for methane combustion and nitric oxide formation. Changes to Chapter 9 reflect advances in both experimentation and modeling related to laminar nonpremixed flames. Chapter 10 has been updated to reflect current practice in the design and operation of gas-turbine combustors; Chapter 10 also cites recent droplet combustion studies conducted in space using the Space Shuttle and the International Space Station. Revisions to Chapter 12 reflect the latest advances in understanding turbulent premixed combustion. Similarly, Chapter 13 has been revised to include recent findings on soot formation and destruction and provides an expanded and updated discussion of flame radiation from turbulent nonpremixed flames. Several new figures and more than 30 new references complement these two chapters.

The title of Chapter 15 has been changed from "Pollutant Emissions" to "Emissions" to reflect that greenhouse gas emissions, as well as pollutant emissions, are both important combustion considerations. Many changes and/or additions have been made to this chapter. These include, but are not limited to, the following: an expanded section on human health effects of particulate matter to reflect new findings; a revised section on $NO_x$ emissions to reflect current understanding of nitrogen chemistry; a discussion of the homogeneous charge compression ignition engine; an interesting development for emission control arising since the previous edition; an updated discussion of catalytic converters for spark-ignition engines; discussion of the emission of particulate matter from both gasoline and diesel engines focusing on ultrafine particles, together with a greatly expanded discussion of particulate matter and its emission control; the introduction of EPA emission factors; additions and revisions to discussions of $NO_x$ and $SO_x$ controls; and the addition of a new section discussing greenhouse gases. Seventy-three new references complement the revisions to Chapter 15.

Concerns over global warming, environmental degradation, and national energy independence, among others, have resulted in a renewed interest in fuels. With this interest comes the need for basic information on fuels. The addition of Chapter 17, Fuels, is intended to fulfill this need. In this chapter, we discuss naming conventions and the molecular structures of hydrocarbons, alcohols, and other organic compounds, followed by a discussion of what properties make a good fuel. Conventional fuels, which include various gasolines, diesel fuels, heating oils, aviation fuels, natural gas, and coal are discussed, as are several alternative fuels. Examples here include biodiesel, ethanol (either corn-based or cellulosic), Fischer-Tropsch liquids from coal or biomass, hydrogen, and others. This new chapter contains eight figures and 22 tables, along with 83 references.

The computer software, previously supplied on a diskette, is now available as a download from the publisher's website at www.mhhe.com/turns3e. The website also contains the instructor's solutions manual and image library.

The author hopes that this new edition will continue to serve well those who have used previous editions and that the changes provided will enhance the usefulness of the book.

*Stephen R. Turns*
*University Park, PA*

# ACKNOWLEDGMENTS

Many people contributed their support, time, and psychic energy to the various editions of this book. First, I would like to thank the many reviewers who contributed along the way. Comments from many reviewers were also very helpful in creating the new chapter on fuels that appears in this third edition. My friend and colleague, Chuck Merkle, continually provided moral support and served as a sounding board for ideas on both content and pedagogy in the first edition. Many students at Penn State contributed in various ways, and I want to acknowledge the particular contributions of Jeff Brown, Jongguen Lee, and Don Michael. Sankaran Venkateswaran deserves special thanks for providing the turbulent jet-flame model calculations, as does Dave Crandall for his assistance with the software. A major debt of thanks is owed to Donn Mueller, who painstakingly solved all of the end-of-chapter problems. I want to acknowledge my friends and colleagues at Auburn University who welcomed me during an extended stay there during a sabbatical leave: Sushil Bhavnani, Roy Knight, Pradeep Lal, Bonnie MacEwan, Tom Manig, P. K. Raju, and Jeff Suhling. I would also like to thank the Gas Research Institute (now Gas Technology Institute) for their support of my research activities through the years, as it was these activities that provided the initial inspiration and impetus to write this book. Cheryl Adams and Mary Newby were instrumental in transcribing hand-written scrawl and modifying various drafts to create the final manuscripts. I owe them both a large debt. The support and assistance from Bill Stenquist and Lora Neyens at McGraw-Hill is much appreciated. Invaluable to my efforts throughout was the unwavering support of my family. They tolerated amazingly well the time spent writing on weekends and holiday breaks—time that I could have spent with them. Joan, my wife and friend of more than forty years, has been unflagging in her support of me and my projects. For this I am eternally grateful. Thank you, Joan.

# CONTENTS

# 1

# Introduction

## MOTIVATION TO STUDY COMBUSTION

Combustion and its control are essential to our existence on this planet as we know it. In 2007, approximately 85 percent of the energy used in the United States came from combustion sources [1] (Fig. 1.1). A quick glance around your local environment shows the importance of combustion in your daily life. More than likely, the heat for your room or home comes directly from combustion (either a gas- or oil-fired furnace or boiler), or, indirectly, through electricity that was generated by burning a fossil fuel. Our nation's electrical needs are met primarily by combustion. In 2006, only 32.7 percent of the electrical generating capability was nuclear or hydroelectric, while more than half was provided by burning coal, as shown in Table 1.1 [2]. Our transportation system relies almost entirely on combustion. Figure 1.2 provides an overview of the energy flow in the production of electricity. In the United States in 2007, ground vehicles and aircraft burned approximately 13.6 million barrels of various petroleum products per day [3], or approximately two-thirds of all of the petroleum imported or produced in the United States. Aircraft are entirely powered by on-board fuel burning, and most trains are diesel-engine powered. Recent times have also seen the rise of gasoline-engine driven appliances such as lawn mowers, leaf blowers, chain saws, weed-whackers, and the like.

Industrial processes rely heavily on combustion. Iron, steel, aluminum, and other metals-refining industries employ furnaces for producing the raw product, and heat-treating and annealing furnaces or ovens are used downstream to add value to the raw material as it is converted into a finished product. Other industrial combustion devices include boilers, refinery and chemical fluid heaters, glass melters, solids dryers, surface-coating curing and drying ovens, and organic fume incinerators [4, 5], to give just a few examples. The cement manufacturing industry is a heavy user of heat energy delivered by combustion. Rotary kilns, in which the cement clinker is produced, use over 0.4 quads[1] of energy, or roughly 1.4 percent of the total industrial

---

[1] quad = 1 quadrillion Btu = $10^{15}$ Btu.

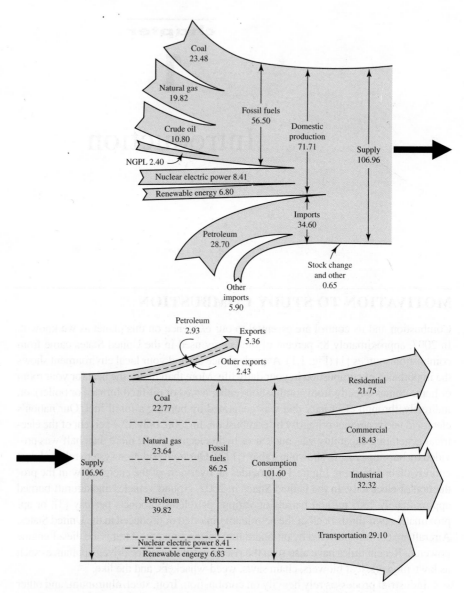

**Figure 1.1**    U.S. energy sources and consumption by end-use sectors, for 2007, in quadrillion Btu (quad). Renewable energy includes conventional hydroelectric power (2.463 quad), biomass (3.584 quad), geothermal (0.353 quad), solar/photovoltaic (0.080 quad), and wind (0.319 quad).

SOURCE: From Ref. [1].

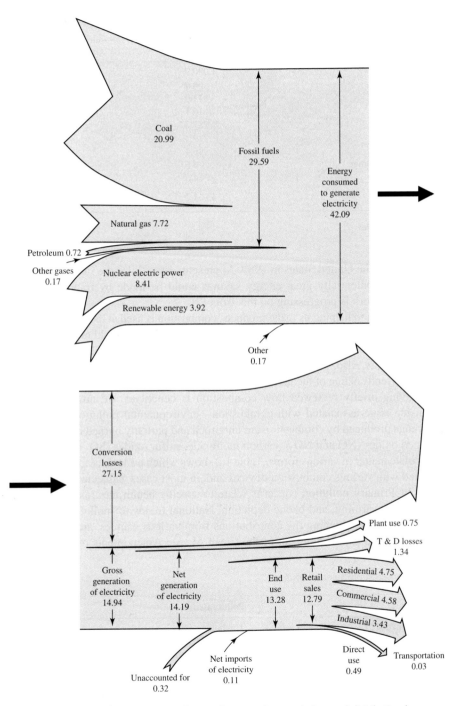

**Figure 1.2**    U.S. electricity generation, end use, and transmission and distribution losses, for 2007, in quadrillion Btu (quad).
SOURCE: From Ref. [2].

**Table 1.1**     2006 U.S. electricity generation

| Source | Billion kW-hr | (%) |
|---|---|---|
| Coal | 1,990.9 | 49.0 |
| Petroleum | 64.4 | 1.6 |
| Natural gas | 813.0 | 20.0 |
| Other gases | 16.1 | 0.4 |
| Nuclear | 787.2 | 19.4 |
| Hydroelectric | 289.2 | 7.1 |
| Other renewables | 96.4 | 2.4 |
| Hydro pumped storage | −6.6 | −0.2 |
| Other | 14.0 | 0.3 |
| Total | 4,064.7 | 100.0 |

SOURCE: From Ref. [2].

energy use in the United States in 1989. At present, rotary kilns are rather inefficient devices, and potentially great energy savings could be made by improving these devices [6]. Work is progressing on this front [7].

In addition to helping us make products, combustion is used at the other end of the product life cycle as a means of waste disposal. Incineration is an old method, but it is receiving renewed interest because of the limited availability of landfill sites in densely populated areas. Also, incineration is attractive for its ability to dispose of toxic wastes safely. Currently, siting of incinerators is a politically controversial and sensitive issue.

Having briefly reviewed how combustion is beneficial, we now look at the downside issue associated with combustion—environmental pollution. The major pollutants produced by combustion are unburned and partially burned hydrocarbons, nitrogen oxides (NO and $NO_2$), carbon monoxide, sulfur oxides ($SO_2$ and $SO_3$), and particulate matter in various forms. Table 1.2 shows which pollutants are typically associated with various combustion devices and, in most cases, subjected to legislated controls. Primary pollution concerns relate to specific health hazards, smogs, acid rain, global warming, and ozone depletion. National trends for pollutant emissions from 1940–1998, showing the contributions from various sources, are presented in Figs. 1.3 to 1.8 [9]. The impact of the Clean Air Act Amendments of 1970 can be

**Table 1.2**     Typical pollutants of concern from selected sources

| Source | Pollutants | | | | |
|---|---|---|---|---|---|
| | Unburned Hydrocarbons | Oxides of Nitrogen | Carbon Monoxide | Sulfur Oxides | Particulate Matter |
| Spark-ignition engines | + | + | + | − | − |
| Diesel engines | + | + | + | − | + |
| Gas-turbine engines | + | + | + | − | + |
| Coal-burning utility boilers | − | + | − | + | + |
| Gas-burning appliances | − | + | + | − | − |

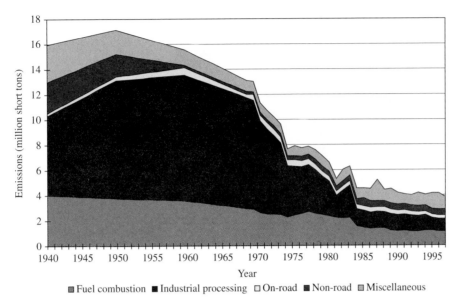

**Figure 1.3**    Trends in emissions of particulate matter (PM$_{10}$) for the United States, 1940–1998, excluding fugitive dust sources. PM$_{10}$ refers to particulate matter smaller than 10 microns. Reading legend left to right corresponds to plotted series bottom to top.

SOURCE: From Ref. [9].

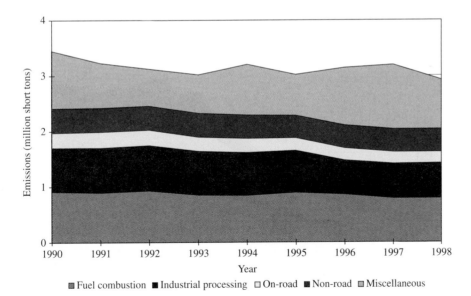

**Figure 1.4**    Trends in directly emitted particulate matter (PM$_{2.5}$) emissions for the United States, 1990 to 1998, excluding fugitive dust sources. PM$_{2.5}$ refers to particulate matter smaller than 2.5 microns. Reading legend left to right corresponds to plotted series bottom to top.

SOURCE: From Ref. [9].

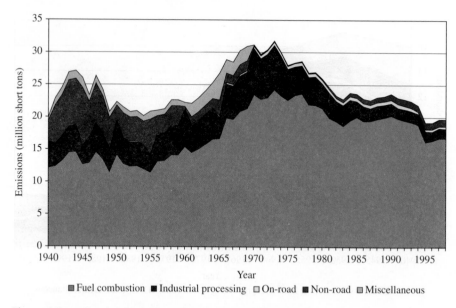

**Figure 1.5**      Trends in emissions of sulfur oxides for the United States, 1940–1998. Reading legend left to right corresponds to plotted series bottom to top.
| SOURCE: From Ref. [9].

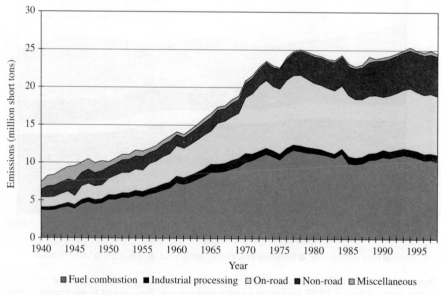

**Figure 1.6**      Trends in emissions of nitrogen oxides for the United States, 1940–1998. Reading legend left to right corresponds to plotted series bottom to top.
| SOURCE: From Ref. [9].

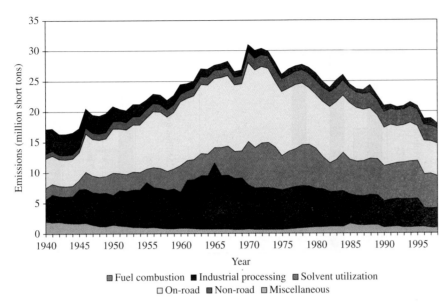

**Figure 1.7**    Trends in emissions of volatile organic compounds for the United States, 1940–1998. Reading legend left to right corresponds to plotted series bottom to top.
| SOURCE: From Ref. [9].

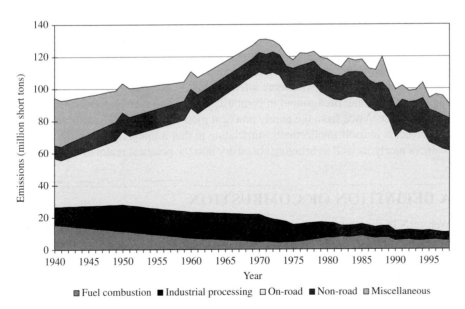

**Figure 1.8**    Trends in emissions of carbon monoxide for the United States, 1940–1998. Reading legend left to right corresponds to plotted series bottom to top.
| SOURCE: From Ref. [9].

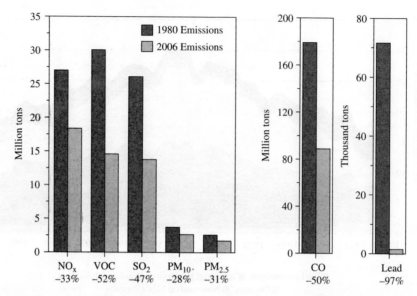

**Figure 1.9** Comparison of U.S. emission rates showing reductions from 1980 to 2006 for oxides of nitrogen (NO$_x$), volatile organic compounds (VOC), sulfur dioxide (SO$_2$), and particulate matter (PM$_{10}$ and PM$_{2.5}$) (left panel), carbon monoxide (CO) (middle panel), and lead (right panel). Particulate matter reductions are referenced to 1990 (PM$_{2.5}$) and 1985 (PM$_{10}$) rather than 1980.
SOURCE: From Ref. [8].

clearly seen in these figures. Reductions in combustion-related pollutant emissions achieved in the past few decades are illustrated in Fig. 1.9.

Considering the importance of combustion in our society, it is somewhat surprising that very few engineers have more than a cursory knowledge of combustion phenomena. However, with an already demanding curriculum, it is unrealistic to expect the subject to be given more attention than it presently receives. Therefore, engineers with some background in combustion may find many opportunities to use their expertise. Aside from the purely practical motivations for studying combustion, the subject is in itself intellectually stimulating in that it integrates all of the thermal sciences nicely, as well as bringing chemistry into the practical realm of engineering.

## A DEFINITION OF COMBUSTION

Webster's Dictionary provides a useful starting point for a definition of **combustion** as *"rapid oxidation generating heat, or both light and heat; also, slow oxidation accompanied by relatively little heat and no light."* For our purposes, we will restrict the definition to include only the rapid oxidation portion, since most practical combustion devices belong in this realm.

This definition emphasizes the intrinsic importance of chemical reactions to combustion. It also emphasizes why combustion is so very important: combustion transforms energy stored in chemical bonds to heat that can be utilized in a variety of ways. Throughout this book, we illustrate the many practical applications of combustion.

# COMBUSTION MODES AND FLAME TYPES

Combustion can occur in either a **flame** or **nonflame** mode, and flames, in turn, are categorized as being either **premixed flames** or **nonpremixed (diffusion) flames.** The difference between flame and nonflame modes of combustion can be illustrated by the processes occurring in a knocking spark-ignition engine (Fig. 1.10). In Fig. 1.10a, we see a thin zone of intense chemical reaction propagating through the unburned fuel–air mixture. The thin reaction zone is what we commonly refer to as a flame.

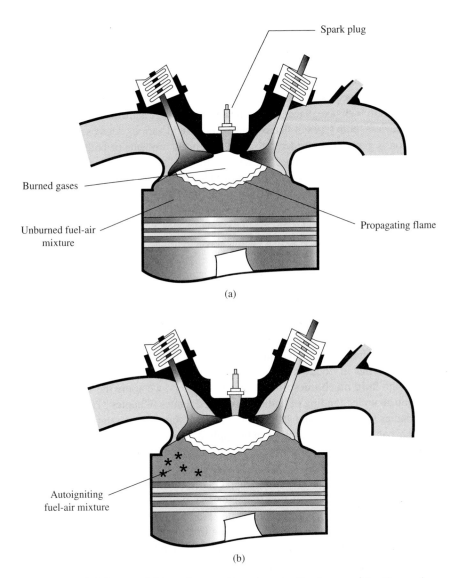

**Figure 1.10**   (a) Flame and (b) nonflame modes of combustion in a spark-ignition engine. Autoignition of the mixture ahead of the propagating flame is responsible for engine knock.

Behind the flame are the hot products of combustion. As the flame moves across the combustion space, the temperature and pressure rise in the unburned gas. Under certain conditions (Fig. 1.10b), rapid oxidation reactions occur at many locations within the unburned gas, leading to very rapid combustion throughout the volume. This essentially volumetric heat release in an engine is called **autoignition,** and the very rapid pressure rise leads to the characteristic sound of engine knock. Knock is undesirable, and a recent challenge to engine designers has been how to minimize the occurrence of knock while operating with lead-free gasolines.[2] In compression-ignition or diesel engines, however, autoignition initiates the combustion process by design.

The two classes of flames, premixed and nonpremixed (or diffusion), are related to the state of mixedness of the reactants, as suggested by their names. In a premixed flame, the fuel and the oxidizer are mixed at the molecular level prior to the occurrence of any significant chemical reaction. The spark-ignition engine is an example where premixed flames occur. Contrarily, in a diffusion flame, the reactants are initially separated, and reaction occurs only at the interface between the fuel and oxidizer, where mixing and reaction both take place. An example of a diffusion flame is a simple candle. In practical devices, both types of flames may be present in various degrees. Diesel-engine combustion is generally considered to have significant amounts of both premixed and nonpremixed or diffusion burning. The term "diffusion" applies strictly to the molecular diffusion of chemical species, i.e., fuel molecules diffuse toward the flame from one direction while oxidizer molecules diffuse toward the flame from the opposite direction. In turbulent nonpremixed flames, turbulent convection mixes the fuel and air together on a macroscopic basis. Molecular mixing at the small scales, i.e., molecular diffusion, then completes the mixing process so that chemical reactions can take place.

## APPROACH TO OUR STUDY

We begin our study of combustion by investigating the key physical processes, or sciences, which form the fundamental framework of combustion science: **thermochemistry** in Chapter 2; **molecular transport of mass (and heat)** in Chapter 3; **chemical kinetics** in Chapters 4 and 5; and, in Chapters 6 and 7, the coupling of all of these with **fluid mechanics.** In subsequent chapters, we apply these fundamentals to develop an understanding of laminar premixed flames (Chapter 8) and laminar diffusion flames (Chapters 9 and 10). In these laminar flames, it is relatively easy to see how basic conservation principles can be applied. Most practical combustion devices operate with turbulent flows, however, and the application of theoretical concepts to these is much more difficult. Chapters 11, 12, and 13 deal with turbulent flames and their practical applications. The final chapters concern the combustion of solids, as exemplified by carbon combustion (Chapter 14); pollutant emissions (Chapter 15); detonations (Chapter 16); and fuels (Chapter 17).

A major goal of this book is to provide a treatment of combustion that is sufficiently simple so that students with no prior introduction to the subject can appreciate

[2]The discovery that tetraethyl lead reduces knock, made by Thomas Midgley in 1921, allowed engine compression ratios to be increased, and thereby improved efficiency and power.

both the fundamental and practical aspects. It is hoped, moreover, that as a result, some may be motivated to learn more about this fascinating field, either through more advanced study, or as a practicing engineers.

## REFERENCES

1. U. S. Energy Information Agency, "Annual Energy Review 2007," DOE/EIA-0384, 2008. (See also http://www.eia.doe.gov/aer/.)

2. U. S. Energy Information Agency, "Electricity," http://www.eia.doe.gov/fuelelectric. html. Accessed 7/30/2008.

3. U. S. Energy Information Agency, "Petroleum," http://www.eia.doe.gov/oil_gas /petroleum/info_glance/petroleum.html. Accessed 7/30/2008.

4. Bluestein, J., "$NO_x$ Controls for Gas-Fired Industrial Boilers and Combustion Equipment: A Survey of Current Practices," Gas Research Institute, GRI-92/0374, October 1992.

5. Baukal, C. E., Jr. (Ed.), *The John Zink Combustion Handbook*, CRC Press, Boca Raton, 2001.

6. Tresouthick, S. W., "The SUBJET Process for Portland Cement Clinker Production," presented at the 1991 Air Products International Combustion Symposium, 24–27 March 1991.

7. U. S. Environmental Protection Agency, "Energy Trends in Selected Manufacturing Sectors: Opportunities for Environmentally Preferable Energy Outcomes," Final Report, March, 2007.

8. U. S. Environmental Protection Agency, "Latest Findings on National Air Quality—Status and Trends through 2006," http://www.epa.gov/air/airtrends/2007/. Accessed 7/30/2008.

9. U. S. Environmental Protection Agency, "National Air Pollutant Emission Trends, 1940–1998," EPA-454/R-00-002, March 2000.

**chapter**

# 2

# Combustion and Thermochemistry

## OVERVIEW

In this chapter, we examine several thermodynamic concepts that are important in the study of combustion. We first briefly review basic property relations for ideal gases and ideal-gas mixtures and the first law of thermodynamics. Although these concepts are likely to be familiar to you from a previous study of thermodynamics, we present them here since they are an integral part of our study of combustion. We next focus on thermodynamic topics related specifically to combustion and reacting systems: concepts and definitions related to element conservation; a definition of enthalpy that accounts for chemical bonds; and first-law concepts defining heat of reaction, heating values, etc., and adiabatic flame temperatures. Chemical equilibrium, a second-law concept, is developed and applied to combustion-product mixtures. We emphasize equilibrium because, in many combustion devices, a knowledge of equilibrium states is sufficient to define many performance parameters of the device; for example, the temperature and major species at the outlet of a steady-flow combustor are likely to be governed by equilibrium considerations. Several examples are presented to illustrate these principles.

## REVIEW OF PROPERTY RELATIONS

### Extensive and Intensive Properties

The numerical value of an **extensive property** depends on the amount (mass or number of moles) of the substance considered. Extensive properties are usually denoted with capital letters; for example, $V$ (m$^3$) for volume, $U$ (J) for internal energy, $H$ (J) ($= U + PV$) for enthalpy, etc. An **intensive property,** on the other hand, is expressed per unit mass (or per mole), and its numerical value is independent of the amount of substance present. Mass-based intensive properties are generally denoted with lowercase letters; for example, $v$ (m³/kg) for specific volume, $u$ (J/kg) for specific internal energy, $h$ (J/kg) ($= u + Pv$) for specific enthalpy, etc. Important exceptions to this lowercase convention are the intensive properties temperature $T$ and pressure $P$. Molar-based intensive properties are indicated in this book with an overbar, e.g., $\bar{u}$ and $\bar{h}$ (J/kmol). Extensive properties are obtained simply from the corresponding intensive properties by multiplying the property value per unit mass (or mole) by the amount of mass (or number of moles); i.e.,

$$V = mv \ (\text{or } N\bar{v})$$

$$U = mu \ (\text{or } N\bar{u}) \tag{2.1}$$

$$H = mh \ (\text{or } N\bar{h}), \text{ etc.}$$

In the following developments, we will use either mass- or molar-based intensive properties, depending on which is most appropriate to a particular situation.

## Equation of State

An **equation of state** provides the relationship among the pressure, $P$, temperature, $T$, and volume $V$ (or specific volume $v$) of a substance. For ideal-gas behavior, i.e., a gas that can be modeled by neglecting intermolecular forces and the volume of the molecules, the following equivalent forms of the equation of state apply:

$$PV = NR_u T, \tag{2.2a}$$

$$PV = mRT, \tag{2.2b}$$

$$Pv = RT, \tag{2.2c}$$

or

$$P = \rho RT, \tag{2.2d}$$

where the specific gas constant $R$ is related to the universal gas constant $R_u$ ($= 8315$ J/kmol-K) and the gas molecular weight $MW$ by

$$R = R_u / MW. \tag{2.3}$$

The density $\rho$ in Eqn. 2.2d is the reciprocal of the specific volume ($\rho = 1/v = m/V$). Throughout this book, we assume ideal-gas behavior for all gaseous species and gas mixtures. This assumption is appropriate for nearly all of the systems we wish to consider because the high temperatures associated with combustion generally result in sufficiently low densities for ideal-gas behavior to be a reasonable approximation.

## Calorific Equations of State

Expressions relating internal energy (or enthalpy) to pressure and temperature are called **calorific equations of state,** i.e.,

$$u = u(T, v) \tag{2.4a}$$

$$h = h(T, P). \tag{2.4b}$$

The word "calorific" relates to expressing energy in units of calories, which has been superseded by the use of joules in the SI system.

General expressions for a differential change in $u$ or $h$ can be expressed by differentiating Eqns. 2.4a and b:

$$du = \left( \frac{\partial u}{\partial T} \right)_v dT + \left( \frac{\partial u}{\partial v} \right)_T dv \qquad (2.5a)$$

$$dh = \left( \frac{\partial h}{\partial T} \right)_P dT + \left( \frac{\partial h}{\partial P} \right)_T dP. \qquad (2.5b)$$

In the above, we recognize the partial derivatives with respect to temperature to be the **constant-volume** and **constant-pressure specific heats,** respectively, i.e.,

$$c_v \equiv \left( \frac{\partial u}{\partial T} \right)_v \qquad (2.6a)$$

$$c_p \equiv \left( \frac{\partial h}{\partial T} \right)_P. \qquad (2.6b)$$

For an ideal gas, the partial derivatives with respect to specific volume, $(\partial u/\partial v)_T$, and pressure, $(\partial h/\partial P)_T$, are zero. Using this knowledge, we integrate Eqn. 2.5, substituting Eqn. 2.6 to provide the following ideal-gas calorific equations of state:

$$u(T) - u_{\text{ref}} = \int_{T_{\text{ref}}}^{T} c_v \, dT \qquad (2.7a)$$

$$h(T) - h_{\text{ref}} = \int_{T_{\text{ref}}}^{T} c_p \, dT. \qquad (2.7b)$$

In a subsequent section, we will define an appropriate reference state that accounts for the different bond energies of various compounds.

For both real **and** ideal gases, the specific heats $c_v$ and $c_p$ are generally functions of temperature. This is a consequence of the internal energy of a molecule consisting of three components: translational, vibrational, and rotational; and the fact that the vibrational and rotational energy storage modes become increasingly active as temperature increases, as described by quantum theory. Figure 2.1 schematically illustrates these three energy storage modes by contrasting a monatomic species, whose internal energy consists solely of translational kinetic energy, and a diatomic molecule, which stores energy in a vibrating chemical bond, represented as a spring between the two nuclei, and by rotation about two orthogonal axes, as well as possessing kinetic energy from translation. With these simple models (Fig. 2.1), we would expect the specific heats of diatomic molecules to be greater than monatomic species. In general, the more complex the molecule, the greater its molar specific heat. This can be seen clearly in Fig. 2.2, where molar specific heats for a number of combustion product species are shown as functions of temperature. As a group, the triatomics have the greatest specific heats, followed by the diatomics, and, lastly, the monatomics. Note that the triatomic molecules also have a greater temperature

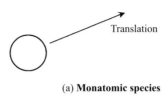

(a) **Monatomic species**

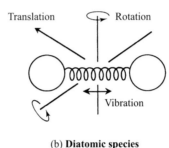

(b) **Diatomic species**

**Figure 2.1**    (a) The internal energy of monatomic species consists only of translational (kinetic) energy, while (b) a diatomic species' internal energy results from translation together with energy from vibration (potential and kinetic) and rotation (kinetic).

dependence than the diatomics, a consequence of the greater number of vibrational and rotational modes that are available to become activated as temperature is increased. In comparison, the monatomic species have nearly constant specific heats over a wide range of temperatures; in fact, the H-atom specific heat is constant ($\bar{c}_p = 20.786$ kJ/kmol-K) from 200 K to 5000 K.

Constant-pressure molar specific heats are tabulated as a function of temperature for various species in Tables A.1 to A.12 in Appendix A. Also provided in Appendix A are the curvefit coefficients, taken from the Chemkin thermodynamic database [1], which were used to generate the tables. These coefficients can be easily used with spreadsheet software to obtain $\bar{c}_p$ values at any temperature within the given temperature range.

## Ideal-Gas Mixtures

Two important and useful concepts used to characterize the composition of a mixture are the constituent mole fractions and mass fractions. Consider a multicomponent mixture of gases composed of $N_1$ moles of species 1, $N_2$ moles of species 2, etc. The **mole fraction of species $i$, $\chi_i$,** is defined as the fraction of the total number of moles in the system that are species $i$; i.e.,

$$\chi_i \equiv \frac{N_i}{N_1 + N_2 + \cdots N_i + \cdots} = \frac{N_i}{N_{tot}}. \tag{2.8}$$

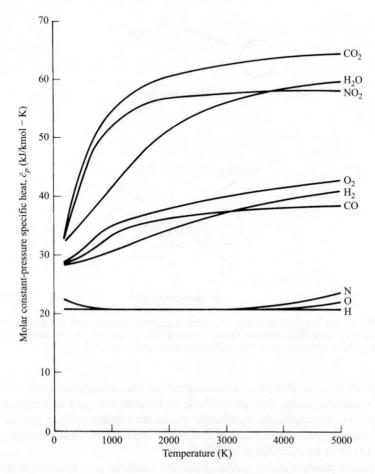

**Figure 2.2**     Molar constant-pressure specific heats as functions of temperature for monatomic (H, N, and O), diatomic (CO, $H_2$, and $O_2$), and triatomic ($CO_2$, $H_2O$, and $NO_2$) species. Values are from Appendix A.

Similarly, the **mass fraction of species $i$, $Y_i$,** is the amount of mass of species $i$ compared with the total mixture mass:

$$Y_i \equiv \frac{m_i}{m_1 + m_2 + \cdots m_i + \cdots} = \frac{m_i}{m_{tot}}. \tag{2.9}$$

Note that, by definition, the sum of all the constituent mole (or mass) fractions must be unity, i.e.,

$$\sum_i \chi_i = 1 \tag{2.10a}$$

$$\sum_i Y_i = 1. \tag{2.10b}$$

Mole fractions and mass fractions are readily converted from one to another using the molecular weights of the species of interest and of the mixture:

$$Y_i = \chi_i MW_i / MW_{\text{mix}} \qquad (2.11a)$$

$$\chi_i = Y_i MW_{\text{mix}} / MW_i. \qquad (2.11b)$$

The **mixture molecular weight, $MW_{\text{mix}}$**, is easily calculated from a knowledge of either the species mole or mass fractions:

$$MW_{\text{mix}} = \sum_i \chi_i MW_i \qquad (2.12a)$$

$$MW_{\text{mix}} = \frac{1}{\sum_i (Y_i / MW_i)}. \qquad (2.12b)$$

Species mole fractions are also used to determine corresponding species partial pressures. The **partial pressure of the $i$th species, $P_i$**, is the pressure of the $i$th species if it were isolated from the mixture at the same temperature and volume as the mixture. For ideal gases, the mixture pressure is the sum of the constituent partial pressures:

$$P = \sum_i P_i. \qquad (2.13)$$

The partial pressure can be related to the mixture composition and total pressure as

$$P_i = \chi_i P. \qquad (2.14)$$

For ideal-gas mixtures, many **mass- (or molar-) specific mixture properties** are calculated simply as mass (or mole) fraction weighted sums of the individual species-specific properties. For example, mixture enthalpies are calculated as

$$h_{\text{mix}} = \sum_i Y_i h_i \qquad (2.15a)$$

$$\bar{h}_{\text{mix}} = \sum_i \chi_i \bar{h}_i. \qquad (2.15b)$$

Other frequently used properties that can be treated in this same manner are internal energies, $u$ and $\bar{u}$. Note that, with our ideal-gas assumption, neither the pure-species properties ($u_i$, $\bar{u}_i$, $h_i$, $\bar{h}_i$) nor the mixture properties depend on pressure.

The mixture entropy also is calculated as a weighted sum of the constituents:

$$s_{\text{mix}}(T, P) = \sum_i Y_i s_i(T, P_i) \qquad (2.16a)$$

$$\bar{s}_{\text{mix}}(T, P) = \sum_i \chi_i \bar{s}_i(T, P_i). \qquad (2.16b)$$

In this case, however, the pure-species entropies ($s_i$ and $\bar{s}_i$) depend on the species partial pressures as indicated in Eqn. 2.16. The constituent entropies in Eqn. 2.16

can be evaluated from standard-state ($P_{ref} \equiv P^o = 1$ atm) values as

$$s_i(T, P_i) = s_i(T, P_{ref}) - R \ln \frac{P_i}{P_{ref}} \tag{2.17a}$$

$$\bar{s}_i(T, P) = \bar{s}_i(T, P_{ref}) - R_u \ln \frac{P_i}{P_{ref}}. \tag{2.17b}$$

Standard-state molar specific entropies are tabulated in Appendix A for many species of interest to combustion.

## Latent Heat of Vaporization

In many combustion processes, a liquid–vapor phase change is important. For example, a liquid fuel droplet must first vaporize before it can burn; and, if cooled sufficiently, water vapor can condense from combustion products. Formally, we define the **latent heat vaporization, $h_{fg}$,** as the heat required in a constant-pressure process to completely vaporize a unit mass of liquid at a given temperature, i.e.,

$$h_{fg}(T, P) \equiv h_{vapor}(T, P) - h_{liquid}(T, P), \tag{2.18}$$

where $T$ and $P$ are the corresponding saturation temperature and pressure, respectively. The latent heat of vaporization is also known as the **enthalpy of vaporization.** Latent heats of vaporization for various fuels at their normal (1 atm) boiling points are tabulated in Table B.1 (Appendix B).

The latent heat of vaporization at a given saturation temperature and pressure is frequently used with the **Clausius–Clapeyron equation** to estimate saturation pressure variation with temperature:

$$\frac{dP_{sat}}{P_{sat}} = \frac{h_{fg}}{R} \frac{dT_{sat}}{T_{sat}^2}. \tag{2.19}$$

This equation assumes that the specific volume of the liquid phase is negligible compared with that of the vapor and that the vapor behaves as an ideal gas. Assuming $h_{fg}$ is constant, Eqn. 2.19 can be integrated from ($P_{sat,1}, T_{sat,1}$) to ($P_{sat,2}, T_{sat,2}$) in order to permit, for example, $P_{sat,2}$ to be estimated from a knowledge of $P_{sat,1}, T_{sat,1}$, and $T_{sat,2}$. We will employ this approach in our discussion of droplet evaporation (Chapter 3) and combustion (Chapter 10).

## FIRST LAW OF THERMODYNAMICS

### First Law—Fixed Mass

Conservation of energy is the fundamental principle embodied in the first law of thermodynamics. For a **fixed mass,** i.e., a **system,** (Fig. 2.3a), energy conservation is

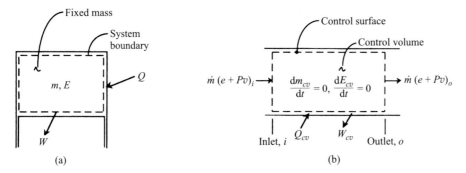

**Figure 2.3** (a) Schematic of fixed-mass system with moving boundary above piston. (b) Control volume with fixed boundaries and steady flow.

expressed for a finite change between two states, 1 and 2, as

$$_1Q_2 \qquad - \qquad _1W_2 \qquad = \qquad \Delta E_{1-2} \qquad (2.20)$$

| Heat added to | Work done by system | Change in total system |
| system in going | on surroundings in going | energy in going from |
| from state 1 to state 2 | from state 1 to state 2 | state 1 to state 2 |

Both $_1Q_2$ and $_1W_2$ are path functions and occur only at the system boundaries; $\Delta E_{1-2}(\equiv E_2 - E_1)$ is the change in the total energy of the system, which is the sum of the internal, kinetic, and potential energies, i.e.,

$$E = \qquad m( \quad u \qquad + \qquad \tfrac{1}{2}v^2 \qquad + \qquad gz \quad ). \qquad (2.21)$$

| Mass-specific system | Mass-specific system | Mass-specific system |
| internal energy | kinetic energy | potential energy |

The system energy is a state variable and, as such, $\Delta E$ does not depend on the path taken to execute a change in state. Equation 2.20 can be converted to unit mass basis or expressed to represent an instant in time. These forms are

$$_1q_2 - _1w_2 = \Delta e_{1-2} = e_2 - e_1 \qquad (2.22)$$

and

$$\dot{Q} \qquad - \qquad \dot{W} \qquad = \qquad dE/dt \qquad (2.23)$$

| Instantaneous rate | Instantaneous rate of | Instantaneous time |
| of heat transferred | work done by system, | rate of change of |
| into system | or power | system energy |

or

$$\dot{q} - \dot{w} = de/dt, \qquad (2.24)$$

where lowercase letters are used to denote mass-specific quantities, e.g., $e \equiv E/m$.

## First Law—Control Volume

We next consider a control volume, illustrated in Fig. 2.3b, in which fluid may flow across the boundaries. The steady-state, steady-flow (SSSF) form of the first law is particularly useful for our purposes and should be reasonably familiar to you from previous studies of thermodynamics [2–4]. Because of its importance, however, we present a brief discussion here. The SSSF first law is expressed as

$$\dot{Q}_{cv} \quad - \quad \dot{W}_{cv} \quad = \quad \dot{m}e_o \quad - \quad \dot{m}e_i \quad + \quad \dot{m}(P_o v_o - P_i v_i),$$

| Rate of heat transferred across the control surface from the surroundings, to the control volume | Rate of all work done by the control volume, including shaft work, but excluding flow work | Rate of energy flowing out of the control volume | Rate of energy flowing into the control volume | Net rate of work associated with pressure forces where fluid crosses the control surface, flow work |

$$(2.25)$$

where the subscripts $o$ and $i$ denote the outlet and inlet, respectively, and $\dot{m}$ is the mass flowrate. Before rewriting Eqn. 2.25 in a more convenient form, it is appropriate to list the principal assumptions embodied in this relation:

1. *The control volume is fixed relative to the coordinate system.* This eliminates any work interactions associated with a moving boundary, as well as eliminating the need to consider changes in the kinetic and potential energies of the control volume itself.

2. *The properties of the fluid at each point within the control volume, or on the control surface, do not vary with time.* This assumption allows us to treat all processes as steady.

3. *Fluid properties are uniform over the inlet and outlet flow areas.* This allows us to use single values, rather than integrating over the area, for the inlet and exit stream properties.

4. *There is only one inlet and one exit stream.* This assumption is invoked to keep the final result in a simple form and can be easily relaxed to allow multiple inlet/exit streams.

The specific energy $e$ of the inlet and outlet streams consists of the specific internal, kinetic, and potential energies, i.e.,

$$e \quad = \quad u \quad + \quad \tfrac{1}{2}v^2 \quad + \quad gz, \qquad (2.26)$$

| Total energy per unit mass | Internal energy per unit mass | Kinetic energy per unit mass | Potential energy per unit mass |

where $v$ and $z$ are the velocity and elevation, respectively, of the stream where it crosses the control surface.

The pressure–specific volume product terms associated with the flow work in Eqn. 2.25 can be combined with the specific internal energy of Eqn. 2.26, which we recognize as the useful property, enthalpy:

$$h \equiv u + Pv = u + P/\rho. \qquad (2.27)$$

Combining Eqns. 2.25–2.27, and rearranging, yields our final form of energy conservation for a control volume:

$$\dot{Q}_{cv} - \dot{W}_{cv} = \dot{m}\left[(h_o - h_i) + \tfrac{1}{2}(v_o^2 - v_i^2) + g(z_o - z_i)\right].$$  (2.28)

The first law can also be expressed on a mass-specific basis by dividing Eqn. 2.28 by the mass flowrate $\dot{m}$, i.e.,

$$q_{cv} - w_{cv} = h_o - h_i + \tfrac{1}{2}(v_o^2 - v_i^2) + g(z_o - z_i).$$  (2.29)

In Chapter 7, we present more complete expressions of energy conservation that are subsequently simplified for our objectives in this book. For the time being, however, Eqn. 2.28 suits our needs.

## REACTANT AND PRODUCT MIXTURES

### Stoichiometry

The **stoichiometric** quantity of oxidizer is just that amount needed to completely burn a quantity of fuel. If more than a stoichiometric quantity of oxidizer is supplied, the mixture is said to be fuel lean, or just **lean;** while supplying less than the stoichiometric oxidizer results in a fuel-rich, or **rich** mixture. The stoichiometric oxidizer– (or air–) fuel ratio (mass) is determined by writing simple atom balances, assuming that the fuel reacts to form an ideal set of products. For a hydrocarbon fuel given by $C_xH_y$, the stoichiometric relation can be expressed as

$$C_xH_y + a(O_2 + 3.76N_2) \rightarrow xCO_2 + (y/2)H_2O + 3.76aN_2,$$  (2.30)

where

$$a = x + y/4.$$  (2.31)

For simplicity, we assume throughout this book that the simplified composition for air is 21 percent $O_2$ and 79 percent $N_2$ (by volume), i.e., that for each mole of $O_2$ in air, there are 3.76 moles of $N_2$.

The **stoichiometric air–fuel ratio** can be found as

$$(A/F)_{stoic} = \left(\frac{m_{air}}{m_{fuel}}\right)_{stoic} = \frac{4.76a}{1}\frac{MW_{air}}{MW_{fuel}},$$  (2.32)

where $MW_{air}$ and $MW_{fuel}$ are the molecular weights of the air and fuel, respectively. Table 2.1 shows stoichiometric air–fuel ratios for methane and solid carbon.

**Table 2.1**    Some combustion properties of methane, hydrogen, and solid carbon for reactants at 298 K

|  | $\Delta h_R$ (kJ/kg$_{fuel}$) | $\Delta h_R$ (kJ/kg$_{mix}$) | $(O/F)_{stoic}$[a] (kg/kg) | $T_{ad,eq}$ (K) |
|---|---|---|---|---|
| CH$_4$ + air | −55,528 | −3,066 | 17.11 | 2226 |
| H$_2$ + O$_2$ | −142,919 | −15,880 | 8.0 | 3079 |
| C(s) + air | −32,794 | −2,645 | 11.4 | 2301 |

[a] O/F is the oxidizer–fuel ratio, where for combustion with air, the air is the oxidizer not just the oxygen in the air.

Also shown is the oxygen–fuel ratio for combusion of $H_2$ in pure $O_2$. For all of these systems, we see that there is many times more oxidizer than fuel.

The **equivalence ratio, $\Phi$**, is commonly used to indicate quantitatively whether a fuel–oxidizer mixture is rich, lean, or stoichometric. The equivalence ratio is defined as

$$\Phi = \frac{(A/F)_{stoic}}{(A/F)} = \frac{(F/A)}{(F/A)_{stoic}} \tag{2.33a}$$

From this definition, we see that for fuel-rich mixtures, $\Phi > 1$, and for fuel-lean mixtures, $\Phi < 1$. For a stoichiometric mixture, $\Phi$ equals unity. In many combustion applications, the equivalence ratio is the single most important factor in determining a system's performance. Other parameters frequently used to define relative stoichiometry are **percent stoichiometric air,** which relates to the equivalence ratio as

$$\% \text{ stoichometric air} = \frac{100\%}{\Phi} \tag{2.33b}$$

and **percent excess air,** or

$$\% \text{ excess air} = \frac{(1-\Phi)}{\Phi} \cdot 100\%. \tag{2.33c}$$

---

**Example 2.1**

A small, low-emission, stationary gas turbine engine (see Fig. 2.4) operates at full load (3950 kW) at an equivalence ratio of 0.286 with an air flowrate of 15.9 kg/s. The equivalent composition of the fuel (natural gas) is $C_{1.16}H_{4.32}$. Determine the fuel mass flowrate and the operating air–fuel ratio for the engine.

**Solution**

Given:   $\Phi = 0.286$,     $MW_{air} = 28.85$,

$\dot{m}_{air} = 15.9$ kg/s,   $MW_{fuel} = 1.16(12.01) + 4.32(1.008) = 18.286$

Find:   $\dot{m}_{fuel}$ and $(A/F)$.

We will proceed by first finding $(A/F)$ and then $\dot{m}_{fuel}$. The solution requires only the application of definitions expressed in Eqns. 2.32 and 2.33, i.e.,

$$(A/F)_{stoic} = 4.76a \frac{MW_{air}}{MW_{fuel}},$$

where $a = x + y/4 = 1.16 + 4.32/4 = 2.24$. Thus,

$$(A/F)_{stoic} = 4.76(2.24)\frac{28.85}{18.286} = 16.82,$$

and, from Eqn. 2.33,

$$\boxed{(A/F)} = \frac{(A/F)_{stoic}}{\Phi} = \frac{16.82}{0.286} = \boxed{58.8}$$

Since $(A/F)$ is the ratio of the air flowrate to the fuel flowrate,

$$\boxed{\dot{m}_{\text{fuel}}} = \frac{\dot{m}_{\text{air}}}{(A/F)} = \frac{15.9 \text{ kg/s}}{58.8} = \boxed{0.270 \text{ kg/s}}$$

**Comment**

Note that even at full power, a large quantity of excess air is supplied to the engine.

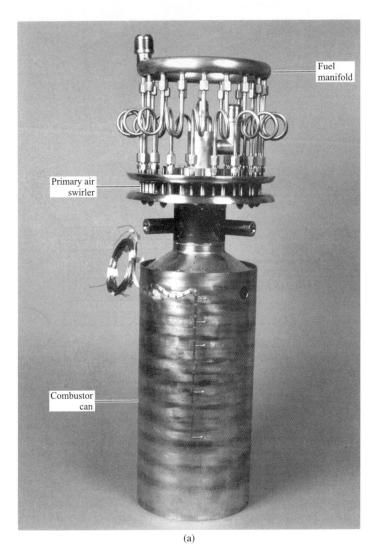

(a)

**Figure 2.4**    (a) Experimental low-NO$_x$ gas-turbine combustor can and (b)(c) fuel and air mixing system. Eight cans are used in a 3950-kW engine.
SOURCE: Copyright © 1987, Electric Power Research Institute, EPRI AP-5347, *NO$_x$ Reduction for Small Gas Turbine Power Plants*; reprinted with permission.

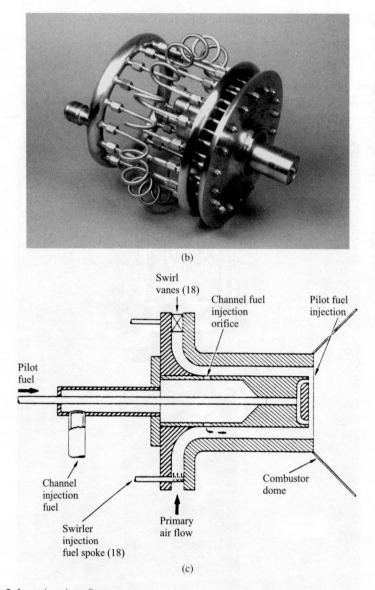

(b)

(c)

**Figure 2.4**     (continued)

---

**Example 2.2**

A natural gas–fired industrial boiler (see Fig. 2.5) operates with an oxygen concentration of 3 mole percent in the flue gases. Determine the operating air–fuel ratio and the equivalence ratio. Treat the natural gas as methane.

**Solution**

Given:     $\chi_{O_2} = 0.03,$     $MW_{fuel} = 16.04,$

$MW_{air} = 28.85.$

Find:     $(A/F)$ and $\Phi$.

**Figure 2.5**    Two 10-MW (34 million Btu/hr) natural-gas burners fire into a boiler combustion chamber 3 m deep. Air enters the burners through the large vertical pipes, and the natural gas enters through the horizontal pipe on the left.
SOURCE: Courtesy of Fives North American Combustion, Inc.

We can use the given $O_2$ mole fraction to find the air–fuel ratio by writing an overall combustion equation assuming "complete combustion," i.e., no dissociation (all fuel C is found in $CO_2$ and all fuel H is found in $H_2O$):

$$CH_4 + a(O_2 + 3.76\,N_2) \rightarrow CO_2 + 2H_2O + bO_2 + 3.76a\,N_2,$$

where $a$ and $b$ are related from conservation of O atoms,

$$2a = 2 + 2 + 2b$$

or

$$b = a - 2.$$

From the definition of a mole fraction (Eqn. 2.8),

$$\chi_{O_2} = \frac{N_{O_2}}{N_{mix}} = \frac{b}{1 + 2 + b + 3.76a} = \frac{a-2}{1+4.76a}.$$

Substituting the known value of $\chi_{O_2}(= 0.03)$ and then solving for $a$ yields

$$0.03 = \frac{a-2}{1+4.76a}$$

or

$$a = 2.368.$$

The mass air–fuel ratio, in general, is expressed as

$$(A/F) = \frac{N_{air}}{N_{fuel}} \frac{MW_{air}}{MW_{fuel}},$$

so

$$(A/F) = \frac{4.76a}{1} \frac{MW_{air}}{MW_{fuel}}$$

$$\boxed{(A/F)} = \frac{4.76(2.368)(28.85)}{16.04} = \boxed{20.3}$$

To find $\Phi$, we need to determine $(A/F)_{stoic}$. From Eq. 2.31, $a = 2$; hence,

$$(A/F)_{stoic} = \frac{4.76(2)28.85}{16.04} = 17.1$$

Applying the definition of $\Phi$ (Eqn. 2.33),

$$\boxed{\Phi} = \frac{(A/F)_{stoic}}{(A/F)} = \frac{17.1}{20.3} = \boxed{0.84}$$

**Comment**

In the solution, we assumed that the $O_2$ mole fraction was on a "wet basis," i.e., moles of $O_2$ per mole of moisture-containing flue gases. Frequently, in the measurement of exhaust species, moisture is removed to prevent condensation in the analyzers; thus, $\chi_{O_2}$ can also be reported on a "dry basis" (see Chapter 15).

## Standardized Enthalpy and Enthalpy of Formation

In dealing with chemically reacting systems, the concept of standardized enthalpies is extremely valuable. For any species, we can define a **standardized enthalpy** that is the sum of an enthalpy that takes into account the energy associated with chemical bonds (or lack thereof), the **enthalpy of formation, $h_f$**, and an enthalpy that is associated only with the temperature, the **sensible enthalpy change, $\Delta h_s$**. Thus, we can write the molar standardized enthalpy for species $i$ as

$$\overline{h}_i(T) \qquad = \qquad \overline{h}^o_{f,i}(T_{ref}) \qquad + \qquad \Delta\overline{h}_{s,i}(T), \qquad (2.34)$$

|Standardized enthalpy at temperature $T$|Enthalpy of formation at standard reference state $(T_{ref}, P^o)$|Sensible enthalpy change in going from $T_{ref}$ to $T$|

where $\Delta\overline{h}_{s,i} \equiv \overline{h}_i(T) - \overline{h}^o_{f,i}(T_{ref})$.

To make practical use of Eqn. 2.34, it is necessary to define a **standard reference state.** We employ a standard-state temperature, $T_{ref} = 25°C$ (298.15 K), and standard-state pressure, $P_{ref} = P^o = 1$ atm (101,325 Pa), consistent with the Chemkin [1] and NASA [5] thermodynamic databases. Furthermore, we adopt the convention that enthalpies of formation are zero for the elements in their naturally occurring state at the reference state temperature and pressure. For example, at 25°C and 1 atm, oxygen exists as diatomic molecules; hence,

$$\left(\bar{h}_{f,O_2}^o\right)_{298} = 0,$$

where the superscript $o$ is used to denote that the value is for the standard-state pressure.

   To form oxygen atoms at the standard state requires the breaking of a rather strong chemical bond. The bond dissociation energy for $O_2$ at 298 K is 498,390 kJ/kmol$_{O_2}$. Breaking this bond creates two O atoms; thus, the enthalpy of formation for atomic oxygen is half the value of the $O_2$ bond dissociation energy, i.e.,

$$\left(\bar{h}_{f,O}^o\right) = 249,195 \text{ kJ/kmol}_O.$$

Thus, enthalpies of formation have a clear physical interpretation as the net change in enthalpy associated with breaking the chemical bonds of the standard-state elements and forming new bonds to create the compound of interest.

   Representing the standardized enthalpy graphically provides a useful way to understand and use this concept. In Fig. 2.6, the standardized enthalpies of atomic oxygen (O) and diatomic oxygen ($O_2$) are plotted versus temperature starting from absolute zero. At 298.15 K, we see that $\bar{h}_{O_2}$ is zero (by definition of the standard-state reference condition) and the standardized enthalpy of atomic oxygen equals its enthalpy of formation, since the sensible enthalpy at 298.15 K is zero. At the temperature indicated (4000 K), we see the additional sensible enthalpy contribution to the standardized enthalpy. In Appendix A, enthalpies of formation at the reference state are given, and sensible enthalpies are tabulated as a function of

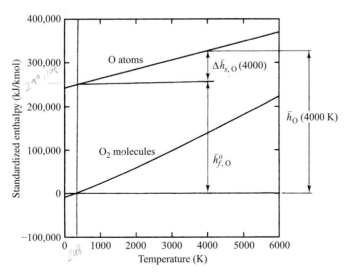

**Figure 2.6**   Graphical interpretation of standardized enthalpy, enthalpy of formation, and sensible enthalpy.

temperature for a number of species of importance in combustion. Enthalpies of formation for reference temperatures other than the standard state 298.15 K are also tabulated.

---

**Example 2.3**

A gas stream at 1 atm contains a mixture of CO, $CO_2$, and $N_2$ in which the CO mole fraction is 0.10 and the $CO_2$ mole fraction is 0.20. The gas-stream temperature is 1200 K. Determine the standardized enthalpy of the mixture on both a mole basis (kJ/kmol) and a mass basis (kJ/kg). Also determine the mass fractions of the three component gases.

**Solution**

Given:    $\chi_{CO} = 0.10$,     $T = 1200$ K,

           $\chi_{CO_2} = 0.20$,     $P = 1$ atm.

Find:    $\bar{h}_{mix}$, $h_{mix}$, $Y_{CO}$, $Y_{CO_2}$, and $Y_{N_2}$.

Finding $\bar{h}_{mix}$ requires the straightforward application of the ideal-gas mixture law, Eqn. 2.15, and, to find $\chi_{N_2}$, the knowledge that $\sum \chi_i = 1$ (Eqn. 2.10). Thus,

$$\chi_{N_2} = 1 - \chi_{CO_2} - \chi_{CO} = 0.70$$

and

$$\bar{h}_{mix} = \sum \chi_i \bar{h}_i$$

$$= \chi_{CO}\left[\bar{h}^o_{f,CO} + \left(\bar{h}(T) - \bar{h}^o_{f,298}\right)_{CO}\right]$$

$$+ \chi_{CO_2}\left[\bar{h}^o_{f,CO_2} + \left(\bar{h}(T) - \bar{h}^o_{f,298}\right)_{CO_2}\right]$$

$$+ \chi_{N_2}\left[\bar{h}^o_{f,N_2} + \left(\bar{h}(T) - \bar{h}^o_{f,298}\right)_{N_2}\right].$$

Substituting values from Appendix A (Table A.1 for CO, Table A.2 for $CO_2$, and Table A.7 for $N_2$):

$$\bar{h}_{mix} = 0.10[-110,541 + 28,440]$$

$$+ 0.20[-393,546 + 44,488]$$

$$+ 0.70[0 + 28,118]$$

$$\boxed{\bar{h}_{mix} = -58,339.1 \text{ kJ/kmol}_{mix}}$$

To find $h_{mix}$, we need to determine the molecular weight of the mixture:

$$MW_{mix} = \sum \chi_i MW_i$$

$$= 0.10(28.01) + 0.20(44.01) + 0.70(28.013)$$

$$= 31.212.$$

Then,

$$\boxed{h_{mix}} = \frac{\bar{h}_{mix}}{MW_{mix}} = \frac{-58,339.1}{31.212} = \boxed{-1869.12 \text{ kJ/kg}_{mix}}$$

Since we have previously found $MW_{mix}$, calculation of the individual mass fractions follows simply from their definitions (Eqn. 2.11):

$$Y_{CO} = 0.10\frac{28.01}{31.212} = 0.0897$$

$$Y_{CO_2} = 0.20\frac{44.01}{31.212} = 0.2820$$

$$Y_{N_2} = 0.70\frac{28.013}{31.212} = 0.6282$$

As a check, we see that $0.0897 + 0.2820 + 0.6282 = 1.000$, as required.

**Comment**

Both molar and mass units are frequently used in combustion. Because of this, you should be quite comfortable with their interconversions.

## Enthalpy of Combustion and Heating Values

Knowing how to express the enthalpy for mixtures of reactants and mixtures of products allows us to define the enthalpy of reaction, or, when dealing specifically with combustion reactions, the enthalpy of combustion. Consider the steady-flow reactor, shown in Fig. 2.7, in which a stoichiometric mixture of reactants enters and products exits, both at standard-state conditions (25°C, 1 atm). The combustion process is assumed to be complete, i.e., all of the fuel carbon is converted to $CO_2$ and all of the fuel hydrogen is converted to $H_2O$. For the products to exit at the same temperature as the entering reactants, heat must be removed from the reactor. The amount of heat removed can be related to the reactant and product standardized enthalpies by applying the steady-flow form of the first law (Eqn. 2.29):

$$q_{cv} = h_o - h_i = h_{prod} - h_{reac}. \tag{2.35}$$

The definition of the **enthalpy of reaction,** or the **enthalpy of combustion, $\Delta h_R$** (per mass of mixture), is

$$\Delta h_R \equiv q_{cv} = h_{prod} - h_{reac}, \tag{2.36a}$$

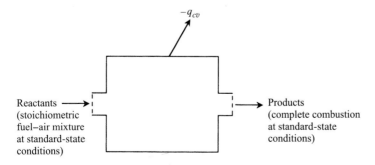

**Figure 2.7**    Steady-flow reactor used to determine enthalpy of combustion.

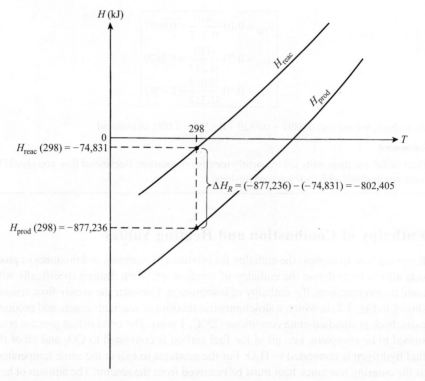

**Figure 2.8** Enthalpy of reaction using representative values for a stoichiometric methane–air mixture. The water in the products is assumed to be in the vapor state.

or, in terms of extensive properties,

$$\Delta H_R = H_{prod} - H_{reac}. \tag{2.36b}$$

The enthalpy of combustion can be illustrated graphically, as shown in Fig. 2.8. Consistent with the heat transfer being negative, the standardized enthalpy of the products lies below that of the reactants. For example, at 25°C and 1 atm, the reactants enthalpy of a stoichiometric mixture of $CH_4$ and air, where 1 kmol of fuel reacts, is −74,831 kJ. At the same conditions (25°C, 1 atm), the combustion products have a standardized enthalpy of −877,236 kJ. Thus,

$$\Delta H_R = -877,236 - (-74,831) = -802,405 \text{ kJ}.$$

This value can be adjusted to a per-mass-of-fuel basis:

$$\Delta h_R \left( \frac{kJ}{kg_{fuel}} \right) = \Delta H_R / MW_{fuel} \tag{2.37}$$

or

$$\Delta h_R \left( \frac{kJ}{kg_{fuel}} \right) = (-802,405/16.043) = -50,016.$$

This value can, in turn, be converted to a per-unit-mass-of-mixture basis:

$$\Delta h_R \left( \frac{\text{kJ}}{\text{kg}_{\text{mix}}} \right) = \Delta h_R \left( \frac{\text{kJ}}{\text{kg}_{\text{fuel}}} \right) \frac{m_{\text{fuel}}}{m_{\text{mix}}}, \tag{2.38}$$

where

$$\frac{m_{\text{fuel}}}{m_{\text{mix}}} = \frac{m_{\text{fuel}}}{m_{\text{air}} + m_{\text{fuel}}} = \frac{1}{(A/F) + 1}. \tag{2.39}$$

From Table 2.1, we see that the stoichiometric air–fuel ratio for $CH_4$ is 17.11; thus,

$$\Delta h_R = \left( \frac{\text{kJ}}{\text{kg}_{\text{mix}}} \right) = \frac{-50,016}{17.11 + 1} = -2761.8.$$

Note that the value of the enthalpy of combustion depends on the temperature chosen for its evaluation since the enthalpies of both the reactants and products vary with temperature; i.e., the distance between the $H_{\text{prod}}$ and $H_{\text{reac}}$ lines in Fig. 2.8 is not constant.

The **heat of combustion,** $\Delta h_c$ (known also as the **heating value**), is numerically equal to the enthalpy of reaction, but with opposite sign. The **upper** or **higher heating value, HHV,** is the heat of combustion calculated assuming that all of the water in the products has condensed to liquid. This scenario liberates the most energy, hence the designation "upper." The **lower heating value, LHV,** corresponds to the case where none of the water is assumed to condense. For $CH_4$, the upper heating value is approximately 11 percent larger than the lower. Standard-state heating values for a variety of hydrocarbon fuels are given in Appendix B.

---

A. Determine the upper and lower heating values at 298 K of gaseous *n*-decane, $C_{10}H_{22}$, per kilomole of fuel and per kilogram of fuel. The molecular weight of *n*-decane is 142.284.

**Example 2.4**

B. If the enthalpy of vaporization of *n*-decane is 359 kJ/$\text{kg}_{\text{fuel}}$ at 298 K, what are the upper and lower heating values of liquid *n*-decane?

**Solution**

A. For 1 mole of $C_{10}H_{22}$, the combustion equation can be written as

$$C_{10}H_{22}(g) + 15.5(O_2 + 3.76N_2) \rightarrow 10CO_2 + 11H_2O(\text{l or g}) + 15.5(3.76)N_2.$$

For either the upper or lower heating value,

$$\Delta H_c = -\Delta H_R = H_{\text{reac}} - H_{\text{prod}},$$

where the numerical value of $H_{\text{prod}}$ depends on whether the $H_2O$ in the products is liquid (determining higher heating value) or gaseous (determining lower heating value). The sensible enthalpies for all species involved are zero since we desire $\Delta H_c$ at the reference state (298 K). Furthermore, the enthalpies of formation of the $O_2$ and $N_2$ are also zero at 298 K. Recognizing that

$$H_{\text{reac}} = \sum_{\text{reac}} N_i \bar{h}_i \quad \text{and} \quad H_{\text{prod}} = \sum_{\text{prod}} N_i \bar{h}_i,$$

we obtain

$$\Delta H_{c,\mathrm{H_2O(l)}} = \mathrm{HHV} = (1)\bar{h}^o_{f,\mathrm{C_{10}H_{22}}} - \left[10\bar{h}^o_{f,\mathrm{CO_2}} + 11\bar{h}^o_{f,\mathrm{H_2O(l)}}\right].$$

Table A.6 (Appendix A) gives the enthalpy of formation for gaseous water and the enthalpy of vaporization. With these values, we can calculate the enthalpy of formation for the liquid water (Eqn. 2.18):

$$\bar{h}^o_{f,\mathrm{H_2O(l)}} = \bar{h}^o_{f,\mathrm{H_2O(g)}} - \bar{h}_{fg} = -241,847 - 44,010 = -285,857 \text{ kJ/kmol}.$$

Using this value, together with enthalpies of formation given in Appendices A and B, we obtain the higher heating value:

$$\Delta H_{c,\mathrm{H_2O,(l)}} = (1)\left(-249,659\,\frac{\text{kJ}}{\text{kmol}}\right)$$
$$-\left[10\left(-393,546\,\frac{\text{kJ}}{\text{kmol}}\right) + 11\left(-285,857\,\frac{\text{kJ}}{\text{kmol}}\right)\right]$$
$$= 6,830,096 \text{ kJ}$$

and

$$\boxed{\Delta \bar{h}_c = \frac{\Delta H_c}{N_{\mathrm{C_{10}H_{22}}}} = \frac{6,830,096 \text{ kJ}}{1\,\text{kmol}} = 6,830,096 \text{ kJ/kmol}_{\mathrm{C_{10}H_{22}}}}$$

or

$$\boxed{\Delta h_c = \frac{\Delta \bar{h}_c}{MW_{\mathrm{C_{10}H_{22}}}} = \frac{6,830,096\,\frac{\text{kJ}}{\text{kmol}}}{142.284\,\frac{\text{kg}}{\text{kmol}}} = 48,003 \text{ kJ/kg}_{\mathrm{C_{10}H_{22}}}}$$

For the lower heating value, we use $\bar{h}^o_{f,\mathrm{H_2O(g)}} = -241,847\,\text{kJ/kmol}$ in place of $\bar{h}^o_{f,\mathrm{H_2O(l)}} = -285,857\,\text{kJ/kmol}$. Thus,

$$\boxed{\Delta \bar{h}_c = 6,345,986 \text{ kJ/kmol}_{\mathrm{C_{10}H_{22}}}}$$

or

$$\boxed{\Delta h_c = 44,601 \text{ kJ/kg}_{\mathrm{C_{10}H_{22}}}}$$

B.    For $\mathrm{C_{10}H_{22}}$, in the liquid state,

$$H_{\text{reac}} = (1)\left(\bar{h}^o_{f,\mathrm{C_{10}H_{22}(g)}} - \bar{h}_{fg}\right),$$

or

$$\Delta h_c\left(\frac{\text{liquid}}{\text{fuel}}\right) = \Delta h_c\left(\frac{\text{gaseous}}{\text{fuel}}\right) - h_{fg}.$$

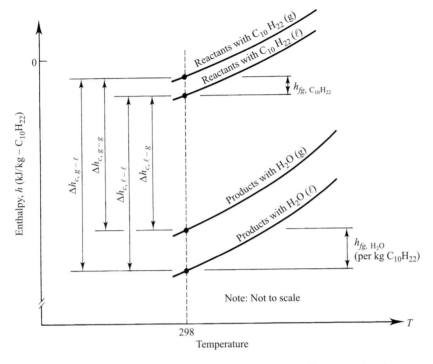

**Figure 2.9**    Enthalpy–temperature plot illustrating calculation of heating values in Example 2.4.

Thus,

$$\boxed{\begin{aligned}
\Delta h_c \text{ (higher)} &= 48{,}003 - 359 \\
&= 47{,}644 \text{ kJ/kg}_{C_{10}H_{22}} \\
\Delta h_c \text{ (lower)} &= 44{,}601 - 359 \\
&= 44{,}242 \text{ kJ/kg}_{C_{10}H_{22}}
\end{aligned}}$$

**Comment**

Graphical representations of the various definitions and/or thermodynamic processes are valuable aids in setting up problems or in checking their solutions. Figure 2.9 illustrates, on $h$–$T$ coordinates, the important quantities used in this example. Note that the enthalpy of vaporization given for $n$-decane is for the standard-state temperature (298.15 K), while the value given in Appendix B is at the boiling point (447.4 K).

## ADIABATIC FLAME TEMPERATURES

We define two adiabatic flame temperatures: one for constant-pressure combustion and one for constant-volume. If a fuel–air mixture burns adiabatically at constant pressure, the standardized enthalpy of the reactants at the initial state (say, $T = 298$ K, $P = 1$ atm) equals the standardized enthalpy of the products at the final state ($T = T_{ad}$, $P = 1$ atm), i.e., application of Eqn. 2.28 results in

$$H_{\text{reac}}(T_i, P) = H_{\text{prod}}(T_{ad}, P), \qquad (2.40a)$$

or, equivalently, on a per-mass-of-mixture basis,

$$h_{\text{reac}}(T_i, P) = h_{\text{prod}}(T_{ad}, P). \qquad (2.40b)$$

This first-law statement, Eqn. 2.40, defines what is called the **constant-pressure adiabatic flame temperature.** This definition is illustrated graphically in Fig. 2.10. Conceptually, the adiabatic flame temperature is simple; however, evaluating this quantity requires knowledge of the composition of the combustion products. At typical flame temperatures, the products dissociate and the mixture comprises many species. As shown in Table 2.1 and Table B.l in Appendix B, flame temperatures are typically several thousand kelvins. Calculating the complex composition by invoking chemical equilibrium is the subject of the next section. The following example illustrates the fundamental concept of constant-pressure adiabatic flame temperatures, while making crude assumptions regarding the product mixture composition and evaluation of the product mixture enthalpy.

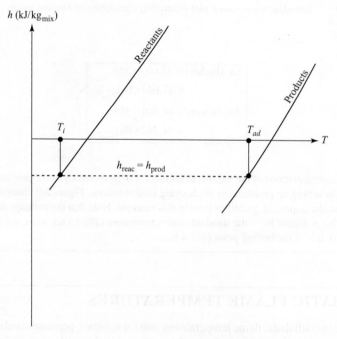

**Figure 2.10**    Illustration of constant-pressure adiabatic flame temperature on *h–T* coordinates.

Example 2.5

Estimate the constant-pressure adiabatic flame temperature for the combustion of a stoichio-metric $CH_4$–air mixture. The pressure is 1 atm and the initial reactant temperature is 298 K.
　　Use the following assumptions:

1. "Complete combustion" (no dissociation), i.e., the product mixture consists of only $CO_2$, $H_2O$, and $N_2$.
2. The product mixture enthalpy is estimated using constant specific heats evaluated at 1200 K ($\approx 0.5(T_i + T_{ad})$), where $T_{ad}$ is guessed to be about 2100 K.

**Solution**

Mixture composition:

$$CH_4 + 2(O_2 + 3.76N_2) \rightarrow 1CO_2 + 2H_2O + 7.52N_2$$
$$N_{CO_2} = 1, N_{H_2O} = 2, N_{N_2} = 7.52.$$

Properties (Appendices A and B):

| Species | Enthalpy of Formation @ 298 K $\bar{h}_{f,i}^o$ (kJ/kmol) | Specific Heat @ 1200 K $\bar{c}_{p,i}$ (kJ/kmol-K) |
|---|---|---|
| $CH_4$ | −74,831 | — |
| $CO_2$ | −393,546 | 56.21 |
| $H_2O$ | −241,845 | 43.87 |
| $N_2$ | 0 | 33.71 |
| $O_2$ | 0 | — |

First law (Eqn. 2.40):

$$H_{react} = \sum_{react} N_i \bar{h}_i = H_{prod} = \sum_{prod} N_i \bar{h}_i$$

$$H_{react} = (1)(-74,831) + 2(0) + 7.52(0)$$
$$= -74,831 \text{ kJ}$$

$$H_{prod} = \sum N_i \left[ \bar{h}_{f,i}^o + \bar{c}_{p,i}(T_{ad} - 298) \right]$$
$$= (1)[-393,546 + 56.21(T_{ad} - 298)]$$
$$+ (2)[-241,845 + 43.87(T_{ad} - 298)]$$
$$+ (7.52)[0 + 33.71(T_{ad} - 298)].$$

Equating $H_{react}$ to $H_{prod}$ and solving for $T_{ad}$ yields

$$\boxed{T_{ad} = 2318\,\text{K}}$$

**Comments**

Comparing the above result with the equilibrium-composition based computation shown in Table 2.1 ($T_{ad,\,eq} = 2226$ K) shows that the simplified approach overestimates $T_{ad}$ by slightly less than 100 K. Considering the crudeness of the assumptions, this appears to be rather surprisingly good agreement. Removing assumption 2 and recalculating $T_{ad}$ using variable specific heats, i.e.,

$$\bar{h}_i = \bar{h}_{f,i}^{o} + \int_{298}^{T} \bar{c}_{p,i} \ dT,$$

yields $T_{ad} = 2328$ K. (Note that Appendix A provides tabulations of these integrated quantities. Similar tabulations are found in the JANAF tables [6].) Since this result is quite close to our constant-$c_p$ solution, we conclude that the ~ 100 K difference is the result of neglecting dissociation. Note that dissociation causes a lowering of $T_{ad}$ since more energy is tied up in chemical bonds (enthalpies of formation) at the expense of the sensible enthalpy.

---

In the above, we dealt with a constant-pressure system, which would be appropriate in dealing with a gas-turbine combustor, or a furnace. Let us look now at **constant-volume adiabatic flame temperatures,** which we might require in an ideal Otto-cycle analysis, for example. The first law of thermodynamics (Eqn. 2.20) requires

$$U_{reac}(T_{init}, P_{init}) = U_{prod}(T_{ad}, P_f),  \tag{2.41}$$

where $U$ is the standardized internal energy of the mixture. Graphically, Eqn. 2.41 resembles the sketch (Fig. 2.10) used to illustrate the constant-pressure adiabatic flame temperature, except the internal energy replaces the enthalpy. Since most compilations or calculations of thermodynamic properties provide values for $H$ (or $h$) rather than $U$ (or $u$) [1, 6], we can rearrange Eqn. 2.41 to the following form:

$$H_{reac} - H_{prod} - V(P_{init} - P_f) = 0.  \tag{2.42}$$

We can apply the ideal-gas law to eliminate the $PV$ terms:

$$P_{init} V = \sum_{reac} N_i R_u T_{init} = N_{reac} R_u T_{init}$$

$$P_f V = \sum_{prod} N_i R_u T_{ad} = N_{prod} R_u T_{ad}.$$

Thus,

$$H_{reac} - H_{prod} - R_u (N_{reac} T_{init} - N_{prod} T_{ad}) = 0.  \tag{2.43}$$

An alternative form of Eqn. 2.43, on a per-mass-of-mixture basis, can be obtained by dividing Eqn. 2.43 by the mass of mixture, $m_{mix}$, and recognizing that

$$m_{mix} / N_{reac} \equiv MW_{reac}$$

or

$$m_{mix} / N_{prod} \equiv MW_{prod}.$$

We thus obtain

$$h_{reac} - h_{prod} - R_u \left( \frac{T_{init}}{MW_{reac}} - \frac{T_{ad}}{MW_{prod}} \right) = 0.  \tag{2.44}$$

Since the equilibrium composition of the product mixture depends upon both temperature and pressure, as we will see in the next section, utilizing Eqn. 2.43 or 2.44 with the ideal-gas law and appropriate calorific equations of state, e.g., $h = h(T, P) = h(T$ only, ideal gas), to find $T_{ad}$ is straightforward, but non-trivial.

---

Estimate the constant-volume adiabatic flame temperature for a stoichiometric $CH_4$–air mixture using the same assumptions as in Example 2.5. Initial conditions are $T_i = 298$ K, $P = 1$ atm (= 101,325 Pa).

**Example 2.6**

### Solution

The same composition and properties used in Example 2.5 apply here. We note, however, that the $c_{p,i}$ values should be evaluated at a temperature somewhat greater than 1200 K, since the constant-volume $T_{ad}$ will be higher than the constant-pressure $T_{ad}$. Nonetheless, we will use the same values as before.

First law (Eqn. 2.43):

$$H_{reac} - H_{prod} - R_u(N_{reac}T_{init} - N_{prod}T_{ad}) = 0$$

or

$$\sum_{reac} N_i \bar{h}_i - \sum_{prod} N_i \bar{h}_i - R_u(N_{reac}T_{init} - N_{prod}T_{ad}) = 0.$$

Substituting numerical values, we have

$$H_{react} = (1)(-74,831) + 2(0) + 7.52(0)$$
$$= -74,831\,kJ$$
$$H_{prod} = (1)[-393,546 + 56.21(T_{ad} - 298)]$$
$$+ (2)[-241,845 + 43.87(T_{ad} - 298)]$$
$$+ (7.52)[0 + 33.71(T_{ad} - 298)]$$
$$= -877,236 + 397.5(T_{ad} - 298)\,kJ$$

and

$$R_u(N_{reac}T_{init} - N_{prod}T_{ad}) = 8.315(10.52)(298 - T_{ad}),$$

where $N_{reac} = N_{prod} = 10.52$ kmol.

Reassembling Eqn. 2.43 and solving for $T_{ad}$ yields

$$\boxed{T_{ad} = 2889\ K}$$

### Comments

(i) For the same initial conditions, constant-volume combustion results in much higher temperatures (571 K higher in this example) than for constant-pressure combustion. This is a consequence of the pressure forces doing no work when the volume is fixed. (ii) Note, also, that the number of moles was conserved in going from the initial to final state. This is a fortuitous result for $CH_4$ and does not occur for other fuels. (iii) The final pressure is well above the initial pressure: $P_f = P_{init}(T_{ad}/T_{init}) = 9.69$ atm.

## CHEMICAL EQUILIBRIUM

In high-temperature combustion processes, the products of combustion are not a simple mixture of ideal products, as may be suggested by the simple atom-balance used to determine stoichiometry (see Eqn. 2.30). Rather, the major species **dissociate,** producing a host of minor species. Under some conditions, what ordinarily might be considered a minor species is actually present in rather large quantities. For example, the ideal combustion products for burning a hydrocarbon with air are $CO_2$, $H_2O$, $O_2$, and $N_2$. Dissociation of these species and reactions among the dissociation products yields the following species: $H_2$, OH, CO, H, O, N, NO, and possibly others. The problem we address in this section is the calculation of the mole fractions of all of the product species at a given temperature and pressure, subject to the constraint of conserving the number of moles of each of the elements present in the initial mixture. This element constraint merely says that the number of C, H, O, and N atoms is constant, regardless of how they are combined in the various species.

There are several ways to approach the calculation of equilibrium composition. To be consistent with the treatment of equilibrium in most undergraduate thermodynamics courses, we focus on the equilibrium-constant approach and limit our discussion to the application of ideal gases. For descriptions of other methods, the interested reader is referred to the literature [5, 7].

### Second-Law Considerations

The concept of chemical equilibrium has its roots in the second law of thermodynamics. Consider a fixed-volume, adiabatic reaction vessel in which a fixed mass of reactants form products. As the reactions proceed, both the temperature and pressure rise until a final equilibrium condition is reached. This final state (temperature, pressure, and composition) is not governed solely by first-law considerations, but necessitates invoking the second law. Consider the combustion reaction

$$CO + \tfrac{1}{2}O_2 \rightarrow CO_2. \tag{2.45}$$

If the final temperature is high enough, the $CO_2$ will dissociate. Assuming the products to consist only of $CO_2$, CO, and $O_2$, we can write

$$\left[ CO + \tfrac{1}{2}O_2 \right]_{\substack{cold \\ reactants}} \rightarrow \left[ (1-\alpha)CO_2 + \alpha CO + \frac{\alpha}{2}O_2 \right]_{\substack{hot \\ products}} \tag{2.46}$$

where $\alpha$ is the fraction of the $CO_2$ dissociated. We can calculate the adiabatic flame temperature as a function of the dissociation fraction, $\alpha$, using Eqn. 2.42. For example, with $\alpha = 1$, no heat is released and the mixture temperature, pressure, and composition remain unchanged; while with $\alpha = 0$, the maximum amount of heat release occurs and the temperature and pressure would be the highest possible allowed by the first law. This variation in temperature with $\alpha$ is plotted in Fig. 2.11.

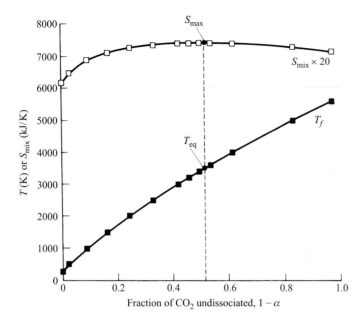

**Figure 2.11**    Illustration of chemical equilibrium for a fixed-mass isolated system.

What constraints are imposed by the second law on this thought experiment where we vary $\alpha$? The entropy of the product mixture can be calculated by summing the product species entropies, i.e.,

$$S_{mix}(T_f, P) = \sum_{i=1}^{3} N_i \bar{s}_i(T_f, P_i) = (1-\alpha)\bar{s}_{CO_2} + \alpha\bar{s}_{CO} + \frac{\alpha}{2}\bar{s}_{O_2}, \qquad (2.47)$$

where $N_i$ is the number of moles of species $i$ in the mixture. The individual species entropies are obtained from

$$\bar{s}_i = \bar{s}_i^o(T_{ref}) + \int_{T_{ref}}^{T_f} \bar{c}_{p,i} \frac{dT}{T} - R_u \ln \frac{P_i}{P^o}, \qquad (2.48)$$

where ideal-gas behavior is assumed, and $P_i$ is the partial pressure of the $i$th species. Plotting the mixture entropy (Eqn. 2.47) as a function of the dissociation fraction, we see that a maximum value is reached at some intermediate value of $\alpha$. For the reaction chosen, $CO + \frac{1}{2}O_2 \rightarrow CO_2$, the maximum entropy occurs near $1 - \alpha = 0.5$.

For our choice of conditions (constant $U$, $V$, and $m$, which implies no heat or work interactions), the second law requires that the entropy change internal to the system

$$dS \geq 0. \qquad (2.49)$$

Thus, we see that the composition of the system will spontaneously shift toward the point of maximum entropy when approaching from either side, since $dS$ is positive. Once the maximum entropy is reached, no further change in composition is allowed, since this would require the system entropy to decrease in violation of the second law (Eqn. 2.49). Formally, the condition for equilibrium can be written

$$(dS)_{U, V, m} = 0. \tag{2.50}$$

In summary, if we fix the internal energy, volume, and mass of an isolated system, the application of Eqn. 2.49 (second law), Eqn. 2.41 (first law), and Eqn. 2.2 (equation of state) define the equilibrium temperature, pressure, and chemical composition.

## Gibbs Function

Although the foregoing was useful in illustrating how the second law comes into play in establishing chemical equilibrium, the use of an isolated (fixed-energy) system of fixed mass and volume is not particularly useful for many of the typical problems involving chemical equilibrium. For example, there is frequently a need to calculate the composition of a mixture at a given temperature, pressure, and stoichiometry. For this problem, the **Gibbs free energy, $G$,** replaces the entropy as the important thermodynamic property.

As you may recall from your previous study of thermodynamics, the Gibbs free energy is defined in terms of other thermodynamic properties as

$$G \equiv H - TS. \tag{2.51}$$

The second law can then be expressed as

$$(dG)_{T, P, m} \leq 0, \tag{2.52}$$

which states that the Gibbs function always decreases for a spontaneous, isothermal, isobaric change of a fixed-mass system in the absence of all work effects except boundary ($P$–$dV$) work. This principle allows us to calculate the equilibrium composition of a mixture at a given temperature and pressure. The Gibbs function attains a minimum in equilibrium, in contrast to the maximum in entropy we saw for the fixed-energy and fixed-volume case (Fig. 2.11). Thus, at equilibrium,

$$(dG)_{T, P, m} = 0. \tag{2.53}$$

For a mixture of ideal gases, the Gibbs function for the $i$th species is given by

$$\bar{g}_{i, T} = \bar{g}_{i, T}^o + R_u T \ln\left(P_i / P^o\right) \tag{2.54}$$

where $\bar{g}_{i, T}^o$ is the Gibbs function of the pure species at the standard-state pressure (i.e., $P_i = P^o$) and $P_i$ is the partial pressure. The standard-state pressure, $P^o$, by convention taken to be 1 atm, appears in the denominator of the logarithm term. In dealing with reacting systems, a **Gibbs function of formation, $\bar{g}_{f, i}^o$,** is frequently employed:

$$\bar{g}_{f, i}^o(T) \equiv \bar{g}_i^o(T) - \sum_{j \text{ elements}} v_j' \bar{g}_j^o(T), \tag{2.55}$$

where the $v_j'$ are the stoichiometric coefficients of the elements required to form one mole of the compound of interest. For example, the coefficients are $v_{O_2}' = \frac{1}{2}$ and

$v'_C = 1$ for forming a mole of CO from $O_2$ and C, respectively. As with enthalpies, the Gibbs functions of formation of the naturally occurring elements are assigned values of zero at the reference state. Appendix A provides tabulations of Gibbs function of formation over a range of temperatures for selected species. Having tabulations of $\bar{g}^o_{f,i}(T)$ as a function of temperature is quite useful. In later calculations, we will need to evaluate differences in $\bar{g}^o_{i,T}$ between different species at the same temperature. These differences can be obtained easily by using the Gibbs function of formation at the temperature of interest, values of which are provided in Appendix A. Tabulations for over 1,000 species can be found in the JANAF tables [6].

The Gibbs function for a mixture of ideal gases can be expressed as

$$G_{mix} = \sum N_i \bar{g}_{i,T} = \sum N_i \left[ \bar{g}^o_{i,T} + R_u T \ln\left(P_i/P^o\right)\right] \tag{2.56}$$

where $N_i$ is the number of moles of the $i$th species.

For fixed temperature and pressure, the equilibrium condition becomes

$$dG_{mix} = 0 \tag{2.57}$$

or

$$\sum dN_i \left[ \bar{g}^o_{i,T} + R_u T \ln\left(P_i/P^o\right)\right] + \sum N_i d\left[ \bar{g}^o_{i,T} + R_u T \ln\left(P_i/P^o\right)\right] = 0. \tag{2.58}$$

The second term in Eqn. 2.58 can be shown to be zero by recognizing that $d(\ln P_i) = dP_i/P_i$ and that $\sum dP_i = 0$, since all changes in the partial pressures must sum to zero because the total pressure is constant. Thus,

$$dG_{mix} = 0 = \sum dN_i \left[ \bar{g}^o_{i,T} + R_u T \ln\left(P_i/P^o\right)\right]. \tag{2.59}$$

For the general system, where

$$aA + bB + \cdots \Leftrightarrow eE + fF + \cdots, \tag{2.60}$$

the change in the number of moles of each species is directly proportional to its stoichiometric coefficient, i.e.,

$$dN_A = -\kappa a \tag{2.61}$$
$$dN_B = -\kappa b$$
$$\vdots \qquad \vdots$$
$$dN_E = +\kappa e$$
$$dN_F = +\kappa f$$
$$\vdots \qquad \vdots$$

Substituting Eqn. 2.61 into Eqn. 2.59 and canceling the proportionality constant $\kappa$, we obtain

$$-a\left[ \bar{g}^o_{A,T} + R_u T \ln\left(P_A/P^o\right)\right] - b\left[ \bar{g}^o_{B,T} + R_u T \ln\left(P_B/P^o\right)\right] - \cdots \tag{2.62}$$
$$+ e\left[ \bar{g}^o_{E,T} + R_u T \ln\left(P_E/P^o\right)\right] + f\left[ \bar{g}^o_{F,T} + R_u T \ln\left(P_F/P^o\right)\right] + \cdots = 0.$$

Equation 2.62 can be rearranged and the log terms grouped together to yield

$$-\left(e\bar{g}^o_{E,T} + f\bar{g}^o_{F,T} + \cdots - a\bar{g}^o_{A,T} - b\bar{g}^o_{B,T} - \cdots\right) \tag{2.63}$$

$$= R_u T \ln \frac{\left(P_E/P^o\right)^e \cdot \left(P_F/P^o\right)^f \cdot \text{etc.}}{\left(P_A/P^o\right)^a \cdot \left(P_B/P^o\right)^b \cdot \text{etc.}}$$

The term in parentheses on the left-hand side of Eqn. 2.63 is called the **standard-state Gibbs function change** $\Delta G^o_T$, i.e.,

$$\Delta G^o_T = \left(e\bar{g}^o_{E,T} + f\bar{g}^o_{F,T} + \cdots - a\bar{g}^o_{A,T} - b\bar{g}^o_{B,T} - \cdots\right) \tag{2.64a}$$

or, alternately,

$$\Delta G^o_T \equiv \left(e\bar{g}^o_{f,E} + f\bar{g}^o_{f,F} + \cdots - a\bar{g}^o_{f,A} - b\bar{g}^o_{f,B} - \cdots\right)_T. \tag{2.64b}$$

The argument of the natural logarithm is defined as the **equilibrium constant** $K_p$ for the reaction expressed in Eqn. 2.60, i.e.,

$$K_p = \frac{\left(P_E/P^o\right)^e \cdot \left(P_F/P^o\right)^f \cdot \text{etc.}}{\left(P_A/P^o\right)^a \cdot \left(P_B/P^o\right)^b \cdot \text{etc.}}. \tag{2.65}$$

With these definitions, Eqn. 2.63, our statement of chemical equilibrium at constant temperature and pressure, is given by

$$\Delta G^o_T = -R_u T \ln K_p, \tag{2.66a}$$

or

$$K_p = \exp\left(-\Delta G^o_T/R_u T\right). \tag{2.66b}$$

From the definition of $K_p$ (Eqn. 2.65) and its relation to $\Delta G^o_T$ (Eqn. 2.66), we can obtain a qualitative indication of whether a particular reaction favors products (goes strongly to completion) or reactants (very little reaction occurs) at equilibrium. If $\Delta G^o_T$ is positive, reactants will be favored since $\ln K_p$ is negative, which requires that $K_p$ itself is less than unity. Similarly, if $\Delta G^o_T$ is negative, the reaction tends to favor products. Physical insight into this behavior can be obtained by appealing to the definition of $\Delta G$ in terms of the enthalpy and entropy changes associated with the reaction. From Eqn. 2.51, we can write

$$\Delta G^o_T = \Delta H^o - T\Delta S^o,$$

which can be substituted into Eqn. 2.66b:

$$K_p = e^{-\Delta H^o/R_u T} \cdot e^{\Delta S^o/R_u}.$$

For $K_p$ to be greater than unity, which favors products, the enthalpy change for the reaction, $\Delta H^o$, should be negative, i.e., the reaction is exothermic and the system energy is lowered. Also, positive changes in entropy, which indicate greater molecular chaos, lead to values of $K_p > 1$.

Consider the dissociation of $CO_2$ as a function of temperature and pressure,

**Example 2.7**

$$CO_2 \Leftrightarrow CO + \tfrac{1}{2}O_2.$$

Find the composition of the mixture, i.e., the mole fractions of $CO_2$, CO, and $O_2$, that results from subjecting originally pure $CO_2$ to various temperatures ($T = 1500$, 2000, 2500, and 3000 K) and pressures (0.1, 1, 10, and 100 atm).

### Solution

To find the three unknown mole fractions, $\chi_{CO_2}$, $\chi_{CO}$, and $\chi_{O_2}$, we will need three equations. The first equation will be an equilibrium expression, Eqn. 2.66. The other two equations will come from element conservation expressions that state that the total amounts of C and O are constant, regardless of how they are distributed among the three species, since the original mixture was pure $CO_2$.

To implement Eqn. 2.66, we recognize that $a = 1$, $b = 1$, and $c = \tfrac{1}{2}$, since

$$(1)CO_2 \Leftrightarrow (1)CO + \left(\tfrac{1}{2}\right)O_2.$$

Thus, we can evaluate the standard-state Gibbs function change. For example, at $T = 2500$ K,

$$\Delta G_T^o = \left[ \left(\tfrac{1}{2}\right)\bar{g}_{f,O_2}^o + (1)\bar{g}_{f,CO}^o - (1)\bar{g}_{f,CO_2}^o \right]_{T=2500}$$
$$= \left(\tfrac{1}{2}\right)0 + (1)(-327,245) - (-396,152)$$
$$= 68,907 \text{ kJ/kmol.}$$

The values above are taken from Appendix Tables A.1, A.2, and A.11.

From the definition of $K_p$, we have

$$K_p = \frac{\left(P_{CO}/P^o\right)^1 \left(P_{O_2}/P^o\right)^{0.5}}{\left(P_{CO_2}/P^o\right)^1}.$$

We can rewrite $K_p$ in terms of the mole fractions by recognizing that $P_i = \chi_i P$. Thus,

$$K_p = \frac{\chi_{CO}\chi_{O_2}^{0.5}}{\chi_{CO_2}} \cdot (P/P^o)^{0.5}$$

Substituting the above into Eqn. 2.66b, we have

$$\frac{\chi_{CO}\chi_{O_2}^{0.5}(P/P^o)^{0.5}}{\chi_{CO_2}} = \exp\left[\frac{-\Delta G_T^o}{R_u T}\right]$$
$$= \exp\left[\frac{-68,907}{(8.315)(2500)}\right]$$
$$\frac{\chi_{CO}\chi_{O_2}^{0.5}(P/P^o)^{0.5}}{\chi_{CO_2}} = 0.03635. \qquad \text{(I)}$$

We create a second equation to express **conservation of elements:**

$$\frac{\text{No. of carbon atoms}}{\text{No. of oxygen atoms}} = \frac{1}{2} = \frac{\chi_{CO} + \chi_{CO_2}}{\chi_{CO} + 2\chi_{CO_2} + 2\chi_{O_2}}.$$

We can make the problem more general by defining the C/O ratio to be a parameter $Z$ that can take on different values depending on the initial composition of the mixture:

$$Z = \frac{\chi_{CO} + \chi_{CO_2}}{\chi_{CO} + 2\chi_{CO_2} + 2\chi_{O_2}}$$

or

$$(Z-1)\chi_{CO} + (2Z-1)\chi_{CO_2} + 2Z\chi_{O_2} = 0. \tag{II}$$

To obtain a third and final equation, we require that all of the mole fractions sum to unity:

$$\sum_i \chi_i = 1$$

or

$$\chi_{CO} + \chi_{CO_2} + \chi_{O_2} = 1. \tag{III}$$

Simultaneous solution of Eqns. I, II, and III for selected values of $P$, $T$, and $Z$ yield values for the mole fractions $\chi_{CO}$, $\chi_{CO_2}$, and $\chi_{O_2}$. Using Eqns. II and III to eliminate $\chi_{CO_2}$ and $\chi_{O_2}$, Eqn. I becomes

$$\chi_{CO}(1 - 2Z + Z\chi_{CO})^{0.5}(P/P^o)^{0.5} - [2Z - (1+Z)\chi_{CO}]\exp\left(-\Delta G_T^o/R_u T\right) = 0.$$

The above expression is easily solved for $\chi_{CO}$ by applying Newton–Raphson iteration, which can be implemented simply using spreadsheet software. The other unknowns, $\chi_{CO}$ and $\chi_{O_2}$, are then recovered using Equations II and III.

Results are shown in Table 2.2 for four levels each of temperature and pressure. Figure 2.12 shows the CO mole fractions over the range of parameters investigated.

**Comments**

Two general observations concerning these results can be made. First, at any fixed temperature, increasing the pressure suppresses the dissociation of $CO_2$ into CO and $O_2$; second, increasing the temperature at a fixed pressure promotes the dissociation. Both of these trends are consistent with the **principle of Le Châtelier** that states that any system initially in a state of equilibrium when subjected to a change (e.g., increasing pressure or temperature) will shift in composition in such a way as to minimize the change. For an increase in pressure, this translates to the equilibrium shifting in the direction to produce fewer moles. For the $CO_2 \Leftrightarrow CO + \frac{1}{2}O_2$ reaction, this means a shift to the left, to the $CO_2$ side. For equimolar reactions, pressure has no effect. When the temperature is increased, the composition shifts in the endothermic direction. Since heat is absorbed when $CO_2$ breaks down into CO and $O_2$, increasing the temperature produces a shift to the right, to the $CO + \frac{1}{2}O_2$ side.

**Table 2.2**    Equilibrium compositions at various temperatures and pressures for $CO_2 \Leftrightarrow CO + \frac{1}{2}O_2$

|  | $P = 0.1$ atm | $P = 1$ atm | $P = 10$ atm | $P = 100$ atm |
|---|---|---|---|---|
| | $T = 1500$ K, $\Delta G_T^o = 1.5268 \cdot 10^8$ J/kmol | | | |
| $\chi_{CO}$ | $7.755 \cdot 10^{-4}$ | $3.601 \cdot 10^{-4}$ | $1.672 \cdot 10^{-4}$ | $7.76 \cdot 10^{-5}$ |
| $\chi_{CO_2}$ | 0.9988 | 0.9994 | 0.9997 | 0.9999 |
| $\chi_{O_2}$ | $3.877 \cdot 10^{-4}$ | $1.801 \cdot 10^{-4}$ | $8.357 \cdot 10^{-5}$ | $3.88 \cdot 10^{-5}$ |
| | $T = 2000$ K, $\Delta G_T^o = 1.10462 \cdot 10^8$ J/kmol | | | |
| $\chi_{CO}$ | 0.0315 | 0.0149 | $6.96 \cdot 10^{-3}$ | $3.243 \cdot 10^{-3}$ |
| $\chi_{CO_2}$ | 0.9527 | 0.9777 | 0.9895 | 0.9951 |
| $\chi_{O_2}$ | 0.0158 | 0.0074 | $3.48 \cdot 10^{-3}$ | $1.622 \cdot 10^{-3}$ |
| | $T = 2500$ K, $\Delta G_T^o = 6.8907 \cdot 10^7$ J/kmol | | | |
| $\chi_{CO}$ | 0.2260 | 0.1210 | 0.0602 | 0.0289 |
| $\chi_{CO_2}$ | 0.6610 | 0.8185 | 0.9096 | 0.9566 |
| $\chi_{O_2}$ | 0.1130 | 0.0605 | 0.0301 | 0.0145 |
| | $T = 3000$ K, $\Delta G_T^o = 2.7878 \cdot 10^7$ J/kmol | | | |
| $\chi_{CO}$ | 0.5038 | 0.3581 | 0.2144 | 0.1138 |
| $\chi_{CO_2}$ | 0.2443 | 0.4629 | 0.6783 | 0.8293 |
| $\chi_{O_2}$ | 0.2519 | 0.1790 | 0.1072 | 0.0569 |

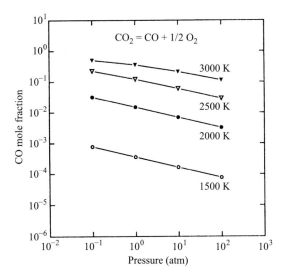

**Figure 2.12**    The CO mole fractions resulting from dissociation of pure $CO_2$ at various pressures and temperatures.

## Complex Systems

The preceding sections focused on simple situations involving a single equilibrium reaction; however, in most combustion systems, many species and several simultaneous equilibrium reactions are important. In principle, the previous example could be extended to include additional reactions. For example, the reaction $O_2 \Leftrightarrow 2O$ is likely to be important at the temperatures considered. Including this reaction introduces only one additional unknown, $\chi_O$. We easily add an additional equation to account for the $O_2$ dissociation:

$$\left(\chi_O^2/\chi_{O_2}\right)P/P^o = \exp\left(-\Delta G_T^{o\,\prime}/R_u T\right),$$

where $\Delta G_T^{o\,\prime}$ is the appropriate standard-state Gibbs function change for the $O_2 \Leftrightarrow 2O$ reaction. The element-conservation expression (Eqn. II) is modified to account for the additional O-containing species,

$$\frac{\text{No. of C atoms}}{\text{No. of O atoms}} = \frac{\chi_{CO} + \chi_{CO_2}}{\chi_{CO} + 2\chi_{CO_2} + 2\chi_{O_2} + \chi_O},$$

and Eqn. III becomes

$$\chi_{CO} + \chi_{CO_2} + \chi_{O_2} + \chi_O = 1.$$

We now have a new set of four equations with four unknowns to solve. Since two of the four equations are nonlinear, it is likely that some method of simultaneously solving nonlinear equations would be applied. Appendix E presents the **generalized Newton's method,** which is easily applied to such systems.

An example of the above approach being applied to the C, H, N, O system is the computer code developed by Olikara and Borman [8]. This code solves for 12 species, invoking seven equilibrium reactions and four atom-conservation relations, one each for C, O, H, and N. This code was developed specifically for internal combustion engine simulations and is readily imbedded as a subroutine in simulation codes. This code is used in the software provided with this book, as explained in Appendix F.

One of the most frequently used general equilibrium codes is the powerful NASA Chemical Equilibrium with Applications [5]. This code is capable of handling over 400 different species, and many special problem features are built into it. For example, rocket nozzle performance and shock calculations can be performed. The theoretical approach to the equilibrium calculation does not employ equilibrium constants, but rather techniques are applied to minimize either the Gibbs or Helmholz energies, subject to atom-balance constraints.

Several equilibrium solvers are available as downloads or for online use, for example, Refs. [9, 10].

## EQUILIBRIUM PRODUCTS OF COMBUSTION

## Full Equilibrium

When we combine the first law with complex chemical equilibrium principles, the adiabatic flame temperature and the detailed composition of the products of combustion can be obtained by solving Eqns. 2.40 (or 2.41) and 2.66 simultaneously, with appropriate atom-conservation constants. An example of such a calculation for constant-pressure (1 atm) combustion of propane with air is shown in Figs. 2.13 and 2.14, where it has been assumed that the only products occurring are $CO_2$, $CO$, $H_2O$, $H_2$, H, OH, $O_2$, O, NO, $N_2$, and N.

In Fig. 2.13, we see the adiabatic flame temperature and the **major species** as functions of equivalence ratio. Major products of lean combustion are $H_2O$, $CO_2$, $O_2$, and $N_2$; while for rich combustion, they are $H_2O$, $CO_2$, $CO$, $H_2$, and $N_2$. It is interesting to note that the maximum flame temperature, 2278.4 K, occurs not at stoichiometric, but, rather, at a slightly rich equivalence ratio ($\Phi \approx 1.05$), as does the water mole fraction ($\Phi \approx 1.15$). That the maximum temperature is at a slightly rich equivalence ratio is a consequence of both the heat of combustion and heat capacity of the products ($N_{prod} \cdot \bar{c}_{p,prod}$) declining beyond $\Phi = 1$. For equivalence ratios between $\Phi = 1$ and $\Phi(T_{max})$, the heat capacity decreases more rapidly with $\Phi$ than $\Delta H_c$; while beyond $\Phi(T_{max})$, $\Delta H_c$ falls more rapidly than does the heat capacity. The decrease in heat capacity is dominated by the decrease in number of product moles formed per mole of fuel burned, with the decrease in the mean specific heat being

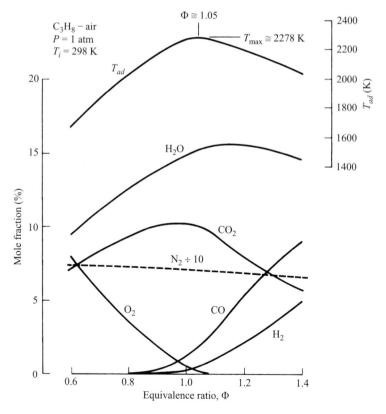

**Figure 2.13**    Equilibrium adiabatic flame temperatures and major product species for propane–air combustion at 1 atm.

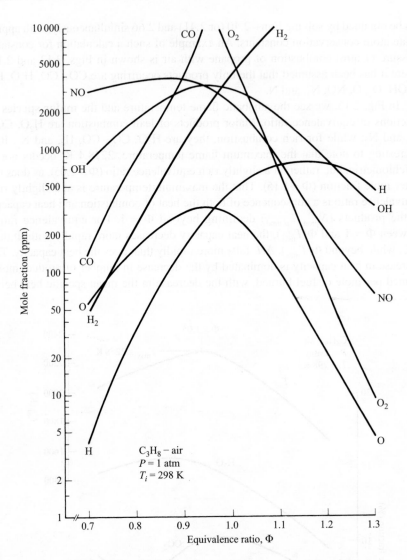

**Figure 2.14**    Minor species distributions for propane-air combustion at 1 atm.

less significant. Also in Fig. 2.13, we see, as a result of dissociation, the simultaneous presence of $O_2$, CO, and $H_2$ at stoichiometric conditions ($\Phi = 1$). Under conditions of "complete combustion," i.e., no dissociation, all three of these species would be zero; thus, we expose the approximate nature of the "complete-combustion" assumption. Later, we will quantify this effect.

Some of the **minor species** of equilibrium combustion of hydrocarbons in air are shown in Fig. 2.14. Here, we see the atoms O and H and the diatomic species OH and NO, all below the 4000 ppm level, and we see that CO is a minor species in lean products, and, conversely, that $O_2$ is a minor product of rich combustion. The CO and $O_2$

concentrations, however, head through the top of the graph on their way to becoming major species in rich and lean products, respectively. It is interesting to note that the level of the hydroxyl radical OH is more than an order of magnitude greater than the O atom, and that both peak slightly lean of stoichiometric conditions. Furthermore, although not shown, N-atom concentrations are several orders of magnitude less than those of the O atom. The lack of dissociation of the $N_2$ molecule is a result of the strong triple covalent bond. The O and OH maxima in the lean region have implications for the kinetics of NO formation. Equilibrium NO concentrations are rather flat and peak in the lean region, falling rapidly in the rich region. In most combustion systems, NO levels are well below the equilibrium concentrations shown, because of the relatively slow formation reactions, as we will see in Chapters 4 and 5.

## Water-Gas Equilibrium

In this section, we will develop simple relations that allow the calculation of the ideal products of combustion (no dissociation producing minor species) for both lean and rich conditions. For lean combustion, nothing new is involved as we need employ only atom balances; for rich combustion, however, we employ a single equilibrium reaction, $CO + H_2O \Leftrightarrow CO_2 + H_2$, the so-called **water-gas shift reaction,** to account for the simultaneous presence of the incomplete products of combustion, CO and $H_2$. This water-gas equilibrium is central to steam reforming of CO in the petroleum industry.

Assuming no dissociation, the combustion of an arbitrary hydrocarbon with our simplified air can be represented as

$$C_xH_y + a(O_2 + 3.76N_2) \rightarrow bCO_2 + cCO + dH_2O + eH_2 + fO_2 + 3.76aN_2, \quad (2.67a)$$

which for *lean or stoichiometric conditions* ($\Phi \leq 1$) becomes

$$C_xH_y + a(O_2 + 3.76N_2) \rightarrow bCO_2 + dH_2O + fO_2 + 3.76aN_2, \quad (2.67b)$$

or for *rich conditions* ($\Phi > 1$) becomes

$$C_xH_y + a(O_2 + 3.76N_2) \rightarrow bCO_2 + cCO + dH_2O + eH_2 + 3.76aN_2. \quad (2.67c)$$

Since the coefficient $a$ represents the ratio of the number of moles of $O_2$ in the reactants to the number of moles of fuel, we can relate $a$ to the equivalence ratio by using Eqn. 2.31, i.e.,

$$a = \frac{x + y/4}{\Phi}; \quad (2.68)$$

thus, given the fuel type and $\Phi$, $a$ is a known quantity.

Our objective is to find the mole fractions of all of the product species. For lean or stoichiometric combustion, the coefficients $c$ and $e$ are zero because there is sufficient $O_2$ to have all the fuel C and H react to form $CO_2$ and $H_2O$, respectively. The coefficients $b$, $d$, and $f$ can be found by C-, H-, and O-atom balances, respectively; thus,

$$b = x, \quad (2.69a)$$

$$c = 0, \quad (2.69b)$$

$$d = y/2, \tag{2.69c}$$

$$e = 0, \tag{2.69d}$$

$$f = \left(\frac{1-\Phi}{\Phi}\right)(x+y/4) \tag{2.69e}$$

The total number of moles of products (per mole of fuel burned) can be found by summing the above coefficients together with the $3.76a$ moles of $N_2$:

$$N_{TOT} = x + y/2 + \left(\frac{x+y/4}{\Phi}\right)(1-\Phi+3.76). \tag{2.70}$$

The mole fractions are then determined by dividing each of the coefficients above by $N_{TOT}$:

*Lean or stoichiometric ($\Phi \le 1$)*

$$\chi_{CO_2} = x/N_{TOT}, \tag{2.71a}$$

$$\chi_{CO} = 0, \tag{2.71b}$$

$$\chi_{H_2O} = (y/2)/N_{TOT}, \tag{2.71c}$$

$$\chi_{H_2} = 0, \tag{2.71d}$$

$$\chi_{O_2} = \left(\frac{1-\Phi}{\Phi}\right)(x+y/4)/N_{TOT}, \tag{2.71e}$$

$$\chi_{N_2} = 3.76(x+y/4)/(\Phi N_{TOT}). \tag{2.71f}$$

For rich combustion ($\Phi > 1$), no oxygen appears, so the coefficient $f$ is zero. That leaves us with four unknowns ($b$, $c$, $d$, and $e$). To solve for these, we employ the three element balances (C, H, and O) and the water-gas shift equilibrium,

$$K_p = \frac{\left(P_{CO_2}/P^o\right)\cdot\left(P_{H_2}/P^o\right)}{\left(P_{CO}/P^o\right)\cdot\left(P_{H_2O}/P^o\right)} = \frac{b\cdot e}{c\cdot d}. \tag{2.72}$$

The use of Eqn. 2.72 causes the system of equations for $b$, $c$, $d$, and $e$ to be nonlinear (quadratic). Solving the element balances in terms of the unknown coefficient $b$ results in

$$c = x - b \tag{2.73a}$$

$$d = 2a - b - x \tag{2.73b}$$

$$e = -2a + b + x + y/2 \tag{2.73c}$$

Substituting Eqns. 2.73a–c into Eqn. 2.72 yields a quadratic equation in $b$, the solution of which is

$$b = \frac{2a(K_p - 1) + x + y/2}{2(K_p - 1)} \tag{2.74}$$

$$- \frac{1}{2(K_p - 1)} \left[ (2a(K_p - 1) + x + y/2)^2 - 4K_p(K_p - 1)(2ax - x^2) \right]^{1/2},$$

where the negative root is selected to yield physically realistic (positive) values of $b$. Again,

$$N_{TOT} = b + c + d + e + 3.76a = x + y/2 + 3.76a, \tag{2.75}$$

and the various mole fractions are expressed in terms of $b$ (Eqn. 2.74):

*Rich* ($\Phi > 1$)

$$\chi_{CO_2} = b/N_{TOT}, \tag{2.76a}$$

$$\chi_{CO} = c/N_{TOT} = (x - b)/N_{TOT}, \tag{2.76b}$$

$$\chi_{H_2O} = d/N_{TOT} = (2a - b - x)/N_{TOT}, \tag{2.76c}$$

$$\chi_{H_2} = e/N_{TOT} = (-2a + b + x + y/2)/N_{TOT}, \tag{2.76d}$$

$$\chi_{O_2} = 0, \tag{2.76e}$$

$$\chi_{N_2} = 3.76a/N_{TOT}, \tag{2.76f}$$

where $a$ is evaluated from Eqn. 2.68. Spreadsheet software can be used conveniently to solve Eqns. 2.76a–f and their ancillary relations for the mole fractions for various fuels (values of $x$ and $y$) and equivalence ratios. Since $K_p$ is a function of temperature, an appropriate temperature must be selected; however, at typical combustion temperatures, say 2000–2400 K, the mole fractions are not strongly dependent on the choice of temperature. Selected values of $K_p$ are shown in Table 2.3.

Table 2.4 shows comparisons between the full-equilibrium calculations and the approximate method above for CO and $H_2$ mole fractions for propane–air combustion products. The equilibrium constant for the water-gas shift was evaluated at

**Table 2.3** Selected values of equilibrium constant $K_p$ for water-gas shift reaction, $CO + H_2O \leftrightarrow CO_2 + H_2$

| $T$ (K) | $K_p$ | $T$ (K) | $K_p$ |
|---|---|---|---|
| 298 | $1.05 \cdot 10^5$ | 2000 | 0.2200 |
| 500 | 138.3 | 2500 | 0.1635 |
| 1000 | 1.443 | 3000 | 0.1378 |
| 1500 | 0.3887 | 3500 | 0.1241 |

**Table 2.4**        CO and $H_2$ mole fractions for rich combustion, $C_3H_8$–air, $P = 1$ atm

| | $\chi_{CO}$ | | | $\chi_{H_2}$ | | |
|---|---|---|---|---|---|---|
| $\Phi$ | Full Equilibrium | Water-Gas Equilibrium[a] | % Difference | Full Equilibrium | Water-Gas Equilibrium[a] | % Difference |
| 1.1 | 0.0317 | 0.0287 | −9.5 | 0.0095 | 0.0091 | −4.2 |
| 1.2 | 0.0537 | 0.0533 | −0.5 | 0.0202 | 0.0203 | +0.5 |
| 1.3 | 0.0735 | 0.0741 | +0.8 | 0.0339 | 0.0333 | −1.8 |
| 1.4 | 0.0903 | 0.0920 | +1.9 | 0.0494 | 0.0478 | −3.4 |

[a]For $K_P = 0.193$ ($T = 2200$ K).

2200 K for all equivalence ratios. Here, we see that for $\Phi \gtrsim 1.2$, the full-equilibrium and approximate methods yield concentrations that differ by only a few percent. As $\Phi$ approaches unity, the simple method becomes increasingly inaccurate, because dissociation was neglected.

To quantify the degree of dissociation at $\Phi = 1$, Table 2.5 shows $CO_2$ and $H_2O$ mole fractions calculated using both full equilibrium and the assumption of no dissociation. Here, we see that, at 1 atm, approximately 12 percent of the $CO_2$ dissociates, whereas just over 4 percent of the $H_2O$ dissociates.

## Pressure Effects

Pressure has a significant effect on dissociation. Table 2.6 shows the decreasing degree of $CO_2$ dissociation with pressure. Since the only other carbon-containing species allowed in the product mixture is CO, the effect shown in Table 2.5 results from the equilibrium reaction $CO_2 \Leftrightarrow CO + \frac{1}{2}O_2$. Since the dissociation of $CO_2$ results in an increase in the total number of moles present, the pressure effect shown is consistent with the principle of Le Châtelier discussed previously. The $H_2O$ dissociation is more complex in that, in addition to $H_2O$, elemental hydrogen is present as OH, $H_2$, and H; thus, we cannot isolate the pressure effect on $H_2O$ in a single equilibrium expression, but need to simultaneously consider other reactions. The net effect of pressure results in a decrease in the $H_2O$ dissociation, as expected. We note also that the temperature increases as the dissociation is suppressed by increased pressure in accord with Le Châtelier's principle.

**Table 2.5**        Degree of dissociation for propane–air combustion products ($P = 1$ atm, $\Phi = 1$)

| Species | Mole Fraction | | |
|---|---|---|---|
| | Full Equilibrium | No Dissociation | % Dissociated |
| $CO_2$ | 0.1027 | 0.1163 | 11.7 |
| $H_2O$ | 0.1484 | 0.1550 | 4.3 |

**Table 2.6**    Effect of pressure on dissociation of propane–air combustion products ($\Phi = 1$)

| Pressure (atm) | $T_{ad}$ (K) | $\chi_{CO_2}$ | % Dissociation | $\chi_{H_2O}$ | % Dissociation |
|---|---|---|---|---|---|
| 0.1 | 2198 | 0.0961 | 17.4 | 0.1444 | 6.8 |
| 1.0 | 2268 | 0.1027 | 11.7 | 0.1484 | 4.3 |
| 10 | 2319 | 0.1080 | 7.1 | 0.1512 | 2.5 |
| 100 | 2353 | 0.1116 | 4.0 | 0.1530 | 1.3 |

## SOME APPLICATIONS

In this section, we present two practical applications: the use of recuperation or regeneration to improve energy utilization and/or increase flame temperatures, and the use of flue (or exhaust) gas recirculation to lower flame temperatures. Our intent here is to apply to "real-world" examples the concepts previously developed in this chapter and to illustrate the use of some of the software included with this book.

### Recuperation and Regeneration

A **recuperator** is a heat exchanger in which energy from a steady flow of hot combustion products, i.e., flue gases, is transferred to the air supplied to the combustion process. A general flow diagram is shown in Fig. 2.15. A wide variety of recuperators is used in practice, many of which employ radiation heat transfer from the flue gases, as well as convection. An example of a recuperator for an indirect-fired application is illustrated in Fig. 2.16.

A **regenerator** also transfers energy from the flue gases to the incoming combustion air, but, in this case, an energy storage medium, such as a corrugated steel or ceramic matrix, is alternately heated by the hot gases and cooled by the air.

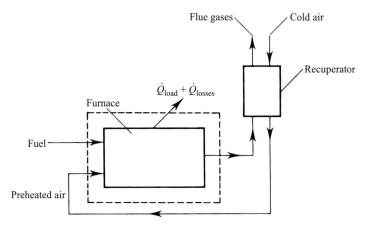

**Figure 2.15**    Schematic of furnace with air preheated by recuperation or regeneration. Dashed line indicates the control volume employed in Example 2.8.

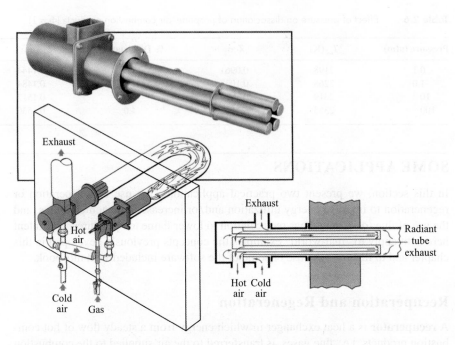

**Figure 2.16**     Radiant-tube burner with coupled recuperator for indirect firing. Note that all the flue gases pass through the recuperator.
SOURCE: Courtesy of Eclipse Combustion.

Figure 2.17 illustrates the application of a spinning-disk regenerator to an automotive gas-turbine engine, and Fig. 2.18 shows a similar concept applied to an industrial furnace. In other regenerator concepts, the flow paths are switched alternately to heat and cool the thermal storage medium.

---

**Example 2.8**

A recuperator, such as shown in Fig. 2.16, is employed in a natural-gas-fired heat-treating furnace. The furnace operates at atmospheric pressure with an equivalence ratio of 0.9. The fuel gas enters the burner at 298 K, and the air is preheated.

A.  Determine the effect of air preheat on the adiabatic temperature of the flame zone for a range of inlet air temperatures from 298 K to 1000 K.
B.  What fuel savings result from preheating the air from 298 K to 600 K? Assume that the temperature of the flue gases at the furnace exit, prior to entering the recuperator, is 1700 K, both with and without air preheat.

**Solution (Part A)**

We will employ the computer program HPFLAME, which incorporates the Olikara and Borman equilibrium routines [8], to solve the first-law problem, $H_{reac} = H_{prod}$. The input file for the program requires the definition of the fuel by providing the number of carbon, hydrogen, oxygen, and nitrogen atoms constituting the fuel molecule, the equivalence ratio, a guess for the adiabatic flame temperature, the pressure, and the reactants' enthalpy. The input file for this

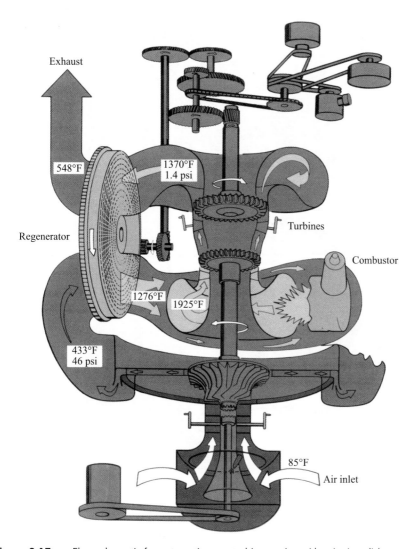

Exhaust

548°F

1370°F
1.4 psi

Regenerator

Turbines

Combustor

1276°F    1925°F

433°F
46 psi

85°F

Air inlet

**Figure 2.17**    Flow schematic for automotive gas-turbine engine with spinning-disk regenerator. Ambient air is compressed to 46 psi and 433°F before passing through the regenerator. The thermal energy given up by the regenerator heats the air to 1276°F prior to combustion. After the products of combustion expand through the two turbines, they enter the opposite side of the regenerator at 1370°F, returning energy to the spinning disk, and exhaust at 548°F.
SOURCE: Courtesy of Chrysler Group LLC.

example, treating natural gas as methane, is shown below:

Adiabatic Flame Calculation for Specified Fuel, Phi, P, & Reactant
Enthalpy Using Olikara & Borman Equilibrium Routines
Problem Title: **EXAMPLE 2.8 Air Preheat at 1000 K**

01          /CARBON ATOMS IN FUEL
04          /HYDROGEN ATOMS IN FUEL

**Figure 2.18** (a) Steel rotary regenerator used in industrial furnace applications. (b) Flow paths are indicated by the arrows in the sketch.

SOURCE: From Ref. [11]. Courtesy of IHEA Combustion Technology Manual.

| | |
|---|---|
| **00** | /OXYGEN ATOMS IN FUEL |
| **00** | /NITROGEN ATOMS IN FUEL |
| **0.900** | /EQUIVALENCE RATIO |
| **2000.** | /TEMPERATURE (K) (Initial Guess) |
| **101325.0** | /PRESSURE (Pa) |
| **155037.0** | /ENTHALPY OF REACTANTS PER KMOL FUEL (kJ/kmol-fuel) |

The only quantity requiring calculation is the reactants' enthalpy, expressed as kJ/kmol of fuel. To find the number of moles of $O_2$ and $N_2$ supplied per mole of fuel, we write our combustion equation as

$$CH_4 + a(O_2 + 3.76N_2) \rightarrow products,$$

where (Eqn. 2.68)

$$a = \frac{x + y/4}{\Phi} = \frac{(1 + 4/4)}{0.9} = 2.22.$$

Thus,

$$CH_4 + 2.22O_2 + 8.35N_2 \rightarrow products.$$

The reactants' enthalpy (per mole of fuel) is then

$$H_{reac} = \bar{h}^o_{f,CH_4} + 2.22\Delta\bar{h}_{s,O_2} + 8.35\Delta\bar{h}_{s,N_2}.$$

Using Tables A.7, A.11, and B.1, the above expression can be evaluated for various air temperatures, as shown in the following table.

| $T$ (K) | $\Delta\bar{h}_{s,O_2}$ (kJ/kmol) | $\Delta\bar{h}_{s,N_2}$ (kJ/kmol) | $H_{reac}$ (kJ/kmol$_{fuel}$) | $T_{ad}$ (K) |
|---|---|---|---|---|
| 298 | 0 | 0 | −74,831 | 2134 |
| 400 | 3,031 | 2,973 | −45,254 | 2183 |
| 600 | 9,254 | 8,905 | +20,140 | 2283 |
| 800 | 15,838 | 15,046 | +86,082 | 2373 |
| 1000 | 22,721 | 21,468 | +155,037 | 2456 |

Using the $H_{reac}$ values from the table, the constant-pressure adiabatic flame temperatures are calculated using HPFLAME. These results also are given in the table and plotted in Fig. 2.19.

### Comment (Part A)

We note from Fig. 2.19 that, for the range of preheat temperatures investigated, a 100 K increase in air temperature results in about a 50 K increase in flame temperature. This effect can be attributed to dissociation and to the larger specific heat of the product gases compared with the air.

### Solution (Part B)

To determine the amount of fuel saved as a result of preheating the air to 600 K, we will write an energy balance for the control volume indicated in Fig. 2.15, assuming both the heat transferred to the load and the heat losses are the same, with and without preheat. Assuming steady flow, we apply Eqn. 2.28, recognizing that heat is transferred out of the control volume:

$$-\dot{Q} = -\dot{Q}_{load} - \dot{Q}_{loss} = \dot{m}(h_{prod} - h_{reac})$$
$$= (\dot{m}_A + \dot{m}_F)h_{prod} - \dot{m}_F h_F - \dot{m}_A h_A.$$

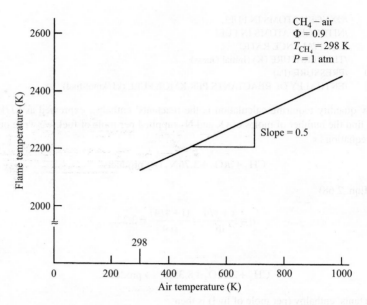

**Figure 2.19**    Effect of combustion air preheat on adiabatic flame temperature for methane combustion ($\Phi = 0.9$, $P = 1$ atm).

For convenience, we define a fuel utilization efficiency as

$$\eta \equiv \frac{\dot{Q}}{\dot{m}_F \text{LHV}} = \frac{-([(A/F)+1]h_{\text{prod}} - (A/F)h_A - h_F)}{\text{LHV}}.$$

To evaluate the above, we require:

$$(A/F) = \frac{(A/F)_{\text{stoic}}}{\Phi} = \frac{17.1}{0.9} = 19.0$$

$$h_F = \bar{h}^o_{f,F}/MW_F = \frac{-74{,}831}{16.043} = -4664.4 \text{ kJ/kg}$$

$$h_{\text{prod}} = -923 \text{ kJ/kg} \qquad \text{(calculated by TPEQUIL code; see Appendix F)}$$

$$h_{A@298\,K} = 0$$

$$h_{A@600\,K} = \left(0.21\Delta\bar{h}_{s,O_2} + 0.79\Delta\bar{h}_{s,N_2}\right)/MW_A$$

$$= \frac{0.21(9254) + 0.79(8905)}{28.85}$$

$$= 311.2 \text{ kJ/kg}.$$

Thus, with air entering at 298 K,

$$\eta_{298} = \frac{-[(19+1)(-923) - 19(0) - (-4664.4)]}{50{,}016}$$

$$= 0.276$$

and, for air entering at 600 K,

$$\eta_{600} = \frac{-1[(19+1)(-923) - 19(311.2) - (-4664.4)]}{50{,}016}$$

$$= 0.394.$$

We now calculate the fuel savings, defined as

$$\text{Savings} = \frac{\dot{m}_{F,600} - \dot{m}_{F,298}}{\dot{m}_{F,298}} = 1 - \frac{\eta_{298}}{\eta_{600}}$$

$$= 1 - \frac{0.276}{0.394} = 0.30$$

or, expressed as a percentage,

$$\boxed{\text{Savings} = 30\%}.$$

### Comment (Part B)

We see that substantial fuel savings can be realized by using recuperators to return some of the energy that would normally go up the stack. Note that the nitric oxide emissions may be affected, since peak temperatures will increase as a result of preheat. With air entering at 600 K, the adiabatic flame temperature increases 150 K (7.1 percent) above the 298 K air case.

## Flue- (or Exhaust-) Gas Recirculation

In one strategy to decrease the amount of oxides of nitrogen ($NO_x$) formed and emitted from certain combustion devices, a portion of the burned product gases is recirculated and introduced with the air and fuel. This emission control strategy, and others, is discussed in Chapter 15. The effect of the recirculated gases is to decrease the maximum temperatures in the flame zone. Decreased flame temperatures result in less $NO_x$ being formed. Figure 2.20a schematically illustrates the application of flue-gas recirculation (FGR) in a boiler or furnace, and Fig. 2.20b shows the exhaust-gas recirculation (EGR) system used in automotive engines. The following example shows how the principle of conservation of energy can be applied to determine the effect of product gas recirculation on flame temperatures.

Consider a spark-ignition engine whose compression and combustion processes have been idealized as a polytropic compression from bottom-dead-center (state 1) to top-dead-center (state 2) and constant-volume combustion (state 2 to state 3), respectively, as shown in the sketch below. Determine the effect of EGR (0–20 percent, expressed as a volume percentage of the air and fuel) on the adiabatic flame temperature and pressure at state 3. The engine compression ratio (CR $\equiv V_1 / V_2$) is 8.0, the polytropic exponent is 1.3, and the initial pressure and temperature (state 1) are fixed at 0.5 atm and 298 K, respectively, regardless of the quantity of recirculated gas. The fuel is isooctane and the equivalence ratio is unity.

**Example 2.9**

$\Phi = 1.0$
$P_1 = 0.5$ atm
$T_1 = 298$ K

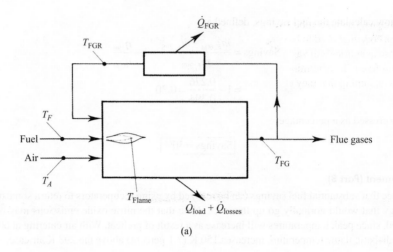

(a)

**ESM EGR System**

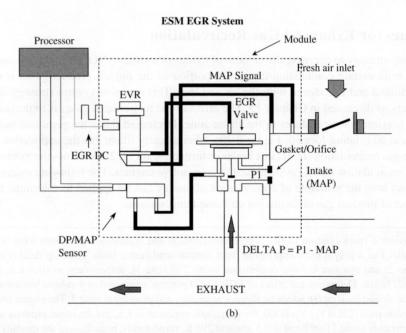

(b)

**Figure 2.20** (a) Schematic of flue-gas recirculation applied to boiler or furnace.
(b) Exhaust-gas recirculation system for spark-ignition engine.
| SOURCE: (b) Courtesy of Ford Motor Company.

**Solution**

To determine the initial temperature and pressure for the start of combustion (state 2), we apply the polytropic relations:

$$T_2 = T_1(V_1/V_2)^{n-1} = 298(8)^{0.3} = 556 \text{ K}$$

$$P_2 = P_1(V_1/V_2)^n = 0.5(8)^{1.3} = 7.46 \text{ atm (755,885 Pa)}.$$

To analyze the combustion process, we will employ the code UVFLAME. Inputs required for the code that need to be calculated are $H_{reac}$ (kJ/kmol—**fuel**), $N_{reac}/N_{fuel}$, and $MW_{reac}$. Each of these quantities will vary as percent EGR is changed, although the temperature and pressure remain fixed. To determine these inputs, we first determine the composition of the recycled gases, assuming that they consist of undissociated products of the reaction

$$C_8H_{18} + 12.5(O_2 + 3.76N_2) \rightarrow 8CO_2 + 9H_2O + 47N_2.$$

Thus,

$$\chi_{CO_2} = \frac{8}{8+9+47} = \frac{8}{64} = 0.1250$$

$$\chi_{H_2O} = \frac{9}{64} = 0.1406$$

$$\chi_{N_2} = \frac{47}{64} = 0.7344.$$

Using Tables A.2, A.6, and A.7, we evaluate the molar specific enthalpy of the recycled gases at $T_2$ (= 556 K):

$$\bar{h}_{EGR} = 0.1250(-382,707) + 0.1406(-232,906) + 0.7344(7588)$$
$$= -75,012.3 \text{ kJ/kmol}_{EGR}.$$

The molar specific enthalpy of the air at $T_2$(= 556 K) is

$$\bar{h}_A = 0.21(7853) + 0.79(7588) = 7643.7 \text{ kJ/kmol}_{air}.$$

The fuel enthalpy at $T_2$ is calculated from the curvefit coefficients given in Table B.2. Note that the enthalpy generated from these coefficients is the sum of both the enthalpy of formation and sensible enthalpy:

$$\bar{h}_F = -161,221 \text{ kJ/kmol}_{fuel}.$$

The enthalpy of the reactants at state 2 can now be calculated:

$$H = N_F\bar{h}_F + N_A\bar{h}_A + N_{EGR}\bar{h}_{EGR},$$

where, by definition,

$$N_{EGR} \equiv (N_A + N_F)\%EGR/100\%.$$

From the stoichiometry given, $N_A = 12.5(4.76) = 59.5$ kmol; thus,

$$H_{reac} = (1)\bar{h}_F + (59.5)\bar{h}_A + 60.5(\%EGR)\bar{h}_{EGR/100\%}.$$

Values of $H_{reac}$ for various percent EGR are shown in the following table.

The reactant mixture molecular weight is

$$MW_{reac} = \frac{N_F MW_F + N_A MW_A + N_{EGR} MW_{EGR}}{N_F + N_A + N_{EGR}},$$

where

$$MW_{EGR} = \sum_{EGR} \chi_i MW_i$$
$$= 0.1250(44.011) + 0.1406(18.016) + 0.7344(28.013)$$
$$= 29.607 \text{ kg/kmol}_{EGR}.$$

Values for $MW_{reac}$ and $N_{tot}(= N_F + N_A + N_{EGR})$ are also shown in the following table.

| %EGR | $N_{EGR}$ | $N_{tot}$ | $MW_{reac}$ | $H_{reac}(kJ/kmol_{fuel})$ |
|---|---|---|---|---|
| 0 | 0 | 60.50 | 30.261 | +293,579 |
| 5 | 3.025 | 63.525 | 30.182 | +66,667 |
| 10 | 6.050 | 66.55 | 30.111 | −160,245 |
| 15 | 9.075 | 69.575 | 30.045 | −387,158 |
| 20 | 12.100 | 72.60 | 29.98 | −614,070 |

Using the above information, we exercise the code UVFLAME and calculate the following adiabatic temperatures and the corresponding pressures at state 3.

| %EGR | $T_{ad}(= T_3)$ (K) | $P_3$ (atm) |
|---|---|---|
| 0 | 2804 | 40.51 |
| 5 | 2742 | 39.41 |
| 10 | 2683 | 38.38 |
| 15 | 2627 | 37.12 |
| 20 | 2573 | 36.51 |

These results are plotted in Fig. 2.21.

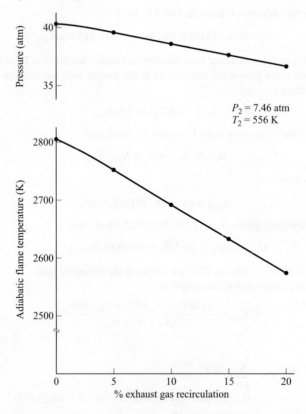

**Figure 2.21**    Calculated adiabatic flame temperatures and peak pressures for constant-volume combustion, with combustion products recycled with the fresh air and fuel (Example 2.9).

**Comments**

From the table and the graph, we see that EGR can have a pronounced effect on peak temperatures, with 20 percent EGR resulting in a drop of about 275 K from the zero recycle condition. As we shall see in Chapters 4, 5, and 15, such temperature decreases can have a dramatic effect on $NO_x$ formation.

We should note that in real applications, the temperature of the recycled gases is likely to vary with the amount recycled, which, in turn, affects the final peak temperature. Furthermore, we did not check to see if the initial (state 1) temperature was below the dew point of the recycled gases, as condensed water would be undesirable in a real EGR system.

---

A package boiler, fueled by natural gas ($CH_4$), operates with a stack-gas $O_2$ concentration of 1.5 percent (by volume). The fuel gas is supplied at 298 K, and the air is preheated to 400 K. Determine the adiabatic flame temperature and the equilibrium NO concentration for operation with 15 percent FGR, where the FGR is expressed as a volume percentage of the fuel and air. The recirculated gases enter the combustion chamber at 600 K. Compare these results with those for operation without FGR.

**Example 2.10**

**Solution**

We start by determining the equivalence ratio. For both cases, the overall stoichiometry is the same and can be written as

$$CH_4 + a(O_2 + 3.76N_2) \rightarrow CO_2 + 2H_2O + xO_2 + 3.76aN_2$$

An O-atom balance yields

$$2a = 2 + 2 + 2x$$

or

$$x = a - 2.$$

Since $\chi_{O_2} = 0.015$, then

$$0.015 = \frac{x}{1 + 2 + x + 3.76a}.$$

Solving the above two relationships simultaneously yields

$$x = 0.1699 \quad \text{and} \quad a = 2.1699.$$

For stoichiometric ($\Phi = 1$) conditions, $a = 2$; thus, the operating equivalence ratio is simply

$$\Phi = \frac{2}{a} = \frac{2}{2.1699} = 0.9217,$$

which follows from Eqns. 2.32 and 2.33a.

We consider first the case without FGR and employ the code HPFLAME to calculate the adiabatic flame temperature and the equilibrium $\chi_{NO}$. To use this code requires the reactants' enthalpy per mole of fuel:

$$H_{reac} = N_F \bar{h}_F + N_{O_2} \bar{h}_{O_2} + N_{N_2} \bar{h}_{N_2},$$

where

$$N_F = 1 \text{ (required by kmol-of-fuel basis)},$$

$$N_{O_2} = a = 2.1699,$$

$$N_{N_2} = 3.76a = 8.1589.$$

Using Appendix Tables B.1, A.11, and A.7, we evaluate the specific molar enthalpies of the $CH_4$, $O_2$, and $N_2$ at their respective temperatures to obtain the reactant mixture enthalpy:

$$H_{reac} = 1(-74,831) + 2.1699(3031) + 8.1589(2973)$$
$$= -43,997 \text{ kJ}.$$

Since the boiler operates at essentially atmospheric pressure, the required input file for HPFLAME is then

Adiabatic Flame Calculation for Specified Fuel, Phi, P, & Reactant
Enthalpy Using Olikara & Borman Equilibrium Routines
Problem Title: **EXAMPLE 2.10 Case without FGR**

| | |
|---|---|
| **01** | /CARBON ATOMS IN FUEL |
| **04** | /HYDROGEN ATOMS IN FUEL |
| **00** | /OXYGEN ATOMS IN FUEL |
| **00** | /NITROGEN ATOMS IN FUEL |
| **0.9217** | /EQUIVALENCE RATIO |
| **2000.** | /TEMPERATURE (K) (Initial Guess) |
| **101325.0** | /PRESSURE (Pa) |
| **−43997.0** | /ENTHALPY OF REACTANTS PER KMOL FUEL (kJ/kmol-fuel) |

Running HPFLAME yields

$$T_{ad} = 2209.8\text{K}$$
$$\chi_{NO} = 0.003497 \quad \text{or} \quad 3497 \text{ ppm}.$$

For the 15 percent FGR case, there are several ways to calculate the adiabatic flame temperature using the codes supplied with this book. Our approach will be first to determine the reactant mixture enthalpy (fuel, air, and FGR) per unit mass of mixture and then to employ Eqn. 2.40b:

$$h_{reac} = h_{prod}(T_{ad}).$$

We will use TPEQUIL to calculate $h_{prod}(T_{ad})$ for assumed values of $T_{ad}$ and iterate to obtain a final result.

As we have previously calculated the reactant fuel and air enthalpies, we need only to add to these the FGR enthalpy to obtain the total reactants' enthalpy; i.e.,

$$H_{reac} = N_F \bar{h}_F + N_{O_2}\bar{h}_{O_2} + N_{N_2}\bar{h}_{N_2} + N_{FGR}\bar{h}_{FGR},$$

Rather than calculating $\bar{h}_{FGR}$ by hand, we employ TPEQUIL with $T = 600$ K, $\Phi = 0.9217$, and $P = 101,325$ Pa. This yields

$$h_{FGR} = -2.499 \cdot 10^3 \text{ kJ/kg}$$

with

$$MW_{FGR} = 27.72 \text{ kg/kmol}.$$

Thus,

$$\bar{h}_{FGR} = h_{FGR}MW_{FGR} = -67,886 \text{ kJ/kmol}_{FGR}.$$

We obtain the number of moles of FGR from the definition

$$N_{FGR} = (N_F + N_A)\%FGR/100\%$$
$$= [1 + (2.1699)4.76]0.15$$
$$= 1.6993 \text{ kmol}.$$

The reactant mixture enthalpy is

$$H_{\text{reac}} = -43,997 + 1.6993(-67,886)$$
$$= -159,356 \,\text{kJ},$$

and the mass-specific enthalpy is

$$h_{\text{reac}} = \frac{H_{\text{reac}}}{m_{\text{reac}}} = \frac{H_{\text{reac}}}{N_F MW_F + N_A MW_A + N_{\text{FGR}} MW_{\text{FGR}}}$$
$$= \frac{-159,356}{1(16.043) + 10.329(28.85) + 1.6993(27.72)}$$
$$= -441.3 \,\text{kJ/kg}.$$

We now employ Eqn. 2.40b using TPEQUIL. Knowing that $T_{ad}$ must be less than the zero-FGR value of 2209.8 K, we guess at 2100 K. With this temperature as input to TPEQUIL, the output value for $h_{\text{prod}}$ (2100 K) is $-348.0$ kJ/kg. A second guess of 2000 K yields $h_{\text{prod}}$ (2000 K) = $-519.2$ kJ/kg. With these two values bracketing the desired value, we linearly interpolate to obtain $T_{ad} = 2045.5$ K. Further iteration using TPEQUIL yields the final value:

$$T_{ad} = 2046.5 \,\text{K}$$

with

$$\chi_{\text{NO}} = 0.002297 \quad \text{or} \quad 2297 \text{ ppm.}$$

Comparing these results with those without FGR, we see that 15 percent FGR results in an adiabatic flame temperature drop of approximately 163 K, and equilibrium NO mole fractions decrease by 34 percent.

## Comments

This problem can be solved directly using HPFLAME, provided the correct value of reactants' enthalpy is used. Since the code assumes that the reactants consist only of fuel and air, the FGR must be treated, on an element basis, to be fuel and air; thus, the number of moles of fuel in the reactants exceeds unity. The number of moles of "fuel" from the FGR is the number of moles of FGR divided by the number of moles of products formed per mole of $CH_4$ burned, as determined by the combustion equation given at the start of the solution; i.e.,

$$N_{\text{"Fuel" from FGR}} = \frac{N_{\text{FGR}}}{N_{\text{prod}}/N_{CH_4}},$$

and the appropriate input enthalpy for HPFLAME is

$$\bar{h}_{\text{reac}} (\text{kJ/kmol}_{\text{"fuel"}}) = \frac{H_{\text{reac}}}{1 + N_{\text{"Fuel" from FGR}}}.$$

For our specific problem, $N_{\text{"Fuel" from FGR}} = 1.6993(0.0883) = 0.15$ and $\bar{h}_{\text{reac}} = -159,356/1.15 = -138,571$ kJ/kmol$_{\text{fuel}}$. Using this value with HPFLAME yields the same result for $T_{ad}$ as obtained above, but without iteration. For the constant-volume combustion process (see Example 2.9), the code UVFLAME works directly on a mass basis using the values input for the total number of moles and the molecular weight of the reactants; therefore, no correction to the reactant enthalpy is required when dealing with FGR or EGR when using UVFLAME.

We note that the NO mole fractions calculated are *equilibrium* values. In a practical application, the NO concentrations in the combustion chamber and the flue are determined by chemical kinetics (Chapters 4 and 5) as, in most cases, insufficient time is available to achieve equilibrium; nevertheless, the differences between the equilibrium NO values with and without FGR indicate the potential for the reduction of kinetically determined NO by FGR (see Chapter 15).

## SUMMARY

All of the concepts presented in this chapter are of fundamental importance to a study of combustion. We began this chapter with a brief review of basic thermodynamic properties of simple substances and ideal-gas mixtures. We also reviewed the conservation of energy principle, the first law of thermodynamics. The first law in its various forms should be an old friend by now. You should be familiar with the equivalence ratio and how it is used to define rich, lean, and stoichiometric mixtures. Other important thermodynamic properties defined include standardized enthalpies, which are used with the first law to define enthalpies of reaction, heating values, and constant-pressure and constant-volume adiabatic flame temperatures. You should be able to illustrate all of these properties graphically on appropriate thermodynamic coordinates ($h$–$T$ or $u$–$T$). In our discussion of chemical equilibrium, we introduced the Gibbs function and demonstrated its utility in calculating the equilibrium composition of ideal-gas mixtures. You should be able to calculate equilibrium compositions for simple mixtures using equilibrium constants ($K_p$), together with element conservation. You should also be able to formulate the more complex problems; however, a familiarity with one or more equilibrium computer codes is useful for their solution. The final topics in this chapter dealt with the variation of combustion product mixture composition with equivalence ratio and the importance of dissociation. You should understand which species may be considered major and which minor, and have an appreciation for the order of magnitude of the mole fractions of the 11 important species we considered. Also shown was the usefulness of water-gas equilibrium for dealing with rich product mixtures in a simplified way. You should be able to calculate the composition of combustion product mixtures assuming no dissociation. Although many seemingly diverse topics were discussed, you should appreciate how the principles of the first and second laws of thermodynamics underpin these topics.

## NOMENCLATURE

| | |
|---|---|
| $a$ | Molar oxygen–fuel ratio (kmol/kmol) |
| $A/F$ | Mass air–fuel ratio (kg/kg) |
| $c_p, \bar{c}_p$ | Constant-pressure specific heat (J/kg-K or J/kmol-K) |
| $c_v, \bar{c}_v$ | Constant-volume specific heat (J/kg-K or J/kmol-K) |
| $E, e$ | Total energy (J or J/kg) |
| $F/A$ | Mass fuel–air ratio (kg/kg) |
| $g$ | Gravitational acceleration (m/s$^2$) |

| $\overline{g}^o$ | Pure-species Gibbs function (J/kmol) |
|---|---|
| $\overline{g}_f^o$ | Gibbs function of formation (J/kmol) |
| $G, \overline{g}$ | Gibbs function or free energy (J or J/kmol) |
| $\Delta G^o$ | Standard-state Gibbs function change, Eqn. 2.64 (J/kmol) |
| $h_f^o, \overline{h}_f^o$ | Enthalpy of formation (J/kg or J/kmol) |
| $H, h, \overline{h}$ | Enthalpy (J or J/kg or J/kmol) |
| $\Delta H_c, \Delta h_c, \Delta \overline{h}_c$ | Heat of combustion (heating value) (J or J/kg or J/kmol) |
| $\Delta H_R, \Delta h, \Delta \overline{h}_R$ | Enthalpy of reaction (J or J/kg or J/kmol) |
| HHV | Higher heating value (J/kg) |
| $K_p$ | Equilibrium constant, Eqn. 2.65 (dimensionless) |
| LHV | Lower heating value (J/kg) |
| $m$ | Mass (kg) |
| $\dot{m}$ | Mass flowrate (kg/s) |
| $MW$ | Molecular weight (kg/kmol) |
| $N$ | Number of moles (kmol) |
| $P$ | Pressure (Pa) |
| $Q, q$ | Heat (J or J/kg) |
| $\dot{Q}, \dot{q}$ | Heat-transfer rate (J/s = W or W/kg) |
| $R$ | Specific gas constant (J/kg-K) |
| $R_u$ | Universal gas constant (J/kmol-K) |
| $S, s, \overline{s}$ | Entropy (J/K or J/kg-K or J/kmol-K) |
| $t$ | Time (s) |
| $T$ | Temperature (K) |
| $U, u, \overline{u}$ | Internal energy (J or J/kg or J/kmol) |
| $v$ | Velocity (m/s) |
| $V, v$ | Volume ($m^3$ or $m^3$/kg) |
| $W, w$ | Work (J or J/kg) |
| $\dot{W}, \dot{w}$ | Rate of work or power (J/s = W or W/kg) |
| $x$ | Number of carbon atoms in fuel |
| $y$ | Number of hydrogen atoms in fuel |
| $Y$ | Mass fraction (kg/kg) |
| $z$ | Elevation (m) |
| $Z$ | Element ratio |

### *Greek Symbols*

| $\alpha$ | Fraction dissociated |
|---|---|
| $\kappa$ | Proportionality constant, Eqn. 2.61 |
| $\rho$ | Density (kg/$m^3$) |
| $\Phi$ | Equivalence ratio |
| $\chi$ | Mole fraction |

### *Subscripts*

| $ad$ | Adiabatic |
|---|---|
| $A$ | Air |
| $cv$ | Control volume |

| | |
|---|---|
| $f$ | Final or formation |
| $F$ | Fuel |
| $g$ | Gas |
| $i$ | $i$th species or inlet |
| init | Initial |
| $l$ | Liquid |
| mix | Mixture |
| $o$ | Outlet |
| prod | Product |
| reac | Reactant |
| ref | Reference |
| $s$ | Sensible |
| sat | Saturation state |
| stoic | Stoichiometric |
| $T$ | At temperature $T$ |

**Superscripts**

| | |
|---|---|
| $o$ | Denotes standard-state pressure ($P^o = 1$ atm) |

## REFERENCES

1. Kee, R. J., Rupley, F. M., and Miller, J. A., "The Chemkin Thermodynamic Data Base," Sandia National Laboratories Report SAND87-8215 B, March 1991.

2. Moran, M. J., and Shapiro, H. N., *Fundamentals of Engineering Thermodynamics,* 5th Ed., Wiley, New York, 2004.

3. Wark, K., Jr., *Thermodynamics,* 6th Ed., McGraw-Hill, New York, 1999.

4. Turns, S. R., *Thermodynamics: Concepts and Applications,* Cambridge University Press, New York, 2006.

5. Gordon, S., and McBride, B. J., "Computer Program for Calculation of Complex Chemical Equilibrium Compositions, Rocket Performance, Incident and Reflected Shocks, and Chapman-Jouguet Detonations," NASA SP-273, 1976. See also Glenn Research Center, "Chemical Equilibrium with Applications," http://www.grc.gov/WWW/CEAWeb /ceaHome.htm. Accessed 1/27/2009.

6. Stull, D. R., and Prophet, H., "JANAF Thermochemical Tables," 2nd Ed., NSRDS-NBS 37, National Bureau of Standards, June 1971. (The 4th Ed. is available from NIST.)

7. Pope, S. B., "Gibbs Function Continuation for the Stable Computation of Chemical Equilibrium," *Combustion and Flame*, 139: 222–226 (2004).

8. Olikara, C., and Borman, G. L., "A Computer Program for Calculating Properties of Equilibrium Combustion Products with Some Applications to I. C. Engines," SAE Paper 750468, 1975.

9. Morley, C., "GASEQ—A Chemical Equilibrium Program for Windows," http://www. gaseq.co.uk/. Accessed 2/3/2009.

10. Dandy, D. S., "Chemical Equilibrium Calculation," http://navier.engr.colostate.edu/tools/ equil.html. Accessed 2/3/2009.

11. Industrial Heating Equipment Association, *Combustion Technology Manual*, 4th Ed., IHEA, Arlington, VA, 1988.

## REVIEW QUESTIONS

**1.** Make a list of all of the boldfaced words in Chapter 2. Make sure you understand the meaning of each.

**2.** Describe the temperature dependence of the specific heats of monatomic and polyatomic gases. What is the underlying cause of this dependence? What implications does the temperature dependence have for combustion?

**3.** Why is the equivalence ratio frequently more meaningful than the air–fuel (or fuel–air) ratio when comparing different fuels?

**4.** What three conditions define the standard reference state?

**5.** Sketch a graph showing $H_{reac}$ and $H_{prod}$ as functions of temperature, taking into account nonconstant specific heats.

**6.** Using your sketch from question 5, illustrate the effect of preheating the reactants on the constant-pressure adiabatic flame temperature.

**7.** Describe the effect of pressure on the equilibrium mole fractions for the following reactions:

$$O_2 \Leftrightarrow 2O$$
$$N_2 + O_2 \Leftrightarrow 2NO$$
$$CO + \tfrac{1}{2}O_2 \Leftrightarrow CO_2$$

What is the effect of temperature?

**8.** Make a list of the major and minor species associated with combustion products at high temperature, ranking them from the highest to the lowest concentrations (mole fraction) and giving an approximate numerical value for each at $\Phi = 0.7$. Repeat for $\Phi = 1.3$. Compare your lists.

**9.** What is the significance of the water-gas shift reaction?

**10.** Describe the effect of increasing temperature on the equilibrium composition of combustion products.

11.  Describe the effect of increasing pressure on the equilibrium composition of combustion products.

12.  Why does flue-gas recirculation decrease flame temperatures? What happens if the flue gas recirculated is at the flame temperature?

## PROBLEMS

**2.1**  Determine the mass fraction of $O_2$ and $N_2$ in air, assuming the molar composition is 21 percent $O_2$ and 79 percent $N_2$.

**2.2**  A mixture is composed of the following number of moles of various species:

| Species | No. of moles |
|---------|--------------|
| CO | 0.095 |
| $CO_2$ | 6 |
| $H_2O$ | 7 |
| $N_2$ | 34 |
| NO | 0.005 |

A.  Determine the mole fraction of nitric oxide (NO) in the mixture. Also, express your result as mole percent, and as parts-per-million.
B.  Determine the molecular weight of the mixture.
C.  Determine the mass fraction of each constituent.

**2.3**  Consider a gaseous mixture consisting of 5 kmol of $H_2$ and 3 kmol of $O_2$. Determine the $H_2$ and $O_2$ mole fractions, the molecular weight of the mixture, and the $H_2$ and $O_2$ mass fractions.

**2.4**  Consider a binary mixture of oxygen and methane. The methane mole fraction is 0.2. The mixture is at 300 K and 100 kPa. Determine the methane mass fraction in the mixture and the methane molar concentration in kmol of methane per $m^3$ of mixture.

**2.5**  Consider a mixture of $N_2$ and Ar in which there are three times as many moles of $N_2$ as there are moles of Ar. Determine the mole fractions of $N_2$ and Ar, the molecular weight of the mixture, the mass fractions of $N_2$ and Ar, and the molar concentration of $N_2$ in $kmol/m^3$ for a temperature of 500 K and a pressure of 250 kPa.

**2.6**  Determine the standardized enthalpy in $J/kmol_{mix}$ of a mixture of $CO_2$ and $O_2$ where $\chi_{CO_2} = 0.10$ and $\chi_{O_2} = 0.90$ at a temperature of 400 K.

**2.7**  Determine the molecular weight of a stoichiometric ($\Phi = 1.0$) methane–air mixture.

**2.8**  Determine the stoichiometric air–fuel ratio (mass) for propane ($C_3H_8$).

**2.9**   Propane burns in a premixed flame at an air–fuel ratio (mass) of 18:1. Determine the equivalence ratio $\Phi$.

**2.10**  For an equivalence ratio of $\Phi = 0.6$, determine the associated air–fuel ratios (mass) for methane, propane, and decane ($C_{10}H_{22}$).

**2.11**  In a propane-fueled truck, 3 percent (by volume) oxygen is measured in the exhaust stream of the running engine. Assuming "complete" combustion without dissociation, determine the air–fuel ratio (mass) supplied to the engine.

**2.12**  Assuming "complete" combustion, write out a stoichiometric balance equation, like Eqn. 2.30, for 1 mol of an arbitrary alcohol $C_xH_yO_z$. Determine the number of moles of air required to burn 1 mol of fuel.

**2.13**  Using the results of problem 2.12, determine the stoichiometric air–fuel ratio (mass) for methanol ($CH_3OH$). Compare your result with the stoichiometric ratio for methane ($CH_4$). What implications does this comparison have?

**2.14**  Consider a stoichiometric mixture of isooctane and air. Calculate the enthalpy of the mixture at the standard-state temperature (298.15 K) on a per-kmol-of-fuel basis ($kJ/kmol_{fuel}$), on a per-kmol-of-mixture basis ($kJ/kmol_{mix}$), and on a per-mass-of-mixture basis ($kJ/kg_{mix}$).

**2.15**  Repeat problem 2.14 for a temperature of 500 K.

**2.16**  Repeat problem 2.15, but now let the equivalence ratio $\Phi = 0.7$. How do these results compare with those of problem 2.15?

**2.17**  Consider a fuel which is an equimolar mixture of propane ($C_3H_8$) and natural gas ($CH_4$). Write out the complete stoichiometric combustion reaction for this fuel burning with air and determine the stoichiometric fuel–air ratio on a molar basis. Also, determine the molar air–fuel ratio for combustion at an equivalence ratio, $\Phi$, of 0.8.

**2.18**  Determine the enthalpy of the products of "ideal" combustion, i.e., no dissociation, resulting from the combustion of an isooctane–air mixture for an equivalence ratio of 0.7. The products are at 1000 K and 1 atm. Express your result using the following three bases: per kmol-of-fuel, per kg-of-fuel, and per kg-of-mixture. *Hint:* You may find Eqns. 2.68 and 2.69 useful; however, you should be able to derive these from atom-conservation considerations.

**2.19**  Butane ($C_4H_{10}$) burns with air at an equivalence ratio of 0.75. Determine the number of **moles** of air required per mole of fuel.

**2.20**  A glass melting furnace is burning ethene ($C_2H_4$) in pure oxygen (not air). The furnace operates at an equivalence ratio of 0.9 and consumes 30 kmol/hr of ethene.

   A.   Determine the energy input rate based on the LHV of the fuel. Express your result in both kW and Btu/hr.

   B.   Determine the $O_2$ consumption rate in kmol/hr and kg/s.

**2.21** Methyl alcohol ($CH_3OH$) burns with excess air at an air–fuel ratio (mass) of 8.0. Determine the equivalence ratio, $\Phi$, and the mole fraction of $CO_2$ in the product mixture assuming complete combustion, i.e., no dissociation.

**2.22** The lower heating value of vapor $n$-decane is 44,597 kJ/kg at $T = 298$ K. The enthalpy of vaporization of $n$-decane is 276.8 kJ/kg of $n$-decane. The enthalpy of vaporization of water at 298 K is 2442.2 kJ/kg of water.

    A. Determine the lower heating value of liquid $n$-decane. Use units of kJ/kg $n$-decane to express your result.

    B. Determine the higher heating value of vapor $n$-decane at 298 K.

**2.23** Determine the enthalpy of formation in kJ/kmol for methane, given the lower heating value of 50,016 kJ/kg at 298 K.

**2.24** Determine the standardized enthalpy of the mixture given in problem 2.2 for a temperature of 1000 K. Express your result in kJ/kmol of mixture.

**2.25** The lower heating value of methane is 50,016 kJ/kg (of methane). Determine the heating value:

    A. per mass of mixture.

    B. per mole of air–fuel mixture.

    C. per cubic meter of air–fuel mixture.

**2.26** The higher heating value for liquid octane ($C_8H_{18}$) at 298 K is 47,893 kJ/kg and the heat of vaporization is 363 kJ/kg. Determine the enthalpy of formation at 298 K for octane vapor.

**2.27** Verify the information in Table 2.1 under the headings $\Delta h_R$ (kJ/kg of fuel), $\Delta h_R$ (kJ/kg of mix), and $(O/F)_{stoic}$ for the following:

    A. $CH_4$–air

    B. $H_2$–$O_2$.

    C. $C(s)$–air.

    Note that any $H_2O$ in the product is assumed to be in the liquid state.

**2.28** Generate the same information requested in problem 2.27 for a stoichiometric mixture of $C_3H_8$ (propane) and air.

**2.29** Consider a liquid fuel. Draw a sketch on $h$–$T$ coordinates illustrating the following quantities: $h_l(T)$; $h_v(T)$; heat of vaporization, $h_{fg}$; heat of formation for fuel vapor; enthalpy of formation for fuel liquid; lower heating value; higher heating value.

**2.30** Determine the adiabatic flame temperature for constant-pressure combustion of a stoichiometric propane–air mixture assuming reactants at 298 K, no dissociation of the products, and constant specific heats evaluated at 298 K.

**2.31** Repeat problem 2.30, but using constant specific heats evaluated at 2000 K. Compare your result with that of problem 2.30 and discuss.

**2.32** Repeat problem 2.30, but now use property tables (Appendix A) to evaluate the sensible enthalpies.

**2.33*** Once more, repeat problem 2.30, but eliminate the unrealistic assumptions, i.e., allow for dissociation of the products and variable specific heats. Use HPFLAME (Appendix F), or other appropriate software. Compare and contrast the results of problems 2.30–2.33. Explain why they differ.

**2.34** Using the data in Appendix A, calculate the adiabatic constant-pressure flame temperature for a boiler operating with the fuel blend and equivalence ratio given in problem 2.17. Assume complete combustion to $CO_2$ and $H_2O$ and neglect any dissociation. Also, assume the heat capacity of the combustion products is constant evaluated at 1200 K. The boiler operates at 1 atm, and both the air and fuel enter at 298 K.

**2.35** Repeat problem 2.30, but for constant-volume combustion. Also, determine the final pressure.

**2.36*** Use the condition given in problem 2.33, but calculate the constant-volume adiabatic flame temperature using UVFLAME (Appendix F), or other appropriate software. Also, determine the final pressure. Compare your results with those of problem 2.35 and discuss.

**2.37** Derive the equivalent system (fixed mass) form of the first law corresponding to Eqn. 2.35, which is used to define the heat of reaction. Treat the system as constant pressure with initial and final temperatures equal.

**2.38** A furnace, operating at 1 atm, uses preheated air to improve its fuel efficiency. Determine the adiabatic flame temperature when the furnace is run at a mass air–fuel ratio of 18 for air preheated to 800 K. The fuel enters at 450 K. Assume the following simplified thermodynamic properties:

$$T_{ref} = 300 \text{ K,}$$

$$MW_{fuel} = MW_{air} = MW_{prod} = 29 \text{ kg/kmol,}$$

$$c_{p,\,fuel} = 3500 \text{ J/kg-K; } c_{p,\,air} = c_{p,\,prod} = 1200 \text{ J/kg-K,}$$

$$\bar{h}^o_{f,\,air} = \bar{h}^o_{f,\,prod} = 0,$$

$$\bar{h}^o_{f,\,fuel} = 1.16 \cdot 10^9 \text{ J/kmol.}$$

**2.39** Consider the constant-pressure, adiabatic combustion of a stoichiometric ($\Phi = 1$) fuel–air mixture where $(A/F)_{stoic} = 15$. Assume the following simplified properties for the fuel, air, and products, with $T_{ref} = 300$ K:

| | Fuel | Air | Products |
|---|---|---|---|
| $c_p$ (J/kg-K) | 3500 | 1200 | 1500 |
| $h^o_{f,300}$ (J/kg) | $2 \cdot 10^7$ | 0 | $-1.25 \cdot 10^6$ |

A. Determine the adiabatic flame temperature for a mixture initially at 600 K.

B. Determine the heating value of the fuel at 600 K. Give units.

**2.40**  Consider the combustion of hydrogen ($H_2$) with oxygen ($O_2$) in a steady-flow reactor as shown in the sketch. The heat loss through the reactor walls per unit mass flow ($\dot{Q}/\dot{m}$) is 187 kJ/kg. The equivalence ratio is 0.5 and the pressure is 5 atm.

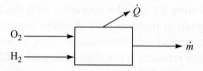

For a zero-Kelvin reference state, approximate enthalpies-of-formation are

$$\bar{h}^o_{f,H_2}(0) = \bar{h}^o_{f,O_2}(0) = 0 \text{ kJ/mol},$$

$$\bar{h}^o_{f,H_2O}(0) = -238{,}000 \text{ kJ/mol},$$

$$\bar{h}^o_{f,OH}(0) = -38{,}600 \text{ kJ/mol}.$$

A.  Determine the mean molecular weight of the combustion product gases in the outlet stream, assuming no dissociation.

B.  For the same assumption as in part A, determine the mass fractions of the species in the outlet stream.

C.  Determine the temperature in the product stream at the reactor outlet, again assuming no dissociation. Furthermore, assume that all species have the same constant molar specific heats, $\bar{c}_{p,i}$, equal to 40 kJ/kmol-K. The $H_2$ enters at 300 K and the $O_2$ at 800 K.

D.  Now assume that dissociation occurs, but that the only minor product is OH. Write out all of the equations necessary to calculate the outlet temperature. List the unknowns in your equation set.

**2.41**  Verify that the results given in Table 2.2 satisfy Eqns. 2.64 and 2.65 for the following conditions:

A.  $T = 2000$ K, $P = 0.1$ atm.

B.  $T = 2500$ K, $P = 100$ atm.

C.  $T = 3000$ K, $P = 1$ atm.

**2.42**  Consider the equilibrium reaction $O_2 \Leftrightarrow 2O$ in a closed vessel. Assume the vessel contains 1 mol of $O_2$ when there is no dissociation. Calculate the mole fractions of $O_2$ and O for the following conditions:

A.  $T = 2500$ K, $P = 1$ atm.

B.  $T = 2500$ K, $P = 3$ atm.

**2.43**  Repeat problem 2.42A, but add 1 mol of an inert diluent to the mixture, e.g., argon. What is the influence of the diluent? Discuss.

**2.44** Consider the equilibrium reaction $CO_2 \Leftrightarrow CO + \frac{1}{2}O_2$. At 10 atm and 3000 K, the equilibrium mole fractions of a particular mixture of $CO_2$, CO, and $O_2$ are 0.6783, 0.2144, and 0.1072, respectively. Determine the equilibrium constant $K_p$ for this situation.

**2.45** Consider the equilibrium reaction $H_2O \Leftrightarrow H_2 + \frac{1}{2}O_2$. At 0.8 atm, the mole fractions are $\chi_{H_2O} = 0.9$, $\chi_{H_2} = 0.03$, and $\chi_{O_2} = 0.07$. Determine the equilibrium constant $K_p$ for this situation.

**2.46** Consider the equilibrium reaction $H_2O + CO \Leftrightarrow CO_2 + H_2$ at a particular temperature $T$. At $T$, the enthalpies-of-formation of each species are as follows:

$$\bar{h}^o_{H_2O} = -251,7000 \text{ kJ/kmol}, \quad \bar{h}^o_{f,CO_2} = -396,600 \text{ kJ/kmol},$$

$$\bar{h}^o_{f,CO} = -118,700 \text{ kJ/kmol}, \quad \bar{h}^o_{f,H_2} = 0.$$

A. What is the effect of pressure on the equilibrium? Explain.

B. What is the effect of temperature on the equilibrium? Explain (calculation required).

**2.47** Calculate the equilibrium composition for the reaction $H_2 + \frac{1}{2}O_2 \Leftrightarrow H_2O$ when the ratio of the number of moles of elemental hydrogen to elemental oxygen is unity. The temperature is 2000 K, and the pressure is 1 atm.

**2.48\*** Calculate the equilibrium composition for the reaction $H_2O \Leftrightarrow H_2 + \frac{1}{2}O_2$ when the ratio of the number of moles of elemental hydrogen to elemental oxygen, Z, is varied. Let $Z = 0.5$, 1.0, and 2.0. The temperature is 2000 K, and the pressure is 1 atm. Plot your results and discuss. *Hint:* Use spreadsheet software to perform your calculations.

**2.49\*** Calculate the equilibrium composition for the reaction $H_2O \Leftrightarrow H_2 + \frac{1}{2}O_2$ when the ratio of the number of moles of elemental hydrogen to elemental oxygen, Z, is fixed at $Z = 2.0$, while the pressure is varied. Let $P = 0.5$, 1.0, and 2.0 atm. The temperature is 2000 K. Plot your results and discuss. *Hint:* Use spreadsheet software to perform your calculations.

**2.50** Reformulate problem 2.47 to include the species OH, O, and H. Identify the number of equations and the number of unknowns. They should, of course, be equal. Do not solve your system.

**2.51\*** Use STANJAN or other appropriate software to calculate the complete equilibrium for the H–O system using the conditions and atom constraints given in problem 2.47.

**2.52\*** For the conditions given below, list from highest to lowest the mole fractions of $CO_2$, CO, $H_2O$, $H_2$, OH, H, $O_2$, O, $N_2$, NO, and N. Also, give approximate values.

A. Propane–air constant-pressure combustion products at their adiabatic flame temperature for $\Phi = 0.8$.

B.   As in part A, but for $\Phi = 1.2$.

C.   Indicate which species may be considered major and which minor in parts A and B.

**2.53\***   Consider the adiabatic, constant-pressure combustion of n-decane ($C_{10}H_{22}$) with air for reactants at 298.15 K. Use HPFLAME (Appendix F) to calculate $T_{ad}$ and species mole fractions for $O_2$, $H_2O$, $CO_2$, $N_2$, CO, $H_2$, OH, and NO. Use equivalence ratios of 0.75, 1.00, and 1.25 and evaluate each condition for three pressure levels: 1, 10, and 100 atm. Construct a table showing your results and discuss the effects of equivalence ratio and pressure on $T_{ad}$ and the product mixture composition.

**2.54\***   Consider the combustion products of decane ($C_{10}H_{22}$) with air at an equivalence ratio of 1.25, pressure of 1 atm, and temperature of 2200 K. Estimate the mixture composition assuming no dissociation except for the water-gas shift equilibrium. Compare with results of TPEQUIL.

**2.55\***   A natural gas–fired industrial boiler operates with excess air such that the $O_2$ concentration in the flue gases is 2 percent (vol.), measured after removal of the moisture in the combustion products. The flue gas temperature is 700 K without air preheat.

A.   Determine the equivalence ratio for the system assuming that the properties of natural gas are the same as methane.

B.   Determine the thermal efficiency of the boiler, assuming that both the air and fuel enter at 298 K.

C.   With air preheat, the flue gases are at 433 K (320°F) after passing through the air preheater. Again, determine the thermal efficiency of the boiler for both air and fuel entering at the preheater and burner, respectively, at 298 K.

D.   Assuming premixed operation of the burners, estimate the maximum temperature in the combustion space ($P = 1$ atm) with air preheat.

**2.56**   The equivalence ratio of a combustion process is often determined by extracting a sample of the exhaust gas and measuring the concentrations of major species. In a combustion experiment using isooctane ($C_8H_{18}$), continuous gas analyzers monitor the exhaust gas and measure a $CO_2$ concentration of 6 percent by volume and a CO concentration of 1 percent by volume. The sample gas is not dried before the measurements are made.

A.   What is the equivalence ratio associated with this combustion process? Assume the process is overall lean.

B.   If an $O_2$ analyzer was monitoring the exhaust gas, what would it be reading?

**2.57**   An inventor has devised an atmospheric-pressure process to manufacture methanol. The inventor claims he has developed a catalyst that promotes the economical reaction of CO and $H_2$ to yield methanol; however, a cheap supply of CO and $H_2$ is needed. The inventor proposes burning natural gas ($CH_4$) in oxygen under fuel-rich conditions to yield a gas mixture of CO, $CO_2$, $H_2O$, and $H_2$.

A.  If methane burns in oxygen at an equivalence ratio $\Phi = 1.5$, and the combustion reactions go to equilibrium, what will be the resulting gas composition? Assume the combustion temperature is controlled to 1500 K.

B.  What would be the composition if the temperature were controlled to 2500 K?

**2.58*** Consider the combustion of 1 kmol of propane with air at 1 atm. Construct a single graph using $H$–$T$ coordinates that shows the following:

A.  Reactants' enthalpy, $H$, in kJ versus temperature, over the range of 298–800 K for $\Phi = 1.0$.

B.  Repeat part A for $\Phi = 0.75$.

C.  Repeat part A for $\Phi = 1.25$.

D.  Products' enthalpy, $H$, for ideal combustion (no dissociation) versus temperature, over the range of 298–3500 K for $\Phi = 1.0$.

E.  Repeat part D for $\Phi = 0.75$.

F.  Repeat part D for $\Phi = 1.25$, using the water-gas equilibrium to account for incomplete combustion.

**2.59** Using the graph constructed in problem 2.58, estimate the constant-pressure adiabatic flame temperatures for the following conditions:

A.  For reactants at 298 K with $\Phi = 0.75$, 1.0, and 1.25.

B.  For $\Phi = 1.0$ with reactants' temperatures of 298 K, 600 K, and 800 K.

C.  Discuss your results from parts A and B.

**2.60*** Repeat problem 2.58, but use the code TPEQUIL (Appendix F) to calculate the products' $H$ versus $T$ curves. Use the same scales as you did in problem 2.58 so that the results can be overlaid for comparison. Discuss the differences associated with the product enthalpy curves for the ideal combustion case compared with the equilibrium case. *Hint:* Make sure the basis for all the enthalpies is per mole of methane. You will have to convert the results from TPEQUIL to this basis.

**2.61** Repeat parts A and B of problem 2.59 using the graph obtained in problem 2.60. Compare your results with those of problem 2.59 and discuss.

**2.62*** Use the code HPFLAME (Appendix F) to determine the adiabatic flame temperature for the conditions given in parts A and B of problem 2.59. Compare your results with those of problems 2.59 and 2.61. Discuss.

**2.63** A furnace uses preheated air to improve its fuel efficiency. Determine the adiabatic flame temperature when the furnace is operating at a mass air–fuel ratio of 16 for air preheated to 600 K. The fuel enters at 300 K. Assume the following simplified thermodynamic properties:

$$T_{ref} = 300 \text{ K},$$

$$MW_{fuel} = MW_{air} = MW_{prod} = 29 \text{ kg/kmol},$$

$$c_{p,\,fuel} = c_{p,\,air} = c_{p,\,prod} = 1200 \text{ J/kg-K},$$

$$h^o_{f,\,air} = h^o_{f,\,prod} = 0,$$

$$h^o_{f,\,fuel} = 4 \cdot 10^7 \text{ J/kg}.$$

**2.64*** In one strategy to decrease the amount of oxides of nitrogen ($NO_x$) formed and emitted from boilers, a portion of the flue gases is recirculated and introduced with the air and fuel, as shown in Fig. 2.20a. The effect of the recirculated gases is to decrease the maximum temperatures in the flame zone. Decreased flame temperatures result in less $NO_x$ being formed. To increase the effectiveness of a given amount of recycled gases, the gases may be cooled. Your job is to determine what combinations of percent FGR and $T_{FGR}$ result in maximum (adiabatic) flame temperatures of approximately 1950 K.

Your design should be based on the following constraints and assumptions: the fuel enters the burner at 298 K and 1 atm; the air enters the burner at 325 K and 1 atm; the oxygen ($O_2$) mole fraction in the cold, i.e., undissociated, flue gases is $\chi_{O_2} = 0.02$; the flue gas composition can be approximated as "complete" combustion products for all conditions, with the equivalence ratio determined from the flue-gas $O_2$ content; the percent FGR is defined as the molar percentage of fuel and air supplied; the natural gas can be treated as methane; and the maximum flue-gas temperature is 1200 K.

Use graphs and tables as appropriate to present your results. Also, discuss the practical ramifications of adding FGR (pumping requirements, capital equipment costs, etc.). How might these considerations affect the choice of operating conditions (%FGR, $T_{EGR}$)?

# 3

# Introduction to Mass Transfer

## OVERVIEW

As indicated in Chapter 1, understanding combustion requires a combined knowledge of thermodynamics (Chapter 2), heat and mass transfer, and chemical reaction rate theory, or chemical kinetics (Chapter 4). Since most readers of this book are unlikely to have had much, if any, exposure to the subject of mass transfer, we present in this chapter a brief introduction to this topic. Mass transfer, a fundamental topic in chemical engineering, is quite complex, much more so than is suggested by the following discussion. We provide here only a rudimentary treatment of the fundamental rate laws and conservation principles governing mass transfer, leaving a more comprehensive treatment of mass transfer until Chapter 7 and to other textbooks [1–4]. To develop some physical insight into mass transfer, we briefly examine the process from a molecular point of view. This has the added advantage of showing the fundamental similarity of mass transfer and heat conduction in gases. Lastly, we illustrate the application of mass transfer concepts to the mathematical descriptions of the evaporation of a liquid layer and a droplet.

## RUDIMENTS OF MASS TRANSFER

Imagine opening a bottle of perfume and placing the opened bottle in the center of a room. Using your nose as a sensor, the presence of perfume molecules in the immediate vicinity of the bottle can be detected shortly after the bottle is opened. At a later time, you will find the perfume odor everywhere in the room. The processes whereby the perfume molecules are transported from a region of high concentration, near the bottle, to a region of low concentration, far from the bottle, are the subject of **mass transfer.** Like heat transfer and momentum transfer, mass may be transported by molecular processes (e.g., collisions in an ideal gas) and/or turbulent processes. The molecular processes are relatively slow and operate on small spatial scales, whereas turbulent transport depends upon the velocity and size of an eddy carrying the transported material. Our focus here is on molecular transport, while Chapters 11, 12, and 13 deal with the turbulent processes.

## Mass Transfer Rate Laws

**Fick's Law of Diffusion**   Let us consider a nonreacting gas mixture comprising just two molecular species: species A and B. **Fick's law** describes the rate at which one species diffuses through the other. For the case of **one-dimensional binary diffusion,** Fick's law on a mass basis is

$$\dot{m}_A'' \qquad = \qquad Y_A(\dot{m}_A'' + \dot{m}_B'') \qquad - \qquad \rho \mathcal{D}_{AB} \frac{dY_A}{dx}, \qquad (3.1)$$

| Mass flow of species A per unit area | Mass flow of species A associated with bulk flow per unit area | Mass flow of species A associated with molecular diffusion per unit area |

where $\dot{m}_A''$ is the mass flux of species A and $Y_A$ is the mass fraction. In this book, the **mass flux** is defined as the mass flowrate of species A per unit area perpendicular to the flow:

$$\dot{m}_A'' = \dot{m}_A / A. \qquad (3.2)$$

The units of $\dot{m}_A''$ are kg/s-m$^2$. The idea of a "flux" should be familiar to you since the "heat flux" is the rate at which energy is transported per unit area; i.e., $\dot{Q}'' = \dot{Q}/A$ with units of J/s-m$^2$ or W/m$^2$. The **binary diffusivity,** $\mathcal{D}_{AB}$, is a property of the mixture and has units of m$^2$/s. Values for some binary diffusion coefficients, i.e., diffusivities, at 1 atm are provided in Appendix D.

Equation 3.1 states that species A is transported by two means: the first term on the right-hand side representing the transport of A resulting from the bulk motion of the fluid, and the second term representing the diffusion of A superimposed on the bulk flow. In the absence of diffusion, we obtain the obvious result that

$$\dot{m}_A'' = Y_A(\dot{m}_A'' + \dot{m}_B'') = Y_A \dot{m}'' \equiv \text{Bulk flux of species A,} \qquad (3.3a)$$

where $\dot{m}''$ is the mixture mass flux. The diffusional flux adds an additional component to the flux of A:

$$-\rho \mathcal{D}_{AB} \frac{dY_A}{dx} \equiv \text{Diffusional flux of species A, } \dot{m}_{A,\text{diff}}''. \qquad (3.3b)$$

This expression says that the **diffusional flux of A,** $\dot{m}_{A,\text{diff}}''$, is proportional to the gradient of the mass fraction, where the constant of proportionality is $-\rho \mathcal{D}_{AB}$. Thus, we see that species A preferentially moves from a region of high concentration to a region of low concentration, analogous to energy traveling in the direction from high temperature to low. Note that the negative sign causes the flux to be positive in the x-direction when the concentration gradient is negative. An analogy between the diffusion of mass and the diffusion of heat (conduction) can be drawn by comparing **Fourier's law of conduction,**

$$\dot{Q}''_x = -k \frac{dT}{dx},$$ (3.4)

with Fick's law of diffusion in the absence of bulk flow, Eqn. 3.3b. Both expressions indicate a flux ($\dot{m}''_{A,\,diff}$ or $\dot{Q}''_x$) being proportional to the gradient of a scalar quantity (($dY_A/dx$) or ($dT/dx$)). We will explore this analogy further when we discuss the physical significance of the **transport properties,** $\rho D$ and $k$, the proportionality constants in Eqns. 3.3b and 3.4, respectively.

Equation 3.1 is the one-dimensional component of the more general expression

$$\dot{m}''_A = Y_A(\dot{m}''_A + \dot{m}''_B) - \rho D_{AB} \nabla Y_A,$$ (3.5)

where the bold symbols represent vector quantities. In many instances, the molar form of Eqn. 3.5 is useful:

$$\dot{N}''_A = \chi_A(\dot{N}''_A + \dot{N}''_B) - c D_{AB} \nabla \chi_A,$$ (3.6)

where $\dot{N}''_A$ is the molar flux (kmol/s-m$^2$) of species A, $\chi_A$ is the mole fraction, and $c$ is the mixture molar concentration (kmol$_{mix}$/m$^3$).

The meanings of bulk flow and diffusional flux become clearer if we express the total mass flux for a binary mixture as the sum of the mass flux of species A and the mass flux of species B:

$$\dot{m}'' \quad = \quad \dot{m}''_A \quad + \quad \dot{m}''_B.$$ (3.7)

|  Mixture mass flux  |  Species A mass flux  |  Species B mass flux  |
|---|---|---|

The mixture mass flux on the left-hand side of Eqn. 3.7 is the total mixture flowrate $\dot{m}$ per unit of area perpendicular to the flow. This is the $\dot{m}$ that you are familiar with from previous studies in thermodynamics and fluid mechanics. Assuming one-dimensional flow for convenience, we now substitute the appropriate expressions for the individual species mass fluxes (Eqn. 3.1) into Eqn. 3.7 and obtain

$$\dot{m}'' = Y_A \dot{m}'' - \rho D_{AB} \frac{dY_A}{dx} + Y_B \dot{m}'' - \rho D_{BA} \frac{dY_B}{dx}$$ (3.8a)

or

$$\dot{m}'' = (Y_A + Y_B)\dot{m}'' - \rho D_{AB} \frac{dY_A}{dx} - \rho D_{BA} \frac{dY_B}{dx}.$$ (3.8b)

For a binary mixture, $Y_A + Y_B = 1$ (Eqn. 2.10); thus,

$$-\rho \mathcal{D}_{AB} \frac{dY_A}{dx} \underbrace{\qquad}_{\substack{\text{Diffusional flux} \\ \text{of species A}}} - \; \rho \mathcal{D}_{BA} \frac{dY_B}{dx} \underbrace{\qquad}_{\substack{\text{Diffusional flux} \\ \text{of species B}}} = 0, \qquad (3.9)$$

i.e., the sum of the diffusional fluxes of the species is zero. In general, overall mass conservation requires that $\Sigma \dot{m}''_{i,\,\text{diff}} = 0$.

It is important at this point to emphasize that we are assuming a binary gas and that species diffusion is a result of concentration gradients only, which is termed **ordinary diffusion.** Real mixtures of interest in combustion contain many components, not just two. Our binary gas assumption, however, allows us to understand the essential physics of many situations without the complications inherent in an analysis of multicomponent diffusion. Also, gradients of temperature and pressure can produce species diffusion, i.e., the **thermal diffusion (Soret)** and **pressure diffusion** effects, respectively. In many systems of interest, these effects are usually small, and, again, their neglect allows us to understand more clearly the essential physics of a problem.

**Molecular Basis of Diffusion**    To obtain insight into the molecular processes that result in the macroscopic laws of mass diffusion (Fick's law) and heat diffusion or conduction (Fourier's law), we apply some concepts from the kinetic theory of gases (e.g., [5, 6]). Consider a stationary (no bulk flow) plane layer of a binary gas mixture consisting of rigid, nonattracting molecules in which the molecular mass of each species A and B is essentially equal. A concentration (mass-fraction) gradient exists in the gas layer in the $x$-direction and is sufficiently small that the mass-fraction distribution can be considered linear over a distance of a few molecular mean free paths, $\lambda$, as illustrated in Fig. 3.1. With these assumptions, we can define the following average molecular properties derived from kinetic theory [1, 5, 6]:

$$\bar{v} \equiv \begin{array}{c} \text{Mean speed of} \\ \text{species A molecules} \end{array} = \left( \frac{8k_B T}{\pi m_A} \right)^{1/2}, \qquad (3.10a)$$

$$Z''_A \equiv \begin{array}{c} \text{Wall collision frequency} \\ \text{of A molecules per unit area} \end{array} = \frac{1}{4} \left( \frac{n_A}{V} \right) \bar{v}, \qquad (3.10b)$$

$$\lambda \equiv \text{Mean free path} = \frac{1}{\sqrt{2}\pi \left( \frac{n_{\text{tot}}}{V} \right) \sigma^2}, \qquad (3.10c)$$

$$a \equiv \begin{array}{c} \text{Average perpendicular distance} \\ \text{from plane of last collision to plane} \\ \text{where next collision occurs} \end{array} = \frac{2}{3}\lambda, \qquad (3.10d)$$

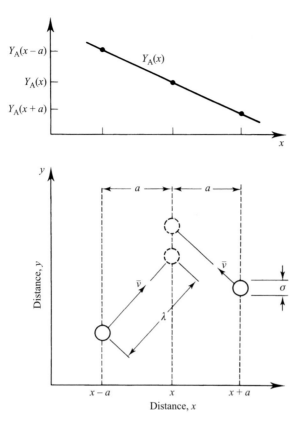

**Figure 3.1**     Schematic diagram illustrating mass diffusion of species A molecules from a region of high concentration to one of low concentration. Mass-fraction distribution is shown at the top.

where $k_B$ is Boltzmann's constant, $m_A$ the mass of a single A molecule, $n_A/V$ the number of A molecules per unit volume, $n_{tot}/V$ the total number of molecules per unit volume, and $\sigma$ the diameter of both A and B molecules.

Assuming no bulk flow for simplicity, the net flux of A molecules at the $x$-plane is the difference between the flux of A molecules in the positive $x$-direction and the flux of A molecules in the negative $x$-direction:

$$\dot{m}_A'' = \dot{m}_{A,(+)x\text{-dir}}'' - \dot{m}_{A,(-)x\text{-dir}}'', \tag{3.11}$$

which, when expressed in terms of the collision frequency, becomes

$$\dot{m}_A'' \quad = \quad (Z_A'')_{x-a} m_A \quad - \quad (Z_A'')_{x+a} m_A. \tag{3.12}$$

$$\begin{pmatrix} \text{Net mass} \\ \text{flux of} \\ \text{species A} \end{pmatrix} \begin{pmatrix} \text{Number of A} \\ \text{molecules crossing} \\ \text{plane at } x \text{ originating} \\ \text{from plane at } x-a \\ \text{per unit time and area} \end{pmatrix} \begin{pmatrix} \text{Mass of} \\ \text{single} \\ \text{molecule} \end{pmatrix} \begin{pmatrix} \text{Number of A} \\ \text{molecules crossing} \\ \text{plane at } x \text{ originating} \\ \text{from plane at } x+a \\ \text{per unit time and area} \end{pmatrix} \begin{pmatrix} \text{Mass of} \\ \text{single} \\ \text{molecule} \end{pmatrix}$$

We can use the definition of density ($\rho \equiv m_{\text{tot}}/V_{\text{tot}}$) to relate $Z_A''$ (Eqn. 3.10b) to the mass fraction of A molecules:

$$Z_A'' m_A = \frac{1}{4} \frac{n_A m_A}{m_{\text{tot}}} \rho \bar{v} = \frac{1}{4} Y_A \rho \bar{v}. \tag{3.13}$$

Substituting Eqn. 3.13 into Eqn. 3.12 and treating the mixture density and mean molecular speeds as constants yields

$$\dot{m}_A'' = \frac{1}{4} \rho \bar{v}(Y_{A,\,x-a} - Y_{A,\,x+a}). \tag{3.14}$$

With our assumption of a linear concentration distribution,

$$\frac{dY_A}{dx} = \frac{Y_{A,\,x+a} - Y_{A,\,x-a}}{2a} = \frac{Y_{A,\,x+a} - Y_{A,\,x-a}}{4\lambda/3}. \tag{3.15}$$

Solving Eqn. 3.15 for the concentration difference and substituting into Eqn. 3.14, we obtain our final result:

$$\dot{m}_A'' = -\rho \frac{\bar{v}\lambda}{3} \frac{dY_A}{dx}. \tag{3.16}$$

Comparing Eqn. 3.16 with Eqn. 3.3b, we identify the binary diffusivity $\mathcal{D}_{AB}$ as

$$\mathcal{D}_{AB} = \frac{\bar{v}\lambda}{3}. \tag{3.17}$$

Using the definitions of the mean molecular speed (Eqn. 3.10a) and mean free path (Eqn. 3.10c), together with the ideal-gas equation of state $PV = nk_B T$, the temperature and pressure dependence of $\mathcal{D}_{AB}$ can easily be determined, i.e.,

$$\mathcal{D}_{AB} = \frac{2}{3}\left(\frac{k_B^3 T}{\pi^3 m_A}\right)^{1/2} \frac{T}{\sigma^2 P} \tag{3.18a}$$

or

$$\mathcal{D}_{AB} \propto T^{3/2}P^{-1}. \tag{3.18b}$$

Thus, we see that the diffusivity depends strongly on temperature (to the $\frac{3}{2}$ power) and varies inversely with pressure. The mass flux of species A, however, depends on the product $\rho \mathcal{D}_{AB}$, which has a square-root temperature dependence and is independent of pressure:

$$\rho \mathcal{D}_{AB} \propto T^{1/2}P^0 = T^{1/2}. \tag{3.18c}$$

In many simplified analyses of combustion processes, the weak temperature dependence is neglected and $\rho \mathcal{D}$ is treated as a constant.

**Comparison with Heat Conduction**    To see clearly the relationship between mass and heat transfer, we now apply kinetic theory to the transport of energy. We assume a homogeneous gas consisting of rigid nonattracting molecules in which a temperature gradient exists. Again, the gradient is sufficiently small that the temperature distribution is essentially linear over several mean free paths, as illustrated in Fig. 3.2. The mean molecular speed and mean free path have the same definitions as given in Eqns. 3.10a and 3.10c, respectively; however, the molecular collision frequency of interest is now based on the total number density of molecules, $n_{tot}/V$, i.e.,

$$Z'' \equiv \frac{\text{Average wall collision}}{\text{frequency per unit area}} = \frac{1}{4}\left(\frac{n_{tot}}{V}\right)\bar{v}. \tag{3.19}$$

In our no-interaction-at-a-distance hard-sphere model of the gas, the only energy storage mode is molecular translational, i.e., kinetic, energy. We write an energy balance at the $x$-plane (see Fig. 3.2) where the net energy flow (per unit area) in the $x$-direction is the difference between the kinetic energy flux associated with molecules traveling from $x - a$ to $x$ and those traveling from $x + a$ to $x$:

$$\dot{Q}''_x = Z''(ke)_{x-a} - Z''(ke)_{x+a}. \tag{3.20}$$

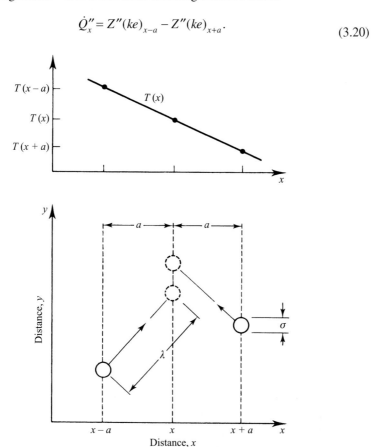

**Figure 3.2** Schematic diagram illustrating energy transfer (heat conduction) associated with molecular motion in a gas. The temperature distribution is shown at the top.

Because the mean kinetic energy of a molecule is given by [5]

$$ke = \frac{1}{2}m\bar{v}^2 = \frac{3}{2}k_B T,$$ (3.21)

the heat flux can be related to the temperature as

$$\dot{Q}''_x = \frac{3}{2}k_B Z''(T_{x-a} - T_{x+a}).$$ (3.22)

The temperature difference in Eqn. 3.22 relates to the temperature gradient following the same form as Eqn. 3.15, i.e.,

$$\frac{dT}{dx} = \frac{T_{x+a} - T_{x-a}}{2a}.$$ (3.23)

Substituting Eqn. 3.23 into Eqn. 3.22, employing the definitions of $Z''$ and $a$, we obtain our final result for the heat flux:

$$\dot{Q}''_x = -\frac{1}{2}k_B\left(\frac{n}{V}\right)\bar{v}\lambda\frac{dT}{dx}.$$ (3.24)

Comparing the above with Fourier's law of heat conduction (Eqn. 3.4), we can identify the thermal conductivity $k$ as

$$k = \frac{1}{2}k_B\left(\frac{n}{V}\right)\bar{v}\lambda.$$ (3.25)

Expressed in terms of $T$ and molecular mass and size, the thermal conductivity is

$$k = \left(\frac{k_B^3}{\pi^3 m\sigma^4}\right)^{1/2} T^{1/2}.$$ (3.26)

The thermal conductivity is thus proportional to the square root of temperature,

$$k \propto T^{1/2},$$ (3.27)

as is the $\rho\mathcal{D}_{AB}$ product. For real gases, the true temperature dependence is greater.

## Species Conservation

In this section, we employ the rate law of species transport (Fick's law) to develop a basic species mass conservation expression. Consider the one-dimensional control volume of Fig. 3.3, a plane layer $\Delta x$ thick. Species A flows into and out of the control volume as a result of the combined action of bulk flow and diffusion. Within the control volume, species A may be created or destroyed as a result of chemical reaction.

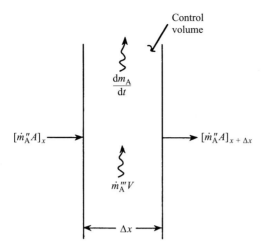

**Figure 3.3**     Control volume for one-dimensional analysis of conservation of species A.

The net rate of increase in the mass of A within the control volume relates to the mass fluxes and reaction rate as follows:

$$\frac{dm_{A,cv}}{dt} \quad = \quad [\dot{m}_A'' A]_x \quad - \quad [\dot{m}_A'' A]_{x+\Delta x} \quad + \quad \dot{m}_A''' V, \quad (3.28)$$

| Rate of increase of mass of A within control volume | Mass flow of A into the control volume | Mass flow of A out of the control volume | Mass production rate of species A by chemical reaction |
|---|---|---|---|

where the species mass flux $\dot{m}_A''$ is given by Eqn. 3.1, and $\dot{m}_A'''$ is the mass production rate of species A per unit volume ($kg_A/m^3$-s). In Chapter 5, we specifically deal with how to determine $\dot{m}_A'''$. Recognizing that the mass of A within the control volume is $m_{A,cv} = Y_A m_{cv} = Y_A \rho V_{cv}$ and that the volume $V_{cv} = A\Delta x$, Eqn. 3.28 can be rewritten:

$$A\Delta x \frac{\partial(\rho Y_A)}{\partial t} = A\left[Y_A \dot{m}'' - \rho \mathcal{D}_{AB} \frac{\partial Y_A}{\partial x}\right]_x$$
$$- A\left[Y_A \dot{m}'' - \rho \mathcal{D}_{AB} \frac{\partial Y_A}{\partial x}\right]_{x+\Delta x} + \dot{m}_A''' A\Delta x. \quad (3.29)$$

Dividing through by $A\Delta x$ and taking the limit as $\Delta x \to 0$, Eqn. 3.29 becomes

$$\frac{\partial(\rho Y_A)}{\partial t} = -\frac{\partial}{\partial x}\left[Y_A \dot{m}'' - \rho \mathcal{D}_{AB} \frac{\partial Y_A}{\partial x}\right] + \dot{m}_A''' \quad (3.30)$$

or, for the case of steady flow where $\partial(\rho Y_A)/\partial t = 0$,

$$\dot{m}_A''' - \frac{d}{dx}\left[Y_A \dot{m}'' - \rho \mathcal{D}_{AB} \frac{dY_A}{dx}\right] = 0. \quad (3.31)$$

Equation 3.31 is the steady-flow, one-dimensional form of species conservation for a binary gas mixture, assuming species diffusion occurs only as a result of concentration gradients; i.e., only ordinary diffusion is considered. For the multidimensional case, Eqn. 3.31 can be generalized as

$$\dot{m}_A''' \qquad - \qquad \nabla \cdot \dot{m}_A'' \qquad = 0. \qquad (3.32)$$

Net rate of production       Net flow of species A
of species A by              out of control volume,
chemical reaction,                per unit volume
per unit volume

In Chapter 7, we will employ Eqns. 3.31 and 3.32 to develop the conservation of energy principle for a reacting system. Chapter 7 also provides a more detailed treatment of mass transfer, extending the present development to multicomponent (nonbinary) mixtures and including thermal diffusion.

# SOME APPLICATIONS OF MASS TRANSFER

## The Stefan Problem

Consider liquid A, maintained at a fixed height in a glass cylinder as illustrated in Fig. 3.4. A mixture of gas A and gas B flow across the top of the cylinder. If the concentration of A in the flowing gas is less than the concentration of A at the liquid–vapor interface, a driving force for mass transfer exists and species A will diffuse from the liquid–gas interface to the open end of the cylinder. If we assume a steady state exists (i.e., the liquid is replenished at a rate to keep the liquid level constant, or the interface regresses so slowly that its movement can be neglected)

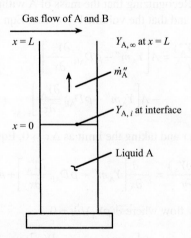

Gas flow of A and B

$x = L$        $Y_{A,\infty}$ at $x = L$

$\dot{m}_A''$

$Y_{A,i}$ at interface

$x = 0$

Liquid A

**Figure 3.4**     Diffusion of vapor A through a stagnant column of gas B, i.e., the Stefan problem.

and, furthermore, assume that B is insoluble in liquid A, then there will be no net transport of B in the tube, producing a stagnant layer of B in the column.

Mathematically, the overall conservation of mass for this system can be expressed as

$$\dot{m}''(x) = \text{constant} = \dot{m}''_A + \dot{m}''_B. \tag{3.33}$$

Since $\dot{m}''_B = 0$, then

$$\dot{m}''_A = \dot{m}''(x) = \text{constant}. \tag{3.34}$$

Equation 3.1 now becomes

$$\dot{m}''_A = Y_A \dot{m}''_A - \rho \mathcal{D}_{AB} \frac{dY_A}{dx}. \tag{3.35}$$

Rearranging and separating variables, we obtain

$$-\frac{\dot{m}''_A}{\rho \mathcal{D}_{AB}} dx = \frac{dY_A}{1 - Y_A}. \tag{3.36}$$

Assuming the product $\rho \mathcal{D}_{AB}$ to be constant, Eqn. 3.36 can be integrated to yield

$$-\frac{\dot{m}''_A}{\rho \mathcal{D}_{AB}} x = -\ln[1 - Y_A] + C, \tag{3.37}$$

where $C$ is the constant of integration. With the boundary condition

$$Y_A(x = 0) = Y_{A,i}, \tag{3.38}$$

we eliminate $C$ and obtain the following mass-fraction distribution after removing the logarithm by exponentiation:

$$Y_A(x) = 1 - (1 - Y_{A,i}) \exp\left[\frac{\dot{m}''_A x}{\rho \mathcal{D}_{AB}}\right]. \tag{3.39}$$

The mass flux of A, $\dot{m}''_A$, can be found by letting $Y_A(x = L) = Y_{A,\infty}$ in Eqn. 3.39. Thus,

$$\dot{m}''_A = \frac{\rho \mathcal{D}_{AB}}{L} \ln\left[\frac{1 - Y_{A,\infty}}{1 - Y_{A,i}}\right]. \tag{3.40}$$

From Eqn. 3.40, we see that the mass flux is directly proportional to the product of the density, $\rho$, and the mass diffusivity, $\mathcal{D}_{AB}$, and inversely proportional to the length, $L$. Larger diffusivities thus produce larger mass fluxes.

To see the effects of the concentrations at the interface and at the top of the tube, let the mass fraction of A in the freestream flow be zero, while arbitrarily

**Table 3.1**    Effect of interface mass fraction on mass flux

| $Y_{A,i}$ | $\dot{m}''_A/(\rho \mathcal{D}_{AB}/L)$ |
|-----------|------------------------------------------|
| 0 | 0 |
| 0.05 | 0.0513 |
| 0.10 | 0.1054 |
| 0.20 | 0.2231 |
| 0.50 | 0.6931 |
| 0.90 | 2.303 |
| 0.999 | 6.908 |

varying $Y_{A,i}$, the interface mass fraction, from zero to unity. Physically, this could correspond to an experiment in which dry nitrogen is blown across the tube outlet and the interface mass fraction is controlled by the partial pressure of the liquid, which, in turn, is varied by changing the temperature. Table 3.1 shows that at small values of $Y_{A,i}$, the dimensionless mass flux is essentially proportional to $Y_{A,i}$. For $Y_{A,i}$ greater than about 0.5, the mass flux increases very rapidly.

## Liquid–Vapor Interface Boundary Conditions

In the example above, we treated the gas-phase mass fraction of the diffusing species at the liquid–vapor interface, $Y_{A,i}$, as a known quantity. Unless this mass fraction is measured, which is unlikely, some means must be found to calculate or estimate its value. This can be done by assuming equilibrium exists between the liquid and vapor phases of species A. With this equilibrium assumption, and the assumption of ideal gases, the partial pressure of species A on the gas side of the interface must equal the saturation pressure associated with the temperature of the liquid, i.e.,

$$P_{A,i} = P_{sat}(T_{liq,i}). \tag{3.41}$$

The partial pressure, $P_{A,i}$, can be related to the mole fraction of species A, $\chi_{A,i} = P_{sat}/P$, and to the mass fraction:

$$Y_{A,i} = \frac{P_{sat}(T_{liq,i})}{P} \frac{MW_A}{MW_{mix,i}}, \tag{3.42}$$

where the molecular weight of the mixture also depends on $\chi_{A,i}$, and, hence, on $P_{sat}$.

The above analysis has transformed the problem of finding the vapor mass fraction at the interface to the problem of finding the temperature at the interface. In some cases, the interface temperature may be given or known, but, in general, the interface temperature must be found by writing energy balances for the liquid and gas phases and solving them with appropriate boundary conditions, including that at the interface. In the following, we will establish this interface boundary condition, but leave the gas-phase and liquid-phase energy balances for later.

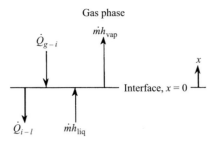

**Figure 3.5**    Energy balance at the surface of evaporating fluid.

In crossing the liquid–vapor boundary, we maintain continuity of temperature, i.e.,

$$T_{\text{liq},i}(x=0^-) = T_{\text{vap},i}(x=0^+) = T(0), \qquad (3.43)$$

and energy is conserved at the interface as illustrated in Fig. 3.5. Heat is transferred from the gas to the liquid surface, $\dot{Q}_{g-i}$. Some of this energy goes into heating the liquid, $\dot{Q}_{i-l}$, while the remainder causes the phase change. This energy balance is expressed as

$$\dot{Q}_{g-i} - \dot{Q}_{i-l} = \dot{m}(h_{\text{vap}} - h_{\text{liq}}) = \dot{m}h_{fg} \qquad (3.44)$$

or

$$\dot{Q}_{\text{net}} = \dot{m}h_{fg}. \qquad (3.45)$$

Equation 3.45 can be used to calculate the net heat transferred to the interface if the evaporation rate, $\dot{m}$, is known. Conversely, if $\dot{Q}_{\text{net}}$ is known, the evaporation rate can be determined.

---

Liquid benzene ($C_6H_6$), at 298 K, is contained in a 1-cm-diameter glass tube and maintained at a level 10 cm below the top of the tube, which is open to the atmosphere. The following properties of benzene are given:

**Example 3.1**

$$T_{\text{boil}} = 353 \text{ K} @ 1 \text{ atm},$$
$$h_{fg} = 393 \text{ kJ/kg} @ T_{\text{boil}},$$
$$MW = 78.108 \text{ kg/kmol},$$
$$\rho_l = 879 \text{ kg/m}^3,$$
$$\mathcal{D}_{C_6H_6-\text{air}} = 0.88 \cdot 10^{-5} \text{ m}^2/\text{s} @ 298 \text{ K}.$$

A. Determine the mass evaporation rate (kg/s) of the benzene.
B. How long does it take to evaporate 1 cm$^3$ of benzene?
C. Compare the evaporation rate of benzene with that of water. Assume
   $\mathcal{D}_{H_2O-\text{air}} = 2.6 \cdot 10^{-5} \text{ m}^2/\text{s}$.

**Solution**

A. Find $\dot{m}_{C_6H_6}$.

Since the configuration given represents the Stefan problem, we can apply Eqn. 3.40:

$$\dot{m}''_{C_6H_6} = \frac{\bar{\rho}\mathcal{D}_{C_6H_6-air}}{L}\ln\left[\frac{1-Y_{C_6H_6,\infty}}{1-Y_{C_6H_6,i}}\right].$$

In the above, $\mathcal{D}$, $L$, and $Y_{C_6H_6,\infty}$ $(=0)$ are all known; however, we need to evaluate the benzene mass fraction at the interface, $Y_{C_6H_6,i}$, and an appropriate mean density, $\bar{\rho}$, before proceeding.

From Eqn. 3.42, we know that

$$Y_{C_6,H_6,i} = \chi_{C_6H_6,i}\frac{MW_{C_6H_6}}{MW_{mix,i}}$$

where

$$\chi_{C_6H_6,i} = \frac{P_{sat}(T_{liq,i})}{P}.$$

To evaluate $P_{sat}/P$, we integrate the Clausius–Clapeyron equation, Eqn. 2.19,

$$\frac{dP}{P} = \frac{h_{fg}}{R_u/MW_{C_6H_6}}\frac{dT}{T^2},$$

from the reference state ($P = 1$ atm, $T = T_{boil} = 353$ K) to the state at 298 K, i.e.,

$$\frac{P_{sat}}{P(=1\,\text{atm})} = \exp\left[-\frac{h_{fg}}{(R_u/MW_{C_6H_6})}\left(\frac{1}{T}-\frac{1}{T_{boil}}\right)\right].$$

Evaluating the above,

$$\frac{P_{sat}}{P(=1\,\text{atm})} = \exp\left[\frac{-393,000}{8315/78.108}\left(\frac{1}{298}-\frac{1}{353}\right)\right]$$

$$= \exp(-1.93) = 0.145.$$

So, $P_{sat} = 0.145$ atm and $\chi_{C_6H_6} = 0.145$. The interface mixture molecular weight is then

$$MW_{mix,i} = 0.145(78.108)+(1-0.145)28.85$$

$$= 35.99\ \text{kg/kmol},$$

where the simplified composition of air is assumed. The interface benzene mass fraction is then

$$Y_{C_6H_6,i} = 0.145\frac{78.108}{35.99} = 0.3147.$$

For isothermal, isobaric conditions, we can estimate the mean gas density in the tube by using the ideal-gas law and the mean mixture molecular weight as follows:

$$\bar{\rho} = \frac{P}{(R_u/\overline{MW})T},$$

where

$$\overline{MW} = \frac{1}{2}(MW_{\text{mix},i} + MW_{\text{mix},\infty})$$

$$= \frac{1}{2}(35.99 + 28.85) = 32.42.$$

Thus,

$$\overline{\rho} = \frac{101,325}{\left(\dfrac{8315}{32.42}\right)298} = 1.326 \text{ kg/m}^3.$$

We can now evaluate the benzene mass flux (Eqn. 3.40):

$$\dot{m}''_{C_6H_6} = \frac{1.326(0.88 \cdot 10^{-5})}{0.1} \ln\left[\frac{1-0}{1-0.3147}\right]$$

$$= 1.167 \cdot 10^{-4} \ln(1.459)$$

$$= 1.167 \cdot 10^{-4}(0.378) = 4.409 \cdot 10^{-5} \text{ kg/s-m}^2$$

and

$$\boxed{\dot{m}_{C_6H_6}} = \dot{m}''_{C_6H_6}\frac{\pi D^2}{4} = 4.409 \cdot 10^{-5}\frac{\pi(0.01)^2}{4} = \boxed{3.46 \cdot 10^{-9} \text{ kg/s}}$$

B. Find the time needed to evaporate 1 cm³ of benzene.

Since the liquid level is maintained, the mass flux is constant during the time to evaporate the 1 cm³; thus,

$$t = \frac{m_{\text{evap}}}{\dot{m}_{C_6H_6}} = \frac{\rho_{\text{liq}}V}{\dot{m}_{C_6H_6}},$$

$$\boxed{t} = \frac{879(\text{kg/m}^3)1 \cdot 10^{-6}(\text{m}^3)}{3.46 \cdot 10^{-9}(\text{kg/s})} = \boxed{2.54 \cdot 10^5 \text{s or 70.6 hr}}$$

C. Find $\dot{m}_{C_6H_6}/\dot{m}_{H_2O}$.

Finding $\dot{m}_{H_2O}$ follows the above; however, the problem is simplified since we can use the steam tables (e.g., [7]), to find $P_{\text{sat}}$ at 298 K rather than invoking the Clausius–Clapeyron approximation.

From the steam tables,

$$P_{\text{sat}}(298 \text{ K}) = 3.169 \text{ kPa}$$

so

$$\chi_{H_2O,i} = \frac{3169}{101,325} = 0.03128$$

and

$$MW_{\text{mix},i} = 0.03128(18.016) + (1 - 0.03128)28.85 = 28.51.$$

Thus,

$$Y_{H_2O,i} = \chi_{H_2O,i} \frac{MW_{H_2O}}{MW_{mix,i}} = 0.03128 \frac{18.016}{28.51} = 0.01977.$$

The mean molecular weight and mean density in the tube are

$$\overline{MW} = \frac{1}{2}(28.51 + 28.85) = 28.68$$

$$\overline{\rho} = \frac{101,325}{\left(\dfrac{8315}{28.68}\right)298} = 1.173 \text{ kg/m}^3,$$

where we assume the air outside of the tube is dry.

The evaporation flux is then

$$\dot{m}''_{H_2O} = \frac{1.173(2.6 \cdot 10^{-5})}{0.1} \ln\left[\frac{1-0}{1-0.01977}\right]$$

$$= 3.050 \cdot 10^{-4} \ \ln(1.020)$$

$$= 3.050 \cdot 10^{-4}(0.01997) = 6.09 \cdot 10^{-6} \text{ kg/s-m}^2$$

so,

$$\boxed{\frac{\dot{m}_{C_6H_6}}{\dot{m}_{H_2O}} = \frac{4.409 \cdot 10^{-5}}{6.09 \cdot 10^{-6}} = \boxed{7.2}}$$

**Comment**

Comparing the details of the calculations in parts A and C, we see that the higher vapor pressure of the benzene dominates over the higher diffusivity of the water, thus causing the benzene to evaporate more than seven times faster than the water.

---

## Droplet Evaporation

The problem of the evaporation of a single liquid droplet in a quiescent environment is just the Stefan problem for a spherically symmetric coordinate system. Our treatment of droplet evaporation illustrates the application of mass-transfer concepts to a problem of practical interest. In Chapter 10, we will deal with the droplet evaporation/combustion problem in greater detail; however, we foreshadow the Chapter 10 development by introducing the concept of an evaporation constant and droplet lifetimes.

Figure 3.6 defines the spherically symmetric coordinate system. The radius $r$ is the only coordinate variable. It has its origin at the center of the droplet, and the droplet radius at the liquid–vapor interface is denoted $r_s$. Very far from the droplet surface ($r \to \infty$), the mass fraction of droplet vapor is $Y_{F,\infty}$.

Physically, heat from the ambient environment supplies the energy necessary to vaporize the liquid, and the vapor then diffuses from the droplet surface into the ambient gas. The mass loss causes the droplet radius to shrink with time until the

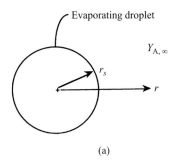

(a)

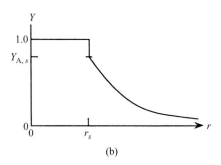

(b)

**Figure 3.6**     Evaporation of a liquid droplet in a quiescent environment.

droplet is completely evaporated ($r_s = 0$). The problem that we wish to solve is the determination of the mass flowrate of the vapor from the surface at any instant in time. Knowledge of this will then enable us to calculate the droplet radius as a function of time and the droplet lifetime.

To mathematically describe this process completely, the following conservation laws are required:

*Droplet:* mass conservation, energy conservation.

*Droplet vapor/ambient gas mixture* ($r_s < r < \infty$): overall mass conservation, droplet vapor (species) conservation, and energy conservation.

Thus, we see that for this complete description we need at least five equations. These equations, in general, take the form of either ordinary or partial differential equations, depending on the simplifying assumptions applied.

**Assumptions**     For our brief treatment here, we can greatly reduce the number of unknowns and, hence, the number of equations, by invoking the same assumptions that we employed in the one-dimensional cartesian problem.

1. The evaporation process is quasi-steady. This means that, at any instant in time, the process can be described as if it were in steady state. This assumption eliminates the need to deal with partial differential equations.

2. The droplet temperature is uniform and, furthermore, the temperature is assumed to be some fixed value below the boiling point of the liquid. In many problems, the transient heating of the liquid does not greatly affect the droplet lifetime. Determination of the temperature at the droplet surface depends on the heat-transfer rate to the droplet. Thus, our assumption of a specified temperature eliminates the need to apply conservation of energy to the gas phase surrounding the liquid droplet and to the droplet itself. As will be shown in Chapter 10, heat-transfer considerations frequently dominate the droplet evaporation problem.

3. The mass fraction of vapor at the droplet surface is determined by liquid–vapor equilibrium at the droplet temperature.

4. We also assume that all thermophysical properties—specifically, the $\rho D$ product—are constant. Although properties may vary greatly as we move through the gas phase from the droplet surface to the faraway surroundings, constant properties allow a simple closed-form solution.

**Evaporation Rate**     With the above assumptions, we can find the mass evaporation rate, $\dot{m}$, and the droplet radius history, $r_s(t)$, by writing a droplet vapor species conservation equation, and a droplet liquid mass conservation equation. From species conservation, we determine the evaporation rate $\dot{m}$, and thus, knowing $\dot{m}(t)$, we can easily find the drop size as a function of time.

As in the cartesian-coordinate Stefan problem, the species originally in the liquid phase is the species transported, while the ambient fluid is stagnant. Thus, our previous analysis (Eqns. 3.33–3.35) needs only to be modified to take into account the change in coordinate system. Overall mass conservation is expressed as

$$\dot{m}(r) = \text{constant} = 4\pi r^2 \dot{m}'', \tag{3.46}$$

where $\dot{m}'' = \dot{m}''_A + \dot{m}''_B = \dot{m}''_A$, since $\dot{m}_B = 0$. Note that it is the mass flowrate, not the mass flux, that is constant. Species conservation for the droplet vapor (Eqn. 3.5) becomes

$$\dot{m}''_A = Y_A \dot{m}''_A - \rho D_{AB} \frac{dY_A}{dr}. \tag{3.47}$$

Substituting Eqn. 3.46 into Eqn. 3.47, and rearranging to solve for the evaporation rate $\dot{m}(=\dot{m}_A)$, yields

$$\dot{m} = -4\pi r^2 \frac{\rho D_{AB}}{1 - Y_A} \frac{dY_A}{dr}. \tag{3.48}$$

Integrating Eqn. 3.48 and applying the boundary condition that at the droplet surface the vapor mass fraction is $Y_{A,s}$, i.e.,

$$Y_A(r = r_s) = Y_{A,s}, \tag{3.49}$$

yields

$$Y_A(r) = 1 - \frac{(1-Y_{A,s})\exp[-\dot{m}/(4\pi\rho\mathcal{D}_{AB}r)]}{\exp[-\dot{m}/(4\pi\rho\mathcal{D}_{AB}r_s)]}. \qquad (3.50)$$

The evaporation rate can be determined from Eqn. 3.50 by letting $Y_A = Y_{A,\infty}$ for $r \to \infty$ and solving for $\dot{m}$:

$$\dot{m} = 4\pi r_s \rho \mathcal{D}_{AB} \ln\left[\frac{(1-Y_{A,\infty})}{(1-Y_{A,s})}\right]. \qquad (3.51)$$

This result (Eqn. 3.51) is analogous to Eqn. 3.40 for the cartesian problem.

To see more conveniently how the vapor mass fractions at the droplet surface, $Y_{A,s}$, and far from the surface, $Y_{A,\infty}$, affect the evaporation rate, the argument of the logarithm in Eqn. 3.51 is used to define the dimensionless **transfer number, $B_Y$:**

$$1 + B_Y \equiv \frac{1-Y_{A,\infty}}{1-Y_{A,s}} \qquad (3.52a)$$

or

$$B_Y = \frac{Y_{A,s}-Y_{A,\infty}}{1-Y_{A,s}}. \qquad (3.52b)$$

Using the transfer number, $B_Y$, the evaporation rate is expressed as

$$\dot{m} = 4\pi r_s \rho \mathcal{D}_{AB} \ln(1+B_Y). \qquad (3.53)$$

From this result, we see that when the transfer number is zero, the evaporation rate is zero; and, correspondingly, as the transfer number increases, so does the evaporation rate. This makes physical sense in that, from the appearance of the mass-fraction difference $Y_{A,s} - Y_{A,\infty}$ in its definition, $B_Y$ can be interpreted as a "driving potential" for mass transfer.

**Droplet Mass Conservation**  We obtain the droplet radius (or diameter) history by writing a mass balance that states that the rate at which the mass of the droplet decreases is equal to the rate at which the liquid is vaporized; i.e.,

$$\frac{dm_d}{dt} = -\dot{m}, \qquad (3.54)$$

where the droplet mass, $m_d$, is given by

$$m_d = \rho_l V = \rho_l \pi D^3/6, \qquad (3.55)$$

and $V$ and $D$ ($= 2r_s$) are the droplet volume and diameter, respectively.

Substituting Eqns. 3.55 and 3.53 into Eqn. 3.54 and performing the differentiation yields

$$\frac{dD}{dt} = -\frac{4\rho \mathcal{D}_{AB}}{\rho_l D} \ln(1 + B_Y).$$  (3.56)

In the combustion literature, however, Eqn. 3.56 is more commonly expressed in terms of $D^2$ rather than $D$. This form is

$$\frac{dD^2}{dt} = -\frac{8\rho \mathcal{D}_{AB}}{\rho_l} \ln(1 + B_Y).$$  (3.57)

Equation 3.57 tells us that the time derivative of the square of the droplet diameter is constant; hence, $D^2$ varies linearly with $t$ with the slope $-(8\rho \mathcal{D}_{AB}/\rho_l)\ln(1 + B_Y)$, as illustrated in Fig. 3.7a. This slope is defined to be the **evaporation constant** $K$:

$$K = \frac{8\rho \mathcal{D}_{AB}}{\rho_l} \ln(1 + B_Y).$$  (3.58)

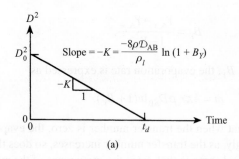

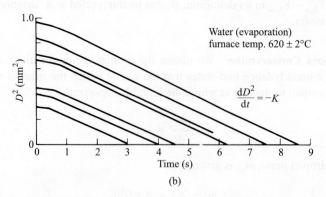

**Figure 3.7**    The $D^2$ law for droplet evaporation. (a) Simplified analysis. (b) Experimental data from Ref. [8] for water droplets with $T_\infty = 620°C$.
Reprinted by permission of The Combustion Institute.

We can use Eqn. 3.57 (or 3.56) to find the time it takes a droplet of given initial size to evaporate completely; i.e., the droplet lifetime, $t_d$. Thus,

$$\int_{D_0^2}^{0} dD^2 = -\int_{0}^{t_d} K\, dt, \tag{3.59}$$

which yields

$$t_d = D_0^2/K. \tag{3.60}$$

We can change the upper limits in Eqn. 3.59 to provide a general relationship expressing the variation of $D$ with time $(t)$:

$$D^2(t) = D_0^2 - Kt. \tag{3.61}$$

Equation 3.61 is referred to as the $D^2$ **law** for droplet evaporation. Experimental data show that the $D^2$ law holds after an initial transient period as shown in Fig. 3.7b. The $D^2$ law is also used to describe burning of fuel droplets, which is discussed in Chapter 10.

---

**Example 3.2**

In mass-diffusion-controlled evaporation of a fuel droplet, the droplet surface temperature is an important parameter. Estimate the droplet lifetime of a 100-$\mu$m-diameter $n$-dodecane droplet evaporating in dry nitrogen at 1 atm if the droplet temperature is 10 K below the dodecane boiling point. Repeat the calculation for a temperature 20 K below the boiling point, and compare the results. For simplicity, assume that, in both cases, the mean gas density is that of nitrogen at a mean temperature of 800 K. Use this same temperature to estimate the fuel vapor diffusivity. The density of liquid dodecane is 749 kg/m³.

Given:   $n$-dodecane droplet

$$D = 100\ \mu\text{m},$$
$$P = 1\ \text{atm},$$
$$\rho_l = 749\ \text{kg/m}^3,$$
$$T_s = T_{\text{boil}} - 10\ (\text{or } 20),$$
$$\rho = \rho_{\text{N}_2}\ @\overline{T} = 800\ \text{K}.$$

Find:   droplet lifetime, $t_d$.

**Solution**

We can estimate the droplet lifetimes by employing Eqn. 3.60, after calculating the evaporation rate constants, $K$, from Eqn. 3.58. Evaluation of properties will be important in our solution.

Properties required:

$$T_{\text{boil}} = 216.3\text{°C} + 273.15 = 489.5\ \text{K (Appendix Table B.1)},$$
$$h_{fg} = 256\ \text{kJ/kg (Appendix Table B.1)},$$
$$MW_{\text{A}} = 170.337\ \text{kg/kmol},$$
$$\mathcal{D}_{\text{AB}} = 8.1 \cdot 10^{-6}\ \text{m}^2/\text{s}\ @\ 399\ \text{K (Appendix Table D.1)}.$$

We start by calculating $B_Y$, which requires knowledge of the fuel mass fraction at the surface. As in Example 3.1, we integrate the Clausius–Clapeyron equation to find the saturation pressure at the given droplet surface temperature. For $T = T_{boil} - 10 \ (= 479.5 \text{ K})$,

$$\frac{P_{sat}}{P(=1 \text{ atm})} = \exp\left[\frac{-256,000}{(8315/170.337)}\left(\frac{1}{479.5} - \frac{1}{489.5}\right)\right] = 0.7998$$

so $P_{sat} = 0.7998$ and $\chi_A (= \chi_{dodecane}) = 0.7998$. We employ Eqn. 2.11 to calculate the fuel mass fraction at the surface:

$$Y_{A,s} = 0.7998 \frac{170.337}{0.7998(170.337) + (1 - 0.7998)28.014} = 0.9605.$$

We now evaluate the transfer number $B_Y$ (Eqn. 3.52b):

$$B_Y = \frac{Y_{A,s} - Y_{A,\infty}}{1 - Y_{A,s}} = \frac{0.9605 - 0}{1 - 0.9605} = 24.32.$$

To evaluate the evaporation constant, we need to estimate $\rho \mathcal{D}_{AB}$, which we treat as $\bar{\rho}_{N_2} \mathcal{D}_{AB} (\bar{T} = 800 \text{ K})$. Extrapolating the tabulated value to 800 K using Eqn. 3.18b gives

$$\mathcal{D}_{AB}(\bar{T}) = 8.1 \cdot 10^{-6}\left(\frac{800}{399}\right)^{3/2} = 23.0 \cdot 10^{-6} \text{ m}^2/\text{s},$$

and using the ideal-gas law to evaluate $\bar{\rho}_{N_2}$:

$$\bar{\rho}_{N_2} = \frac{101,325}{(8315/28.014)800} = 0.4267 \text{ kg/m}^3.$$

Thus,

$$K = \frac{8\bar{\rho}\mathcal{D}_{AB}}{\rho_l}\ln(1 + B_Y)$$

$$= \frac{8(0.4267)23.0 \cdot 10^{-6}}{749}\ln(1 + 24.32)$$

$$= 3.39 \cdot 10^{-7} \text{ m}^2/\text{s},$$

and the droplet lifetime is

$$t_d = D^2/K = \frac{(100 \cdot 10^{-6})^2}{3.39 \cdot 10^{-7}}$$

$$\boxed{t_d = 0.030 \text{ s}}$$

Repeating the calculations for $T = T_{boil} - 20 = 469.5 \text{ K}$, we can compare the various parameters:

| $\Delta T$ (K) | $T$ (K) | $P_{sat}$ (atm) | $Y_s$ | $B_Y$ | $K$ (m²/s) | $t_d$ (s) |
|---|---|---|---|---|---|---|
| 10 | 479.5 | 0.7998 | 0.9605 | 24.32 | $3.39 \cdot 10^{-7}$ | 0.030 |
| 20 | 469.5 | 0.6336 | 0.9132 | 10.52 | $2.56 \cdot 10^{-7}$ | 0.039 |

From the above table, we see that about a 2 percent change in the droplet surface temperature results in a 30 percent increase in the droplet lifetime. This large effect of temperature manifests itself in the $B_Y$ term, where the denominator, $1 - Y_{A,s}$, is clearly sensitive to temperature when $Y_{A,s}$ is large.

**Comments**

Evaporation of fuel droplets at elevated temperatures is important for many practical devices, particularly gas-turbine engines and diesel engines. In these devices, evaporation occurs when droplets are injected into hot compressed air or into zones where combustion is already in progress. In spark-ignition engines employing fuel injection, temperatures in the intake system are typically much closer to ambient levels and pressures are subatmospheric. In most cases, forced-convection effects are important in the evaporation process. These effects will be discussed in Chapter 10, which focuses on droplet combustion.

# SUMMARY

In this chapter, you were introduced to the concept of mass diffusion or mass transfer, and Fick's law governing the rate of species mass transfer in a binary mixture was presented. A new transport property, the mass diffusivity, provides the proportionality between the species diffusional flux and the species concentration gradient, analogous to the kinematic viscosity and thermal diffusivity in momentum and heat transfer, respectively. You should be familiar with the physical interpretation of Fick's law and how to apply it to simple binary systems. As an example of simple binary diffusion, the Stefan problem was developed. From this example, you saw how overall mass and species mass conservation equations apply to a single species diffusing through a stagnant layer of a second species. With application of appropriate boundary conditions, we were able to solve for the liquid evaporation rate. A similar analysis was performed for the spherically symmetric problem of droplet evaporation. You should have some physical appreciation for this problem, as well as being familiar with how to calculate droplet evaporation rates and droplet lifetimes.

# NOMENCLATURE

| | |
|---|---|
| $a$ | Defined in Eqn. 3.10d |
| $A$ | Area (m$^2$) |
| $B_Y$ | Dimensionless transfer number or Spalding number, Eqn. 3.52 |
| $c$ | Molar concentration (kmol/m$^3$) |
| $D$ | Diameter (m) |
| $\mathcal{D}_{AB}$ | Binary diffusivity or diffusion coefficient (m$^2$/s) |
| $h$ | Enthalpy (J/kg) |
| $h_{fg}$ | Heat of vaporization (J/kg) |
| $k$ | Thermal conductivity (W/m-K) |
| $k_B$ | Boltzmann constant (J/K) |
| $ke$ | Kinetic energy (J) |
| $K$ | Evaporation constant, Eqn. 3.58 (m$^2$/s) |

| | |
|---|---|
| $L$ | Tube height (m) |
| $m$ | Mass (kg) |
| $\dot{m}$ | Mass flowrate (kg/s) |
| $\dot{m}''$ | Mass flux (kg/s-m$^2$) |
| $MW$ | Molecular weight (kg/kmol) |
| $n$ | Number of molecules |
| $N$ | Number of moles |
| $\dot{N}$ | Molar flowrate (kmol/s) |
| $\dot{N}''$ | Molar flux (kmol/s-m$^2$) |
| $P$ | Pressure (Pa) |
| $\dot{Q}$ | Heat-transfer rate (W) |
| $\dot{Q}''$ | Heat flux (W/m$^2$) |
| $r$ | Radius (m) |
| $R_u$ | Universal gas constant (J/kmol-K) |
| $t$ | Time (s) |
| $T$ | Temperature (K) |
| $\overline{v}$ | Average molecular speed (m/s) |
| $V$ | Volume (m$^3$) |
| $x$ | Cartesian coordinate (m) |
| $y$ | Cartesian coordinate (m) |
| $Y$ | Mass fraction (kg/kg) |
| $Z''$ | Molecular collision frequency per unit area (no./m$^2$-s) |

### Greek Symbols

| | |
|---|---|
| $\lambda$ | Molecular mean free path (m) |
| $\rho$ | Density (kg/m$^3$) |
| $\sigma$ | Molecular diameter (m) |
| $\chi$ | Mole fraction (kmol/kmol) |

### Subscripts

| | |
|---|---|
| A | Species A |
| B | Species B |
| boil | Boiling point |
| cv | Control volume |
| d | Droplet |
| diff | Diffusion |
| evap | Evaporation |
| g | Gas |
| i | Interface |
| l, liq | Liquid |
| mix | Mixture |
| s | Surface |
| sat | Saturation |
| tot | Total |
| vap | Vapor |
| $\infty$ | Freestream or far removed from surface |

# REFERENCES

1. Bird, R. B., Stewart, W. E., and Lightfoot, E. N., *Transport Phenomena,* John Wiley & Sons, New York, 1960.

2. Thomas, L. C., *Heat Transfer–Mass Transfer Supplement,* Prentice-Hall, Englewood Cliffs, NJ, 1991.

3. Williams, F. A., *Combustion Theory,* 2nd Ed., Addison-Wesley, Redwood City, CA, 1985.

4. Kuo, K. K., *Principles of Combustion,* 2nd Ed., John Wiley & Sons, Hoboken, NJ, 2005.

5. Pierce, F. J., *Microscopic Thermodynamics,* International, Scranton, PA, 1968.

6. Daniels, F., and Alberty, R. A., *Physical Chemistry,* 4th Ed., John Wiley & Sons, New York, 1975.

7. Irvine, T. F., Jr., and Hartnett, J. P. (eds.), *Steam and Air Tables in SI Units,* Hemisphere, Washington, 1976.

8. Nishiwaki, N., "Kinetics of Liquid Combustion Processes: Evaporation and Ignition Lag of Fuel Droplets," *Fifth Symposium (International) on Combustion,* Reinhold, New York, pp. 148–158, 1955.

# REVIEW QUESTIONS

**1.** Make a list of all of the boldfaced words in Chapter 3. Make sure you understand the meaning of each.

**2.** Assuming constant properties, recast Eqn. 3.4 into a form where the constant of proportionality in Fourier's law is the thermal diffusivity, $\alpha = k/\rho c_p$, rather than $k$. The gradient (or spatial derivative) of what property now appears?

**3.** Starting with the sketch in Fig. 3.3, derive the species conservation equation (Eqn. 3.30) without reference to the text derivation.

**4.** Using words only, explain what is happening in the Stefan problem.

**5.** How is the evaporating droplet problem similar to the Stefan problem?

**6.** If no heat is supplied to the liquid A (Fig. 3.4), what happens to the temperature of the liquid? Can you write a simple energy conservation equation to justify your answer?

**7.** Define a transfer number, $B_Y$, that can be used with Eqn. 3.40. Furthermore, assume the flow area is $A$, and write an expression for $\dot{m}$ in terms of fluid properties, geometrical parameters, and $B_Y$. Compare your result with Eqn. 3.53 and discuss.

**8.** Why, and in what way, does the presence of species A in the free stream (Fig. 3.4) affect the evaporation rate?

9. How does the mass average velocity vary with distance from the surface of an evaporating droplet at any instant in time?

10. Explain what is meant by "quasi-steady" flow.

11. Starting with Eqns. 3.53 and 3.54, derive the "$D^2$ law" for droplet evaporation.

## PROBLEMS

**3.1** Consider an equimolar mixture of oxygen ($O_2$) and nitrogen ($N_2$) at 400 K and 1 atm. Calculate the mixture density $\rho$ and the mixture molar concentration $c$.

**3.2** Calculate the mass fractions of $O_2$ and $N_2$ in the mixture given in problem 3.1.

**3.3** Estimate the value of the binary diffusivity of $n$-octane in air at a temperature of 400 K and pressure of 3.5 atm utilizing the value given in Appendix D. Also, compare the ratio $\mathcal{D}_{ref}/\mathcal{D}$ ($T = 400$ K, $P = 3.5$ atm) with the ratio of the $\rho\mathcal{D}$ products, i.e., $(\rho\mathcal{D})_{ref}/(\rho\mathcal{D})$ ($T = 400$ K, $P = 3.5$ atm).

**3.4** Equation 3.18a was derived assuming that molecules A and B are essentially the same mass and size. Reference [1] indicates that Eqn. 3.18a can be generalized for the case where $m_A \neq m_B$ and $\sigma_A \neq \sigma_B$ by using

$$m = \frac{m_A m_B}{(m_A + m_B)/2} \quad \text{and} \quad \sigma = \frac{1}{2}(\sigma_A + \sigma_B).$$

With this information, estimate the binary diffusivity for $O_2$ in $N_2$ at 273 K for $\sigma_{O_2} = 3.467$ and $\sigma_{N_2} = 3.798$ Å. Compare your simple estimate with the handbook value of $1.8 \cdot 10^{-5}$ m$^2$/s. Should you expect good agreement? Why not? *Note:* You will need to use Avogadro's number $6.022 \cdot 10^{26}$ molecules/kmol to calculate the mass of a molecule.

**3.5** Consider liquid $n$-hexane in a 50-mm-diameter graduated cylinder. The distance from the liquid–gas interface to the top of the cylinder is 20 cm. The steady-state $n$-hexane evaporation rate is $8.2 \cdot 10^{-8}$ kg/s and the $n$-hexane mass fraction at the liquid–air interface is 0.482. The diffusivity of $n$-hexane in air is $8.0 \cdot 10^{-4}$ m$^2$/s.

A. Determine the mass flux of $n$-hexane vapor. Give units.

B. Determine the bulk flux of $n$-hexane vapor, i.e., that portion associated with the bulk flow, at the liquid–gas interface.

C. Determine the diffusional flux of $n$-hexane vapor at the liquid–gas interface.

**3.6** Consider water in a 25-mm-diameter test tube evaporating into dry air at 1 atm. The distance from the water–air interface to the top of the tube is $L = 15$ cm. The mass fraction of the water vapor at the water–air interface is 0.0235, and the binary diffusivity for water vapor in air is $2.6 \cdot 10^{-5}$ m$^2$/s.

A.  Determine the mass evaporation rate of the water.

B.  Determine the water vapor mass fraction at $x = L/2$.

C.  Determine the fraction of the water mass flow that is contributed by bulk flow and the fraction contributed by diffusion at $x = L/2$.

D.  Repeat part C for $x = 0$ and $x = L$. Plot your results. Discuss.

**3.7**  Consider the physical situation described in problem 3.6, except that the interface vapor mass fraction is unknown. Find the mass evaporation rate of the water when the liquid water is at 21°C. Assume equilibrium at the interface, i.e., $P_{H_2O}(x = 0) = P_{sat}(T)$. The air outside the tube is dry.

**3.8**  Repeat problem 3.6 when the air outside the tube is at 21°C and has a relative humidity of 50 percent. Also, determine the rate of heat transfer required to maintain the liquid water at 21°C.

**3.9**  Consider liquid *n*-hexane in a 50-mm-diameter graduated cylinder. Air blows across the top of the cylinder. The distance from the liquid–air interface to the open end of the cylinder is 20 cm. Assume the diffusivity of *n*-hexane is $8.8 \cdot 10^{-6}$ m$^2$/s. The liquid *n*-hexane is at 25 °C. Estimate the evaporation rate of the *n*-hexane. (*Hint:* Review the Clausius–Clapeyron relation as applied in Example 3.1.)

**3.10**  Calculate the evaporation rate constant for a 1-mm-diameter water droplet at 75 °C evaporating into dry, hot air at 500 K and 1 atm.

**3.11**  Determine the influence of the ambient water vapor mole fraction on the lifetimes of 50-*μ*m-diameter water droplets. The droplets are evaporating in air at 1 atm. Assume the droplet temperature is 75 °C and the mean air temperature is 200 °C. Use values of $\chi_{H_2O,\infty} = 0.1, 0.2,$ and $0.3$.

**3.12**  Consider the two general forms of Fick's law presented in this chapter:

$$\dot{m}_A'' = Y_A(\dot{m}_A'' + \dot{m}_B'') - \rho\mathcal{D}_{AB}\nabla Y_A \qquad (3.5)$$

$$\dot{N}_A'' = \chi_A(\dot{N}_A'' + \dot{N}_B'') - c\mathcal{D}_{AB}\nabla\chi_A, \qquad (3.6)$$

where $c$ is the molar concentration ($= \rho/MW_{mix}$).

The first expression (Eqn. 3.5) represents the **mass flux** of species A (kg$_A$/s-m$^2$) relative to a stationary laboratory reference frame, and the second is an equivalent expression for the **molar flux** of species A (kmol$_A$/s-m$^2$), also relative to a stationary laboratory reference frame.

The first term on the right-hand-side of Eqns. 3.5 and 3.6, respectively, is the bulk transport of species A at the **mass-average velocity, $V$** (for Eqn. 3.5) and the **molar-average velocity, $V^*$** (for Eqn. 3.6). You are used to working with the mass-average velocity as the velocity of choice in fluid mechanics. Both of these velocities are relative to our fixed laboratory reference frame. Useful relationships associated with these two velocities are the following [1]:

$$\dot{m}_A'' + \dot{m}_B'' = \rho V \tag{I}$$

or

$$\rho Y_A v_A + \rho Y_B v_B = \rho V \tag{II}$$

and

$$\dot{N}_A'' + \dot{N}_B'' = cV^* \tag{III}$$

or

$$c\chi_A v_A + c\chi_B v_B = cV^*, \tag{IV}$$

where $v_A$ and $v_B$ are the species velocities relative to the fixed frame.

The second term on the right-hand side of Eqns. 3.5 and 3.6, respectively, expresses the diffusion flux of species A relative to the mass-average velocity $V$ (for Eqn. 3.5) and relative to the molar-average velocity $V^*$ (for Eqn. 3.6).

Use the above relations (Eqns. I–IV), and others that you may need, to transform the 1-D planar form of Eqn. 3.6 to the 1-D planar form of Eqn. 3.5. *Hint:* There are a few elegant ways to do this, as well as tedious brute-force approaches.

**3.13**  Rederive the simple Stefan problem result (Eqn. 3.40) using the **molar flux** form of Fick's law (Eqn. 3.6) with boundary conditions expressed in mole fractions. Note that the molar concentration $c(= P/R_u T)$ is constant for an ideal gas, unlike the density $\rho$, which depends on the mixture molecular weight. Leave the final result in terms of the mole fractions.

**3.14**  Use the result from problem 3.13 to solve part A of Example 3.1 in the text. How does this result for $\dot{m}_{C_6H_6}$ compare with the text's result? Which is more accurate? Why?

# 4

# Chemical Kinetics

## OVERVIEW

Understanding the underlying chemical processes is essential to our study of combustion. In many combustion processes, chemical reaction rates control the rate of combustion, and, in essentially all combustion processes, chemical rates determine pollutant formation and destruction. Also, ignition and flame extinction are intimately related to chemical processes. The study of the elementary reactions and their rates, **chemical kinetics,** is a specialized field of physical chemistry. In the past few decades, much progress has been made in combustion because chemists have been able to define the detailed chemical pathways leading from reactants to products, and to measure or calculate their associated rates. With this knowledge, combustion scientists and engineers are able to construct computer models that simulate reacting systems. Although a tremendous amount of progress has been made, the problem of predicting the details of combustion in a complex flow field, where both fluid mechanics and chemistry are treated from first principles, has not yet been solved. In general, the fluid mechanical problem alone still taxes the largest computers (Chapter 11), and the addition of detailed chemistry makes the solution impossible.

In this chapter, we will look at basic chemical kinetics concepts. The subsequent chapter outlines the most important, or at least most well known, chemical mechanisms of importance to combustion; and, in Chapter 6, we will see also how models of chemical processes can be coupled to simple thermodynamic models of some reacting systems of interest to combustion engineers.

## GLOBAL VERSUS ELEMENTARY REACTIONS

The overall reaction of a mole of fuel with $a$ moles of an oxidizer to form $b$ moles of combustion products can be expressed by the **global reaction mechanism**

$$F + aOx \rightarrow bPr. \tag{4.1}$$

From experimental measurements, the rate at which the fuel is consumed can be expressed as

$$\frac{d[X_F]}{dt} = -k_G(T)[X_F]^n[X_{Ox}]^m,$$ (4.2)

where the notation $[X_i]$ is used to denote the molar concentration ($kmol/m^3$ in SI units or $gmol/cm^3$ in CGS units) of the $i$th species in the mixture. Equation 4.2 states that the rate of disappearance of the fuel is proportional to each of the reactants raised to a power. The constant of proportionality, $k_G$, is called the **global rate coefficient,** and, in general, is not constant, but rather is a strong function of temperature. The minus sign indicates that the fuel concentration decreases with time. The exponents $n$ and $m$ relate to the **reaction order.** Equation 4.2 says that the reaction is $n$th order with respect to the fuel, $m$th order with respect to the oxidizer, and $(n + m)$th order overall. For global reactions, $n$ and $m$ are not necessarily integers and arise from curvefitting experimental data. Later, we will see that for elementary reactions, reaction orders will always be integers. In general, a particular global expression in the form of Eqn. 4.2 holds only over a limited range of temperatures and pressures, and may depend on the details of the apparatus used to define the rate parameters. For example, different expressions for $k_G(T)$ and different values for $n$ and $m$ must be applied to cover a wide range of temperatures.

The use of global reactions to express the chemistry in a specific problem is frequently a "black box" approach. Although this approach may be useful in solving some problems, it does not provide a basis for understanding what is actually happening chemically in a system. For example, it is totally unrealistic to believe that $a$ oxidizer molecules simultaneously collide with a single fuel molecule to form $b$ product molecules, since this would require breaking several bonds and subsequently forming many new bonds. In reality, many sequential processes can occur involving many **intermediate species.** For example, consider the global reaction

$$2H_2 + O_2 \rightarrow 2H_2O.$$ (4.3)

To effect this global conversion of hydrogen and oxygen to water, the following **elementary reactions** are important:

$$H_2 + O_2 \rightarrow HO_2 + H,$$ (4.4)

$$H + O_2 \rightarrow OH + O,$$ (4.5)

$$OH + H_2 \rightarrow H_2O + H,$$ (4.6)

$$H + O_2 + M \rightarrow HO_2 + M,$$ (4.7)

among others.

In this partial mechanism for hydrogen combustion, we see from reaction 4.4 that when oxygen and hydrogen molecules collide and react, they do not yield water, but, instead, form the intermediate species $HO_2$, the hydroperoxy radical, and a hydrogen atom, H, another radical. **Radicals** or **free radicals** are reactive molecules, or atoms, that have unpaired electrons. To form $HO_2$ from $H_2$ and $O_2$, only one bond is broken and one bond formed. Alternatively, one might consider that $H_2$ and $O_2$ would react to form two hydroxyl radicals (OH); however, such a reaction is unlikely since it requires the breaking of two bonds and the creation of two new bonds. The hydrogen atom created in reaction 4.4 then reacts with $O_2$ to form two additional radicals, OH and O (reaction 4.5). It is the subsequent reaction (4.6) of the hydroxyl radical (OH) with molecular hydrogen that forms water. To have a complete picture of the combustion of $H_2$ and $O_2$, more than 20 elementary reactions can be considered [1, 2]. These we consider in Chapter 5. The collection of elementary reactions necessary to describe an overall reaction is called a reaction **mechanism.** Reaction mechanisms may involve only a few steps (i.e., elementary reactions) or as many as several hundred. A field of active research involves selecting the minimum number of elementary steps necessary to describe a particular global reaction.

## ELEMENTARY REACTION RATES

### Bimolecular Reactions and Collision Theory

Most elementary reactions of interest in combustion are **bimolecular;** that is, two molecules collide and react to form two different molecules. For an arbitrary bimolecular reaction, this is expressed as

$$A + B \rightarrow C + D. \tag{4.8}$$

Reactions 4.4–4.6 are examples of bimolecular elementary reactions.

The rate at which the reaction proceeds is directly proportional to the concentrations ($kmol/m^3$) of the two reactant species, i.e.,

$$\frac{d[A]}{dt} = -k_{bimolec}[A][B]. \tag{4.9}$$

All elementary bimolecular reactions are overall second order, being first order with respect to each of the reacting species. The rate coefficient, $k_{bimolec}$, again is a function of temperature, but unlike the global rate coefficient, this rate coefficient has a theoretical basis. The SI units for $k_{bimolec}$ are $m^3/kmol\text{-}s$; however, much of the chemistry and combustion literature still uses CGS units.

Molecular collision theory can be used to provide insight into the form of Eqn. 4.9 and to suggest the temperature dependence of the bimolecular rate coefficient. As we will see, the collision theory for bimolecular reactions has many shortcomings; nevertheless, the approach is important for historical reasons and provides a way to visualize bimolecular reactions. In our discussion of molecular transport in Chapter 3, we introduced the concepts of wall collision frequency, mean molecular speed, and mean free path (Eqn. 3.10). These same concepts are important in our discussion of molecular collision rates. To determine the collision frequency of a pair of molecules, we start with the simpler case of a single molecule of diameter $\sigma$ traveling with constant speed $v$ and experiencing collisions with identical, but

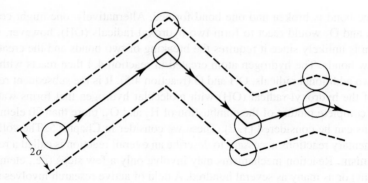

**Figure 4.1**      Collision volume swept out by a molecule, with diameter $\sigma$, striking like molecules.

stationary, molecules. The random path of the molecule is illustrated in Fig. 4.1. If the distance traveled between collisions, i.e., the mean free path, is large, then the moving molecule sweeps out a cylindrical volume in which collisions are possible, equal to $v\pi\sigma^2\Delta t$, in the time interval $\Delta t$. If the stationary molecules are distributed randomly and have a number density $n/V$, the number of collisions experienced by the traveling molecule per unit time can be expressed as

$$Z \equiv \frac{\text{collisions}}{\text{per unit time}} = (n/V)v\pi\sigma^2. \tag{4.10}$$

In an actual gas, all of the molecules are moving. If we assume Maxwellian velocity distributions for all the molecules, the collision frequency for identical molecules is given by [2, 3]

$$Z_c = \sqrt{2}(n/V)\pi\sigma^2\bar{v}, \tag{4.11}$$

where $\bar{v}$ is the mean speed whose value depends on temperature (Eqn. 3.10a).

Equation 4.11 applies to identical molecules. We can extend our analysis to collisions between unlike molecules having hard-sphere diameters of $\sigma_A$ and $\sigma_B$. The diameter of the collision volume (Fig. 4.1) is then $\sigma_A + \sigma_B \equiv 2\sigma_{AB}$. Thus, Eqn. 4.11 becomes

$$Z_c = \sqrt{2}(n_B/V)\pi\sigma_{AB}^2\bar{v}_A, \tag{4.12}$$

which expresses the frequency of collision of a single A molecule with all B molecules. We are interested, however, in the collision frequency associated with all A and B molecules. Thus, the total number of collisions per unit volume and per unit time is obtained by multiplying the collision frequency of a single A molecule (Eqn. 4.12) by the number of A molecules per unit volume, and using the appropriate mean molecular speeds, i.e.,

$$Z_{AB}/V = \frac{\text{No. of collisions between all A and all B}}{\text{Unit volume} \cdot \text{unit time}} \tag{4.13}$$

$$= (n_A/V)(n_B/V)\pi\sigma_{AB}^2\left(\bar{v}_A^2 + \bar{v}_B^2\right)^{1/2},$$

which can be expressed in terms of temperature as [2, 3]

$$Z_{AB}/V = (n_A/V)(n_B/V)\pi\sigma_{AB}^2\left(\frac{8k_BT}{\pi\mu}\right)^{1/2},$$  (4.14)

where

$k_B$ = Boltzmann constant = $1.381 \cdot 10^{-23}$ J/K;

$\mu = \dfrac{m_A m_B}{m_A + m_B} = $ reduced mass where $m_A$ and $m_B$ are the masses of species A and B, respectively, in kilograms;

$T$ = absolute temperature (K).

Note that the average speed is obtained by replacing the mass of the single molecule in Eqn. 3.10a with the reduced mass $\mu$.

To relate the above to the problem of reaction rates, we write

$$-\frac{d[A]}{dt} = \left[\frac{\text{No. of collisions A and B molecules}}{\text{Unit volume} \cdot \text{unit time}}\right] \cdot \left[\begin{array}{c}\text{Probability that a}\\ \text{collision leads to}\\ \text{reaction}\end{array}\right] \cdot \left[\frac{\text{kmol of A}}{\text{No. of molecules of A}}\right],$$

(4.15a)

or

$$-\frac{d[A]}{dt} = (Z_{AB}/V)\mathcal{P}N_{AV}^{-1},$$  (4.15b)

where $N_{AV}$ is the Avogadro number ($6.022 \cdot 10^{26}$ molecules/kmol). The probability that a collision leads to reaction can be expressed as a product of two factors: an energy factor, $\exp[-E_A/R_uT]$, which expresses the fraction of collisions that occur with an energy above the threshold level necessary for reaction, $E_A$, or **activation energy;** and a geometrical or **steric factor, $p$,** that takes into account the geometry of collisions between A and B. For example, in the reaction of OH and H to form $H_2O$, intuitively one expects a reaction to be more likely if the H atom strikes the O side of the hydroxyl rather than the H side, since the product has bonds of the form H–O–H. In general, steric factors are much less than unity; however, there are exceptions. Thus, Eqn. 4.15b becomes

$$-\frac{d[A]}{dt} = pN_{AV}\sigma_{AB}^2\left[\frac{8\pi k_BT}{\mu}\right]^{1/2}\exp[-E_A/R_uT][A][B],$$  (4.16)

where the substitutions $n_A/V = [A]N_{AV}$ and $n_B/V = [B]N_{AV}$ have been employed. Comparing Eqn. 4.9 with Eqn. 4.16, we see that the bimolecular rate coefficient, based on collision theory, is

$$k(T) = pN_{AV}\sigma_{AB}^2\left[\frac{8\pi k_BT}{\mu}\right]^{1/2}\exp[-E_A/R_uT].$$  (4.17)

Unfortunately, collision theory provides no means to determine the activation energy or the steric factor. More-advanced theories, which postulate the structure of the molecule in the process of breaking and forming bonds, i.e., an **activated complex,** do allow calculation of $k_{bimolec}$ from first principles. A discussion of such theories is beyond the scope of this book and the interested reader is referred to Refs. [2] and [3].

If the temperature range of interest is not too great, the bimolecular rate coefficient can be expressed by the empirical **Arrhenius form,**

$$k(T) = A \exp(-E_A / R_u T), \tag{4.18}$$

where $A$ is a constant termed the **pre-exponential factor** or the **frequency factor.** Comparing Eqns. 4.17 and 4.18, we see that $A$ is not strictly constant but depends, based on collision theory, on $T^{1/2}$. **Arrhenius plots** of log $k$ versus $1/T$ for experimental data are used to extract values for the activation energy, since the slope of such plots is $-E_A/R_u$.

Although tabulation of experimental values for rate coefficients in Arrhenius form is common, current practice frequently utilizes the three-parameter functional form:

$$k(T) = AT^b \exp(-E_A / R_u T), \tag{4.19}$$

where $A$, $b$, and $E_A$ are the three empirical parameters. Table 4.1 illustrates the three-parameter form showing the recommendations of Warnatz [4] for the $H_2$–$O_2$ system.

**Table 4.1**     Recommended rate coefficients for $H_2$–$O_2$ reactions from Ref. [4]

| Reaction | $A$ $((cm^3/gmol)^{n-1}/s)^a$ | $b$ | $E_A$ (kJ/gmol) | Temperature Range (K) |
|---|---|---|---|---|
| $H + O_2 \rightarrow OH + O$ | $1.2 \cdot 10^{17}$ | $-0.91$ | 69.1 | 300–2500 |
| $OH + O \rightarrow O_2 + H$ | $1.8 \cdot 10^{13}$ | 0 | 0 | 300–2500 |
| $O + H_2 \rightarrow OH + H$ | $1.5 \cdot 10^{7}$ | 2.0 | 31.6 | 300–2500 |
| $OH + H_2 \rightarrow H_2O + H$ | $1.5 \cdot 10^{8}$ | 1.6 | 13.8 | 300–2500 |
| $H + H_2O \rightarrow OH + H_2$ | $4.6 \cdot 10^{8}$ | 1.6 | 77.7 | 300–2500 |
| $O + H_2O \rightarrow OH + OH$ | $1.5 \cdot 10^{10}$ | 1.14 | 72.2 | 300–2500 |
| $H + H + M \rightarrow H_2 + M$ | | | | |
| M = Ar (low $P$) | $6.4 \cdot 10^{17}$ | $-1.0$ | 0 | 300–5000 |
| M = $H_2$ (low $P$) | $0.7 \cdot 10^{16}$ | $-0.6$ | 0 | 100–5000 |
| $H_2 + M \rightarrow H + H + M$ | | | | |
| M = Ar (low $P$) | $2.2 \cdot 10^{14}$ | 0 | 402 | 2500–8000 |
| M = $H_2$ (low $P$) | $8.8 \cdot 10^{14}$ | 0 | 402 | 2500–8000 |
| $H + OH + M \rightarrow H_2O + M$ | | | | |
| M = $H_2O$ (low $P$) | $1.4 \cdot 10^{23}$ | $-2.0$ | 0 | 1000–3000 |
| $H_2O + M \rightarrow H + OH + M$ | | | | |
| M = $H_2O$ (low $P$) | $1.6 \cdot 10^{17}$ | 0 | 478 | 2000–5000 |
| $O + O + M \rightarrow O_2 + M$ | | | | |
| M = Ar (low $P$) | $1.0 \cdot 10^{17}$ | $-1.0$ | 0 | 300–5000 |
| $O_2 + M \rightarrow O + O + M$ | | | | |
| M = Ar (low $P$) | $1.2 \cdot 10^{14}$ | 0 | 451 | 2000–10,000 |

$^a n$ is the reaction order.

Determine the collision-theory steric factor for the reaction    **Example 4.1**

$$O + H_2 \rightarrow OH + H$$

at 2000 K given the hard-sphere diameters $\sigma_O = 3.050$ and $\sigma_{H_2} = 2.827$ Å and the experimental parameters given in Table 4.1.

### Solution

Equating the collision-theory rate coefficient (Eqn. 4.17) with the three-parameter experimental rate coefficient (Eqn. 4.19) yields

$$k(T) = pN_{AV}\sigma_{AB}^2 \left[\frac{8\pi k_B T}{\mu}\right]^{1/2} \exp[-E_A/R_u T] = AT^b \exp[-E_A/R_u T],$$

where we assume the activation energy $E_A$ is the same for both expressions. Solving for the steric factor, $p$, is straightforward, although care must be exercised in treating the units:

$$p = \frac{AT^b}{N_{AV}\left(\dfrac{8\pi k_B T}{\mu}\right)^{1/2}\sigma_{AB}^2}.$$

To evaluate the above, we employ

$$A = 1.5\cdot 10^7 \text{ cm}^3/\text{gmol-s} \quad \text{(Table 4.1)},$$

$$b = 2.0 \quad \text{(Table 4.1)},$$

$$\sigma_{AB} = \left(\sigma_O + \sigma_{H_2}\right)/2$$
$$= (3.050 + 2.827)/2 = 2.939 \text{ Å}$$
$$= 2.939\cdot 10^{-8} \text{ cm},$$

$$m_O = \frac{16 \text{ g/gmol}}{6.022\cdot 10^{23} \text{ molecules/gmol}} = 2.66\cdot 10^{-23} \text{ g},$$

$$m_{H_2} = \frac{2.008}{6.022\cdot 10^{23}} = 0.332\cdot 10^{-23} \text{ g},$$

$$\mu = \frac{m_O m_{H_2}}{m_O + m_{H_2}} = \frac{2.66(0.332)}{2.66 + 0.332}\cdot 10^{-23} = 2.95\cdot 10^{-24} \text{ g},$$

$$k_B = 1.381\cdot 10^{-23} \text{ J/K} = 1.381\cdot 10^{-16} \text{ g-cm}^2/\text{s}^2\text{-K}.$$

Thus,

$$p = \frac{1.5\cdot 10^7 (2000)^2}{6.022\cdot 10^{23}\left(\dfrac{8\pi(1.381\cdot 10^{-16})2000}{2.95\cdot 10^{-24}}\right)^{1/2}(2.939\cdot 10^{-8})^2}$$

$$= 0.075.$$

Units check:

$$p[=]\frac{\text{cm}^3}{\text{gmol-s}}\frac{1}{\dfrac{1}{\text{gmol}}\left(\dfrac{\text{g-cm}^2}{\text{s}^2\text{-K}}\dfrac{K}{g}\right)^{1/2}\text{cm}^2} = 1$$

$$\boxed{p = 0.075 \text{ (dimensionless)}}$$

**Comment**

The value of $p = 0.075$ is less than unity, as expected, and its small value points out the shortcomings of the simple theory. Note that CGS units were employed and note the use of the Avogadro number to calculate the species mass in grams. Note also that all of the units for $k(T)$ are in the pre-exponential factor $A$; thus, the factor $T^b$ is dimensionless by definition.

## Other Elementary Reactions

As the name suggests, **unimolecular** reactions involve a single species undergoing a rearrangement (isomerization or decomposition) to form one or two product species, i.e.,

$$A \rightarrow B \tag{4.20}$$

or

$$A \rightarrow B + C. \tag{4.21}$$

Examples of unimolecular reactions include the typical dissociation reactions important to combustion: e.g., $O_2 \rightarrow O + O$, $H_2 \rightarrow H + H$, etc.

Unimolecular reactions are first order at high pressures:

$$\frac{d[A]}{dt} = -k_{uni}[A], \tag{4.22}$$

whereas at low pressures, the reaction rate also depends on the concentration of any molecules, M, with which the reacting species may collide. In this case,

$$\frac{d[A]}{dt} = -k[A][M]. \tag{4.23}$$

To explain this interesting behavior requires postulating a three-step mechanism. Since we have yet to explore some of the concepts required to deal with this, we conclude our treatment of unimolecular reactions for the time being.

**Termolecular** reactions involve three reactant species and correspond to the reverse of the unimolecular reaction at low pressures. The general form of a termolecular reaction is

$$A + B + M \rightarrow C + M. \tag{4.24}$$

Recombination reactions such as $H + H + M \rightarrow H_2 + M$ and $H + OH + M \rightarrow H_2O + M$ are important examples of termolecular reactions in combustion. Termolecular reactions are third order, and their rates can be expressed as

$$\frac{d[A]}{dt} = -k_{ter}[A][B][M], \tag{4.25}$$

where, again, M may be any molecule and is frequently referred to as a **third body.** When A and B are the same species, as in $H + H + M$, a factor of two must multiply the right-hand side of Eqn. 4.25 since two A molecules disappear to form C. In radical–radical reactions, the third body is required to carry away the energy liberated in forming the stable species. During the collision, the internal energy of the newly formed molecule is transferred to the third body, M, and is manifest as kinetic energy of M. Without this energy transfer, the newly formed molecule would dissociate back to its constituent atoms.

# RATES OF REACTION FOR MULTISTEP MECHANISMS

## Net Production Rates

In the previous sections, we introduced the idea of a sequence of elementary reactions that leads from reactants to products, which we termed the reaction mechanism. Knowing how to express the rates of elementary reactions, we can now mathematically express the net rates of production or destruction for any species participating in a series of elementary steps. For example, let us return to the $H_2$–$O_2$ reaction mechanism, which is incompletely given by Eqns. 4.4–4.7, and include both forward and reverse reactions as indicated by the $\Leftrightarrow$ symbol:

$$H_2 + O_2 \underset{k_{r1}}{\overset{k_{f1}}{\Leftrightarrow}} HO_2 + H, \tag{R.1}$$

$$H + O_2 \underset{k_{r2}}{\overset{k_{f2}}{\Leftrightarrow}} OH + O, \tag{R.2}$$

$$OH + H_2 \underset{k_{r3}}{\overset{k_{f3}}{\Leftrightarrow}} H_2O + H, \tag{R.3}$$

$$H + O_2 + M \underset{k_{r4}}{\overset{k_{f4}}{\Leftrightarrow}} HO_2 + M, \tag{R.4}$$

$$\vdots$$

where $k_{fi}$ and $k_{ri}$ are the elementary forward and reverse rate coefficients, respectively, for the $i$th reaction. The net rate of production of $O_2$, for example, is the sum of all of the individual elementary rates producing $O_2$ minus the sum of all of the rates destroying $O_2$; i.e.,

$$\frac{d[O_2]}{dt} = k_{r1}[HO_2][H] + k_{r2}[OH][O] \tag{4.26}$$
$$+ k_{r4}[HO_2][M] + \cdots$$
$$- k_{f1}[H_2][O_2] - k_{f2}[H][O_2]$$
$$- k_{f4}[H][O_2][M] - \cdots$$

and for H atoms,

$$\frac{d[H]}{dt} = k_{f1}[H_2][O_2] + k_{r2}[OH][O]$$

$$+ k_{f3}[OH][H_2] + k_{r4}[HO_2][M] + \cdots$$

$$- k_{r1}[HO_2][H] - k_{f2}[H][O_2]$$

$$- k_{r3}[H_2O][H] - k_{f4}[H][O_2][M] - \cdots$$

(4.27)

We can write similar expressions for each species participating in the mechanism, which yields a system of first-order ordinary differential equations that describes the evolution of the chemical system starting from given initial conditions, i.e.,

$$\frac{d[X_i](t)}{dt} = f_i([X_1](t), [X_2](t),\ldots, [X_n](t))$$

(4.28a)

with

$$[X_i](0) = [X_i]_0.$$

(4.28b)

For a particular system, this set of equations (4.28a) couples with any necessary statements of conservation of mass, momentum, or energy, and state equations, which can be integrated numerically using a computer. Packaged routines, such as IMSL's DGEAR [5], efficiently integrate the **stiff** system of equations that arise in chemical systems. A set of equations is considered stiff when one or more variables change very rapidly, while others change very slowly. This disparity in time scales is common in chemical systems where radical reactions are very fast compared with reactions involving stable species. Several numerical integration routines have been used or developed specifically for reacting chemical systems [1, 6–8].

## Compact Notation

Since mechanisms may involve many elementary steps and many species, a compact notation has been developed to represent both the mechanism, e.g., R.1–R.4 . . . , and the individual species production rates, e.g., Eqns. 4.26 and 4.27. For the mechanism, one can write

$$\sum_{j=1}^{N} v'_{ji} X_j \Leftrightarrow \sum_{j=1}^{N} v''_{ji} X_j \quad \text{for} \quad i = 1, 2,\ldots, L,$$

(4.29)

where $v'_{ji}$ and $v''_{ji}$ are the **stoichiometric coefficients** on the reactants and products side of the equation, respectively, for the $j$th species in the $i$th reaction. For example, consider the four reactions R.1–R.4, involving the eight species $O_2$, $H_2$, $H_2O$, $HO_2$, $O$, $H$, $OH$, and $M$. Defining $j$ and $i$ as follows:

| $j$ | Species | $i$ | Reaction |
|---|---|---|---|
| 1 | $O_2$ | 1 | R.1 |
| 2 | $H_2$ | 2 | R.2 |
| 3 | $H_2O$ | 3 | R.3 |
| 4 | $HO_2$ | 4 | R.4 |
| 5 | $O$ | | |
| 6 | $H$ | | |
| 7 | $OH$ | | |
| 8 | $M$ | | |

and using $j$ as the column index and $i$ the row index, we can write the stoichiometric coefficient matrices as

$$v'_{ji} = \begin{bmatrix} 1 & 1 & 0 & 0 & 0 & 0 & 0 & 0 \\ 1 & 0 & 0 & 0 & 0 & 1 & 0 & 0 \\ 0 & 1 & 0 & 0 & 0 & 0 & 1 & 0 \\ 1 & 0 & 0 & 0 & 0 & 1 & 0 & 1 \end{bmatrix} \tag{4.30a}$$

and

$$v''_{ji} = \begin{bmatrix} 0 & 0 & 0 & 1 & 0 & 1 & 0 & 0 \\ 0 & 0 & 0 & 0 & 1 & 0 & 1 & 0 \\ 0 & 0 & 1 & 0 & 0 & 1 & 0 & 0 \\ 0 & 0 & 0 & 1 & 0 & 0 & 0 & 1 \end{bmatrix}. \tag{4.30b}$$

Since elementary reactions involve, at most, three reactant species, the coefficient matrices will always be sparse (many more zero than nonzero elements) when the number of species involved is large.

The following three relations compactly express the net production rate of each species in a multistep mechanism:

$$\dot{\omega}_j = \sum_{i=1}^{L} v_{ji} q_i \quad \text{for} \quad j = 1, 2, \ldots, N, \tag{4.31}$$

where

$$v_{ji} = (v''_{ji} - v'_{ji}) \tag{4.32}$$

and

$$q_i = k_{fi} \prod_{j=1}^{N} [X_j]^{v'_{ji}} - k_{ri} \prod_{j=1}^{N} [X_j]^{v''_{ji}}. \tag{4.33}$$

The **production rates,** $\dot{\omega}_j$, correspond to the left-hand sides of Eqns. 4.26 and 4.27, for example. For systems in which the species concentrations are only affected by chemical transformations, $\dot{\omega}_j \equiv d[X_j]/dt$ for the complete mechanism. Equation 4.33 defines the **rate-of-progress variable,** $q_i$, for the $i$th elementary reaction. The symbol $\prod$ is used to denote a product of terms in the same sense that $\sum$ is used to represent a summation of terms. For example, $q_i$ ($= q_1$) for reaction R.1 is expressed

$$q_i = k_{f1}[O_2]^1[H_2]^1[H_2O]^0[HO_2]^0[O]^0[H]^0[OH]^0[M]^0$$
$$- k_{r1}[O_2]^0[H_2]^0[H_2O]^0[HO_2]^1[O]^0[H]^1[OH]^0[M]^0 \qquad (4.34)$$
$$= k_{f1}[O_2][H_2] - k_{r1}[HO_2][H].$$

Writing similar expressions for $i = 2$, 3, and 4 and summing (Eqn. 4.31), taking into account whether the $j$th species is created, destroyed, or does not participate in the $i$th step (Eqn. 4.32), completes the total rate expression for $\dot{\omega}_j$. Chapter 6 illustrates how these rate expressions are applied to various reacting systems.

The compact notation embodied in Eqns. 4.29 and 4.31–4.34 is particularly useful in solving chemical kinetics problems using a computer. The software CHEMKIN [1] is a widely used general-purpose package for solving chemical kinetics problems developed at Sandia (Livermore) National Laboratories.

## Relation Between Rate Coefficients and Equilibrium Constants

Measuring rate coefficients of elementary reactions is a difficult task that frequently leads to results that possess a rather large degree of uncertainty. The more reliable rate coefficients often are known no better than within a factor of two, whereas others may be uncertain by an order of magnitude, or more. On the other hand, equilibrium constants, which are based on thermodynamic measurements or calculations, are very accurate and precise in most cases. We can take advantage of accurate thermodynamic data in solving chemical kinetics problems by recognizing that, at equilibrium, the forward and reverse reaction rates must be equal. For example, consider both the forward and reverse rates for the arbitrary bimolecular reaction

$$A + B \underset{k_r}{\overset{k_f}{\Longleftrightarrow}} C + D. \qquad (4.35)$$

For species A, we can write

$$\frac{d[A]}{dt} = -k_f[A][B] + k_r[C][D]. \qquad (4.36)$$

For the equilibrium condition, $A + B = C + D$, the time rate of change of [A] must be zero, as must also be the time rates of change of species B, C, and D. Thus, we can express equilibrium as

$$0 = -k_f[A][B] + k_r[C][D], \qquad (4.37)$$

which, upon rearranging, becomes

$$\frac{[C][D]}{[A][B]} = \frac{k_f(T)}{k_r(T)}. \tag{4.38}$$

In Chapter 2, we defined the equilibrium constant based on partial pressures for an arbitrary equilibrium reaction to be

$$K_p = \frac{\left(P_C/P^o\right)^c \left(P_D/P^o\right)^d \cdots}{\left(P_A/P^o\right)^a \left(P_B/P^o\right)^b \cdots} \tag{4.39}$$

where the superscripts are the stoichiometric coefficients, i.e., $v_i' = a, b, \ldots$ and $v_i'' = c, d, \ldots$. Since the molar concentrations relate to the mole fractions and partial pressures as follows,

$$[X_i] = \chi_i P/R_u T = P_i/R_u T, \tag{4.40}$$

we can define an equilibrium constant based on molar concentrations, $K_c$. The relationship between $K_c$ and $K_p$ is

$$K_p = K_c \left(R_u T/P^o\right)^{c+d+\cdots-a-b-\cdots} \tag{4.41a}$$

or

$$K_p = K_c \left(R_u T/P^o\right)^{\Sigma v'' - \Sigma v'}, \tag{4.41b}$$

where

$$K_c = \frac{[C]^c [D]^d \cdots}{[A]^a [B]^b \cdots} = \frac{\displaystyle\prod_{\text{prod}} [X_i]^{v_i''}}{\displaystyle\prod_{\text{react}} [X_i]^{v_i'}}. \tag{4.42}$$

From the above, we see that the ratio of the forward and reverse rate coefficients equals the equilibrium constant $K_c$,

$$\frac{k_f(T)}{k_r(T)} = K_c(T), \tag{4.43}$$

and that for a bimolecular reaction, $K_c = K_p$. Using Eqn. 4.43, one can compute a reverse reaction rate from a knowledge of the forward rate and the equilibrium constant for the reaction; or, conversely, one can calculate the forward rate knowing the reverse rate. In performing kinetic calculations, the most accurate experimental rate coefficient over the temperature range of interest should be used, and the rate in the opposite direction calculated from the equilibrium constant. Over a wide range of temperatures, different choices may have to be made. The National Institute of Standards and Technology (formerly the U.S. Bureau of Standards) maintains a chemical kinetics database providing rate information on well over 6000 reactions [9].

**Example 4.2**

In their survey of experimental determinations of rate coefficients for the N–H–O system, Hanson and Salimian [10] recommend the following rate coefficient for the reaction $NO + O \rightarrow N + O_2$:

$$k_f = 3.80 \cdot 10^9 \, T^{1.0} \exp(-20,820/T) [=] \, cm^3/gmol\text{-}s.$$

Determine the rate coefficient $k_r$ for the reverse reaction, i.e., $N + O_2 \rightarrow NO + O$, at 2300 K.

**Solution**

The forward and reverse rate coefficients are related to the equilibrium constant, $K_p$, via Eqn. 4.43:

$$\frac{k_f(T)}{k_r(T)} = K_c(T) = K_p(T).$$

Thus, to find $k_r$, we need to evaluate $k_f$ and $K_p$ at 2300 K. From Eqns. 2.66 and 2.64, we can evaluate $K_p$:

$$K_p = \exp\left(\frac{-\Delta G_T^o}{R_u T}\right),$$

where

$$\Delta G_{2300\,K}^o = \left[ \overline{g}_{f,N}^o + \overline{g}_{f,O_2}^o - \overline{g}_{f,NO}^o - \overline{g}_{f,O}^o \right]_{2300\,K}$$

$$= 326,331 + 0 - 61,243 - 101,627 \quad \text{(Appendix Tables A.8, A.9, A.11, A.12)}$$

$$= 163,461 \, kJ/kmol$$

$$K_p = \exp\left(\frac{-163,461}{8.315(2300)}\right) = 1.94 \cdot 10^{-4} \quad \text{(dimensionless)}.$$

The forward rate coefficient at 2300 K is

$$k_f = 3.8 \cdot 10^9 (2300) \exp\left(\frac{-20,820}{2300}\right)$$

$$= 1.024 \cdot 10^9 \, cm^3/gmol\text{-}s.$$

So,

$$\boxed{k_r} = k_f / K_p = \frac{1.024 \cdot 10^9}{1.94 \cdot 10^{-4}} = \boxed{5.28 \cdot 10^{12} \, cm^3/gmol\text{-}s}$$

**Comment**

The reactions used in this example are part of the important Zeldovich or thermal NO mechanism: $O + N_2 \Leftrightarrow NO + O$ and $N + O_2 \Leftrightarrow NO + O$. We will explore this mechanism further in a subsequent example, as well as in Chapters 5 and 15.

## Steady-State Approximation

In many chemical systems of interest to combustion, highly reactive intermediate species, e.g., radicals, are formed. Analyses of such systems can sometimes be simplified by applying the **steady-state approximation** to these reactive intermediates or radicals.

Physically, what occurs is that, after a rapid initial buildup in concentration, the radical is destroyed as rapidly as it is formed, so that its rate of formation and its rate of destruction are equal [11]. This situation typically occurs when the reaction forming the intermediate species is slow, while the reaction destroying the intermediate is very fast. As a result, the concentrations of the radical are quite small in comparison with those of the reactants and products. A good example of this is the Zeldovich mechanism for the formation of nitric oxide, where the reactive intermediate of interest is the N atom:

$$O + N_2 \xrightarrow{k_1} NO + N$$

$$N + O_2 \xrightarrow{k_2} NO + O.$$

The first reaction in this pair is slow and, hence, rate limiting; while the second is extremely fast. We can write the net production rate of N atoms as

$$\frac{d[N]}{dt} = k_1[O][N_2] - k_2[N][O_2]. \tag{4.44}$$

After a rapid transient allowing the buildup of N atoms to some low concentration, the two terms on the right-hand-side of Eqn. 4.44 become equal, and $d[N]/dt$ approaches zero. With $d[N]/dt \rightarrow 0$, we are able to determine the steady concentration of N atoms:

$$0 = k_1[O][N_2] - k_2[N]_{ss}[O_2] \tag{4.45}$$

or

$$[N]_{ss} = \frac{k_1[O][N_2]}{k_2[O_2]}. \tag{4.46}$$

Although invoking the steady-state approximation suggests that $[N]_{ss}$ does not change with time, $[N]_{ss}$ may change as it rapidly readjusts according to Eqn. 4.46. To determine the time rate of change requires differentiating Eqn. 4.46, rather than applying Eqn. 4.44:

$$\frac{d[N]_{ss}}{dt} = \frac{d}{dt} \left[ \frac{k_1[O][N_2]}{k_2[O_2]} \right]. \tag{4.47}$$

In the next section, we apply the steady-state approximation to the mechanism for unimolecular reactions.

## The Mechanism for Unimolecular Reactions

In a previous section, we deferred discussing the pressure dependence of unimolecular reactions until we had a grasp of the multistep mechanism concept. To explain the pressure dependence requires a three-step mechanism:

$$A + M \xrightarrow{k_e} A^* + M, \tag{4.48a}$$

$$A^* + M \xrightarrow{k_{de}} A + M, \tag{4.48b}$$

$$A^* \xrightarrow{k_{uni}} \text{products.} \tag{4.48c}$$

In the first step, Eqn. 4.48a, molecule A collides with the "third body," M. As a result of the collision, some of the translational kinetic energy from M is transferred to molecule A, resulting in an increase in A's internal vibrational and rotational energies. The high-internal-energy A molecule is referred to as the *energized* A molecule, $A^*$. After the energizing of A, two things may happen: $A^*$ may collide with another molecule resulting in $A^*$'s internal energy being converted back to translational energy in the reverse of the energizing process (Eqn. 4.48b), or the energized $A^*$ may fly apart in a true unimolecular process (Eqn. 4.48c). To observe the pressure dependence, we can write an expression that describes the rate at which products are formed:

$$\frac{d[\text{products}]}{dt} = k_{uni}[A^*]. \tag{4.49}$$

To evaluate $[A^*]$, we will employ the steady-state approximation discussed above. The net production of $A^*$ can be expressed as

$$\frac{d[A^*]}{dt} = k_e[A][M] - k_{de}[A^*][M] - k_{uni}[A^*]. \tag{4.50}$$

If we assume that $d[A^*]/dt = 0$, following some initial rapid transient during which $[A^*]$ reaches a steady state, then we can solve for $[A^*]$:

$$[A^*] = \frac{k_e[A][M]}{k_{de}[M] + k_{uni}}. \tag{4.51}$$

Substituting Eqn. 4.51 into Eqn. 4.49 yields

$$\frac{d[\text{products}]}{dt} = \frac{k_{uni}k_e[A][M]}{k_{de}[M] + k_{uni}} = \frac{k_e[A][M]}{(k_{de}/k_{uni})[M] + 1}. \tag{4.52}$$

For the overall reaction,

$$A \xrightarrow{k_{app}} \text{products,} \tag{4.53}$$

we write

$$-\frac{d[A]}{dt} = \frac{d[\text{products}]}{dt} = k_{app}[A], \tag{4.54}$$

where $k_{app}$ is defined as the apparent unimolecular rate coefficient. Equating Eqns. 4.52 and 4.54, we find that the apparent rate coefficient is of the form

$$k_{app} = \frac{k_e[M]}{(k_{de}/k_{uni})[M] + 1}. \tag{4.55}$$

Analyzing Eqn. 4.55, we can explain the interesting pressure dependence of unimolecular reactions. When the pressure is increased, [M] ($kmol/m^3$) increases. At high enough pressures, the term $k_{de}[M]/k_{uni}$ becomes much larger than unity, and the [M]s in the numerator and denominator of Eqn. 4.55 cancel. Thus,

$$k_{app}(P \to \infty) = k_{uni}k_e/k_{de}. \qquad (4.56)$$

At sufficiently low pressures, $k_{de}[M]/k_{uni}$ is much less than unity and can be neglected in comparison; hence,

$$k_{app}(P \to 0) = k_e[M], \qquad (4.57)$$

and the dependence of the reaction rate on [M] becomes apparent. Thus, we see that the three-step mechanism indeed provides a logical explanation for the high- and low-pressure limits of unimolecular reactions. Procedures for obtaining rate coefficients for pressures between the two limits are discussed by Gardiner and Troe [12].

## Chain and Chain-Branching Reactions

**Chain reactions** involve the production of a radical species that subsequently reacts to produce another radical. This radical, in turn, reacts to produce yet another radical. This sequence of events, or chain reaction, continues until a reaction involving the formation of a stable species from two radicals breaks the chain. Chain reactions can occur in many chemical processes of importance to combustion, as you will see in Chapter 5.

In the following, we illustrate some of the features of chain reactions by exploring a hypothetical chain mechanism, which is globally represented as

$$A_2 + B_2 \to 2AB.$$

The **chain-initiation reaction** is

$$A_2 + M \xrightarrow{k_1} A + A + M \qquad (C.1)$$

and the **chain-propagating reactions** involving the free radicals A and B are

$$A + B_2 \xrightarrow{k_2} AB + B \qquad (C.2)$$

$$B + A_2 \xrightarrow{k_3} AB + A. \qquad (C.3)$$

The **chain-terminating reaction** is

$$A + B + M \xrightarrow{k_4} AB + M. \qquad (C.4)$$

In the early stages of reaction, the concentration of the product AB is small, as are the concentrations of A and B throughout the course of the reaction; thus, we can neglect the reverse reactions to determine the reaction rates for the stable species at this reaction stage:

$$\frac{d[A_2]}{dt} = -k_1[A_2][M] - k_3[A_2][B], \tag{4.58}$$

$$\frac{d[B_2]}{dt} = -k_2[B_2][A], \tag{4.59}$$

$$\frac{d[AB]}{dt} = k_2[A][B_2] + k_3[B][A_2] + k_4[A][B][M]. \tag{4.60}$$

For the radicals A and B, the steady-state approximation is invoked. Thus,

$$2k_1[A_2][M] - k_2[A][B_2] + k_3[B][A_2] - k_4[A][B][M] = 0 \tag{4.61}$$

and

$$k_2[A][B_2] - k_3[B][A_2] - k_4[A][B][M] = 0. \tag{4.62}$$

Solving Eqns. 4.61 and 4.62 simultaneously for [A] yields

$$[A] = \frac{k_1}{2k_2}\frac{[A_2][M]}{[B_2]}\left\{1 + \left[1 + \frac{4k_2k_3}{k_1k_4}\frac{[B_2]}{[M]^2}\right]^{1/2}\right\}. \tag{4.63}$$

A similarly complicated expression results for [B]. Knowing the steady-state values for [A] and [B], the initial reaction rates $d[A_2]/dt$ and $d[B_2]/dt$, and $d[AB]/dt$, can be determined for some initial concentrations of $[A_2]$ and $[B_2]$. Of the three reaction rates, $d[B_2]/dt$ is the simplest and is expressed as

$$\frac{d[B_2]}{dt} = -\frac{k_1}{2}[A_2][M]\left\{1 + \left[1 + \frac{4k_2k_3}{k_1k_4}\frac{[B_2]}{[M]^2}\right]^{1/2}\right\}. \tag{4.64}$$

Equations 4.63 and 4.64 can be simplified further by recognizing that the term within the brackets $4k_2k_3[B_2]/(k_1k_4[M]^2) \gg 1$. This inequality holds because the rate coefficients for the radical reactions, $k_2$ and $k_3$, must be much larger than $k_1$ and $k_4$ for the steady-state approximation to apply. If we assume that the square root of $4k_2k_3[B_2]/(k_1k_4[M]^2)$ is also much greater than unity, we can write approximate expressions for [A] and $d[B_2]/dt$ that can be easily analyzed:

$$[A] \approx \frac{[A_2]}{[B_2]^{1/2}}\left(\frac{k_1k_3}{k_2k_4}\right)^{1/2} \tag{4.65}$$

and

$$\frac{d[B_2]}{dt} \approx -[A_2][B_2]^{1/2}\left(\frac{k_1k_2k_3}{k_4}\right)^{1/2}. \tag{4.66}$$

From Eqn. 4.65, we see that the radical concentration depends on the square root of the ratio of the initiation-step (C.1) rate coefficient $k_1$ to the termolecular chain-terminating step (C.4) rate coefficient $k_4$; the greater the initiation rate, the greater the radical concentration, and conversely, the greater the termination rate, the lesser the radical concentration. The response of the disappearance rate of $[B_2]$ to $k_1$ and $k_4$ is similar.

We also see that increasing the rate coefficients $k_2$ and $k_3$ of the chain-propagating steps (C.2 and C.3) results in an increased rate of disappearance of $[B_2]$. This effect is in direct proportion to the propagating-reaction rate coefficient, i.e., $k_{prop} \equiv (k_2 k_3)^{1/2}$. The rate coefficients of the chain-propagating steps are likely to have little effect upon the radical concentrations. Because $k_2$ and $k_3$ appear as a ratio in Eqn. 4.65, their influence on the radical concentrations would be small for rate coefficients of similar magnitude.

These simple scalings break down at pressures sufficiently high to cause our assumption that $4k_2 k_3 [B_2]/(k_1 k_4 [M]^2) \gg 1$ to break down. Recall that the molar concentration $(kmol/m^3)$ scales directly with pressure if the mole fractions and temperature are fixed.

**Chain-branching reactions** involve the formation of two radical species from a reaction that consumes only one radical. The reaction, $O + H_2O \rightarrow OH + OH$, is an example of a chain-branching reaction. The existence of a chain-branching step in a chain mechanism can have an explosive effect, literally, and the interesting explosion behavior of mixtures of $H_2$ and $O_2$ (Chapter 5) is a result of chain-branching steps.

In a system with chain branching, it is possible for the concentration of a radical species to build up geometrically, causing the rapid formation of products. Unlike in the previous hypothetical example, the rate of the chain-initiation step does not control the overall reaction rate. With chain branching, the rates of radical reactions dominate. Chain-branching reactions are responsible for a flame being self-propagating and are an essential ingredient in combustion chemistry.

---

As mentioned previously, a famous chain mechanism is the Zeldovich, or thermal, mechanism for the formation of nitric oxide from atmospheric nitrogen:

**Example 4.3**

$$N_2 + O \xrightarrow{k_{1f}} NO + N$$

$$N + O_2 \xrightarrow{k_{2f}} NO + O.$$

Because the second reaction is much faster than the first, the steady-state approximation can be used to evaluate the N-atom concentration. Furthermore, in high-temperature systems, the NO formation reaction is typically much slower than other reactions involving $O_2$ and $O$. Thus, $O_2$ and $O$ can be assumed to be in equilibrium:

$$O_2 \underset{K_p}{\Leftrightarrow} 2O.$$

Construct a global mechanism

$$N_2 + O_2 \xrightarrow{k_G} 2NO$$

represented as

$$\frac{d[NO]}{dt} = k_G [N_2]^m [O_2]^n,$$

i.e., determine $k_G$, $m$, and $n$ using the elementary rate coefficients, etc., from the detailed mechanism.

**Solution**

From the elementary reactions, we can write

$$\frac{d[NO]}{dt} = k_{1f}[N_2][O] + k_{2f}[N][O_2]$$

$$\frac{d[N]}{dt} = k_{1f}[N_2][O] - k_{2f}[N][O_2],$$

where we assume reverse reaction rates are negligible.

With the steady-state approximation, $d[N]/dt = 0$; thus,

$$[N]_{ss} = \frac{k_{1f}[N_2][O]}{k_{2f}[O_2]}.$$

Substituting $[N]_{ss}$ into the above expression for $d[NO]/dt$, yields

$$\frac{d[NO]}{dt} = k_{1f}[N_2][O] + k_{2f}[O_2]\left(\frac{k_{1f}[N_2][O]}{k_{2f}[O_2]}\right)$$

$$= 2k_{1f}[N_2][O].$$

We now eliminate [O] through our equilibrium approximation,

$$K_p = \frac{P_O^2}{P_{O_2}P^o} = \frac{[O]^2(R_uT)^2}{[O_2](R_uT)P^o} = \frac{[O]^2}{[O_2]}\frac{R_uT}{P^o}$$

or

$$[O] = \left[[O_2]\frac{K_pP^o}{R_uT}\right]^{1/2}.$$

Thus,

$$\frac{d[NO]}{dt} = 2k_{1f}\left(\frac{K_pP^o}{R_uT}\right)^{1/2}[N_2][O_2]^{1/2}.$$

From the above, we can identify the global parameters:

$$\boxed{k_G \equiv 2k_{1f}\left(\frac{K_pP^o}{R_uT}\right)^{1/2}}$$

$$\boxed{\begin{array}{l} m = 1 \\ n = 1/2 \end{array}}$$

**Comment**

In many cases where global reactions are invoked, the detailed kinetics are not known. This example shows that global parameters can be interpreted or inferred from a knowledge of detailed chemistry. This allows the testing or verification of elementary mechanisms in terms of global measurements. Note, also, that the global mechanism developed above applies only to the initial formation rate of NO. This is because we have ignored the reverse reactions, which would become important as NO concentrations rise.

---

Consider the shock-heating of air to 2500 K and 3 atm. Use the results of Example 4.3 to determine:

**Example 4.4**

A. The initial nitric oxide formation rate in ppm/s.
B. The amount of nitric oxide formed (in ppm) in 0.25 ms.

The rate coefficient, $k_{1f}$, is [10]:

$$k_{1f} = 1.82 \cdot 10^{14} \exp[-38,370/T(\mathrm{K})]$$

$$[=] \mathrm{cm}^3/\mathrm{gmol\text{-}s}.$$

**Solution**

A. Find $d\chi_{\mathrm{NO}}/dt$. We can evaluate $d[\mathrm{NO}]/dt$ from

$$\frac{d[\mathrm{NO}]}{dt} = 2k_{1f} \left( \frac{K_p P^o}{R_u T} \right)^{1/2} [\mathrm{N}_2][\mathrm{O}_2]^{1/2},$$

where we assume that

$$\chi_{\mathrm{N}_2} \cong \chi_{\mathrm{N}_2, i} = 0.79$$

$$\chi_{\mathrm{O}_2, e} \cong \chi_{\mathrm{O}_2, i} = 0.21$$

since $\chi_{\mathrm{NO}}$ and $\chi_{\mathrm{O}}$ are both likely to be quite small and, hence, can be neglected in estimating the values of $\chi_{\mathrm{N}_2}$ and $\chi_{\mathrm{O}_2, e}$.

We can convert the mole fractions to molar concentrations (Eqn. 4.40):

$$[\mathrm{N}_2] = \chi_{\mathrm{N}_2} \frac{P}{R_u T} = 0.79 \frac{3(101,325)}{8315(2500)}$$

$$= 1.155 \cdot 10^{-2} \ \mathrm{kmol/m}^3$$

$$[\mathrm{O}_2] = \chi_{\mathrm{O}_2} \frac{P}{R_u T} = 0.21 \frac{3(101,325)}{8315(2500)}$$

$$= 3.071 \cdot 10^{-3} \ \mathrm{kmol/m}^3$$

and evaluate the rate coefficient,

$$k_{1f} = 1.82 \cdot 10^{14} \exp[-38,370/2500]$$

$$= 3.93 \cdot 10^{7} \ \mathrm{cm}^3/\mathrm{gmol\text{-}s}$$

$$= 3.93 \cdot 10^{4} \ \mathrm{m}^3/\mathrm{kmol\text{-}s}.$$

To find the equilibrium constant, we employ Eqns. 2.64 and 2.66:

$$\Delta G_T^o = \left[ (2) \, \overline{g}_{f,\mathrm{O}}^o - (1)\overline{g}_{f,\mathrm{O}_2}^o \right]_{2500 \ \mathrm{K}}$$

$$= 2(88,203) - (1)0 = 176,406 \ \mathrm{kJ/kmol} \quad \text{(Appendix Tables A.11 and A.12),}$$

$$K_p = \frac{P_O^2}{P_{O_2} P^o} = \exp\left(\frac{-\Delta G_T^o}{R_u T}\right),$$

$$K_p P^o = \exp\left(\frac{-176,406}{8.315(2500)}\right) 1 \text{ atm} = 2.063 \cdot 10^{-4} \text{ atm} = 20.90 \text{ Pa}.$$

We can now numerically evaluate $d[NO]/dt$:

$$\frac{d[NO]}{dt} = 2(3.93 \cdot 10^4)\left(\frac{20.90}{8315(2500)}\right)^{1/2} 1.155 \cdot 10^{-2}(3.071 \cdot 10^{-3})^{1/2}$$

$$= 0.0505 \text{ kmol/m}^3\text{-s}$$

or, in terms of parts per million,

$$\frac{d\chi_{NO}}{dt} = \frac{R_u T}{P} \frac{d[NO]}{dt}$$

$$= \frac{8315(2500)}{3(101,325)} 0.0505 = 3.45 \text{ (kmol/kmol)/s}$$

$$\boxed{\frac{d\chi_{NO}}{dt} = 3.45 \cdot 10^6 \text{ ppm/s}}$$

The reader should verify that the units are correct in the above calculations of $d[NO]/dt$ and $d\chi_{NO}/dt$.

B. Find $\chi_{NO}(t = 0.25 \text{ ms})$. If we assume that the $N_2$ and $O_2$ concentrations do not change with time and that reverse reactions are unimportant over the 0.25-ms interval, we can quite simply integrate $d[NO]/dt$ or $\chi_{NO}/dt$, i.e.,

$$\int_0^{[NO](t)} d[NO] = \int_0^t k_G[N_2][O_2]^{1/2} \, dt$$

so

$$[NO](t) = k_G[N_2][O_2]^{1/2}t$$

$$= 0.0505(0.25 \cdot 10^{-3}) = 1.263 \cdot 10^{-5} \text{ kmol/m}^3$$

or

$$\chi_{NO} = [NO]\frac{R_u T}{P}$$

$$= 1.263 \cdot 10^{-5}\left(\frac{8315(2500)}{3(101,325)}\right) = 8.64 \cdot 10^{-4} \text{ kmol/kmol}$$

$$\boxed{\chi_{NO} = 864 \text{ ppm}}$$

**Comment**

We could use the above value of NO together with the appropriate reverse reaction rate coefficients to see the relative importance of the reverse reactions in the Zeldovich mechanism (Example 4.3) to determine if our approach to part B was valid.

The next chapter outlines several chemical mechanisms that have importance to combustion and shows the usefulness of the theoretical concepts presented in this chapter.

## Chemical Time Scales

In analyzing combustion processes, insight can be gained from a knowledge of chemical time scales. More precisely, knowing the magnitudes of chemical times relative to either convective or mixing times is of importance. For example, we will see in Chapter 12 how the Damköhler number, the ratio of flow times to chemical times, defines the various regimes of premixed combustion. In this section, we develop expressions that permit estimation of characteristic chemical time scales for elementary reactions.

**Unimolecular Reactions**  Consider the unimolecular reaction represented by Eqn. 4.53 and its corresponding reaction rate expression, Eqn. 4.54. This rate equation can be integrated, assuming constant temperature, to yield the following expression for the time history of [A]:

$$[A](t) = [A]_0 \exp(-k_{app}t) \qquad (4.67)$$

where $[A]_0$ is the initial concentration of species A.

In the same manner that a characteristic time, or time constant, is defined for a simple resistor–capacitor electrical network, we define a characteristic chemical time, $\tau_{chem}$, to be the time required for the concentration of A to fall from its initial value to a value equal to $1/e$ times the initial value, i.e.,

$$\frac{[A](\tau_{chem})}{[A]_0} = 1/e. \qquad (4.68)$$

Thus, combining Eqns. 4.67 to 4.68,

$$1/e = \exp(-k_{app}\tau_{chem}) \qquad (4.69)$$

or

$$\tau_{chem} = 1/k_{app}. \qquad (4.70)$$

Equation 4.70 shows that to estimate the time scale of simple unimolecular reactions all that is required is the apparent rate coefficient, $k_{app}$.

**Bimolecular Reactions**  Consider now the bimolecular reaction

$$A + B \rightarrow C + D \qquad (4.8)$$

and its rate expression

$$\frac{d[A]}{dt} = -k_{bimolec}[A][B]. \qquad (4.9)$$

For this single reaction taking place in the absence of all other reactions, the concentrations of A and B are related simply through stoichiometry. From Eqn. 4.8, we see that for every mole of A destroyed, one mole of B is similarly destroyed; thus, any change in [A] has a corresponding change in [B]:

$$x \equiv [A]_0 - [A] = [B]_0 - [B]. \tag{4.71}$$

The concentration of B then relates to the concentration of A simply as

$$[B] = [A] + [B]_0 - [A]_0. \tag{4.72}$$

Substituting Eqn. 4.71 into Eqn. 4.9 and then integrating yields the following:

$$\frac{[A](t)}{[B](t)} = \frac{[A]_0}{[B]_0} \exp[([A]_0 - [B]_0)k_{bimolec} t]. \tag{4.73}$$

Substituting Eqn. 4.72 into the above, and setting $[A]/[A]_0 = 1/e$ when $t = \tau_{chem}$ to define our characteristic time, results in

$$\tau_{chem} = \frac{\ln[e + (1 - e)([A]_0/[B]_0)]}{([B]_0 - [A]_0)k_{bimolec}}, \tag{4.74}$$

where $e = 2.718$.

Frequently, one of the reactants is in much greater abundance than the other. For the case when $[B]_0 \gg [A]_0$, Eqn. 4.74 simplifies to

$$\tau_{chem} = \frac{1}{[B]_0 k_{bimolec}}. \tag{4.75}$$

From Eqns. 4.74 and 4.75, we see that characteristic times for simple bimolecular reactions depend only on the concentrations of the initial reactants and the value of the rate coefficient.

**Termolecular Reactions**     We can deal quite simply with the termolecular reaction

$$A + B + M \rightarrow C + M, \tag{4.24}$$

since, for a simple system at constant temperature, the third body concentration [M] is constant. The reaction rate expression, Eqn. 4.25, is mathematically identical to the bimolecular rate expression where $k_{ter}[M]$ plays the same role as $k_{bimolec}$:

$$\frac{d[A]}{dt} = (-k_{ter}[M])[A][B]. \tag{4.76}$$

The characteristic time for the termolecular reaction is then expressed as

$$\tau_{chem} = \frac{\ln[e + (1 - e)([A]_0/[B]_0)]}{([B]_0 - [A]_0)k_{ter}[M]}, \tag{4.77}$$

and, when $[B]_0 \gg [A]_0$,

$$\tau_{chem} = \frac{1}{[B]_0 [M] k_{ter}}. \qquad (4.78)$$

The following example illustrates the application of the above to a few selected reactions of interest in combustion.

---

Consider the following combustion reactions:                                    **Example 4.5**

| Reaction | Rate coefficient |
|----------|------------------|
| i. $CH_4 + OH \rightarrow CH_3 + H_2O$ | $k(cm^3/gmol\text{-}s) = 1.00 \cdot 10^8\, T\,(K)^{1.6} \exp[-1570/T\,(K)]$ |
| ii. $CO + OH \rightarrow CO_2 + H$ | $k(cm^3/gmol\text{-}s) = 4.76 \cdot 10^7\, T\,(K)^{1.23} \exp[-35.2/T\,(K)]$ |
| iii. $CH + N_2 \rightarrow HCN + N$ | $k(cm^3/gmol\text{-}s) = 2.86 \cdot 10^8\, T\,(K)^{1.1} \exp[-10{,}267/T\,(K)]$ |
| iv. $H + OH + M \rightarrow H_2O + M$ | $k(cm^6/gmol^2\text{-}s) = 2.20 \cdot 10^{22}\, T\,(K)^{-2.0}$ |

Reaction i is an important step in the oxidation of $CH_4$, and reaction ii is the key step in CO oxidation. Reaction iii is a rate-limiting step in the prompt-NO mechanism, and reaction iv is a typical radical recombination reaction. (The importance of these reactions, and others, is discussed more fully in Chapter 5.)

Estimate the characteristic chemical times associated with the least abundant reactant in each of these reactions for the following two conditions:

| Condition I (low temperature) | Condition II (high temperature) |
|:---:|:---:|
| $T = 1344.3$ K | $T = 2199.2$ K |
| $P = 1$ atm | $P = 1$ atm |
| $\chi_{CH_4} = 2.012 \cdot 10^{-4}$ | $\chi_{CH_4} = 3.773 \cdot 10^{-6}$ |
| $\chi_{N_2} = 0.7125$ | $\chi_{N_2} = 0.7077$ |
| $\chi_{CO} = 4.083 \cdot 10^{-3}$ | $\chi_{CO} = 1.106 \cdot 10^{-2}$ |
| $\chi_{OH} = 1.818 \cdot 10^{-4}$ | $\chi_{OH} = 3.678 \cdot 10^{-3}$ |
| $\chi_{H} = 1.418 \cdot 10^{-4}$ | $\chi_{H} = 6.634 \cdot 10^{-4}$ |
| $\chi_{CH} = 2.082 \cdot 10^{-9}$ | $\chi_{CH} = 9.148 \cdot 10^{-9}$ |
| $\chi_{H_2O} = 0.1864$ | $\chi_{H_2O} = 0.1815$ |

Assume that each of the four reactions is uncoupled from all other reactions and that the third-body collision partner concentration is the sum of the $N_2$ and $H_2O$ concentrations.

**Solution**

To determine the characteristic chemical time associated with each reaction, we employ Eqn. 4.74 (or Eqn. 4.75) for the bimolecular reactions i, ii, and iii, and Eqn. 4.77 (or Eqn. 4.78) for the termolecular reaction iv. For reaction i at condition I, we treat OH as species A, for its mole fraction is just slightly smaller than that of $CH_4$. Converting the mole fractions to molar concentrations, we have

$$[OH] = \chi_{OH} \frac{P}{R_u T}$$

$$= 1.818 \cdot 10^{-4} \frac{101{,}325}{8315(1344.3)}$$

$$= 1.648 \cdot 10^{-6} \text{ kmol/m}^3 \text{ or } 1.648 \cdot 10^{-9} \text{ gmol/cm}^3$$

and

$$[CH_4] = \chi_{CH_4} \frac{P}{R_u T}$$
$$= 2.012 \cdot 10^{-4} \frac{101{,}325}{8315(1344.3)}$$
$$= 1.824 \cdot 10^{-6} \text{ kmol/m}^3 \quad \text{or} \quad 1.824 \cdot 10^{-9} \text{ gmol/cm}^3.$$

Evaluating the rate coefficient, given in CGS units, we obtain

$$k_1 = 1.00 \cdot 10^8 (1344.3)^{1.6} \exp[-1507/1344.3]$$
$$= 3.15 \cdot 10^{12} \text{ cm}^3/\text{gmol-s}.$$

Since $[CH_4]$ and $[OH]$ are of the same order of magnitude, we employ Eqn. 4.74 to evalute $\tau_{chem}$:

$$\tau_{OH} = \frac{\ln[2.718 - 1.718([OH]/[CH_4])]}{([CH_4] - [OH])k_i}$$
$$= \frac{\ln[2.718 - 1.718(1.648 \cdot 10^{-9}/1.824 \cdot 10^{-9})]}{(1.824 \cdot 10^{-9} - 1.648 \cdot 10^{-9})3.15 \cdot 10^{12}} = \frac{0.1534}{554.4}$$

$$\boxed{\tau_{OH} = 2.8 \cdot 10^{-4} \text{ s or 0.28 ms}}$$

All of the characteristic times for the least abundant reactants for reactions i–iii are computed similarly, and the results summarized in the table below. For the termolecular reaction iv at condition I,

$$[M] = \left(\chi_{N_2} + \chi_{H_2O}\right) \frac{P}{R_u T}$$
$$= (0.7125 + 0.1864) \frac{101{,}325}{8315(1344.3)}$$
$$= 8.148 \cdot 10^{-3} \text{ kmol/m}^3 \quad \text{or} \quad 8.148 \cdot 10^{-6} \text{ gmol/cm}^3$$

and, from Eqn. 4.77,

$$\tau_H = \frac{\ln[2.718 - 1.718([H]/[OH])]}{([OH] - [H])[M]k_{iv}}.$$

Numerically evaluating $[H]$, $[OH]$, and $k_{iv}$ and substituting into the above yields

$$\tau_H = \frac{\ln[2.718 - 1.718(1.285 \cdot 10^{-9})/(1.648 \cdot 10^{-9})]}{(1.648 \cdot 10^{-9} - 1.285 \cdot 10^{-9})8.149 \cdot 10^{-6}(1.217 \cdot 10^{16})}$$

$$\boxed{\tau_H = 8.9 \cdot 10^{-3} \text{ s or 8.9 ms}}$$

A similar calculation is performed for condition II, and the results are shown in the following table:

| Condition | Reaction | Species A | $k_i(T)$ | $\tau_A$ (ms) |
|-----------|----------|-----------|----------|----------------|
| I | i | OH | $3.15 \cdot 10^{12}$ | 0.28 |
| I | ii | OH | $3.27 \cdot 10^{11}$ | 0.084 |
| I | iii | CH | $3.81 \cdot 10^{8}$ | 0.41 |
| I | iv | H | $1.22 \cdot 10^{16}$ | 8.9 |
| II | i | $CH_4$ | $1.09 \cdot 10^{13}$ | 0.0045 |
| II | ii | OH | $6.05 \cdot 10^{11}$ | 0.031 |
| II | iii | CH | $1.27 \cdot 10^{10}$ | 0.020 |
| II | iv | H | $4.55 \cdot 10^{15}$ | 2.3 |

### Comment

We make several observations from these calculations. First, the increase in temperature from 1344 K to 2199 K shortens the chemical time scales for all of the species, but most dramatically for the $CH_4 + OH$ and the $CH + N_2$ reactions (i and iii). For the $CH_4 + OH$ reaction, the dominant factor is the decrease in the given $CH_4$ concentration; and for the $CH + N_2$ reaction, the large increase in the rate coefficient dominates. A second observation is that the characteristic times at either condition vary widely among the various reactions. It is particularly interesting to note the very long characteristic times associated with recombination reaction $H + OH + M \rightarrow H_2O + M$, compared with the bimolecular reactions. That recombination reactions are relatively slow plays a key role in using partial equilibrium assumptions to simplifiy complex chemical mechanisms (see next section). The last comment is that the simplified relation, Eqn. 4.75, could have been used with good accuracy for reaction iii at both the low- and high-temperature conditions, as well as reaction i at the high temperature, since in these cases one of the reactants was much more abundant than the other.

## Partial Equilibrium

Many combustion processes simultaneously involve both fast and slow reactions such that the fast reactions are rapid in both the forward and reverse directions. These fast reactions are usually chain-propagating or chain-branching steps, whereas the slow reactions are termolecular recombination reactions. Treating the fast reactions as if they were equilibrated simplifies the chemical kinetics by eliminating the need to write rate equations for the radical species involved. This treatment is called the **partial-equilibrium approximation.** We now illustrate these ideas using the following hypothetical mechanism:

$$A + B_2 \rightarrow AB + B, \qquad (P.1f)$$

$$AB + B \rightarrow A + B_2, \qquad (P.1r)$$

$$B + A_2 \rightarrow AB + A, \qquad (P.2f)$$

$$AB + A \rightarrow B + A_2, \qquad (P.2r)$$

$$AB + A_2 \rightarrow A_2B + A, \qquad (P.3f)$$

$$A_2B + A \rightarrow AB + A_2, \qquad (P.3r)$$

$$A + AB + M \rightarrow A_2B + M. \qquad (P.4f)$$

In this mechanism, the reactive intermediates are A, B, and AB, and the stable species are $A_2$, $B_2$ and $A_2B$. Note that the bimolecular reactions are grouped as forward and reverse reaction pairs (e.g., P.1$f$ and P.1$r$). The reaction rates of each reaction in the three pairs of bimolecular reactions are assumed to be much faster than the rate of the recombination reaction P.4$f$. Bimolecular pairs behaving in this manner are often referred to as **shuffle** reactions, since the radical species shuffle between being reactants and products. We further assume that the forward and reverse reaction rates in each pair are equal, i.e.,

$$k_{P.1f}[A][B_2] = k_{P.1r}[AB][B], \tag{4.79a}$$

$$k_{P.2f}[B][A_2] = k_{P.2r}[AB][A], \tag{4.79b}$$

and

$$k_{P.3f}[AB][A_2] = k_{P.3r}[A_2B][A], \tag{4.79c}$$

or alternatively,

$$\frac{[AB][B]}{[A][B_2]} = K_{p,1}, \tag{4.80a}$$

$$\frac{[AB][A]}{[B][A_2]} = K_{p,2}, \tag{4.80b}$$

and

$$\frac{[A_2B][A]}{[AB][A_2]} = K_{p,3}. \tag{4.80c}$$

Simultaneous solution of Eqns. 4.80a, b, and c allows the radical species A, B, and AB to be expressed in terms of the stable species $A_2$, $B_2$, $A_2B$, eliminating the need for rate equations for the radicals, i.e.,

$$[A] = K_{p,3}(K_{p,1}K_{p,2}[B_2])^{1/2} \frac{[A_2]^{3/2}}{[A_2B]}, \tag{4.81a}$$

$$[B] = K_{p,3}K_{p,1} \frac{[A_2][B_2]}{[A_2B]}, \tag{4.81b}$$

and

$$[AB] = (K_{p,1}K_{p,2}[A_2][B_2])^{1/2}. \tag{4.81c}$$

Knowing the radical concentrations from Eqns. 4.81a, b, and c allows the product formation rate to be calculated from reaction P.4$f$:

$$\frac{d[A_2B]}{dt} = k_{P.4f}[A][AB][M]. \tag{4.82}$$

Of course, to evaluate and integrate Eqn. 4.82, $[A_2]$ and $[B_2]$ must be known or calculated from simultaneous integration of similar expressions.

It is interesting to note that the net result of invoking either the partial equilibrium assumption or the steady-state approximation is the same: a radical concentration is determined by an algebraic equation rather than the integration of an ordinary differential equation. It is important, however, to keep in mind that, physically, the two approximations are quite different, with the partial equilibrium approximation forcing a reaction, or sets of reactions, to be essentially equilibrated; whereas the steady-state approximation forces the individual net production rate of one or more species to be essentially zero.

There are many examples in the combustion literature where the partial equilibrium approximation is invoked to simplify a problem. Two particular interesting cases are the calculation of carbon monoxide concentrations during the expansion stroke of a spark-ignition engine [13, 14] and the calculation of nitric oxide emissions in turbulent jet flames [15, 16]. In both of these problems, the slow recombination reactions cause radical concentrations to build up to levels in excess of their values for full equilibrium.

# REDUCED MECHANISMS

Complex reaction mechanisms of interest in combustion may involve many elementary reactions. For example, the methane oxidation mechanism presented in Chapter 5 considers 279 elementary steps. In computational models of complex systems, such as spark-ignition engines, the simultaneous solution of the time-dependent conservation equations that determine the flowfield, the temperature field, and the distribution of a large number of species is a formidable task. Several days of computer time may be required for a single simulation. To decrease the computational effort in such simulations, researchers and practitioners frequently employ **reduced chemical mechanisms.** These reduced mechanisms are designed to capture the most important aspects of the combustion process without capturing unnecessary details inherent in the complex mechanisms.

Mechanism reduction techniques include the elimination of both relatively unimportant species and elementary reactions using sensitivity analyses [18], or other means, to create a skeletal mechanism. The skeletal mechanism resembles the complex mechanism in that it consists solely of elementary steps, although many fewer are involved. The skeletal mechanism can be reduced further by the use of the steady-state approximation for selected radical species, and invoking the partial equilibrium approximation for a set or sets of elementary reactions [23]. These two approximations are discussed in previous sections. The result is a set of global reactions involving yet fewer species. Criteria for eliminating species and reactions are usually tied to specific applications. For example, one reduced mechanism may be based on accurate predictions of ignition times, whereas another reduced mechanism for the same fuel may be based on accurate prediction of the propagation speed of premixed flames.

Many different tools have been developed for mechanism reduction. References [18–26] provide an entry to this literature. We revisit reduced mechanisms in our discussion of some chemical mechanisms of importance to combustion in Chapter 5.

# CATALYSIS AND HETEROGENEOUS REACTIONS

Our discussion of chemical kinetics has thus far focused on reactions that occur in the gas phase. Because all participating species, both reactants and products, exist in a single (gaseous) phase, such reactions are referred to as **homogeneous** reactions. We now introduce the topic of **heterogeneous** reactions. Here we consider systems in which the participating substances do not exist in a single phase. Such heterogeneous systems comprise various combinations of gas, liquid, and solid phases. Gas-solid systems of importance to combustion include coal-char combustion (see Chapter 14); catalytic cleanup of automotive exhaust gases (see Chapter 15); catalytic combustion of natural gas, an advanced technology with application to gas-turbine engines [27]; and others.

## Surface Reactions

The physics of heterogeneous reactions differs significantly from that of homogeneous, gas-phase reactions. In the latter, reactant molecules collide and interact, rearranging the bonds among atoms to form products. For reaction to take place, the collision must occur with sufficient energy and favorable orientation of the reactant molecules, as we discussed at the beginning of this chapter. Although this description is overly simplistic, the basic physics of homogeneous reactions is straightforward. In contrast, heterogeneous reactions require us to consider the additional processes of absorption of molecules from the gas phase to the solid surface and the converse process of desorption. Two types of adsorption processes exist: physisorption and chemisorption. In **physisorption,** gas molecules are held at the solid surface by van der Waals forces, whereas in **chemisorption** chemical bonds hold the gas molecule to the surface. In many systems, a spectrum of binding forces exists between these extremes. Physisorption is reversible; hence, equilibrium is readily attained between molecules in the gas phase and those adsorbed into the surface. In contrast, chemisorption is frequently irreversible. In such cases, the chemisorbed molecule is strongly bound to the solid and is unable to return to the gas phase. This attribute of chemisorption is important to our discussion of heterogeneous reactions and catalysis in that the chemisorbed molecule can react with other molecules in adjacent sites to form a product molecule that is free to be desorbed. In the adsorption process, adsorbed molecules may reside at different locations on the solid substrate. For example, some sites may be edges or ledges of the crystalline substrate, whereas others may be planar sites. Specific site types that promote reaction are known as **active sites.** Understanding the nature of active sites for coal char and soot oxidation is an active research area.

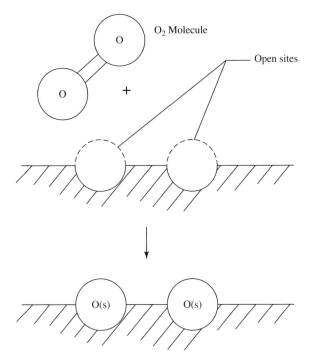

**Figure 4.2**    An oxygen molecule is adsorbed onto the surface of a solid, filling two open sites. This results in two oxygen atoms available to react heterogeneously. For example, in the catalytic oxidation of CO on platinum, an adsorbed O reacts with an adsorbed CO to form $CO_2$ and provide an open site on the platinum, i.e., $O(s) + CO(s) \rightarrow CO_2(s) + Pt(s)$.

To illustrate these ideas, we consider the oxidation of CO to $CO_2$ over a platinum surface. Figure 4.2 shows the chemisorption of an oxygen molecule into two open sites. We represent this reaction as

$$O_2 + 2Pt(s) \rightarrow 2O(s), \qquad (HR.1)$$

where the notation (s) represents an open site for the platinum, i.e., Pt(s); and an adsorbed atom for the oxygen, i.e., O(s). Using this same notation for adsorbed CO and $CO_2$ molecules, we write the remaining steps in the CO oxidation process:

$$CO + Pt(s) \rightarrow CO(s) \qquad (HR.2)$$

$$CO(s) + O(s) \rightarrow CO_2(s) + Pt(s) \qquad (HR.3)$$

$$CO_2(s) \rightarrow CO_2 + Pt(s) \qquad (HR.4)$$

In Eqn. HR.2, a gas-phase CO molecule is absorbed into an open site on the platinum. Step HR.3 represents the reaction of an adsorbed CO molecule and an adsorbed O atom to produce an adsorbed $CO_2$ molecule. For this reaction to occur, the adsorbed species must be in close proximity to allow the rearrangement of chemical bonds. The last step, HR.4, represents desorption of the $CO_2$ molecule into the gas phase and the resulting creation of an open site on the platinum surface. The net effect of these four steps is

$$O_2 + 2CO \rightarrow 2CO_2,$$   (HR.5)

where we have multiplied Eqns. HR.2–HR.4 by two and summed the resulting reactions with HR.1.

We note from Eqn. HR.5 that the platinum is not consumed, yet the platinum is essential for the reaction to occur (Eqns. HR.1–HR.4). This result illustrates the traditional definition of a catalyst [28]: A **catalyst** is a substance which increases the speed of a chemical reaction without itself undergoing change. For our specific example of CO oxidation, we see that the pathway expressed by Eqns. HR.1–HR.5 offer an alternative to the gas-phase homogeneous oxidation of CO, which we will discuss in Chapter 5. In the practical application of automotive exhaust after-treatment, the use of noble metal catalysts (Pt, Pd, and Rh) permit oxidation of the CO at typical exhaust temperatures, temperatures that are much too low for any significant gas-phase CO oxidation to occur.

We also note that the presence of a catalyst does not alter the equilibrium composition of a mixture. In fact, a catalyst can be used to take a slowly reacting, non-equilibrium system to its equilibrium state.

## Complex Mechanisms

Similar to our treatment of homogeneous chemistry (see Eqns. 4.29–4.34), complex mechanisms for heterogeneous chemistry can be expressed in a compact form. To define the reaction set analogous to that expressed by Eqn. 4.29, we note that active (open) sites are treated as a species. Furthermore, species that exist both in the gas phase and adsorbed on a solid are treated as a separate species. For example, the simple mechanism expressed by Eqns. HR.1–HR.5 involves seven species: $O_2$, $O(s)$, $Pt(s)$, CO, $CO(s)$, $CO_2$, and $CO_2(s)$.

In heterogeneous systems, the production rate of species $j$ associated with surface reactions is denoted $\dot{s}_j$ and has units of kmol/m$^2$-s. Thus, for a multistep heterogeneous mechanism involving $N$ species and $L$ reaction steps

$$\dot{s}_j = \sum_{i=1}^{L} v_{ji} \dot{q}_i^s \quad \text{for} \quad j = 1, 2, \ldots N$$   (4.83)

where

$$v_{ji} = (v_{ji}'' - v_{ji}')$$   (4.84)

and

$$\dot{q}_i^s = k_{fi} \prod_{j=1}^{N} [X_j]^{v_{ji}'} - k_{ri} \prod_{j=1}^{N} [X_j]^{v_{ji}''}.$$   (4.85)

We emphasize that Eqns. 4.83–4.85 apply only to the set of reactions comprising the heterogeneous system mechanism. Note that in applying the rate-of-progress

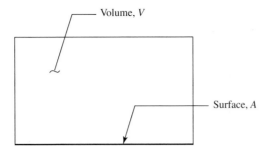

Volume, *V*

Surface, *A*

**Figure 4.3**    Gas-phase reactions occur uniformly throughout the volume *V* simultaneously with heterogeneous reactions occurring uniformly at the catalytically active surface A.

variable $\dot{q}_i^s$ to the heterogeneous system, the units associated with the species concentrations $[X_j]$ depend on whether the species is in the gas phase, e.g., CO, or associated with the surface, e.g., CO(s). For example, consider the simple heterogeneous mechanism expressed by Eqns. HR.1–HR4. For this mechanism, the net rate of CO production would simply be

$$\dot{s}_{CO} = -k_1[CO][Pt(s)]. \tag{4.86}$$

Here the associated units are $\dot{s}_{CO}$ [=] kmol/m$^2$-s, [CO] [=] kmol/m$^3$, and [Pt(s)] [=] kmol/m$^2$. The units associated with the rate constant $k_1$ are m$^3$/kmol for this particular case. In general, the rate coefficient units depend on the specific reaction.

Figure 4.3 illustrates a system in which both homogeneous and heterogeneous chemistry are important. Here a gas mixture is contained in a vessel of volume. One of the walls of the vessel is catalytically active and has an area *A*. Of interest is the oxidation of CO. We assume that the gas mixture is homogeneous, as might occur if the mixture is stirred with a fan; furthermore, we assume that surface reactions occur uniformly over the area *A*. With these assumptions, we can express the net mass production rate of CO from chemical reaction as the combined rates from the gas phase and at the surface as follows:

$$(\dot{m}_{CO})_{chem} = \dot{\omega}_{CO}MW_{CO}V + \dot{s}_{CO}MW_{CO}A, \tag{4.87}$$

where $MW_{CO}$ is the molecular weight of CO, $\dot{\omega}_{CO}$ is the volumetric molar rate of CO production (kmol/m$^3$) defined by Eqn. 4.31, and $\dot{s}_{CO}$ is the production rate of CO per unit area of the catalytic surface (kmol/m$^2$) defined by Eqn. 4.83. Of course, if CO is being oxidized this net production rate would be negative.

## SUMMARY

In this chapter, we explored many concepts essential to the understanding of combustion chemistry. You should be able to distinguish between global and elementary reactions and mechanisms and have an appreciation for the types of elementary reactions, i.e., bimolecular, termolecular, and unimolecular. You should also have an understanding of the relation between molecular collision theory and reaction rates. In particular, you should understand the physical meaning of the steric factor, pre-exponential factor, and activation energy—components of the rate coefficient that have their origins in collision theory. We also developed the concept of net species production rates and procedures to formulate appropriate equations from complex mechanisms. A compact notation was introduced to facilitate numerical calculations using a computer. You should understand chain mechanisms and the concepts of chain initiation, chain propagation, and chain termination. The approximation of a steady state for highly reactive species, such as atoms and other radicals, was introduced and used to simplify chain mechanisms. The partial equilibrium approximation was also discussed. We introduced the concept of a reduced mechanism in which a complex mechanism is reduced by eliminating unimportant species and reactions and then further simplified using steady-state and partial-equilibrium approximations. The Zeldovich nitric oxide formation mechanism, introduced in the examples, was used to illustrate many of the concepts in concrete ways.

## NOMENCLATURE

| | |
|---|---|
| $A$ | Pre-exponential factor (various units) |
| $b$ | Temperature exponent |
| $E_A$ | Activation energy (J/kmol) |
| $k$ | Rate coefficient (various units) |
| $k_B$ | Boltzmann constant, $1.381 \cdot 10^{-23}$ (J/K) |
| $K_c$ | Equilibrium constant based on concentrations |
| $K_p$ | Equilibrium constant based on partial pressures |
| $m$ | Mass (kg) or reaction order |
| M | Third-body collision partner |
| $MW$ | Molecular weight (kg/kmol) |
| $n$ | Reaction order |
| $n/V$ | Number density (molecules/m$^3$ or 1/m$^3$) |
| $N_{AV}$ | Avogadro number, $6.022 \cdot 10^{26}$ (molecules/kmol) |
| $p$ | Steric factor |
| $P$ | Pressure |
| $\mathcal{P}$ | Probability |
| $q$ | Rate-of-progress variable, Eqn. 4.33 |
| $R_u$ | Universal gas constant (J/kmol-K) |
| $\dot{s}$ | Net species production rate for surface reactions (kmol/m$^2$-s) |
| $t$ | Time (s) |
| $T$ | Temperature (K) |
| $v$ | Velocity (m/s) |
| $\bar{v}$ | Maxwellian mean velocity (m/s) |
| $V$ | Volume (m$^3$) |

| | |
|---|---|
| $X_j$ | Chemical formula of species $j$, Eqn. 4.29 |
| $Z_c$ | Collision frequency (1/s) |

### *Greek Symbols*

| | |
|---|---|
| $\Delta t$ | Time interval (s) |
| $\mu$ | Reduced mass (kg) |
| $\nu'$ | Reactant stoichiometric coefficient |
| $\nu''$ | Product stoichiometric coefficient |
| $\nu_{ji}$ | $\nu''_{ji} - \nu'_{ji}$, Eqn. 4.32 |
| $\chi$ | Mole fraction |
| $\dot{\omega}$ | Net species volumetric production rate (kmol/m$^3$-s) |

### *Subscripts*

| | |
|---|---|
| app | Apparent |
| bimolec | Bimolecular |
| *de* | De-energized |
| *e* | Energized |
| *f* | Forward |
| *F* | Fuel |
| *G* | Global or overall |
| *i* | *i*th species |
| *Ox* | Oxidizer |
| *Pr* | Products |
| *r* | Reverse |
| *ss* | Steady state |
| ter | Termolecular |
| uni | Unimolecular |
| 0 | Initial |

### *Other Notation*

| | |
|---|---|
| $[X]$ | Molar concentration of species $X$ (kmol/m$^3$) |
| $\Pi$ | Multiplication |

## REFERENCES

1. Kee, R. J., Rupley, F. M., and Miller, J. A., "Chemkin-II: A Fortran Chemical Kinetics Package for the Analysis of Gas-Phase Chemical Kinetics," Sandia National Laboratories Report SAND89-8009, March 1991.

2. Gardiner, W. C., Jr., *Rates and Mechanisms of Chemical Reactions,* Benjamin, Menlo Park, CA, 1972.

3. Benson, S. W., *The Foundations of Chemical Kinetics,* McGraw-Hill, New York, 1960.

4. Warnatz, J., "Rate Coefficients in the C/H/O System," Chapter 5 in *Combustion Chemistry* (W. C. Gardiner, Jr., ed.), Springer-Verlag, New York, pp. 197–360, 1984.

5. IMSL, Inc., "DGEAR," IMSL Library, Houston, TX.

6. Hindmarsh, A. C., "ODEPACK, A Systematic Collection of ODE Solvers," *Scientific Computing—Applications of Mathematics and Computing to the Physical Sciences* (R. S. Stapleman, ed.), North-Holland, Amsterdam, p. 55, 1983.

7. Bittker, D. A., and Soullin, V. J., "GCKP-84-General Chemical Kinetics Code for Gas Flow and Batch Processing Including Heat Transfer," NASA TP-2320, 1984.

8. Pratt, D. T., and Radhakrishnan, K., "CREK-ID: A Computer Code for Transient, Gas-Phase Combustion Kinetics," NASA Technical Memorandum TM-83806, 1984.

9. National Institute of Standards and Technology, *NIST Chemical Kinetics Database*, NIST, Gaithersburg, MD, published annually.

10. Hanson, R. K., and Salimian, S., "Survey of Rate Constants in the N/H/O System," Chapter 6 in *Combustion Chemistry* (W. C. Gardiner, Jr., ed.), Springer-Verlag, New York, pp. 361–421, 1984.

11. Williams, F. A., *Combustion Theory,* 2nd Ed., Addison-Wesley, Redwood City, CA, p. 565, 1985.

12. Gardiner, W. C., Jr., and Troe, J., "Rate Coefficients of Thermal Dissociation, Isomerization, and Recombination Reactions," Chapter 4 in *Combustion Chemistry* (W. C. Gardiner, Jr., ed.), Springer-Verlag, New York, pp. 173–196, 1984.

13. Keck, J. C., and Gillespie, D., "Rate-Controlled Partial-Equilibrium Method for Treating Reactive Gas Mixtures," *Combustion and Flame,* 17: 237–241 (1971).

14. Delichatsios, M. M., "The Kinetics of CO Emissions from an Internal Combustion Engine," S.M. Thesis, Massachusetts Institute of Technology, Cambridge, MA, June 1972.

15. Chen, C.-S., Chang, K.-C., and Chen, J.-Y., "Application of a Robust $\beta$-pdf Treatment to Analysis of Thermal NO Formation in Nonpremixed Hydrogen–Air Flame," *Combustion and Flame,* 98: 375–390 (1994).

16. Janicka, J., and Kollmann, W., "A Two-Variables Formalism for the Treatment of Chemical Reactions in Turbulent $H_2$–Air Diffusion Flames," *Seventeenth Symposium (International) on Combustion,* The Combustion Institute, Pittsburgh, PA, p. 421, 1979.

17. Svehla, R. A., "Estimated Viscosities and Thermal Conductivities of Gases at High Temperature," NASA Technical Report R-132, 1962.

18. Tomlin, A. S., Pilling, M. J., Turanyi, T., Merkin, J. H., and Brindley, J., "Mechanism Reduction for the Oscillatory Oxidation of Hydrogen: Sensitivity and Quasi-Steady State Analyses," *Combustion and Flame,* 91: 107–130 (1992).

19. Peters, N. and Rogg, B. (eds.), *Reduced Kinetic Mechanisms for Applications in Combustion Systems,* Lecture Notes in Physics, m 15, Springer-Verlag, Berlin, 1993.

20. Sung, C. J., Law, C. K., and Chen, J.-Y., "Augmented Reduced Mechanisms for NO Emission in Methane Oxidation," *Combustion and Flame,* 125: 906–919 (2001).

21. Montgomery, C. J., Cremer, M. A., Chen, J.-Y., Westbrook, C. K., and Maurice, L. Q., "Reduced Chemical Kinetic Mechanisms for Hydrocarbon Fuels," *Journal of Propulsion and Power,* 18: 192–198 (2002).

22. Bhattacharjee, B., Schwer, D. A., Barton, P. I., and Green, W. H., Jr., "Optimally-Reduced Kinetic Models: Reaction Elimination in Large-Scale Kinetic Mechanisms," *Combustion and Flame,* 135: 191–208 (2003).

23. Lu, T., and Law, C. K., "A Directed Relation Graph Method for Mechanism Reduction," *Proceedings of the Combustion Institute,* 30: 1333–1341 (2005).

24. Brad, R. B., Tomlin, A. S., Fairweather, M., and Griffiths, J. F., "The Application of Chemical Reduction Methods to a Combustion System Exhibiting Complex Dynamics," *Proceedings of the Combustion Institute,* 31: 455–463 (2007).

25. Lu, T., and Law, C. K., "Towards Accommodating Realistic Fuel Chemistry in Large-Scale Computation," *Progress in Combustion Science and Technology,* 35: 192–215 (2009).

26. Law, C. K., *Combustion Physics,* Cambridge University Press, New York, 2006.

27. Dalla Betta, R. A., "Catalytic Combustion Gas Turbine Systems: The Preferred Technology for Low Emissions Electric Power Production and Co-generation," *Catalysis Today,* 35: 129–135 (1997).

28. Brown, T. L., Lemay, H. E., Bursten, B. E., Murphy, C. J., and Woodward, P. M., *Chemistry: The Central Science,* 11th Ed., Prentice Hall, Upper Saddle River, NJ, 2008.

## QUESTIONS AND PROBLEMS

**4.1** Make a list of all of the boldfaced words in Chapter 4 and discuss their meanings.

**4.2** Several species and their structural forms are given below. Using sketches of colliding molecules, show that the reaction $2H_2 + O_2 \rightarrow 2H_2O$ is highly improbable based on simple collisions and the given structures.

$$H_2 : \quad H-H$$
$$O_2 : \quad O=O$$
$$H_2O: \quad H \diagup^{O}\diagdown H$$

**4.3** Consider the reaction $H_2 + O_2 \rightarrow HO_2 + H$. Show that this is likely to be an elementary reaction. Use a sketch, as in problem 4.2. The structure of the hydroperoxy radical is H–O–O.

**4.4** Consider the reaction $CH_4 + O_2 \rightarrow CH_3 + HO_2$. Although a $CH_4$ molecule may collide with an $O_2$ molecule, a chemical reaction may not necessarily occur. List two factors important in determining whether or not a reaction occurs during a collision.

**4.5** Consider the overall oxidation reaction of propane:

$$C_3H_8 + 5O_2 \rightarrow 3CO_2 + 4H_2O.$$

The following global mechanism has been proposed for this reaction:

$$\text{Reaction rate} = 8.6 \cdot 10^{11} \exp(-30/R_u T)[C_3H_8]^{0.1}[O_2]^{1.65},$$

where CGS units (cm, s, gmol, kcal, K) are employed.

A. Identify the order of the reaction with respect to propane.

B. Identify the order of the reaction with respect to $O_2$.

C. What is the overall order of the global reaction?

D. Identify the activation energy for the reaction.

**4.6** In a global, single-step mechanism for butane combustion, the reaction order with respect to butane is 0.15 and with respect to oxygen is 1.6. The rate coefficient can be expressed in Arrhenius form: the pre-exponential factor is $4.16 \cdot 10^9$ [$(kmol/m^3)^{-0.75}/s$] and the activation energy is 125,000 kJ/kmol. Write out an expression for the rate of butane destruction, $d[C_4H_{10}]/dt$.

**4.7** Using the results of problem 4.6, determine the volumetric mass oxidation rate of butane, in kg/s-m$^3$, for a fuel–air mixture with an equivalence ratio of 0.9, temperature of 1200 K, and pressure of 1 atm.

**4.8** Classify the following reactions as being either global or elementary. For those identified as elementary, further classify them as unimolecular, bimolecular, or termolecular. Give reasons for your classification.

A. $CO + OH \rightarrow CO_2 + H.$

B. $2CO + O_2 \rightarrow 2CO_2.$

C. $H_2 + O_2 \rightarrow H + H + O_2.$

D. $HOCO \rightarrow H + CO_2.$

E. $CH_4 + 2O_2 \rightarrow CO_2 + 2H_2O.$

F. $OH + H + M \rightarrow H_2O + M.$

**4.9** The following hard-sphere collision diameters, $\sigma$, are taken from Svehla [17]:

| Molecule | $\sigma(\text{Å})$ |
|----------|--------------------|
| H        | 2.708              |
| $H_2$    | 2.827              |
| OH       | 3.147              |
| $H_2O$   | 2.641              |
| O        | 3.050              |
| $O_2$    | 3.467              |

Using the above data with Eqn. 4.17 and Table 4.1, determine a temperature-dependent expression for the steric factors for the rate coefficients of the following reactions. Evaluate the steric factor at a temperature of 2500 K. Be careful with units and remember that the reduced mass should be in grams or kilograms.

$$H + O_2 \rightarrow OH + O$$

$$OH + O \rightarrow O_2 + H.$$

**4.10**  In methane combustion, the following reaction pair is important:

$$CH_4 + M \underset{k_r}{\overset{k_f}{\Longleftrightarrow}} CH_3 + H + M,$$

where the reverse reaction coefficient is given by

$$k_r \, (m^6/kmol^2\text{-s}) = 2.82 \cdot 10^5 T \exp[-9835/T].$$

At 1500 K, the equilibrium constant $K_p$ has a value of 0.003691 based on a reference-state pressure of 1 atm (101,325 Pa). Derive an algebraic expression for the forward rate coefficient $k_f$. Evaluate your expression for a temperature of 1500 K. Give units.

**4.11***  Plot the forward rate coefficient for the reaction $O + N_2 \rightarrow NO + N$ versus temperature for the temperature range 1500 K $< T <$ 2500 K. The rate coefficient is given as [10] $k(T) = 1.82 \cdot 10^{14} \exp[-38, 370/T(K)]$ in units of $cm^3/gmol\text{-s}$. What conclusions can you draw?

**4.12**  Consider the following mechanism for the production of ozone from the heating of oxygen:

$$O_3 \underset{k_{1r}}{\overset{k_{1f}}{\Longleftrightarrow}} O_2 + O \qquad\qquad (R.1)$$

$$O + O_3 \underset{k_{2r}}{\overset{k_{2f}}{\Longleftrightarrow}} 2O_2 \qquad\qquad (R.2)$$

A.  Write out the coefficient matrices for $\nu'_{ji}$ and $\nu''_{ji}$. Use the convention that species 1 is $O_3$, species 2 is $O_2$, and species 3 is O.

B.  Starting with the compact notation defined by Eqns. 4.31–4.33, express the rate-of-progress variables, $\dot{\omega}_j$, for the three species involved in the above mechanism. Retain all terms without simplifying, i.e., retain terms raised to the zero power.

**4.13**  Generate the coefficient matrices $\nu'_{ji}$ and $\nu''_{ji}$ for the $H_2$–$O_2$ reaction mechanism (H.1–H.20) given in Chapter 5. Consider both forward and reverse reactions. Do not attempt to include the radical destruction at the walls (Eqn. H.21).

**4.14**  Consider the following elementary reaction mechanism where both the forward and reverse reactions are important:

$$CO + O_2 \overset{1}{\Longleftrightarrow} CO_2 + O,$$

$$O + H_2O \overset{2}{\Longleftrightarrow} OH + OH,$$

$$CO + OH \overset{3}{\Longleftrightarrow} CO_2 + H,$$

$$H + O_2 \overset{4}{\Longleftrightarrow} OH + O.$$

*Indicates required use of computer.

How many chemical rate equations are needed to determine the chemical evolution of a system defined by this mechanism? Write out the rate equation for the hydroxyl radical.

**4.15**  Consider the production of the stable product, HBr, from $H_2$ and $Br_2$. The following complex reaction mechanism is thought to apply:

$$M + Br_2 \rightarrow Br + Br + M, \tag{R.1}$$

$$M + Br + Br \rightarrow Br_2 + M, \tag{R.2}$$

$$Br + H_2 \rightarrow HBr + H, \tag{R.3}$$

$$H + HBr \rightarrow Br + H_2. \tag{R.4}$$

A.  For each reaction, identify the type of elementary reaction, e.g., unimolecular, etc., and indicate its role in the chain mechanism, e.g., chain initiation.

B.  Write out a complete expression for the Br-atom reaction rate, $d[Br]/dt$.

C.  Write out an expression that can be used to determine the steady-state concentration of the hydrogen atom, $[H]$.

**4.16**  Consider the following chain-reaction mechanism for the high-temperature formation of nitric oxide, i.e., the Zeldovich mechanism:

$$O + N_2 \xrightarrow{k_{1f}} NO + N \quad \text{Reaction 1}$$

$$N + O_2 \xrightarrow{k_{2f}} NO + O \quad \text{Reaction 2}$$

A.  Write out expressions for $d[NO]/dt$ and $d[N]/dt$.

B.  Assuming N atoms exist in steady state and that the concentrations of O, $O_2$, and $N_2$ are at their equilibrium values for a specified temperature and composition, simplify your expression obtained above for $d[NO]/dt$ for the case of negligible reverse reactions. (*Answer:* $d[NO]/dt = 2k_{1f}[O]_{eq}[N_2]_{eq}$.)

C.  Write out the expression for the steady-state N-atom concentration used in part B.

D.  For the conditions given below and using the assumptions of part B, how long does it take to form 50 ppm (mole fraction $\cdot 10^6$) of NO?

$$T = 2100 \text{ K},$$

$$\rho = 0.167 \text{ kg/m}^3,$$

$$MW = 28.778 \text{ kg/kmol},$$

$$\chi_{O,eq} = 7.6 \cdot 10^{-5} \text{(mole fraction)},$$

$$\chi_{O_2,eq} = 3.025 \cdot 10^{-3} \text{(mole fraction)},$$

$$\chi_{N_2,eq} = 0.726 \text{ (mole fraction)},$$

$$k_{1f} = 1.82 \cdot 10^{14} \exp[-38,370/T(\text{K})] \text{ with units of cm}^3/\text{gmol-s}.$$

E.  Calculate the value of the reverse reaction rate coefficient for the first reaction, i.e., $O + N_2 \leftarrow NO + N$, for a temperature of 2100 K.

F.  For your computations in part D, how good is the assumption that reverse reactions are negligible? Be quantitative.

G.  For the conditions of part D, determine numerical values for [N] and $\chi_N$. (*Note:* $k_{2f} = 1.8 \cdot 10^{10} \, T \exp(-4680/T)$ with units of $cm^3/gmol\text{-}s$.)

**4.17**  When the following reaction is added to the two reactions in problem 4.16, the NO formation mechanism is called the extended Zeldovich mechanism:

$$N + OH \overset{k_{3f}}{\rightarrow} NO + H.$$

With the assumption of equilibrium concentrations of O, $O_2$, $N_2$, H, and OH applied to this three-step mechanism, find expressions for:

A.  The steady-state N-atom concentration, neglecting reverse reactions.

B.  The NO formation rate $d[NO]/dt$, again neglecting reverse reactions.

**4.18**  Consider the following CO oxidation reactions:

$$CO + OH \overset{k_1}{\rightarrow} CO_2 + H$$

$$CO + O_2 \overset{k_2}{\rightarrow} CO_2 + O,$$

where $k_1 \, (cm^3/gmol\text{-}s) = 1.17 \cdot 10^7 \, T \, (K)^{1.35} \exp[+3000/R_u T \, (K)]$, $k_2 \, (cm^3/gmol\text{-}s) = 2.50 \cdot 10^{12} \exp[-200{,}000/R_u T \, (K)]$, and $R_u = 8.315 \, J/gmol\text{-}K$. Calculate and compare the characteristic times associated with these two reactions for $T = 2000$ K and $P = 1$ atm. The CO mole fraction is 0.011 and the OH and $O_2$ mole fractions are $3.68 \cdot 10^{-3}$ and $6.43 \cdot 10^{-3}$, respectively.

**4.19***  Consider the production of nitric oxide via the Zeldovich mechanism. For methane–air combustion at an equivalence ratio of 0.9 and 1 atm, the following amounts of NO are calculated to be produced after 10 ms:

| $T(K)$ | $\chi_{NO}(ppm)$ |
|--------|------------------|
| 1600   | 0.0015           |
| 1800   | 0.150            |
| 2000   | 6.58             |
| 2200   | 139              |
| 2400   | 1823             |

Use TPEQUIL to calculate the **equilibrium** NO mole fractions. Construct a table showing the equilibrium values and the ratio of the kinetically determined NO mole fractions to the equilibrium values. Also, compute a characteristic NO-formation time, defined as the time required to reach the equilibrium value using the initial rate of formation. For example, at 1600 K, the initial formation rate is $0.0015 \, ppm/0.010 s = 0.15 \, ppm/s$. Discuss the temperature dependence of your findings. What is the significance of this temperature dependence?

**4.20**   In the combustion of hydrogen, the following reactions involving radicals are fast in both the forward and reverse directions:

$$H + O_2 \underset{k_{1r}}{\overset{k_{1f}}{\Longleftrightarrow}} OH + O, \tag{R.1}$$

$$O + H_2 \underset{k_{2r}}{\overset{k_{2f}}{\Longleftrightarrow}} OH + H, \tag{R.2}$$

$$OH + H_2 \underset{k_{3r}}{\overset{k_{3f}}{\Longleftrightarrow}} H_2O + H. \tag{R.3}$$

Use the assumption of partial equilibrium to derive algebraic expressions for the molar concentrations of the three radical species, O, H, and OH, in terms of the kinetic rate coefficients and the molar concentrations of reactant and product species, $H_2$, $O_2$, and $H_2O$.

# 5

# Some Important Chemical Mechanisms

## OVERVIEW

Our approach in this chapter will be to present, or outline, the elementary steps involved in a number of chemical mechanisms of major importance to combustion and combustion-generated air pollution. For detailed discussions of these mechanisms, the reader is referred to the original literature, reviews, and more advanced textbooks that emphasize kinetics [1, 2]. Our purpose is simply to illustrate real systems, which are generally complex, and to show that the fundamentals discussed in Chapter 4 are indeed important to understanding such systems.

It is important to point out that complex mechanisms are evolutionary products of chemists' thoughts and experiments, and, as such, they may change with time as new insights are developed. Thus, when we discuss a particular mechanism, we are not referring to *the* mechanism in the same sense that we refer to *the* first law of thermodynamics, or other well-known conservation principles.

## THE $H_2$–$O_2$ SYSTEM

The hydrogen–oxygen system is important in its own right as, for example, in rocket propulsion. This system is also important as a subsystem in the oxidation of hydrocarbons and moist carbon monoxide. Detailed reviews of $H_2$–$O_2$ kinetics can be found in Refs. [3]–[5]. Relying heavily on Glassman [1], we outline the oxidation of hydrogen as follows.

The initiation reactions are

$$H_2 + M \rightarrow H + H + M \quad \text{(very high temperatures)} \qquad \text{(H.1)}$$

$$H_2 + O_2 \rightarrow HO_2 + H \quad \text{(other temperatures).} \qquad \text{(H.2)}$$

Chain-reaction steps involving O, H, and OH radicals are

$$H + O_2 \rightarrow O + OH, \tag{H.3}$$

$$O + H_2 \rightarrow H + OH, \tag{H.4}$$

$$H_2 + OH \rightarrow H_2O + H, \tag{H.5}$$

$$O + H_2O \rightarrow OH + OH. \tag{H.6}$$

Chain-terminating steps involving O, H, and OH radicals are the three-body recombination reactions:

$$H + H + M \rightarrow H_2 + M, \tag{H.7}$$

$$O + O + M \rightarrow O_2 + M, \tag{H.8}$$

$$H + O + M \rightarrow OH + M, \tag{H.9}$$

$$H + OH + M \rightarrow H_2O + M. \tag{H.10}$$

To complete the mechanism, we need to include reactions involving $HO_2$, the hydroperoxy radical, and $H_2O_2$, hydrogen peroxide. When

$$H + O_2 + M \rightarrow HO_2 + M \tag{H.11}$$

becomes active, then the following reactions, and the reverse of H.2, come into play:

$$HO_2 + H \rightarrow OH + OH, \tag{H.12}$$

$$HO_2 + H \rightarrow H_2O + O, \tag{H.13}$$

$$HO_2 + O \rightarrow O_2 + OH, \tag{H.14}$$

and

$$HO_2 + HO_2 \rightarrow H_2O_2 + O_2, \tag{H.15}$$

$$HO_2 + H_2 \rightarrow H_2O_2 + H, \tag{H.16}$$

with

$$H_2O_2 + OH \rightarrow H_2O + HO_2, \tag{H.17}$$

$$H_2O_2 + H \rightarrow H_2O + OH, \tag{H.18}$$

$$H_2O_2 + H \rightarrow HO_2 + H_2, \tag{H.19}$$

$$H_2O_2 + M \rightarrow OH + OH + M. \tag{H.20}$$

Depending upon the temperature, pressure, and extent of reaction, the reverse reactions of all of the above can be important; therefore, in modeling the $H_2$–$O_2$ system as many as 40 reactions can be taken into account involving the eight species: $H_2$, $O_2$, $H_2O$, OH, O, H, $HO_2$, and $H_2O_2$.

The $H_2$–$O_2$ system has interesting explosion characteristics (Fig. 5.1) that can be explained using the above mechanism. Figure 5.1 shows that there are distinct

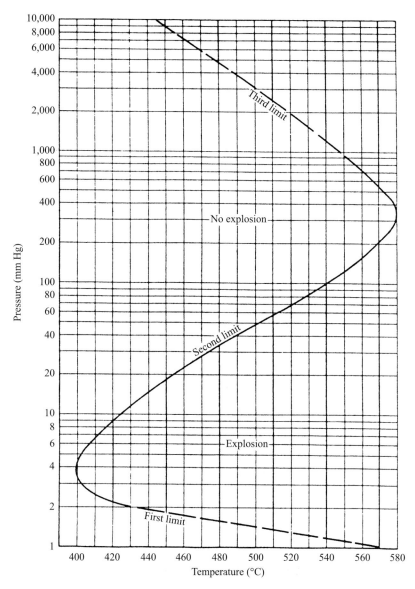

**Figure 5.1**    Explosion limits for a stoichiometric hydrogen–oxygen mixture in a spherical vessel.
SOURCE: From Ref. [2]. Reprinted by permission of Academic Press.

regions in temperature–pressure coordinates where a stoichiometric mixture of H$_2$ and O$_2$ will and will not explode. The temperatures and pressures correspond to the initial charging conditions of the spherical vessel containing the reactants. To explore the explosion behavior, let us follow a vertical line, say at 500 °C, from the lowest pressure shown (1 mm Hg) upwards to pressures of several atmospheres. Up until about 1.5 mm Hg, there is no explosion. This lack of explosion is the result of the free radicals produced in the initiation step (H.2) and chain sequence (H.3–H.6) being destroyed by reactions on the walls of the vessel. These wall reactions break the chain, preventing the rapid buildup of radicals that leads to explosion. The wall reactions are not explicitly included in the mechanism since they are not strictly gas-phase reactions. We can symbolically write a first-order reaction for wall destruction of radicals as

$$\text{Radical} \xrightarrow{k_{wall}} \text{absorbed products,} \qquad \text{(H.21)}$$

where $k_{wall}$ is a function of both diffusion (transport) and chemical processes, as well as the nature of the wall surface. Heterogeneous (surface) reactions are discussed in Chapters 4 and 14.

When the initial pressure is set above approximately 1.5 mm Hg, the mixture explodes. This is a direct result of the gas-phase chain sequence H.3–H.6 prevailing over the radical destruction at the wall. Recall from our examination of a generic chain mechanism that increasing pressure increased the radical concentration linearly while increasing the reaction rate geometrically.

Continuing our journey up the 500 °C isotherm, we remain in an explosive regime until about 50 mm Hg; at this point, the mixture ceases to be explosive. The cessation of explosive behavior can be explained by the competition for H atoms between the chain-branching reaction, H.3, and what is effectively a chain-terminating step at low temperatures, reaction H.11 [1, 2]. Reaction H.11 is chain-terminating because the hydroperoxy radical, $HO_2$, is relatively unreactive at these conditions, and, because of this, it can diffuse to the wall where it is destroyed (H.21).

At the third limit, we cross over again into an explosive regime at about 3,000 mm Hg. At these conditions, reaction H.16 adds a chain-branching step which opens up the $H_2O_2$ chain sequence [1, 2].

From this brief discussion of the explosion limits of the $H_2$–$O_2$ system, it is clear how useful an understanding of the *detailed* chemistry of a system is in explaining experimental observations. It is also clear that such understanding is essential to the development of predictive models of combustion phenomena when chemical effects are important.

## CARBON MONOXIDE OXIDATION

Although the oxidation of carbon monoxide has importance in its own right, it is extremely important to the oxidation of hydrocarbons. Hydrocarbon combustion simplistically can be characterized as a two-step process: the first step involves the breakdown of the fuel to carbon monoxide, with the second step being the final oxidation of carbon monoxide to carbon dioxide.

It is well known that CO is slow to oxidize unless there are some hydrogen-containing species present; small quantities of $H_2O$ or $H_2$ can have a tremendous effect on the oxidation rate. This is because the CO oxidation step involving the hydroxyl radical is much faster than the steps involving $O_2$ and O.

Assuming water is the primary hydrogen-containing species, the following four steps describe the oxidation of CO [1]:

$$CO + O_2 \rightarrow CO_2 + O, \tag{CO.1}$$

$$O + H_2O \rightarrow OH + OH, \tag{CO.2}$$

$$CO + OH \rightarrow CO_2 + H, \tag{CO.3}$$

$$H + O_2 \rightarrow OH + O. \tag{CO.4}$$

The reaction CO.1 is slow and does not contribute significantly to the formation of $CO_2$, but rather serves as the initiator of the chain sequence. The actual CO oxidation step, CO.3, is also a chain-propagating step, producing H atoms that react with $O_2$ to form OH and O (reaction CO.4). These radicals, in turn, feed back into the oxidation step (CO.3) and the first chain-branching step (CO.2). The $CO + OH \rightarrow CO_2 + H$ (CO.3) step is the key reaction in the overall scheme.

If hydrogen is the catalyst instead of water, the following steps are involved [1]:

$$O + H_2 \rightarrow OH + H \tag{CO.5}$$

$$OH + H_2 \rightarrow H_2O + H. \tag{CO.6}$$

With hydrogen present, the entire $H_2$–$O_2$ reaction system (H.1–H.21) needs to be included to describe the CO oxidation. Glassman [1] indicates that with $HO_2$ present, another route for CO oxidation opens up:

$$CO + HO_2 \rightarrow CO_2 + OH, \tag{CO.7}$$

although this reaction is not nearly as important as the OH attack on CO (i.e., reaction CO.3). A comprehensive $CO/H_2$ oxidation mechanism is presented in Ref. [5].

# OXIDATION OF HYDROCARBONS

## General Scheme for Alkanes

**Alkanes**, or **paraffins**, are saturated, straight-chain or branched-chain, single-bonded hydrocarbons with the general molecular formula $C_nH_{2n+2}$. In this section, we briefly discuss the generic oxidation of the higher alkanes, that is, paraffins with $n > 2$. The oxidation of methane (and ethane) exhibits unique characteristics and will be discussed in the next section.

Our discussion of higher alkanes is different from the preceding sections in that no attempt is made to explore or list the many elementary reactions involved; rather, we will present an overview of the oxidation process, indicating key reaction steps, and then briefly discuss multistep global approaches that have been applied with some success. Reviews of detailed mechanisms for alkanes and other hydrocarbons can be found in Refs. [6] and [7].

The oxidation of alkanes can be characterized by three sequential processes [1], given below, and illustrated by the species and temperature distributions shown in Fig. 5.2:

   I.  The fuel molecule is attacked by O and H atoms and breaks down, primarily forming alkenes and hydrogen. The hydrogen oxidizes to water, subject to available oxygen.

  II.  The unsaturated alkenes further oxidize to CO and $H_2$. Essentially, all of the $H_2$ is converted to water.

 III.  The CO burns out via reaction CO.3, $CO + OH \rightarrow CO_2 + H$. Nearly all of the heat release associated with the overall combustion process occurs in this step.

We will now flesh out these three processes in steps 1–8, following Glassman [1], and illustrate them with the example of propane ($C_3H_8$) oxidation.

*Step 1.* A carbon–carbon (C–C) bond is broken in the original fuel molecule. The C–C bonds are preferentially broken over hydrogen–carbon bonds because the C–C bonds are weaker.

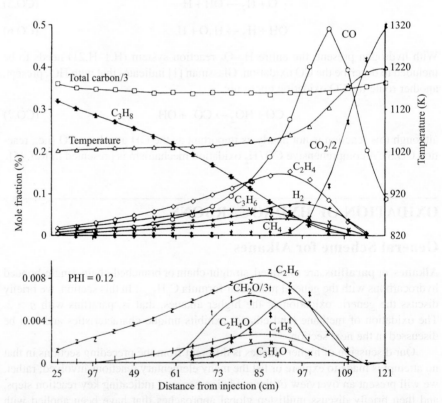

**Figure 5.2**   Species mole fractions and temperature as functions of distance (time) for the oxidation of propane in a steady-flow reactor.

SOURCE: From Ref. [8]. Reprinted by permission of Gordon & Breach Science Publishers.

$$Example: \ C_3H_8 + M \rightarrow C_2H_5 + CH_3 + M. \qquad (P.1)$$

*Step 2.* The two resulting hydrocarbon radicals break down further, creating alkenes (hydrocarbons with double carbon bonds) and hydrogen atoms. The removal of an H atom from the hydrocarbon is termed **H-atom abstraction.** In the example for this step, ethylene and methylene are produced.

$$Example: \ C_2H_5 + M \rightarrow C_2H_4 + H + M \qquad (P.2a)$$

$$CH_3 + M \rightarrow CH_2 + H + M. \qquad (P.2b)$$

*Step 3.* The creation of H atoms from step 2 starts the development of a pool of radicals.

$$Example: \ H + O_2 \rightarrow O + OH. \qquad (P.3)$$

*Step 4.* With the establishment of the radicals, new fuel-molecule attack pathways open up.

$$Example: \ C_3H_8 + OH \rightarrow C_3H_7 + H_2O, \qquad (P.4a)$$

$$C_3H_8 + H \rightarrow C_3H_7 + H_2, \qquad (P.4b)$$

$$C_3H_8 + O \rightarrow C_3H_7 + OH. \qquad (P.4c)$$

*Step 5.* As in step 2, the hydrocarbon radicals again decay into alkenes and H atoms via H-atom abstraction,

$$Example: \ C_3H_7 + M \rightarrow C_3H_6 + H + M, \qquad (P.5)$$

and following the **$\beta$-scission rule** [1]. This rule states that the C–C or C–H bond broken will be the one that is one place removed from the radical site, i.e., the site of the unpaired electron. The unpaired electron at the radical site strengthens the adjacent bonds at the expense of those one place removed from the site. For the $C_3H_7$ radical created in step 4, we have two possible paths:

$$Example: \ C_3H_7 + M \ \Big\langle {\nearrow \atop \searrow} \ \begin{matrix} C_3H_6 + H + M \\ \\ C_2H_4 + CH_3 + M. \end{matrix} \qquad (P.6)$$

The application of the $\beta$-scission rule to the $C_3H_7$ radical breakdown (P.6) is illustrated in Fig. 5.3.

*Step 6.* The oxidation of the olefins created in steps 2 and 5 is initiated by O-atom attack, which produces formyl radicals (HCO) and formaldehyde ($H_2CO$).

$$Example: \ C_3H_6 + O \rightarrow C_2H_5 + HCO \qquad (P.7a)$$

$$C_3H_6 + O \rightarrow C_2H_4 + H_2CO. \qquad (P.7b)$$

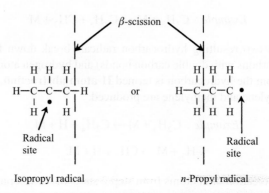

**Figure 5.3**    The β-scission rule applied to bond breaking in $C_3H_7$ where the radical site occurs at different locations. Note the intervening C–C bond between the radical site and the broken bond.

*Step 7a.* Methyl radicals ($CH_3$) oxidize.

*Step 7b.* Formaldehyde ($H_2CO$) oxidizes.

*Step 7c.* Methylene ($CH_2$) oxidizes.

The details of steps 7a–7c can be found in Ref. [1]; however, each of these steps produces carbon monoxide, the oxidation of which is the final step (step 8).

*Step 8.* The carbon monoxide oxidizes following the moist CO mechanism defined by CO.1–CO.7.

As can be seen from the above, the mechanism of higher-alkane oxidation is quite complex indeed. The details of such mechanisms are still the subject of research [6, 7].

## Global and Quasi-Global Mechanisms

The sequential nature of the processes I–III above has led to empirical global models that capture the overall behavior in a sequence of global or quasi-global steps. Westbrook and Dryer [9] present and evaluate one-step, two-step, and multistep global kinetics for a wide variety of hydrocarbons. Global models, by definition, do not capture the details of hydrocarbon oxidation; they may, however, be useful for engineering approximations, if their limitations are recognized. The following single-step expression [9] is suggested for engineering approximations for the global reaction

$$C_xH_y + (x + y/4)O_2 \xrightarrow{k_G} xCO_2 + (y/2)H_2O \qquad (5.1)$$

$$\frac{d[C_xH_y]}{dt} = -A \exp(-E_a/R_uT)[C_xH_y]^m[O_2]^n \qquad (5.2)$$

$$[=] \text{ gmol/cm}^3\text{-s,}$$

**Table 5.1**    Single-step reaction rate parameters for use with Eqn. 5.2 (Adapted from Ref. [9])

| Fuel | Pre-exponential Factor, $A$[a] | Activation Temperature, $E_a/R_u$ (K) | $m$ | $n$ |
|------|------|------|------|------|
| $CH_4$ | $1.3 \cdot 10^8$ | $24{,}358$[b] | $-0.3$ | 1.3 |
| $CH_4$ | $8.3 \cdot 10^5$ | $15{,}098$[c] | $-0.3$ | 1.3 |
| $C_2H_6$ | $1.1 \cdot 10^{12}$ | $15{,}098$ | 0.1 | 1.65 |
| $C_3H_8$ | $8.6 \cdot 10^{11}$ | $15{,}098$ | 0.1 | 1.65 |
| $C_4H_{10}$ | $7.4 \cdot 10^{11}$ | $15{,}098$ | 0.15 | 1.6 |
| $C_5H_{12}$ | $6.4 \cdot 10^{11}$ | $15{,}098$ | 0.25 | 1.5 |
| $C_6H_{14}$ | $5.7 \cdot 10^{11}$ | $15{,}098$ | 0.25 | 1.5 |
| $C_7H_{16}$ | $5.1 \cdot 10^{11}$ | $15{,}098$ | 0.25 | 1.5 |
| $C_8H_{18}$ | $4.6 \cdot 10^{11}$ | $15{,}098$ | 0.25 | 1.5 |
| $C_8H_{18}$ | $7.2 \cdot 10^{12}$ | $20{,}131$[d] | 0.25 | 1.5 |
| $C_9H_{20}$ | $4.2 \cdot 10^{11}$ | $15{,}098$ | 0.25 | 1.5 |
| $C_{10}H_{22}$ | $3.8 \cdot 10^{11}$ | $15{,}098$ | 0.25 | 1.5 |
| $CH_3OH$ | $3.2 \cdot 10^{12}$ | $15{,}098$ | 0.25 | 1.5 |
| $C_2H_5OH$ | $1.5 \cdot 10^{12}$ | $15{,}098$ | 0.15 | 1.6 |
| $C_6H_6$ | $2.0 \cdot 10^{11}$ | $15{,}098$ | $-0.1$ | 1.85 |
| $C_7H_8$ | $1.6 \cdot 10^{11}$ | $15{,}098$ | $-0.1$ | 1.85 |
| $C_2H_4$ | $2.0 \cdot 10^{12}$ | $15{,}098$ | 0.1 | 1.65 |
| $C_3H_6$ | $4.2 \cdot 10^{11}$ | $15{,}098$ | $-0.1$ | 1.85 |
| $C_2H_2$ | $6.5 \cdot 10^{12}$ | $15{,}098$ | 0.5 | 1.25 |

[a]Units of A are consistent with concentrations in Eqn. 5.2 expressed in units of $gmol/cm^3$, i.e., $A[=]$ $(gmol/cm^3)^{1-m-n}/s$.

[b]$E_a = 48.4$ kcal/gmol.

[c]$E_a = 30$ kcal/gmol.

[d]$E_a = 40$ kcal/gmol.

where the parameters $A$, $E_a/R_u$, $m$, and $n$, shown in Table 5.1, have been chosen to provide best agreement between experimental and predicted flame speeds and flammability limits (see Chapter 8). Note the treatment of units for A in the footnote.

An example of a multistep, quasi-global mechanism is that of Hautman *et al.* [8], modeling propane oxidation using a four-step scheme:

$$C_nH_{2n+2} \rightarrow (n/2)C_2H_4 + H_2, \tag{HC.1}$$

$$C_2H_4 + O_2 \rightarrow 2CO + 2H_2, \tag{HC.2}$$

$$CO + \tfrac{1}{2}O_2 \rightarrow CO_2, \tag{HC.3}$$

$$H_2 + \tfrac{1}{2}O_2 \rightarrow H_2O, \tag{HC.4}$$

**Table 5.2** Constants[a] for multistep global mechanism for $C_nH_{2n+2}$ oxidation [8]

|  | **Propane ($n = 3$)** | | | |
| --- | --- | --- | --- | --- |
| Rate eqn. | 5.3 | 5.4 | 5.5 | 5.6 |
| $x$ | 17.32 | 14.7 | 14.6 | 13.52 |
| $[E_A/R_u]$ (K) | 24,962 | 25,164 | 20,131 | 20,634 |
| $a$ | 0.50 | 0.90 | 1.0 | 0.85 |
| $b$ | 1.07 | 1.18 | 0.25 | 1.42 |
| $c$ | 0.40 | −0.37 | 0.50 | −0.56 |

[a]Initial conditions: $T$ (K): 960–1145; $[C_3H_8]_i$ (gmol/cm³): $1 \cdot 10^{-8} - 1 \cdot 10^{-7}$; $[O_2]_i$ (gmol/cm³): $1 \cdot 10^{-7} - 5 \cdot 10^{-6}$; Φ: 0.03–2.0.

where the reaction rates (in gmol/cm³-s) are expressed as

$$\frac{d[C_nH_{2n+2}]}{dt} = -10^x \exp(-E_A/R_uT)[C_nH_{2n+2}]^a[O_2]^b[C_2H_4]^c, \tag{5.3}$$

$$\frac{d[C_2H_4]}{dt} = -10^x \exp(-E_A/R_uT)[C_2H_4]^a[O_2]^b[C_nH_{2n+2}]^c, \tag{5.4}$$

$$\frac{d[CO]}{dt} = -10^x \exp(-E_A/R_uT)[CO]^a[O_2]^b[H_2O]^c 7.93 \exp(-2.48\Phi), \tag{5.5}$$

$$\frac{d[H_2]}{dt} = -10^x \exp(-E_A/R_uT)[H_2]^a[O_2]^b[C_2H_4]^c, \tag{5.6}$$

where Φ is the equivalence ratio. The exponents $x$, $a$, $b$, and $c$ for each reaction are given in Table 5.2.

Notice that in the mechanism it is assumed that ethene ($C_2H_4$) is the intermediate hydrocarbon; and that in the rate equations, $C_3H_8$ and $C_2H_4$ inhibit the oxidation of $C_2H_4$ and $H_2$, respectively, since the $C_3H_8$ and $C_2H_4$ exponents are negative. Note also that the rate equations do not directly follow from the global steps, since Eqns. 5.3–5.6 involve the product of three species, not just two, as suggested by the individual global steps HC.1–HC.4. Other approaches to simplified hydrocarbon oxidation are discussed in Ref. [10].

## Real Fuels and Their Surrogates

Most real fuels, such as gasoline and diesel fuel, contain many different hydrocarbons, as discussed in Chapter 17. One approach to modeling the combustion of a real fuel is to consider a single component to be representative of the real fuel blend or to employ a mixture of a small number of hydrocarbon components such that many of the important characteristics of the fuel are matched. Chemical kinetic mechanisms have been developed to model the combustion of representative higher molecular weight hydrocarbons and blends of these as surrogates to real fuels. Table 5.3 presents several studies that have addressed this problem of kinetic modeling.

**Table 5.3**    Chemical kinetic studies targeting real fuel combustion

| Target Fuel | Surrogate Blend[1] | Reference | Comment |
|---|---|---|---|
| Natural gas | Methane ($CH_4$)<br>Ethane ($C_2H_6$)<br>Propane ($C_3H_8$) | Dagaut [11] | — |
| Kerosene (Jet A-1) | n-Decane ($C_{10}H_{22}$) | Dagaut [11] | Single-component model fuel |
| Kerosene (Jet A-1) | 74% n-Decane ($C_{10}H_{22}$)<br>15% n-Propylbenzene<br>11% n-Propylcyclohexane | Dagaut [11] | 207 species and 1,592 reactions |
| Diesel fuel | 36.5% n-Hexadecane ($C_{16}H_{24}$)<br>24.5% Isooctane ($C_8H_{18}$)<br>20.4% n-Propylcyclohexane<br>18.2% n-Propylbenzene | Dagaut [11] | 298 species and 2,352 reactions |
| JP-8 (Jet fuel) | 10% Isooctane<br>20% Methylcyclohexane ($C_7H_{14}$)<br>15% m-Xylene ($C_8H_{10}$)<br>30% n-Dodecane ($C_7H_{14}$)<br>5% Tetralin ($C_7H_{14}$)<br>20% Tetradecane ($C_{14}H_{30}$) | Cooke et al. [12]<br>Violi et al. [13]<br>Ranzi et al. [14]<br>Ranzi et al. [15]<br>Ranzi et al. [16] | 221 species and 5,032 reactions |
| Gasoline | Isooctane (neat) ($C_8H_{18}$)<br>Isooctane ($C_8H_{18}$) – n-Heptane ($C_7H_{16}$) | Curran et al. [17]<br>Curran et al. [18] | Single-component model fuel and two-component surrogates; 860–990 species and 3,600–4,060 reactions |
| Gasoline | 63–69% (liq. vol.) Isooctane ($C_8H_{18}$)<br>14–20% (liq. vol.) Toluene ($C_7H_8$)<br>17% (liq. vol.) n-Heptane ($C_7H_{16}$)<br>and<br>62% (liq. vol.) Isooctane ($C_8H_{18}$)<br>20% (liq. vol.) Ethanol ($C_2H_5OH$)<br>18% (liq. vol.) n-Heptane ($C_7H_{16}$)<br>and<br>45% (liq. vol.) Toluene ($C_7H_8$)<br>25% (liq. vol.) Isooctane ($C_8H_{18}$)<br>20% (liq. vol.) n-Heptane ($C_7H_{16}$)<br>10% (liq. vol.) Diisobutylene ($C_8H_{16}$) | Andrae et al. [19]<br>Andrae [20] | Octane numbers of blends match standard European gasoline. |
| Biodiesel | Methyl decaoate ($C_{10}H_{22}O_2$, i.e., $CH_3(CH_2)_8COOCH_3$) | Herbinet et al. [21] | 3,012 species and 8,820 reactions |

[1]Compositions given in mole percent unless otherwise noted.

# METHANE COMBUSTION

## Complex Mechanism

Because of its unique tetrahedral molecular structure with large C–H bond energies, methane exhibits some unique combustion characteristics. For example, it has a high ignition temperature, low flame speed, and low reactivity in photochemical smog chemistry compared to other hydrocarbons.

Methane chemical kinetics are perhaps the most widely researched and, hence, most well understood. Kaufman [22], in a review of combustion kinetics indicated that the methane combustion mechanism evolved in the period 1970–1982 from less than 15 elementary steps with 12 species to 75 elementary steps, plus the 75 reverse reactions, with 25 species. More recently, several research groups have collaborated in the creation of an optimized methane kinetic mechanism [23]. This mechanism, designated GRI Mech, is based on the optimization techniques of Frenklach *et al.* [24]. GRI Mech [23] is available on the Internet and is continually updated. Version 3.0, shown in Table 5.4, considers 325 elementary reactions involving 53 species. Many of these steps we have seen before as part of the $H_2$ and CO oxidation mechanisms.

To make some sense of this complex system, we present reaction pathway analyses for both high-temperature and low-temperature combustion of $CH_4$ with air in a well-stirred reactor [25] using GRI Mech 2.11. A detailed discussion of the well-stirred reactor is presented in Chapter 6; however, for our purposes here, we need only recognize that reactions take place in a homogeneous, iso-thermal environment. The choice of a well-stirred reactor eliminates the need to account for a spatial distribution of species as would be encountered in a flame, for example.

**Table 5.4**     Complex methane combustion mechanism (GRI Mech 3.0) [23]

| | | Forward Rate Coefficient[a] | | |
|---|---|---|---|---|
| **No.** | **Reaction** | **A** | **b** | **E** |
| *C–H–O Reactions* | | | | |
| 1 | $O + O + M \rightarrow O_2 + M$ | 1.20E + 17 | −1.0 | 0.0 |
| 2 | $O + H + M \rightarrow OH + M$ | 5.00E + 17 | −1.0 | 0.0 |
| 3 | $O + H_2 \rightarrow H + OH$ | 3.87E + 04 | 2.7 | 6,260 |
| 4 | $O + HO_2 \rightarrow OH + O_2$ | 2.00E + 13 | 0.0 | 0.0 |
| 5 | $O + H_2O_2 \rightarrow OH + HO_2$ | 9.63E + 06 | 2.0 | 4,000 |
| 6 | $O + CH \rightarrow H + CO$ | 5.70E + 13 | 0.0 | 0.0 |
| 7 | $O + CH_2 \rightarrow H + HCO$ | 8.00E + 13 | 0.0 | 0.0 |
| 8[b] | $O + CH_2(S) \rightarrow H_2 + CO$ | 1.50E + 13 | 0.0 | 0.0 |
| 9[b] | $O + CH_2(S) \rightarrow H + HCO$ | 1.50E + 13 | 0.0 | 0.0 |
| 10 | $O + CH_3 \rightarrow H + CH_2O$ | 5.06E + 13 | 0.0 | 0.0 |
| 11 | $O + CH_4 \rightarrow OH + CH_3$ | 1.02E + 09 | 1.5 | 8,600 |
| 12 | $O + CO + M \rightarrow CO_2 + M$ | 1.8E + 10 | 0.0 | 2,385 |
| 13 | $O + HCO \rightarrow OH + CO$ | 3.00E + 13 | 0.0 | 0.0 |
| 14 | $O + HCO \rightarrow H + CO_2$ | 3.00E + 13 | 0.0 | 0.0 |
| 15 | $O + CH_2O \rightarrow OH + HCO$ | 3.90E + 13 | 0.0 | 3,540 |
| 16 | $O + CH_2OH \rightarrow OH + CH_2O$ | 1.00E + 13 | 0.0 | 0.0 |
| 17 | $O + CH_3O \rightarrow OH + CH_2O$ | 1.00E + 13 | 0.0 | 0.0 |
| 18 | $O + CH_3OH \rightarrow OH + CH_2OH$ | 3.88E + 05 | 2.5 | 3,100 |
| 19 | $O + CH_3OH \rightarrow OH + CH_3O$ | 1.30E + 05 | 2.5 | 5,000 |
| 20 | $O + C_2H \rightarrow CH + CO$ | 5.00E + 13 | 0.0 | 0.0 |
| 21 | $O + C_2H_2 \rightarrow H + HCCO$ | 1.35E + 07 | 2.0 | 1,900 |
| 22 | $O + C_2H_2 \rightarrow OH + C_2H$ | 4.60E + 19 | −1.4 | 28,950 |
| 23 | $O + C_2H_2 \rightarrow CO + CH_2$ | 9.64E + 06 | 2.0 | 1,900 |
| 24 | $O + C_2H_3 \rightarrow H + CH_2CO$ | 3.00E + 13 | 0.0 | 0.0 |

**Table 5.4**      (continued)

| No. | Reaction | Forward Rate Coefficient[a] | | |
|---|---|---|---|---|
| | | $A$ | $b$ | $E$ |
| *C–H–O Reactions (continued)* | | | | |
| 25 | $O + C_2H_4 \rightarrow CH_3 + HCO$ | 1.25E + 07 | 1.83 | 220 |
| 26 | $O + C_2H_5 \rightarrow CH_3 + CH_2O$ | 2.24E + 13 | 0.0 | 0.0 |
| 27 | $O + C_2H_6 \rightarrow OH + C_2H_5$ | 8.98E + 07 | 1.9 | 5,690 |
| 28 | $O + HCCO \rightarrow H + CO + CO$ | 1.00E + 14 | 0.0 | 0.0 |
| 29 | $O + CH_2CO \rightarrow OH + HCCO$ | 1.00E + 13 | 0.0 | 8,000 |
| 30 | $O + CH_2CO \rightarrow CH_2 + CO_2$ | 1.75E + 12 | 0.0 | 1,350 |
| 31 | $O_2 + CO \rightarrow O + CO_2$ | 2.50E + 12 | 0.0 | 47,800 |
| 32 | $O_2 + CH_2O \rightarrow HO_2 + HCO$ | 1.00E + 14 | 0.0 | 40,000 |
| 33 | $H + O_2 + M \rightarrow HO_2 + M$ | 2.80E + 18 | −0.9 | 0.0 |
| 34 | $H + O_2 + O_2 \rightarrow HO_2 + O_2$ | 2.08E + 19 | −1.2 | 0.0 |
| 35 | $H + O_2 + H_2O \rightarrow HO_2 + H_2O$ | 1.13E + 19 | −0.8 | 0.0 |
| 36 | $H + O_2 + N_2 \rightarrow HO_2 + N_2$ | 2.60E + 19 | −1.2 | 0.0 |
| 37 | $H + O_2 + Ar \rightarrow HO_2 + Ar$ | 7.00E + 17 | −0.8 | 0.0 |
| 38 | $H + O_2 \rightarrow O + OH$ | 2.65E + 16 | −0.7 | 17,041 |
| 39 | $H + H + M \rightarrow H_2 + M$ | 1.00E + 18 | −1.0 | 0.0 |
| 40 | $H + H + H_2 \rightarrow H_2 + H_2$ | 9.00E + 16 | −0.6 | 0.0 |
| 41 | $H + H + H_2O \rightarrow H_2 + H_2O$ | 6.00E + 19 | −1.2 | 0.0 |
| 42 | $H + H + CO_2 \rightarrow H_2 + CO_2$ | 5.50E + 20 | −2.0 | 0.0 |
| 43 | $H + OH + M \rightarrow H_2O + M$ | 2.20E + 22 | −2.0 | 0.0 |
| 44 | $H + HO_2 \rightarrow O + H_2O$ | 3.97E + 12 | 0.0 | 671 |
| 45 | $H + HO_2 \rightarrow O_2 + H_2$ | 4.48E + 13 | 0.0 | 1,068 |
| 46 | $H + HO_2 \rightarrow OH + OH$ | 8.4E + 13 | 0.0 | 635 |
| 47 | $H + H_2O_2 \rightarrow HO_2 + H_2$ | 1.21E + 07 | 2.0 | 5,200 |
| 48 | $H + H_2O_2 \rightarrow OH + H_2O$ | 1.00E + 13 | 0.0 | 3,600 |
| 49 | $H + CH \rightarrow C + H_2$ | 1.65E + 14 | 0.0 | 0.0 |
| 50 | $H + CH_2 (+ M) \rightarrow CH_3 (+ M)$ | pressure dependent | | |
| 51[b] | $H + CH_2(S) \rightarrow CH + H_2$ | 3.00E + 13 | 0.0 | 0.0 |
| 52 | $H + CH_3 (+ M) \rightarrow CH_4 (+ M)$ | pressure dependent | | |
| 53 | $H + CH_4 \rightarrow CH_3 + H_2$ | 6.60E + 08 | 1.6 | 10,840 |
| 54 | $H + HCO (+ M) \rightarrow CH_2O (+ M)$ | pressure dependent | | |
| 55 | $H + HCO \rightarrow H_2 + CO$ | 7.34E + 13 | 0.0 | 0.0 |
| 56 | $H + CH_2O (+ M) \rightarrow CH_2OH (+ M)$ | pressure dependent | | |
| 57 | $H + CH_2O (+ M) \rightarrow CH_3O (+ M)$ | pressure dependent | | |
| 58 | $H + CH_2O \rightarrow HCO + H_2$ | 5.74E + 07 | 1.9 | 2,742 |
| 59 | $H + CH_2OH (+ M) \rightarrow CH_3OH (+ M)$ | pressure dependent | | |
| 60 | $H + CH_2OH \rightarrow H_2 + CH_2O$ | 2.00E + 13 | 0.0 | 0.0 |
| 61 | $H + CH_2OH \rightarrow OH + CH_3$ | 1.65E + 11 | 0.7 | −284 |
| 62[b] | $H + CH_2OH \rightarrow CH_2(S) + H_2O$ | 3.28E + 13 | −0.1 | 610 |
| 63 | $H + CH_3O (+ M) \rightarrow CH_3OH (+ M)$ | pressure dependent | | |
| 64[b] | $H + CH_2OH \rightarrow CH_2(S) + H_2O$ | 4.15E + 07 | 1.6 | 1,924 |
| 65 | $H + CH_3O \rightarrow H_2 + CH_2O$ | 2.00E + 13 | 0.0 | 0.0 |
| 66 | $H + CH_3O \rightarrow OH + CH_3$ | 1.50E + 12 | 0.5 | −110 |
| 67[b] | $H + CH_3O \rightarrow CH_2(S) + H_2O$ | 2.62E + 14 | −0.2 | 1,070 |
| 68 | $H + CH_3OH \rightarrow CH_2OH + H_2$ | 1.70E + 07 | 2.1 | 4,870 |
| 69 | $H + CH_3OH \rightarrow CH_3O + H_2$ | 4.20E + 06 | 2.1 | 4,870 |
| 70 | $H + C_2H (+ M) \rightarrow C_2H_2 (+ M)$ | pressure dependent | | |

**Table 5.4**     (continued)

| | | Forward Rate Coefficient[a] | | |
|---|---|---|---|---|
| No. | Reaction | $A$ | $b$ | $E$ |
| ***C–H–O Reactions (continued)*** | | | | |
| 71 | $H + C_2H_2 (+ M) \rightarrow C_2H_3 (+ M)$ | pressure dependent | | |
| 72 | $H + C_2H_3 (+ M) \rightarrow C_2H_4 (+ M)$ | pressure dependent | | |
| 73 | $H + C_2H_3 \rightarrow H_2 + C_2H_2$ | 3.00E + 13 | 0.0 | 0.0 |
| 74 | $H + C_2H_4 (+ M) \rightarrow C_2H_5 (+ M)$ | pressure dependent | | |
| 75 | $H + C_2H_4 \rightarrow C_2H_3 + H_2$ | 1.32E + 06 | 2.5 | 12,240 |
| 76 | $H + C_2H_5 (+ M) \rightarrow C_2H_6 (+ M)$ | pressure dependent | | |
| 77 | $H + C_2H_5 \rightarrow C_2H_4 + H_2$ | 2.00E + 12 | 0.0 | 0.0 |
| 78 | $H + C_2H_6 \rightarrow C_2H_5 + H_2$ | 1.15E + 08 | 1.9 | 7,530 |
| 79[b] | $H + HCCO \rightarrow CH_2(S) + CO$ | 1.00E + 14 | 0.0 | 0.0 |
| 80 | $H + CH_2CO \rightarrow HCCO + H_2$ | 5.00E + 13 | 0.0 | 8,000 |
| 81 | $H + CH_2CO \rightarrow CH_3 + CO$ | 1.13E + 13 | 0.0 | 3,428 |
| 82 | $H + HCCOH \rightarrow H + CH_2CO$ | 1.00E + 13 | 0.0 | 0.0 |
| 83 | $H_2 + CO (+ M) \rightarrow CH_2O (+ M)$ | pressure dependent | | |
| 84 | $OH + H_2 \rightarrow H + H_2O$ | 2.16E + 08 | 1.5 | 3,430 |
| 85 | $OH + OH (+ M) \rightarrow H_2O_2 (+ M)$ | pressure dependent | | |
| 86 | $OH + OH \rightarrow O + H_2O$ | 3.57E + 04 | 2.4 | −2,110 |
| 87 | $OH + HO_2 \rightarrow O_2 + H_2O$ | 1.45E + 13 | 0.0 | −500 |
| 88 | $OH + H_2O_2 \rightarrow HO_2 + H_2O$ | 2.00E + 12 | 0.0 | 427 |
| 89 | $OH + H_2O_2 \rightarrow HO_2 + H_2O$ | 1.70E + 18 | 0.0 | 29,410 |
| 90 | $OH + C \rightarrow H + CO$ | 5.00E + 13 | 0.0 | 0.0 |
| 91 | $OH + CH \rightarrow H + HCO$ | 3.00E + 13 | 0.0 | 0.0 |
| 92 | $OH + CH_2 \rightarrow H + CH_2O$ | 2.00E + 13 | 0.0 | 0.0 |
| 93 | $OH + CH_2 \rightarrow CH + H_2O$ | 1.13E + 07 | 2.0 | 3,000 |
| 94[b] | $OH + CH_2(S) \rightarrow H + CH_2O$ | 3.00E + 13 | 0.0 | 0.0 |
| 95 | $OH + CH_3 (+ M) \rightarrow CH_3OH (+ M)$ | pressure dependent | | |
| 96 | $OH + CH_3 \rightarrow CH_2 + H_2O$ | 5.60E + 07 | 1.6 | 5,420 |
| 97[b] | $OH + CH_3 \rightarrow CH_2 (S) + H_2O$ | 6.44E + 17 | −1.3 | 1,417 |
| 98 | $OH + CH_4 \rightarrow CH_3 + H_2O$ | 1.00E + 08 | 1.6 | 3,120 |
| 99 | $OH + CO \rightarrow H + CO_2$ | 4.76E + 07 | 1.2 | 70 |
| 100 | $OH + HCO \rightarrow H_2O + CO$ | 5.00E + 13 | 0.0 | 0.0 |
| 101 | $OH + CH_2O \rightarrow HCO + H_2O$ | 3.43E + 09 | 1.2 | −447 |
| 102 | $OH + CH_2OH \rightarrow H_2O + CH_2O$ | 5.00E + 12 | 0.0 | 0.0 |
| 103 | $OH + CH_3O \rightarrow H_2O + CH_2O$ | 5.00E + 12 | 0.0 | 0.0 |
| 104 | $OH + CH_3OH \rightarrow CH_2OH + H_2O$ | 1.44E + 06 | 2.0 | −840 |
| 105 | $OH + CH_3OH \rightarrow CH_3O + H_2O$ | 6.30E + 06 | 2.0 | 1,500 |
| 106 | $OH + C_2H \rightarrow H + HCCO$ | 2.00E + 13 | 0.0 | 0.0 |
| 107 | $OH + C_2H_2 \rightarrow H + CH_2CO$ | 2.18E − 04 | 4.5 | −1,000 |
| 108 | $OH + C_2H_2 \rightarrow H + HCCOH$ | 5.04E + 05 | 2.3 | 13,500 |
| 109 | $OH + C_2H_2 \rightarrow C_2H + H_2O$ | 3.37E + 07 | 2.0 | 14,000 |
| 110 | $OH + C_2H_2 \rightarrow CH_3 + CO$ | 4.83E − 04 | 4.0 | −2,000 |
| 111 | $OH + C_2H_3 \rightarrow H_2O + C_2H_2$ | 5.00E + 12 | 0.0 | 0.0 |
| 112 | $OH + C_2H_4 \rightarrow C_2H_3 + H_2O$ | 3.60E + 06 | 2.0 | 2,500 |
| 113 | $OH + C_2H_6 \rightarrow C_2H_5 + H_2O$ | 3.54E + 06 | 2.1 | 870 |
| 114 | $OH + CH_2CO \rightarrow HCCO + H_2O$ | 7.50E + 12 | 0.0 | 2,000 |
| 115 | $HO_2 + HO_2 \rightarrow O_2 + H_2O_2$ | 1.30E + 11 | 0.0 | −1,630 |
| 116 | $HO_2 + HO_2 \rightarrow O_2 + H_2O_2$ | 4.20E + 14 | 0.0 | 12,000 |

**Table 5.4**    (continued)

| No. | Reaction | Forward Rate Coefficient[a] | | |
|-----|----------|------|------|------|
| | | *A* | *b* | *E* |
| **C–H–O Reactions (continued)** | | | | |
| 117 | $HO_2 + CH_2 \rightarrow OH + CH_2O$ | 2.00E + 13 | 0.0 | 0.0 |
| 118 | $HO_2 + CH_3 \rightarrow O_2 + CH_4$ | 1.00E + 12 | 0.0 | 0.0 |
| 119 | $HO_2 + CH_3 \rightarrow OH + CH_3O$ | 3.78E + 13 | 0.0 | 0.0 |
| 120 | $HO_2 + CO \rightarrow OH + CO_2$ | 1.50E + 14 | 0.0 | 23,600 |
| 121 | $HO_2 + CH_2O \rightarrow HCO + H_2O_2$ | 5.60E + 06 | 2.0 | 12,000 |
| 122 | $C + O_2 \rightarrow O + CO$ | 5.80E + 13 | 0.0 | 576 |
| 123 | $C + CH_2 \rightarrow H + C_2H$ | 5.00E + 13 | 0.0 | 0.0 |
| 124 | $C + CH_3 \rightarrow H + C_2H_2$ | 5.00E + 13 | 0.0 | 0.0 |
| 125 | $CH + O_2 \rightarrow O + HCO$ | 6.71E + 13 | 0.0 | 0.0 |
| 126 | $CH + H_2 \rightarrow H + CH_2$ | 1.08E + 14 | 0.0 | 3,110 |
| 127 | $CH + H_2O \rightarrow H + CH_2O$ | 5.71E + 12 | 0.0 | −755 |
| 128 | $CH + CH_2 \rightarrow H + C_2H_2$ | 4.00E + 13 | 0.0 | 0.0 |
| 129 | $CH + CH_3 \rightarrow H + C_2H_3$ | 3.00E + 13 | 0.0 | 0.0 |
| 130 | $CH + CH_4 \rightarrow H + C_2H_4$ | 6.00E + 13 | 0.0 | 0.0 |
| 131 | $CH + CO (+ M) \rightarrow HCCO (+ M)$ | pressure dependent | | |
| 132 | $CH + CO_2 \rightarrow HCO + CO$ | 1.90E + 14 | 0.0 | 15,792 |
| 133 | $CH + CH_2O \rightarrow H + CH_2CO$ | 9.46E + 13 | 0.0 | −515 |
| 134 | $CH + HCCO \rightarrow CO + C_2H_2$ | 5.00E + 13 | 0.0 | 0.0 |
| 135 | $CH_2 + O_2 \rightarrow OH + HCO$ | 5.00E + 12 | 0.0 | 1,500 |
| 136 | $CH_2 + H_2 \rightarrow H + CH_3$ | 5.00E + 05 | 2.0 | 7,230 |
| 137 | $CH_2 + CH_2 \rightarrow H_2 + C_2H_2$ | 1.60E + 15 | 0.0 | 11,944 |
| 138 | $CH_2 + CH_3 \rightarrow H + C_2H_4$ | 4.00E + 13 | 0.0 | 0.0 |
| 139 | $CH_2 + CH_4 \rightarrow CH_3 + CH_3$ | 2.46E + 06 | 2.0 | 8,270 |
| 140 | $CH_2 + CO (+ M) \rightarrow CH_2CO (+ M)$ | pressure dependent | | |
| 141 | $CH_2 + HCCO \rightarrow C_2H_3 + CO$ | 3.00E + 13 | 0.0 | 0.0 |
| 142[b] | $CH_2(S) + N_2 \rightarrow CH_2 + N_2$ | 1.50E + 13 | 0.0 | 600 |
| 143[b] | $CH_2(S) + Ar \rightarrow CH_2 + Ar$ | 9.00E + 12 | 0.0 | 600 |
| 144[b] | $CH_2(S) + O_2 \rightarrow H + OH + CO$ | 2.80E + 13 | 0.0 | 0.0 |
| 145[b] | $CH_2(S) + O_2 \rightarrow CO + H_2O$ | 1.20E + 13 | 0.0 | 0.0 |
| 146[b] | $CH_2(S) + H_2 \rightarrow CH_3 + H$ | 7.00E + 13 | 0.0 | 0.0 |
| 147[b] | $CH_2(S) + H_2O (+ M) \rightarrow CH_3OH (+ M)$ | pressure dependent | | |
| 148[b] | $CH_2(S) + H_2O \rightarrow CH_2 + H_2O$ | 3.00E + 13 | 0.0 | 0.0 |
| 149[b] | $CH_2(S) + CH_3 \rightarrow H + C_2H_4$ | 1.20E + 13 | 0.0 | −570 |
| 150[b] | $CH_2(S) + CH_4 \rightarrow CH_3 + CH_3$ | 1.60E + 13 | 0.0 | −570 |
| 151[b] | $CH_2(S) + CO \rightarrow CH_2 + CO$ | 9.00E + 12 | 0.0 | 0.0 |
| 152[b] | $CH_2(S) + CO_2 \rightarrow CH_2 + CO_2$ | 7.00E + 12 | 0.0 | 0.0 |
| 153[b] | $CH_2(S) + CO_2 \rightarrow CO + CH_2O$ | 1.40E + 13 | 0.0 | 0.0 |
| 154[b] | $CH_2(S) + C_2H_6 \rightarrow CH_3 + C_2H_5$ | 4.00E + 13 | 0.0 | −550 |
| 155 | $CH_3 + O_2 \rightarrow O + CH_3O$ | 3.56E + 13 | 0.0 | 30,480 |
| 156 | $CH_3 + O_2 \rightarrow OH + CH_2O$ | 2.31E + 12 | 0.0 | 20,315 |
| 157 | $CH_3 + H_2O_2 \rightarrow HO_2 + CH_4$ | 2.45E + 04 | 2.47 | 5,180 |
| 158 | $CH_3 + CH_3 (+ M) \rightarrow C_2H_6 (+ M)$ | pressure dependent | | |
| 159 | $CH_3 + CH_3 \rightarrow H + C_2H_5$ | 6.48E + 12 | 0.1 | 10,600 |
| 160 | $CH_3 + HCO \rightarrow CH_4 + CO$ | 2.65E + 13 | 0.0 | 0.0 |
| 161 | $CH_3 + CH_2O \rightarrow HCO + CH_4$ | 3.32E + 03 | 2.8 | 5,860 |

**Table 5.4**     (continued)

| No. | Reaction | Forward Rate Coefficient[a] | | |
|---|---|---|---|---|
| | | $A$ | $b$ | $E$ |
| *C–H–O Reactions (continued)* | | | | |
| 162 | $CH_3 + CH_3OH \rightarrow CH_2OH + CH_4$ | 3.00E + 07 | 1.5 | 9,940 |
| 163 | $CH_3 + CH_3OH \rightarrow CH_3O + CH_4$ | 1.00E + 07 | 1.5 | 9,940 |
| 164 | $CH_3 + C_2H_4 \rightarrow C_2H_3 + CH_4$ | 2.27E + 05 | 2.0 | 9,200 |
| 165 | $CH_3 + C_2H_6 \rightarrow C_2H_5 + CH_4$ | 6.14E + 06 | 1.7 | 10,450 |
| 166 | $HCO + H_2O \rightarrow H + CO + H_2O$ | 1.55E + 18 | −1.0 | 17,000 |
| 167 | $HCO + M \rightarrow H + CO + M$ | 1.87E + 17 | −1.0 | 17,000 |
| 168 | $HCO + O_2 \rightarrow HO_2 + CO$ | 1.35E + 13 | 0.0 | 400 |
| 169 | $CH_2OH + O_2 \rightarrow HO_2 + CH_2O$ | 1.80E + 13 | 0.0 | 900 |
| 170 | $CH_3O + O_2 \rightarrow HO_2 + CH_2O$ | 4.28E − 13 | 7.6 | −3,530 |
| 171 | $C_2H + O_2 \rightarrow HCO + CO$ | 1.00E + 13 | 0.0 | −755 |
| 172 | $C_2H + H_2 \rightarrow H + C_2H_2$ | 5.68E + 10 | 0.9 | 1,993 |
| 173 | $C_2H_3 + O_2 \rightarrow HCO + CH_2O$ | 4.58E + 16 | −1.4 | 1,015 |
| 174 | $C_2H_4 (+ M) \rightarrow H_2 + C_2H_2 (+ M)$ | pressure dependent | | |
| 175 | $C_2H_5 + O_2 \rightarrow HO_2 + C_2H_4$ | 8.40E + 11 | 0.0 | 3,875 |
| 176 | $HCCO + O_2 \rightarrow OH + CO + CO$ | 3.20E + 12 | 0.0 | 854 |
| 177 | $HCCO + HCCO \rightarrow CO + CO + C_2H_2$ | 1.00E + 13 | 0.0 | 0.0 |
| *N-Containing Reactions* | | | | |
| 178 | $N + NO \rightarrow N_2 + O$ | 2.70E + 13 | 0.0 | 355 |
| 179 | $N + O_2 \rightarrow NO + O$ | 9.00E + 09 | 1.0 | 6,500 |
| 180 | $N + OH \rightarrow NO + H$ | 3.36E + 13 | 0.0 | 385 |
| 181 | $N_2O + O \rightarrow N_2 + O_2$ | 1.40E + 12 | 0.0 | 10,810 |
| 182 | $N_2O + O \rightarrow NO + NO$ | 2.90E + 13 | 0.0 | 23,150 |
| 183 | $N_2O + H \rightarrow N_2 + OH$ | 3.87E + 14 | 0.0 | 18,880 |
| 184 | $N_2O + OH \rightarrow N_2 + HO_2$ | 2.00E + 12 | 0.0 | 21,060 |
| 185 | $N_2O (+ M) \rightarrow N_2 + O (+ M)$ | pressure dependent | | |
| 186 | $HO_2 + NO \rightarrow NO_2 + OH$ | 2.11E + 12 | 0.0 | −480 |
| 187 | $NO + O + M \rightarrow NO_2 + M$ | 1.06E + 20 | −1.4 | 0.0 |
| 188 | $NO_2 + O \rightarrow NO + O_2$ | 3.90E + 12 | 0.0 | −240 |
| 189 | $NO_2 + H \rightarrow NO + OH$ | 1.32E + 14 | 0.0 | 360 |
| 190 | $NH + O \rightarrow NO + H$ | 4.00E + 13 | 0.0 | 0.0 |
| 191 | $NH + H \rightarrow N + H_2$ | 3.20E + 13 | 0.0 | 330 |
| 192 | $NH + OH \rightarrow HNO + H$ | 2.00E + 13 | 0.0 | 0.0 |
| 193 | $NH + OH \rightarrow N + H_2O$ | 2.00E + 09 | 1.2 | 0.0 |
| 194 | $NH + O_2 \rightarrow HNO + O$ | 4.61E + 05 | 2.0 | 6,500 |
| 195 | $NH + O_2 \rightarrow NO + OH$ | 1.28E + 06 | 1.5 | 100 |
| 196 | $NH + N \rightarrow N_2 + H$ | 1.50E + 13 | 0.0 | 0.0 |
| 197 | $NH + H_2O \rightarrow HNO + H_2$ | 2.00E + 13 | 0.0 | 13,850 |
| 198 | $NH + NO \rightarrow N_2 + OH$ | 2.16E + 13 | −0.2 | 0.0 |
| 199 | $NH + NO \rightarrow N_2O + H$ | 3.65E + 14 | −0.5 | 0.0 |
| 200 | $NH_2 + O \rightarrow OH + NH$ | 3.00E + 12 | 0.0 | 0.0 |

**Table 5.4**    (continued)

| No. | Reaction | Forward Rate Coefficient[a] | | |
|---|---|---|---|---|
| | | $A$ | $b$ | $E$ |
| ***N-Containing Reactions (continued)*** | | | | |
| 201 | $NH_2 + O \rightarrow H + HNO$ | 3.9E + 13 | 0.0 | 0.0 |
| 202 | $NH_2 + H \rightarrow NH + H_2$ | 4.00E + 13 | 0.0 | 3,650 |
| 203 | $NH_2 + OH \rightarrow NH + H_2O$ | 9.00E + 07 | 1.5 | −460 |
| 204 | $NNH \rightarrow N_2 + H$ | 3.30E + 08 | 0.0 | 0.0 |
| 205 | $NNH + M \rightarrow N_2 + H + M$ | 1.30E + 14 | −0.1 | 4,980 |
| 206 | $NNH + O_2 \rightarrow HO_2 + N_2$ | 5.00E + 12 | 0.0 | 0.0 |
| 207 | $NNH + O \rightarrow OH + N_2$ | 2.50E + 13 | 0.0 | 0.0 |
| 208 | $NNH + O \rightarrow NH + NO$ | 7.00E + 13 | 0.0 | 0.0 |
| 209 | $NNH + H \rightarrow H_2 + N_2$ | 5.00E + 13 | 0.0 | 0.0 |
| 210 | $NNH + OH \rightarrow H_2O + N_2$ | 2.00E + 13 | 0.0 | 0.0 |
| 211 | $NNH + CH_3 \rightarrow CH_4 + N_2$ | 2.50E + 13 | 0.0 | 0.0 |
| 212 | $H + NO + M \rightarrow HNO + M$ | 4.48E + 19 | −1.3 | 740 |
| 213 | $HNO + O \rightarrow NO + OH$ | 2.50E + 13 | 0.0 | 0.0 |
| 214 | $HNO + H \rightarrow H_2 + NO$ | 9.00E + 11 | 0.7 | 660 |
| 215 | $HNO + OH \rightarrow NO + H_2O$ | 1.30E + 07 | 1.9 | −950 |
| 216 | $HNO + O_2 \rightarrow HO_2 + NO$ | 1.00E + 13 | 0.0 | 13,000 |
| 217 | $CN + O \rightarrow CO + N$ | 7.70E + 13 | 0.0 | 0.0 |
| 218 | $CN + OH \rightarrow NCO + H$ | 4.00E + 13 | 0.0 | 0.0 |
| 219 | $CN + H_2O \rightarrow HCN + OH$ | 8.00E + 12 | 0.0 | 7,460 |
| 220 | $CN + O_2 \rightarrow NCO + O$ | 6.14E + 12 | 0.0 | −440 |
| 221 | $CN + H_2 \rightarrow HCN + H$ | 2.95E + 05 | 2.5 | 2,240 |
| 222 | $NCO + O \rightarrow NO + CO$ | 2.35E + 13 | 0.0 | 0.0 |
| 223 | $NCO + H \rightarrow NH + CO$ | 5.40E + 13 | 0.0 | 0.0 |
| 224 | $NCO + OH \rightarrow NO + H + CO$ | 2.50E + 12 | 0.0 | 0.0 |
| 225 | $NCO + N \rightarrow N_2 + CO$ | 2.00E + 13 | 0.0 | 0.0 |
| 226 | $NCO + O_2 \rightarrow NO + CO_2$ | 2.00E + 12 | 0.0 | 20,000 |
| 227 | $NCO + M \rightarrow N + CO + M$ | 3.10E + 14 | 0.0 | 54,050 |
| 228 | $NCO + NO \rightarrow N_2O + CO$ | 1.90E + 17 | −1.5 | 740 |
| 229 | $NCO + NO \rightarrow N_2 + CO_2$ | 3.80E + 18 | −2.0 | 800 |
| 230 | $HCN + M \rightarrow H + CN + M$ | 1.04E + 29 | −3.3 | 126,600 |
| 231 | $HCN + O \rightarrow NCO + H$ | 2.03E + 04 | 2.6 | 4,980 |
| 232 | $HCN + O \rightarrow NH + CO$ | 5.07E + 03 | 2.6 | 4,980 |
| 233 | $HCN + O \rightarrow CN + OH$ | 3.91E + 09 | 1.6 | 26,600 |
| 234 | $HCN + OH \rightarrow HOCN + H$ | 1.10E + 06 | 2.0 | 13,370 |
| 235 | $HCN + OH \rightarrow HNCO + H$ | 4.40E + 03 | 2.3 | 6,400 |
| 236 | $HCN + OH \rightarrow NH_2 + CO$ | 1.60E + 02 | 2.6 | 9,000 |
| 237 | $H + HCN + M \rightarrow H_2CN + M$ | pressure dependent | | |
| 238 | $H_2CN + N \rightarrow N_2 + CH_2$ | 6.00E + 13 | 0.0 | 400 |
| 239 | $C + N_2 \rightarrow CN + N$ | 6.30E + 13 | 0.0 | 46,020 |
| 240 | $CH + N_2 \rightarrow HCN + N$ | 3.12E + 09 | 0.9 | 20,130 |
| 241 | $CH + N_2 (+ M) \rightarrow HCNN (+ M)$ | pressure dependent | | |
| 242 | $CH_2 + N_2 \rightarrow HCN + NH$ | 1.00E + 13 | 0.0 | 74,000 |
| 243[b] | $CH_2(S) + N_2 \rightarrow NH + HCN$ | 1.00E + 11 | 0.0 | 65,000 |

**Table 5.4**   (continued)

| No. | Reaction | Forward Rate Coefficient[a] | | |
|-----|----------|-----|-----|-----|
| | | *A* | *b* | *E* |
| *N-Containing Reactions (continued)* | | | | |
| 244 | $C + NO \rightarrow CN + O$ | 1.90E + 13 | 0.0 | 0.0 |
| 245 | $C + NO \rightarrow CO + N$ | 2.90E + 13 | 0.0 | 0.0 |
| 246 | $CH + NO \rightarrow HCN + O$ | 4.10E + 13 | 0.0 | 0.0 |
| 247 | $CH + NO \rightarrow H + NCO$ | 1.62E + 13 | 0.0 | 0.0 |
| 248 | $CH + NO \rightarrow N + HCO$ | 2.46E + 13 | 0.0 | 0.0 |
| 249 | $CH_2 + NO \rightarrow H + HNCO$ | 3.10E + 17 | −1.4 | 1,270 |
| 250 | $CH_2 + NO \rightarrow OH + HCN$ | 2.90E + 14 | −0.7 | 760 |
| 251 | $CH_2 + NO \rightarrow H + HCNO$ | 3.80E + 13 | −0.4 | 580 |
| 252[b] | $CH_2(S) + NO \rightarrow H + HNCO$ | 3.10E + 17 | −1.4 | 1,270 |
| 253[b] | $CH_2(S) + NO \rightarrow OH + HCN$ | 2.90E + 14 | −0.7 | 760 |
| 254[b] | $CH_2(S) + NO \rightarrow H + HCNO$ | 3.80E + 13 | −0.4 | 580 |
| 255 | $CH_3 + NO \rightarrow HCN + H_2O$ | 9.60E + 13 | 0.0 | 28,800 |
| 256 | $CH_3 + NO \rightarrow H_2CN + OH$ | 1.00E + 12 | 0.0 | 21,750 |
| 257 | $HCNN + O \rightarrow CO + H + N_2$ | 2.20E + 13 | 0.0 | 0.0 |
| 258 | $HCNN + O \rightarrow HCN + NO$ | 2.00E + 12 | 0.0 | 0.0 |
| 259 | $HCNN + O_2 \rightarrow O + HCO + N_2$ | 1.20E + 13 | 0.0 | 0.0 |
| 260 | $HCNN + OH \rightarrow H + HCO + N_2$ | 1.20E + 13 | 0.0 | 0.0 |
| 261 | $HCNN + H \rightarrow CH_2 + N_2$ | 1.00E + 14 | 0.0 | 0.0 |
| 262 | $HNCO + O \rightarrow NH + CO_2$ | 9.80E + 07 | 1.4 | 8,500 |
| 263 | $HNCO + O \rightarrow HNO + CO$ | 1.50E + 08 | 1.6 | 44,000 |
| 264 | $HNCO + O \rightarrow NCO + OH$ | 2.20E + 06 | 2.1 | 11,400 |
| 265 | $HNCO + H \rightarrow NH_2 + CO$ | 2.25E + 07 | 1.7 | 3,800 |
| 266 | $HNCO + H \rightarrow H_2 + NCO$ | 1.05E + 05 | 2.5 | 13,300 |
| 267 | $HNCO + OH \rightarrow NCO + H_2O$ | 3.30E + 07 | 1.5 | 3,600 |
| 268 | $HNCO + OH \rightarrow NH_2 + CO_2$ | 3.30E + 06 | 1.5 | 3,600 |
| 269 | $HNCO + M \rightarrow NH + CO + M$ | 1.18E + 16 | 0.0 | 84,720 |
| 270 | $HCNO + H \rightarrow H + HNCO$ | 2.10E + 15 | −0.7 | 2,850 |
| 271 | $HCNO + H \rightarrow OH + HCN$ | 2.70E + 11 | 0.2 | 2,120 |
| 272 | $HCNO + H \rightarrow NH_2 + CO$ | 1.70E + 14 | −0.8 | 2,890 |
| 273 | $HOCN + H \rightarrow H + HNCO$ | 2.00E + 07 | 2.0 | 2,000 |
| 274 | $HCCO + NO \rightarrow HCNO + CO$ | 9.00E + 12 | 0.0 | 0.0 |
| 275 | $CH_3 + N \rightarrow H_2CN + H$ | 6.10E + 14 | −0.3 | 290 |
| 276 | $CH_3 + N \rightarrow HCN + H_2$ | 3.70E + 12 | 0.1 | −90 |
| 277 | $NH_3 + H \rightarrow NH_2 + H_2$ | 5.40E + 05 | 2.4 | 9,915 |
| 278 | $NH_3 + OH \rightarrow NH_2 + H_2O$ | 5.00E + 07 | 1.6 | 955 |
| 279 | $NH_3 + O \rightarrow NH_2 + OH$ | 9.40E + 06 | 1.9 | 6,460 |
| *Reactions Added in Update from Version 2.11 to Version 3.0* | | | | |
| 280 | $NH + CO_2 \rightarrow HNO + CO$ | 1.00E + 13 | 0.0 | 14,350 |
| 281 | $CN + NO_2 \rightarrow NCO + NO$ | 6.16E + 15 | −0.8 | 345 |
| 282 | $NCO + NO_2 \rightarrow N_2O + CO_2$ | 3.25E + 12 | 0.0 | −705 |
| 283 | $N + CO_2 \rightarrow NO + CO$ | 3.00E + 12 | 0.0 | 11,300 |
| 284 | $O + CH_3 \rightarrow H + H_2 + CO$ | 3.37E + 13 | 0.0 | 0.0 |
| 285 | $O + C_2H_4 \rightarrow CH_2CHO$ | 6.70E + 06 | 1.8 | 220 |
| 286 | $O + C_2H_5 \rightarrow H + CH_3CHO$ | 1.10E + 14 | 0.0 | 0.0 |

**Table 5.4**      (continued)

| No. | Reaction | Forward Rate Coefficient[a] | | |
|---|---|---|---|---|
| | | *A* | *b* | *E* |
| *Reactions Added in Update from Version 2.11 to Version 3.0 (continued)* | | | | |
| 287 | $OH + HO_2 \rightarrow O_2 + H_2O$ | 5.00E + 15 | 0.0 | 17,330 |
| 288 | $OH + CH_3 \rightarrow H_2 + CH_2O$ | 8.00E + 09 | 0.5 | −1,755 |
| 289 | $CH + H_2 + M \rightarrow CH_3 + M$ | pressure dependent | | |
| 290 | $CH_2 + O_2 \rightarrow H + H + CO_2$ | 5.80E + 12 | 0.0 | 1,500 |
| 291 | $CH_2 + O_2 \rightarrow O + CH_2O$ | 2.40E + 12 | 0.0 | 1,500 |
| 292 | $CH_2 + CH_2 \rightarrow H + H + C_2H_2$ | 2.00E + 14 | 0.0 | 10,989 |
| 293[b] | $CH_2(S) + H_2O \rightarrow H_2 + CH_2O$ | 6.82E + 10 | 0.2 | −935 |
| 294 | $C_2H_3 + O_2 \rightarrow O + CH_2CHO$ | 3.03E + 11 | 0.3 | 11 |
| 295 | $C_2H_3 + O_2 \rightarrow HO_2 + C_2H_2$ | 1.34E + 06 | 1.6 | −384 |
| 296 | $O + CH_3CHO \rightarrow OH + CH_2CHO$ | 2.92E + 12 | 0.0 | 1,808 |
| 297 | $O + CH_3CHO \rightarrow OH + CH_3 + CO$ | 2.92E + 12 | 0.0 | 1,808 |
| 298 | $O_2 + CH_3CHO \rightarrow HO_2 + CH_3 + CO$ | 3.01E + 13 | 0.0 | 39,150 |
| 299 | $H + CH_3CHO \rightarrow CH_2CHO + H_2$ | 2.05E + 09 | 1.2 | 2,405 |
| 300 | $H + CH_3CHO \rightarrow CH_3 + H_2 + CO$ | 2.05E + 09 | 1.2 | 2,405 |
| 301 | $OH + CH_3CHO \rightarrow CH_3 + H_2O + CO$ | 2.34E + 10 | 0.7 | −1,113 |
| 302 | $HO_2 + CH_3CHO \rightarrow CH_3 + H_2O_2 + CO$ | 3.01E + 12 | 0.0 | 11,923 |
| 303 | $CH_3 + CH_3CHO \rightarrow CH_3 + CH_4 + CO$ | 2.72E + 06 | 1.8 | 5,920 |
| 304 | $H + CH_2CO + M \rightarrow CH_3CHO + M$ | pressure dependent | | |
| 305 | $O + CH_2CHO \rightarrow H + CH_2 + CO_2$ | 1.50E + 14 | 0.0 | 0.0 |
| 306 | $O_2 + CH_2CHO \rightarrow OH + CO + CH_2O$ | 1.81E + 10 | 0.0 | 0.0 |
| 307 | $O_2 + CH_2CHO \rightarrow OH + HCO + HCO$ | 2.35E + 10 | 0.0 | 0.0 |
| 308 | $H + CH_2CHO \rightarrow CH_3 + HCO$ | 2.20E + 13 | 0.0 | 0.0 |
| 309 | $H + CH_2CHO \rightarrow CH_2CO + H_2$ | 1.10E + 13 | 0.0 | 0.0 |
| 310 | $OH + CH_2CHO \rightarrow H_2O + CH_2CO$ | 1.20E + 13 | 0.0 | 0.0 |
| 311 | $OH + CH_2CHO \rightarrow HCO + CH_2OH$ | 3.01E + 13 | 0.0 | 0.0 |
| 312 | $CH_3 + C_2H_5 + M \rightarrow C_3H_8 + M$ | pressure dependent | | |
| 313 | $O + C_3H_8 \rightarrow OH + C_3H_7$ | 1.93E + 05 | 2.7 | 3,716 |
| 314 | $H + C_3H_8 \rightarrow C_3H_7 + H_2$ | 1.32E + 06 | 2.5 | 6,756 |
| 315 | $OH + C_3H_8 \rightarrow C_3H_7 + H_2O$ | 3.16E + 07 | 1.8 | 934 |
| 316 | $C_3H_7 + H_2O_2 \rightarrow HO_2 + C_3H_8$ | 3.78E + 02 | 2.7 | 1,500 |
| 317 | $CH_3 + C_3H_8 \rightarrow C_3H_7 + CH_4$ | 9.03E − 01 | 3.6 | 7,154 |
| 318 | $CH_3 + C_2H_4 + M \rightarrow C_3H_7 + M$ | pressure dependent | | |
| 319 | $O + C_3H_7 \rightarrow C_2H_5 + CH_2O$ | 9.64E + 13 | 0.0 | 0.0 |
| 320 | $H + C_3H_7 + M \rightarrow C_3H_8 + M$ | pressure dependent | | |
| 321 | $H + C_3H_7 \rightarrow CH_3 + C_2H_5$ | 4.06E + 06 | 2.2 | 890 |
| 322 | $OH + C_3H_7 \rightarrow C_2H_5 + CH_2OH$ | 2.41E + 13 | 0.0 | 0.0 |
| 323 | $HO_2 + C_3H_7 \rightarrow O_2 + C_3H_8$ | 2.55E + 10 | 0.3 | −943 |
| 324 | $HO_2 + C_3H_7 \rightarrow OH + C_2H_5 + CH_2O$ | 2.41E + 13 | 0.0 | 0.0 |
| 325 | $CH_3 + C_3H_7 \rightarrow C_2H_5 + C_2H_5$ | 1.93E + 13 | −0.3 | 0.0 |

[a]The forward rate coefficient $k = AT^b \exp(-E/RT)$. $R$ is the universal gas constant, $T$ is the temperature in K. The units of $A$ involve $gmol/cm^3$ and s, and those of $E$, $cal/gmol$.

[b]$CH_2(S)$ designates the singlet state of $CH_2$.

## High-Temperature Reaction Pathway Analysis

Major chemical pathways in the conversion of methane to carbon dioxide at a high temperature (2200 K) are illustrated in Fig. 5.4. Each arrow represents an elementary reaction, or set of reactions, with the primary reactant species at the tail and the primary product species at the head. Additional reactant species are shown along the length of the arrow, and the corresponding reaction number (see Table 5.4) is indicated. The width of the arrow gives a visual indication of the relative importance of a particular reaction path, while the parenthetical numerical values quantify the destruction rate of the reactant.

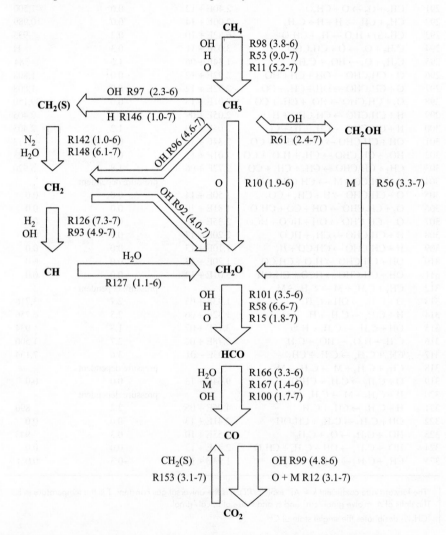

**Figure 5.4**     High-temperature reaction pathway diagram for combustion of methane in a well-stirred reactor at $T = 2200$ K and $P = 1$ atm for a 0.1-s residence time. Reaction numbers refer to Table 5.4, and reaction rates are shown in parentheses. For example, 2.6–7 implies $2.6 \cdot 10^{-7}$ (gmol/cm³-s). Results shown are for GRI Mech 2.11.

From Fig. 5.4, we see a linear progression of $CH_4$ to $CO_2$, together with several side loops originating from the methyl ($CH_3$) radical. The linear progression, or backbone, starts with an attack on the $CH_4$ molecule by OH, O, and H radicals to produce the methyl radical; the methyl radical then combines with an oxygen atom to form formaldehyde ($CH_2O$); the formaldehyde, in turn, is attacked by OH, H, and O radicals to produce the formyl radical (HCO); the formyl radical is converted to CO by a trio of reactions; and, finally, CO is converted to $CO_2$, primarily by reaction with OH, as discussed earlier. Molecular structure diagrams illuminate the elegance

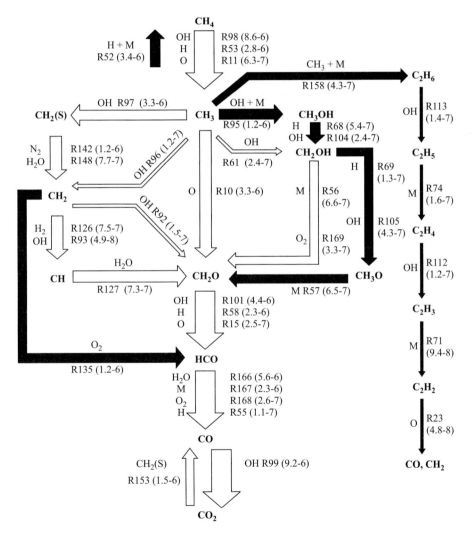

**Figure 5.5** Low-temperature (< 1500 K) reaction pathway diagram for combustion of methane in a well-stirred reactor at $T = 1345$ K and $P = 1$ atm for a 0.1-s residence time. Reaction numbers refer to Table 5.4, and reaction rates are shown in parentheses. For example, 2.6–7 implies $2.6 \cdot 10^{-7}$ (gmol/cm³-s). Results shown are for GRI Mech 2.11.

of this sequence of elementary reactions to explain, in part, the oxidation of methane. Drawing such is left as an exercise for the reader.

In addition to the direct pathway from the methyl radical to formaldehyde ($CH_3 + O \rightarrow CH_2O + H$), the $CH_2$ radicals also react to form $CH_2$ radicals in two possible electronic configurations. The singlet electronic state of $CH_2$ is designated $CH_2(S)$ and should not be confused with similar notation used to indicate a solid. There is also a side loop whereby $CH_3$ is first converted to $CH_2OH$, which, in turn, is converted to $CH_2O$. Other less important pathways complete the mechanism. Those pathways with reaction rates less than $1 \cdot 10^{-7}$ gmol/cm$^3$-s are not shown in Fig. 5.4.

## Low-Temperature Reaction Pathway Analysis

At low temperatures (< 1500 K, say), pathways that were unimportant at higher temperatures now become prominent. Figure 5.5 illustrates the scenario at 1345 K. Black arrows show the new pathways that now complement all of the high-temperature pathways. Several interesting features appear: first, there is a strong recombination of $CH_3$ back to $CH_4$; second, an alternate route from $CH_3$ to $CH_2O$ appears through the intermediate production of methanol ($CH_3OH$); and, most interestingly, $CH_3$ radicals combine to form ethane ($C_2H_6$), a higher hydrocarbon than the original reactant methane. The $C_2H_6$ is ultimately converted to CO (and $CH_2$) through $C_2H_4$ (ethene) and $C_2H_2$ (acetylene). The appearance of hydrocarbons higher than the initial reactant hydrocarbon is a common feature of low-temperature oxidation processes.

Because of the importance of methane oxidation, much effort has been directed to formulating reduced (simplified) mechanisms. We address this topic following our discussion of oxides of nitrogen kinetics.

## OXIDES OF NITROGEN FORMATION

Nitric oxide, introduced in Chapter 1 and more fully discussed in Chapter 4, is an important minor species in combustion because of its contribution to air pollution. The detailed nitrogen chemistry involved in the combustion of methane is presented in the second half of Table 5.4. In the combustion of fuels that contain no nitrogen, nitric oxide is formed by four chemical mechanisms or routes that involve nitrogen from the air: the **thermal** or **Zeldovich mechanism,** the **Fenimore** or **prompt mechanism,** the **N₂O-intermediate mechanism,** and the **NNH mechanism.** The thermal mechanism dominates in high-temperature combustion over a fairly wide range of equivalence ratios, whereas the Fenimore mechanism is particularly important in rich combustion. It appears that the $N_2O$-intermediate mechanism plays an important role in the production of NO in very lean, low-temperature combustion processes. The NNH mechanism is a relative newcomer to the named NO formation routes. Various studies show the relative contributions of the first three mechanisms in premixed [26] and diffusion flames [27, 28] and the relative contributions of all four pathways for lean premixed combustion in jet-stirred reactors [29]. For more information on the chemistry involved in forming and controlling oxides of nitrogen in combustion processes than is provided in this section, the reader is referred to the reviews found in Dagaut *et al.* [30], Dean and Bozzelli [31], and Miller and Bowman [32]. Additional information and references also are provided in Chapter 15.

The **thermal** or **Zeldovich mechanism** consists of two chain reactions:

$$O + N_2 \Leftrightarrow NO + N \tag{N.1}$$

$$N + O_2 \Leftrightarrow NO + O, \tag{N.2}$$

which can be extended by adding the reaction

$$N + OH \Leftrightarrow NO + H. \tag{N.3}$$

The rate coefficients for N.1–N.3 are [33]

$$k_{N.1f} = 1.8 \cdot 10^{11} \exp[-38{,}370/T(K)] \qquad [=] \, m^3/kmol\text{-}s,$$

$$k_{N.1r} = 3.8 \cdot 10^{10} \exp[-425/T(K)] \qquad [=] \, m^3/kmol\text{-}s,$$

$$k_{N.2f} = 1.8 \cdot 10^{7} \, T \exp[-4680/T(K)] \qquad [=] \, m^3/kmol\text{-}s,$$

$$k_{N.2r} = 3.8 \cdot 10^{6} \, T \exp[-20{,}820/T(K)] \qquad [=] \, m^3/kmol\text{-}s,$$

$$k_{N.3f} = 7.1 \cdot 10^{10} \exp[-450/T(K)] \qquad [=] \, m^3/kmol\text{-}s,$$

$$k_{N.3r} = 1.7 \cdot 10^{11} \exp[-24{,}560/T(K)] \qquad [=] \, m^3/kmol\text{-}s.$$

This three-reaction set is referred to as the **extended Zeldovich mechanism.** In general, this mechanism is coupled to the fuel combustion chemistry through the $O_2$, O, and OH species; however, in processes where the fuel combustion is complete before NO formation becomes significant, the two processes can be uncoupled. In this case, if the relevant timescales are sufficiently long, one can assume that the $N_2, O_2$, O, and OH concentrations are at their equilibrium values and N atoms are in steady state. These assumptions greatly simplify the problem of calculating the NO formation. If we make the additional assumption that the NO concentrations are much less than their equilibrium values, the reverse reactions can be neglected. This yields the following rather simple rate expression:

$$\frac{d[NO]}{dt} = 2k_{N.1f}[O]_{eq}[N_2]_{eq}. \tag{5.7}$$

In Chapter 4 (Example 4.3), we showed that Eqn. 5.7 obtains for the above assumptions. Within flame zones proper and in short-timescale, postflame processes, the equilibrium assumption is not valid. Superequilibrium concentrations of O atoms, up to several orders of magnitude greater than equilibrium, greatly increase NO formation rates. This **superequilibrium O** (and OH) atom contribution to NO production rates is sometimes classified as part of the so-called prompt-NO mechanism; however, for historical reasons, we refer to prompt-NO only in conjunction with the Fenimore mechanism.

The activation energy for (N.1) is relatively large (319,050 kJ/kmol); thus, this reaction has a very strong temperature dependence (see problem 4.11). As a rule of thumb, the thermal mechanism is usually unimportant at temperatures below 1800 K. Compared with the timescales of fuel oxidation processes, NO is formed rather slowly by the thermal mechanism; thus, thermal NO is generally considered to be formed in the postflame gases.

The **N₂O-intermediate mechanism** is important in fuel-lean ($\Phi < 0.8$), low-temperature conditions. The three steps of this mechanism are

$$O + N_2 + M \Leftrightarrow N_2O + M, \qquad\qquad (N.4)$$

$$H + N_2O \Leftrightarrow NO + NH, \qquad\qquad (N.5)$$

$$O + N_2O \Leftrightarrow NO + NO. \qquad\qquad (N.6)$$

This mechanism becomes important in NO control strategies that involve lean premixed combustion, which are used by gas-turbine manufacturers [34].

The **Fenimore mechanism** is intimately linked to the combustion chemistry of hydrocarbons. Fenimore [35] discovered that some NO was rapidly produced in the flame zone of laminar premixed flames long before there would be time to form NO by the thermal mechanism, and he gave this rapidly formed NO the appellation **prompt NO.** The general scheme of the Fenimore mechanism is that hydrocarbon radicals react with molecular nitrogen to form amines or cyano compounds. The amines and cyano compounds are then converted to intermediate compounds that ultimately form NO. Ignoring the processes that form CH radicals to initiate the mechanism, the Fenimore mechanism can be written

$$CH + N_2 \Leftrightarrow HCN + N \qquad\qquad (N.7)$$

$$C + N_2 \Leftrightarrow CN + N, \qquad\qquad (N.8)$$

where (N.7) is the primary path and is the rate-limiting step in the sequence. For equivalence ratios less than about 1.2, the conversion of hydrogen cyanide, HCN, to form NO follows the following chain sequence:

$$HCN + O \Leftrightarrow NCO + H, \qquad\qquad (N.9)$$

$$NCO + H \Leftrightarrow NH + CO, \qquad\qquad (N.10)$$

$$NH + H \Leftrightarrow N + H_2, \qquad\qquad (N.11)$$

$$N + OH \Leftrightarrow NO + H. \qquad\qquad (N.3)$$

For equivalence ratios richer than 1.2, other routes open up and the chemistry becomes much more complex. Miller and Bowman [32] point out that the above scheme is no longer rapid and that NO is recycled to HCN, inhibiting NO production. Furthermore, the Zeldovich reaction that couples with the prompt mechanism actually destroys rather than forms NO, i.e., N + NO → N₂ + O. Figure 5.6 schematically

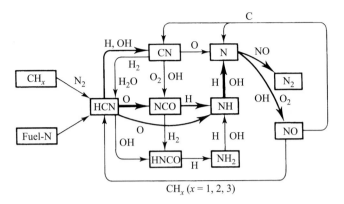

**Figure 5.6**     Production of NO associated with the Fenimore prompt mechanism.
SOURCE: Reprinted from Ref. [36] by permission of The Combustion Institute.

illustrates the processes involved, and the reader is referred to Refs. [32] and [36] for further details.

The **NNH mechanism** for NO formation is a recently discovered reaction pathway [31, 37–39]. The two key steps in this mechanism are

$$N_2 + H \rightarrow NNH \tag{N.12}$$

and

$$NNH + O \rightarrow NO + NH. \tag{N.13}$$

This route has been shown to be particularly important in the combustion of hydrogen [40] and for hydrocarbon fuels with large carbon-to-hydrogen ratios [29].

Certain fuels contain nitrogen in their molecular structure. The NO formed from this nitrogen is frequently designated **fuel nitrogen,** and is considered to be another NO formation pathway adding to those already discussed. Coal, in particular, may contain bound nitrogen up to about 2 percent by mass. In the combustion of fuels with bound nitrogen, the nitrogen in the parent fuel is rapidly converted to hydrogen cyanide, HCN, or ammonia, $NH_3$. The remaining steps follow the prompt-NO mechanism discussed above and outlined in Fig. 5.6.

In the atmosphere, nitric oxide ultimately oxidizes to form **nitrogen dioxide,** which is important to the production of acid rain and photochemical smog. Many combustion processes, however, emit significant fractions of their total oxides of nitrogen ($NO_x = NO + NO_2$) as $NO_2$. The elementary reactions responsible for forming $NO_2$ prior to the exhausting of the combustion products into the atmosphere are the following:

$$NO + HO_2 \Leftrightarrow NO_2 + OH \quad \text{(formation)}, \tag{N.14}$$

$$NO_2 + H \Leftrightarrow NO + OH \quad \text{(destruction)}, \tag{N.15}$$

$$NO_2 + O \Leftrightarrow NO + O_2 \quad \text{(destruction)}, \tag{N.16}$$

where the $HO_2$ radical is formed by the three-body reaction

$$H + O_2 + M \Leftrightarrow HO_2 + M. \tag{N.17}$$

The $HO_2$ radicals are formed in relatively low-temperature regions; hence, $NO_2$ formation occurs when NO molecules from high-temperature regions diffuse or are transported by fluid mixing into the $HO_2$-rich regions. The $NO_2$ destruction reactions (N.15) and (N.16) are active at high temperatures, thus preventing the formation of $NO_2$ in high-temperature zones.

## METHANE COMBUSTION AND OXIDES OF NITROGEN FORMATION—A REDUCED MECHANISM

Many different reduced mechanisms have been developed for methane oxidation. See, for example, Refs. [41–43]. Here we present the mechanism of Sung *et al.* [43] based on a reduction of GRI Mech 3.0 (Table 5.4). This reduced mechanism also includes nitrogen chemistry that allows predictions of NO, $NO_2$, and $N_2O$. The mechanism includes 21 species (see Table 5.5) and 17 steps (see Table 5.6). Performance constraints employed in the creation of this reduced model include well-stirred reactor performance (see Chapter 6); the evolution of autoignition (see Chapter 6); the propagation speed and detailed structure of 1-D, laminar, premixed flames (see Chapter 8); and the properties of counterflow non-premixed flames (see Chapter 9). In nearly all situations, the reduced mechanism produced results in excellent agreement with those of the full mechanism.

Implementation of the reduced mechanism, although straightforward, is nontrivial. We refer the reader to the Sung *et al.* [43] paper for details.

**Table 5.5**   Species retained in the reduced mechanism for methane oxidation and $NO_x$ formation of Sung *et al.* [43]

| *C–H–O Species: 15* | |
|---|---|
| $CH_3$ | $H_2$ |
| $CH_4$ | $H$ |
| $CO$ | $O_2$ |
| $CO_2$ | $OH$ |
| $CH_2O$ | $H_2O$ |
| $C_2H_2$ | $HO_2$ |
| $C_2H_4$ | $H_2O_2$ |

| *Nitrogen-Containing Species: 6* | |
|---|---|
| $N_2$ | $NO$ |
| $HCN$ | $NO_2$ |
| $NH_3$ | $N_2O$ |

**Table 5.6**    Reaction steps in the reduced mechanism for methane oxidation and $NO_x$ formation of Sung et al. [43]

---

*$CH_4$ to $CO_2$ and $H_2O$ Steps:*

| | |
|---|---|
| $CH_4 + H \leftrightarrow CH_3 + H_2O$ | (RM.1) |
| $CH_3 + OH \leftrightarrow CH_2O + H_2$ | (RM.2) |
| $CH_2O \leftrightarrow CO + H_2$ | (RM.3) |
| $C_2H_6 \leftrightarrow C_2H_4 + H_2$ | (RM.4) |
| $C_2H_4 + OH \leftrightarrow CH_3 + CO + H_2$ | (RM.5) |
| $C_2H_2 + O_2 \leftrightarrow 2CO + H_2$ | (RM.6) |
| $CO + OH + H \leftrightarrow CO_2 + H_2$ | (RM.7) |
| $H + OH \leftrightarrow H_2O$ | (RM.8) |
| $2H_2 + O_2 \leftrightarrow 2H + 2OH$ | (RM.9) |
| $2H \leftrightarrow H_2$ | (RM.10) |
| $HO_2 + H \leftrightarrow H_2 + O_2$ | (RM.11) |
| $H_2O_2 + H \leftrightarrow H_2 + HO_2$ | (RM.12) |

*NO Formation via Thermal, Prompt, and $N_2O$-Intermediate Mechanisms:*

| | |
|---|---|
| $N_2 + O_2 \leftrightarrow 2NO$ | (RM.13) |
| $HCN + H + O_2 \leftrightarrow H_2 + CO + NO$ | (RM.14) |

*$NH_3$ Conversion to $NH_2$:*

| | |
|---|---|
| $NH_3 + 3H + H_2O \leftrightarrow 4H_2 + NO$ | (RM.15) |

*$NO_2$ and $N_2O$ Formation:*

| | |
|---|---|
| $HO_2 + NO \leftrightarrow HO + NO_2$ | (RM.16) |
| $H_2 + O_2 + N_2 \leftrightarrow H + OH + N_2O$ | (RM.17) |

---

# SUMMARY

In this chapter, we outlined the kinetic mechanisms for five chemical systems of importance to combustion: $H_2$ oxidation, CO oxidation, higher alkane oxidation, $CH_4$ oxidation, and oxides of nitrogen (NO and $NO_2$) formation. In the $H_2$–$O_2$ system, we discovered how the changes in reaction pathways with temperature and pressure result in various regimes of explosive and nonexplosive behavior. The idea of heterogeneous, or wall, reactions was introduced to explain the destruction of radicals at low pressures where mean free paths are long. We also saw the importance of moisture, or other $H_2$-containing species, in the oxidation of CO. Without a trace of these species, CO oxidation is very slow. The $CO + OH \rightarrow CO_2 + H$ reaction was found to be a key step in CO oxidation. The oxidation of higher alkanes ($C_nH_{2n+2}$ with $n > 2$) can be characterized as a three-step process: first, the fuel molecule is attacked by radicals producing intermediates (alkenes and $H_2$); second, the intermediates oxidize to CO and $H_2O$; and, third, the CO and any remaining $H_2$ are oxidized to $CO_2$ and $H_2O$, respectively. Global and quasi-global mechanisms were presented for use in engineering approximations. Methane was shown to be unique among hydrocarbons in that it is significantly less reactive. The mechanism for $CH_4$ oxidation is perhaps the best elaborated of all hydrocarbons, and an example was presented that consists of 325 elementary steps involving 53 species. Reaction pathways were outlined for low- and high-temperature $CH_4$ oxidation in a well-stirred reactor. This chapter introduced the various NO formation pathways: the extended Zeldovich or thermal route, together with the superequilibrium O (and OH) contributions; the Fenimore prompt mechanism; the $N_2O$-intermediate path; the NNH route; and, lastly, the fuel-N mechanism. The concept of mechanism reduction was also introduced. Although an overwhelming number of reactions were discussed in this chapter, the reader should still be able to grasp certain key features of each system, without getting bogged down in a myriad of elementary reactions. Moreover, this chapter should provide a heightened awareness of the importance of chemistry in combustion.

## REFERENCES

1. Glassman, I., *Combustion*, 2nd Ed., Academic Press, Orlando, FL, 1987.

2. Lewis, B., and von Elbe, G., *Combustion, Flames and Explosions of Gases,* 3rd Ed., Academic Press, Orlando, FL, 1987.

3. Gardiner, W. C., Jr., and Olson, D. B., "Chemical Kinetics of High Temperature Combustion," *Annual Review of Physical Chemistry,* 31: 377–399 (1980).

4. Westbrook, C. K., and Dryer, F. L., "Chemical Kinetic Modeling of Hydrocarbon Combustion," *Progress in Energy and Combustion Science,* 10: 1–57 (1984).

5. Davis, S. G., Joshi, A. V., Wang, H., and Egolfopoulos, F., "An Optimized Kinetic Model of $H_2$/CO Combustion," *Proceedings of the Combustion Institute,* 30: 1283–1292 (2005).

6. Simmie, J. M., "Detailed Chemical Kinetic Models for Combustion of Hydrocarbon Fuels," *Progress in Energy and Combustion Science,* 29: 599–634 (2003).

7. Battin-Leclerc, F., "Detailed Chemical Kinetic Models for the Low-Temperature Combustion of Hydrocarbons with Application to Gasoline and Diesel Fuel Surrogates," *Progress in Energy and Combustion Science,* 34: 440–498 (2008).

8. Hautman, D. J., Dryer, F. L., Schug, K. P., and Glassman, I., "A Multiple-Step Overall Kinetic Mechanism for the Oxidation of Hydrocarbons," *Combustion Science and Technology,* 25: 219–235 (1981).

9. Westbrook, C. K., and Dryer, F. L., "Simplified Reaction Mechanisms for the Oxidation of Hydrocarbon Fuels in Flames," *Combustion Science and Technology,* 27: 31–43 (1981).

10. Card, J. M., and Williams, F. A., "Asymptotic Analysis with Reduced Chemistry for the Burning of *n*-Heptane Droplets," *Combustion and Flame,* 91: 187–199 (1992).

11. Dagaut, P., "On the Kinetics of Hydrocarbons Oxidation from Natural Gas to Kerosene and Diesel Fuel," *Physical Chemistry and Chemical Physics,* 4: 2079–2094 (2002).

12. Cooke, J. A., *et al.*, "Computational and Experimental Study of JP-8, a Surrogate, and Its Components in Counterflow Diffusion Flames," *Proceedings of the Combustion Institute,* 30: 439–446 (2005).

13. Violi, A., *et al.*, "Experimental Formulation and Kinetic Model for JP-8 Surrogate Mixtures," *Combustion Science and Technology,* 174: 399–417 (2002).

14. Ranzi, E., Dente, M., Goldaniga, A., Bozzano, G., and Faravelli, T., "Lumping Procedures in Detailed Kinetic Modeling of Gasification, Pyrolysis, Partial Oxidation and Combustion of Hydrocarbon Mixtures," *Progress in Energy and Combustion Science,* 27: 99–139 (2001).

15. Ranzi, E., *et al.*, "A Wide-Range Modeling Study of Iso-Octane Oxidation," *Combustion and Flame,* 108: 24–42 (1997).

16. Ranzi, E., Gaffuri, P., Faravelli, T., and Dagaut, P., "A Wide-Range Modeling Study of *n*-Heptane Oxidation," *Combustion and Flame,* 103: 91–106 (1995).

17. Curran, H. J., Gaffuri, P., Pitz, W. J., and Westbrook, C. K., "A Comprehensive Modeling Study of Iso-Octane Oxidation," *Combustion and Flame,* 129: 253–280 (2002).

18. Curran, H. J., Pitz, W. J., Westbrook, C. K., Callahan, C. V., and Dryer, F. L., "Oxidation of Automotive Primary Reference Fuels at Elevated Pressures," *Twenty-Seventh Symposium (International) on Combustion,* The Combustion Institute, Pittsburgh, PA, pp. 379–387, 1998.

19. Andrae, J. C. G., Björnbom, P., Cracknell, R. F., and Kalghatgi, G. T., "Autoignition of Toluene Reference Fuels at High Pressures Modeled with Detailed Chemical Kinetics," *Combustion and Flame,* 149: 2–24 (2007).

20. Andrae, J. C. G., "Development of a Detailed Chemical Kinetic Model for Gasoline Surrogate Fuels," *Fuel,* 87: 2013–2022 (2008).

21. Herbinet, O., Pitz, W. J., and Westbrook, C. K., "Detailed Chemical Kinetic Oxidation Mechanism for a Biodiesel Surrogate," *Combustion and Flame,* 154: 507–528 (2008).

22. Kaufman, F., "Chemical Kinetics and Combustion: Intricate Paths and Simple Steps," *Nineteenth Symposium (International) on Combustion,* The Combustion Institute, Pittsburgh, PA, pp. 1–10, 1982.

23. Smith, G. P., Golden, D. M., Frenklach, M., Moriarity, N. M., Eiteneer, B., Goldenberg, M., Bowman, C. T., Hanson, R. K., Song, S., Gardiner, W. C., Jr., Lissianski, V. V., and Qin, Z., *GRI-Mech Home Page,* http://www.me.berkeley.edu/gri_mech/.

24. Frenklach, M., Wang, H., and Rabinowitz, M. J., "Optimization and Analysis of Large Chemical Kinetic Mechanisms Using the Solution Mapping Method—Combustion of Methane," *Progress in Energy and Combustion Science,* 18: 47–73 (1992).

25. Glarborg, P., Kee, R. J., Grcar, J. F., and Miller, J. A., "PSR: A Fortran Program for Modeling Well-Stirred Reactors," Sandia National Laboratories Report SAND86-8209, 1986.

26. Drake, M. C., and Blint, R. J., "Calculations of $NO_x$ Formation Pathways in Propagating Laminar, High Pressure Premixed $CH_4$/Air Flames," *Combustion Science and Technology,* 75: 261–285 (1991).

27. Drake, M. C., and Blint, R. J., "Relative Importance of Nitric Oxide Formation Mechanisms in Laminar Opposed-Flow Diffusion Flames," *Combustion and Flame,* 83: 185–203 (1991).

28. Nishioka, M., Nakagawa, S., Ishikawa, Y., and Takeno, T., "NO Emission Characteristics of Methane-Air Double Flame," *Combustion and Flame,* 98: 127–138 (1994).

29. Rutar, T., Lee, J. C. Y., Dagaut, P., Malte, P. C., and Byrne, A. A., "NO$_x$ Formation Pathways in Lean-Premixed-Prevapourized Combustion of Fuels with Carbon-to-Hydrogen Ratios between 0.25 and 0.88," *Proceedings of the Institution of Mechanical Engineers, Part A: Journal of Power and Energy,* 221: 387–398 (2007).

30. Dagaut, P., Glarborg, P., and Alzueta, M. U., "The Oxidation of Hydrogen Cyanide and Related Chemistry," *Progress in Energy and Combustion Science,* 34: 1–46 (2008).

31. Dean, A., and Bozzelli, J., "Combustion Chemistry of Nitrogen," in *Gas-phase Combustion Chemistry* (Gardiner, W. C., Jr., ed.), Springer, New York, pp. 125–341, 2000.

32. Miller, J. A., and Bowman, C. T., "Mechanism and Modeling of Nitrogen Chemistry in Combustion," *Progress in Energy and Combustion Science,* 15: 287–338 (1989).

33. Hanson, R. K., and Salimian, S., "Survey of Rate Constants in the N/H/O System," Chapter 6 in *Combustion Chemistry* (W. C. Gardiner, Jr., ed.), Springer-Verlag, New York, pp. 361–421, 1984.

34. Correa, S. M., "A Review of NO$_x$ Formation under Gas-Turbine Combustion Conditions," *Combustion Science and Technology,* 87: 329–362 (1992).

35. Fenimore, C. P., "Formation of Nitric Oxide in Premixed Hydrocarbon Flames," *Thirteenth Symposium (International) on Combustion,* The Combustion Institute, Pittsburgh, PA, pp. 373–380, 1970.

36. Bowman, C. T., "Control of Combustion-Generated Nitrogen Oxide Emissions: Technology Driven by Regulations," *Twenty-Fourth Symposium (International) on Combustion,* The Combustion Institute, Pittsburgh, PA, pp. 859–878, 1992.

37. Bozzelli, J. W., and Dean, A. M., "O + NNH: A Possible New Route for NO$_x$ Formation in Flames," *International Journal of Chemical Kinetics,* 27: 1097–1109 (1995).

38. Harrington, J. E., *et al.,* "Evidence for a New NO Production Mechanism in Flames," *Twenty-Sixth Symposium (International) on Combustion,* The Combustion Institute, Pittsburgh, PA, pp. 2133–2138, 1996.

39. Hayhurst, A. N., and Hutchinson, E. M., "Evidence for a New Way of Producing NO via NNH in Fuel-Rich Flames at Atmospheric Pressure," *Combustion and Flame,* 114: 274–279 (1998).

40. Konnov, A. A., Colson, G., and De Ruyck, J., "The New Route Forming NO via NNH," *Combustion and Flame,* 121: 548–550 (2000).

41. Peters, N., and Kee, R. J., "The Computation of Stretched Laminar Methane-Air Diffusion Flames Using a Reduced Four-Step Mechanism," *Combustion and Flame,* 68: 17–29 (1987).

42. Smooke, M. D. (ed.), *Reduced Kinetic Mechanisms and Asymptotic Approximations for Methane-Air Flames,* Lecture Notes in Physics, 384, Springer-Verlag, New York, 1991.

43. Sung, C. J., Law, C. K., and Chen, J.-Y., "Augmented Reduced Mechanism for NO Emission in Methane Oxidation, *Combustion and Flame,* 125: 906–919 (2001).

44. Lide, D. R., (ed.), *Handbook of Chemistry and Physics,* 77th Ed., CRC Press, Boca Raton, 1996.

## QUESTIONS AND PROBLEMS

**5.1** Identify and discuss the processes involved in the $H_2$–$O_2$ system that result in:

    A.  The first explosion limit (see Fig. 5.1).

    B.  The second explosion limit.

    C.  The third explosion limit.

**5.2** What is the difference between a homogeneous reaction and a heterogeneous reaction? Give examples of each.

**5.3** Why is moisture, or other $H_2$-containing species, important for the rapid oxidation of CO?

**5.4** Identify the primary elementary reaction step in which CO is converted to $CO_2$.

**5.5** The oxidation of higher alkanes can be treated as three major sequential steps. What are the key features of each step?

**5.6** Show how the $\beta$-scission rule would apply to the breaking of C–C bonds in the following hydrocarbon radicals. The line indicates a C–H bond and the dot the radical site.

    A.  *n*-butyl radical—$C_4H_9$.

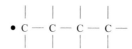

    B.  *sec*-butyl radical—$C_4H_9$

**5.7** The oxidation of $C_3H_8$ has been broken down into eight semidetailed steps (P.1 through P.7 plus other steps). Following the example of $C_3H_8$, show the first five steps in the oxidation of butane, $C_4H_{10}$.

**5.8** Using the single-step global mechanism given by Eqns. 5.1 and 5.2 for combustion of a hydrocarbon with air, compare the mass rates of fuel carbon conversion to $CO_2$ for $\Phi = 1$, $P = 1$ atm, and $T = 1600$ K for the following fuels:

    A.  $CH_4$—methane.

    B.  $C_3H_8$—propane.

    C.  $C_8H_{18}$—octane.

*Hint*: Make sure that you include the nitrogen from the air in determining your reactants' concentrations. Also note the treatment of units given in the footnote to Table 5.1.

**5.9**  What features of the methane molecule contribute to its low reactivity? Use data from the *CRC Handbook of Chemistry and Physics* [44] (or other reference book) to support your statements.

**5.10**  The production of nitric oxide from the combustion of nitrogen-free fuels with air occurs by several mechanisms. List and discuss these mechanisms.

**5.11**  Many experiments have shown that nitric oxide is formed very rapidly within flame zones, and more slowly in postflame gases. What factors contribute to this rapid formation of NO in flame zones?

**5.12**  Identify the key radical in the conversion of NO to $NO_2$ in combustion systems. Why does $NO_2$ not appear in high-temperature flame regions?

**5.13**  Consider the production of nitric oxide (NO) in the following combustion systems using the Zeldovich mechanism given in Eqn. 5.7. (See also Chapter 4, Examples 4.3 and 4.4.) In each case, assume that the environment is well mixed; O, $O_2$, and $N_2$ are at their equilibrium values (given); N atoms are in steady state; the temperature is fixed (given); and reverse reactions are negligible. Calculate the NO concentration in parts per million and the ratio of the kinetically formed NO to the given equilibrium NO concentration. Comment on the validity of neglecting reverse reactions. Do your results make sense?

A.  Consider the operation of a stationary power-generation, gas-turbine engine operating without any emission controls. The primary zone of this gas-turbine combustor has an equivalence ratio of 1.0. The mean residence time of the combustion products at the conditions below is 7 ms.

$$n\text{-Decane}/\text{air}, \qquad \chi_{O,eq} = 7.93 \cdot 10^{-5},$$
$$T = 2300 \text{ K}, \qquad \chi_{O_2,eq} = 3.62 \cdot 10^{-3},$$
$$P = 14 \text{ atm}, \qquad \chi_{N_2,eq} = 0.7295,$$
$$MW = 28.47, \qquad \chi_{NO,eq} = 2.09 \cdot 10^{-3}.$$

B.  The residence time in the primary combustion zone of a gas-fired (methane/air) furnace is 200 ms for conditions given below.

$$\Phi = 1.0, \qquad \chi_{O,eq} = 1.99 \cdot 10^{-4},$$
$$T = 2216 \text{ K}, \qquad \chi_{O_2,eq} = 4.46 \cdot 10^{-3},$$
$$P = 1 \text{ atm}, \qquad \chi_{N_2,eq} = 0.7088,$$
$$MW = 27.44, \qquad \chi_{NO,eq} = 1.91 \cdot 10^{-3}.$$

C.  Additional air is added after the primary zone combustion of part A to form a secondary combustion zone. How much additional NO is formed in the secondary zone of the combustor, assuming a step change to the conditions given below? The mean residence time of the gases in the secondary zone is 10 ms. Use the NO formed in part A as the initial condition for your calculation.

$$\Phi = 0.55, \qquad \chi_{O_2, eq} = 0.0890,$$

$$P = 14 \text{ atm}, \qquad \chi_{O, eq} = 2.77 \cdot 10^{-5},$$

$$T = 1848 \text{ K}, \qquad \chi_{N_2, eq} = 0.7570,$$

$$MW = 28.73, \qquad \chi_{NO, eq} = 3.32 \cdot 10^{-3}.$$

D. How much additional NO is formed in the secondary combustion zone of the furnace of part B? The residence time in the secondary zone is 0.5 s. Again, assume instantaneous mixing.

$$\Phi = 0.8958, \qquad \chi_{O, eq} = 9.149 \cdot 10^{-5},$$

$$T = 2000 \text{ K}, \qquad \chi_{O_2, eq} = 0.0190,$$

$$P = 1 \text{ atm}, \qquad \chi_{N_2, eq} = 0.720,$$

$$MW = 27.72, \qquad \chi_{NO, eq} = 2.34 \cdot 10^{-3}.$$

**5.14** Determine the units conversion factor required to express the pre-exponential factors in Table 5.1 in SI units (i.e., kmol, m, and s). Perform the conversion for the first and last entries in the table.

**5.15** An experimenter finds that she is unable to establish a premixed carbon monoxide flame when using an oxidizer composed of high-purity $O_2$ (21 percent) and high-purity $N_2$ (79 percent). When room air is used as the oxidizer, however, she has no trouble maintaining a flame! Explain why the carbon monoxide burns with the room air but not with the synthetic bottled air. How do the results of problem 4.18 relate to this problem?

**5.16** Draw molecular structure diagrams illustrating the transformation of a $CH_4$ molecule through the various intermediate compounds to the final oxidation product $CO_2$. Follow the linear "backbone" of the high-temperature oxidation pathway illustrated in Fig. 5.4.

**5.17** A researcher measures trace quantities of ethane and methanol in an experiment involving the extinction of methane–air flames at a cold wall. To explain the presence of the ethane and the methanol, a colleague argues that the samples were contaminated, while the researcher maintains that contamination did not occur. Is the researcher's position tenable? Explain.

**5.18\*** Methane is burned with air in an adiabatic, constant-volume vessel. The equivalence ratio is 0.9 and the initial temperature and pressure prior to combustion are 298 K and 1 atm, respectively. Consider the formation of nitric oxide by the thermal mechanism (N.1 and N.2) in the postflame gases.

| *Indicates required use of computer.

A.  Derive an expression for the NO production rate $d[NO]/dt$. Do not neglect reverse reactions. Assume $[N]$ is in steady state and that $[O]$, $[O_2]$, and $[N_2]$ are equilibrium values unaffected by the NO concentration.

B.  Determine the NO mole fraction as a function of time. Use numerical integration to accomplish this. Use UVFLAME to establish the initial conditions ($T$, $P$, $[O]$, $[O_2]$, and $[N_2]$).

C.  Plot $\chi_{NO}$ versus time. Indicate the equilibrium mole fraction from UVFLAME on your graph. Discuss.

D.  At what time do reverse reactions become important in this analysis? Discuss.

# 6

# Coupling Chemical and Thermal Analyses of Reacting Systems

## OVERVIEW

In Chapter 2, we reviewed the thermodynamics of reacting systems, considering only the initial and final states. For example, the concept of an adiabatic flame temperature was derived based on the knowledge of the initial state of the reactants and the final composition of the products, as determined by equilibrium. Performing an adiabatic flame temperature calculation required no knowledge of chemical rate processes. In this chapter, we couple the knowledge gained of chemical kinetics in Chapter 4 with fundamental conservation principles (e.g., mass and energy conservation) for various archetypal thermodynamic systems. This coupling allows us to describe the detailed evolution of the system from its initial reactant state to its final product state, which may or may not be in chemical equilibrium. In other words, we will be able to calculate the system temperature and the various species concentrations as functions of time as the system proceeds from reactants to products.

Our analyses in this chapter will be simple, without the complication of mass diffusion. The systems that are chosen for study in this chapter, shown in Fig. 6.1, make bold assumptions about the mixedness of the system. Three of the four systems assume that the systems are perfectly mixed and homogeneous in composition; the fourth system, the plug-flow reactor, totally ignores mixing and diffusion in the flow (axial) direction, while assuming perfect mixedness in the radial direction perpendicular to the flow. Although the concepts developed here can be used as building blocks for modeling more complex flows, perhaps more importantly, they are pedagogically useful for developing a very basic understanding of the interrelationships among thermodynamics, chemical kinetics, and fluid mechanics. In the next chapter, we will extend our simple analysis to include the effects of mass diffusion.

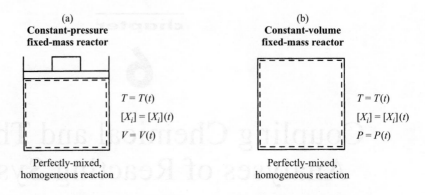

(a)
**Constant-pressure**
**fixed-mass reactor**

$T = T(t)$
$[X_i] = [X_i](t)$
$V = V(t)$

Perfectly-mixed,
homogeneous reaction

(b)
**Constant-volume**
**fixed-mass reactor**

$T = T(t)$
$[X_i] = [X_i](t)$
$P = P(t)$

Perfectly-mixed,
homogeneous reaction

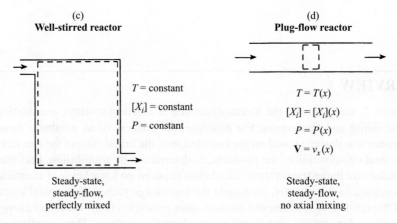

(c)
**Well-stirred reactor**

$T$ = constant
$[X_i]$ = constant
$P$ = constant

Steady-state,
steady-flow,
perfectly mixed

(d)
**Plug-flow reactor**

$T = T(x)$
$[X_i] = [X_i](x)$
$P = P(x)$
$\mathbf{V} = v_x(x)$

Steady-state,
steady-flow,
no axial mixing

**Figure 6.1**     Simple chemically reacting systems: (a) constant-pressure, fixed mass;
(b) constant-volume, fixed mass; (c) well-stirred reactor; (d) plug-flow reactor.

## CONSTANT-PRESSURE, FIXED-MASS REACTOR

### Application of Conservation Laws

Consider reactants contained in a piston–cylinder arrangement (Fig. 6.1a) that react at each and every location within the gas volume at the same rate. Thus, there are no temperature or composition gradients within the mixture, and a single temperature and set of species concentrations suffice to describe the evolution of this system. For exothermic combustion reactions, both the temperature and volume will increase with time, and there may be heat transfer through the reaction vessel walls.

In the following, we will develop a system of first-order ordinary differential equations whose solution describes the desired temperature and species evolution.

These equations and their initial conditions define an **initial-value problem.** Starting with the rate form of the conservation of energy for a fixed-mass system, we write

$$\dot{Q} - \dot{W} = m\frac{\mathrm{d}u}{\mathrm{d}t}. \qquad (6.1)$$

Applying the definition of enthalpy, $h \equiv u + Pv$, and differentiating gives

$$\frac{\mathrm{d}u}{\mathrm{d}t} = \frac{\mathrm{d}h}{\mathrm{d}t} - P\frac{\mathrm{d}v}{\mathrm{d}t}. \qquad (6.2)$$

Assuming the only work is the $P$–$\mathrm{d}v$ work at the piston,

$$\frac{\dot{W}}{m} = P\frac{\mathrm{d}v}{\mathrm{d}t}. \qquad (6.3)$$

Substituting Eqns. 6.2 and 6.3 into Eqn. 6.1, the $P\,\mathrm{d}v/\mathrm{d}t$ terms cancel leaving

$$\frac{\dot{Q}}{m} = \frac{\mathrm{d}h}{\mathrm{d}t}. \qquad (6.4)$$

We can express the system enthalpy in terms of the system chemical composition as

$$h = \frac{H}{m} = \frac{\displaystyle\sum_{i=1}^{N} N_i \bar{h}_i}{m}, \qquad (6.5)$$

where $N_i$ and $\bar{h}_i$ are the number of moles and molar enthalpy of species $i$, respectively. Differentiation of Eqn. 6.5 yields

$$\frac{\mathrm{d}h}{\mathrm{d}t} = \frac{1}{m}\left[\sum_i\left(\bar{h}_i\frac{\mathrm{d}N_i}{\mathrm{d}t}\right) + \sum_i\left(N_i\frac{\mathrm{d}\bar{h}_i}{\mathrm{d}t}\right)\right]. \qquad (6.6)$$

Assuming ideal-gas behavior, i.e., $\bar{h}_i = \bar{h}_i(T \text{ only})$,

$$\frac{\mathrm{d}\bar{h}_i}{\mathrm{d}t} = \frac{\partial \bar{h}_i}{\partial T}\frac{\mathrm{d}T}{\mathrm{d}t} = \bar{c}_{p,i}\frac{\mathrm{d}T}{\mathrm{d}t}, \qquad (6.7)$$

where $\bar{c}_{p,i}$ is the molar constant-pressure specific heat of species $i$. Equation 6.7 provides the desired link to the system temperature, and the definition of the molar concentration $[X_i]$ and the mass-action expressions, $\dot{\omega}_i = \ldots$, provide the necessary link to the system composition, $N_i$, and chemical dynamics, $\mathrm{d}N_i/\mathrm{d}t$. These expressions are

$$N_i = V[X_i] \qquad (6.8)$$

$$\frac{\mathrm{d}N_i}{\mathrm{d}t} \equiv V\dot{\omega}_i, \qquad (6.9)$$

where the $\dot{\omega}_i$ values are calculated from the detailed chemical mechanism as discussed in Chapter 4 (see Eqns. 4.31–4.33).

Substituting Eqns. 6.7–6.9 into Eqn. 6.6, our statement of energy conservation (Eqn. 6.4) becomes, after rearrangement,

$$\frac{dT}{dt} = \frac{(\dot{Q}/V) - \sum_i(\bar{h}_i \dot{\omega}_i)}{\sum_i([X_i]\bar{c}_{p,i})}, \tag{6.10}$$

where we use the following calorific equation of state to evaluate the enthalpies:

$$\bar{h}_i = \bar{h}_{f,i}^o + \int_{T_{\text{ref}}}^{T} \bar{c}_{p,i}\, dT. \tag{6.11}$$

To obtain the volume, we apply mass conservation and the definition of $[X_i]$ in Eqn. 6.8:

$$V = \frac{m}{\sum_i([X_i]MW_i)}. \tag{6.12}$$

The species molar concentrations, $[X_i]$, change with time as a result of both chemical reactions and changing volume, i.e.,

$$\frac{d[X_i]}{dt} = \frac{d(N_i/V)}{dt} = \frac{1}{V}\frac{dN_i}{dt} - N_i\frac{1}{V^2}\frac{dV}{dt} \tag{6.13a}$$

or

$$\frac{d[X_i]}{dt} = \dot{\omega}_i - [X_i]\frac{1}{V}\frac{dV}{dt}, \tag{6.13b}$$

where the first term on the right-hand side is the chemical production term and the second term accounts for the changing volume.

The ideal-gas law can be used to eliminate the $dV/dt$ term. Differentiating

$$PV = \sum_i N_i R_u T \tag{6.14a}$$

for the case of constant pressure, and rearranging, yields

$$\frac{1}{V}\frac{dV}{dt} = \frac{1}{\sum_i N_i}\sum_i \frac{dN_i}{dt} + \frac{1}{T}\frac{dT}{dt}. \tag{6.14b}$$

First substituting Eqn. 6.9 into Eqn. 6.14b and then substituting the result into Eqn. 6.13b, provides, after rearrangement, our final expression for the rate of change of the species molar concentrations:

$$\frac{d[X_i]}{dt} = \dot{\omega}_i - [X_i]\left[\frac{\sum_i \dot{\omega}_i}{\sum_j [X_j]} + \frac{1}{T}\frac{dT}{dt}\right]. \tag{6.15}$$

## Reactor Model Summary

Succinctly stated, our problem is to find the solution to

$$\frac{dT}{dt} = f([X_i], T) \tag{6.16a}$$

$$\frac{d[X_i]}{dt} = f([X_i], T) \qquad i = 1, 2, \ldots, N \tag{6.16b}$$

with initial conditions

$$T(t = 0) = T_0 \tag{6.17a}$$

and

$$[X_i](t = 0) = [X_i]_0. \tag{6.17b}$$

The functional expressions for Eqns. 6.16a and 6.16b are obtained from Eqn. 6.10 and Eqn. 6.15, respectively. Enthalpies are calculated using Eqn. 6.11, and the volume is obtained from Eqn. 6.12.

To carry out the solution of the above system, an integration routine capable of handling stiff equations should be employed, as discussed in Chapter 4.

## CONSTANT-VOLUME, FIXED-MASS REACTOR

### Application of Conservation Laws

The application of energy conservation to the constant-volume reactor follows closely that of the constant-pressure reactor, with the major difference being the absence of work in the former. Starting with Eqn. 6.1, with $\dot{W} = 0$, the first law takes the following form:

$$\frac{du}{dt} = \frac{\dot{Q}}{m}. \tag{6.18}$$

Recognizing that the specific internal energy, $u$, now plays the same mathematical role as the specific enthalpy, $h$, in our previous analysis, expressions equivalent to Eqns. 6.5–6.7 are developed and substituted into Eqn. 6.18. This yields, after rearrangement,

$$\frac{dT}{dt} = \frac{(\dot{Q}/V) - \sum_i (\bar{u}_i \dot{\omega}_i)}{\sum_i ([X_i] \bar{c}_{v,i})}. \tag{6.19}$$

Recognizing that, for ideal gases, $\bar{u}_i = \bar{h}_i - R_u T$ and $\bar{c}_{v,i} = \bar{c}_{p,i} - R_u$, we can express Eqn. 6.19 using enthalpies and constant-pressure specific heats:

$$\frac{dT}{dt} = \frac{(\dot{Q}/V) + R_u T \sum_i \dot{\omega}_i - \sum_i (\bar{h}_i \dot{\omega}_i)}{\sum_i [[X_i](\bar{c}_{p,i} - R_u)]}. \tag{6.20}$$

In constant-volume explosion problems, the time-rate-of-change of the pressure is of interest. To calculate $dP/dt$, we differentiate the ideal-gas law, subject to the constant volume constraint, i.e.,

$$PV = \sum_i N_i R_u T \tag{6.21}$$

and

$$V \frac{dP}{dt} = R_u T \frac{d \sum_i N_i}{dt} + R_u \sum_i N_i \frac{dT}{dt}. \tag{6.22}$$

Applying the definitions of $[X_i]$ and $\dot{\omega}_i$ (see Eqns. 6.8 and 6.9), Eqns. 6.21 and 6.22 become

$$P = \sum_i [X_i] R_u T \tag{6.23}$$

and

$$\frac{dP}{dt} = R_u T \sum_i \dot{\omega}_i + R_u \sum_i [X_i] \frac{dT}{dt}, \tag{6.24}$$

which completes our simple analysis of homogeneous constant-volume combustion.

## Reactor Model Summary

Equation 6.20 can be integrated simultaneously with the chemical rate expressions to determine $T(t)$ and $[X_i](t)$, i.e.,

$$\frac{dT}{dt} = f([X_i], T) \tag{6.25a}$$

$$\frac{d[X_i]}{dt} = \dot{\omega}_i = f([X_i], T) \quad i = 1, 2, \ldots, N \tag{6.25b}$$

with initial conditions

$$T(t = 0) = T_0 \tag{6.26a}$$

and

$$[X_i](t = 0) = [X_i]_0. \tag{6.26b}$$

The required enthalpies are evaluated using Eqn. 6.11, and the pressure from Eqn. 6.23. Again, a stiff equation solver should be used to carry out the integration.

**Example 6.1**

In spark-ignition engines, knock occurs when the unburned fuel–air mixture ahead of the flame reacts homogeneously, i.e., it autoignites. The rate-of-pressure rise is a key parameter in determining knock intensity and propensity for mechanical damage to the piston–crank assembly. Pressure-versus-time traces for normal and knocking combustion in a spark-ignition engine are illustrated in Fig. 6.2. Note the very rapid pressure rise in the case of heavy knock. Figure 6.3 shows schlieren (index-of-refraction gradient) photographs of flame propagation for normal and knocking combustion.

Create a simple constant-volume model of the autoignition process and determine the temperature and the fuel and product concentration histories. Also determine $dP/dt$ as a function of time. Assume initial conditions corresponding to compression of a fuel–air mixture from 300 K and 1 atm to top-dead-center for a compression ratio of 10:1. The initial volume before compression is $3.68 \cdot 10^{-4}$ m³, which corresponds to an engine with both a bore and a stroke of 75 mm. Use ethane as fuel.

### Solution

We will make some bold and sweeping assumptions about the thermodynamics and the chemical kinetics to keep the computational complexity to a minimum. Our solution, however, will still retain the strong coupling between the thermochemistry and chemical kinetics. Our assumptions are as follows:

i. One-step global kinetics using the rate parameters for ethane $C_2H_6$ (see Table 5.1).
ii. The fuel, air, and products all have equal molecular weights; i.e., $MW_F = MW_{Ox} = MW_{Pr} = 29$.
iii. The specific heats of the fuel, air, and products are constants and equal; that is, $c_{p,F} = c_{p,Ox} = c_{p,Pr} = 1200$ J/kg-K.
iv. The enthalpy of formation of the air and products are zero; the enthalpy of formation of the fuel is $4 \cdot 10^7$ J/kg.
v. We assume that the stoichiometric air–fuel ratio is 16.0 and restrict combustion to stoichiometric or lean conditions.

The use of global kinetics is hard to justify for a problem like engine knock where detailed chemistry is important [3]. Our only justification is that we are trying to illustrate principles, recognizing that our answers may be inaccurate in detail. Assumptions ii–iv provide values

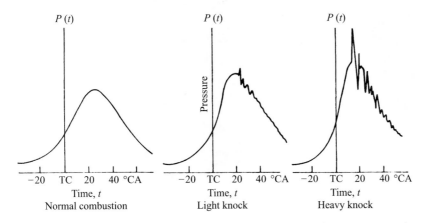

**Figure 6.2**   Cylinder pressure-versus-time measurements in a spark-ignition engine for normal combustion, light knock, and heavy knock cycles. The crank angle interval of 40° corresponds to 1.67 ms.

SOURCE: Adapted from Refs. [1] and [17] by permission of McGraw-Hill, Inc.

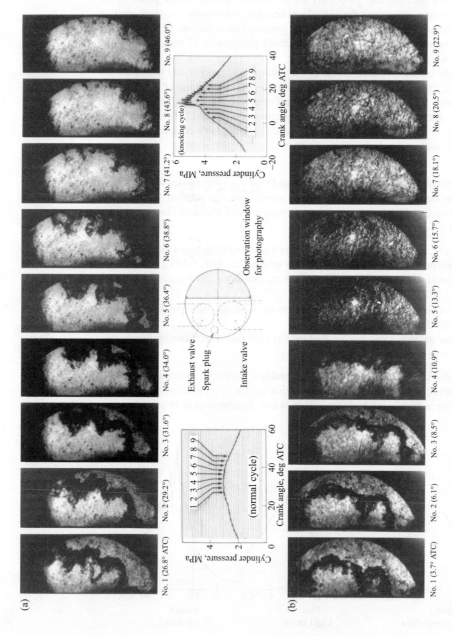

**Figure 6.3** Schlieren photographs from high-speed movies of (a) normal combustion and (b) knocking combustion. Pressure–time traces corresponding to the photographs are also shown.

SOURCE: From Refs. [2] and [17]. Reprinted by permission of McGraw-Hill, Inc.

that give reasonable estimates of flame temperatures, yet trivialize the problem of obtaining thermodynamic properties [4].

With these assumptions, we can now formulate our model. From Eqn. 5.2 and Table 5.1, the fuel (ethane) reaction rate is

$$\frac{d[F]}{dt} = -6.19 \cdot 10^9 \exp\left(\frac{-15,098}{T}\right)[F]^{0.1}[O_2]^{1.65} \tag{I}$$

$$[=] \text{ kmol/m}^3\text{-s},$$

where, assuming 21% percent $O_2$ in the air,

$$[O_2] = 0.21[Ox].$$

Note the conversion of units from a gmol-cm$^3$ to a kmol-m$^3$ basis for the pre-exponential factor $(1.1 \cdot 10^{12} \cdot [1000]^{1-0.1-1.65} = 6.19 \cdot 10^9)$.

We can simply relate the oxidizer and product reaction rates to the fuel rate through the stoichiometry (assumptions ii and v):

$$\frac{d[Ox]}{dt} = (A/F)_s \frac{MW_F}{MW_{Ox}} \frac{d[F]}{dt} = 16 \frac{d[F]}{dt} \tag{II}$$

and

$$\frac{d[Pr]}{dt} = -[(A/F)_s + 1]\frac{MW_F}{MW_{Pr}} \frac{d[F]}{dt} = -17 \frac{d[F]}{dt}. \tag{III}$$

We complete our model by applying Eqn. 6.20:

$$\frac{dT}{dt} = \frac{(\dot{Q}/V) + R_u T \sum \dot{\omega}_i - \sum(\bar{h}_i \dot{\omega}_i)}{\sum[[X_i](\bar{c}_{p,i} - R_u)]}.$$

This simplifies, by noting that

$$\dot{Q}/V = 0 \quad \text{(adiabatic)},$$

$$\sum \dot{\omega}_i = 0 \quad \text{(assumptions ii and v)},$$

$$\sum \bar{h}_i \dot{\omega}_i = \dot{\omega}_F \bar{h}^o_{f,F} \quad \text{(assumptions ii–v)},$$

and

$$\sum [X_i](\bar{c}_{p,i} - R_u) = (\bar{c}_p - R_u)\sum[X_i] = (\bar{c}_p - R_u)\sum \chi_i \frac{P}{R_u T} = (\bar{c}_p - R_u)\frac{P}{R_u T},$$

to be

$$\frac{dT}{dt} = \frac{-\dot{\omega}_F \bar{h}^o_{f,F}}{(\bar{c}_p - R_u)P/(R_u T)}. \tag{IV}$$

Although our basic model is complete, we can add ancillary relations for the pressure and pressure-derivative. From Eqn. 6.23 and 6.24,

$$P = R_u T([F] + [Ox] + [Pr]),$$

or

$$P = P_0 \frac{T}{T_0}$$

and

$$\frac{dP}{dt} = \frac{P}{T}\frac{dT}{dt} = \frac{P_0}{T_0}\frac{dT}{dt}.$$

Before we can integrate our system of first-order ordinary differential equations (Eqns. I–IV), we need to determine initial conditions for each of the variables: $[F]$, $[Ox]$, $[Pr]$, and $T$. Assuming isentropic compression from bottom-dead-center to top-dead-center and a specific heat ratio of 1.4, the initial temperature and pressure can be found:

$$T_0 = T_{TDC} = T_{BDC}\left(\frac{V_{BDC}}{V_{TDC}}\right)^{\gamma-1} = 300\left(\frac{10}{1}\right)^{1.4-1} = 753 \text{ K}$$

and

$$P_0 = P_{TDC} = P_{BDC}\left(\frac{V_{BDC}}{V_{TDC}}\right)^{\gamma} = (1)\left(\frac{10}{1}\right)^{1.4} = 25.12 \text{ atm.}$$

The initial concentrations can be found by employing the given stoichiometry. The oxidizer and fuel mole fractions are

$$\chi_{Ox,0} = \frac{(A/F)_s/\Phi}{[(A/F)_s/\Phi]+1},$$

$$\chi_{Pr,0} = 0,$$

$$\chi_{F,0} = 1 - \chi_{Ox,0}.$$

The molar concentrations, $[X_i] = \chi_i P/(R_u T)$, are

$$[Ox]_0 = \left[\frac{(A/F)_s/\Phi}{[(A/F)_s/\Phi]+1}\right]\frac{P_0}{R_u T_0},$$

$$[F]_0 = \left[1 - \frac{(A/F)_s/\Phi}{[(A/F)_s/\Phi]+1}\right]\frac{P_0}{R_u T_0},$$

$$[Pr]_0 = 0.$$

Equations I–IV were integrated numerically, and the results are shown in Fig. 6.4. From this figure, we see that the temperature increases only about 200 K in the first 3 ms, while, thereafter, it rises to the adiabatic flame temperature ($c$. 3300 K) in less than 0.1 ms. This rapid temperature rise and concomitant rapid consumption of the fuel is characteristic of a **thermal explosion,** where the energy released and the temperature rise from reaction feeds back to produce ever-increasing reaction rates because of the $[-E_a/R_u T]$ temperature dependence of the reaction rate. From Fig. 6.4, we also see the huge pressure derivative in the explosive stage, with a peak value of about $1.9 \cdot 10^{13}$ Pa/s.

**Comments**

Although this model predicted the explosive combustion of the mixture after an initial period of slow combustion, as is observed in real knocking combustion, the single-step kinetics mechanism

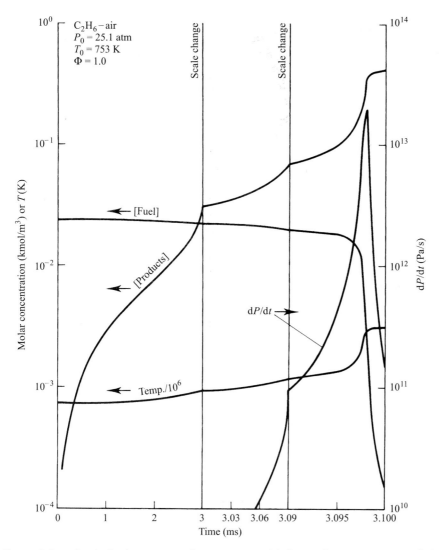

**Figure 6.4**   Results for the constant-volume reactor model of Example 6.1. Temperature, fuel, and products concentrations, and rate-of-pressure rise (d$P$/d$t$) are shown. Note the expansion of the time scale at 3 ms and again at 3.09 ms allows the explosion to be resolved.

does not model the true behavior of autoigniting mixtures. In reality, the **induction period,** or **ignition delay,** is controlled by the formation of intermediate species, which subsequently react. Recall the three basic stages of hydrocarbon oxidation presented in Chapter 5. To accurately model knock, a more detailed mechanism would be required. Ongoing research efforts aim at elucidating the details of the "low-temperature" kinetics of the induction period [3].

Control of engine knock has always been important to performance improvements, and, more recently, has received attention because of legislated requirements to remove lead-based antiknock compounds from gasoline.

## WELL-STIRRED REACTOR

The well-stirred, or perfectly-stirred, reactor is an ideal reactor in which perfect mixing is achieved inside the control volume, as shown in Fig. 6.5. Experimental reactors employing high-velocity inlet jets approach this ideal and have been used to study many aspects of combustion, such as flame stabilization [5] and $NO_x$ formation [6–8] (Fig. 6.6). Well-stirred reactors have also been used to obtain values for global reaction parameters [9]. The well-stirred reactor is sometimes called a Longwell reactor in recognition of the early work of Longwell and Weiss [5]. Chomiak [10] cites that Zeldovich [11] described the operation of the well-stirred reactor nearly a decade earlier.

## Application of Conservation Laws

To develop the theory of well-stirred reactors, we review the concept of mass conservation of individual species. In Chapter 3, we developed a species conservation equation for a differential control volume. We now write mass conservation for an arbitrary species $i$, for an integral control volume (see Fig. 6.5), as

$$\frac{\mathrm{d}m_{i,cv}}{\mathrm{d}t} = \dot{m}_i''' V + \dot{m}_{i,\mathrm{in}} - \dot{m}_{i,\mathrm{out}}. \qquad (6.27)$$

| Rate at which mass of $i$ accumulates within control volume | Rate at which mass of $i$ is generated within control volume | Mass flow of $i$ into control volume | Mass flow of $i$ out of control volume |

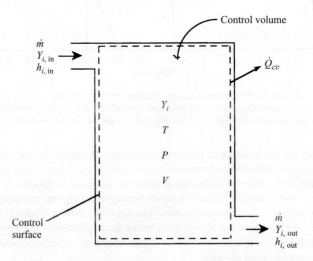

**Figure 6.5**    Schematic of a well-stirred reactor.

**Figure 6.6**    Longwell reactor with one hemisphere removed. Fuel–air mixture enters through small holes in central hollow steel sphere, and products exit through larger holes in firebrick lining. Scale shown is in inches.
SOURCE: From Ref. [5]. Reprinted by permission, © The American Chemical Society.

What distinguishes Eqn. 6.27 from the overall continuity equation is the presence of the generation term $\dot{m}_i''' V$. This term arises because chemical reactions transform one species into another; hence, a positive generation rate indicates the formation of a species, and a negative generation rate signifies that the species is being destroyed during the reaction. In the combustion literature, this generation term is frequently referred to as a **source** or **sink.** When the appropriate form of Eqn. 6.27 is written for each of the species in the reactor ($i = 1, 2, \ldots, N$), the sum of these equations yields the familiar form of the continuity equation,

$$\frac{dm_{cv}}{dt} = \dot{m}_{in} - \dot{m}_{out}. \tag{6.28}$$

The mass generation rate of a species, $\dot{m}_i'''$, is easily related to the net production rate, $\dot{\omega}_i$, developed in Chapter 4:

$$\dot{m}_i''' = \dot{\omega}_i MW_i. \tag{6.29}$$

Ignoring any diffusional flux, the individual species mass flowrate is simply the product of the total mass flowrate and that species mass fraction, i.e.,

$$\dot{m}_i = \dot{m} Y_i. \tag{6.30}$$

When we apply Eqn. 6.27 to the well-stirred reactor, assuming steady-state operation, the time derivative of the left-hand side disappears. With this assumption, and substituting Eqns. 6.29 and 6.30, Eqn. 6.27 becomes

$$\dot{\omega}_i MW_i V + \dot{m}(Y_{i,\text{in}} - Y_{i,\text{out}}) = 0 \quad \text{for} \quad i = 1, 2, \ldots, N \text{ species.} \tag{6.31}$$

Furthermore, we can identify the outlet mass fractions, $Y_{i,\text{out}}$, as being equal to the mass fractions within the reactor. Since the composition within the reactor is everywhere the same, the composition at the outlet of the control volume must be the same as in the interior. With this knowledge, the species production rates are of the form

$$\dot{\omega}_i = f([X_i]_{cv}, T) = f([X_i]_{\text{out}}, T), \tag{6.32}$$

where the mass fractions and molar concentrations are related by

$$Y_i = \frac{[X_i]MW_i}{\sum\limits_{j=1}^{N}[X_j]MW_j}. \tag{6.33}$$

Equation 6.31, when written for each species, provides $N$ equations with $N + 1$ unknowns, with the assumed known parameters $\dot{m}$ and $V$. An energy balance provides the additional equation needed for closure.

The steady-state, steady-flow conservation of energy equation (Eqn. 2.28) applied to the well-stirred reactor is

$$\dot{Q} = \dot{m}(h_{\text{out}} - h_{\text{in}}), \tag{6.34}$$

where we neglect changes in kinetic and potential energies. Rewriting Eqn. 6.34 in terms of the individual species, we obtain

$$\dot{Q} = \dot{m}\left(\sum\limits_{i=1}^{N}Y_{i,\text{out}}h_i(T) - \sum\limits_{i=1}^{N}Y_{i,\text{in}}h_i(T_{\text{in}})\right), \tag{6.35}$$

where

$$h_i(T) = h_{f,i}^o + \int\limits_{T_{\text{ref}}}^{T} c_{p,i}\, dT. \tag{6.36}$$

Solving for the temperature, $T$, and species mass fractions, $Y_{i,\text{out}}$, is quite similar to our computation of equilibrium flame temperatures in Chapter 2; however, now the product composition is constrained by chemical kinetics, rather than by chemical equilibrium.

It is common in the discussion of well-stirred reactors to define a mean **residence time** for the gases in the reactor:

$$t_R = \rho V / \dot{m} \tag{6.37}$$

where the mixture density is calculated from the ideal-gas law,

$$\rho = P MW_{\text{mix}} / R_u T. \qquad (6.38)$$

The mixture molecular weight is readily calculated from a knowledge of the mixture composition. Appendix 6A provides relationships between $MW_{\text{mix}}$ and $Y_i$, $\chi_i$, and $[X_i]$.

## Reactor Model Summary

Because the well-stirred reactor is assumed to be operating at steady state, there is no time dependence in the mathematical model. The equations describing the reactor are a set of coupled nonlinear algebraic equations, rather than a system of ordinary differential equations (ODEs), which was the result for the previous two examples. Thus, the $\dot{\omega}_i$ appearing in Eqn. 6.31 depends only on the $Y_i$ (or $[X_i]$) and temperature, not time. To solve this system of $N + 1$ equations, Eqns. 6.31 and 6.35, the generalized Newton's method (Appendix E) can be employed. Depending on the chemical system under study, it may be difficult to achieve convergence with Newton's method and more sophisticated numerical techniques may be necessary [12].

---

Develop a simplified model of a well-stirred reactor using the same simplified chemistry and thermodynamics used in Example 6.1 (equal and constant $c_p$s and $MWs$, and one-step global kinetics). Use the model to determine the blowout characteristics of a spherical (80-mm-diameter) reactor with premixed reactants ($C_2H_6$–air) entering at 298 K. Plot the equivalence ratio at blowout as a function of mass flowrate for $\Phi \leq 1.0$. Assume the reactor is adiabatic.

**Example 6.2**

**Solution**

Noting that the molar concentrations relate to mass fractions as

$$[X_i] = \frac{P MW_{\text{mix}}}{R_u T} \frac{Y_i}{MW_i},$$

our global reaction rate, $\dot{\omega}_F$, can be expressed as

$$\dot{\omega}_F = \frac{d[F]}{dt} = -k_G \left( \frac{P MW_{\text{mix}}}{R_u T} \right)^{m+n} \left( \frac{Y_F}{MW_F} \right)^m \left( \frac{0.233 Y_{Ox}}{MW_{Ox}} \right)^n,$$

where $m = 0.1$ and $n = 1.65$, the factor 0.233 is the mass fraction of $O_2$ in the oxidizer (air), and the mixture molecular weight is given by

$$MW_{\text{mix}} = \left[ \frac{Y_F}{MW_F} + \frac{Y_{Ox}}{MW_{Ox}} + \frac{Y_{Pr}}{MW_{Pr}} \right]^{-1}.$$

The global rate coefficient is, as in Example 6.1,

$$k_G = 6.19 \cdot 10^9 \exp\left( \frac{-15,098}{T} \right).$$

We can now write species conservation equations for the fuel by applying Eqn. 6.31 to give

$$f_1 \equiv \dot{m}(Y_{F,\text{in}} - Y_F) - k_G MW_F V \left( \frac{P}{R_u T} \right)^{1.75} \frac{\left( \dfrac{Y_F}{MW_F} \right)^{0.1} \left( \dfrac{0.233 Y_{Ox}}{MW_{Ox}} \right)^{1.65}}{\left[ \dfrac{Y_F}{MW_F} + \dfrac{Y_{Ox}}{MW_{Ox}} + \dfrac{Y_{Pr}}{MW_{Pr}} \right]^{1.75}} = 0,$$

which further simplifies by applying our assumption of equal molecular weights and noting that $\Sigma Y_i = 1$:

$$f_1 \equiv \dot{m}(Y_{F,\text{in}} - Y_F) - k_G MW \, V \left( \frac{P}{R_u T} \right)^{1.75} \frac{Y_F^{0.1}(0.233 Y_{Ox})^{1.65}}{1} = 0. \tag{I}$$

For the oxidizer (air),

$$f_2 \equiv \dot{m}(Y_{Ox,\text{in}} - Y_{Ox}) - (A/F)_s k_G MW \, V \left( \frac{P}{R_u T} \right)^{1.75} \frac{Y_F^{0.1}(0.233 Y_{Ox})^{1.65}}{1} = 0. \tag{II}$$

For the product mass fraction, we write

$$f_3 \equiv 1 - Y_F - Y_{Ox} - Y_{Pr} = 0. \tag{III}$$

Our final equation in the model results from the application of Eqn. 6.35:

$$\begin{aligned}
f_4 \equiv \; & Y_F \left[ h_{f,F}^o + c_{p,F}(T - T_{\text{ref}}) \right] \\
& + Y_{Ox} \left[ h_{f,Ox}^o + c_{p,Ox}(T - T_{\text{ref}}) \right] \\
& + Y_{Pr} \left[ h_{f,Pr}^o + c_{p,Pr}(T - T_{\text{ref}}) \right] \\
& - Y_{F,\text{in}} \left[ h_{f,F}^o + c_{p,F}(T_{\text{in}} - T_{\text{ref}}) \right] \\
& - Y_{Ox,\text{in}} \left[ h_{f,Ox}^o + c_{p,Ox}(T_{\text{in}} - T_{\text{ref}}) \right] = 0,
\end{aligned}$$

which also further simplifies by the assumptions of equal specific heats and $h_{f,Ox}^o = h_{f,Pr}^o = 0$ to give

$$f_4 \equiv (Y_F - Y_{F,\text{in}}) h_{f,F}^o + c_p (T - T_{\text{in}}) = 0. \tag{IV}$$

Equations I–IV constitute our reactor model and involve the four unknowns $Y_F$, $Y_{Ox}$, $Y_{Pr}$, and $T$ and the parameter $\dot{m}$. To determine the reactor blowout characteristic, we solve the non-linear algebraic equation set (I–IV) for a sufficiently small value of $\dot{m}$ that allows combustion for a given equivalence ratio. We then increase $\dot{m}$ until we fail to achieve a solution, or until the solution yields the input values. Figure 6.7 illustrates the results of such a procedure for $\Phi = 1$. The generalized Newton's method (Appendix E) was used to solve the equation set.

In Fig. 6.7, we see the decreasing conversion of fuel to products and decreased temperature as the flowrate increases to the blowout condition ($\dot{m} > 0.193$ kg/s). The ratio of the temperature at blowout to the adiabatic flame temperature is given by (1738 K/2381 K = 0.73), which is in agreement with the results in Ref. [5].

Repeating the calculations at various equivalence ratios generates the blowout characteristics shown in Fig. 6.8. Note that the reactor is more easily blown out as the fuel–air mixture

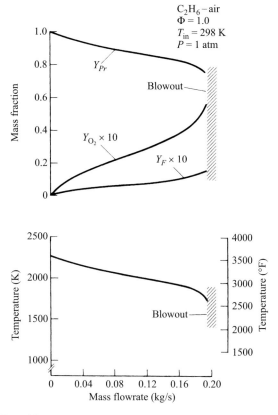

**Figure 6.7**    Effect of flowrate on conditions inside a model well-stirred reactor. For flowrates greater than 0.193 kg/s, combustion cannot be sustained within the reactor (blowout).

becomes leaner. The shape of the blowout curve in Fig. 6.8 is similar to those determined for experimental reactors and turbine combustors.

**Comment**

Well-stirred-reactor theory and experiments were used in the 1950s as a guide to the development of high-intensity combustors for gas turbines and ramjets. This example provides a good illustration of how reactor theory can be applied to the problem of blowout. The blowout condition, plus some margin of safety, determines the maximum-load condition for continuous-flow combustors. Although well-stirred-reactor theory captures some of the characteristics of blowout, other theories also have been proposed to explain flameholding. We explore this topic further in Chapter 12.

---

Use the following detailed kinetic mechanism for $H_2$ combustion to investigate the behavior of an adiabatic, well-stirred reactor operating at 1 atm. The reactant stream is a stoichiometric ($\Phi = 1$) mixture of $H_2$ and air at 298 K, and the reactor volume is 67.4 cm³. Vary the residence time between the long-time (equilibrium) and the short-time (blowout) limits. Plot the temperature and the $H_2O$, $H_2$, OH, $O_2$, O, and NO mole fractions versus residence time.

**Example 6.3**

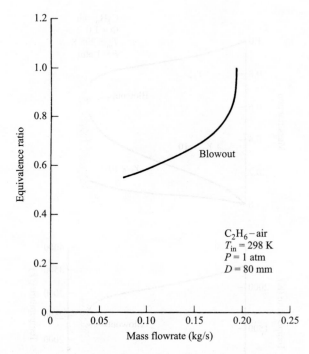

**Figure 6.8**  Blowout characteristics of a model well-stirred reactor.

### H–O–N mechanism

| No. | Reaction | Forward Rate Coefficient[a] | | |
|---|---|---|---|---|
| | | $A$ | $b$ | $E_a$ |
| 1[b] | $H + O_2 + M \leftrightarrow HO_2 + M$ | $3.61 \cdot 10^{17}$ | $-0.72$ | 0 |
| 2 | $H + H + M \leftrightarrow H_2 + M$ | $1.0 \cdot 10^{18}$ | $-1.0$ | 0 |
| 3 | $H + H + H_2 \leftrightarrow H_2 + H_2$ | $9.2 \cdot 10^{16}$ | $-0.6$ | 0 |
| 4 | $H + H + H_2O \leftrightarrow H_2 + H_2O$ | $6.0 \cdot 10^{19}$ | $-1.25$ | 0 |
| 5[c] | $H + OH + M \leftrightarrow H_2O + M$ | $1.6 \cdot 10^{22}$ | $-2.0$ | 0 |
| 6[c] | $H + O + M \leftrightarrow OH + M$ | $6.2 \cdot 10^{16}$ | $-0.6$ | 0 |
| 7 | $O + O + M \leftrightarrow O_2 + M$ | $1.89 \cdot 10^{13}$ | 0 | $-1,788$ |
| 8 | $H_2O_2 + M \leftrightarrow OH + OH + M$ | $1.3 \cdot 10^{17}$ | 0 | 45,500 |
| 9 | $H_2 + O_2 \leftrightarrow OH + OH$ | $1.7 \cdot 10^{13}$ | 0 | 47,780 |
| 10 | $OH + H_2 \leftrightarrow H_2O + H$ | $1.17 \cdot 10^{9}$ | 1.3 | 3,626 |
| 11 | $O + OH \leftrightarrow O_2 + H$ | $3.61 \cdot 10^{14}$ | $-0.5$ | 0 |
| 12 | $O + H_2 \leftrightarrow OH + H$ | $5.06 \cdot 10^{4}$ | 2.67 | 6,290 |
| 13 | $OH + HO_2 \leftrightarrow H_2O + O_2$ | $7.5 \cdot 10^{12}$ | 0 | 0 |
| 14 | $H + HO_2 \leftrightarrow OH + OH$ | $1.4 \cdot 10^{14}$ | 0 | 1,073 |
| 15 | $O + HO_2 \leftrightarrow O_2 + OH$ | $1.4 \cdot 10^{13}$ | 0 | 1,073 |
| 16 | $OH + OH \leftrightarrow O + H_2O$ | $6.0 \cdot 10^{8}$ | 1.3 | 0 |
| 17 | $H + HO_2 \leftrightarrow H_2 + O_2$ | $1.25 \cdot 10^{13}$ | 0 | 0 |
| 18 | $HO_2 + HO_2 \leftrightarrow H_2O_2 + O_2$ | $2.0 \cdot 10^{12}$ | 0 | 0 |

**H–O–N mechanism** *(continued)*

| | | Forward Rate Coefficient[a] | | |
|---|---|---|---|---|
| No. | Reaction | A | b | $E_a$ |
| 19 | $H_2O_2 + H \leftrightarrow HO_2 + H_2$ | $1.6 \cdot 10^{12}$ | 0 | 3,800 |
| 20 | $H_2O_2 + OH \leftrightarrow H_2O + HO_2$ | $1.0 \cdot 10^{13}$ | 0 | 1,800 |
| 21 | $O + N_2 \leftrightarrow NO + N$ | $1.4 \cdot 10^{14}$ | 0 | 75,800 |
| 22 | $N + O_2 \leftrightarrow NO + O$ | $6.40 \cdot 10^9$ | 1.0 | 6,280 |
| 23 | $OH + N \leftrightarrow NO + H$ | $4.0 \cdot 10^{13}$ | 0 | 0 |

[a] The forward rate coefficient is of the form $k_f = AT^b \exp(-E_a / R_u T)$, where A is expressed in CGS units (cm, s, K, gmol), b is dimensionless with T in Kelvins, and $E_a$ is given in cal/gmol.

[b] When $H_2O$ and $H_2$ act as collision partners M, the rate coefficient is multiplied by 18.6 and 2.86, respectively.

[c] When $H_2O$ acts as a collision partner M, the rate coefficient is multiplied by 5.

**Solution**

Although we will use CHEMKIN software to solve this problem, it is instructive to outline the solution steps as if we were solving the problem from scratch. Examination of the chemical mechanism above shows that there are 11 species involved: $H_2$, H, $O_2$, O, OH, $HO_2$, $H_2O_2$, $H_2O$, $N_2$, N, and NO.

Thus, we need to write 11 equations of the form of Eqn. 6.31 (or 10 equations of this form if we choose to invoke $\Sigma Y_{i,out} = 1$). The $Y_{i,in}$ values are readily calculated from the given stoichiometry, assuming that the only species present in the inlet stream are $H_2$, $O_2$, and $N_2$. For $\Phi = 1$, $a = 0.5$ in the combustion reaction, $H_2 + a(O_2 + 3.76N_2) \rightarrow H_2O + 3.76aN_2$; thus,

$$\chi_{H_2, in} = 1/3.38 = 0.2959,$$

$$\chi_{O_2, in} = 0.5/3.38 = 0.1479,$$

$$\chi_{N_2, in} = 1.88/3.38 = 0.5562,$$

$$\chi_{H, in} = \chi_{O_2, in} = \chi_{OH, in} = \ldots = 0.$$

Using these mole fractions, the reactant mixture molecular weight, $\Sigma \chi_i MW_i$, is calculated to be 20.91. The corresponding mass fractions, $\chi_i MW_i / MW_{mix}$, are

$$Y_{H_2, in} = 0.0285,$$

$$Y_{O_2, in} = 0.2263,$$

$$Y_{N_2, in} = 0.7451,$$

$$Y_{H, in} = Y_{O, in} = Y_{OH, in} = \ldots = 0.$$

Choosing the equation for O atoms to illustrate one of the equations in the set (Eqn. 6.31), we write

$$\dot{\omega}_O MW_O V - \frac{PMW_{mix} V}{R_u T t_R} Y_{O, out} = 0,$$

where Eqns. 6.37 and 6.38 have been combined to eliminate $\dot{m}$, and

$$
\begin{aligned}
\dot{\omega}_O = & -k_{6f}[H][O](P/(R_uT)+4[H_2O]) \\
& +k_{6r}[OH](P/(R_uT)+4[H_2O]) \\
& -2k_{7f}[O]^2P/(R_uT)+2k_{7r}[O_2]P/(R_uT) \\
& -k_{11f}[O][OH]+k_{11r}[O_2][H]-k_{12f}[O][H_2] \\
& +k_{12r}[OH][H]-k_{15f}[O][HO_2] \\
& +k_{15r}[O_2][OH]+k_{16f}[OH]^2-k_{16r}[O][H_2O] \\
& -k_{21f}[O][N_2]+k_{21r}[NO][N] \\
& +k_{22f}[N][O_2]-k_{22r}[NO][O].
\end{aligned}
$$

In the above expression, note that $P/(R_uT)$ has been substituted for $[M]$ and how the enhanced third-body efficiency for $H_2O$ in reaction 6 has been treated. Note also that outlet quantities are implied for all $[X_i]$s. Since both $Y_{O,out}$ and $[O]$ appear in the O-atom conservation expression, Eqn. 6.33 is needed to express one in terms of the other. As expanding this expression is straightforward, this step is left to the reader. Again, Eqn. 6.33 would be employed for each of the 11 species. Closure is obtained to our system of 23 unknowns (11 $[X_i]$s, 11 $Y_{i,out}$s, and $T_{ad}$) by applying a single conservation-of-energy expression, Eqn. 6.35:

$$
\begin{aligned}
Y_{H_2,in}h_{H_2}(298)+&Y_{O_2,in}h_{O_2}(298)+Y_{N_2,in}h_{N_2}(298) \\
= Y_{H_2}h_{H_2}(T_{ad})+&Y_Hh_H(T_{ad})+Y_{O_2}h_{O_2}(T_{ad}) \\
+&Y_{OH}h_{OH}(T_{ad})+\cdots+Y_{NO}h_{NO}(T_{ad}).
\end{aligned}
$$

Of course, a thermodynamic database is needed to relate the $h_i$ values and $T_{ad}$.

The above framework for our solution is embodied in the Fortran computer code PSR [12]. This code, together with the subroutines contained in the CHEMKIN library [16], is used to generate numerical results for our particular problem. Input quantities include the chemical mechanism, reactant-stream constitutents, equivalence ratio, inlet temperature, and reactor volume and pressure. The residence time is also an input variable. With some experimentation, a residence time of 1 s was found to yield essentially equilibrium conditions within the reactor, while blowout occurs near $t_R = 1.7 \cdot 10^{-5}$ s. Figures 6.9 and 6.10 show predicted mole fractions and gas temperatures within this range of residence times. As expected, the adiabatic temperature and $H_2O$ product concentration drop as residence times become shorter; conversely, the $H_2$ and $O_2$ concentrations rise. The behavior of the O and OH radicals is more complicated, with maxima exhibited between the two residence-time limits. The NO concentrations fall rapidly as residence times fall below $10^{-2}$ s.

**Comment**

This example provides a concrete illustration of how relatively complex chemistry couples with a simple thermodynamic process. A similar analysis was used to generate the reaction pathway diagrams for $CH_4$ combustion shown in Chapter 5.

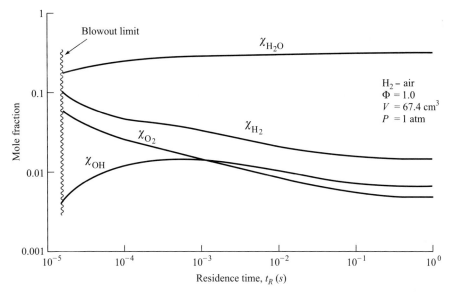

**Figure 6.9**    Predicted mole fractions of $H_2O$, $H_2$, $O_2$, and OH in a well-stirred reactor for conditions between blowout ($t_R \approx 1.75 \cdot 10^{-5}$ s) and near equilibrium ($t_R = 1$ s).

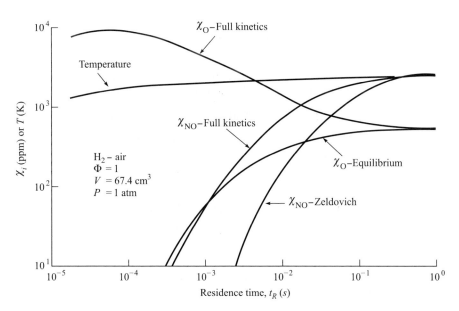

**Figure 6.10**    Predicted temperatures and O-atom and NO mole fractions in a well-stirred reactor for conditions between blowout ($t_R \approx 1.75 \cdot 10^{-5}$ s) and near equilibrium ($t_R = 1$ s). Also shown are O-atom concentrations associated with equilibrium at the kinetically derived temperature together with NO concentrations based on these equilibrium O atoms.

**Example 6.4**

Explore the degree to which nonequilibrium O atoms affect NO formation in the well-stirred reactor of Example 6.3. Assume O atoms, $O_2$, and $N_2$ are at their equilibrium values for the reactor temperatures predicted by the full kinetics. Use the following corresponding temperatures and residence times from Example 6.3: $T_{ad} = 2378$ K for $t_R = 1$ s, $T_{ad} = 2366.3$ K for $t_R = 0.1$ s, and $T_{ad} = 2298.5$ K for $t_R = 0.01$ s. Also assume that N atoms are in steady state and that NO formation is controlled by the simple Zeldovich chain-reaction pair (Chapter 5, Eqns. N.1 and N.2):

$$N_2 + O \overset{k_{1f}}{\underset{k_{1r}}{\Leftrightarrow}} NO + N$$

$$N + O_2 \overset{k_{2f}}{\underset{k_{2r}}{\Leftrightarrow}} NO + O.$$

**Solution**

We first formulate the simplified reactor model. Since the temperature and equilibrium values for $\chi_O$, $\chi_{O_2}$, and $\chi_{N_2}$ will be inputs to the model, we need only write the NO species conservation equation (Eqn. 6.31):

$$\dot{\omega}_{NO} MW_{NO} V - \dot{m} Y_{NO} = 0,$$

where $Y_{NO}$ is the NO mass fraction within the reactor. Rearranging the above and converting $Y_{NO}$ to a molar concentration basis (Eqn. 6A.3) yields

$$\dot{\omega}_{NO} - \frac{\dot{m}}{\rho V}[NO] = 0,$$

or, more simply,

$$\dot{\omega}_{NO} - [NO]/t_R = 0. \tag{I}$$

We now need only to apply the simple Zeldovich mechanism to express $\dot{\omega}_{NO}$ in Eqn. I in terms of presumably known quantities ($T$, $[O_2]_e$, $[O]_e$) and the unknown [NO]. The result is a complex transcendental equation with [NO] as the only unknown variable. Accordingly,

$$\dot{\omega}_{NO} = k_{1f}[O]_e[N_2]_e - k_{2r}[NO][O]_e + [N]_{ss}(k_{2f}[O_2]_e - k_{1r}[NO]). \tag{II}$$

Applying the steady-state approximation for the N atom, we obtain

$$[N]_{ss} = \frac{k_{1f}[O]_e[N_2]_e + k_{2r}[NO][O]_e}{k_{1r}[NO] + k_{2f}[O_2]_e}. \tag{III}$$

Substituting this result for $[N]_{ss}$ back into our expression for $\dot{\omega}_{NO}$ yields

$$\dot{\omega}_{NO} = k_{1f}[O]_e[N_2]_e(Z+1) + k_{2r}[NO][O]_e(Z-1), \tag{IV}$$

where

$$Z = \frac{k_{2f}[O_2]_e - k_{1r}[NO]}{k_{1r}[NO] + k_{2f}[O_2]_e}.$$

Before we can solve Eqn. I for [NO], values for $[O_2]_e$, $[O]_e$, and $[N_2]_e$ are needed. Since several equilibrium expressions are involved, finding these values is nontrivial; however, we can make this job simple by employing the code TPEQUIL (Appendix F). Although the code is set up to deal with fuels of the type $C_xH_yO_z$, neither $x$ nor $y$ may be identically zero, so dealing with pure $H_2$ does not appear possible. We can still utilize the code, however, by setting $x$ equal to unity and $y$ to some very large integer, say $10^6$. The small amount of carbon allows the code to avoid division by zero, without introducing any significant error in the evaluation of the species mole fractions. For example, CO and $CO_2$ concentrations are less than 1 ppm for all calculations performed for this example. The following table shows the equilibrium O, $O_2$, and $N_2$ mole fractions calculated using TPEQUIL:

| $t_R$ (s) | $T_{ad}$ (K) | $\chi_{O,e}$ | $\chi_{O_2,e}$ | $\chi_{N_2,e}$ |
|-----------|--------------|--------------|----------------|----------------|
| 1.0 | 2378 | $5.28 \cdot 10^{-4}$ | $4.78 \cdot 10^{-3}$ | 0.6445 |
| 0.1 | 2366.3 | $4.86 \cdot 10^{-4}$ | $4.60 \cdot 10^{-3}$ | 0.6449 |
| 0.01 | 2298.5 | $2.95 \cdot 10^{-4}$ | $3.65 \cdot 10^{-3}$ | 0.6468 |

Equation I was solved using a Newton–Raphson algorithm, i.e., $[NO]^{new} = [NO]^{old} - f([NO]^{old})/f'([NO]^{old})$, where $f([NO]) (= 0)$ represents Eqn. I, implemented by spreadsheet software. Within the spreadsheet, the input mole fractions were converted to molar concentrations, for example, $[O]_e = \chi_{O,e} P/(R_u T)$, and rate coefficients were evaluated from the following expressions presented in Chapter 5:

$$k_{1f} = 1.8 \cdot 10^{11} \exp[-38{,}370/T(K)],$$

$$k_{1r} = 3.8 \cdot 10^{10} \exp[-425/T(K)],$$

$$k_{2f} = 1.8 \cdot 10^7 T(K) \exp[-4680/T(K)],$$

$$k_{2r} = 3.8 \cdot 10^6 T(K) \exp[-20{,}820/T(K)].$$

The results of this iterative solution for the three residence times are shown below, along with the full-chemistry results generated in Example 6.3 for comparison. These results also are plotted in Fig. 6.10.

| $t_R$ (s) | Equilibrium O-atom Assumption | | Full Chemistry | |
|-----------|----------------------|-----------------|-----------------|-----------------|
| | $\chi_{O,e}$ (ppm) | $\chi_{NO}$ (ppm) | $\chi_O$ (ppm) | $\chi_{NO}$ (ppm) |
| 1.0 | 528 | 2,473 | 549 | 2,459 |
| 0.1 | 486 | 1,403 | 665 | 2,044 |
| 0.01 | 295 | 162 | 1,419 | 744 |

**Comments**

Figure 6.10 shows the dramatic departure of the kinetically derived O-atom concentration from the corresponding equilibrium values as residence times are reduced below 1 s. This is a result of there being insufficient time for the recombination reactions to form stable species from the radicals. In turn, the superequilibrium O-atom concentration results in greater NO production; for example, at $t_R = 0.01$ s, the NO concentration predicted by the full kinetics is nearly five times that based on the assumption of equilibrium O atoms.

# PLUG-FLOW REACTOR

## Assumptions

A plug-flow reactor represents an ideal reactor that has the following attributes:

1.  Steady-state, steady flow.

2.  No mixing in the axial direction. This implies that molecular and/or turbulent mass diffusion is negligible in the flow direction.

3.  Uniform properties in the direction perpendicular to the flow, i.e., one-dimensional flow. This means that at any cross section, a single velocity, temperature, composition, etc., completely characterize the flow.

4.  Ideal frictionless flow. This assumption allows the use of the simple Euler equation to relate pressure and velocity.

5.  Ideal-gas behavior. This assumption allows simple state relations to be employed to relate $T$, $P$, $\rho$, $Y_i$, and $h$.

## Application of Conservation Laws

Our goal here is to develop a system of first-order ODEs whose solution describes the reactor flow properties, including composition, as functions of distance, $x$. The geometry and coordinate definition are schematically illustrated at the top of Fig. 6.11. Table 6.1 provides an overview of the analysis listing the physical and chemical principles that generate $6 + 2N$ equations and a like number of unknown variables and functions. The number of unknowns could be easily reduced by $N$, by recognizing that the species production rates, $\dot{\omega}_i$, can be immediately expressed in terms of the mass fractions (see Appendix 6A) without the need to explicitly involve the $\dot{\omega}_1$. Explicitly

**Table 6.1**      Overview of relationships and variables for plug-flow reactor with $N$ species

| Source of Equations | Number of Equations | Variables or Derivatives Involved |
| --- | --- | --- |
| Fundamental conservation principles: mass, $x$-momentum, energy, species | $3 + N$ | $\dfrac{d\rho}{dx}, \dfrac{dv_x}{dx}, \dfrac{dP}{dx}, \dfrac{dh}{dx}, \dfrac{dY_i}{dx}\ (i = 1, 2, \ldots, N), \dot{\omega}_i\ (i = 1, 2, \ldots, N)$ |
| Mass action laws | $N$ | $\dot{\omega}_i\ (i = 1, 2, \ldots, N)$ |
| Equation of state | $1$ | $\dfrac{d\rho}{dx}, \dfrac{dP}{dx}, \dfrac{dT}{dx}, \dfrac{dMW_{\text{mix}}}{dx}$ |
| Calorific equation of state | $1$ | $\dfrac{dh}{dx}, \dfrac{dT}{dx}, \dfrac{dY_i}{dx}\ (i = 1, 2, \ldots, N)$ |
| Definition of mixture molecular weight | $1$ | $\dfrac{dMW_{\text{mix}}}{dx}, \dfrac{dY_i}{dx}\ (i = 1, 2, \ldots, N)$ |

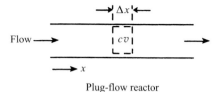

Plug-flow reactor

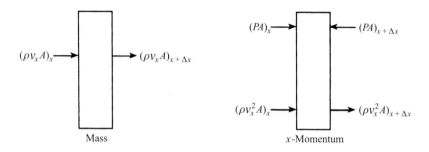

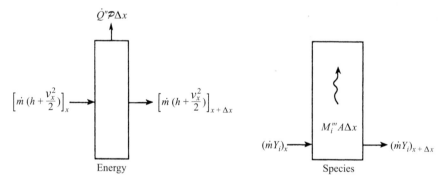

**Figure 6.11**    Control volumes showing fluxes of mass, x-momentum, energy, and species for a plug-flow reactor.

retaining them, however, clearly reminds us of the importance of chemical reactions in our analysis. Although not shown in Table 6.1, the following parameters are treated as known quantities, or functions, and are necessary to obtain a solution: $\dot{m}$, $k_i(T)$, $A(x)$, and $\dot{Q}''(x)$. The area function $A(x)$ defines the cross-sectional area of the reactor as a function of $x$; thus, our model reactor could represent a nozzle, or a diffuser, or any particular one-dimensional geometry, and not just a constant cross-sectional device as suggested by the top sketch in Fig. 6.11. The heat flux function $\dot{Q}''(x)$, although explicitly indicating that the wall heat flux is known, is also intended to indicate that the heat flux may be calculated from a given wall-temperature distribution.

With reference to the fluxes and control volumes illustrated in Fig. 6.11, we can easily derive the following conservation relationships:

**Mass Conservation**

$$\frac{d(\rho v_x A)}{dx} = 0. \tag{6.39}$$

**x-Momentum Conservation**

$$\frac{dP}{dx} + \rho v_x \frac{dv_x}{dx} = 0. \tag{6.40}$$

**Energy Conservation**

$$\frac{d\left(h + v_x^2/2\right)}{dx} + \frac{\dot{Q}''\mathcal{P}}{\dot{m}} = 0. \tag{6.41}$$

**Species Conservation**

$$\frac{dY_i}{dx} - \frac{\dot{\omega}_i MW_i}{\rho v_x} = 0. \tag{6.42}$$

The symbols $v_x$ and $\mathcal{P}$ represent the axial velocity and local perimeter of the reactor, respectively. All of the other quantities have been defined previously. The derivation of these equations is left as an exercise for the reader (see problem 6.1).

To obtain a useful form of the equations where the individual variable derivatives can be isolated, Eqns. 6.39 and 6.41 can be expanded and rearranged to yield the following:

$$\frac{1}{\rho}\frac{d\rho}{dx} + \frac{1}{v_x}\frac{dv_x}{dx} + \frac{1}{A}\frac{dA}{dx} = 0 \tag{6.43}$$

$$\frac{dh}{dx} + v_x \frac{dv_x}{dx} + \frac{\dot{Q}''\mathcal{P}}{\dot{m}} = 0. \tag{6.44}$$

The $\dot{\omega}_i$s appearing in Eqn. 6.42 can be expressed using Eqn. 4.31, with the $[X_i]$s transformed to $Y_i$s.

The functional relationship of the ideal-gas calorific equation of state,

$$h = h(T, Y_i), \tag{6.45}$$

can be exploited using the chain rule to relate $dh/dx$ and $dT/dx$, yielding

$$\frac{dh}{dx} = c_p \frac{dT}{dx} + \sum_{i=1}^{N} h_i \frac{dY_i}{dx}. \tag{6.46}$$

To complete our mathematical description of the plug-flow reactor, we differentiate the ideal-gas equation of state,

$$P = \rho R_u T / MW_{mix}, \tag{6.47}$$

to yield

$$\frac{1}{P}\frac{\mathrm{d}P}{\mathrm{d}x} = \frac{1}{\rho}\frac{\mathrm{d}\rho}{\mathrm{d}x} + \frac{1}{T}\frac{\mathrm{d}T}{\mathrm{d}x} - \frac{1}{MW_{\mathrm{mix}}}\frac{\mathrm{d}MW_{\mathrm{mix}}}{\mathrm{d}x}, \tag{6.48}$$

where the mixture molecular weight derivative follows simply from its definition expressed in terms of species mass fractions, i.e.,

$$MW_{\mathrm{mix}} = \left[\sum_{i=1}^{N} Y_i/MW_i\right]^{-1} \tag{6.49}$$

and

$$\frac{\mathrm{d}MW_{\mathrm{mix}}}{\mathrm{d}x} = -MW_{\mathrm{mix}}^2 \sum_{i=1}^{N} \frac{1}{MW_i}\frac{\mathrm{d}Y_i}{\mathrm{d}x}. \tag{6.50}$$

Equations 6.40, 6.42, 6.43, 6.44, 6.46, 6.48, and 6.49 contain in a linear fashion the derivatives $\mathrm{d}\rho/\mathrm{d}x$, $\mathrm{d}v_x/\mathrm{d}x$, $\mathrm{d}P/\mathrm{d}x$, $\mathrm{d}h/\mathrm{d}x$, $\mathrm{d}Y_i/\mathrm{d}x$ ($i = 1, 2, \ldots, N$), $\mathrm{d}T/\mathrm{d}x$, and $\mathrm{d}MW_{\mathrm{mix}}/\mathrm{d}x$. The number of equations can be reduced by eliminating some of the derivatives by substitution. One logical choice is to retain the derivatives $\mathrm{d}T/\mathrm{d}x$, $\mathrm{d}\rho/\mathrm{d}x$, and $\mathrm{d}Y_i/\mathrm{d}x$ ($i = 1, 2, \ldots, N$). With this choice, the following equations constitute the system of ODEs that must be integrated starting from an appropriate set of initial conditions:

$$\frac{\mathrm{d}\rho}{\mathrm{d}x} = \frac{\left(1 - \dfrac{R_u}{c_p MW_{\mathrm{mix}}}\right)\rho^2 v_x^2 \left(\dfrac{1}{A}\dfrac{\mathrm{d}A}{\mathrm{d}x}\right) + \dfrac{\rho R_u}{v_x c_p MW_{\mathrm{mix}}}\displaystyle\sum_{i=1}^{N} MW_i\dot{\omega}_i\left(h_i - \dfrac{MW_{\mathrm{mix}}}{MW_i}c_p T\right)}{P\left(1 + \dfrac{v_x^2}{c_p T}\right) - \rho v_x^2}, \tag{6.51}$$

$$\frac{\mathrm{d}T}{\mathrm{d}x} = \frac{v_x^2}{\rho c_p}\frac{\mathrm{d}\rho}{\mathrm{d}x} + \frac{v_x^2}{c_p}\left(\frac{1}{A}\frac{\mathrm{d}A}{\mathrm{d}x}\right) - \frac{1}{v_x \rho c_p}\sum_{i=1}^{N} h_i\dot{\omega}_i MW_i, \tag{6.52}$$

$$\frac{\mathrm{d}Y_i}{\mathrm{d}x} = \frac{\dot{\omega}_i MW_i}{\rho v_x} \tag{6.53}$$

Note that in Eqns. 6.41 and 6.52, $\dot{Q}''$ has been set to zero for simplicity.

A residence time, $t_R$, can also be defined, and one more equation added to the set:

$$\frac{\mathrm{d}t_R}{\mathrm{d}x} = \frac{1}{v_x}. \tag{6.54}$$

Initial conditions necessary to solve Eqns. 6.51–6.54 are

$$T(0) = T_0, \tag{6.55a}$$

$$\rho(0) = \rho_0, \tag{6.55b}$$

$$Y_i(0) = Y_{i0} \qquad i = 1, 2, \ldots, N, \tag{6.55c}$$

$$t_R(0) = 0. \tag{6.55d}$$

In summary, we see that the mathematical description of the plug-flow reactor is similar to the constant-pressure and constant-volume reactor models in that all three result in a coupled set of ordinary differential equations; the plug-flow reactor variables, however, are expressed as functions of a spatial coordinate rather than time.

## APPLICATIONS TO COMBUSTION SYSTEM MODELING

Various combinations of well-stirred reactors and plug-flow reactors are frequently used to approximate more complex combustion systems. A simple illustration of this approach is shown in Fig. 6.12. Here we see a gas-turbine combustor modeled as two well-stirred reactors and a plug-flow reactor, all in series, with provisions for some recycle (recirculation) of combustion products in the first reactor, which represents the primary zone (see Fig. 10.4a in Chapter 10). The secondary zone and dilution zones are modeled by the second well-stirred reactor and the plug-flow reactor, respectively. To accurately model a real combustion device, many reactors may be required, with judicious selection of the proportioning of the various flows into each reactor. This approach relies much on the art and craft of an experienced designer to achieve useful results. Reactor modeling approaches are often used to complement more sophisticated finite-difference or finite-element numerical models of turbine combustors, furnaces, and boilers, etc.

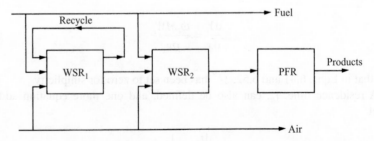

**Figure 6.12**     Conceptual model of gas-turbine combustor using a combination of well-stirred and plug-flow reactors.
SOURCE: After Ref. [13].

## SUMMARY

In this chapter, four model reactors were explored: a constant-pressure reactor, a constant-volume reactor, a well-stirred reactor, and a plug-flow reactor. A description of each of these systems was developed from fundamental conservation principles and linked to chemical kinetics. You should be familiar with these principles and be able to apply them to the model reactors. A numerical example of a constant-volume reactor was developed employing three species (fuel, oxidizer, and products) with one-step global kinetics and simplified thermochemistry. With the model, some characteristics of thermal explosions were elucidated and related to autoignition (knock) in reciprocating engines. As a second example, an equally simple numerical model of a well-stirred reactor was developed. This model was exercised to demonstrate the concept of blowout and the dependence of the blowout mass flowrate on equivalence ratio. With a firm grasp of these simple models, you should be in a good position to understand more complex and more rigorous analyses of combustion systems. Moreover, these simple models frequently are useful as a first step in analyzing many real devices.

## NOMENCLATURE

| | |
|---|---|
| $A$ | Area ($m^2$) |
| $A/F$ | Mass air–fuel ratio (kg/kg) |
| $c_p, \bar{c}_p$ | Constant-pressure specific heat (J/kg-K or J/kmol-K) |
| $c_v, \bar{c}_v$ | Constant-volume specific heat (J/kg-K or J/kmol-K) |
| $h_f^o, \bar{h}_f^o$ | Enthalpy of formation (J/kg or J/kmol) |
| $h, \bar{h}, H$ | Enthalpy (J/kmol or J/kg or J) |
| $k$ | Chemical kinetic rate coefficient (various units) |
| $m$ | Mass (kg) or reaction order with respect to fuel |
| $\dot{m}$ | Mass flowrate (kg/s) |
| $\dot{m}'''$ | Volumetric mass production rate (kg/s-$m^3$) |
| $MW$ | Molecular weight (kg/kmol) |
| $n$ | Reaction order with respect to oxygen |
| $N$ | Number of moles |
| $P$ | Pressure (Pa) |
| $\mathcal{P}$ | Perimeter (m) |
| $\dot{Q}$ | Heat transfer rate (W) |
| $\dot{Q}''$ | Heat flux (W/$m^2$) |
| $R_u$ | Universal gas constant (J/kmol-K) |
| $t$ | Time (s) |
| $T$ | Temperature (K) |
| $u, \bar{u}, U$ | Internal energy (J/kmol or J/kg or J) |
| $v$ | Velocity (m/s) |
| $v$ | Specific volume ($m^3$/kg) |
| $V$ | Volume ($m^3$) |
| $\mathbf{V}$ | Velocity vector (m/s) |
| $\dot{W}$ | Power (W) |

| $x$ | Distance (m) |
|---|---|
| $Y$ | Mass fraction (kg/kg) |

**Greek Symbols**

| $\gamma$ | Specific heat ratio, $c_p/c_v$ |
|---|---|
| $\rho$ | Density (kg/m$^3$) |
| $\Phi$ | Equivalence ratio |
| $\chi$ | Mole fraction (kmol/kmol) |
| $\dot{\omega}$ | Species production rate (kmol/s-m$^3$) |

**Subscripts**

| $ad$ | Adiabatic |
|---|---|
| BDC | Bottom-dead-center |
| $cv$ | Control volume |
| $e$ | Equilibrium |
| $f$ | Forward |
| $F$ | Fuel |
| $G$ | Global |
| $i$ | $i$th species |
| in | Inlet condition |
| mix | Mixture |
| out | Outlet condition |
| $Ox$ | Oxidizer |
| $Pr$ | Product |
| $r$ | Reverse |
| ref | Reference state |
| $R$ | Residence |
| $s$ | Stoichiometric |
| $ss$ | Steady state |
| TDC | Top-dead-center |
| $x$ | $x$-Direction |
| 0 | Initial |

**Other**

| $[X]$ | Molar concentration of species $X$ (kmol/m$^3$) |
|---|---|

# REFERENCES

1. Douaud, A., and Eyzat, P., "DIGITAP—An On-Line Acquisition and Processing System for Instantaneous Engine Data—Applications," SAE Paper 770218, 1977.

2. Nakajima, Y., *et al.*, "Analysis of Combustion Patterns Effective in Improving Anti-Knock Performance of a Spark-Ignition Engine," *Japan Society of Automotive Engineers Review,* 13: 9–17 (1984).

3. Litzinger, T. A., "A Review of Experimental Studies of Knock Chemistry in Engines," *Progress in Energy and Combustion Science,* 16: 155–167 (1990).

4. Spalding, D. B., *Combustion and Mass Transfer,* Pergamon, New York, 1979.

5. Longwell, J. P., and Weiss, M. A., "High Temperature Reaction Rates in Hydrocarbon Combustion," *Industrial & Engineering Chemistry,* 47: 1634–1643 (1955).

6. Glarborg, P., Miller, J. A., and Kee, R. J., "Kinetic Modeling and Sensitivity Analysis of Nitrogen Oxide Formation in Well-Stirred Reactors," *Combustion and Flame,* 65: 177–202 (1986).

7. Duterque, J., Avezard, N., and Borghi, R., "Further Results on Nitrogen Oxides Production in Combustion Zones," *Combustion Science and Technology,* 25: 85–95 (1981).

8. Malte, P. C., Schmidt, S. C., and Pratt, D. T., "Hydroxyl Radical and Atomic Oxygen Concentrations in High-Intensity Turbulent Combustion," *Sixteenth Symposium (International) on Combustion,* The Combustion Institute, Pittsburgh, PA, p. 145, 1977.

9. Bradley, D., Chin, S. B., and Hankinson, G., "Aerodynamic and Flame Structure within a Jet-Stirred Reactor," *Sixteenth Symposium (International) on Combustion,* The Combustion Institute, Pittsburgh, PA, p. 1571, 1977.

10. Chomiak, J., *Combustion: A Study in Theory, Fact and Application,* Gordon & Breach, New York, p. 334, 1990.

11. Zeldovich, Y. B., and Voyevodzkii, V. V., *Thermal Explosion and Flame Propagation in Gases,* Izd. MMI, Moscow, 1947.

12. Glarborg, P., Kee, R. J., Grcar, J. F., and Miller, J. A., "PSR: A Fortran Program for Modeling Well-Stirred Reactors," Sandia National Laboratories Report SAND86-8209, 1986.

13. Swithenbank, J., Poll, I., Vincent, M. W., and Wright, D. D., "Combustion Design Fundamentals," *Fourteenth Symposium (International) on Combustion,* The Combustion Institute, Pittsburgh, PA, p. 627, 1973.

14. Dryer, F. L., and Glassman, I., "High-Temperature Oxidation of CO and $H_2$," *Fourteenth Symposium (International) on Combustion,* The Combustion Institute, Pittsburgh, PA, p. 987, 1972.

15. Westbrook, C. K., and Dryer, F. L., "Simplified Reaction Mechanisms for the Oxidation of Hydrocarbon Fuels in Flames," *Combustion Science and Technology,* 27: 31–43 (1981).

16. Kee, R. J., Rupley, F. M., and Miller, J. A., "Chemkin-II: A Fortran Chemical Kinetics Package for the Analysis of Gas-Phase Chemical Kinetics," Sandia National Laboratories Report SAND89-8009, March 1991.

17. Heywood, J. B., *Internal Combustion Engine Fundamentals,* McGraw-Hill, New York, 1988.

18. Incropera, F. P., and DeWitt, D. P., *Fundamentals of Heat and Mass Transfer,* 3rd Ed., John Wiley & Sons, New York, p. 496, 1990.

## PROBLEMS AND PROJECTS

**6.1**  Derive the basic differential conservation equations for the plug-flow reactor (Eqns. 6.39–6.42) using Fig. 6.11 as a guide. *Hint:* This is relatively straightforward and does not involve much manipulation.

**6.2**    Show that

$$\frac{d(\rho v_x A)}{dx} = 0 = \frac{1}{\rho}\frac{d\rho}{dx} + \cdots, \text{etc.} \qquad \text{(see Eqn. 6.43).}$$

**6.3**    Show that

$$\frac{d}{dx}\left(P = \rho\frac{R_u T}{MW_{mix}}\right) \Rightarrow \frac{1}{P}\frac{dP}{dx} = \frac{1}{\rho}\frac{d\rho}{dx} + \cdots, \text{etc.} \qquad \text{(see Eqn. 6.48).}$$

**6.4**    Show that

$$\frac{dMW_{mix}}{dx} = -MW_{mix}^2 \sum_i (dY_i/dx)MW_i^{-1}.$$

**6.5\***    A.   Use MATHEMATICA or other symbolic manipulation software to verify Eqns. 6.51–6.53.

  B.   Add the heat flux distribution, $\dot{Q}''(x)$, to the problem defined by Eqns. 6.51–6.53.

**6.6**    In the well-stirred-reactor literature, a "reactor loading parameter" is frequently encountered. This single parameter lumps together the effect of pressure, mass flowrate, and reactor volume. Can you identify (create) such a parameter for the well-stirred-reactor model developed in Example 6.2? *Hint:* The parameter is expressed as $P^a \dot{m}^b V^c$. Find the exponents $a$, $b$, and $c$.

**6.7**    Consider a nonadiabatic well-stirred reactor with simplified chemistry, i.e., fuel, oxidizer, and a single product species. The reactants, consisting of fuel ($Y_F = 0.2$) and oxidizer ($Y_{Ox} = 0.8$) at 298 K, flow into the 0.003-m³ reactor at 0.5 kg/s. The reactor operates at 1 atm and has a heat loss of 2000 W. Assume the following simplified thermodynamic properties: $c_p$ = 1100 J/kg-K (all species), $MW = 29$ kg/kmol (all species), $h_{f,F}^o = -2000$ kJ/kg, $h_{f,Ox}^o = 0$, and $h_{f,Pr}^o = -4000$ kJ/kg. The fuel and oxidizer mass fractions in the outlet stream are 0.001 and 0.003, respectively. Determine the temperature in the reactor and the residence time.

**6.8**    Consider the combustion of a fuel and oxidizer in an adiabatic plug-flow reactor of constant cross-sectional area. Assume that the reaction is a single-step reaction with the following stoichiometry and kinetics: $1 \text{ kg}_F + v\text{kg}_{Ox} \rightarrow (1 + v)\text{kg}_{Pr}$, and $\dot{\omega}_F = -A\exp(-E_a/R_u T)[F][Ox]$. Assume also the following simplified thermodynamic properties: $MW_F = MW_{Ox} = MW_{Pr}$, $c_{p,F} = c_{p,Ox} = c_{p,Pr}$ = constant, $h_{f,Ox}^o = h_{f,Pr}^o = 0$, and $h_{f,F}^o = \Delta h_c$.

  Develop a conservation-of-energy relationship in which the temperature, $T$, is the principal dependent variable. Express all concentration variables, or parameters that depend on concentrations, in terms of mass fractions, $Y_i$. The only unknown functions that should appear in your final

---

˙Indicates required use of computer.

result are $T$, $Y_i$, and the axial velocity, $v_x$, each being a function of the axial coordinate $x$. For simplicity, neglect kinetic energy changes and assume that the pressure is essentially constant. *Hint*: You may find species conservation useful.

**6.9*** Create a computer code embodying the simple constant-volume reactor developed in Example 6.1. Verify that it reproduces the results shown in Fig. 6.4 and then use the model to explore the effects of $P_0$, $T_0$, and $\Phi$ on combustion times and maximum rates-of-pressure rise. Discuss your results. *Hint:* You will need to decrease the time interval between output printings when combustion rates are rapid.

**6.10*** Develop a constant-pressure-reactor model using the same chemistry and thermodynamics as in Example 6.1. Using an initial volume of 0.008 m³, explore the effects of $P$ and $T_0$ on combustion durations. Use $\Phi = 1$ and assume the reactor is adiabatic.

**6.11*** Develop a plug-flow-reactor model using the same chemistry and thermodynamics as in Example 6.1. Assume the reactor is adiabatic. Use the model to

    A. Determine the mass flowrate such that the reaction is 99 percent complete in a flow length of 10 cm for $T_{in} = 1000$ K, $P_{in} = 0.2$ atm, and $\Phi = 0.2$. The circular duct has a diameter of 3 cm.

    B. Explore the effects of $P_{in}$, $T_{in}$, and $\Phi$ on the flow length required for 99 percent complete combustion using the flowrate determined in Part A.

**6.12*** Develop a model of the combustion of carbon monoxide with moist air in a constant-volume adiabatic reactor. Assume the following global mechanism of Dryer and Glassman [14] applies:

$$CO + \tfrac{1}{2}O_2 \underset{k_r}{\overset{k_f}{\Longleftrightarrow}} CO_2,$$

where the forward and reverse reaction rates are expressed as

$$\frac{d[CO]}{dt} = -k_f [CO][H_2O]^{0.5}[O_2]^{0.25}$$

$$\frac{d[CO_2]}{dt} = -k_r [CO_2],$$

where

$$k_f = 2.24 \cdot 10^{12} \left[ \left( \frac{kmol}{m^3} \right)^{-0.75} \frac{1}{s} \right] \exp\left[ \frac{-1.674 \cdot 10^8 \, (J/kmol)}{R_u T(K)} \right]$$

$$k_r = 5.0 \cdot 10^8 \left( \frac{1}{s} \right) \exp\left[ \frac{-1.674 \cdot 10^8 \, (J/kmol)}{R_u T(K)} \right].$$

In your model, assume constant (but not equal) specific heats evaluated at 2000 K.

A. Write out all the necessary equations to describe your model, explicitly expressing them in terms of the molar concentrations of CO, $CO_2$, $H_2O$, $O_2$, and $N_2$; individual $\bar{c}_p$ values, $\bar{c}_{p,CO}, \bar{c}_{p,CO_2}$, etc.; and individual enthalpies of formation. Note that the $H_2O$ is a catalyst so its mass fraction is preserved.

B. Exercise your model to determine the influence of the initial $H_2O$ mole fraction (0.1–3.0 percent) on the combustion process. Use maximum rates-of-pressure rise and combustion durations to characterize the process. Use the following initial conditions: $T_0 = 1000$ K, $P = 1$ atm, and $\Phi = 0.25$.

**6.13*** Incorporate the global CO oxidation kinetics given in problem 6.12 in a model of a well-stirred reactor. Assume constant (but not equal) specific heats evaluated at 2000 K.

A. Write out all of the necessary equations to describe your model, explicitly expressing them in terms of the molar concentrations of CO, $CO_2$, $H_2O$, $O_2$, and $N_2$; individual $\bar{c}_p$ values, $\bar{c}_{p,CO}, \bar{c}_{p,CO_2}$, etc.; and individual enthalpies of formation.

B. Exercise your model to determine the influence of the initial $H_2O$ mole fraction (0.1–3.0 percent) on the blowout-limit mass flow-rate. The incoming gases are a stoichiometric mixture of CO and air (plus moisture) at 298 K. The reactor operates at atmospheric pressure.

**6.14*** Incorporate Zeldovich NO formation kinetics in the well-stirred-reactor model presented in Example 6.2. Assume that the NO formation kinetics are *uncoupled* from the combustion process, i.e., the heat lost or evolved from the NO reactions can be neglected, as can the small amount of mass. Assume equilibrium O-atom concentrations.

A. Write out explicitly all of the equations required by your model.

B. Determine the NO mass fraction as a function of $\Phi(0.8–1.1)$ for $\dot{m} = 0.1$ kg/s, $T_{in} = 298$ K, and $P = 1$ atm. The equilibrium constant for

$$\tfrac{1}{2}O_2 \overset{K_p}{\Leftrightarrow} O$$

is given by

$$K_p = 3030\exp(-30{,}790/T).$$

**6.15*** Develop a model of a gas-turbine combustor as two well-stirred reactors in series, where the first reactor represents the primary zone and the second reactor represents the secondary zone. Assume the fuel is decane. Use the following two-step hydrocarbon oxidation mechanism [15]:

$$C_xH_y + \left(\frac{x}{2} + \frac{y}{4}\right)O_2 \xrightarrow{k_F} x\,CO + \frac{y}{2}H_2O$$

$$CO + \tfrac{1}{2}O_2 \underset{k_{CO,r}}{\overset{k_{CO,f}}{\Longleftrightarrow}} CO_2.$$

The rate expressions for the CO oxidation step are given in problem 6.12, and the rate expression for decane conversion to CO is given by

$$\frac{d[C_{10}H_{22}]}{dt} = -k_F[C_{10}H_{22}]^{0.25}[O_2]^{1.5},$$

where

$$k_F = 2.64 \cdot 10^9 \exp\left[\frac{-15{,}098}{T}\right] \quad \text{(SI units)}.$$

A.  Write out all of the governing equations treating the equivalence ratios, $\Phi_1$ and $\Phi_2$, of the two zones as known parameters. Assume constant (but not equal) specific heats.

B.  Write a computer code embodying your model from part A. Perform a design exercise with objectives and constraints provided by your instructor.

**6.16\*** Use CHEMKIN [16] subroutines to model $H_2$–air combustion and thermal NO formation, including superequilibrium O-atom contributions, for the following systems:

A.  A constant-volume reactor.

B.  A plug-flow reactor.

C.  Exercise your models using initial and/or flow conditions provided by your instructor.

**6.17\*** Consider a simple tube furnace as shown on next page in the sketch. A natural gas (methane)–air mixture ($\Phi = 0.9$) is burned in a rapid mixing burner at the inlet of the furnace, and heat is transferred from the hot products of combustion to the constant-temperature wall ($T_w = 600$ K). Assume that the nitric oxide produced by the burner is negligible compared with that produced in the postflame gases in the furnace. The natural gas flowrate is 0.0147 kg/s. Assuming that the hot products enter the furnace at 2350 K and exit the furnace at 1700 K, determine the following, taking into account the constraints given below:

A.  For a furnace diameter of 0.30 m, what is the length of the furnace?

B.  What is the nitric oxide (NO) mole fraction at the exit?

C.  For the same fuel flowrate and stoichiometry, can the NO emissions be reduced by changing the diameter ($D$) of the furnace? (Note that the length ($L$) of the furnace will have to be changed accordingly to maintain the same outlet temperature of 1700 K). What are the values of $D$ and $L$ that result in lower NO if the mean velocity at the entrance to the furnace is constrained to be within a factor of two (up or down) from the original design point? On a single graph, plot $\chi_{NO}$ versus pipe length for each case.

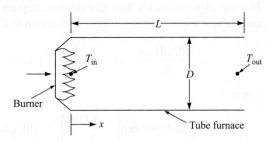

Additional constraints and assumptions are the following:

1.  The pressure is constant at 1 atm.

2.  The flow is fully developed, both hydraulically and thermally, within the tube furnace.

3.  Use the Dittus–Boelter equation [18] $Nu_D = 0.023 \, Re_D^{0.8} \, Pr^{0.3}$ for your heat-transfer analysis, where the properties are evaluated at $(T_{in} + T_{out})/2$. Use air properties to simplify your calculations.

4.  Assume Zeldovich NO kinetics apply. Do not neglect reverse reactions. Use the steady-state approximation for N atoms and assume that $O_2$ and O are in simple equilibrium. Furthermore, assume that the $O_2$ mole fraction is constant at 0.02, even though it changes slightly with temperature. Use the equilibrium constant $K_p = 3.6 \cdot 10^3 \exp[-31{,}090/T \, (K)]$ for the $\frac{1}{2}O_2 \Leftrightarrow O$ equilibrium.

# APPENDIX 6A
# SOME USEFUL RELATIONSHIPS AMONG MASS FRACTIONS, MOLE FRACTIONS, MOLAR CONCENTRATIONS, AND MIXTURE MOLECULAR WEIGHTS

Mole fraction / mass fractions:

$$\chi_i = Y_i MW_{mix}/MW_i \qquad (6A.1)$$

$$Y_i = \chi_i MW_i/MW_{mix}. \qquad (6A.2)$$

Mass fraction / molar concentration:

$$[X_i] = PMW_{\text{mix}}Y_i /(R_u T MW_i) = Y_i \rho / MW_i \tag{6A.3}$$

$$Y_i = \frac{[X_i]MW_i}{\sum\limits_{j}[X_j]MW_j}. \tag{6A.4}$$

Mole fraction / molar concentration:

$$[X_i] = \chi_i P/R_u T = \chi_i \rho / MW_{\text{mix}} \tag{6A.5}$$

$$\chi_i = [X_i]/\sum\limits_{j}[X_j]. \tag{6A.6}$$

Mass concentration:

$$\rho_i = \rho Y_i = [X_i]MW_i. \tag{6A.7}$$

$MW_{\text{mix}}$ defined in terms of mass fractions:

$$MW_{\text{mix}} = \frac{1}{\sum\limits_{i} Y_i / MW_i}. \tag{6A.8}$$

$MW_{\text{mix}}$ defined in terms of mole fractions:

$$MW_{\text{mix}} = \sum\limits_{i} \chi_i MW_i. \tag{6A.9}$$

$MW_{\text{mix}}$ defined in terms of molar concentrations:

$$MW_{\text{mix}} = \frac{\sum\limits_{i}[X_i]MW_i}{\sum\limits_{i}[X_i]}. \tag{6A.10}$$

# Simplified Conservation Equations for Reacting Flows[1]

## OVERVIEW

One of the objectives of this book is to present, in as simple as possible a manner, the essential physics and chemistry of combustion. When one considers the details of multicomponent reacting mixtures, a complex situation arises, both physically and mathematically, which can be somewhat intimidating to a newcomer to the field. The primary objective of this chapter is to present the simplified governing equations expressing the conservation of mass, species, momentum, and energy for reacting flows. In particular, we wish to treat the following three situations:

1. Steady flow for a one-dimensional *planar* (*x*-coordinate only) geometry.
2. Steady flow for a one-dimensional *spherical* (*r*-coordinate only) geometry.
3. Steady flow for a two-dimensional *axisymmetric* (*r*- and *x*-coordinates) geometry.

From the first of these, we will develop an analysis of premixed laminar flames in Chapter 8; from the second, analyses of evaporation and combustion of fuel droplets in Chapter 10; and from the third, analyses of laminar (Chapter 9) and turbulent (Chapter 13) axisymmetric jet flames. These systems and the coordinate geometries are illustrated in Fig. 7.1.

Our approach is first to develop quite simple forms of the conservation equations, usually focusing on one-dimensional cartesian systems, to illustrate the essential physics of each conservation principle. We then present more general relationships from which the basic conservation equations are obtained for the radial and axisymmetric geometries of interest. Although much of the physics can be captured in such simple analyses, it is important to caution that certain interesting and important phenomena will be excluded with this approach. For example, recent research [1] shows that the

---

[1]This chapter may be skipped in its entirety without any loss of continuity. It is recommended that the chapter be treated as a reference when dealing with the simplified fundamental conservation relations employed in various subsequent chapters.

Planar
flame

Spherical
flame

Axisymmetric
flame

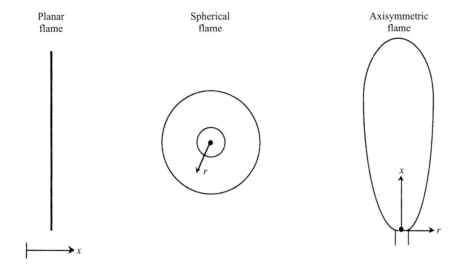

**Figure 7.1**    Coordinate systems for planar flames, spherically symmetric flames (droplet burning), and axisymmetric flames (jet flames).

unequal rates of temperature-gradient-induced diffusion among species (thermal diffusion) in premixed flames have a profound effect on turbulent flame propagation. Therefore, for those desiring a more comprehensive treatment, multicomponent diffusion, including thermal diffusion, is also introduced in this chapter. Furthermore, we extend the development of the energy conservation equation to a form that can be used as a starting point for detailed numerical models of flames.

The use of "conserved scalars" to simplify and analyze certain combustion problems is quite common in the literature. To introduce this concept, we discuss and develop equations for the conserved scalars of mixture fraction and mixture enthalpy.

## OVERALL MASS CONSERVATION (CONTINUITY)

Consider the one-dimensional control volume shown in Fig. 7.2, a plane layer $\Delta x$ thick. Mass enters at $x$ and exits at $x + \Delta x$, with the difference between the flow in and out being the rate at which mass accumulates within the control volume; i.e.,

$$\underset{\substack{\text{Rate of increase} \\ \text{of mass within} \\ \text{control volume}}}{\frac{\mathrm{d}m_{cv}}{\mathrm{d}t}} = \underset{\substack{\text{Mass flow} \\ \text{into the} \\ \text{control volume}}}{[\dot{m}]_x} - \underset{\substack{\text{Mass flow} \\ \text{out of the} \\ \text{control volume}}}{[\dot{m}]_{x+\Delta x}}. \qquad (7.1)$$

Recognizing that the mass within the control volume is $m_{cv} = \rho V_{cv}$, where the volume $V_{cv} = A\Delta x$, and that the mass flowrate is $\dot{m} = \rho v_x A$, we rewrite Eqn. 7.1 as

$$\frac{\mathrm{d}(\rho A\, \Delta x)}{\mathrm{d}t} = [\rho v_x A]_x - [\rho v_x A]_{x+\Delta x}. \qquad (7.2)$$

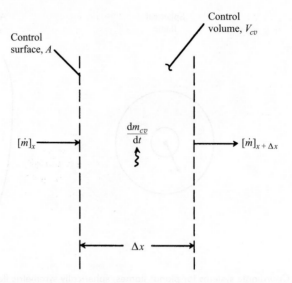

**Figure 7.2**    Control volume for one-dimensional analysis of mass conservation.

Dividing through by $A\Delta x$ and taking the limit as $\Delta x \rightarrow 0$, Eqn. 7.2 becomes

$$\frac{\partial \rho}{\partial t} = -\frac{\partial(\rho v_x)}{\partial x}. \tag{7.3}$$

In the case of steady flow, where $\partial \rho / \partial t = 0$,

$$\boxed{\frac{d(\rho v_x)}{dx} = 0} \tag{7.4a}$$

or

$$\boxed{\rho v_x = \text{constant}} \tag{7.4b}$$

In combustion systems, the density will vary greatly with position in the flow; thus, we see from Eqn. 7.4 that the velocity must also vary with position such that the product $\rho v_x$, the mass flux $\dot{m}''$, remains constant. In its most general form, conservation of mass associated with a fixed point in a flow can be expressed as

$$\frac{\partial \rho}{\partial t} + \boldsymbol{\nabla} \cdot (\rho \mathbf{V}) = 0. \tag{7.5}$$

Rate of gain        Net rate of
of mass per        mass flow out
unit volume        per unit volume

Assuming steady flow and applying the vector operations appropriate for the coordinate system of interest (e.g., see Ref. [2] for a complete compilation), we obtain, first for the spherical system,

$$\frac{1}{r^2}\frac{\partial}{\partial r}\left(r^2\rho v_r\right) + \frac{1}{r\sin\theta}\frac{\partial}{\partial \theta}(\rho v_\theta \sin\theta) + \frac{1}{r\sin\theta}\frac{\partial(\rho v_\phi)}{\partial \phi} = 0,$$

which simplifies for our 1-D spherically symmetric system, where $v_\theta = v_\phi = 0$, and $\partial(\ )/\partial\theta = \partial(\ )/\partial\phi = 0$, to be

$$\frac{1}{r^2}\frac{d}{dr}\left(r^2\rho v_r\right) = 0 \tag{7.6a}$$

or

$$r^2\rho v_r = \text{constant} \tag{7.6b}$$

Equation 7.6b is equivalent to writing $\dot{m} = \text{constant} = \rho v_r A(r)$ where $A(r) = 4\pi r^2$.

For our axisymmetric system with steady flow, the general continuity equation (Eqn. 7.5) yields

$$\frac{1}{r}\frac{\partial}{\partial r}(r\rho v_r) + \frac{\partial}{\partial x}(\rho v_x) = 0 \tag{7.7}$$

which results from setting $v_\theta = 0$ in the complete cylindrical formulation. Note that now, for the first time, two velocity components, $v_r$ and $v_x$, appear, rather than just a single component as in previous analyses.

# SPECIES MASS CONSERVATION (SPECIES CONTINUITY)

In Chapter 3, we derived the one-dimensional species conservation equation with the assumptions that species diffused only as a result of concentration gradients and that the mixture comprised only two species, i.e., a binary mixture. We will not repeat that development here, but rather will restate our final result (Eqn. 3.31), which for steady flow is written as

$$\frac{d}{dx}\left[\dot{m}''Y_A - \rho\mathcal{D}_{AB}\frac{dY_A}{dx}\right] = \dot{m}'''_A$$

or

$$\frac{d}{dx}(\dot{m}''Y_A) - \frac{d}{dx}\left(\rho\mathcal{D}_{AB}\frac{dY_A}{dx}\right) = \dot{m}'''_A \tag{7.8}$$

| Mass flow of species A due to convection (advection by bulk flow) per unit volume (kg/s-m³) | Mass flow of species A due to molecular diffusion per unit volume (kg/s-m³) | Net mass production rate of species A by chemical reaction per unit volume (kg/s-m³) |

where $\dot{m}''$ is the mass flux $\rho v_x$, and $\dot{m}_A'''$ is the net production rate of species A per unit volume associated with chemical reaction. A more general one-dimensional form of species continuity can be expressed as

$$\boxed{\frac{d\dot{m}_i''}{dx} = \dot{m}_i''' \quad i = 1, 2, \ldots, N,}$$  (7.9)

where the subscript $i$ represents the $i$th species. In this relation, no restrictions, such as binary diffusion governed by Fick's law, have been imposed to describe the species flux $\dot{m}_i''$.

The general vector form for mass conservation of the $i$th species is expressed as

$$\underset{\substack{\text{Rate of gain of mass} \\ \text{of species } i \text{ per} \\ \text{unit volume}}}{\frac{\partial(\rho Y_i)}{\partial t}} \quad + \quad \underset{\substack{\text{Net rate of mass flow} \\ \text{of species } i \text{ out by} \\ \text{diffusion and bulk} \\ \text{flow per unit} \\ \text{volume}}}{\nabla \cdot \dot{m}_i''} \quad = \quad \underset{\substack{\text{Net rate of mass} \\ \text{production of} \\ \text{species } i \text{ per} \\ \text{unit volume}}}{\dot{m}_i'''} \quad \text{for} \quad i = 1, \quad (7.10)$$

At this juncture, it is worthwhile to digress somewhat to discuss the species mass flux in a bit more detail. The mass flux of $i$, $\dot{m}_i''$, is defined by the mass average velocity of $i$, $\mathbf{v}_i$, as follows:

$$\dot{\mathbf{m}}_i'' \equiv \rho Y_i \mathbf{v}_i,$$  (7.11)

where the **species velocity** $\mathbf{v}_i$ is, in general, a quite complicated expression that takes into account mass diffusion associated with concentration gradients (**ordinary diffusion**), as well as other modes (see next section). The sum of all of the individual species mass fluxes is the mixture mass flux, i.e.,

$$\sum \dot{\mathbf{m}}_i'' = \sum \rho Y_i \mathbf{v}_i = \dot{\mathbf{m}}''.$$  (7.12)

Thus, we see that, since $\dot{\mathbf{m}}'' \equiv \rho \mathbf{V}$, the mass average velocity $\mathbf{V}$ is

$$\mathbf{V} = \sum Y_i \mathbf{v}_i.$$  (7.13)

This is the fluid velocity with which you are familiar and is referred to as the mass average **bulk velocity.** The difference between the species velocity and the bulk velocity is defined as the **diffusional velocity,** $\mathbf{v}_{i,\,\text{diff}} \equiv \mathbf{v}_i - \mathbf{V}$, i.e., the velocity of an individual species relative to the bulk velocity. The diffusional mass flux can be expressed in terms of the diffusional velocity:

$$\dot{\mathbf{m}}_{i,\,\text{diff}}'' \equiv \rho Y_i (\mathbf{v}_i - \mathbf{V}) = \rho Y_i \mathbf{v}_{i,\,\text{diff}}.$$  (7.14)

As discussed in Chapter 3, the total species mass flux is the sum of the bulk flow and diffusion contributions, i.e.,

$$\dot{m}''_i = \dot{m}'' Y_i + \dot{m}''_{i,\,\text{diff}} \qquad (7.15a)$$

or, in terms of velocities,

$$\rho Y_i \mathbf{v}_i = \rho Y_i \mathbf{V} + \rho Y_i \mathbf{v}_{i,\,\text{diff}}. \qquad (7.15b)$$

Depending on the direction of the species concentration gradients, the diffusional flux, or velocity, can be directed either against or with the bulk flow. For example, a high concentration of a species downstream creates a diffusional flux upstream against the bulk flow. Using the above definitions (Eqns. 7.11 and 7.14), our general species conservation equation (Eqn. 7.10) can be rewritten in terms of species diffusion velocities, $\mathbf{v}_{i,\,\text{diff}}$, and mass fractions, $Y_i$:

$$\frac{\partial(\rho Y_i)}{\partial t} + \nabla \cdot [\rho Y_i (\mathbf{V} + \mathbf{v}_{i,\,\text{diff}})] = \dot{m}'''_i \quad \text{for} \quad i = 1, 2, \ldots, N. \qquad (7.16)$$

This form frequently appears in the literature and is the formulation employed in the various combustion codes developed at Sandia National Laboratories (e.g., Refs. [3] and [4]). We end our digression here and return to the development of the simplified species conservation equations for the spherical and axisymmetrical geometries.

For the case of ordinary diffusion only (no thermal or pressure diffusion) in a binary mixture, the general form of Fick's law given below can be used to evaluate the species mass flux, $\dot{m}''_i$, which appears in our general species conservation relation (Eqn. 7.10):

$$\dot{m}''_A = \dot{m}'' Y_A - \rho \mathcal{D}_{AB} \nabla Y_A. \qquad (7.17)$$

For the spherically symmetric system with steady flow, Eqn. 7.10 becomes

$$\frac{1}{r^2} \frac{d}{dr}\left(r^2 \dot{m}''_i\right) = \dot{m}'''_i \quad i = 1, 2, \ldots, N \qquad (7.18)$$

or, with the assumption of binary diffusion, Eqn. 7.17,

$$\boxed{\frac{1}{r^2} \frac{d}{dr}\left[r^2\left(\rho v_r Y_A - \rho \mathcal{D}_{AB} \frac{dY_A}{dr}\right)\right] = \dot{m}'''_A} \qquad (7.19)$$

The physical interpretation of the above relation is the same as previously shown (Eqn. 7.8) except that the mass flow of species A is directed in the radial direction, rather than in the $x$-direction.

For the axisymmetric geometry ($r$-, and $x$-coordinates), the corresponding species conservation equation for a binary mixture is

$$
\underbrace{\frac{1}{r}\frac{\partial}{\partial r}(r\rho v_r Y_A)}_{\substack{\text{Mass flow of species A} \\ \text{due to radial convection} \\ \text{(radial advection by bulk flow)} \\ \text{per unit volume } (kg_A/s\text{-}m^3)}}
+
\underbrace{\frac{1}{r}\frac{\partial}{\partial x}(r\rho v_x Y_A)}_{\substack{\text{Mass flow of species A} \\ \text{due to axial convection} \\ \text{(axial advection by bulk flow)} \\ \text{per unit volume } (kg_A/s\text{-}m^3)}}
$$

$$
\underbrace{-\frac{1}{r}\frac{\partial}{\partial r}\left[r\rho \mathcal{D}_{AB}\frac{\partial Y_A}{\partial r}\right]}_{\substack{\text{Mass flow of species A} \\ \text{due to molecular diffusion} \\ \text{in radial direction per unit volume} \\ (kg_A/s\text{-}m^3)}}
=
\underbrace{\dot{m}_A'''}_{\substack{\text{Net mass production rate} \\ \text{of species A by chemical} \\ \text{reaction per unit volume} \\ (kg_A/s\text{-}m^3)}}
\tag{7.20}
$$

In the above equation, we have assumed that axial diffusion is negligible in comparison with radial diffusion and with both axial and radial convection (advection).

## MULTICOMPONENT DIFFUSION

In modeling and understanding the *details* of many combustion systems, particularly the structure of laminar premixed and nonpremixed flames, the problem cannot be reduced to a simple representation involving a binary mixture. In these cases, the formulation of the species transport laws must account for the fact that there are a large number of species present and that the properties of the individual species may be greatly different. For example, we might expect heavy fuel molecules to diffuse much less rapidly than lightweight H atoms. Furthermore, large temperature gradients, typically found in flames, produce a second driving force for mass transfer in addition to that produced by concentration gradients. This temperature-gradient-driven mass diffusion, referred to as **thermal diffusion** or the **Soret effect,** results in light molecules diffusing from low- to high-temperature regions, and, conversely, heavy molecules diffusing from high- to low-temperature regions.

We begin our discussion of multicomponent diffusion by presenting some of the most general relations used to express individual species mass diffusion fluxes and/or diffusion velocities. These relations are then simplified to the forms most pertinent to combustion applications. With yet more restrictive assumptions, we obtain relatively simple approximate expressions that are sometimes used to model multicomponent diffusion in flame systems.

### General Formulations

The general problem of species diffusion in multicomponent mixtures allows for four distinct modes of mass diffusion: **ordinary diffusion,** resulting from concentration gradients; **thermal (or Soret) diffusion,** resulting from temperature gradients; **pressure diffusion,** resulting from pressure gradients; and **forced diffusion,** resulting from unequal body forces per unit mass among the species. The mass fluxes associated with each diffusion mode are additive; thus,

$$\dot{m}''_{i,\,\text{diff}} = \dot{m}''_{i,\,\text{diff},\,\chi} + \dot{m}''_{i,\,\text{diff},\,T} + \dot{m}''_{i,\,\text{diff},\,P} + \dot{m}''_{i,\,\text{diff},\,f}, \qquad (7.21a)$$

where the subscripts $\chi$, $T$, $P$, and $f$ refer, respectively, to ordinary, thermal, pressure, and forced diffusion. Similarly, the diffusion velocities add vectorially:

$$\mathbf{v}_{i,\,\text{diff}} = \mathbf{v}_{i,\,\text{diff},\,\chi} + \mathbf{v}_{i,\,\text{diff},\,T} + \mathbf{v}_{i,\,\text{diff},\,P} + \mathbf{v}_{i,\,\text{diff},\,f}. \qquad (7.21b)$$

In typical combustion systems, pressure gradients are not sufficiently large to induce pressure diffusion, so we can neglect this effect. Forced diffusion results primarily from charged species (ions) interacting with an electric field; although ions do exist in small concentrations in flames, forced diffusion is not thought to be significant. In the following development, therefore, we retain only the ordinary and thermal contributions to the diffusion flux. For a thorough treatment of all aspects of diffusion, the reader is referred to Refs. [2] and [5–7].

With the assumption of ideal-gas behavior, the most general expressions for ordinary diffusion simplify to the following [2]:

$$\dot{m}''_{i,\,\text{diff},\,\chi} = \frac{P}{R_u T} \frac{MW_i}{MW_{\text{mix}}} \sum_{j=1}^{N} MW_j \, \mathrm{D}_{ij} \boldsymbol{\nabla} \chi_j \qquad i = 1, 2, \ldots, N, \qquad (7.22)$$

where $MW_{\text{mix}}$ is the mixture molecular weight and the $\mathrm{D}_{ij}$s are the **ordinary multicomponent diffusion coefficients.** It is important to note that the multicomponent diffusivity, $\mathrm{D}_{ij}$, is not identical to the binary diffusivity, $\mathcal{D}_{ij}$, for the same pair of species. We will discuss the $\mathrm{D}_{ij}$s more fully in a section below. The corresponding expression to Eqn. 7.22 for the diffusion velocity of species $i$ is given by

$$\mathbf{v}_{i,\,\text{diff},\,\chi} = \frac{1}{\chi_i MW_{\text{mix}}} \sum_{j=1}^{N} MW_j \mathrm{D}_{ij} \boldsymbol{\nabla} \chi_j \qquad i = 1, 2, \ldots, N. \qquad (7.23)$$

The **Stefan–Maxwell equation,** an alternative to Eqn. 7.23, eliminates the need to determine the concentration-dependent $\mathrm{D}_{ij}$s by coupling the various species diffusion velocities [2]:

$$\boldsymbol{\nabla} \chi_i = \sum_{j=1}^{N} \left[ \frac{\chi_i \chi_j}{\mathcal{D}_{ij}} (\mathbf{v}_{j,\,\text{diff},\,\chi} - \mathbf{v}_{i,\,\text{diff},\,\chi}) \right] \qquad i = 1, 2, \ldots, N. \qquad (7.24)$$

In Eqn. 7.23, all of the species mole-fraction gradients appear in each of the $N$ equations ($i = 1, 2, \ldots, N$), whereas only the $i$th diffusion velocity appears; conversely, all of the diffusion velocities appear in each of the $N$ equations represented by Eqn. 7.24, whereas only the $i$th mole-fraction gradient appears in each.

The **thermal diffusion velocity** for the $i$th species is expressed as [2]

$$\mathbf{v}_{i,\,\mathrm{diff},\,T} = -\frac{D_i^T}{\rho Y_i}\frac{1}{T}\nabla T, \tag{7.25}$$

where $D_i^T$ is the **thermal diffusion coefficient.** This coefficient may be positive or negative, indicating diffusion towards colder or hotter regions, respectively. For an interesting discussion of thermal diffusion, the reader is referred to Ref. [8].

## Calculation of Multicomponent Diffusion Coefficients

The following, somewhat complex, expressions for the multicomponent diffusion coefficients $D_{ij}$ for ordinary diffusion have been derived from kinetic theory [7, 9]:

$$D_{ij} = \chi_i\,\frac{MW_{\mathrm{mix}}}{MW_j}(F_{ij} - F_{ii}), \tag{7.26}$$

where $F_{ij}$ and $F_{ii}$ are components of the matrix $[F_{ij}]$. The $[F_{ij}]$ matrix is the inverse of $[L_{ij}]$; i.e.,

$$[F_{ij}] = [L_{ij}]^{-1}. \tag{7.27}$$

The components of $[L_{ij}]$ are determined from

$$[L_{ij}] = \sum_{k=1}^{K}\frac{\chi_k}{MW_i\mathcal{D}_{ik}}[MW_j\chi_j(1-\delta_{ik}) - MW_i\chi_i(\delta_{ij}-\delta_{jk})], \tag{7.28}$$

where $\delta_{mn}$ is the Kronecker delta function, which takes on the value of unity for $m = n$ and is otherwise zero. The summation $k = 1$ to $K$ is taken over all species. The multicomponent diffusion coefficients have the following properties [2]:

$$D_{ii} = 0 \tag{7.29a}$$

$$\sum_{i=1}^{N}(MW_i MW_h D_{ih} - MW_i MW_k D_{ik}) = 0. \tag{7.29b}$$

We note in the above that the multicomponent diffusion coefficients depend in a complex way on both the mixture composition, through the appearance of the component mole fractions $\chi_i$, and on the individual binary diffusion coefficients for pairs of species, $\mathcal{D}_{ij}$. Note that, generally, only in the case of a two-component (i.e., binary) mixture do the $D_{ij}$s equal the $\mathcal{D}_{ij}$s. Numerical values on the $\mathcal{D}_{ij}$s can be estimated using the methods presented in Appendix D, whereas software [10] is available, as part of the CHEMKIN library, for the determination of the multicomponent diffusion coefficients $D_{ij}$, along with the other mixture transport properties.

Example 7.1

Determine numerical values for all of the multicomponent diffusion coefficients $D_{ij}$ associated with a mixture of $H_2$, $O_2$, and $N_2$ having the following properties: $\chi_{H_2} = 0.15$, $\chi_{O_2} = 0.20$, $\chi_{N_2} = 0.65$, $T = 600$ K, and $P = 1$ atm.

### Solution

To determine the $D_{ij}$s requires the straightforward application of Eqn. 7.26. This is simple, in principle, but considerable computation is involved. First, we write out the nine components of the $L_{ij}$ matrix (Eqn. 7.28), designating $i$ (and $j$) = 1, 2, and 3 to represent $H_2$, $O_2$, and $N_2$, respectively:

$$[L_{ij}] = \begin{bmatrix} L_{11} & L_{12} & L_{13} \\ L_{21} & L_{22} & L_{23} \\ L_{31} & L_{32} & L_{33} \end{bmatrix},$$

where, taking advantage of the fact that $\mathcal{D}_{ij} = \mathcal{D}_{ji}$ (see Chapter 3),

$$L_{11} = L_{22} = L_{33} = 0,$$
$$L_{12} = \chi_2(MW_2\chi_2 + MW_1\chi_1)/(MW_1\mathcal{D}_{12}) + \chi_3(MW_2\chi_2)/(MW_1\mathcal{D}_{13}),$$
$$L_{13} = \chi_2(MW_3\chi_3)/(MW_1\mathcal{D}_{12}) + \chi_3(MW_3\chi_3 + MW_1\chi_1)/(MW_1\mathcal{D}_{13}),$$
$$L_{21} = \chi_1(MW_1\chi_1 + MW_2\chi_2)/(MW_2\mathcal{D}_{21}) + \chi_3(MW_1\chi_1)/(MW_2\mathcal{D}_{23}),$$
$$L_{23} = \chi_1(MW_3\chi_3)/(MW_2\mathcal{D}_{21}) + \chi_3(MW_3\chi_3 + MW_2\chi_2)/(MW_2\mathcal{D}_{23}),$$
$$L_{31} = \chi_1(MW_1\chi_1 + MW_3\chi_3)/(MW_3\mathcal{D}_{31}) + \chi_2(MW_1\chi_1)/(MW_3\mathcal{D}_{32}),$$
$$L_{32} = \chi_1(MW_2\chi_2)/(MW_3\mathcal{D}_{31}) + \chi_2(MW_2\chi_2 + MW_3\chi_3)/(MW_3\mathcal{D}_{32}).$$

To obtain numerical values for the $L_{ij}$s above, we need to evaluate the binary diffusion coefficients, $\mathcal{D}_{12}$, $\mathcal{D}_{13}$, and $\mathcal{D}_{23}$, which can be accomplished using the method described in Appendix D. The values for the characteristic Lennard–Jones length $\sigma_i$ and energy $\varepsilon_i$ from Appendix Table D.2 are listed below:

| $i$ | Species | $\chi_i$ | $MW_i$ | $\sigma_i$(Å) | $\varepsilon_i/k_B$ (K) |
|---|---|---|---|---|---|
| 1 | $H_2$ | 0.15 | 2.016 | 2.827 | 59.17 |
| 2 | $O_2$ | 0.20 | 32.000 | 3.467 | 106.7 |
| 3 | $N_2$ | 0.65 | 28.014 | 3.798 | 71.4 |

To calculate $\mathcal{D}_{H_2-O_2}$ from Appendix Eqn. D.2 requires us first to evaluate the collision integral $\Omega_D$ for $H_2$ and $O_2$, which, in turn, requires us to determine $\varepsilon_{H_2-O_2}/k_B$ and $T^*$:

$$\varepsilon_{H_2-O_2}/k_B = \left[ (\varepsilon_{H_2}/k_B)(\varepsilon_{O_2}/k_B) \right]^{1/2} = (59.7 \cdot 106.7)^{1/2} = 79.8 \text{ K}$$

$$T^* = k_B T/\varepsilon_{H_2-O_2} = 600/79.8 = 7.519.$$

The collision integral $\Omega_D$ (Appendix Eqn. D.3) is

$$\Omega_D = \frac{1.06036}{(7.519)^{0.15610}} + \frac{0.19300}{\exp(0.47635 \cdot 7.519)}$$
$$+ \frac{1.03587}{\exp(1.52996 \cdot 7.519)} + \frac{1.76474}{\exp(3.89411 \cdot 7.519)} = 0.7793.$$

Other parameters required are

$$\sigma_{H_2-O_2} = \frac{\sigma_{H_2} + \sigma_{O_2}}{2} = \frac{2.827 + 3.467}{2} = 3.147 \text{ Å}$$

$$MW_{H_2-O_2} = 2\left[\left(1/MW_{H_2}\right) + \left(1/M_{O_2}\right)\right]^{-1} = 2[(1/2.016) + (1/32.00)]^{-1} = 3.793.$$

Thus,

$$\mathcal{D}_{H_2-O_2} = \frac{0.0266 T^{3/2}}{P \, MW_{H_2-O_2}^{1/2} \sigma_{H_2-O_2}^2 \Omega_D}$$

$$= \frac{0.0266 \, (600)^{3/2}}{101,325 \, (3.793)^{1/2} \, (3.147)^2 \, 0.7793}$$

$$= 2.5668 \cdot 10^{-4} \text{ m}^2/\text{s}$$

or    $2.5668 \text{ cm}^2/\text{s}.$

The binary diffusion coefficients $\mathcal{D}_{H_2-N_2}$ and $\mathcal{D}_{O_2-N_2}$ are similarly evaluated (spreadsheet software is helpful):

$$\mathcal{D}_{H_2-N_2} = 2.4095 \text{ cm}^2/\text{s} \quad \text{and} \quad \mathcal{D}_{O_2-N_2} = 0.6753 \text{ cm}^2/\text{s}.$$

We calculate the value of the $L_{12}$ element of the $[L]$ matrix:

$$L_{12} = \chi_2 (MW_2\chi_2 + MW_1\chi_1)/(MW_1\mathcal{D}_{12}) + \chi_3(MW_2\chi_2)/(MW_1\mathcal{D}_{13})$$

$$= 0.20 \, (32.000 \cdot 0.20 + 2.016 \cdot 0.15)/(2.016 \cdot 2.5668)$$

$$+ 0.65 \, (32.000 \cdot 0.20)/(2.016 \cdot 2.4095) = 1.1154.$$

The other elements are similarly evaluated to yield

$$[L] = \begin{bmatrix} 0 & 1.1154 & 3.1808 \\ 0.0213 & 0 & 0.7735 \\ 0.0443 & 0.2744 & 0 \end{bmatrix}.$$

The inverse of $[L]$ is obtained with the aid of a computer or calculator:

$$[L_{ij}]^{-1} = [F_{ij}] = \begin{bmatrix} -3.7319 & 15.3469 & 15.1707 \\ 0.6030 & -2.4796 & 1.1933 \\ 0.1029 & 0.8695 & -0.4184 \end{bmatrix}.$$

With the above result, we are finally in a position to calculate the multicomponent diffusion coefficients $D_{ij}$ from Eqn. 7.26, noting that the molecular weight of the mixture $MW_{mix}$ is $24.9115[= 0.15(2.016) + 0.20(32.000) + 0.65(28.014)]$. For example,

$$D_{12} = D_{H_2-O_2} = \chi_1 \frac{MW_{mix}}{MW_2}(F_{12} - F_{11})$$

$$= 0.15 \frac{24.9115}{32.000}(15.3469 + 3.7319)$$

$$= 2.228 \text{ cm}^2/\text{s}.$$

Similarly, we can evaluate the other $D_{ij}$s to complete the matrix:

$$[D_{ij}] = \begin{bmatrix} 0 & 2.228 & 2.521 \\ 7.618 & 0 & 0.653 \\ 4.188 & 0.652 & 0 \end{bmatrix} [=] \, cm^2/s.$$

**Comments**

In this example, we dealt with only three species and, with a bit of algebra, could have solved for each $D_{ij}$ analytically. For example,

$$D_{12} = \mathcal{D}_{12} \left[ 1 + \chi_3 \frac{(MW_3/MW_2)\mathcal{D}_{13} - \mathcal{D}_{12}}{\chi_1 \mathcal{D}_{23} + \chi_2 \mathcal{D}_{13} + \chi_3 \mathcal{D}_{12}} \right].$$

For most systems of interest in combustion, many species are involved, and all aspects of the approach presented above are executed using a computer. See, for example, Ref. [10].

To obtain a better appreciation for the differences between the multicomponent diffusion coefficients $D_{ij}$ and the binary diffusion coefficients $\mathcal{D}_{ij}$, we can compare the results above with the complete $\mathcal{D}_{ij}$ matrix:

$$[\mathcal{D}_{ij}] = \begin{bmatrix} 4.587 & 2.567 & 2.410 \\ 2.567 & 0.689 & 0.675 \\ 2.410 & 0.675 & 0.661 \end{bmatrix} [=] \, cm^2/s.$$

We note, first, that the $\mathcal{D}_{ij}$s are all nonzero, in contrast to the $D_{ij}$s. Furthermore, we see that the $[\mathcal{D}_{ij}]$ matrix is symmetric, whereas all of the nonzero $D_{ij}$s take on unique values.

---

Calculation of the multicomponent thermal diffusion coefficients is not nearly as straightforward as the determination of the ordinary diffusion coefficients discussed above. For thermal diffusion, an additional six matrices of greater complexity than the $[L_{ij}]$ matrix given above are involved. The CHEMKIN software [10], previously referenced, also calculates thermal diffusion coefficients, along with the other multi-component transport properties, i.e., thermal conductivity and viscosity.

## Simplified Approach

A commonly used approximate method for treating diffusion in multicomponent mix-tures is to cast the species diffusion flux or diffusion velocity equations (Eqns. 7.22 and 7.23) in forms analogous to Fick's law for binary diffusion for all but one spe-cies; i.e.,

$$\dot{m}''_{i, \, diff, \, \chi} = -\rho \mathcal{D}_{im} \nabla Y_i \qquad i = 1, 2, \ldots, N-1 \tag{7.30}$$

and

$$\mathbf{v}_{i,\,\text{diff},\,\chi} = -\frac{\mathcal{D}_{im}}{Y_i}\nabla Y_i \quad i = 1, 2, \ldots, N-1, \tag{7.31}$$

where $\mathcal{D}_{im}$ is the **effective binary diffusion coefficient** for species $i$ in the mixture $m$, as defined below. Since overall mass conservation does not necessarily obtain when Eqns. 7.30 and 7.31 are used consistently for all $N$ species, we use the fact that the sum of all of the species diffusion fluxes must be zero to obtain the diffusion velocity for the $N$th species:

$$\sum_{i=1}^{N}\rho Y_i \mathbf{v}_{i,\,\text{diff},\chi} = 0 \tag{7.32}$$

or

$$\mathbf{v}_{N,\,\text{diff},\,\chi} = -\frac{1}{Y_N}\sum_{i=1}^{N-1} Y_i \mathbf{v}_{i,\,\text{diff},\,\chi}. \tag{7.33}$$

Kee *et al.* [10] suggest that Eqn. 7.33 be applied to the species in excess, which in many combustion systems is $N_2$. This method effects a simplification because the effective binary diffusion coefficients $\mathcal{D}_{im}$ are easy to calculate. The following expression for $\mathcal{D}_{im}$ [5], although rigorously true only for the special case of all species except $i$ having the same velocities [2], can be employed:

$$\mathcal{D}_{im} = \frac{1-\chi_i}{\sum_{j\neq i}^{N}(\chi_j/\mathcal{D}_{ij})} \quad \text{for} \quad i = 1, 2, \ldots, N-1. \tag{7.34}$$

One particularly useful simplification obtains when all species but one, the $N$th, are in trace quantities. For this special case, then,

$$\mathcal{D}_{im} = \mathcal{D}_{iN}. \tag{7.35}$$

Here we see that one needs only to compute the $N-1$ binary diffusion coefficients to determine all of the species diffusion velocities.

---

**Example 7.2**

Consider the same mixture of $H_2$, $O_2$, and $N_2$ presented in Example 7.1.

A. Calculate the effective binary diffusion coefficients for each of the three species.
B. Write out expressions for the diffusion velocities for each of the species using effective binary diffusion coefficients. Assume a 1-D planar geometry. Compare this representation with the exact formulation employing multicomponent diffusion coefficients.

**Solution**

To determine the three $\mathcal{D}_{im}$s, we straightforwardly apply Eqn. 7.34 using the same convention as in Example 7.1 to designate the species. The required binary diffusivities, $\mathcal{D}_{H_2-O_2}$, $\mathcal{D}_{H_2-N_2}$ and $\mathcal{D}_{O_2-N_2}$, have the same values previously determined. Thus,

$$\mathcal{D}_{H_2-m} = \frac{1-\chi_{H_2}}{\dfrac{\chi_{O_2}}{\mathcal{D}_{H_2-O_2}} + \dfrac{\chi_{N_2}}{\mathcal{D}_{H_2-N_2}}} = \frac{1-0.15}{\dfrac{0.20}{2.567} + \dfrac{0.65}{2.410}} = 2.445\ \text{cm}^2/\text{s},$$

$$\mathcal{D}_{O_2-m} = \frac{1-\chi_{O_2}}{\dfrac{\chi_{H_2}}{\mathcal{D}_{O_2-H_2}} + \dfrac{\chi_{N_2}}{\mathcal{D}_{O_2-N_2}}} = \frac{1-0.20}{\dfrac{0.15}{2.567} + \dfrac{0.65}{0.675}} = 0.783\ \text{cm}^2/\text{s},$$

$$\mathcal{D}_{N_2-m} = \frac{1-\chi_{N_2}}{\dfrac{\chi_{H_2}}{\mathcal{D}_{N_2-H_2}} + \dfrac{\chi_{O_2}}{\mathcal{D}_{N_2-O_2}}} = \frac{1-0.65}{\dfrac{0.15}{2.410} + \dfrac{0.20}{0.675}} = 0.976\ \text{cm}^2/\text{s}.$$

With a mole fraction of 0.65, $N_2$ is the species in excess; thus, we apply Eqn. 7.31 for $H_2$ and $O_2$, and Eqn. 7.33 for $N_2$ to obtain expressions for the species diffusion velocities:

$$v_{H_2,\text{diff}} = -\frac{\mathcal{D}_{H_2-m}}{Y_{H_2}}\frac{dY_{H_2}}{dx},$$

$$v_{O_2,\text{diff}} = -\frac{\mathcal{D}_{O_2-m}}{Y_{O_2}}\frac{dY_{O_2}}{dx},$$

$$v_{N_2,\text{diff}} = -\frac{\mathcal{D}_{H_2-m}}{Y_{N_2}}\frac{dY_{H_2}}{dx} + \frac{\mathcal{D}_{O_2-m}}{Y_{N_2}}\frac{dY_{O_2}}{dx}.$$

To compare with the above, the exact multicomponent expressions for the diffusion velocities are obtained by applying Eqn. 7.23 for each species:

$$v_{H_2,\text{diff}} = \frac{1}{\chi_{H_2} MW_{\text{mix}}}\left[ MW_{O_2}D_{H_2-O_2}\frac{d\chi_{O_2}}{dx} + MW_{N_2}D_{H_2-N_2}\frac{d\chi_{N_2}}{dx}\right],$$

$$v_{O_2,\text{diff}} = \frac{1}{\chi_{O_2} MW_{\text{mix}}}\left[ MW_{H_2}D_{O_2-H_2}\frac{d\chi_{H_2}}{dx} + MW_{N_2}D_{O_2-N_2}\frac{d\chi_{N_2}}{dx}\right],$$

$$v_{N_2,\text{diff}} = \frac{1}{\chi_{N_2} MW_{\text{mix}}}\left[ MW_{H_2}D_{N_2-H_2}\frac{d\chi_{H_2}}{dx} + MW_{O_2}D_{N_2-O_2}\frac{d\chi_{O_2}}{dx}\right].$$

**Comment**

Note the computational simplicity of the effective-binary method: first, the computation of the $\mathcal{D}_{im}$s is nearly trivial in comparison with calculating the $D_{ij}$s; second, all of the diffusion velocities but one are expressed as single terms containing a single concentration gradient. In contrast, each species diffusion velocity in the exact multicomponent is expressed by $N-1$ terms, each term containing a different concentration gradient.

# MOMENTUM CONSERVATION

## One-Dimensional Forms

Momentum conservation for our 1-D planar and spherical systems is exceedingly simple because we neglect both viscous forces and the gravitational body force.

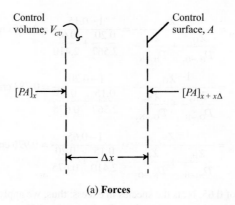

**(a) Forces**

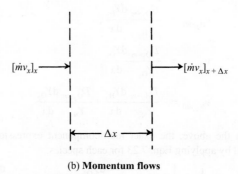

**(b) Momentum flows**

**Figure 7.3**    Control volume for one-dimensional analysis of momentum conservation, neglecting all effects of viscosity.

Figure 7.3 illustrates that the only forces acting on our planar control volume are those due to pressure. Also, there is only a single momentum flow in and a single momentum flow out of the control volume because of the simple geometry. For steady state, the general statement of momentum conservation is that the sum of all forces acting in a given direction on a control volume equals the net flow of momentum out of the control volume in the same direction, i.e.,

$$\sum \mathbf{F} = \dot{m}\mathbf{v}_{\text{out}} - \dot{m}\mathbf{v}_{\text{in}}.$$
(7.36)

For the 1-D system shown in Fig. 7.3, Eqn. 7.36 is written

$$[PA]_x - [PA]_{x+\Delta x} = \dot{m}([v_x]_{x+\Delta x} - [v_x]_x).$$
(7.37)

Dividing the left- and right-hand sides of the above equation by $\Delta x$, recognizing that both $A$ and $\dot{m}$ are constant, and taking the limit $\Delta x \to 0$, we recover the following ordinary differential equation:

$$-\frac{dP}{dx} = \dot{m}''\frac{dv_x}{dx}.$$
(7.38a)

Expressing the mass flux in terms of the velocity ($\dot{m}'' = \rho v_x$), Eqn. 7.38a becomes

$$-\frac{dP}{dx} = \rho v_x \frac{dv_x}{dx}.$$

(7.38b)

This is the 1-D Euler equation with which you are probably familiar. For the spherically symmetric flow, a similar result is found where $r$ and $v_r$ replace $x$ and $v_x$, respectively.

For the 1-D laminar premixed flame (Chapter 8), and droplet combustion (Chapter 10), we will assume that kinetic energy change across the flame is small; i.e.,

$$\frac{d\left(v_x^2/2\right)}{dx} = v_x \frac{dv_x}{dx} \approx 0.$$

Hence, the momentum equation simplifies to the trivial result that

$$\frac{dP}{dx} = 0,$$

(7.39)

which implies that the pressure is constant throughout the flowfield. The same result obtains for the spherical system.

## Two-Dimensional Forms

Rather than proceeding directly with the axisymmetric problem, we first illustrate the essential elements of momentum conservation for a two-dimensional viscous flow in cartesian ($x$, $y$) coordinates. Working in the cartesian system allows us to visualize and assemble the various terms in the momentum equation in a more straightforward manner than is possible in cylindrical coordinates. Following this development, we present the analogous axisymmetric formulation and simplify this for the boundary-layer-like jet flow.

Figure 7.4 illustrates the various forces acting in the $x$-direction on a control volume having a width $\Delta x$, height $\Delta y$, and unit depth in a steady two-dimensional flow. Acting normal to the $x$-faces are the normal viscous stresses $\tau_{xx}$, and the pressure $P$, each multiplied by the area over which they act, $\Delta y(1)$. Acting on the $y$-faces, but generating a force in the $x$-direction, are the viscous shear stresses $\tau_{yx}$, also multiplied by the area over which they act, $\Delta x(1)$. Acting at the center of mass of the control volume is the body force associated with gravity, $m_{cv}g_x(= \rho\,\Delta x\,\Delta y(1)\,g_x)$. The various $x$-direction momentum flows associated with the same control volume are shown in Fig. 7.5. Each of these corresponds to the product of the mass flowrate through the control-volume face of interest and the $x$-component of the velocity at that face. Applying the principle of momentum conservation, which states that the

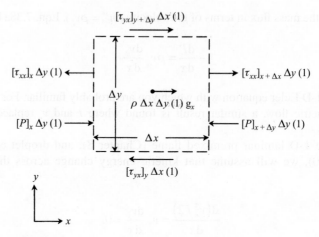

**Figure 7.4**    Forces in the x-direction acting on x- and y-faces of a two-dimensional control volume of unit depth (perpendicular to page).

sum of the forces in the x-direction must equal the net momentum flow out of the control volume, we write

$$([\tau_{xx}]_{x+\Delta x} - [\tau_{xx}]_x)\Delta y(1) + ([\tau_{yx}]_{y+\Delta y} - [\tau_{yx}]_y)\Delta x(1)$$
$$+ ([P]_x - [P]_{x+\Delta x})\Delta y(1) + \rho\,\Delta x\,\Delta y(1)g_x$$
$$= ([\rho v_x v_x]_{x+\Delta x} - [\rho v_x v_x]_x)\,\Delta y(1)$$
$$+ ([\rho v_y v_x]_{y+\Delta y} - [\rho v_y v_x]_y)\Delta x(1). \tag{7.40}$$

Dividing each term above by $\Delta x\,\Delta y$, taking the limits $\Delta x \to 0$ and $\Delta y \to 0$, and recognizing the definitions of the various partial derivatives, Eqn. 7.40 becomes

$$\frac{\partial(\rho v_x v_x)}{\partial x} + \frac{\partial(\rho v_y v_x)}{\partial y} = \frac{\partial \tau_{xx}}{\partial x} + \frac{\partial \tau_{yx}}{\partial y} - \frac{\partial P}{\partial x} + \rho g_x, \tag{7.41}$$

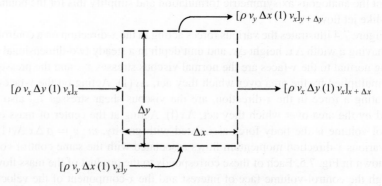

**Figure 7.5**    Momentum flows through x- and y-faces of a two-dimensional control volume of unit depth (perpendicular to page).

where the momentum flows have been placed on the left-hand side, and the forces on the right.

A similar procedure yields the steady-flow $y$-component of the momentum equation:

$$\frac{\partial(\rho v_x v_y)}{\partial x} + \frac{\partial(\rho v_y v_y)}{\partial y} = \frac{\partial \tau_{xy}}{\partial x} + \frac{\partial \tau_{yy}}{\partial y} - \frac{\partial P}{\partial y} + \rho g_y. \tag{7.42}$$

The corresponding equations for the axial and radial components of the momentum equation for the axisymmetric flow expressed in cylindrical coordinates are

*Axial (x) component*

$$\frac{\partial}{\partial x}(r\rho v_x v_x) + \frac{\partial}{\partial r}(r\rho v_x v_r) = \frac{\partial}{\partial r}(r\tau_{rx}) + r\frac{\partial \tau_{xx}}{\partial x} - r\frac{\partial P}{\partial x} + \rho g_x r \tag{7.43}$$

*Radial (r) component*

$$\frac{\partial}{\partial x}(r\rho v_r v_x) + \frac{\partial}{\partial r}(r\rho v_r v_r) = \frac{\partial}{\partial r}(r\tau_{rr}) + r\frac{\partial \tau_{rx}}{\partial x} - r\frac{\partial P}{\partial r}. \tag{7.44}$$

To preserve axisymmetry in the presence of a gravitational field, we assume that the gravitational acceleration is aligned with the $x$-direction.

For a Newtonian fluid, the viscous stresses appearing in the above equations are given by

$$\tau_{xx} = \mu \left[ 2\frac{\partial v_x}{\partial x} - \frac{2}{3}(\boldsymbol{\nabla} \cdot \mathbf{V}) \right], \tag{7.45a}$$

$$\tau_{rr} = \mu \left[ 2\frac{\partial v_r}{\partial r} - \frac{2}{3}(\boldsymbol{\nabla} \cdot \mathbf{V}) \right], \tag{7.45b}$$

$$\tau_{rx} = \mu \left[ 2\frac{\partial v_x}{\partial r} + \frac{\partial v_r}{\partial x} \right], \tag{7.45c}$$

where $\mu$ is the fluid viscosity and

$$(\boldsymbol{\nabla} \cdot \mathbf{V}) = \frac{1}{r}\frac{\partial}{\partial r}(rv_r) + \frac{\partial v_x}{\partial x}.$$

Our purpose in developing the axisymmetric flow momentum equations is to apply them to jet flames in subsequent chapters. Jets have characteristics very similar to boundary layers that develop in flows adjoining a solid surface. First, the jet width is typically small in comparison with its length in the same sense that a boundary layer is thin in comparison with its length. Second, velocities change much more rapidly in the direction transverse to the flow than they do in the axial direction, i.e., $\partial(\ )/\partial r \gg \partial(\ )/\partial x$. And lastly, axial velocities are much larger than the transverse velocities, i.e., $v_x \gg v_r$. Using these properties of a jet (boundary layer) flow, the axial component of the momentum equation (Eqn. 7.43) can be simplified using dimensional (order-of-magnitude) analysis. Specifically, for the axial component, $r(\partial \tau_{xx}/\partial x)$ is neglected, since

$$\frac{\partial}{\partial r}(r\tau_{rx}) \gg r\frac{\partial \tau_{xx}}{\partial x},$$

and $\tau_{rx}$ simplifies to

$$\tau_{rx} = \mu \frac{\partial v_x}{\partial r},$$

because

$$\frac{\partial v_x}{\partial r} \gg \frac{\partial v_r}{\partial x}.$$

With these simplifications, the axial momentum equation becomes

$$\frac{\partial}{\partial x}(r\rho v_x v_x) + \frac{\partial}{\partial r}(r\rho v_x v_r) = \frac{\partial}{\partial r}\left(r\mu \frac{\partial v_x}{\partial r}\right) - r\frac{\partial P}{\partial x} + \rho g_x r. \qquad (7.46)$$

From a similar order-of-magnitude analysis of the radial momentum equation, one comes to the conclusion that $\partial P/\partial r$ is very small (see, e.g., Schlichting [11]). This implies that the pressure inside the jet at any axial station is essentially the same as the pressure in the ambient fluid, outside of the jet, at the same axial position. With this knowledge, we can equate $\partial P/\partial x$ appearing in the axial momentum equation to the hydrostatic pressure gradient in the ambient fluid; and, furthermore, the velocity components $v_x$ and $v_r$ can be determined by simultaneously solving the overall continuity (Eqn. 7.7) and axial momentum (Eqn. 7.46) equations, with no need to include the radial momentum equation explicitly.

In our future developments (Chapters 9 and 13), we will assume that the jet is oriented vertically upwards with gravity acting downwards, thus providing a positive buoyancy effect, or, in some cases, we will neglect gravity altogether. For the first case, then, we recognize, as mentioned above, that

$$\frac{\partial P}{\partial x} \approx \frac{\partial P_\infty}{\partial x} = -\rho_\infty g \qquad (7.47)$$

where $g(= -g_x)$ is the scalar gravitational acceleration (9.81 m/s$^2$), and $P_\infty$ and $\rho_\infty$ are the ambient fluid pressure and density, respectively. Combining Eqn. 7.47 with Eqn. 7.46 yields our final form of the axial momentum equation:

$$
\begin{array}{ccc}
\dfrac{1}{r}\dfrac{\partial}{\partial x}(r\rho v_x v_x) & + & \dfrac{1}{r}\dfrac{\partial}{\partial r}(r\rho v_x v_r) \\[4pt]
\text{\small\it x-Momentum flow} & & \text{\small\it x-Momentum flow} \\
\text{\small by axial convection} & & \text{\small by radial convection} \\
\text{\small per unit volume} & & \text{\small per unit volume} \\[10pt]
= \quad \dfrac{1}{r}\dfrac{\partial}{\partial r}\!\left(r\mu\dfrac{\partial v_x}{\partial r}\right) & + & (\rho_\infty - \rho)g \\[4pt]
\text{\small Viscous force} & & \text{\small Buoyant force} \\
\text{\small per unit volume} & & \text{\small per unit volume}
\end{array}
\tag{7.48}
$$

Note that the above relation allows for variable density, an inherent characteristic of combusting flows, and variable (temperature-dependent) viscosity.

## ENERGY CONSERVATION

### General One-Dimensional Form

Starting with a one-dimensional cartesian system, we consider the control volume shown in Fig. 7.6, where the various energy flows into and out of a plane layer, $\Delta x$ in length, are shown. Following Eqn. 2.28, the first law of thermodynamics can be expressed as follows:

$$
(\dot{Q}''_x - \dot{Q}''_{x+\Delta x})A - \dot{W}_{cv} = \dot{m}''A\left[\left(h+\frac{v_x^2}{2}+gz\right)_{x+\Delta x} - \left(h+\frac{v_x^2}{2}+gz\right)_x\right].
\tag{7.49}
$$

From the outset, we assume steady state; thus, no energy accumulation within the control volume is indicated. We also assume that no work is done by the control volume and that there is no change in potential energies of the inlet and outlet streams. With these assumptions, dividing through by the area $A$, and rearranging, Eqn. 7.49 becomes

$$
-(\dot{Q}''_{x+\Delta x} - \dot{Q}''_x) = \dot{m}''\left[\left(h+\frac{v_x^2}{2}\right)_{x+\Delta x} - \left(h+\frac{v_x^2}{2}\right)_x\right].
\tag{7.50}
$$

Dividing both sides of Eqn. 7.50 by $\Delta x$, taking the limit $\Delta x \to 0$, and recognizing the definition of a derivative, we obtain the following differential equation:

$$
-\frac{d\dot{Q}''_x}{dx} = \dot{m}''\left(\frac{dh}{dx} + v_x\frac{dv_x}{dx}\right).
\tag{7.51}
$$

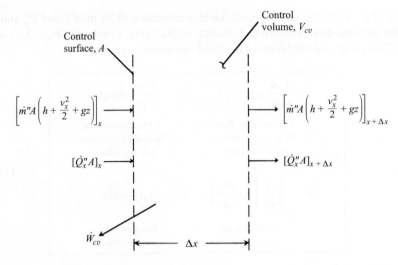

**Figure 7.6**    Control volume for one-dimensional, steady-state analysis of energy conservation.

If we were dealing with a system where there was no species diffusion, we would simply replace the heat flux $\dot{Q}''_x$ with Fourier's law for conduction; however, in our system, which assumes that species are diffusing, the heat flux consists of both conduction and an additional enthalpy flux resulting from the diffusing species. Assuming there is no radiation, the general vector form of the heat flux is given by

$$\dot{\mathbf{Q}}'' = -k\nabla T + \sum \dot{m}''_{i,\,\text{diff}} h_i, \qquad (7.52a)$$

| Heat flux | Conduction | Species diffusion |
| vector | contribution | contribution |

where $\dot{m}''_{i,\,\text{diff}}$ is the diffusional flux of the $i$th species, which was introduced in our discussion of species conservation. For our 1-D plane layer, the heat flux is

$$\dot{Q}''_x = -k\frac{dT}{dx} + \sum \rho Y_i (v_{ix} - v_x)h_i, \qquad (7.52b)$$

where we have related the diffusional flux to the diffusion velocity (Eqn. 7.14). At this juncture, we have all the physics we wish to consider defined by Eqn. 7.51 with Eqn. 7.52b. The following development is primarily mathematical manipulation and coupling with concepts and definitions from our discussion of species conservation.

Before substituting $\dot{Q}''_x$ back into our expression for overall energy conservation, we rewrite Eqn. 7.52b in terms of the bulk and species mass fluxes, i.e.,

$$\dot{Q}''_x = -k\frac{dT}{dx} + \sum \rho v_{ix} Y_i h_i - \rho v_x \sum Y_i h_i = -k\frac{dT}{dx} + \sum \dot{m}''_i h_i - \dot{m}''h, \qquad (7.53)$$

where we recognize that $\dot{m}_i'' = \rho v_{ix} Y_i$, $\rho v_x = \dot{m}''$, and $\sum Y_i h_i = h$. Now substituting Eqn. 7.53 into Eqn. 7.51, canceling the $\dot{m}'' dh/dx$ terms which appear on each side, and rearranging, results in

$$\frac{d}{dx}\left(\sum h_i \dot{m}_i''\right) + \frac{d}{dx}\left(-k\frac{dT}{dx}\right) + \dot{m}'' v_x \frac{dv_x}{dx} = 0. \tag{7.54}$$

We now expand the first term in Eqn. 7.54; i.e.,

$$\frac{d}{dx}\left(\sum h_i \dot{m}_i''\right) = \sum \dot{m}_i'' \frac{dh_i}{dx} + \sum h_i \frac{d\dot{m}_i''}{dx}.$$

The term $d\dot{m}_i''/dx$, which now appears, is key in that from species conservation (Eqn. 7.9),

$$\frac{d\dot{m}_i''}{dx} = \dot{m}_i'''.$$

With the substitution of the above, energy conservation (Eqn. 7.54) is explicitly linked to the species production rates associated with chemical reactions. Our final 1-D energy conservation equation is

$$\boxed{\sum \dot{m}_i'' \frac{dh_i}{dx} + \frac{d}{dx}\left(-k\frac{dT}{dx}\right) + \dot{m}'' v_x \frac{dv_x}{dx} = -\sum h_i \dot{m}_i'''} \tag{7.55}$$

Equation 7.55 is often the starting point for further simplification and, as such, applies equally well to multicomponent (nonbinary) and binary systems. It is also important to note that, so far, no assumptions have been made with regard to the thermophysical properties $(k, \rho, c_p, \mathcal{D})$; we have, however, assumed that there is no radiation, no viscous dissipation, and no potential energy changes (see Table 7.1).

## Shvab–Zeldovich Forms

The **Shvab–Zeldovich energy equation,** so named in recognition of the researchers who first developed it, is particularly useful in that the species mass fluxes and enthalpies are eliminated from the left-hand side of Eqn. 7.55 and replaced with terms having temperature as the only dependent variable. A key assumption in the development of the Shvab–Zeldovich equation is that the **Lewis number** $(Le = k/\rho c_p \mathcal{D})$ is unity. The **unity $Le$ assumption** is frequently invoked in the analysis of combustion problems, and the great simplification that it affords will be emphasized here. Another key assumption in our development is that Fick's law can be used to describe the species mass fluxes.

We begin by examining the heat flux for a reacting flow defined in Eqn. 7.52a:

$$\dot{Q}_x'' = -k\frac{dT}{dx} + \sum \dot{m}_{i,\,\mathrm{diff}}'' h_i. \tag{7.56}$$

**Table 7.1**    Energy equation assumptions

| Equation | Basic Assumptions/Neglected Effects | Properties | Mass Transfer Laws | Geometry |
|---|---|---|---|---|
| Eqn. 7.55 | i.  Steady state<br>ii.  No effect of gravity<br>iii.  No shaft work or viscous dissipation<br>iv.  No radiation heat transfer | Variable, temperature-dependent properties | Ordinary (concentration gradient) diffusion only | Constant-area plane layer (1-D cartesian) |
| Eqn. 7.62 | i–iv above<br>+<br>Thermal diffusivity ($\alpha$) equals mass diffusivity ($\mathcal{D}$), i.e., the Lewis number ($Le$) is unity | As above | As above<br>+<br>Binary (or effective binary) diffusion governed by Fick's law | As above |
| Eqn. 7.63 | As above<br>+<br>Kinetic energy changes are negligible, which implies that the pressure is constant | As above | As above | As above |
| Eqn. 7.65 | As above | As above | As above | One-dimensional, spherically symmetric |
| Eqn. 7.66 | As above<br>+<br>Negligible diffusion in the axial ($x$) direction | As above | As above | Two-dimensional, axisymmetric |
| Eqn. 7.67 | As in Eqn. 7.55<br>+<br>Kinetic energy changes negligible | As above | Multicomponent diffusion | As in Eqn. 7.55 |

Using the definition of the species flux (Eqn. 7.15a) and Fick's law (Eqn. 3.1 or 7.17), Eqn. 7.56 becomes

$$\dot{Q}''_x = -k\frac{dT}{dx} - \sum \rho \mathcal{D}\frac{dY_i}{dx}h_i \qquad (7.57a)$$

or, assuming a single diffusivity characterizes the mixture,

$$\dot{Q}''_x = -k\frac{dT}{dx} - \rho \mathcal{D}\sum h_i\frac{dY_i}{dx}. \qquad (7.57b)$$

By applying the definition of the derivative of the product,

$$\frac{d\sum h_i Y_i}{dx} = \sum h_i\frac{dY_i}{dx} + \sum Y_i\frac{dh_i}{dx},$$

the heat flux is now expressed as

$$\dot{Q}''_x = -k\frac{dT}{dx} - \rho \mathcal{D}\frac{d\sum h_i Y_i}{dx} + \rho \mathcal{D}\sum Y_i\frac{dh_i}{dx}. \qquad (7.57c)$$

We use the definition of $h \equiv \sum h_i Y_i$ to simplify the second term on the right-hand side; and the third term can be expressed in terms of $c_p$ and $T$ by recognizing that

$$\sum Y_i \frac{dh_i}{dx} = \sum Y_i c_{p,i} \frac{dT}{dx} = c_p \frac{dT}{dx}.$$

The three terms that constitute the heat flux are now

$$\dot{Q}_x'' = -k \frac{dT}{dx} - \rho D \frac{dh}{dx} + \rho D c_p \frac{dT}{dx}. \tag{7.57d}$$

Using the definition of the thermal diffusivity, $\alpha \equiv k/\rho c_p$, to express the thermal conductivity appearing in the above, results in

$$\dot{Q}_x'' = \underbrace{-\rho \alpha c_p \frac{dT}{dx}}_{\substack{\text{Flux of sensible} \\ \text{enthalpy due to} \\ \text{conduction}}} - \underbrace{\rho D \frac{dh}{dx}}_{\substack{\text{Flux of standardized} \\ \text{enthalpy due to} \\ \text{species diffusion}}} + \underbrace{\rho D c_p \frac{dT}{dx}}_{\substack{\text{Flux of sensible} \\ \text{enthalpy due to} \\ \text{species diffusion}}}, \tag{7.58}$$

where a physical interpretation of each of the terms is shown. In general, all three terms contribute to the total heat flux; however, for the special case where $\alpha = D$, we see that the sensible enthalpy flux due to conduction cancels with the sensible enthalpy flux arising from species diffusion. Since the Lewis number is defined as the ratio of $\alpha$ to $D$,

$$Le \equiv \frac{\alpha}{D} = 1 \tag{7.59}$$

for this special case. For many species of interest in combustion, Lewis numbers are of the order of unity, thus providing some physical justification for equating $\alpha$ and $D$. With this assumption, the heat flux is simply

$$\dot{Q}_x'' = -\rho D \frac{dh}{dx}. \tag{7.60}$$

We now employ the above in our basic energy conservation equation (Eqn. 7.51); thus,

$$\frac{d}{dx}\left(\rho D \frac{dh}{dx}\right) = \dot{m}'' \frac{dh}{dx} + \dot{m}'' v_x \frac{dv_x}{dx}. \tag{7.61}$$

Employing the definition of the standardized enthalpy,

$$h = \sum Y_i h_{f,i}^o + \int_{T_{\text{ref}}}^{T} c_p dT,$$

Eqn. 7.61 becomes

$$\frac{d}{dx}\left[\rho D \sum h_{f,i}^{o}\frac{dY_i}{dx} + \rho D \frac{d\int c_p dT}{dx}\right] = \dot{m}''\sum h_{f,i}^{o}\frac{dY_i}{dx} + \dot{m}''\frac{d\int c_p dT}{dx} + \dot{m}''v_x\frac{dv_x}{dx}.$$

Rearranging the above yields

$$\dot{m}''\frac{d\int c_p dT}{dx} - \frac{d}{dx}\left[\rho D\frac{d\int c_p dT}{dx}\right] + \dot{m}''v_x\frac{dv_x}{dx} = -\frac{d}{dx}\left[\sum h_{f,i}^{o}\left(\dot{m}''Y_i - \rho D\frac{dY_i}{dx}\right)\right].$$

The right-hand side of the above is simplified by employing both Fick's law and species conservation (Eqn. 7.9):

$$-\frac{d}{dx}\left[\sum h_{f,i}^{o}\left(\dot{m}''Y_i - \rho D\frac{dY_i}{dx}\right)\right] = -\frac{d}{dx}\left[\sum h_{f,i}^{o}\dot{m}_i''\right] = -\sum h_{f,i}^{o}\dot{m}_i'''.$$

We now reassemble our final result:

$$\dot{m}''\frac{d\int c_p dT}{dx} - \frac{d}{dx}\left[\rho D\frac{d\int c_p dT}{dx}\right] + \dot{m}''v_x\frac{dv_x}{dx} = -\sum h_{f,i}^{o}\dot{m}_i'''. \tag{7.62}$$

Thus far, we have retained the kinetic energy change term for completeness; however, this term is usually small and generally neglected in most developments of the Shvab–Zeldovich energy equation. Dropping this term, we have the following result, which has the simple physical interpretation that the combined rates of convection (advection) and diffusion of sensible enthalpy (thermal energy) equal the rate at which chemical energy is converted to thermal energy by chemical reaction:

$$\begin{array}{ccc} \underbrace{\dot{m}''\dfrac{d\int c_p dT}{dx}} & + & \underbrace{\dfrac{d}{dx}\left[-\rho D\dfrac{d\int c_p dT}{dx}\right]} & = & \underbrace{-\sum h_{f,i}^{o}\dot{m}_i'''} \end{array} \tag{7.63}$$

| Rate of sensible enthalpy transport by convection (advection) per unit volume (W/m³) | Rate of sensible enthalpy transport by diffusion per unit volume (W/m³) | Rate of sensible enthalpy production by chemical reaction per unit volume (W/m³) |
|---|---|---|

The general form of the Shvab–Zeldovich energy equation is

$$\nabla\cdot\left[\dot{m}''\int c_p dT - \rho D\nabla\left(\int c_p dT\right)\right] = -\sum h_{f,i}^{o}\dot{m}_i'''. \tag{7.64}$$

We can apply the definitions of the vector operators to obtain the Shvab–Zeldovich energy equation for our spherical and axisymmetric geometries. The 1-D spherical form is

$$\frac{1}{r^2}\frac{d}{dr}\left[r^2\left(\rho v_r \int c_p \, dT - \rho \mathcal{D}\frac{d\int c_p \, dT}{dr}\right)\right] = -\sum h_{f,i}^o \dot{m}_i''' \qquad (7.65)$$

and the axisymmetric form is

$$\frac{1}{r}\frac{\partial}{\partial x}\left(r\rho v_x \int c_p \, dT\right) + \frac{1}{r}\frac{\partial}{\partial r}\left(r\rho v_r \int c_p \, dT\right) - \frac{1}{r}\frac{\partial}{\partial r}\left(r\rho\mathcal{D}\frac{\partial\int c_p \, dT}{\partial r}\right) = -\sum h_{f,i}^o \dot{m}_i'''$$

| Rate of sensible enthalpy transport by axial convection (advection) per unit volume (W/m³) | Rate of sensible enthalpy transport by radial convection (advection) per unit volume (W/m³) | Rate of sensible enthalpy transport by radial diffusion per unit volume (W/m³) | Rate of sensible enthalpy production by chemical reaction per unit volume (W/m³) |
|---|---|---|---|

$$(7.66)$$

## Useful Form for Flame Calculations

In numerical models of both premixed [3] and nonpremixed [4], steady, laminar flames, it is convenient for the species diffusion velocity $v_{i,\text{diff}} (= v_{i,\text{diff},\chi} + v_{i,\text{diff},T})$ to appear explicitly in the energy equation. For the 1-D problem, we can start with Eqn. 7.55, neglect the kinetic-energy change term, and obtain the following:

$$\dot{m}''c_p\frac{dT}{dx} + \frac{d}{dx}\left(-k\frac{dT}{dx}\right) + \sum_{i=1}^{N}\rho Y_i v_{i,\text{diff}}\, c_{p,i}\frac{dT}{dx} = -\sum_{i=1}^{N} h_i \dot{m}_i''', \qquad (7.67)$$

where $c_p \equiv \sum Y_i c_{p,i}$ and $N$ is the number of species considered.

To arrive at this result requires the application of the ideal-gas calorific equation of state (Eqn. 6.11), steady-state species conservation from Eqn. 7.16, the definition of mixture properties in terms of the components (Eqn. 2.15), and the product rule for derivatives. With these hints, the derivation of Eqn. 7.67 is left as an exercise for the reader.

Note that, in all of the forms of the energy equation given above, we have made no assumptions that the properties are constant; however, in many simplified analyses of combustion systems, as we will employ in subsequent chapters, it is useful to treat $c_p$ and the product $\rho\mathcal{D}$ as constants. Table 7.1 summarizes the assumptions we have employed in developing our various energy conservation equations.

## THE CONCEPT OF A CONSERVED SCALAR

The **conserved scalar** concept greatly simplifies the solution of reacting flow problems (i.e., the determination of the fields of velocity, species, and temperature), particularly those involving nonpremixed flames. A somewhat tautological definition of a conserved scalar is any scalar property that is conserved throughout the flowfield.

For example, with certain restrictions, the standardized enthalpy is conserved at every point in a flow when there are no sources (or sinks) of thermal energy, such as radiation into or out of the flow or viscous dissipation. In this case, then, the standardized enthalpy would qualify as being a conserved scalar. Element mass fractions are also conserved scalars, since elements are neither created nor destroyed by chemical reaction. There are many conserved scalars that can be defined [12]; however, in the development here we choose to deal with only two: the mixture fraction, defined below, and the previously mentioned mixture standardized enthalpy.

## Definition of Mixture Fraction

If we restrict our flow system to consist of a single inlet stream of pure fuel together with a single stream of pure oxidizer, which react to form a single product, we can define the **mixture fraction,** $f$, a conserved scalar, as

$$f \equiv \frac{\text{Mass of material having its origin in the fuel stream}}{\text{Mass of mixture}}. \qquad (7.68)$$

Since Eqn. 7.68 applies to an infinitesimally small volume, $f$ is just a special kind of mass fraction, formed as a combination of fuel, oxidizer, and product mass fractions, as shown below. For example, $f$ is unity in the fuel stream and zero in the oxidizer stream; and within the flowfield, $f$ takes on values between unity and zero.

For our three-"species" system, we can define $f$ in terms of the fuel, oxidizer, and product mass fractions at any point in the flow, when

$$1 \text{ kg fuel} + \nu \text{ kg oxidizer} \rightarrow (\nu + 1)\text{kg products}. \qquad (7.69)$$

That is,

$$
\underset{\substack{\text{Mass fraction of} \\ \text{material having} \\ \text{its origin in the} \\ \text{fuel stream}}}{f} = \underset{\left(\frac{\text{kg fuel stuff}}{\text{kg fuel}}\right)}{(1)} \quad \underset{\left(\frac{\text{kg fuel}}{\text{kg mixture}}\right)}{Y_F}
$$

$$
+ \underset{\left(\frac{\text{kg fuel stuff}}{\text{kg products}}\right)}{\left(\frac{1}{\nu+1}\right)} \underset{\left(\frac{\text{kg products}}{\text{kg mixture}}\right)}{Y_{Pr}} + \underset{\left(\frac{\text{kg fuel stuff}}{\text{kg oxidizer}}\right)}{(0)} \underset{\left(\frac{\text{kg oxidizer}}{\text{kg mixture}}\right)}{Y_{Ox}}, \qquad (7.70)
$$

where "fuel stuff" is that material originating in the fuel stream. For a hydrocarbon fuel, fuel stuff is carbon and hydrogen. Equation 7.70 is more simply written

$$\boxed{f = Y_F + \left(\frac{1}{\nu+1}\right)Y_{Pr}} \qquad (7.71)$$

This conserved scalar is particularly useful in dealing with diffusion flames where the fuel and oxidizer streams are initially segregated. For premixed combustion, the mixture fraction is everywhere uniform, assuming all species have the same diffusivities; hence, a conservation equation for $f$ provides no new information.

## Conservation of Mixture Fraction

The utility of the conserved scalar, $f$, is that it can be used to generate a species conservation equation that has no reaction rate terms appearing, i.e., the equation is "sourceless." We can develop this idea simply using the 1-D cartesian species equations as an example. Writing out Eqn. 7.8 for the fuel and product species, we have

$$\dot{m}''\frac{dY_F}{dx} - \frac{d}{dx}\left(\rho D \frac{dY_F}{dx}\right) = \dot{m}'''_F \tag{7.72}$$

and

$$\dot{m}''\frac{dY_{Pr}}{dx} - \frac{d}{dx}\left(\rho D \frac{dY_{Pr}}{dx}\right) = \dot{m}'''_{Pr}. \tag{7.73}$$

Dividing Eqn. 7.73 by $(v+1)$ results in

$$\dot{m}''\frac{d(Y_{Pr}/(v+1))}{dx} - \frac{d}{dx}[\rho D d(Y_{Pr}/(v+1))] = \frac{1}{v+1}\dot{m}'''_{Pr}. \tag{7.74}$$

Mass conservation (Eqn. 7.69) also implies that

$$\dot{m}'''_{Pr}/(v+1) = -\dot{m}'''_F, \tag{7.75}$$

where the minus sign reflects that fuel is being consumed and products are being generated. Substituting Eqn. 7.75 into Eqn. 7.74 and adding the resultant equation to Eqn. 7.72 yields

$$\dot{m}''\frac{d(Y_F + Y_{Pr}/(v+1))}{dx} - \frac{d}{dx}\left[\rho D \frac{d}{dx}(Y_F + Y_{Pr}/(v+1))\right] = 0. \tag{7.76}$$

We note that Eqn. 7.76 is "sourceless," i.e., the right-hand side is zero, and that the quantity in the derivatives is our conserved scalar, the mixture fraction, $f$. Recognizing this, Eqn. 7.76 is rewritten as

$$\boxed{\dot{m}''\frac{df}{dx} - \frac{d}{dx}\left(\rho D \frac{df}{dx}\right) = 0} \tag{7.77}$$

Similar manipulations can be performed on the 1-D spherical and the 2-D axisymmetric system equations. For the spherical system, we obtain

$$\boxed{\frac{d}{dr}\left[r^2\left(\rho v_r f - \rho D \frac{df}{dr}\right)\right] = 0} \tag{7.78}$$

and, for the axisymmetric geometry, we obtain

$$\boxed{\frac{\partial}{\partial x}(r\rho v_x f) + \frac{\partial}{\partial r}(r\rho v_r f) - \frac{\partial}{\partial r}\left(r\rho D \frac{\partial f}{\partial r}\right) = 0} \tag{7.79}$$

| | |
|---|---|
| **Example 7.3** | Consider a nonpremixed, ethane ($C_2H_6$)–air flame in which the mole fractions of the following species are measured using various techniques: $C_2H_6$, CO, $CO_2$, $H_2$, $H_2O$, $N_2$, $O_2$, and OH. The mole fractions of all other species are assumed to be negligible. Define a mixture fraction $f$ expressed in terms of the mole fractions of the measured species. |

**Solution**

Our approach will be to first express $f$ in terms of the known species mass fractions by exploiting the definition of $f$ (Eqn. 7.68), and then express the mass fractions in terms of the mole fractions (Eqn. 2.11). Thus,

$$f \equiv \frac{\text{Mass of material originating in fuel stream}}{\text{Mass of mixture}}$$

$$= \frac{[m_C + m_H]_{\text{mix}}}{m_{\text{mix}}},$$

since the fuel stream consists only of carbon and hydrogen, and assuming that no carbon or hydrogen are present in the oxidizer stream; i.e., the air consists solely of $N_2$ and $O_2$.

In the flame gases, carbon is present in any unburned fuel and in the CO and $CO_2$; and hydrogen is present in unburned fuel, $H_2$, $H_2O$, and OH. Summing the mass fractions of carbon and hydrogen associated with each species yields

$$f = Y_{C_2H_6} \frac{2MW_C}{MW_{C_2H_6}} + Y_{CO} \frac{MW_C}{MW_{CO}} + Y_{CO_2} \frac{MW_C}{MW_{CO_2}}$$

$$+ Y_{C_2H_6} \frac{3MW_{H_2}}{MW_{C_2H_6}} + Y_{H_2} + Y_{H_2O} \frac{MW_{H_2}}{MW_{H_2O}} + Y_{OH} \frac{0.5MW_{H_2}}{MW_{OH}}$$

where the weighted ratios of molecular weights are the fractions of the element (C or $H_2$) in each of the species. Substituting for the mass fractions, $Y_i = \chi_i MW_i / MW_{\text{mix}}$, yields

$$f = \chi_{C_2H_6} \frac{MW_{C_2H_6}}{MW_{\text{mix}}} \frac{2MW_C}{MW_{C_2H_6}} + \chi_{CO} \frac{MW_{CO}}{MW_{\text{mix}}} \frac{MW_C}{MW_{CO}} + \cdots$$

$$= \frac{\left(2\chi_{C_2H_6} + \chi_{CO} + \chi_{CO_2}\right)MW_C + \left(3\chi_{C_2H_6} + \chi_{H_2} + \chi_{H_2O} + \frac{1}{2}\chi_{OH}\right)MW_{H_2}}{MW_{\text{mix}}},$$

where

$$MW_{\text{mix}} = \sum \chi_i MW_i$$

$$= \chi_{C_2H_6} MW_{C_2H_6} + \chi_{CO} MW_{CO} + \chi_{CO_2} MW_{CO_2}$$

$$+ \chi_{H_2} MW_{H_2} + \chi_{H_2O} MW_{H_2O} + \chi_{N_2} MW_{N_2}$$

$$+ \chi_{O_2} MW_{O_2} + \chi_{OH} MW_{OH}.$$

**Comment**

Here we see that although the mixture fraction is simple in concept, experimental determinations of $f$ may require the measurement of many species. Approximate values, of course, can be obtained by neglecting the minor species that are difficult to measure.

For the experiment discussed in Example 7.3, determine a numerical value for the mixture fraction, where the mole fractions measured at a point in the flame are the following:

**Example 7.4**

$$\chi_{CO} = 949 \text{ ppm,} \qquad \chi_{H_2O} = 0.1488,$$
$$\chi_{CO_2} = 0.0989, \qquad \chi_{O_2} = 0.0185,$$
$$\chi_{H_2} = 315 \text{ ppm,} \qquad \chi_{OH} = 1350 \text{ ppm.}$$

Assume the balance of the mixture is $N_2$. Also, determine the equivalence ratio for the mixture using the calculated mixture fraction.

**Solution**

Calculation of the mixture fraction is a straightforward application of the result obtained in Example 7.3. We start by calculating the $N_2$ mole fraction:

$$\chi_{N_2} = 1 - \sum \chi_i$$
$$= 1 - 0.0989 - 0.1488 - 0.0185 - (949 + 315 + 1350)1 \cdot 10^{-6}$$
$$= 0.7312.$$

The mixture molecular weight can now be determined:

$$MW_{mix} = \sum \chi_i MW_i$$
$$= 28.16 \text{ kg}_{mix} / \text{kmol}_{mix}.$$

Substituting numerical values into the expression for $f$ from Example 7.3 yields

$$f = \frac{(949 \cdot 10^{-6} + 0.0989)12.011 + (315 \cdot 10^{-6} + 0.1488 + (0.5)1350 \cdot 10^{-6})2.016}{28.16}$$

$$\boxed{f = 0.0533}$$

To calculate the equivalence ratio, we first recognize that, through its definition, the mixture fraction is related to the fuel–air ratio, $(F/A)$, by

$$(F/A) = f/(1-f),$$

and that

$$\Phi = (F/A)/(F/A)_{stoic}.$$

For an arbitrary hydrocarbon $C_xH_y$, the stoichiometric fuel–air ratio is calculated from Eqn. 2.32,

$$(F/A)_{stoic} = \left[ 4.76\,(x + y/4)\,\frac{MW_{air}}{MW_{C_xH_y}} \right]^{-1}$$
$$= \left[ 4.76\,(2 + 6/4)\,\frac{28.85}{30.07} \right]^{-1}$$
$$= 0.0626.$$

Thus,

$$\Phi = \frac{f/(1-f)}{(F/A)_{stoic}} = \frac{0.0533/(1-0.0533)}{0.0626}$$

$$\boxed{\Phi = 0.90}$$

**Comment**

Here we see how the mixture fraction relates to previously defined measures of stoichiometry. You should be able to develop relationships among these various measures, $f$, $(A/F)$, $(F/A)$, and $\Phi$, through a knowledge of their fundamental definitions.

---

**Example 7.5**

Consider a nonpremixed jet flame (see Figs 9.6 and 13.3) in which the fuel is $C_3H_8$ and the oxidizer is an equimolar mixture of $O_2$ and $CO_2$. The species existing within the flame are $C_3H_8$, CO, $CO_2$, $O_2$, $H_2$, $H_2O$, and OH. Determine the numerical value of the stoichiometric mixture fraction $f_{stoic}$ for this system. Also, write out an expression for the local mixture fraction, $f$, at any location within the flame in terms of the flame-species mass fractions, $Y_i$. Assume all pairs of binary diffusion coefficients are equal, i.e., no differential diffusion.

**Solution**

To determine the stoichiometric mixture fraction, we need only calculate the fuel mass fraction $Y_{C_3H_8}$ in a reactants mixture of stoichiometric proportions:

$$C_3H_8 + a(O_2 + CO_2) \rightarrow bCO_2 + cH_2O.$$

From H-, C-, and O-atom conservation, we obtain

$$H: \; 8 = 2c,$$
$$C: \; 3 + a = b,$$
$$O: \; 2a + 2a = 2b + c,$$

which yields

$$a = 5, b = 8, c = 4.$$

Thus,

$$f_{stoic} = Y_F = \frac{MW_{C_3H_8}}{MW_{C_3H_8} + 5(MW_{O_2} + MW_{CO_2})}$$

$$= \frac{44.096}{44.096 + 5(32.000 + 44.011)}$$

$$\boxed{f_{stoic} = 0.1040}$$

To determine the local mixture fraction requires us to take into account that not all of the carbon atoms in the flame originate with the fuel since the oxidizer contains $CO_2$. We note, however, that the only source of elemental H within the flame is $C_3H_8$; thus, we can determine the local mixture fraction knowing that it must be proportional to the local elemental H mass fraction. More explicitly,

$$f = \left( \frac{\text{mass of fuel}}{\text{mass of H}} \right) \left( \frac{\text{mass of H}}{\text{mass of mixture}} \right) = \frac{44.096}{8(1.008)} Y_{\text{H}} = 5.468 Y_{\text{H}}.$$

where $Y_H$ is given by the following weighted sum of the mass fractions of species containing hydrogen:

$$Y_H = \frac{8(1.008)}{44.096} Y_{C_3H_8} + Y_{H_2} + \frac{2.016}{18.016} Y_{H_2O} + \frac{1.008}{17.008} Y_{OH}$$

$$= 0.1829 Y_{C_3H_8} + Y_{H_2} + 0.1119 Y_{H_2O} + 0.0593 Y_{OH}.$$

Thus, our final result is expressed as

$$\boxed{f = Y_{C_3H_8} + 5.468 Y_{H_2} + 0.6119 Y_{H_2O} + 0.3243 Y_{OH}}$$

**Comment**

Note that even though the carbon in the fuel may be converted to CO and $CO_2$, we did not need to explicitly account for this. Note, also that if the H-containing species were to diffuse differentially, the local H:fuel-C ratio would not be the same everywhere within the flame, thus causing our final result to be approximate rather than exact. Although solid carbon (soot) was not considered in this problem, most real, hydrocarbon–air, nonpremixed flames will contain soot, which can complicate both flame-species measurements and the mixture-fraction determination.

## Conserved Scalar Energy Equation

Subject to all of the assumptions underlying the Shvab–Zeldovich energy equation (Eqns. 7.63–7.66), the mixture enthalpy, $h$, is also a conserved scalar:

$$h \equiv \sum_i Y_i h_{f,i}^o + \int_{T_{\text{ref}}}^{T} c_p \, dT. \tag{7.80}$$

This can be seen directly from energy conservation expressed by Eqn. 7.61 when the kinetic energy term $\dot{m}'' v_x dv_x / dx$ is assumed to be negligible. For our three geometries of interest—the 1-D planar, 1-D spherical, and 2-D axisymmetric—the conserved scalar forms of energy conservation are expressed, respectively, as

$$\boxed{\dot{m}'' \frac{dh}{dx} - \frac{d}{dx} \left( \rho \mathcal{D} \frac{dh}{dx} \right) = 0} \tag{7.81}$$

$$\boxed{\frac{d}{dr} \left[ r^2 \left( \rho v_r h - \rho \mathcal{D} \frac{dh}{dr} \right) \right] = 0} \tag{7.82}$$

**Table 7.2**     Summary of conservation equations for reacting flows

| Conserved Quantity | General Forms | 1-D Planar | 1-D Spherical | 2-D Axisymmetric |
|---|---|---|---|---|
| Mass | Eqn. 7.5 | Eqn. 7.4 | Eqn. 7.6 | Eqn. 7.7 |
| Species | Eqn. 7.10 | Eqns. 7.8 and 7.9 | Eqn. 7.19 | Eqn. 7.20 |
| Momentum | N/A | Eqn. 7.38 and 7.39 | See discussion of Eqn. 7.39 | Eqn. 7.48 |
| Energy | Eqn. 7.64 (Shvab–Zeldovich form) | Eqn. 7.55 and Eqn. 7.63 (Shvab–Zeldovich form) | Eqn. 7.65 (Shvab–Zeldovich form) | Eqn. 7.66 (Shvab–Zeldovich form) |
| Mixture fraction[a] | — | Eqn. 7.77 | Eqn. 7.78 | Eqn. 7.79 |
| Enthalpy[a] | — | Eqn. 7.81 | Eqn. 7.82 | Eqn. 7.83 |

[a]These are just alternate conserved scalar forms of species and energy conservation, respectively, and are not separate conservation principles.

and

$$\frac{\partial}{\partial x}(r\rho v_x h) + \frac{\partial}{\partial r}(r\rho v_r h) - \frac{\partial}{\partial r}\left(r\rho \mathcal{D}\frac{\partial h}{\partial r}\right) = 0 \qquad (7.83)$$

The derivations of Eqns. 7.82 and 7.83 are left as exercises for the reader.

## SUMMARY

In this chapter, the general forms of the conservation equations for mass, species, momentum, and energy have been presented along with their brief description. You should be able to recognize these equations and have some appreciation of the physical meaning of each term. Simplified equations were developed for three different geometries and are summarized in Table 7.2 for easy reference. These equations will be the starting point for subsequent developments throughout the text. The concept of a conserved scalar was also introduced in this chapter. Conservation equations for mixture fraction and mixture enthalpy, two conserved scalars, were developed, and are also cited in Table 7.2.

## NOMENCLATURE

| | |
|---|---|
| $A$ | Area (m²) |
| $c_p$ | Constant-pressure specific heat (J/kg-K) |
| $D_{ij}$ | Multicomponent diffusion coefficient (m²/s) |
| $D_j^T$ | Thermal diffusion coefficient (kg/m-s) |
| $\mathcal{D}_{ij}$ | Binary diffusion coefficient (m²/s) |
| $f$ | Mixture fraction, Eqn. 7.68 |
| $F$ | Force (N) |
| $F_{ij}$ | Components of matrix defined by Eqn.7.27 |

| | |
|---|---|
| $g$ | Gravitational acceleration $(m/s^2)$ |
| $h$ | Enthalpy $(J/kg)$ |
| $h_f^o$ | Enthalpy of formation $(J/kg)$ |
| $k$ | Thermal conductivity $(W/m\text{-}K)$ |
| $k_B$ | Boltzmann constant, $1.381 \cdot 10^{-23}$ $(J/K)$ |
| $L_{ij}$ | Components of matrix defined by Eqn. 7.28 |
| $Le$ | Lewis number, Eqn. 7.59 |
| $m$ | Mass $(kg)$ |
| $\dot{m}$ | Mass flowrate $(kg/s)$ |
| $\dot{m}''$ | Mass flux $(kg/s\text{-}m^2)$ |
| $\dot{m}'''$ | Mass production rate per unit volume $(kg/s\text{-}m^3)$ |
| $MW$ | Molecular weight $(kg/kmol)$ |
| $P$ | Pressure $(Pa)$ |
| $\dot{\mathbf{Q}}''$ | Heat flux vector, Eqn. 7.52 $(W/m^2)$ |
| $r$ | Radial coordinate $(m)$ |
| $R_u$ | Universal gas constant $(J/kmol\text{-}K)$ |
| $t$ | Time $(s)$ |
| $T$ | Temperature $(K)$ |
| $\mathbf{v}_i$ | Velocity of species $i$, Eqn. 7.12 $(m/s)$ |
| $v_r, v_\theta, v_x$ | Cylindrical coordinate velocity components $(m/s)$ |
| $v_r, v_\theta, v_\phi$ | Spherical coordinate velocity components $(m/s)$ |
| $v_x, v_y, v_z$ | Cartesian coordinate velocity components $(m/s)$ |
| $V$ | Volume $(m^3)$ |
| $\mathbf{V}$ | Velocity vector $(m/s)$ |
| $x$ | Cartesian or cylindrical axial coordinate $(m)$ |
| $y$ | Cartesian coordinate $(m)$ |
| $Y$ | Mass fraction $(kg/kg)$ |
| $z$ | Cartesian coordinate $(m)$ |

**Greek Symbols**

| | |
|---|---|
| $\alpha$ | Thermal diffusivity $(m^2/s)$ |
| $\delta_{ij}$ | Kronecker delta function |
| $\varepsilon_i$ | Characteristic Lennard–Jones energy |
| $\theta$ | Spherical coordinate circumferential angle $(rad)$ |
| $\mu$ | Absolute or dynamic viscosity $(N\text{-}s/m^2)$ |
| $\nu$ | Kinematic viscosity, $\mu/\rho$ $(m^2/s)$, or stoichiometric oxidizer-to-fuel ratio $(kg/kg)$ |
| $\rho$ | Density $(kg/m^3)$ |
| $\tau$ | Viscous stress $(N/m^2)$ |
| $\phi$ | Spherical coordinate azimuthal angle $(rad)$ |
| $\chi$ | Mole fraction $(kmol/kmol)$ |

**Subscripts**

| | |
|---|---|
| A | Species A |
| B | Species B |
| $cv$ | Control volume |

| diff | Diffusion |
|------|-----------|
| $f$ | Forced |
| $F$ | Fuel |
| $i$ | Species $i$ |
| $m$, mix | Mixture |
| $Ox$ | Oxidizer |
| $P$ | Pressure |
| $Pr$ | Products |
| ref | Reference state |
| stoic | Stoichiometric |
| $T$ | Thermal |
| $\chi$ | Ordinary |
| $\infty$ | Ambient condition |

# REFERENCES

1. Tseng, L.-K., Ismail, M. A., and Faeth, G. M., "Laminar Burning Velocities and Markstein Numbers of Hydrocarbon/Air Flames," *Combustion and Flame,* 95: 410–426 (1993).

2. Bird, R. B., Stewart, W. E., and Lightfoot, E. N., *Transport Phenomena,* John Wiley & Sons, New York, 1960.

3. Kee, R. J., Grcar, J. F., Smooke, M. D., and Miller, J. A., "A Fortran Program for Modeling Steady Laminar One-Dimensional Premixed Flames," Sandia National Laboratories Report SAND85-8240, 1991.

4. Lutz, A. E., Kee, R. J., Grcar, J. F., and Rupley, F. M., "OPPDIF: A Fortran Program for Computing Opposed-Flow Diffusion Flames," Sandia National Laboratories Report SAND96-8243, 1997.

5. Williams, F. A., *Combustion Theory,* 2nd Ed., Addison-Wesley, Redwood City, CA, 1985.

6. Kuo, K. K., *Principles of Combustion,* 2nd Ed., John Wiley & Sons, Hoboken, NJ, 2005.

7. Hirschfelder, J. O., Curtis, C. F., and Bird, R. B., *Molecular Theory of Gases and Liquids,* John Wiley & Sons, New York, 1954.

8. Grew, K. E., and Ibbs, T. L., *Thermal Diffusion in Gases,* Cambridge University Press, Cambridge, 1952.

9. Dixon-Lewis, G., "Flame Structure and Flame Reaction Kinetics, II. Transport Phenomena in Multicomponent Systems," *Proceedings of the Royal Society of London, Series A,* 307: 111–135 (1968).

10. Kee, R. J., Dixon-Lewis, G., Warnatz, J., Coltrin, M. E., and Miller, J. A., "A Fortran Computer Code Package for the Evaluation of Gas-Phase Multicomponent Transport Properties," Sandia National Laboratories Report SAND86-8246, 1990.

11. Schlichting, H., *Boundary-Layer Theory,* 6th Ed., McGraw-Hill, New York, 1968.

12. Bilger, R. W., "Turbulent Flows with Nonpremixed Reactants," in *Turbulent Reacting Flows* (P. A. Libby and F. A. Williams, eds.), Springer-Verlag, New York, 1980.

# REVIEW QUESTIONS

1. What are the three types of mass diffusion? Which one(s) do we neglect?

2. Discuss how the heat flux vector, $\dot{Q}''$, in a multicomponent mixture with diffusion differs from that used in a single-component gaseous system.

3. Define the Lewis number, $Le$, and discuss its physical significance. What role does the assumption that the Lewis number is unity play in simplifying the conservation of energy equation?

4. We considered three mass-average velocities in our discussion of species conservation: the bulk flow velocity, the individual species velocities, and the individual species diffusion velocities. Define each and discuss their physical significance. How do the various velocities relate to each other?

5. What does it mean for a conservation equation to be "sourceless." Give examples of governing equations that contain sources and those that are sourceless.

6. Discuss what is meant by a conserved scalar.

7. Define the mixture fraction.

# PROBLEMS

**7.1** With the aid of overall continuity (Eqn. 7.7), transform the left-hand side of the axial momentum equation, Eqn. 7.48, to a form involving the substantial (or material) derivative, where the cylindrical-coordinate substantial derivative operator is given by

$$\frac{D(\ )}{Dt} \equiv \frac{\partial(\ )}{\partial t} + v_r \frac{\partial(\ )}{\partial r} + \frac{v_\theta}{r} \frac{\partial(\ )}{\partial \theta} + v_x \frac{\partial(\ )}{\partial x}.$$

**7.2** Equation 7.55 expresses conservation of energy for a 1-D (cartesian) reacting flow where no assumptions are made regarding the form of the species transport law (i.e., Fick's law has not been invoked) or the relationship among properties (i.e., $Le = 1$ has not been assumed). Starting with this equation, derive the Shvab–Zeldovich energy equation (Eqn. 7.63) by applying Fick's law with effective binary diffusion and assuming $Le = 1$.

**7.3** Derive the conserved scalar equation for enthalpy for a 1-D spherical flow (Eqn. 7.82) starting with Eqn. 7.65.

**7.4** Derive the conserved scalar equation for enthalpy for an axisymmetric flow (Eqn. 7.83) starting with Eqn. 7.66.

**7.5** Consider the combustion of propane with air giving products CO, $CO_2$, $H_2O$, $H_2$, $O_2$, and $N_2$. Define the mixture fraction in terms of the various product species mole fractions, $\chi_i$.

**7.6\*** So-called "state relations" are frequently employed in the analysis of diffu-sion flames. These state relations relate various mixture properties to the mixture fraction, or other appropriate conserved scalars. Construct state relations for the adiabatic flame temperature $T_{ad}$ and mixture density $\rho$ for ideal combustion (no dissociation) of propane with air at 1 atm. Plot $T_{ad}$ and $\rho$ as functions of the mixture fraction $f$ for the range $0 \leq f \leq 0.12$.

**7.7** Laser-based techniques are used to measure the major species, $N_2$, $O_2$, $H_2$, and $H_2O$ (spontaneous Raman scattering), and the minor species, OH and NO (laser-induced fluorescence), in turbulent hydrogen jet flames burning in air. The hydrogen fuel is diluted, in some cases, with helium. A mixture fraction is defined as follows:

$$f = \frac{\left(MW_{H_2} + \alpha MW_{He}\right)([H_2O] + [H_2]) + (MW_H + \frac{\alpha}{2} MW_{He})[OH]}{A + B}$$

where

$$A = MW_{N_2}[N_2] + MW_{O_2}[O_2] + \left(MW_{H_2O} + \alpha MW_{He}\right)[H_2O]$$

and

$$B = \left(MW_{H_2} + \alpha MW_{He}\right)[H_2] + (MW_{OH} + \frac{\alpha}{2} MW_{He})[OH],$$

and where the concentrations $[X_i]$ are expressed as $kmol/m^3$ and $\alpha$ is the mole ratio of helium to hydrogen in the fuel stream. Here, it is assumed that there are no effects of differential diffusion and that the NO concen-tration is negligible.

Show that the mixture fraction defined above is the same as mass of fuel stuff per mass of mixture defined by Eqn. 7.68.

**7.8** Starting with Eqn. 7.51, derive the expanded form of energy conservation expressed by Eqn. 7.67. Neglect changes in kinetic energy.

**7.9** Consider the heat flux for a one-dimensional reacting flow:

$$\dot{Q}''_x = \dot{Q}''_{cond} + \dot{Q}''_{\substack{species \\ diff}}$$

Express the right-hand side of the heat flux expression in terms of the temperature and appropriate mass fluxes. Using this result, show that $\dot{Q}''_x$ simplifies to

$$\dot{Q}''_x = \rho \mathcal{D} c_p (1 - Le) \, dT/dx - \rho \mathcal{D} \, dh/dx,$$

---

subject to the assumption that a single binary diffusivity, $\mathcal{D}$, characterizes the mixture.

**7.10** Consider a jet diffusion flame in which the fuel is methanol vapor ($CH_3OH$) and the oxidizer is air. The species existing within the flame are $CH_3OH$, $CO$, $CO_2$, $O_2$, $H_2$, $H_2O$, $N_2$, and $OH$.

    A. Determine the numerical value of the **stoichiometric** mixture fraction.

    B. Write out an expression for the mixture fraction, $f$, in terms of the flame-species mass fractions, $Y_i$. Assume all pairs of binary diffusion coefficients are equal, i.e., no differential diffusion.

**7.11\*** Consider the Stefan problem (Fig. 3.4) where liquid normal hexane ($n\text{-}C_6H_{14}$) evaporates through a mixture of combustion products. The products can be treated as the ideal (no dissociation) products from combustion of $C_6H_{14}$ with simple air (21 percent $O_2$/79 percent $N_2$) at an equivalence ratio of 0.30. The temperature is fixed at 32°C and the pressure is 1 atm. The length of the tube is 20 cm and $Y_{C_6H_{14}}(L) = 0$.

    A. Determine the rate of hexane evaporation per unit area, $\dot{m}''_{C_6H_{14}}$, assuming an effective binary diffusivity where $\mathcal{D}_{im} \approx \mathcal{D}_{C_6H_{14}-N_2}$. Use the mass (not molar) flux solution to the Stefan problem. Evaluate the density as in Example 3.1.

    B. Determine $\dot{m}''_{C_6H_{14}}$ using an effective binary diffusivity that considers all five species present. Use average mole fractions, i.e., $\chi_i = 0.5\,(\chi_i(0) + \chi_i(L))$, for your computation of $\mathcal{D}_{im}$. Again, use the mass (not molar) flux solution. How does the result compare with that in part A?

**7.12** Consider a mixture containing an equal number of moles of He, $O_2$, and $CH_4$. Determine the multicomponent diffusion coefficients associated with this mixture at 500 K and 1 atm.

# chapter

# 8

# Laminar Premixed Flames

## OVERVIEW

In previous chapters, we introduced the concepts of mass transfer (Chapter 3) and chemical kinetics (Chapters 4 and 5) and linked them with familiar thermodynamic and heat transfer concepts in Chapters 6 and 7. Understanding premixed laminar flames requires us to utilize all of these concepts. Our development in Chapter 7 of the one-dimensional conservation equations for a reacting flow will be the starting point for analyzing laminar flames.

Laminar premixed flames, frequently in conjunction with diffusion flames, have application in many residential, commercial, and industrial devices and processes. Examples include gas ranges and ovens, heating appliances, and Bunsen burners. An advanced cooktop burner for a gas range is illustrated in Fig. 8.1. Laminar, premixed, natural-gas flames also are frequently employed in the manufacturing of glass products. As suggested by the examples given above, laminar premixed flames are by themselves important; but, perhaps more importantly, understanding laminar flames is a necessary prerequisite to the study of turbulent flames. In both laminar and turbulent flows, the same physical processes are active, and many turbulent flame theories are based on an underlying laminar flame structure. In this chapter, we will qualitatively describe the essential characteristics of laminar premixed flames and develop a simplified analysis of these flames that allows us to see what factors influence the laminar flame speed and the flame thickness. A detailed analysis using state-of-the-art methods will illustrate the power of numerical simulations in understanding flame structure. We will also examine experimental data that illustrate how equivalence ratio, temperature, pressure, and fuel type affect flame speed and flame thickness. Flame speed is emphasized because it is this property that dictates flame shape and important flame-stability characteristics, such as blowoff and flashback. The chapter concludes with discussion of flammability limits and ignition and extinction phenomena.

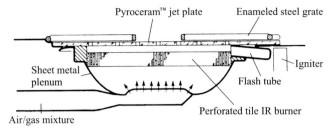

**Figure 8.1**    Advanced residential cooktop burner for gas ranges.
SOURCE: Courtesy of the Gas Research Institute.

## PHYSICAL DESCRIPTION

### Definition

Before proceeding, it is useful to define what we mean by a flame. A **flame** is a self-sustaining propagation of a localized combustion zone at subsonic velocities. There are several key words in this definition. First, we require a flame to be localized; that is, the flame occupies only a small portion of the combustible mixture at any one time. This is in contrast to the various homogeneous reactors we studied in Chapter 6, where reaction was assumed to occur uniformly throughout the reaction vessel. The second key word is subsonic. A discrete combustion wave that travels subsonically is termed a **deflagration.** It is also possible for combustion waves to propagate at supersonic velocities. Such a wave is called a **detonation.** The fundamental propagation mechanisms are different in deflagrations and detonations, and, because of this, these are distinct phenomena. Detonations are discussed in Chapter 16.

### Principal Characteristics

The temperature profile through a flame is perhaps its most important characteristic. Figure 8.2 illustrates a typical flame temperature profile, together with other essential flame features.

   To understand this figure, we need to establish a reference frame for our coordinate system. A flame may be freely propagating, as occurs when a flame is initiated in a tube containing a combustible gas mixture. The appropriate coordinate system would be fixed to the propagating combustion wave. An observer riding with the flame would experience the unburned mixture approaching at the **flame speed, $S_L$.** This is equivalent to a flat flame stabilized on a burner. Here, the flame is stationary relative to the laboratory reference frame and, once again, the reactants enter the flame with a velocity equal to the flame propagation velocity, $S_L$. In both examples, we assume that the flame is one dimensional and that the unburned gas enters the flame in a direction normal to the flame sheet. Since a flame creates hot products, the product density is less than the reactant density. Continuity thus requires that the burned gas velocity be greater than the velocity of the unburned gas:

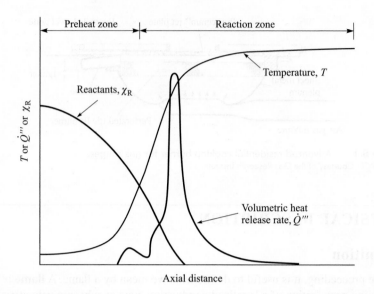

**Figure 8.2**    Laminar flame structure. Temperature and heat-release-rate profiles based on experiments of Friedman and Burke [1].

$$\rho_u S_L A \equiv \rho_u v_u A = \rho_b v_b A, \tag{8.1}$$

where the subscripts $u$ and $b$ refer to the unburned and burned gases, respectively. For a typical hydrocarbon–air flame at atmospheric pressure, the density ratio is approximately seven. Thus, there is considerable acceleration of the gas flow across the flame.

It is convenient to divide a flame into two zones: the **preheat zone,** where little heat is released; and the **reaction zone,** where the bulk of the chemical energy is released. At atmospheric pressure, the flame thickness is quite thin, on the order of a millimeter. It is useful to divide the reaction zone further into a thin region of very fast chemistry followed by a much wider region of slow chemistry. The destruction of the fuel molecules and the creation of many intermediate species occur in the fast-chemistry region. This region is dominated by bimolecular reactions. At atmospheric pressure, the fast-reaction zone is quite thin, typically less than a millimeter. Because this zone is thin, temperature gradients and species concentration gradients are very large. These gradients provide the driving forces that cause the flame to be self-sustaining: the diffusion of heat and radical species from the reaction zone to the preheat zone. In the secondary reaction zone, the chemistry is dominated by three-body radical recombination reactions, which are much slower than typical bimolecular reactions, and the final burnout of CO via $CO + OH \rightarrow CO_2 + H$. This secondary reaction zone may extend several millimeters in a 1-atm flame. Later in this chapter, we present a more detailed description of flame structure illustrating these ideas. Additional information may also be found in Fristrom [2].

Hydrocarbon flames are also characterized by their visible radiation. With an excess of air, the fast-reaction zone appears blue. This blue radiation results from excited CH radicals in the high-temperature zone. When the air is decreased to less than stoichiometric proportions, the zone appears blue-green, now as a result of radiation from excited $C_2$. In both flames, OH radicals also contribute to the visible radiation, and to a lesser degree, chemiluminescence from the reaction $CO + O \rightarrow CO_2 + h\nu$ [3]. If the flame is made richer still, soot will form, with its consequent blackbody continuum radiation. Although the soot radiation has its maximum intensity in the infrared (recall Wien's law?), the spectral sensitivity of the human eye causes us to see a bright yellow (nearly white) to dull orange emission, depending on the flame temperature. References [4] and [5] provide a wealth of information on radiation from flames.

## Typical Laboratory Flames

The Bunsen-burner flame provides an interesting example of laminar premixed flames with which most students have some familiarity and that can be easily used in classroom demonstrations. Figure 8.3a schematically illustrates a Bunsen burner and the flame it produces. A jet of fuel at the base induces a flow of air through the variable area port, and the air and fuel mix as they flow up through the tube. The typical Bunsen-burner flame is a dual flame: a fuel-rich premixed inner flame surrounded by a diffusion flame. The secondary diffusion flame results when the carbon monoxide and hydrogen products from the rich inner flame encounter the ambient air. The shape of the flame is determined by the combined effects of the velocity profile and heat losses to the tube wall. For the flame to remain stationary, the flame speed must equal the speed of the normal component of unburned gas at each location, as illustrated in the vector diagram in Fig. 8.3b. Thus,

$$S_L = v_u \sin \alpha, \tag{8.2}$$

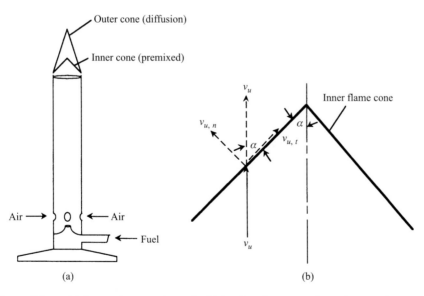

(a)                                            (b)

**Figure 8.3**     (a) Bunsen-burner schematic. (b) Laminar flame speed equals normal component of unburned gas velocity, $v_{u,n}$.

where $S_L$ is the laminar burning velocity. This principle causes the essential conical character of the flame.

One-dimensional flat flames are frequently studied in the laboratory and are also used in some radiant heating burners (Fig. 8.4). Figure 8.5 illustrates the laboratory

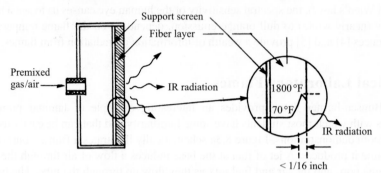

Fiber matrix burner (ceramic or metal)

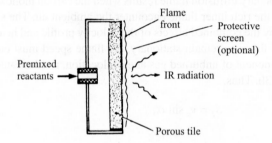

Porous ceramic foam burner

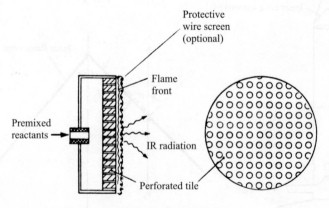

Ported ceramic tile burner

**Figure 8.4**    Direct-fired radiant burners provide uniform heat flux and high efficiency.
SOURCE: Reprinted with permission from the Center for Advanced Materials, *Newsletter*, (1), 1990, Penn State University.

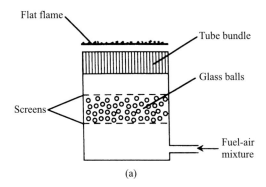

(a)

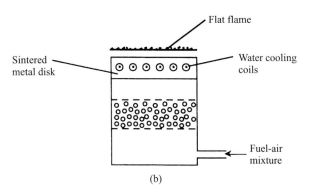

(b)

**Figure 8.5**     (a) Adiabatic flat-flame burner. (b) Nonadiabatic flat-flame burner.

genre. In the adiabatic burner, a flame is stabilized over a bundle of small tubes through which the fuel–air mixture passes laminarly [6]. Over a narrow range of conditions, a stable flat flame is produced. The nonadiabatic burner utilizes a water-cooled face that allows heat to be extracted from the flame, which, in turn, decreases the flame speed, allowing flames to be stabilized over a relatively wide range of flow conditions [7].

---

A premixed laminar flame is stabilized in a one-dimensional gas flow where the vertical velocity of the unburned mixture, $v_u$, varies linearly with the horizontal coordinate, $x$, as shown in the lower half of Fig. 8.6. Determine the flame shape and the distribution of the local angle of the flame surface from vertical. Assume the flame speed is independent of position and equal to 0.4 m/s, a nominal value for a stoichiometric methane–air flame.

**Example 8.1**

**Solution**

From Fig. 8.7, we see that the local angle, $\alpha$, which the flame sheet makes with a vertical plane is (Eqn. 8.2),

$$\alpha = \sin^{-1}(S_L / v_u),$$

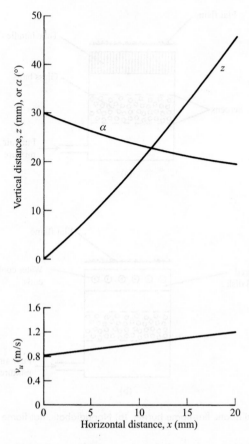

**Figure 8.6**     Flow velocity, flame position, and angle from vertical of line tangent to flame, for Example 8.1.

where, from Fig. 8.6,

$$v_u(\text{mm/s}) = 800 + \frac{1200 - 800}{20} x \,(\text{mm}).$$

So,

$$\alpha = \sin^{-1}\left(\frac{400}{800 + 20x \,(\text{mm})}\right),$$

and has values ranging from 30° at $x = 0$ to 19.5° at $x = 20$ mm, as shown in the top part of Fig. 8.6.

To calculate the flame position, we first obtain an expression for the local slope of the flame sheet ($dz/dx$) in the $x$–$z$ plane, and then integrate this expression with respect to $x$ to find $z(x)$. From Fig. 8.7, we see that

$$\frac{dz}{dx} = \tan \beta = \left(\frac{v_u^2(x) - S_L^2}{S_L^2}\right)^{1/2},$$

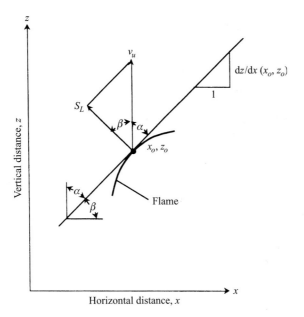

**Figure 8.7**    Definition of flame geometry for Example 8.1.

which, for $v_u \equiv A + Bx$, becomes

$$\frac{dz}{dx} = \left[ \left( \frac{A}{S_L} + \frac{Bx}{S_L} \right)^2 - 1 \right]^{1/2}.$$

Integrating the above with $A/S_L = 2$ and $B/S_L = 0.05$ yields

$$z(x) = \int_0^x \left( \frac{dz}{dx} \right) dx$$

$$= (x^2 + 80x + 1200)^{1/2} \left( \frac{x}{40} + 1 \right)$$

$$- 10 \ln[(x^2 + 80x + 1200)^{1/2} + (x + 40)]$$

$$- 20\sqrt{3} + 10 \ln(20\sqrt{3} + 40).$$

The flame position $z(x)$ is plotted in the upper half of Fig. 8.6. Here we see that the flame sheet is quite steeply inclined. (Note that the horizontal scale is twice that of the vertical.)

**Comment**

From this example, we see how the flame shape is intimately linked to the velocity distribution of the oncoming unburned gas.

In the next section, we turn our attention to establishing some theoretical basis for how various parameters, such as pressure, temperature, and fuel type, affect laminar flame speeds.

## SIMPLIFIED ANALYSIS

Theories of laminar flames abound and have occupied many researchers for many decades. For example, Kuo [8] cites more than a dozen major papers dealing with laminar flame theory published between 1940 and 1980. Various approaches have assumed that either heat diffusion or mass diffusion effects dominate, while other detailed theories include both effects, assuming that both phenomena are important. The earliest description of a laminar flame is that of Mallard and Le Chatelier [9] in 1883. Our simplified approach here follows that of Spalding [10], which lays bare the essential physics of the problem without a great deal of mathematics. The analysis couples principles of heat transfer, mass transfer, chemical kinetics, and thermodynamics to understanding the factors governing flame speed and thickness. The simplified analysis presented below relies on the one-dimensional conservation relations developed in the previous chapter, with additional simplifying assumptions for thermodynamic and transport properties applied. Our objective is to find a simple analytical expression for the laminar flame speed.

## Assumptions

1. One-dimensional, constant-area, steady flow.

2. Kinetic and potential energies, viscous shear work, and thermal radiation are all neglected.

3. The small pressure difference across the flame is neglected; thus, the pressure is constant.

4. The diffusion of heat and mass are governed by Fourier's and Fick's laws, respectively. Binary diffusion is assumed.

5. The **Lewis number, *Le*,** which expresses the ratio of thermal diffusivity to mass diffusivity, i.e.,

$$Le \equiv \frac{\alpha}{\mathcal{D}} = \frac{k}{\rho c_p \mathcal{D}}, \tag{8.3}$$

is unity. This has the result that $k/c_p = \rho \mathcal{D}$, which greatly simplifies the energy equation.

6. The mixture specific heat depends neither on temperature nor on the mixture composition. This is equivalent to assuming that the individual species specific heats are all equal and constant.

7. Fuel and oxidizer form products in a single-step exothermic reaction.

8. The oxidizer is present in stoichiometric or excess proportions; thus, the fuel is completely consumed at the flame.

## Conservation Laws

To understand flame propagation, we apply conservation of mass, species, and energy to the differential control volume illustrated in Fig. 8.8. Using the

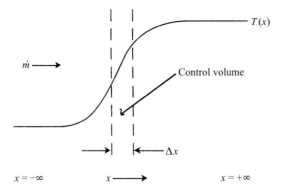

**Figure 8.8**     Control volume for flame analysis.

relationships from Chapter 7, these conservation principles are expressed as follows:

**Mass Conservation**

$$\frac{d(\rho v_x)}{dx} = 0 \tag{7.4a}$$

or

$$\dot{m}'' = \rho v_x = \text{constant.} \tag{7.4b}$$

**Species Conservation**

$$\frac{d\dot{m}_i''}{dx} = \dot{m}_i''' \tag{7.9}$$

or, with the application of Fick's law,

$$\frac{d\left[\dot{m}''Y_i - \rho\mathcal{D}\,\dfrac{dY_i}{dx}\right]}{dx} = \dot{m}_i''', \tag{7.8}$$

where $\dot{m}_i'''$ is the mass production rate of species $i$ per unit volume (kg/s-m$^3$).

Equation 7.8 can be written for each of the three species, where the mass production rates of the oxidizer and products are related to the fuel production rate. Obviously, the production rates of the fuel, $\dot{m}_F'''$, and oxidizer, $\dot{m}_{Ox}'''$, are negative since these species are being consumed, not produced. For our simple reaction, the overall stoichiometry is

$$1 \text{ kg fuel} + \nu \text{ kg oxidizer} \rightarrow (\nu+1) \text{ kg products.} \tag{8.4}$$

Thus,

$$\dot{m}_F''' = \frac{1}{v}\dot{m}_{Ox}''' = -\frac{1}{v+1}\dot{m}_{Pr}'''. \tag{8.5}$$

Equation 7.8 becomes for each species:

*Fuel*

$$\dot{m}''\frac{dY_F}{dx} - \frac{d\left(\rho D\dfrac{dY_F}{dx}\right)}{dx} = \dot{m}_F''', \tag{8.6a}$$

*Oxidizer*

$$\dot{m}''\frac{dY_{Ox}}{dx} - \frac{d\left(\rho D\dfrac{dY_{Ox}}{dx}\right)}{dx} = v\dot{m}_F''', \tag{8.6b}$$

*Products*

$$\dot{m}''\frac{dY_{Pr}}{dx} - \frac{d\left(\rho D\dfrac{dY_{Pr}}{dx}\right)}{dx} = -(v+1)\dot{m}_F'''. \tag{8.6c}$$

In this analysis, the species conservation relations are used only to simplify the energy equation. Because of the assumptions of binary diffusion governed by Fick's law and unity Lewis number, there will be no need to solve the species equations.

**Energy Conservation**    The assumptions we have adopted for our analysis are consistent with those embodied in the Shvab–Zeldovich form of energy conservation (Eqn. 7.63),

$$\dot{m}''c_p\frac{dT}{dx} - \frac{d}{dx}\left[(\rho D c_p)\frac{dT}{dx}\right] = -\sum h_{f,i}^o\dot{m}_i'''. \tag{8.7a}$$

With the overall stoichiometry expressed by Eqns. 8.4 and 8.5, the right-hand side of the above becomes

$$-\sum h_{f,i}^o\dot{m}_i''' = -\left[h_{f,F}^o\dot{m}_F''' + h_{f,Ox}^o v\dot{m}_F''' - h_{f,Pr}^o(v+1)\dot{m}_F'''\right]$$

or

$$-\sum h_{f,i}^o\dot{m}_i''' = -\dot{m}_F'''\Delta h_c,$$

where $\Delta h_c$ is the heat of combustion of the fuel, $\Delta h_c \equiv h_{f,F}^o + vh_{f,Ox}^o - (v+1)h_{f,Pr}^o$, based on the given stoichiometry. Because of the unity Lewis number approximation, we also can replace $\rho D c_p$ with $k$. With these two substitutions, Eqn. 8.7a becomes

$$\dot{m}''\frac{dT}{dx} - \frac{1}{c_p}\frac{d\left(k\dfrac{dT}{dx}\right)}{dx} = -\frac{\dot{m}_F'''\Delta h_c}{c_p}. \tag{8.7b}$$

Recall that our objective is to find a useful expression for the laminar flame speed, which is related simply to the mass flux, $\dot{m}''$, appearing in Eqn. 8.7b, by

$$\dot{m}'' = \rho_u S_L. \tag{8.8}$$

To achieve this objective, we again follow Spalding's [10] approach.

## Solution

To find the mass burning rate, we will assume a temperature profile that satisfies the boundary conditions given below and then integrate Eqn. 8.7b using the assumed temperature distribution. The boundary conditions far upstream of the flame are

$$T(x \rightarrow -\infty) = T_u, \tag{8.9a}$$

$$\frac{\mathrm{d}T}{\mathrm{d}x}(\rightarrow -\infty) = 0, \tag{8.9b}$$

and far downstream of the flame,

$$T(x \rightarrow +\infty) = T_b, \tag{8.9c}$$

$$\frac{\mathrm{d}T}{\mathrm{d}x}(x \rightarrow +\infty) = 0. \tag{8.9d}$$

For simplicity, we assume a simple linear temperature profile that goes from $T_u$ to $T_b$ over the small distance, $\delta$, as shown in Fig. 8.9. We define $\delta$ to be the flame thickness. Mathematically, we have a second-order ordinary differential equation (Eqn. 8.7b) with two unknown parameters, which are referred to in the combustion literature as the **eigenvalues,** $\dot{m}''$ and $\delta$. The specification of *four* boundary conditions, rather than just *two*, allows us to determine the eigenvalues. (It is interesting to note the similarity of the present analysis to von Karman's integral analysis of the flat-plate boundary layer that you may have studied in fluid mechanics. In the fluid mechanics problem, reasonable estimates of boundary layer thickness and shear stresses were obtained using an assumed velocity profile.)

Integrating Eqn. 8.7b over $x$, noting that the contributions of the discontinuities in $\mathrm{d}T/\mathrm{d}x$ at $x = 0$ and $\delta$ cancel, and applying the conditions at $-\infty$ and $+\infty$, yields

$$\dot{m}''[T]_{T=T_u}^{T=T_b} - \frac{k}{c_p}\left[\frac{\mathrm{d}T}{\mathrm{d}x}\right]_{\mathrm{d}T/\mathrm{d}x=0}^{\mathrm{d}T/\mathrm{d}x=0} = \frac{-\Delta h_c}{c_p}\int\limits_{-\infty}^{\infty} \dot{m}_F''' \, \mathrm{d}x \tag{8.10}$$

which, by evaluating the limits, simplifies to

$$\dot{m}''(T_b - T_u) = -\frac{\Delta h_c}{c_p}\int\limits_{-\infty}^{\infty} \dot{m}_F''' \, \mathrm{d}x. \tag{8.11}$$

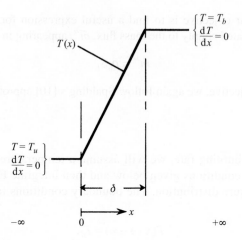

**Figure 8.9**   Assumed temperature profile for laminar premixed flame analysis.

We can change the limits on the reaction rate integral appearing on the right-hand side of Eqn. 8.11 from space to temperature, since $\dot{m}_F'''$ is only nonzero between $T_u$ and $T_b$ over the region $\delta$, i.e.,

$$\frac{dT}{dx} = \frac{T_b - T_u}{\delta} \quad \text{or} \quad dx = \frac{\delta}{T_b - T_u} dT. \tag{8.12}$$

With the change of variables,

$$\dot{m}''(T_b - T_u) = -\frac{\Delta h_c}{c_p} \frac{\delta}{(T_b - T_u)} \int_{T_u}^{T_b} \dot{m}_F''' \, dT, \tag{8.13}$$

and, recognizing the definition of the average reaction rate,

$$\bar{\dot{m}}_F''' \equiv \frac{1}{(T_b - T_u)} \int_{T_u}^{T_b} \dot{m}_F''' \, dT, \tag{8.14}$$

we obtain the simple result that

$$\dot{m}''(T_b - T_u) = -\frac{\Delta h_c}{c_p} \delta \bar{\dot{m}}_F'''. \tag{8.15}$$

This result, Eqn. 8.15, is a simple algebraic equation involving the two unknowns $\dot{m}''$ and $\delta$; thus, we need to find another equation to complete the solution. This can be done by following the same procedure as before, but now integrating from $x = -\infty$ to $x = \delta/2$. Since the reaction zone of a flame lies in the high-temperature region, it is reasonable to assume that $\dot{m}_F'''$ is zero in the interval $-\infty < x \leq \delta/2$. Noting that at $x = \delta/2$,

$$T = \frac{T_b + T_u}{2} \tag{8.16}$$

and

$$\frac{dT}{dx} = \frac{T_b - T_u}{\delta},$$

(8.12)

we obtain from Eqn. 8.10, with the modified limits,

$$\dot{m}''\delta/2 - k/c_p = 0.$$

(8.17)

Solving Eqns. 8.15 and 8.17 simultaneously yields

$$\dot{m}'' = \left[ 2\frac{k}{c_p^2} \frac{(-\Delta h_c)}{(T_b - T_u)} \bar{\dot{m}}_F''' \right]^{1/2}$$

(8.18)

and

$$\delta = 2k/(c_p \dot{m}'').$$

(8.19)

Applying the definitions of flame speed, $S_L \equiv \dot{m}''/\rho_u$, and thermal diffusivity, $\alpha \equiv k/\rho_u c_p$, and recognizing that $\Delta h_c = (\nu + 1)c_p(T_b - T_u)$, we obtain the final results:

$$S_L = \left[ -2\alpha(\nu+1)\frac{\bar{\dot{m}}_F'''}{\rho_u} \right]^{1/2}$$

(8.20)

$$\delta = \left[ \frac{-2\rho_u \alpha}{(\nu+1)\bar{\dot{m}}_F'''} \right]^{1/2}$$

(8.21a)

or, in terms of $S_L$,

$$\delta = 2\alpha/S_L.$$

(8.21b)

We can now use Eqns. 8.20 and 8.21 to analyze how $S_L$ and $\delta$ theoretically are affected by the fuel–air mixture properties. This is done in the next section, where, also, comparisons are made with experimental observations.

---

Estimate the laminar flame speed of a stoichiometric propane–air mixture using the simplified-theory result (Eqn. 8.20). Use the global one-step reaction mechanism (Eqn. 5.2, Table 5.1) to estimate the mean reaction rate.

**Example 8.2**

**Solution**

To find the laminar flame speed, we evaluate Eqn. 8.20,

$$S_L = \left[ -2\alpha(\nu+1)\frac{\bar{\dot{m}}_F'''}{\rho_u} \right]^{1/2}.$$

The essence of this problem is how to evaluate $\bar{\dot{m}}_F'''$ and $\alpha$. Since the simple theory assumed that the reaction was confined to the second half of the flame thickness $(\delta/2 < x < \delta)$, we will choose a mean temperature to evaluate the reaction rate as

$$\bar{T} = \tfrac{1}{2}\left(\tfrac{1}{2}(T_b + T_u) + T_b\right)$$
$$= 1770 \text{ K},$$

where we assume $T_b = T_{ad} = 2260$ K (Chapter 2) and $T_u = 300$ K. Assuming there is neither fuel nor oxygen in the burned gas, the mean concentrations used in the rate equation are

$$\bar{Y}_F = \tfrac{1}{2}(Y_{F,u} + 0)$$
$$= 0.06015/2 = 0.0301$$

and

$$\bar{Y}_{O_2} = \tfrac{1}{2}[0.2331(1 - Y_{F,u}) + 0]$$
$$= 0.1095,$$

where the $A/F$ of a stoichiometric propane–air mixture is $15.625$ ($= \nu$) and the mass fraction of $O_2$ in the air is $0.233$.

The reaction rate, given by

$$\dot{\omega}_F \equiv \frac{d[\text{C}_3\text{H}_8]}{dt} = -k_G[\text{C}_3\text{H}_8]^{0.1}[\text{O}_2]^{1.65},$$

with

$$k_G = 4.836 \cdot 10^9 \exp\left(\frac{-15,098}{T}\right) [=] \left(\frac{\text{kmol}}{\text{m}^3}\right)^{-0.75} \frac{1}{\text{s}},$$

can be transformed to

$$\bar{\dot{\omega}}_F = -k_G(\bar{T})\bar{\rho}^{1.75} \left(\frac{\bar{Y}_F}{MW_F}\right)^{0.1} \left(\frac{\bar{Y}_{O_2}}{MW_{O_2}}\right)^{1.65},$$

where we now use our judiciously selected mean values. Evaluating, using the appropriate units conversion for Table 5.1 (see footnote $a$ in this table and Problem 5.14), gives

$$k_G = 4.836 \cdot 10^9 \exp\left(\frac{-15,098}{1770}\right) = 9.55 \cdot 10^5 \left(\frac{\text{kmol}}{\text{m}^3}\right)^{-0.75} \frac{1}{\text{s}},$$

$$\bar{\rho} = \frac{P}{(R_u/MW)\,\bar{T}} = \frac{101,325}{(8315/29)1770} = 0.1997 \text{ kg/m}^3,$$

$$\bar{\dot{\omega}}_F = -9.55 \cdot 10^5 (0.1997)^{1.75} \left(\frac{0.0301}{44}\right)^{0.1} \left(\frac{0.1095}{32}\right)^{1.65}$$

$$= -2.439 \text{ kmol/s-m}^3.$$

Then, from Eqn. 6.29,

$$\bar{m}_F''' = \bar{\omega}_F M W_F = -2.439(44)$$
$$= -107.3 \, \text{kg/s-m}^3.$$

The thermal diffusivity employed in Eqn. 8.20 is defined as

$$\alpha = \frac{k(\bar{T})}{\rho_u c_p(\bar{T})}.$$

The appropriate mean temperature, however, is now the average over the entire flame thickness $(0 \le x \le \delta)$ since conduction occurs over this interval, not just the half-interval as assumed for the reaction. Thus,

$$\bar{T} = \tfrac{1}{2}(T_b + T_u)$$
$$= 1280 \, \text{K}$$

and

$$\alpha = \frac{0.0809}{1.16(1186)} = 5.89 \cdot 10^{-5} \, \text{m}^2/\text{s},$$

where air properties were used to evaluate $k$, $c_p$, and $\rho$.

We can now substitute numerical values into Eqn. 8.20:

$$S_L = \left[ \frac{-2(5.89 \cdot 10^{-5})(15.625 + 1)(-107.3)}{1.16} \right]^{1/2}$$

$$\boxed{S_L = 0.425 \, \text{m/s or } 42.5 \, \text{cm/s}}$$

**Comment**

From correlations yet to be discussed [11], the experimental value of $S_L$ for this mixture is 38.9 cm/s. Considering the crude nature of our theoretical analysis, this is fortuitous agreement with the calculated value of 42.5 cm/s. Of course, rigorous theory with detailed kinetics can be used to obtain much more accurate predictions. We also note that, with the assumptions embodied in our analysis, the fuel and oxidizer concentrations can be related linearly to the temperature and $\bar{m}_F'''$ can be evaluated exactly by integrating the single-step, irreversible global reaction rate, rather than using estimated mean concentrations and temperature as was done in the example. We also note that the numerical values for the pre-exponential factors from Table 5.1 are not used directly, but first were converted to SI units.

# DETAILED ANALYSIS[1]

Numerical simulations of laminar premixed flames employing both detailed chemical kinetics and mixture transport properties have become standard tools for combustion engineers and scientists. Many simulations are based on the CHEMKIN library

[1]The mathematical development in this section may be omitted without loss of continuity and the reader may wish to skip to the subsection discussing flame structure after reading the brief introductory paragraph.

of Fortran main programs and subroutines, which, for our purposes here, include those described in Refs. [12–16].

## Governing Equations

The basic conservation equations describing steady, one-dimensional flames were developed in Chapter 7 and are repeated below:

**Continuity**

$$\frac{d\dot{m}''}{dx} = 0 \tag{7.4a}$$

**Species Conservation**   Simplifying for steady one-dimensional flow and substituting $\dot{\omega}_i M W_i$ for $\dot{m}_i'''$, Eqn. 7.16 becomes

$$\dot{m}'' \frac{dY_i}{dx} + \frac{d}{dx}(\rho Y_i v_{i,\,\text{diff}}) = \dot{\omega}_i M W_i \quad \text{for} = 1, 2, \ldots, N \text{ species,} \tag{8.22}$$

where the species molar production rates $\dot{\omega}_i$ are defined by Eqns. 4.31–4.33.

**Energy Conservation**   Again, replacing $\dot{m}_i'''$ with $\dot{\omega}_i M W_i$, Eqn. 7.67 becomes

$$\dot{m}'' c_p \frac{dT}{dx} + \frac{d}{dx}\left(-k\frac{dT}{dx}\right) + \sum_{i=1}^{N} \rho Y_i v_{i,\,\text{diff}} c_{p,\,i} \frac{dT}{dx} = -\sum_{i=1}^{N} h_i \dot{\omega}_i M W_i. \tag{8.23}$$

Note that momentum conservation is not explicitly required since we assume that the pressure is constant, as in our simplified flame analysis. In addition to these conservation equations, the following ancillary relations or data are required:

- Ideal-gas equation of state (Eqn. 2.2).
- Constitutive relations for diffusion velocities (Eqns. 7.23 and 7.25 or Eqn. 7.31).
- Temperature-dependent species properties: $h_i(T)$, $c_{p,\,i}(T)$, $k_i(T)$, and $\mathcal{D}_{ij}(T)$.
- Mixture property relations to calculate $MW_{\text{mix}}$, $k$, $D_{ij}$, and $D_i^T$ from individual species properties and mole (or mass) fractions (e.g., for the $D_{ij}$s, Eqn. 7.26).
- A detailed chemical kinetic mechanism to obtain the $\dot{\omega}_i$s (e.g., Table 5.3).
- Interconversion relations for $\chi_i$s, $Y_i$s, and $[X_i]$s (Eqns. 6A.1–6A.10).

## Boundary Conditions

The conservation relations (Eqns. 7.4, 8.22, and 8.23) described a **boundary value problem;** i.e., given information about the unknown functions ($T$, $Y_i$) at an upstream location (boundary) and a downstream location (boundary), the problem at hand is to determine the functions $T(x)$ and $Y_i(x)$ between these boundaries. To complete our detailed mathematical description of the flame now requires that we specify appropriate boundary conditions for the species and energy conservation equations. In our analysis, we assume that the flame is freely propagating and, thus, attach the coordinate system to the flame.

Equation 8.23 is simply, and clearly, second-order in only $T$ and, thus, requires two boundary conditions:

$$T(x \to -\infty) = T_u \qquad (8.24a)$$

$$\frac{dT}{dx}(x \to +\infty) = 0. \qquad (8.24b)$$

Of course, in a numerical solution, the domain $-\infty < x < \infty$ is severely truncated, with the boundaries separated by only a few centimeters.

The appearance of both $Y_i$ and $v_{i,\,diff}$ in derivatives results in the species conservation equation (Eqn. 8.22) being first order in both $Y_i$ and $v_{i,\,diff}$. We note, however, that the constitutive relations defining $v_{i,\,diff}$ (e.g., Eqns. 7.23 and 7.31) relate $v_{i,\,diff}$ to a concentration gradient, either $d\chi_i/dx$ or $dY_i/dx$; thus, one might alternatively consider Eqn. 8.22 to be second order in $Y_i$. Appropriate conditions are that the $Y_i$s are known values far upstream and that the mass-fraction gradients become vanishingly small far downstream:

$$Y_i(x \to -\infty) = Y_{i,o} \qquad (8.25a)$$

$$\frac{dY_i}{dx}(x \to +\infty) = 0. \qquad (8.25b)$$

As we discussed earlier in our simplified analysis of the premixed flame, the mass flux $\dot{m}''$ is not known a priori but is an eigenvalue to the problem—its value is a part of the solution to the problem. To determine the value of $\dot{m}''$ simultaneously with the $T(x)$ and $Y_i(x)$ functions, overall continuity, Eqn. 7.4, is explicitly required to be solved along with the species and energy conservation equations. This additional first-order equation necessitates the specification of an additional boundary condition. In the Sandia code [16], this is achieved for a freely propagating flame by specifying the temperature at a fixed location; i.e., one explicitly fixes the coordinate system to move with the flame:[2]

$$T(x_1) = T_1. \qquad (8.26)$$

---

[2]The Sandia code also allows the modeling of burner-attached flames. For this, and other details, the reader is referred to Ref. [16].

With this, our model of a freely propagating 1-D flame is complete. A discussion of the numerical techniques employed to obtain a solution to this model can be found in Ref. [16].

Before we look at an application of this analysis to the simulation of a particular flame, it should be pointed out that the problem may be alternatively cast as an unsteady one, whose steady-state solution is the desideratum. In this formulation, the unsteady terms, $\partial \rho / \partial t$, $\partial (\rho Y_i) / \partial t$, and $c_p \partial (\rho T) / \partial t$, are added to mass conservation (Eqn. 7.4), species conservation (Eqn. 8.22), and energy conservation (Eqn. 8.23), respectively. For further information on this approach, the reader is referred to Ref. [17].

## Structure of CH₄–Air Flame

We now use the above analysis to obtain an understanding of the detailed structure of a premixed flame. Figure 8.10 shows the temperature distribution and selected species mole-fraction profiles through a 1-atm, stoichiometric, CH₄–air flame, simulated

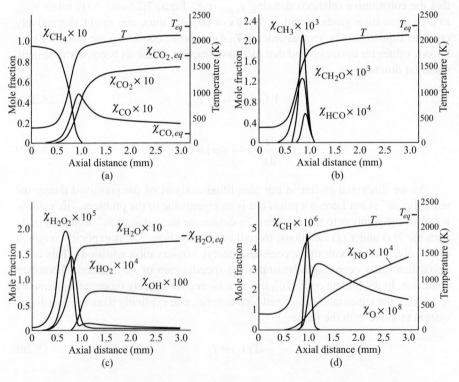

**Figure 8.10**   Calculated species mole-fraction and temperature profiles for laminar, stoichiometric, CH₄–air premixed flame. (a) $T$, $\chi_{CH_4}$, $\chi_{CO}$, and $\chi_{CO_2}$; (b) $T$, $\chi_{CH_3}$, $\chi_{CH_2O}$, and $\chi_{HCO}$; (c) $\chi_{H_2O}$, $\chi_{OH}$, $\chi_{H_2O_2}$, and $\chi_{HO_2}$; (d) $T$, $\chi_{CH}$, $\chi_O$, and $\chi_{NO}$.

using CHEMKIN library codes [14–16] with GRI-Mech 2.11 methane combustion kinetics [18]. The first panel in Fig. 8.10 shows the principal C-containing species $CH_4$, CO, and $CO_2$. Here we see the disappearance of the fuel, the appearance of the intermediate species CO, and burnout of the CO to form $CO_2$. The CO concentration has its peak value at approximately the same location where the $CH_4$ concentration goes to zero, whereas the $CO_2$ concentration at first lags the CO concentration but then continues to rise as the CO is oxidized. Figure 8.11 provides additional insight into the $CH_4 \rightarrow CO \rightarrow CO_2$ sequence by showing the local molar production (destruction) rates for these species. We see that the peak fuel destruction rate nominally corresponds with the peak CO production rate and that the $CO_2$ production rate initially lags that of the CO. Even before the location where there is no longer any $CH_4$ to produce additional CO, the net CO production rate becomes negative, i.e., CO is destroyed. The maximum rate of CO destruction occurs just downstream of the peak $CO_2$ production rate. Note that the bulk of the chemical activity is contained in an interval extending from about 0.5 mm to 1.5 mm. Figure 8.10b shows, similarly, that the C-containing intermediate species, $CH_3$, $CH_2O$, and HCO, are produced and destroyed in a narrow interval (0.4–1.1 mm), as is the CH radical (Fig. 8.10d). The H-intermediates, $HO_2$ and $H_2O_2$, have somewhat broader profiles

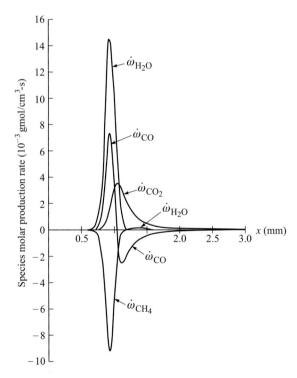

**Figure 8.11**    Calculated volumetric species production rate profiles for laminar, stoichiometric, $CH_4$–air premixed flame. Corresponds to the same conditions as in Fig. 8.10.

than the C-intermediates, and their peak concentrations appear slightly earlier in the flame (Fig. 8.10c). Note also the $H_2O$ reaches its 80-percent-of-equilibrium mole fraction much sooner than does the $CO_2$, i.e., approximately 0.9 mm versus 2 mm.

While the fuel is completely gone in approximately 1 mm and the major portion of the total temperature rise (73 percent) occurs in the same interval, the approach to equilibrium conditions is relatively slow beyond this point. In fact, we see that equilibrium has yet to be reached even at the 3-mm location. This slow approach to equilibrium is primarily a consequence of the dominance of three-body recombination reactions in this region, as mentioned in the beginning of this chapter. Plotting mole fractions as a function of distance, rather than time, exaggerates the slow approach to equilibrium somewhat as a consequence of the stretching of the distance-time relation ($dx = v_x dt$) through continuity ($\rho v_x = $ constant). For example, given the same time interval, a fluid particle in the hot, high-velocity region of the flame travels a distance much greater than a fluid particle in the cold, low-velocity region.

Figure 8.10d focuses on nitric oxide production. Here we see a somewhat rapid rise in the NO mole fraction in the same region where the CH radical is present in the flame. This is followed by a continual, almost linear, increase in the NO mole fraction. In this latter region, NO formation is dominated by Zeldovich kinetics (see Chapter 5). Of course, the NO mole-fraction curve must bend over at some point downstream, as reverse reactions become more important, and approach the equilibrium condition asymptotically. We can better understand the NO mole-fraction profile by examining the NO production rate, $\dot{\omega}_{NO}$, through the flame (Fig. 8.12). From Fig. 8.12, it is apparent that the early appearance of NO within the flame (0.5–0.8 mm in Fig. 8.10d) is the result of passive diffusion since the production rate is essentially zero in that region. It is interesting to observe (Fig. 8.12) that the first chemical

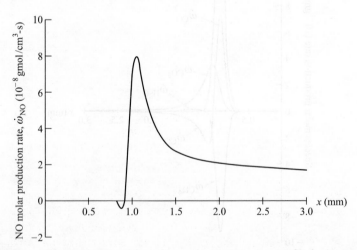

**Figure 8.12**     Calculated nitric oxide molar production rate, $\dot{\omega}_{NO}$, profile for laminar, stoichiometric, $CH_4$–air premixed flame. Corresponds to the same conditions as in Figs. 8.10 and 8.11.

activity associated with NO is a destruction process (0.8–0.9 mm). Nitric oxide production reaches a maximum at an axial location between the peak CH and O-atom concentrations. It is likely that Fenimore and Zeldovich pathways are both important here. Beyond the O-atom peak at $x = 1.2$ mm (Fig. 8.10d), the NO production rate falls rapidly at first, then more slowly (Fig. 8.12). Since the temperature continues to rise in this region, the decline in the net NO production rate must be a consequence of the decaying O-atom concentration and the building strength of reverse reactions.

# FACTORS INFLUENCING FLAME VELOCITY AND THICKNESS

## Temperature

The temperature dependencies of $S_L$ and $\delta$ can be inferred from Eqns. 8.20 and 8.21, recognizing the following approximate temperature scalings. For simplicity, we use $T_b$ to estimate $\bar{m}_F'''$. The pressure dependencies are also indicated.

$$\alpha \propto T_u \bar{T}^{0.75} P^{-1} \tag{8.27}$$

$$\bar{m}_F'''/\rho_u \propto T_u T_b^{-n} P^{n-1} \exp(-E_A/R_u T_b), \tag{8.28}$$

where the exponent $n$ is the overall reaction order, and $\bar{T} \equiv 0.5(T_b + T_u)$. Combining the above scalings yields

$$S_L \propto \bar{T}^{0.375} T_u T_b^{-n/2} \exp(-E_A/2R_u T_b) P^{(n-2)/2} \tag{8.29}$$

and

$$\delta \propto \bar{T}^{0.375} T_b^{n/2} \exp(+E_A/2R_u T_b) P^{-n/2}. \tag{8.30}$$

We see that the laminar flame speed has a strong temperature dependence, since global reaction orders for hydrocarbons are about 2, and apparent activation energies are approximately $1.67 \cdot 10^8$ J/kmol (40 kcal/gmol) (see Table 5.1). For example, Eqn. 8.29 predicts the flame speed to increase by a factor of 3.64 when the unburned gas temperature is increased from 300 K to 600 K. Increasing the unburned gas temperature will also increase the burned gas temperature by about the same amount, if we neglect dissociation and temperature-dependent specific heats. Table 8.1 shows comparisons of flame speeds and flame thicknesses for the case just mentioned (case B), and for the use of fixed unburned gas temperature but a decreased burned gas temperature (case C). Case A is the reference condition. Case C captures the effect of heat transfer or changing the equivalence ratio, either rich or lean, from the maximum-temperature condition. In this case, we see flame speeds decrease, whereas flame thicknesses increase significantly.

**Table 8.1**   Estimate of effects of unburned and burned gas temperature on laminar flame speeds and thickness using Eqns. 8.29 and 8.30

| Case | A | B | C |
|---|---|---|---|
| $T_u$ (K) | 300 | 600 | 300 |
| $T_b$ (K) | 2000 | 2300 | 1700 |
| $S_L/S_{L,A}$ | 1 | 3.64 | 0.46 |
| $\delta/\delta_A$ | 1 | 0.65 | 1.95 |

We can compare our simple estimates of the influence of temperature on flame speeds using the empirical correlation of Andrews and Bradley [19] for stoichiometric methane–air flames,

$$S_L(\text{cm/s}) = 10 + 3.71 \cdot 10^{-4}[T_u(\text{K})]^2, \tag{8.31}$$

which is shown in Fig. 8.13, along with data from several experimenters. Using Eqn. 8.31, an increase in $T_u$ from 300 K to 600 K results in $S_L$ increasing by a factor of 3.3, which compares quite favorably with our estimate of 3.64 (Table 8.1).

Useful correlations of flame speed with unburned gas temperature have been developed by Metghalchi and Keck [11] and are presented in the next section.

## Pressure

Equation 8.30 shows that $S_L \propto P^{(n-2)/2}$. If, again, we assume a global reaction order of 2, flame speed should be independent of pressure. Experimental measurements generally show a negative dependence of pressure. Andrews and Bradley [19] found that

$$S_L(\text{cm/s}) = 43[P(\text{atm})]^{-0.5} \tag{8.32}$$

fits their data for $P > 5$ atm for methane–air flames (Fig. 8.14). Law [20] provides a summary of flame speed data for a range of pressure (up to 5 atm or less) for the following fuels: $H_2$, $CH_4$, $C_2H_2$, $C_2H_4$, $C_2H_6$, and $C_3H_8$. The previously cited work by Metghalchi and Keck [11] also provides flame speed–pressure correlations for selected fuels.

## Equivalence Ratio

Except for very rich mixtures, the primary effect of equivalence ratio on flame speed for similar fuels is a result of how this parameter affects flame temperatures; thus, we would expect flame speeds to be a maximum at a slightly rich mixture and fall off on either side (see Fig. 2.13). Indeed, Fig. 8.15 shows this behavior for methane. Flame thickness shows the inverse trend, having a minimum near stoichiometric (Fig. 8.16). Note that many definitions of $\delta$ are applied in experimental measurements, so caution should be exercised when comparing numerical values from different investigations.

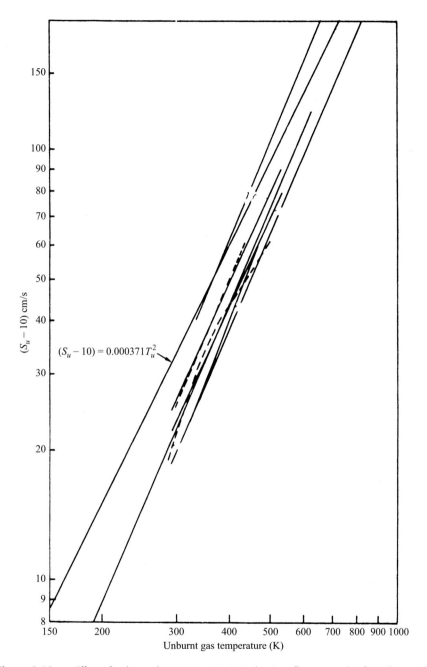

**Figure 8.13** Effect of unburned gas temperature on laminar flame speeds of stoichiometric methane–air mixtures at 1 atm. Various lines are data from various investigators.

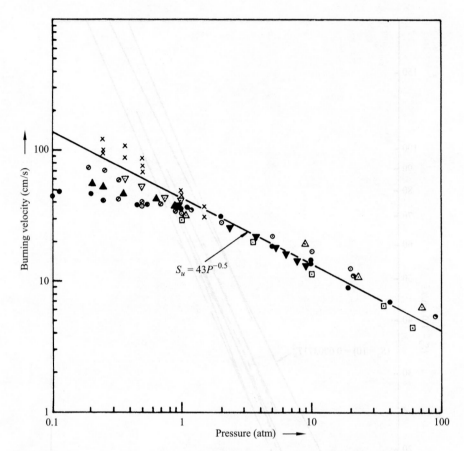

**Figure 8.14**    Effect of pressure on laminar flame speeds of stoichiometric methane–air mixtures for $T_u = 16$–$27°C$.

SOURCE: Reprinted with permission, Elsevier Science, Inc., from Ref. [19], © 1972, The Combustion Institute.

## Fuel Type

An extensive, but somewhat dated, summary of flame speed measurements for a wide variety of fuels is contained in Ref. [21]. Data from this report are shown in Fig. 8.17 for $C_2$–$C_6$ alkanes (single bonds), alkenes (double bonds), and alkynes (triple bonds). Also shown are $CH_4$ and $H_2$. The flame velocity of propane is used as a reference. Roughly speaking the $C_3$–$C_6$ hydrocarbons all follow the same trend as a function of flame temperature. Ethylene ($C_2H_4$) and acetylene ($C_2H_2$) have velocities greater than the $C_3$–$C_6$ group, whereas methane lies somewhat below. Hydrogen's maximum flame speed is many times greater than that of propane. Several factors combine to give $H_2$ its high flame speed: first, the thermal diffusivity of pure $H_2$ is many times greater than the hydrocarbon fuels; second, the mass diffusivity of hydrogen likewise is much greater

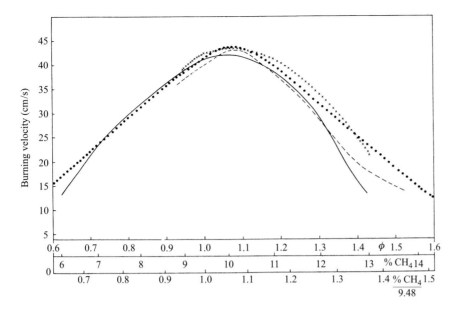

**Figure 8.15** Effect of equivalence ratio on the laminar flame speed of methane–air mixtures at atmospheric pressure.

SOURCE: Reprinted with permission, Elsevier Science, Inc., from Ref. [19], © 1972, The Combustion Institute.

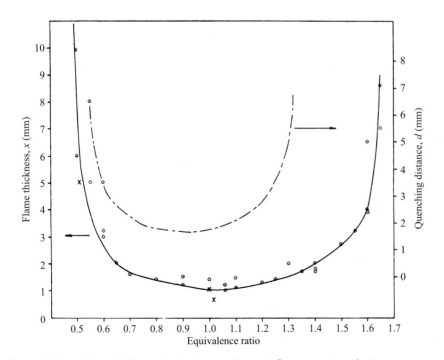

**Figure 8.16** Flame thickness for laminar methane–air flames at atmospheric pressure. Also shown is the quenching distance.

SOURCE: Reprinted with permission, Elsevier Science, Inc., from Ref. [19], © 1972, The Combustion Institute.

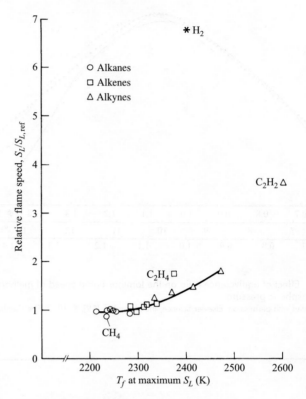

**Figure 8.17**     Relative flame speeds for $C_1$–$C_6$ hydrocarbon fuels. The reference flame speed is based on propane using the tube method [21].

than for the hydrocarbons; and third, the reaction kinetics for $H_2$ are very rapid since the relatively slow $CO \rightarrow CO_2$ step that is a major factor in hydrocarbon combustion is absent. Law [20] presents a compilation of laminar flame-speed data for various pure fuels and mixtures that are considered to be some of the more reliable data obtained to date. Table 8.2 shows a subset of these data.

**Table 8.2**     Laminar flame speeds for various pure fuels burning in air for $\Phi = 1.0$ and at 1 atm ($T_u$ = room temperature) from Ref. [20]

| Fuel | Formula | Laminar Flame Speed, $S_L$ (cm/s) |
|------|---------|-----------------------------------|
| Methane | $CH_4$ | 40 |
| Acetylene | $C_2H_2$ | 136 |
| Ethylene | $C_2H_4$ | 67 |
| Ethane | $C_2H_6$ | 43 |
| Propane | $C_3H_8$ | 44 |
| Hydrogen | $H_2$ | 210 |

# FLAME SPEED CORRELATIONS FOR SELECTED FUELS

Metghalchi and Keck [11] experimentally determined laminar flame speeds for various fuel–air mixtures over a range of temperatures and pressures typical of conditions associated with reciprocating internal combustion engines and gas-turbine combustors. Several forms for correlations were tried [11], including one similar to Eqn. 8.29, with the most useful being

$$S_L = S_{L,\text{ref}} \left( \frac{T_u}{T_{u,\text{ref}}} \right)^{\gamma} \left( \frac{P}{P_{\text{ref}}} \right)^{\beta} (1 - 2.1 Y_{\text{dil}}), \qquad (8.33)$$

for $T_u \gtrsim 350$ K. The subscript ref refers to reference conditions defined by

$$T_{u,\text{ref}} = 298 \text{ K},$$
$$P_{\text{ref}} = 1 \text{ atm},$$

and

$$S_{L,\text{ref}} = B_M + B_2 (\Phi - \Phi_M)^2$$

where the constants $B_M$, $B_2$, and $\Phi_M$ depend on fuel type and are given in Table 8.3. The temperature and pressure exponents, $\gamma$ and $\beta$, are functions of the equivalence ratio, expressed as

$$\gamma = 2.18 - 0.8(\Phi - 1)$$

and

$$\beta = -0.16 + 0.22(\Phi - 1).$$

The term $Y_{\text{dil}}$ is the mass fraction of diluent present in the air–fuel mixture, specifically included in Eqn. 8.33 to account for any recirculated combustion products. Recirculation of exhaust or flue gases is a common technique used to control oxides of nitrogen in many combustion systems (Chapter 15); and in internal combustion engines, residual combustion products mix with the incoming charge under most operating conditions.

**Table 8.3**   Values for $B_M$, $B_2$, and $\Phi_M$ used with Eqn. 8.33 [11]

| Fuel | $\Phi_M$ | $B_M$ (cm/s) | $B_2$ (cm/s) |
|------|---------|--------------|--------------|
| Methanol | 1.11 | 36.92 | −140.51 |
| Propane | 1.08 | 34.22 | −138.65 |
| Isooctane | 1.13 | 26.32 | −84.72 |
| RMFD-303 | 1.13 | 27.58 | −78.34 |

| **Example 8.3** | Compare the laminar flame speeds of gasoline–air mixtures with $\Phi = 0.8$ for the following three cases: |

i. At reference conditions of $T = 298$ K and $P = 1$ atm.
ii. At conditions typical of a spark-ignition engine operating at wide-open throttle: $T = 685$ K and $P = 18.38$ atm.
iii. Same as condition ii above, but with 15 percent (by mass) exhaust-gas recirculation.

### Solution

We will employ the correlation of Metghalchi and Keck, Eqn. 8.33, for RMFD-303. This research fuel (also called indolene) has a controlled composition simulating typical gasolines. The flame speed at 298 K and 1 atm is given by

$$S_{L,\,ref} = B_M + B_2(\Phi - \Phi_M)^2$$

where, from Table 8.3,

$$B_M = 27.58 \text{ cm/s},$$
$$B_2 = -78.38 \text{ cm/s},$$
$$\phi_M = 1.13.$$

Thus,

$$S_{L,\,ref} = 27.58 - 78.34(0.8 - 1.13)^2,$$

$$\boxed{S_{L,\,ref} = 19.05 \text{ cm/s}}$$

To find the flame speed at temperatures and pressures other than the reference state, we employ (Eqn. 8.33)

$$S_L(T_u, P) = S_{L,\,ref} \left( \frac{T_u}{T_{u,\,ref}} \right)^{\gamma} \left( \frac{P}{P_{ref}} \right)^{\beta}$$

where

$$\gamma = 2.18 - 0.8(\Phi - 1)$$
$$= 2.34$$
$$\beta = -0.16 + 0.22(\Phi - 1)$$
$$= -0.204.$$

Thus,

$$S_L(685 \text{ K}, 18.38 \text{ atm}) = 19.05 \left( \frac{685}{298} \right)^{2.34} \left( \frac{18.38}{1} \right)^{-0.204}$$

$$= 19.05(7.012)(0.552)$$

$$\boxed{S_L = 73.8 \text{ cm/s}}$$

With dilution by exhaust-gas recirculation, the flame speed above is reduced by the factor $(1 - 2.1\ Y_{dil})$:

$$S_L(685\ \text{K}, 18.38\ \text{atm}, 15\%\ \text{EGR}) = 73.8[1 - 2.1(0.15)]$$

$$\boxed{S_L = 50.6\ \text{cm/s}}$$

**Comments**

We see that the laminar flame velocity is much greater at engine conditions than at the reference state, with the dominant influence being the temperature. In Chapter 12, we will learn that the laminar flame speed is an important factor in determining the turbulent flame speed, which controls the burning rate in spark-ignition engines. Results of this example also show that dilution decreases the flame speed, which can have a detrimental effect on engine performance if too much gas is recirculated. Note that we have employed a value for $T_u$ less than the recommended minimum (350 K) for accurate use of Eqn. 8.33, thus underestimating the true value of $S_L$ at 298 K.

# QUENCHING, FLAMMABILITY, AND IGNITION

So far in this chapter we have considered only the steady propagation of premixed laminar flames. We now turn our attention to what are essentially transient processes: flame quenching and ignition. Although the processes are transient, we will confine our attention to examining limit behavior, i.e., conditions under which a flame will either extinguish or not, or ignite or not, and ignore the time-dependent details of the extinction and ignition processes.

There are many ways in which a flame can be extinguished. For example, flames will extinguish while passing through narrow passageways. This phenomenon is the basis for many practical flame-arresting devices in use today, and was first put into practice by Sir Humphrey Davey in his invention of the miner's safety lamp in 1815. Other techniques for extinguishing premixed flames are the addition of diluents, such as water, which have primarily a thermal effect, or chemical suppressants, such as halogens, which alter the chemical kinetics. Blowing the flame away from the reactants is also effective in extinguishing flames, as is easily demonstrated with a weakly burning Bunsen-burner flame. A more practical application is the blowout of oil-well fires with explosive charges; although, in this case, the flames may have a strongly nonpremixed character, rather than being premixed.

In the following, we discuss three concepts: quenching distances, flammability limits, and minimum ignition energies. In all of these, heat losses are assumed to control the phenomena. For more detailed analyses and discussions, the reader is referred to the literature [8, 22–32].

## Quenching by a Cold Wall

As mentioned above, flames extinguish upon entering a sufficiently small passageway. If the passageway is not too small, the flame will propagate through it.

The critical diameter of a circular tube where a flame extinguishes, rather than propagates, is referred to as the **quenching distance.** Experimentally, quenching distances are determined by observing whether a flame stabilized above a tube does or does not **flashback** for a particular tube diameter when the reactant flow is rapidly shut off. Quenching distances are also determined using high-aspect-ratio rectangular-slot burners. In this case, the quenching distance is the distance between the long sides, i.e., the slit width. Tube-based quenching distances are somewhat larger (~20–50 percent) than slit-based ones [21].

**Ignition and Quenching Criteria**    Williams [22] provides the following two rules of thumb governing ignition and its converse, flame extinction. The second criterion is the one applicable to the problem of flame quenching by a cold wall.

> Criterion I—Ignition will only occur if enough energy is added to the gas to heat a slab about as thick as a steadily propagating laminar flame to the adiabatic flame temperature.
> Criterion II—The rate of liberation of heat by chemical reactions inside the slab must approximately balance the rate of heat loss from the slab by thermal conduction.

In the following section, we employ these criteria to develop a greatly simplified analysis of flame quenching.

**Simplified Quenching Analysis**    Consider a flame that has just entered a slot formed by two plane-parallel plates as shown in Fig. 8.18. Applying Williams' second criterion and following the approach of Friedman [28], we write an energy balance equating the heat produced by reaction to the heat lost by conduction to the walls, i.e.,

$$\dot{Q}'''V = \dot{Q}_{\text{cond, tot}}, \tag{8.34}$$

where the volumetric heat release rate $\dot{Q}'''$ is related to the previously defined $\bar{\dot{m}}_F'''$ as

$$\dot{Q}''' = -\bar{\dot{m}}_F''' \Delta h_c. \tag{8.35}$$

Before proceeding, it is important to note that the thickness of the slab of gas analyzed (Fig. 8.18) has been taken to be $\delta$, the adiabatic laminar flame thickness as expressed in Eqn. 8.21. Our objective now is to determine the distance $d$, the quenching distance, that satisfies the quenching criterion expressed by Eqn. 8.34.

The heat loss from the flame slab to the wall can be expressed using Fourier's law as

$$\dot{Q}_{\text{cond}} = -kA \frac{dT}{dx}\bigg|_{\substack{\text{In gas} \\ \text{at wall}}}, \tag{8.36}$$

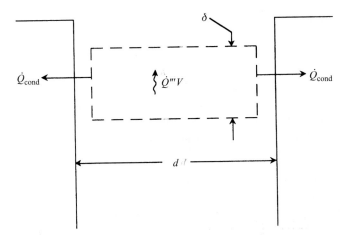

**Figure 8.18**    Schematic of flame quenching between two parallel walls.

where both the conductivity, $k$, and the temperature gradient are evaluated in the gas at the wall. The area $A$ is easily expressed as $2\delta L$, where $L$ is the slot width (perpendicular to the page) and the factor 2 accounts for the flame being in contact with walls on each side. The temperature gradient $dT/dx$, however, is much more difficult to approximate. A reasonable lower bound for the magnitude of $dT/dx$ is $(T_b - T_w)/(d/2)$, where we assume a linear distribution from the centerline plane at $T_b$ to the wall at $T_w$. Since $dT/dx$ is likely to be much greater than this, we introduce an arbitrary constant $b$, defined by

$$\left|\frac{dT}{dx}\right| \equiv \frac{T_b - T_w}{d/b}, \tag{8.37}$$

where $b$ is a number generally much greater than 2. Utilizing Eqns. 8.35–8.37, the quenching criterion (Eqn. 8.34) becomes

$$\left(-\bar{\dot{m}}_F''' \Delta h_c\right)(\delta dL) = k(2\delta L)\frac{T_b - T_w}{d/b} \tag{8.38a}$$

or

$$d^2 = \frac{2kb(T_b - T_w)}{-\bar{\dot{m}}_F''' \Delta h_c}. \tag{8.38b}$$

Assuming $T_w = T_u$, using the previously developed relationships between $\bar{\dot{m}}_F'''$ and $S_L$ (Eqn. 8.20), and relating $\Delta h_c = (\nu + 1)c_p(T_b - T_u)$, Eqn. 8.38b becomes

$$d = 2\sqrt{b}\,\alpha/S_L \tag{8.39a}$$

or, in terms of $\delta$,

$$d = \sqrt{b}\delta. \tag{8.39b}$$

Equation 8.39b is in agreement with the experimental results shown in Fig. 8.16 for methane, which show quenching distances to be greater than the flame thickness $\delta$. Quenching distances for a wide variety of fuels are shown in Table 8.4. It should be pointed out that the temperature and pressure dependencies of the quenching distance can be estimated using Eqn. 8.30.

| | |
|---|---|
| **Example 8.4** | Consider the design of a laminar-flow, adiabatic, flat-flame burner consisting of a square arrangement of thin-walled tubes as illustrated in the sketch below. Fuel–air mixture flows through both the tubes and the interstices between the tubes. It is desired to operate the burner with a stoichiometric methane–air mixture exiting the burner tubes at 300 K and 5 atm. |

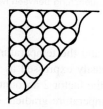

Burner tube layout

A. Determine the mixture mass flowrate per unit cross-sectional area at the design condition.

B. Estimate the maximum tube diameter allowed so that flashback will be prevented.

**Solution**

A. To establish a flat flame, the mean flow velocity must equal the laminar flame speed at the design temperature and pressure. From Fig. 8.14,

$$S_L (300 \text{ K, } 5 \text{ atm}) = 43/\sqrt{P(\text{atm})}$$
$$= 43/\sqrt{5} = 19.2 \text{ cm/s.}$$

The mass flux, $\dot{m}''$, is

$$\dot{m}'' = \dot{m}/A = \rho_u S_L.$$

We can approximate the density by assuming an ideal-gas mixture, where

$$MW_{mix} = \chi_{CH_4} MW_{CH_4} + (1 - \chi_{CH_4}) MW_{air}$$
$$= 0.095(16.04) + 0.905(28.85)$$
$$= 27.6 \text{ kg/kmol}$$

**Table 8.4**   Flammability limits, quenching distances, and minimum ignition energies for various fuels[a]

| Fuel | Flammability Limits | | | Quenching Distance | | Minimum Ignition Energy | |
| --- | --- | --- | --- | --- | --- | --- | --- |
| | $\Phi_{min}$ (Lean or Lower Limit) | $\Phi_{max}$ (Rich or Upper Limit) | Stoichiometric Mass Air–Fuel Ratio | For $\Phi = 1$ (mm) | Absolute Minimum (mm) | For $\Phi = 1$ ($10^{-5}$ J) | Absolute Minimum ($10^{-5}$ J) |
| Acetylene, $C_2H_2$ | 0.19[b] | $\infty$[b] | 13.3 | 2.3 | — | 3 | — |
| Carbon monoxide, CO | 0.34 | 6.76 | 2.46 | — | — | — | — |
| n-Decane, $C_{10}H_{22}$ | 0.36 | 3.92 | 15.0 | 2.1[c] | — | — | — |
| Ethane, $C_2H_6$ | 0.50 | 2.72 | 16.0 | 2.3 | 1.8 | 42 | 24 |
| Ethylene, $C_2H_4$ | 0.41 | >6.1 | 14.8 | 1.3 | — | 9.6 | — |
| Hydrogen, $H_2$ | 0.14[b] | 2.54[b] | 34.5 | 0.64 | 0.61 | 2.0 | 1.8 |
| Methane, $CH_4$ | 0.46 | 1.64 | 17.2 | 2.5 | 2.0 | 33 | 29 |
| Methanol, $CH_3OH$ | 0.48 | 4.08 | 6.46 | 1.8 | 1.5 | 21.5 | 14 |
| n-Octane, $C_8H_{18}$ | 0.51 | 4.25 | 15.1 | — | — | — | — |
| Propane, $C_3H_8$ | 0.51 | 2.83 | 15.6 | 2.0 | 1.8 | 30.5 | 26 |

[a]SOURCE: Data from Ref. [21] unless otherwise noted.

[b]Zabetakis (U.S. Bureau of Mines, Bulletin 627, 1965).

[c]Chomiak [25].

and

$$\rho_u = \frac{P}{(R_u/MW_{mix})T_u} = \frac{5(101,325)}{(8315/27.6)(300)}$$
$$= 5.61 \, \text{kg/m}^3.$$

Thus, the mass flux is

$$\dot{m}'' = \rho_u S_L = 5.61(0.192)$$

$$\boxed{\dot{m}'' = 1.08 \, \text{kg/s-m}^2}.$$

B. We assume that if the tube diameter is less than the quenching distance, with some factor of safety applied, the burner will operate without danger of flashback. Thus, we need to find the quench distance at the design conditions. From Fig. 8.16, we see that the 1-atm quench distance for a slit is approximately 1.7 mm. Since slit quenching distances are 20–50 percent smaller than for tubes, we will use this value outright, with the difference being our factor of safety. We now need to correct this value to the 5-atm condition. From Eqn. 8.39a, we see that

$$d \propto \alpha/S_L,$$

and, from Eqn. 8.27,

$$\alpha \propto T^{1.75}/P.$$

Combining the pressure effects on $\alpha$ and $S_L$, we have

$$d_2 = d_1 \frac{\alpha_2}{\alpha_1} \frac{S_{L,1}}{S_{L,2}} = d_1 \frac{P_1}{P_2} \frac{S_{L,1}}{S_{L,2}}$$

$$d(5 \, \text{atm}) = 1.7 \, \text{mm} \frac{1 \, \text{atm}}{5 \, \text{atm}} \frac{43 \, \text{cm/s}}{19.2 \, \text{cm/s}},$$

so,

$$\boxed{d_{design} \le 0.76 \, \text{mm}}$$

We need to verify that, when using this diameter, laminar flow is maintained in the tube; i.e., $Re_d < 2300$. Using air properties for the viscosity,

$$Re_d = \frac{\rho_u d_{design} S_L}{\mu}$$
$$= \frac{5.61(0.00076)(0.192)}{15.89 \cdot 10^{-6}}$$
$$= 51.5.$$

This value is well below the transitional value; hence, the quenching criteria control the design.

**Comment**

The final design should be based on the worst-case scenario where the quenching distance is a minimum. From Fig. 8.16, we see that the minimum quenching distance is close to the value used above for $\Phi = 0.8$.

## Flammability Limits

Experiments show that a flame will propagate only within a range of mixture strengths between the so-called lower and upper limits of flammability. The **lower limit** is the leanest mixture ($\Phi < 1$) that will allow steady flame propagation, whereas the **upper limit** represents the richest mixture ($\Phi > 1$). Flammability limits are frequently quoted as percent fuel by volume in the mixture, or as a percentage of the stoichiometric fuel requirement, i.e., $\Phi \times 100$ percent.

Table 8.4 shows flammability limits for a number of fuel–air mixtures at atmospheric pressure obtained from experiments employing the "tube method." In this method, it is ascertained whether or not a flame initiated at the bottom of a vertical tube (approximately 50-mm diameter by 1.2-m long) propagates the length of the tube. A mixture that sustains the flame is said to be flammable. By adjusting the mixture strength, the flammability limit can be ascertained. The effect of pressure on the lower flammability limit is relatively weak. Figure 8.19 illustrates this behavior showing measurements obtained using methane–air mixtures in a closed combustion vessel [33].

Although flammability limits can be defined that are physiochemical properties of the fuel–air mixture, experimental flammability limits are related to heat losses from the system, in addition to the mixture properties, and, hence, are generally apparatus dependent [31].

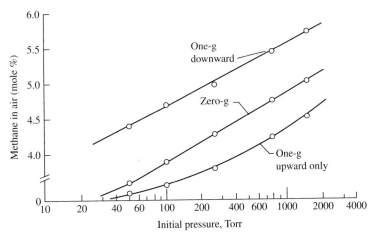

**Figure 8.19** Lower (lean) flammability limits for methane–air mixtures for various pressures. Note that a 5 percent methane mole percentage corresponds to an equivalence ratio of 0.476. Experiments were conducted in normal gravity and zero-gravity conditions. Adapted from Ref. [33]. Reprinted with permission from Elsevier.

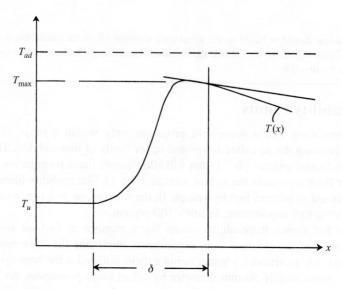

**Figure 8.20**    Temperature profile through flame with heat losses.

Even if conduction losses are minimal, radiation losses can account for the existence of flammability limits. Figure 8.20 illustrates an instantaneous axial temperature profile along the centerline of a tube in which a flame is propagating. Because the high-temperature product gases radiate to a lower-temperature environment, they cool. Their cooling creates a negative temperature gradient at the rear of the flame zone; hence, heat is lost by conduction from the flame proper. When sufficient heat is removed, such that Williams' criteria are not met, the flame ceases to propagate. Williams [22] provides a theoretical analysis for the situation described in Fig. 8.20, a discussion of which is beyond the scope of this book.

---

**Example 8.5**

A full propane cylinder from a camp stove leaks its contents of 1.02 lb (0.464 kg) into a 12′ × 14′ × 8′ (3.66 m × 4.27 m × 2.44 m) room at 20°C and 1 atm. After a long time, the fuel gas and room air are well mixed. Is the mixture in the room flammable?

**Solution**

From Table 8.4, we see that propane–air mixtures are flammable for $0.51 < \Phi < 2.83$. Our problem, thus, is to determine the equivalence ratio of the mixture filling the room. We can determine the partial pressure of the propane by assuming ideal-gas behavior:

$$P_F = \frac{m_F(R_u/MW_F)T}{V_{room}}$$

$$= \frac{0.464(8315/44.094)(20+273)}{3.66(4.27)(2.44)}$$

$$= 672.3 \text{ Pa.}$$

The propane mole fraction is

$$\chi_F = \frac{P_F}{P} = \frac{672.3}{101,325} = 0.00664$$

and

$$\chi_{\text{air}} = 1 - \chi_F = 0.99336.$$

The air–fuel ratio of the mixture in the room is

$$(A/F) = \frac{\chi_{\text{air}} MW_{\text{air}}}{\chi_F MW_F} = \frac{0.99336(28.85)}{0.00664(44.094)}$$

$$= 97.88.$$

From the definition of $\Phi$ and the value of $(A/F)_{\text{stoic}}$ from Table 8.4, we have

$$\Phi = \frac{15.6}{97.88} = 0.159.$$

Since $\Phi = 0.159$ is less than the lower limit of flammability ($\Phi_{\text{min}} = 0.51$), the mixture in the room is not capable of supporting a flame.

**Comment**

Although our calculations show that in the fully mixed state the mixture is not flammable, it is quite possible that, during the transient leaking process, a flammable mixture can exist somewhere within the room. Propane is heavier than air and would tend to accumulate near the floor until it is mixed by bulk motion and molecular diffusion. In environments employing flammable gases, monitors should be located at both low and high positions to detect leakage of heavy and light fuels, respectively.

# Ignition

In this section, we limit our discussion to ignition by electrical sparks and, in particular, focus on the concept of **minimum ignition energy.** Spark ignition is perhaps the most frequently employed means of ignition in practical devices: for example, spark-ignition and gas-turbine engines, and various industrial, commercial, and residential burners. Spark ignition is highly reliable and does not require a preexisting flame, as required by pilot ignition. In the following, a simple analysis is presented from which the pressure, temperature, and equivalence ratio dependencies of the minimum ignition energy are determined. Experimental data are also presented and compared with the predictions of the simple theory.

**Simplified Ignition Analysis**   Consider Williams' second criterion, applied now to a spherical volume of gas, which represents the incipient propagating flame created by a point spark. Using this criterion, we can define a critical gas-volume radius such that a flame will not propagate if the actual radius is smaller than the critical value. The second step of the analysis is to assume that the minimum ignition energy to be supplied by the spark is the energy required to heat the critical gas volume from its initial state to the flame temperature.

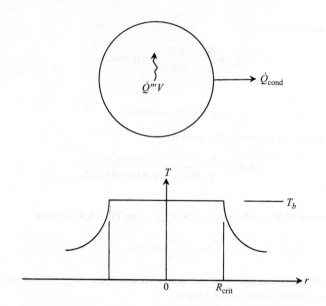

**Figure 8.21**    Critical volume of gas for spark ignition.

To determine the critical radius, $R_{crit}$, we equate the rate of heat liberated by reaction to the rate of heat lost to the cold gas by conduction, as illustrated in Fig. 8.21, i.e.,

$$\dot{Q}'''V = \dot{Q}_{cond} \tag{8.40}$$

or

$$-\bar{m}_F''' \Delta h_c\, 4\pi R_{crit}^3 /3 = -k 4\pi R_{crit}^2 \left.\frac{dT}{dr}\right|_{R_{crit}}, \tag{8.41}$$

where Eqn. 8.35 has been substituted for $\dot{Q}'''$, Fourier's law employed, and the volume and surface area of the sphere expressed in terms of the critical radius, $R_{crit}$.

The temperature gradient in the cold gas at the sphere boundary, $(dT/dr)_{crit}$, can be evaluated by determining the temperature distribution in the gas beyond the sphere ($R_{crit} \leq r \leq \infty$) with the boundary conditions $T(R_{crit}) = T_b$ and $T(\infty) = T_u$. (The well-known result that $Nu = 2$ follows from this analysis, where $Nu$ is the Nusselt number.) This results in

$$\left.\frac{dT}{dr}\right|_{R_{crit}} = -\frac{(T_b - T_u)}{R_{crit}}. \tag{8.42}$$

Substituting Eqn. 8.42 into Eqn. 8.41 yields

$$R_{crit}^2 = \frac{3k(T_b - T_u)}{-\bar{m}_F''' \Delta h_c}. \tag{8.43}$$

The critical radius can be related to the laminar flame speed, $S_L$, and flame thickness, $\delta$, by solving Eqn. 8.20 for $\bar{m}_F'''$ and substituting the result into Eqn. 8.43. This substitution, together with the recognition that $\Delta h_c = (v + 1)c_p(T_b - T_u)$, yields

$$R_{crit} = \sqrt{6}\,\frac{\alpha}{S_L}, \tag{8.44a}$$

where $\alpha = k/\rho_u c_p$ with $k$ and $c_p$ evaluated at some appropriate mean temperature. The critical radius can also be expressed in terms of $\delta$ (Eqn. 8.21b):

$$R_{crit} = (\sqrt{6}/2)\delta. \tag{8.44b}$$

Considering the crudeness of our analysis, the constant $\sqrt{6}/2$ should not be construed to be in any way precise but, rather, expresses an order of magnitude. Thus, from Eqn. 8.44b, we see that the critical radius is roughly equal to—or, at most, a few times larger than—the laminar flame thickness. In contrast, the quenching distance, $d$, as expressed by Eqn. 8.39b, can be many times larger than the flame thickness.

Knowing the critical radius, we now turn our attention to determining the minimum ignition energy, $E_{ign}$. This is done simply by assuming the energy added by the spark heats the critical volume to the burned-gas temperature, i.e.,

$$E_{ign} = m_{crit}c_p(T_b - T_u), \tag{8.45}$$

where the mass of the critical sphere, $m_{crit}$, is $\rho_b 4\pi R_{crit}^3/3$, or

$$E_{ign} = 61.6\rho_b c_p(T_b - T_u)(\alpha/S_L)^3. \tag{8.46}$$

Eliminating $\rho_b$ using the ideal-gas mixture law yields our final result:

$$E_{ign} = 61.6 P\left(\frac{c_p}{R_b}\right)\left(\frac{T_b - T_u}{T_b}\right)\left(\frac{\alpha}{S_L}\right)^3, \tag{8.47}$$

where $R_b = R_u/MW_b$.

**Pressure, Temperature, and Equivalence Ratio Dependencies**    The effect of pressure on minimum ignition energy is the result of the obvious direct influence seen in Eqn. 8.47, and the indirect influences buried in the thermal diffusivity, $\alpha$, and the flame speed, $S_L$. Using Eqns. 8.27 and 8.29 ($n \approx 2$), with Eqn. 8.47, shows the combined pressure effects to be

$$E_{ign} \propto P^{-2}, \tag{8.48}$$

which agrees remarkably well with experimental results. Figure 8.22 shows experimental ignition energies as a function of pressure [29], together with the power law of Eqn. 8.48 fitted to match the central datapoint.

**Table 8.5**    Temperature influence on spark-ignition energy [30]

| Fuel | Initial Temperature (K) | $E_{ign}$ (mJ)[a] |
|---|---|---|
| n-Heptane | 298 | 14.5 |
| | 373 | 6.7 |
| | 444 | 3.2 |
| Isooctane | 298 | 27.0 |
| | 373 | 11.0 |
| | 444 | 4.8 |
| n-Pentane | 243 | 45.0 |
| | 253 | 14.5 |
| | 298 | 7.8 |
| | 373 | 4.2 |
| | 444 | 2.3 |
| Propane | 233 | 11.7 |
| | 243 | 9.7 |
| | 253 | 8.4 |
| | 298 | 5.5 |
| | 331 | 4.2 |
| | 356 | 3.6 |
| | 373 | 3.5 |
| | 477 | 1.4 |

[a] $P = 1$ atm.

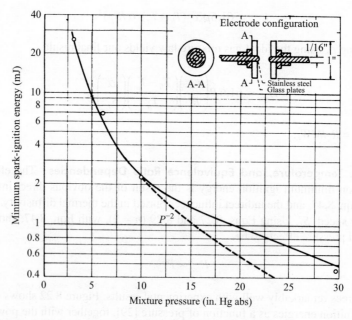

**Figure 8.22**    Effect of pressure on minimum spark-ignition energy.
SOURCE: From Ref. [29]. Reprinted by permission of the American Institute of Physics.

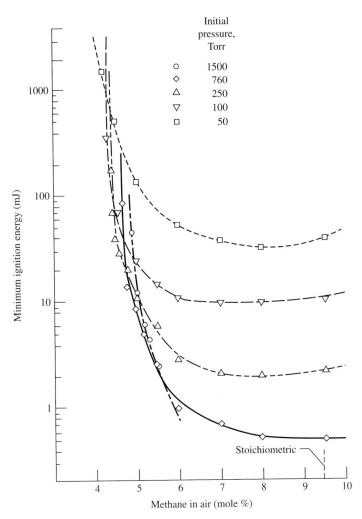

**Figure 8.23**     Minimum ignition energies increase greatly as the lean flammability limit is approached. Reprinted from Ref. [34] Reprinted with permission from Elsevier.

In general, increasing the initial mixture temperature results in decreased ignition energies, as shown in Table 8.5. Determining the theoretical influence from the simplified analyses presented in this chapter is left as an exercise for the reader.

At sufficiently lean equivalence ratios, the minimum energy required to ignite the mixture increases. This effect is illustrated in Fig. 8.23 [34]. Near the lean flammability limit, minimum ignition energies increase more than an order of magnitude from their values at stoichiometric conditions. This behavior is consistent with Eqn. 8.47, which shows a strong reciprocal dependence on laminar flame speed, i.e., $S_L^{-3}$. As shown in Fig. 8.15, the laminar flame speed decreases as the lean limit is approached.

## FLAME STABILIZATION

Important design criteria for gas burners are the avoidance of **flashback** and **liftoff**. Flashback occurs when the flame enters and propagates through the burner tube or port without quenching; liftoff is the condition where the flame is not attached to the burner tube or port but, rather, is stabilized at some distance from the port. Flashback is not only a nuisance, but is a safety hazard as well. In a gas appliance, propagation of a flame through a port can ignite the relatively large volume of gas in the mixer leading to the port, which might result in an explosion. Conversely, flame propagation through a "flash tube" from the pilot flame to the burner proper is used for ignition. In practical burners, flame lifting is generally undesirable for several reasons [35]. First, it can contribute to some escape of unburned gas or incomplete combustion. Also, ignition is difficult to achieve above the lifting limit. Accurate control of the position of a lifted flame is hard to achieve, so that poor heat-transfer characteristics can result. Lifted flames can also be noisy.

The phenomena of flashback and liftoff are both related to matching the local laminar flame speed to the local flow velocity. These matching conditions are schematically illustrated by the velocity vector diagrams of Fig. 8.24. Flashback is generally a transient event, occurring as the fuel flow is decreased or turned off. When the local flame speed exceeds the local flow velocity, the flame propagates upstream through the tube or port (Fig. 8.24a). When the fuel flow is stopped, flames will flashback through any tubes or ports that are larger than the quenching distance. Thus, we expect the controlling parameters for flashback to be the same as those affecting quenching, e.g., fuel type, equivalence ratio, flow velocity, and burner geometry.

Figure 8.25 illustrates flashback stability for a fixed burner geometry (a straight row of 2.7-mm-diameter ports with 6.35-mm spacing) for two different fuels: natural gas (Fig. 8.25a), and a manufactured gas that contains hydrogen (Fig. 8.25b). For a fixed gas and port size, the abscissa is proportional to the port exit velocity. Operation in the region to the left of the flashback zone line results in flashback, whereas flashback-free operation occurs to the right, where velocities are higher. We see that for both fuels, slightly rich stoichiometries provide the least tolerance to flashback, as expected, since maximum laminar flame speeds generally occur with slightly rich mixtures (see Fig. 8.15). We also observe from Fig. 8.25 that the flashback stability of natural gas, which is primarily methane, is much greater than for the manufactured gas. This is primarily a result of the high flame speed associated with the hydrogen in the manufactured gas. (Referring to Table 8.2, we see that the flame speed of $H_2$ is more than five times that of $CH_4$.)

Flame lifting depends on local flame and flow properties near the edges of the burner port. Consider a flame stabilized on a circular tube. At low flow velocities, the edge of the flame lies quite close to the burner lip and is said to be **attached.**

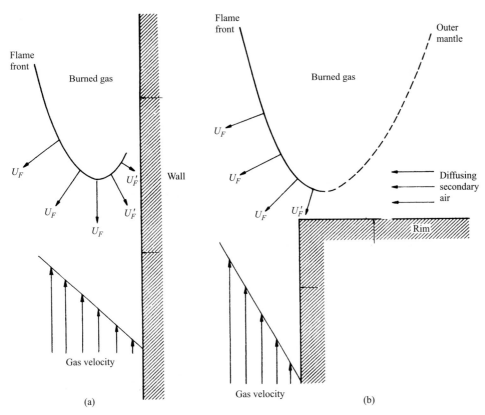

**Figure 8.24**    Velocity vectors showing flow velocities and local flame velocity for (a) flashback and (b) liftoff.
SOURCE: Reprinted with permission from Ref. [36]. © 1955, American Chemical Society.

When the velocity is increased, the cone angle of the flame decreases in accordance with the condition $\alpha = \sin^{-1}(S_L/v_u)$, Eqn. 8.2, and the edge of the flame is displaced a small distance downstream. With further increases in flow velocity, a critical velocity is reached where the flame edge jumps to a downstream position far from the burner lip. When the flame jumps to this position, it is said to be lifted. Increasing the velocity beyond the liftoff value results in increasing the liftoff distance until the flame abruptly blows off the tube altogether, an obviously undesirable condition.

Liftoff and blowoff can be explained by the countervailing effects of decreased heat and radical loss to the burner tube and increased dilution with ambient fluid, which occur as the flow velocity is increased. Consider a flame that is stabilized quite close to the burner rim. The local flow velocity at the stabilization location is small as a result of the boundary layer that develops in the tube. Inside the tube, the velocity at the wall is zero. Because of the close proximity of the flame to the cold

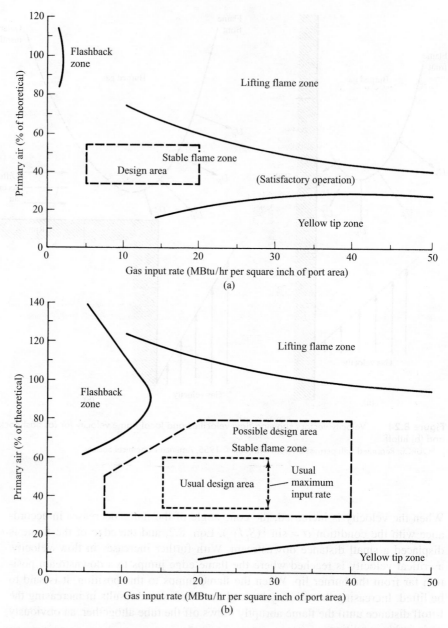

**Figure 8.25**     Stability diagrams for flashback, liftoff, and yellow-tipping for (a) natural gas and (b) manufactured gas for a burner with a single row of 2.7–mm-diameter ports with 6.35–mm spacing. Yellow-tipping indicates soot formation within the flame.
| SOURCE: From Ref. [35]. Reprinted with permission of Industrial Press.

wall, both heat and reactive species diffuse to the wall, causing the local laminar flame speed at the stabilization point also to be small. With the flame speed and flow velocities equal and of relatively small magnitude, the flame edge lies close to the burner tube. When the flow velocity is increased, the flame anchor point moves downstream; however, $S_L$ increases since the heat/radical losses are less because the flame is now not as close to the cold wall. This increase in the burning velocity results in only a small downstream adjustment; and, hence, the flame remains attached. With yet further increases in flow velocity, another effect becomes important—dilution of the mixture with ambient fluid as a result of diffusion. Since dilution tends to offset the effects of less heat loss, the flame lifts, as shown in Fig. 8.24b. With further increases in flow velocity, a point is reached at which there is no location across the flow at which the local flame speed matches the flow velocity and the flame blows off the tube.

Figure 8.25 shows how the region of lift-free operation varies as a function of stoichiometry (percent primary air) and flow velocity (gas input rate per unit port area). Here we see that the natural gas flame is more prone to lifting than the manufactured gas flame. Again, the high flame speed of the hydrogen in the manufactured gas explains the greater stability of this fuel. For more detailed information on laminar flame stability, the reader is referred to Refs. [3] and [21].

## SUMMARY

In this chapter, we considered the properties of laminar flames: their propagation velocity and thickness, quenching distances, flammability limits, and minimum ignition energies. Simplified theories were presented to illuminate the underlying physics and chemistry of these flame properties. We used the results of one such analysis to explore the pressure and temperature dependencies of flame speed and flame thickness, drawing from our conclusion that $S_L \propto (\alpha \bar{\dot{m}}_F''' / \rho_u)^{1/2}$, and finding that $S_L$ increases rapidly with temperature and has a weak inverse dependence on pressure. We saw that maximum laminar flame speeds for hydrocarbons typically occur in slightly rich mixtures, as do peak adiabatic flame temperatures. Correlations of laminar flame speeds for several fuels were presented. These correlations are useful for estimating flame properties in practical devices, such as engines, and you should be familiar with their use. Simple criteria for flame quenching and ignition were presented and used to develop simplistic models of these phenomena. Our analyses showed that the quenching distance is directly proportional to the thermal diffusivity and inversely proportional to the flame speed. We defined the upper and lower flammability limits and presented data for several fuels. In an analysis of the ignition process, the ideas of a critical radius and minimum energy input required to produce a self-propagating flame were developed. We saw that the minimum ignition energy exhibits a strong inverse pressure dependence, which has strong implications for engine applications where reliable ignition must be assured over a wide range of pressures. The chapter concluded with a discussion of the stabilization of laminar flames, i.e., their flashback and liftoff behavior—topics of great practical importance.

## NOMENCLATURE

| | |
|---|---|
| $A$ | Area (m$^2$) |
| $b$ | Dimensionless parameter defined in Eqn. 8.37 |
| $B_M, B_2$ | Parameters defined in Table 8.3 |
| $c_p$ | Specific heat (J/kg-K) |
| $d$ | Quenching distance (m) |
| $D_{ij}$ | Multicomponent diffusion coefficient (m$^2$/s) |
| $D_i^T$ | Thermal diffusion coefficient (kg/m-s) |
| $\mathcal{D}_{ij}$ | Binary diffusion coefficient (m$^2$/s) |
| $E_A$ | Activation energy (J/kmol) |
| $E_{ign}$ | Minimum ignition energy (J) |
| $h$ | Enthalpy (J/kg) |
| $k$ | Thermal conductivity (W/m-K) |
| $L$ | Slot width (m) |
| $Le$ | Lewis number, $\alpha/\mathcal{D}$ |
| $m$ | Mass (kg) |
| $\dot{m}$ | Mass flowrate (kg/s) |
| $\dot{m}''$ | Mass flux (kg/s-m$^2$) |
| $\dot{m}'''$ | Volumetric mass production rate (kg/s-m$^3$) |
| $MW$ | Molecular weight (kg/kmol) |
| $Nu$ | Nusselt number |
| $P$ | Pressure (Pa) |
| $\dot{Q}$ | Heat-transfer rate (W) |
| $\dot{Q}'''$ | Volumetric energy generation rate (W/m$^3$) |
| $r$ | Radial coordinate (m) |
| $R$ | Radius (m) or specific gas constant (J/kg-K) |
| $R_u$ | Universal gas constant (J/kmol-K) |
| $Re_d$ | Reynolds number |
| $S_L$ | Laminar flame speed (m/s) |
| $T$ | Temperature (K) |
| $v$ | Velocity (m/s) |
| $x$ | Distance (m) |
| $Y$ | Mass fraction (kg/kg) |

### Greek Symbols

| | |
|---|---|
| $\alpha$ | Angle (rad) or thermal diffusivity (m$^2$/s) |
| $\beta$ | Pressure exponent, Eqn. 8.33 |
| $\gamma$ | Temperature exponent, Eqn. 8.33 |
| $\delta$ | Laminar flame thickness (m) |

| | |
|---|---|
| $\Delta h_c$ | Heat of combustion (J/kg) |
| $\nu$ | Mass oxidizer-to-fuel ratio (kg/kg) |
| $\rho$ | Density (kg/m$^3$) |
| $\Phi$ | Equivalence ratio |
| $\Phi_M$ | Parameter defined in Table 8.3 |
| $\dot{\omega}$ | Species production rate (kmol/m$^3$-s) |

*Subscripts*

| | |
|---|---|
| *ad* | Adiabatic |
| *b* | Burned gas |
| cond | Conduction |
| crit | Critical |
| dil | Diluent |
| *F* | Fuel |
| *i* | *i*th species |
| max | Maximum |
| mix | Mixture |
| *Ox* | Oxidizer |
| *Pr* | Product |
| ref | Reference state |
| *u* | Unburned gas |

*Other Notation*

| | |
|---|---|
| $\overline{(\ )}$ | Average over reaction zone |
| $[X]$ | Molar concentration of species $X$ (kmol/m$^3$) |

# REFERENCES

1. Friedman, R., and Burke, E., "Measurement of Temperature Distribution in a Low-Pressure Flat Flame," *Journal of Chemical Physics,* 22: 824–830 (1954).

2. Fristrom, R. M., *Flame Structure and Processes,* Oxford University Press, New York, 1995.

3. Lewis, B., and Von Elbe, G., *Combustion, Flames and Explosions of Gases,* 3rd Ed., Academic Press, Orlando, FL, 1987.

4. Gordon, A. G., *The Spectroscopy of Flames,* 2nd Ed., Halsted Press, New York, 1974.

5. Gordon, A. G., and Wolfhard, H. G., *Flames: Their Structure, Radiation and Temperature,* 4th Ed., Halsted Press, New York, 1979.

6. Powling, J., "A New Burner Method for the Determination of Low Burning Velocities and Limits of Inflammability," *Fuel,* 28: 25–28 (1949).

7. Botha, J. P., and Spalding, D. B., "The Laminar Flame Speed of Propane–Air Mixtures with Heat Extraction from the Flame," *Proceedings of the Royal Society of London Series A,* 225: 71–96 (1954).

8. Kuo, K. K., *Principles of Combustion,* 2nd Ed., John Wiley & Sons, Hoboken, NJ, 2005.

9. Mallard, E., and Le Chatelier, H. L., *Annals of Mines,* 4: 379–568 (1883).

10. Spalding, D. B., *Combustion and Mass Transfer,* Pergamon, New York, 1979.

11. Metghalchi, M., and Keck, J. C., "Burning Velocities of Mixtures of Air with Methanol, Isooctane, and Indolene at High Pressures and Temperatures," *Combustion and Flame,* 48: 191–210 (1982).

12. Kee, R. J., and Miller, J. A., "A Structured Approach to the Computational Modeling of Chemical Kinetics and Molecular Transport in Flowing Systems," Sandia National Laboratories Report SAND86-8841, February 1991.

13. Kee, R. J., Rupley, F. M., and Miller, J. A., "Chemkin-II: A Fortran Chemical Kinetics Package for the Analysis of Gas-Phase Chemical Kinetics," Sandia National Laboratories Report SAND89-8009/UC-401, March 1991.

14. Kee, R. J., Dixon-Lewis, G., Warnatz, J., Coltrin, M. E., and Miller, J. A., "A Fortran Computer Code Package for the Evaluation of Gas-Phase Multicomponent Transport Properties," Sandia National Laboratories Report SAND86-8246/UC-401, December 1990.

15. Kee, R. J., Rupley, F. M., and Miller, J. A., "The Chemkin Thermodynamic Data Base," Sandia National Laboratories Report SAND87-8215B/UC-4, March 1991 (supersedes SAND87-8215).

16. Kee, R. J., Grcar, J. F., Smooke, M. D., and Miller, J. A., "A Fortran Program for Modeling Steady Laminar One-Dimensional Premixed Flames," Sandia National Laboratories Report SAND85-8240/UC-401, March 1991.

17. Warnatz, J., Maas, U., and Dibble, R. W., *Combustion,* Springer-Verlag, Berlin, 1996.

18. Bowman, C. T., Hanson, R. K., Davidson, D. F., Gardiner, W. C., Jr., Lissianski, V., Smith, G. P., Golden, D. M., Frenklach, M., Wang, H., and Goldenberg, M., *GRI-Mech 2.11 Home Page,* access via http://www.me.berkeley.edu/gri_mech/, 1995.

19. Andrews, G. E., and Bradley, D., "The Burning Velocity of Methane–Air Mixtures," *Combustion and Flame,* 19: 275–288 (1972).

20. Law, C. K., "A Compilation of Experimental Data on Laminar Burning Velocities," in *Reduced Kinetic Mechanisms for Applications in Combustion Systems* (N. Peters and B. Rogg, eds.), Springer-Verlag, New York, pp. 15–26, 1993.

21. Barnett, H. C., and Hibbard, R. R. (eds.), "Basic Considerations in the Combustion of Hydrocarbon Fuels with Air," NACA Report 1300, 1959.

22. Williams, F. A., *Combustion Theory,* 2nd Ed., Addison-Wesley, Redwood City, CA, 1985.

23. Glassman, I., *Combustion,* 3rd Ed., Academic Press, San Diego, CA, 1996.

24. Strehlow, R. A., *Fundamentals of Combustion,* Krieger, Huntington, NY, 1979.

25. Chomiak, J., *Combustion: A Study in Theory, Fact and Application,* Gordon & Breach, New York, 1990.

26. Frendi, A., and Sibulkin, M., "Dependence of Minimum Ignition Energy on Ignition Parameters," *Combustion Science and Technology,* 73: 395–413 (1990).

27. Lovachev, L. A., *et al.,* "Flammability Limits: An Invited Review," *Combustion and Flame,* 20: 259–289 (1973).

28. Friedman, R., "The Quenching of Laminar Oxyhydrogen Flames by Solid Surfaces," *Third Symposium on Combustion and Flame and Explosion Phenomena,* Williams & Wilkins, Baltimore, p. 110, 1949.

29. Blanc, M. V., Guest, P. G., von Elbe, G., and Lewis, B., "Ignition of Explosive Gas Mixture by Electric Sparks. I. Minimum Ignition Energies and Quenching Distances of Mixtures of Methane, Oxygen, and Inert Gases," *Journal of Chemical Physics,* 15(11): 798–802 (1947).

30. Fenn, J. B., "Lean Flammability Limit and Minimum Spark Ignition Energy," *Industrial & Engineering Chemistry,* 43(12): 2865–2868 (1951).

31. Law, C. K., and Egolfopoulos, F. N., "A Unified Chain-Thermal Theory of Fundamental Flammability Limits," *Twenty-Fourth Symposium (International) on Combustion,* The Combustion Institute, Pittsburgh, PA, p. 137, 1992.

32. Andrews, G. E., and Bradley, D., "Limits of Flammability and Natural Convection for Methane–Air Mixtures," *Fourteenth Symposium (International) on Combustion,* The Combustion Institute, Pittsburgh, PA, p. 1119, 1973.

33. Ronney, P. D., and Wachman, H. Y., "Effect of Gravity on Laminar Premixed Gas Combustion I: Flammability Limits and Burning Velocities," *Combustion and Flame,* 62: 107–119 (1985).

34. Ronney, P. D., "Effect of Gravity on Laminar Premixed Gas Combustion II: Ignition and Extinction Phenomena," *Combustion and Flame,* 62: 121–133 (1985).

35. Weber, E. J., and Vandaveer, F. E., "Gas Burner Design," *Gas Engineers Handbook,* Industrial Press, New York, pp. 12/193–12/210, 1965.

36. Dugger, G. L., "Flame Stability of Preheated Propane–Air Mixtures," *Industrial & Engineering Chemistry,* 47(1): 109–114, 1955.

37. Wu, C. K., and Law, C. K., "On the Determination of Laminar Flame Speeds from Stretched Flames," *Twentieth Symposium (International) on Combustion,* The Combustion Institute, Pittsburgh, PA, p. 1941, 1984.

## REVIEW QUESTIONS

**1.** Make a list of all the boldfaced words in Chapter 8 and discuss each.

**2.** Distinguish between deflagration and detonation.

**3.** Discuss the structure/appearance of a Bunsen-burner flame where the air–fuel mixture in the tube is rich.

**4.** What is the physical significance of the Lewis number? What role does the $Le = 1$ assumption play in the analysis of laminar flame propagation?

**5.** In the context of laminar flame theory, what is an eigenvalue? Discuss.

6.  Discuss the origins of the pressure and temperature dependencies of the laminar flame speed. *Hint:* Refer to the global hydrocarbon oxidation mechanism in Chapter 5.

7.  What are the basic criteria for ignition and quenching?

## PROBLEMS

**8.1**   Consider the outward propagation of a spherical laminar flame into an infinite medium of unburned gas. Assuming that $S_L$, $T_u$, and $T_b$ are all constants, determine an expression for the radial velocity of the flame front for a fixed coordinate system with its origin at the center of the sphere. *Hint:* Use mass conservation for an integral control volume.

**8.2**   Prove that, for the simplified thermodynamics employed in the Chapter 8 flame speed theory, $\Delta h_c = (\nu + 1)c_p(T_b - T_u)$.

**8.3**   Using the simplified theory, estimate the laminar flame speed of $CH_4$ for $\Phi = 1$ and $T_u = 300$ K. Use the global, single-step kinetics given in Chapter 5. Compare your estimate with experimental results. Also, compare your results for $CH_4$ with those for $C_3H_8$ from Example 8.2. Note the need to convert the pre-exponential factor to SI units.

**8.4**   Consider a one-dimensional, adiabatic, laminar, flat flame stabilized on a burner such as in Fig. 8.5a. The fuel is propane and the mixture ratio is stoichiometric. Determine the velocity of the burned gases for operation at atmospheric pressure and an unburned gas temperature of 300 K.

**8.5**   Consider a premixed flame stabilized above a circular tube. For the flame to be conical (constant angle $\alpha$), what is the shape of the velocity profile at the tube exit? Explain.

**8.6**   A premixed propane–air mixture emerges from a round nozzle with a uniform velocity of 75 cm/s. The laminar flame speed of the propane–air mixture is 35 cm/s. A flame is lit at the nozzle exit. What is the cone angle of this flame? What principle determines the cone angle?

**8.7***   Derive the theoretical flame shape for a premixed laminar flame stabilized over a circular tube, assuming the unburned mixture velocity profile is parabolic: $v(r) = v_o(1 - r^2/R^2)$, where $v_o$ is the centerline velocity and $R$ is the burner-tube radius. Ignore the region close to the tube wall where $S_L$ is greater than $v(r)$. Discuss your results.

**8.8**   Derive the pressure and temperature dependencies of the laminar flame speed, as shown in Eqn. 8.29, starting with Eqn. 8.20 and utilizing Eqn. 8.27 as given.

---

*Indicates required or optional use of computer.

**8.9\*** Using the correlations of Metghalchi and Keck [11], calculate laminar flame speeds for the stoichiometric mixtures of the following fuels for $P = 1$ atm and $T_u = 400$ K: propane; isooctane; and indolene (RMFD-303), a gasoline reference fuel.

**8.10\*** Use Eqn. 8.33 to estimate the laminar flame speed of a laminar flamelet in a spark-ignition internal combustion engine for conditions shortly after the spark is fired and a flame is established. The conditions are as follows:

Fuel: indolene (gasoline),     $\Phi = 1.0$,

$P = 13.4$ atm,                       $T = 560$ K.

**8.11\*** Repeat problem 8.10 for an equivalence ratio of $\Phi = 0.80$ with all other conditions the same. Compare your results with those of problem 8.10 and discuss the practical implications of your comparison.

**8.12\*** Use Eqn. 8.33 to calculate the laminar flame speed of propane for the following sets of conditions:

|            | Set 1 | Set 2 | Set 3 | Set 4 |
|------------|-------|-------|-------|-------|
| P (atm)    | 1     | 1     | 1     | 10    |
| $T_u$ (K)  | 350   | 700   | 350   | 350   |
| $\Phi$     | 0.9   | 0.9   | 1.2   | 0.9   |

Using your results, discuss the effects of $T_u$, $P$, and $\Phi$ on $S_L$.

**8.13** Using a combination of correlations and simplified theory, estimate propane–air flame thicknesses for $P = 1$, 10, and 100 atm for $\Phi = 1$ and $T_u = 300$ K. What are the implications of your calculations for the design of explosion-proof housings for electrical devices?

**8.14** Estimate values of the parameter $b$ employed in quenching theory for methane–air mixtures over the range of $0.6 \leq \Phi \leq 1.2$. Plot your results and discuss.

**8.15** Determine the critical radius for ignition of a stoichiometric propane–air mixture at 1 atm.

**8.16** How many times more ignition energy is required to achieve ignition of a fuel–air mixture at sea level with $T = 298$ K than at altitude of 6000 m (19,685 ft) where $P = 47,166$ Pa and $T = 248$ K? Discuss the implications of your calculations for high-altitude relight in aircraft gas-turbine engines.

**8.17\*** Use the CHEMKIN library codes to explore the structure of a freely propagating, laminar, premixed $H_2$–air flame for a reactants temperature of 298 K, $P = 1$ atm, and $\Phi = 1.0$.

A. Plot the temperature as a function of axial distance.

B. Plot the mole fractions of the following species as functions of the axial distance: O, OH, $O_2$, H, $H_2$, $HO_2$, $H_2O_2$, $H_2O$, and NO. Design and group your plots to be usefully communicative.

C. Plot the molar production rates $\dot{\omega}_i$ for the following species as functions of the axial distance: OH, $O_2$, $H_2$, $H_2O$, and NO.

D. Discuss the results from parts A–C.

**8.18*** Use the CHEMKIN library codes to explore the effects of equivalence ratio on the flame speed of freely propagating, laminar, premixed $H_2$–air flames. The reactants temperature is 298 K and $P = 1$ atm. Consider the range of equivalence ratios from 0.7 to 3.0. Compare your results with the measurements presented in Fig. 8 of Ref. [37].

# 9

# Laminar Diffusion Flames

## OVERVIEW

In this chapter, we begin our study of laminar diffusion flames by focusing on burning jets of fuel. Laminar jet flames have been the subject of much fundamental research and, more recently, have been used to develop an understanding of how soot is formed in diffusion burning, e.g., Refs. [1–9]. A familiar example of a nonpremixed jet flame is the Bunsen-burner outer-cone flame, which was briefly mentioned in Chapter 8. Many residential gas appliances, e.g., cooking ranges and ovens, employ laminar jet flames. In these applications, the fuel stream is usually partially premixed with air, which is essential to provide nonsooting operation. Although many analytical [10–17] and numerical [6, 18–22] analyses of laminar jet flames have been performed, current design practices rely heavily on the art and craft of experienced burner designers [23, 24]; however, recent concerns with indoor air quality and pollutant emissions have resulted in the use of more sophisticated design methods. Of particular concern are the emissions of nitrogen dioxide ($NO_2$) and carbon monoxide (CO), both of which are toxic gases.

A primary concern in the design of any system utilizing laminar jet flames is flame geometry, with short flames frequently being desired. Also of interest is the effect of fuel type. For example, some appliances may be designed to operate on natural gas (primarily methane) or propane. In the following sections, we will develop analyses that allow us to see what parameters control flame size and shape. We will also examine the factors controlling soot from laminar jet flames. The chapter concludes with an analysis and discussion of counterflow flames.

# NONREACTING CONSTANT-DENSITY LAMINAR JET

## Physical Description

Before discussing jet flames proper, let us consider the simpler case of a non-reacting laminar jet of fluid (fuel) flowing into an infinite reservoir of quiescent fluid (oxidizer). This simpler case allows us to develop an understanding of the basic flow and diffusional processes that occur in laminar jets uncomplicated by the effects of chemical reaction.

Figure 9.1 illustrates the essential features of a fuel jet issuing from a nozzle of radius $R$ into still air. For simplicity, we assume that the velocity profile is uniform at the tube exit. Close to the nozzle, there exists a region called the **potential core** of the jet. In the potential core, the effects of viscous shear and diffusion have yet to be felt; hence, both the velocity and nozzle-fluid mass fraction remain unchanged from their nozzle-exit values and are uniform in this region. This situation is quite similar to the developing flow in a pipe, except that, in the pipe, conservation of mass requires the uniform core flow to accelerate.

In the region between the potential core and the jet edge, both the velocity and fuel concentration (mass fraction) decrease monotonically to zero at the edge of the jet. Beyond the potential core ($x > x_c$), the effects of viscous shear and mass diffusion are active across the whole width of the jet.

Throughout the entire flowfield, the initial jet momentum is conserved. As the jet issues into the surrounding air, some of its momentum is transferred to the air. Thus, the velocity of the jet decreases, while greater and greater quantities of air are **entrained** into the jet as it proceeds downstream. This idea can be expressed mathematically using an integral form of momentum conservation:

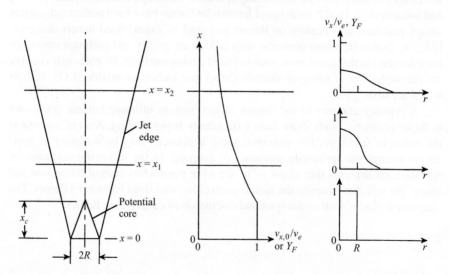

**Figure 9.1**     Nonreacting, laminar fuel jet issuing into an infinite reservoir of quiescent air.

|  Momentum flow of the jet at any $x$, $J$ | $=$ | Momentum flow issuing from the nozzle, $J_e$ |

or

$$2\pi \int_0^\infty \rho(r,\, x) v_x^2(r,\, x) r\, \mathrm{d}r \quad = \quad \rho_e v_e^2 \pi R^2, \qquad (9.1)$$

where $\rho_e$ and $v_e$ are the density and velocity of the fuel at the nozzle exit, respectively. The central graph of Fig. 9.1 illustrates the decay of the centerline velocity with distance beyond the potential core, whereas the right-hand graphs of Fig. 9.1 show the radial velocity decay from the maximum, which occurs at the centerline, to zero at the edge of the jet.

The processes that control the velocity field, i.e., the convection and diffusion of momentum, are similar to the processes that control the fuel concentration field, i.e., the convection and diffusion of mass. Thus, we expect that the distribution of the fuel mass fraction, $Y_F(r,\, x)$, should be similar to the dimensionless velocity distribution, $v_x(r,\, x)/v_e$, as indicated in Fig. 9.1. Because of the high concentration of fuel in the center of the jet, fuel molecules diffuse radially outward in accordance with Fick's law. The effect of moving downstream is to increase the time available for diffusion to take place; hence, the width of the region containing fuel molecules grows with axial distance, $x$, and the centerline fuel concentration decays. Similar to the initial jet momentum, the mass of the fluid issuing from the nozzle is conserved, i.e.,

$$2\pi \int_0^\infty \rho(r,\, x) v_x(r,\, x) Y_F(r,\, x) r\, \mathrm{d}r = \rho_e v_e \pi R^2 Y_{F,\, e}, \qquad (9.2)$$

where $Y_{F,\,e} = 1$. The problem now before us is the determination of the detailed velocity and fuel mass-fraction fields.

## Assumptions

In order to provide a very simple analysis of a nonreacting laminar jet, we employ the following assumptions:

1. The molecular weights of the jet and reservoir fluids are equal. This assumption, combined with ideal-gas behavior and the further assumptions of constant pressure and temperature, provide a uniform fluid density throughout the flowfield.

2. Species molecular transport is by simple binary diffusion governed by Fick's law.

3. Momentum and species diffusivities are constant and equal; thus, the **Schmidt number**, $Sc \equiv \nu/\mathcal{D}$, which expresses the ratio of these two quantities, is unity.

4. Only radial diffusion of momentum and species is important; axial diffusion is neglected. This implies that our solution only applies some distance downstream of the nozzle exit, since near the exit, axial diffusion is quite important.

## Conservation Laws

The basic governing equations expressing conservation of mass, momentum, and species appropriate for our purposes are the so-called **boundary-layer equations,** which are obtained by simplifying the more general equations of motion and species conservation, as discussed in Chapter 7. The pertinent equations (Eqns. 7.7, 7.48, and 7.20) are simplified for our assumptions of constant density, viscosity, and mass diffusivity to yield the following.

### Mass Conservation

$$\frac{\partial v_x}{\partial x} + \frac{1}{r}\frac{\partial (v_r r)}{\partial r} = 0. \tag{9.3}$$

### Axial Momentum Conservation

$$v_x \frac{\partial v_x}{\partial x} + v_r \frac{\partial v_x}{\partial r} = v\frac{1}{r}\frac{\partial}{\partial r}\left( r\frac{\partial v_x}{\partial r} \right). \tag{9.4}$$

### Species Conservation    For the jet fluid, i.e., the fuel,

$$v_x \frac{\partial Y_F}{\partial x} + v_r \frac{\partial Y_F}{\partial r} = \mathcal{D}\frac{1}{r}\frac{\partial}{\partial r}\left( r\frac{\partial Y_F}{\partial r} \right), \tag{9.5}$$

and since there are only two species, fuel and oxidizer, the mass fractions of the two must sum to unity, i.e.,

$$Y_{Ox} = 1 - Y_F. \tag{9.6}$$

## Boundary Conditions

Solving the above equations for the unknown functions, $v_x(r, x)$, $v_r(r, x)$, and $Y_F(r, x)$, requires three boundary conditions each for $v_x$ and $Y_F$ (two as functions of $x$ at specified $r$, and one as a function of $r$ at specified $x$) and one boundary condition for $v_r$ (as a function of $x$ at specified $r$). These are as follows:
Along the jet centerline ($r = 0$),

$$v_r(0, x) = 0, \tag{9.7a}$$

$$\frac{\partial v_x}{\partial r}(0, x) = 0, \tag{9.7b}$$

$$\frac{\partial Y_F}{\partial r}(0, x) = 0, \tag{9.7c}$$

where the first condition (Eqn. 9.7a) implies that there is no source or sink of fluid along the jet axis, whereas the second two (Eqns. 9.7b and c) result from symmetry. At large radii ($r \rightarrow \infty$), the fluid is stagnant and no fuel is present, i.e.,

$$v_x(\infty, x) = 0, \tag{9.7d}$$

$$Y_F(\infty, x) = 0. \tag{9.7e}$$

At the jet exit ($x = 0$), we assume the axial velocity and fuel mass fraction are uniform at the mouth of the nozzle ($r \leq R$) and zero elsewhere:

$$v_x(r \leq R, \ 0) = v_e,$$
$$v_x(r > R, \ 0) = 0, \tag{9.7f}$$

$$Y_F(r \leq R, \ 0) = Y_{F, e} = 1,$$
$$Y_F(r > R, \ 0) = 0. \tag{9.7g}$$

## Solution

The velocity field can be obtained by assuming the profiles to be **similar.** The idea of similarity is that the intrinsic shape of the velocity profiles is the same everywhere in the flowfield. For the present problem, this implies that the radial distribution of $v_x(r, x)$, when normalized by the local centerline velocity $v_x(0, x)$, is a universal function that depends only on the **similarity variable,** $r/x$. The solution for the axial and radial velocities is [25]

$$v_x = \frac{3}{8\pi} \frac{J_e}{\mu x} \left[ 1 + \frac{\xi^2}{4} \right]^{-2}, \tag{9.8}$$

$$v_r = \left( \frac{3J_e}{16\pi\rho_e} \right)^{1/2} \frac{1}{x} \frac{\xi - \dfrac{\xi^3}{4}}{\left( 1 + \dfrac{\xi^2}{4} \right)^2}, \tag{9.9}$$

where $J_e$ is the jet initial momentum flow,

$$J_e = \rho_e v_e^2 \pi R^2, \tag{9.10}$$

and $\xi$ contains the similarity variable $r/x$,

$$\xi = \left( \frac{3\rho_e J_e}{16\pi} \right)^{1/2} \frac{1}{\mu} \frac{r}{x}. \tag{9.11}$$

The axial velocity distribution in dimensionless form can be obtained by substituting Eqn. 9.10 into Eqn. 9.8 and rearranging:

$$v_x/v_e = 0.375(\rho_e v_e R/\mu)(x/R)^{-1}[1 + \xi^2/4]^{-2}, \tag{9.12}$$

and the dimensionless centerline velocity decay is obtained by setting $r = 0$ ($\xi = 0$):

$$v_{x,0}/v_e = 0.375(\rho_e v_e R/\mu)(x/R)^{-1}. \tag{9.13}$$

Equation 9.12 shows that the velocity decays inversely with axial distance and is directly proportional to the jet Reynolds number ($Re_j \equiv \rho_e v_e R/\mu$). Furthermore, Eqn. 9.13 reminds us that the solution is not valid near the nozzle, since $v_{x,0}/v_e$ should not exceed unity. The centerline velocity decay pattern predicted by Eqn. 9.13 is shown in Fig. 9.2. Here we see that the decay is more rapid with the lower-$Re_j$ jets. This behavior occurs because as the Reynolds number is decreased, the relative importance of the initial jet momentum becomes smaller in comparison with the viscous shearing action that slows the jet.

Other parameters frequently used to characterize jets are the **spreading rate** and **spreading angle, $\alpha$**. To define these parameters, we need to introduce the idea of the **jet half-width, $r_{1/2}$**. The jet half-width is simply the radial location where the jet velocity has decayed to one half of its centerline value (Fig. 9.3). An expression for $r_{1/2}$ can be derived by setting $v_x/v_{x,0}$, obtained by taking the ratio of Eqns. 9.12 and 9.13, to be one-half and solving for $r (= r_{1/2})$. The ratio of the jet half-width to the axial distance $x$ is termed the jet spreading rate, and the spreading angle is the angle whose tangent is the spreading rate. Thus,

$$r_{1/2}/x = 2.97\left(\frac{\mu}{\rho v_e R}\right) = 2.97 Re_j^{-1} \tag{9.14}$$

and

$$\alpha \equiv \tan^{-1}(r_{1/2}/x). \tag{9.15}$$

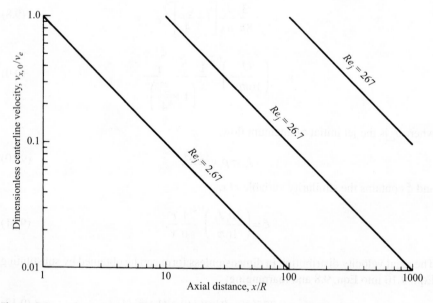

**Figure 9.2**    Centerline velocity decay for laminar jets.

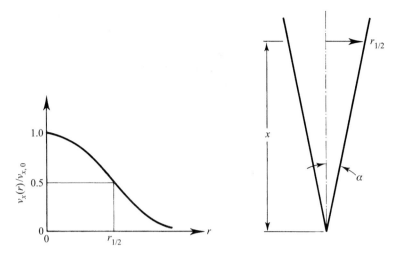

**Figure 9.3**    Definitions of jet half-width, $r_{1/2}$, and jet spreading angle, $\alpha$.

Equations 9.14 and 9.15 reveal that high-$Re_j$ jets are narrow, whereas low-$Re_j$ jets are wide. This result is consistent with the Reynolds number dependence of the velocity decay discussed above.

Let us now examine the solution for the concentration field. Referring to the governing equations for momentum conservation (Eqn. 9.4) and species conservation (Eqn. 9.5), we see that the fuel mass fraction, $Y_F$, plays the same mathematical role as the dimensionless axial velocity, $v_x/v_e$, if $\nu$ and $\mathcal{D}$ are equal. Since the equality of $\nu$ and $\mathcal{D}$ was one of our original assumptions, $Sc = \nu/\mathcal{D} = 1$, the functional form of the solution for $Y_F$ is identical to that for $v_x/v_e$, i.e.,

$$Y_F = \frac{3}{8\pi} \frac{Q_F}{\mathcal{D}x}[1+\xi^2/4]^{-2}, \tag{9.16}$$

where $Q_F$ is the volumetric flowrate of fuel from the nozzle ($Q_F = v_e \pi R^2$).

By applying $Sc = 1(\nu = \mathcal{D})$ to Eqn. 9.16, we can identify again the jet Reynolds number as a controlling parameter, i.e.,

$$Y_F = 0.375 Re_j (x/R)^{-1}[1+\xi^2/4]^{-2}, \tag{9.17}$$

and, for the centerline values of the mass fraction,

$$Y_{F,0} = 0.375 Re_j (x/R)^{-1}. \tag{9.18}$$

Again, we see that the solutions above can only be applied far from the nozzle, that is, the dimensionless distance downstream where the solution is valid must exceed the jet Reynolds number, i.e.,

$$(x/R) \gtrsim 0.375 Re_j. \tag{9.19}$$

Note that Fig. 9.2 also represents the decay of the centerline mass fractions, since Eqns. 9.18 and 9.13 are identical.

---

**Example 9.1**

A jet of ethylene ($C_2H_4$) exits a 10-mm-diameter nozzle into still air at 300 K and 1 atm. Compare the spreading angles and axial locations where the jet centerline mass fraction drops to the stoichiometric value for initial jet velocities of 10 cm/s and 1.0 cm/s. The viscosity of ethylene at 300 K is $102.3 \cdot 10^{-7}$ N-s/m$^2$.

**Solution**

Since the molecular weights of $C_2H_4$ and air are nearly the same ($MW = 28.05$ and 28.85, respectively), we assume that the constant-density jet solutions (Eqns. 9.8–9.15) can be applied to this problem. Designating the 10 cm/s case as case I and the 1 cm/s as case II, we determine the jet Reynolds numbers to be

$$Re_{j,\text{I}} = \frac{\rho v_{e,\text{I}} R}{\mu} = \frac{1.14(0.10)0.005}{102.3 \cdot 10^{-7}} = 55.7$$

and

$$Re_{j,\text{II}} = \frac{\rho v_{e,\text{II}} R}{\mu} = \frac{1.14(0.01)0.005}{102.3 \cdot 10^{-7}} = 5.57,$$

where the density has been estimated from the ideal-gas law, i.e.,

$$\rho = \frac{P}{(R_u/MW)T} = \frac{101,325}{(8315/28.05)300} = 1.14 \text{ kg/m}^3.$$

A. The spreading angle is determined by combining Eqns. 9.14 and 9.15 to yield

$$\alpha = \tan^{-1}[2.97/Re_j]$$

where

$$\alpha_\text{I} = \tan^{-1}[2.97/55.7]$$

$$\boxed{\alpha_\text{I} = 3.05°}$$

and

$$\alpha_\text{II} = \tan^{-1}[2.97/5.57]$$

$$\boxed{\alpha_\text{II} = 28.1°}$$

**Comment**

From these calculations, we see that the low-velocity jet is much wider, with a spreading angle about nine times larger than for the high-velocity jet.

B. The stoichiometric fuel mass fraction can be calculated as

$$Y_{F,\text{stoic}} = \frac{m_F}{m_A + m_F} = \frac{1}{(A/F)_{\text{stoic}} + 1},$$

where

$$(A/F)_{\text{stoic}} = (x + (y/4))4.76\frac{MW_A}{MW_F}$$

$$= (2 + (4/4))4.76\frac{28.85}{28.05} = 14.7.$$

Thus,

$$Y_{F,\,\text{stoic}} = \frac{1}{14.7+1} = 0.0637.$$

To find the axial location where the centerline fuel mass fraction takes on the stoichiometric value, we set $Y_{F,0} = Y_{F,\,\text{stoic}}$ in Eqn. 9.18 and solve for $x$:

$$x = \left( \frac{0.375\,Re_j}{Y_{F,\,\text{stoic}}} \right) R,$$

which, when evaluated for the two cases, becomes

$$\boxed{x_{\mathrm{I}}} = \left( \frac{0.375(55.7)0.005}{0.0637} \right) = \boxed{1.64 \text{ m}}$$

and

$$\boxed{x_{\mathrm{II}}} = \left( \frac{0.375(5.57)0.005}{0.0637} \right) = \boxed{0.164 \text{ m}}$$

**Comment**

We see that the fuel concentration of the low-velocity jet decays to the same value as the high-velocity jet in $1/10$ the distance.

---

Using the case II ($v_e = 1.0$ cm/s, $R = 5$ mm) jet from Example 9.1 as the base case, determine what nozzle exit radius is required to maintain the same flowrate if the exit velocity is increased by a factor of ten to 10 cm/s. Also, determine the axial location for $Y_{F,\,0} = Y_{F,\,\text{stoic}}$ for this condition and compare it with the base case.

**Example 9.2**

**Solution**

A.  We can relate the velocities and exit radii to the flowrate as

$$Q = v_{e,1} A_1 = v_{e,2} A_2$$

$$Q = v_{e,1} \pi R_1^2 = v_{e,2} \pi R_2^2.$$

Thus,

$$R_2 = \left( \frac{v_{e,1}}{v_{e,2}} \right)^{1/2} R_1 = \left( \frac{1}{10} \right)^{1/2} 5 \text{ mm}$$

$$\boxed{R_2 = 1.58 \text{ mm}}$$

B.  For the high-velocity, small-diameter jet, the Reynolds number is

$$Re_j = \frac{\rho v_{e,2} R}{\mu} = \frac{1.14(0.1)0.00158}{102.3 \cdot 10^{-7}} = 17.6,$$

and, from Eqn. 9.18,

$$x = \left( \frac{0.375\,Re_j}{Y_{F,\,\text{stoic}}} \right) R = \frac{0.375(17.6)0.00158}{0.0637}$$

$$\boxed{x = 0.164 \text{ m}}$$

**Comment**

The distance calculated in part B is identical to the case II value from Example 9.1. Thus, we see that the spatial fuel mass-fraction distribution depends only on the initial volumetric flowrate for a given fuel ($\mu/\rho$ constant).

## JET FLAME PHYSICAL DESCRIPTION

The burning laminar fuel jet has much in common with our previous discussion of the isothermal jet. Some essential features of the jet flame are illustrated in Fig. 9.4. As the fuel flows along the flame axis, it diffuses radially outward, while the oxidizer (e.g., air) diffuses radially inward. The flame surface is nominally defined to exist where the fuel and oxidizer meet in stoichiometric proportions, i.e.,

$$\text{Flame surface} \equiv \begin{array}{l}\text{Locus of points where the}\\ \text{equivalence ratio, } \Phi, \text{ equals unity.}\end{array} \qquad (9.20)$$

Note that, although the fuel and oxidizer are consumed at the flame, the equivalence ratio still has meaning since the products composition relates to a unique value of $\Phi$. The products formed at the flame surface diffuse radially both inward and outward. For an **overventilated** flame, where there is more than enough oxidizer in the surroundings to continuously burn the fuel, the flame length, $L_f$, is simply determined by the axial location where

$$\Phi(r = 0, x = L_f) = 1. \qquad (9.21)$$

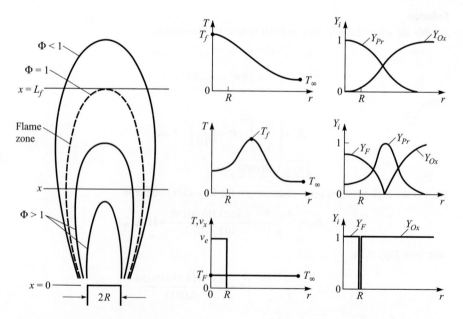

**Figure 9.4**    Laminar diffusion flame structure.

The region where chemical reactions occur is generally quite narrow. As seen in Fig. 9.4, the high-temperature reaction zone occurs in an annular region until the flame tip is reached. That the flame zone is an annulus can be demonstrated in a simple experiment where a metal screen is placed perpendicular to the axis of an unaerated Bunsen-burner flame. In the flame zone, the screen becomes hot and glows, showing the annular structure.

In the upper regions of a vertical flame, there is a sufficient quantity of hot gases that buoyant forces become important. Buoyancy accelerates the flow and causes a narrowing of the flame, since conservation of mass requires streamlines to come closer together as the velocity increases. The narrowing of the flow increases the fuel concentration gradients, $dY_F/dr$, thus enhancing diffusion. The effects of these two phenomena on the length of flames issuing from circular nozzles tend to cancel [12, 13]; thus, simple theories that neglect buoyancy fortuitously are able to predict flame lengths reasonably well for the circular- (and square-) port geometry.

For hydrocarbon flames, soot is frequently present, giving the flame its typical orange or yellow appearance. Given sufficient time, soot is formed on the fuel side of the reaction zone and is consumed when it flows into an oxidizing region; for example, the flame tip. Figure 9.5 illustrates soot formation and destruction zones in a simple jet flame. Depending on the fuel type and flame residence times, not all of the soot that is formed may be oxidized on its journey through high-temperature oxidizing regions. In this case, soot "wings" may appear, with the soot breaking through the flame. This soot that breaks through is generally referred to as **smoke.** Figure 9.6 shows a photograph of an ethylene flame where a soot wing is apparent on the right hand side of the flame tip. We will discuss the formation and destruction of soot in greater detail in a later section of this chapter.

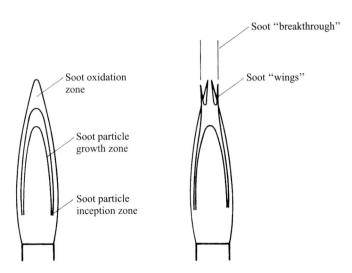

**Figure 9.5**     Soot formation and destruction zones in laminar jet flames.

**Figure 9.6** Laminar ethylene jet diffusion flame. Note soot "wings" at sides of the flame near the tip.

SOURCE: Photograph courtesy of R. J. Santoro.

The last feature of laminar jet diffusion flames we wish to highlight is the relationship between flame length and initial conditions. For circular-port flames, the flame length does not depend on initial velocity or diameter but, rather, on the initial volumetric flowrate, $Q_F$. Since $Q_F = v_e \pi R^2$, various combinations of $v_e$ and $R$ can yield the same flame length. We can show that this is reasonable by appealing to our previous analysis of the nonreacting laminar jet (see Example 9.2). By ignoring the effects of heat released by reaction, Eqn. 9.16 provides a crude description of the flame boundaries when $Y_F$ is set equal to $Y_{F,\text{stoic}}$. The flame length is then obtained when $r$ equals zero, i.e.,

$$L_f \approx \frac{3}{8\pi} \frac{Q_F}{\mathcal{D} Y_{F,\text{stoic}}}. \tag{9.22}$$

Thus, we see that the flame length is indeed proportional to the volumetric flowrate; furthermore, we see that the flame length is inversely proportional to the stoichiometric fuel mass fraction. This implies that fuels that require less air for complete combustion produce shorter flames, as one might expect. Faeth and coworkers [15, 16] have used this model (Eqn. 9.22) with reasonable results for nonbuoyant (zero-gravity) flames in still air with a length adjustment accounting for the nonsimilar behavior close to the jet exit and an overall empirical correction factor. Lin and Faeth [17] applied this approach to nonbuoyant flames in coflowing air. In a subsequent section, we present other approximations that can be used in engineering calculations of flame lengths.

## SIMPLIFIED THEORETICAL DESCRIPTIONS

The earliest theoretical description of the laminar jet diffusion flame is that of Burke and Schumann [10] published in 1928. Although many simplifying assumptions were employed, e.g., the velocity field was everywhere constant and parallel to the flame axis, their theory predicted flame lengths reasonably well for axisymmetric (circular-port) flames. This good agreement led other investigators to refine the original theory, yet retaining the constant-velocity assumption. In 1977, Roper [12–14] published a new theory that retained the essential simplicity of the Burke–Schumann analysis, but relaxed the requirement of a single constant velocity. Roper's approach provides reasonable estimates of flame lengths for both circular and noncircular nozzles. Subsequently, we will present his results, which are useful for engineering calculations. Before doing so, however, let us describe the problem mathematically so that we can see the inherent difficulty of the problem and thus appreciate the elegance of the simplified theories. We begin by first presenting a somewhat general formulation using the more familiar variables: velocities, mass fractions, and temperature; then, with a few additional assumptions, we develop the conserved scalar approach that requires the solution of only *two* partial differential equations. More complete mathematical descriptions can be found in advanced texts [27] and original references [21].

### Primary Assumptions

In the same spirit as Chapter 8, we can greatly simplify the basic conservation equations, yet retain the essential physics, by invoking the following assumptions:

1. The flow is laminar, steady, and axisymmetric, produced by a jet of fuel emerging from a circular nozzle of radius $R$, which burns in a quiescent, infinite reservoir of oxidizer.
2. Only three "species" are considered: fuel, oxidizer, and products. Inside the flame zone, only fuel and products exist; and beyond the flame, only oxidizer and products exist.

3. Fuel and oxidizer react in stoichiometric proportions at the flame. Chemical kinetics are assumed to be infinitely fast, resulting in the flame being represented as an infinitesimally thin sheet. This is commonly referred to as the **flame-sheet approximation.**

4. Species molecular transport is by simple binary diffusion governed by Fick's law.

5. Thermal energy and species diffusitivities are equal; thus, the Lewis number ($Le = \alpha / \mathcal{D}$) is unity.

6. Radiation heat transfer is negligible.

7. Only radial diffusion of momentum, thermal energy, and species is important; axial diffusion is neglected.

8. The flame axis is oriented vertically upward.

## Basic Conservation Equations

With these assumptions, the differential equations that govern the velocity, temperature, and species distributions throughout the flowfield are developed as follows:

**Mass Conservation**    The appropriate mass conservation equation is identical to the axisymmetric continuity equation (Eqn. 7.7) developed in Chapter 7, since none of the flame assumptions above result in any further simplification:

$$\frac{1}{r}\frac{\partial(r\rho v_r)}{\partial r} + \frac{\partial(\rho v_x)}{\partial x} = 0. \tag{9.23}$$

**Axial Momentum Conservation**    Similarly, the axial momentum equation (Eqn. 7.48) also stands as previously developed:

$$\frac{1}{r}\frac{\partial}{\partial x}(r\rho v_x v_x) + \frac{1}{r}\frac{\partial}{\partial r}(r\rho v_x v_r) - \frac{1}{r}\frac{\partial}{\partial r}\left(r\mu \frac{\partial v_x}{\partial r}\right) = (\rho_\infty - \rho)g. \tag{9.24}$$

This equation applies throughout the entire domain, i.e., both inside and outside of the flame boundary, with no discontinuities at the flame sheet.

**Species Conservation**    With the flame-sheet assumption, the chemical production rates ($\dot{m}_i'''$) become zero in the species conservation equation (Eqn. 7.20), with all chemical reaction phenomena embodied in the boundary conditions. Thus,

$$\frac{1}{r}\frac{\partial}{\partial x}(r\rho v_x Y_i) + \frac{1}{r}\frac{\partial}{\partial r}(r\rho v_r Y_i) - \frac{1}{r}\frac{\partial}{\partial r}\left(r\rho \mathcal{D}\frac{\partial Y_i}{\partial r}\right) = 0, \tag{9.25}$$

where $i$ represents fuel, and Eqn. 9.25 applies inside the flame boundary, or $i$ represents oxidizer, and Eqn. 9.25 applies outside of the flame boundary. Since there are only three species, the product mass fraction can be found from

$$Y_{Pr} = 1 - Y_F - Y_{Ox}. \tag{9.26}$$

**Energy Conservation**    The Shvab–Zeldovich form of energy conservation (Eqn. 7.66) is consistent with all of our assumptions and simplifies further in the same manner as the species equation, in that the production term ($\sum h^o_{f,i} \dot{m}'''_i$) becomes zero everywhere except at the flame boundary. Thus, Eqn. 7.66 becomes

$$\frac{\partial}{\partial x}\left(r\rho v_x \int c_p \, dT\right) + \frac{\partial}{\partial r}\left(r\rho v_r \int c_p \, dT\right) - \frac{\partial}{\partial r}\left(r\rho D \frac{\partial \int c_p \, dT}{\partial r}\right) = 0, \qquad (9.27)$$

and applies both inside and outside of the flame, but with discontinuities at the flame itself. Thus, the heat release from reaction must enter the problem formulation as a boundary condition, where the boundary is the flame surface.

## Additional Relations

To completely define our problem, an equation of state is needed to relate density and temperature:

$$\rho = \frac{P \, MW_{\text{mix}}}{R_u T}, \qquad (9.28)$$

where the mixture molecular weight is determined from the species mass fractions (Eqn. 2.12b) as

$$MW_{\text{mix}} = (\sum Y_i / MW_i)^{-1}. \qquad (9.29)$$

Before proceeding, it is worthwhile to summarize our jet flame model equations and point out the intrinsic difficulties associated with obtaining their solutions, for it is these difficulties that motivate reformulating the problem in terms of conserved scalars. We have a total of five conservation equations: mass, axial momentum, energy, fuel species, and oxidizer species; involving five unknown functions: $v_r(r, x)$, $v_x(r, x)$, $T(r, x)$, $Y_F(r, x)$, and $Y_{Ox}(r, x)$. Determining the five functions that simultaneously satisfy all five equations, subject to appropriate boundary conditions, defines our problem. Simultaneously solving five partial differential equations is itself a formidable task. The problem becomes even more complicated when we realize that some of the boundary conditions necessary to solve the fuel and oxidizer species and energy equations must be specified at the flame, the location of which is not known *a priori*. To eliminate this problem of the unknown location of the flame sheet, we seek to recast the governing equations, i.e., to employ conserved scalars, which require boundary conditions only along the flame axis ($r = 0, x$), far from the flame ($r \rightarrow \infty, x$), and at the nozzle exit plane ($r, x = 0$).

## Conserved Scalar Approach

**Mixture Fraction**    We can eliminate our boundary-condition dilemma, without reducing the complexity of the problem much, by replacing the two species conservation equations with the single mixture-fraction equation (Eqn. 7.79) developed in Chapter 7:

$$\frac{\partial}{\partial x}(r\rho v_x f) + \frac{\partial}{\partial r}(r\rho v_r f) - \frac{\partial}{\partial r}\left(r\rho D \frac{\partial f}{\partial r}\right) = 0. \tag{9.30}$$

This equation involves no discontinuities at the flame and requires no additional assumptions beyond those previously listed. Recall that the definition of the mixture fraction, $f$, is the mass fraction of material having its origin in the fuel system (see Eqns. 7.68 and 7.70), and, as such, has a maximum value of unity at the nozzle exit and a value of zero far from the flame. The appropriate boundary conditions for $f$ can be given by

$$\frac{\partial f}{\partial r}(0, x) = 0 \quad \text{(symmetry)}, \tag{9.31a}$$

$$f(\infty, x) = 0 \quad \text{(no fuel in oxidizer)}, \tag{9.31b}$$

$$\begin{aligned} f(r \le R, 0) &= 1 \\ f(r > R, 0) &= 0 \end{aligned} \quad \text{(top-hat exit profile)}. \tag{9.31c}$$

Once the function $f(r, x)$ is known, the location of the flame is easily found, because $f = f_{\text{stoic}}$ at this location.

**Standardized Enthalpy**    We continue our development by turning our attention now to the energy equation. Here, again with no additional simplifying assumptions, we can replace the Shvab–Zeldovich energy equation, which explicitly involves $T(r, x)$, with the conserved scalar form, Eqn. 7.83, which involves the standardized enthalpy function, $h(r, x)$, with temperature no longer appearing explicitly:

$$\frac{\partial}{\partial x}(r\rho v_x h) + \frac{\partial}{\partial r}(r\rho v_r h) - \frac{\partial}{\partial r}\left(r\rho D \frac{\partial h}{\partial r}\right) = 0. \tag{9.32}$$

As with the mixture fraction, no discontinuities in $h$ occur at the flame, and the boundary conditions are given by

$$\frac{\partial h}{\partial r}(0, x) = 0, \tag{9.33a}$$

$$h(\infty, x) = h_{Ox}, \tag{9.33b}$$

$$h(r \le R, 0) = h_F \tag{9.33c}$$

$$h(r > R, 0) = h_{Ox}.$$

The continuity and axial momentum equations remain as given above, i.e., Eqns. 9.23 and 9.24, respectively, unaffected by our desire to use the conserved scalars to replace the species and energy equations. The boundary conditions for the velocities are the same as given previously for the nonreacting jet, i.e.,

$$v_r(0, x) = 0, \tag{9.34a}$$

$$\frac{\partial v_x}{\partial r}(0, x) = 0, \tag{9.34b}$$

$$v_x(\infty, x) = 0, \tag{9.34c}$$

$$v_x(r \leq R,\ 0) = v_e$$

$$v_x(r > R,\ 0) = 0. \tag{9.34d}$$

Before our system of equations (Eqns. 9.23, 9.24, 9.30, and 9.32) can be solved, we need to be able to determine the density, $\rho(r, x)$, which appears in each conservation equation, using some appropriate state relationship. Before we do so, however, we can effect further simplification of our problem by recasting the governing equations to a nondimensional form.

**Nondimensional Equations** Frequently, valuable insights can be obtained by defining dimensionless variables and substituting these into the governing equations. Such procedures result in the identification of important dimensionless parameters, such as the Reynolds number, with which you are well acquainted. We start by using the nozzle radius $R$ as the characteristic length scale, and the nozzle exit velocity $v_e$ as the characteristic velocity, to define the following dimensionless spatial coordinates and velocities:

$$x^* \equiv \begin{array}{c} \text{dimensionless axial} \\ \text{distance} \end{array} = x/R, \tag{9.35a}$$

$$r^* \equiv \begin{array}{c} \text{dimensionless radial} \\ \text{distance} \end{array} = r/R, \tag{9.35b}$$

$$v_x^* \equiv \begin{array}{c} \text{dimensionless axial} \\ \text{velocity} \end{array} = v_x/v_e, \tag{9.35c}$$

$$v_r^* \equiv \begin{array}{c} \text{dimensionless radial} \\ \text{velocity} \end{array} = v_r/v_e. \tag{9.35d}$$

Since the mixture fraction $f$ is already a dimensionless variable with the desired property that $0 \leq f \leq 1$, we will use it directly. The mixture standardized enthalpy $h$, however, is not dimensionless; thus, we define

$$h^* \equiv \begin{array}{c} \text{dimensionless} \\ \text{standardized enthalpy} \end{array} = \frac{h - h_{Ox,\infty}}{h_{F,e} - h_{Ox,\infty}}. \tag{9.35e}$$

Note that at the nozzle exit, $h = h_{F,e}$ and, thus, $h^* = 1$; and in the ambient ($r \to \infty$), $h = h_{Ox,\infty}$, and $h^* = 0$.

To make our governing equations completely dimensionless, we also define the density ratio:

$$\rho^* \equiv \begin{array}{c} \text{dimensionless} \\ \text{density} \end{array} = \frac{\rho}{\rho_e}, \tag{9.35f}$$

where $\rho_e$ is the fuel density at the nozzle exit.

Relating each of the dimensional variables or parameters to its dimensionless counterpart, and substituting these back into the basic conservation equations, results in the following dimensionless governing equations:

*Continuity*

$$\frac{\partial}{\partial x^*}\left(\rho^* v_x^*\right) + \frac{1}{r^*}\frac{\partial}{\partial r^*}\left(r^* \rho^* v_r^*\right) = 0. \tag{9.36}$$

*Axial momentum*

$$\frac{\partial}{\partial x^*}\left(r^*\rho^* v_x^* v_x^*\right)+\frac{\partial}{\partial r^*}\left(r^*\rho^* v_r^* v_x^*\right)-\frac{\partial}{\partial r^*}\left[\left(\frac{\mu}{\rho_e v_e R}\right)r^*\frac{\partial v_x^*}{\partial r^*}\right]=\frac{gR}{v_e^2}\left(\frac{\rho_\infty}{\rho_e}-\rho^*\right)r^*.$$

(9.37)

*Mixture fraction*

$$\frac{\partial}{\partial x^*}\left(r^*\rho^* v_x^* f\right)+\frac{\partial}{\partial r^*}\left(r^*\rho^* v_r^* f\right)-\frac{\partial}{\partial r^*}\left[\left(\frac{\rho D}{\rho_e v_e R}\right)r^*\frac{\partial f}{\partial r^*}\right]=0.$$

(9.38)

*Dimensionless enthalpy*

$$\frac{\partial}{\partial x^*}\left(r^*\rho^* v_x^* h^*\right)+\frac{\partial}{\partial r^*}\left(r^*\rho^* v_r^* h^*\right)-\frac{\partial}{\partial r^*}\left[\left(\frac{\rho D}{\rho_e v_e R}\right)r^*\frac{\partial h^*}{\partial r^*}\right]=0.$$

(9.39)

The dimensionless boundary conditions for the above are

$$v_r^*(0, x^*)=0,$$

(9.40a)

$$v_x^*(\infty, x^*)=f(\infty, x^*)=h^*(\infty, x^*)=0,$$

(9.40b)

$$\frac{\partial v_x^*}{\partial r^*}(0, x^*)=\frac{\partial f}{\partial r^*}(0, x^*)=\frac{\partial h^*}{\partial r^*}(0, x^*)=0,$$

(9.40c)

$$v_x^*(r^*\leq1, 0)=f(r^*\leq1, 0)=h^*(r^*\leq1, 0)=1$$
$$v_x^*(r^*>1, 0)=f(r^*>1, 0)=h^*(r^*>1, 0)=0.$$

(9.40d)

Inspection of the dimensionless governing equation and boundary conditions shows some very interesting features. First, we see that the mixture fraction and dimensionless enthalpy equations, and their boundary conditions, are of identical form; that is, $f$ and $h^*$ play the same role in their respective governing equations. Therefore, we do not need to solve both Eqn. 9.38 and Eqn. 9.39, but, rather, one or the other. For example, if we were to solve for $f(r^*, x^*)$, then simply, $h^*(r^*, x^*)=f(r^*, x^*)$.

**Additional Assumptions**     We also note that, if buoyancy were neglected, the right-hand side of the axial momentum equation (Eqn. 9.37) becomes zero and the general form of this equation is the same as the mixture fraction and dimensionless enthalpy equations except that $\mu$ appears in the former where $\rho D$ appears in the latter. Our problem can be simplified even further if we assume that viscosity $\mu$ equals the $\rho D$ product. Since the **Schmidt number, *Sc*,** is defined as

$$Sc\equiv\frac{\text{momentum diffusivity}}{\text{mass diffusivity}}=\frac{\nu}{D}=\frac{\mu}{\rho D},$$

(9.41)

we see that with $\mu=\rho D$, the Schmidt number is unity ($Sc=1$). Assuming that the momentum diffusivity and mass diffusivity are equal ($Sc=1$) is analogous to our previous assumption of equal thermal and mass diffusivities ($Le=1$).

With the assumptions of negligible buoyancy and $Sc = 1$, the following single conservation equation replaces the individual axial momentum, mixture fraction (species mass), and enthalpy (energy) equations (Eqns. 9.37–9.39):

$$\frac{\partial}{\partial x^*}\left(r^*\rho^*v_x^*\zeta\right)+\frac{\partial}{\partial r^*}\left(r^*\rho^*v_r^*\zeta\right)-\frac{\partial}{\partial r^*}\left(\frac{1}{Re}r^*\frac{\partial\zeta}{\partial r^*}\right)=0, \qquad (9.42)$$

where our generic variable $\zeta = v_x^* = f = h^*$ and $Re = \rho_e v_e R/\mu$. Although $v_x^*$, $f$, and $h^*$ each satisfy Eqn. 9.42, $\rho^*$ and $v_x^*$ are coupled through continuity (Eqn. 9.36), whereas $\rho^*$ and $f$ (or $h^*$) are coupled through state relationships, as discussed below.

**State Relationships**  To be able to solve the jet flame problem as developed above requires that the density $\rho^*(=\rho/\rho_e)$ be related to the mixture fraction, $f$, or any of the other conserved scalars. To do this, we employ the ideal-gas equation of state (Eqn. 9.28); however, this requires a knowledge of the species mass fractions and temperature. Our immediate problem then is to relate the $Y_i$s and $T$ as functions of the mixture fraction. Knowing these primary relationships, they can be combined to obtain the needed relationship $\rho = \rho(f)$. For our simple system, which consists of only products and fuel inside the flame sheet, and only products and oxidizer outside the flame sheet (see assumption 2), our task is to find the following **state relationships:**

$$Y_F = Y_F(f), \qquad (9.43a)$$

$$Y_{Pr} = Y_{Pr}(f), \qquad (9.43b)$$

$$Y_{Ox} = Y_{Ox}(f), \qquad (9.43c)$$

$$T = T(f), \qquad (9.43d)$$

$$\rho = \rho(f). \qquad (9.43e)$$

With our flame-sheet approximation (assumption 3), the definition of the mixture fraction (Eqn. 7.70) can be used to relate the species mass fractions, $Y_F$, $Y_{Ox}$, and $Y_{Pr}$, to $f$ in the region inside the flame sheet, at the flame, and beyond the flame sheet, as illustrated in Fig. 9.7, as follows:

*Inside the flame ($f_{stoic} < f \le 1$)*

$$Y_F = \frac{f - f_{stoic}}{1 - f_{stoic}}, \qquad (9.44a)$$

$$Y_{Ox} = 0, \qquad (9.44b)$$

$$Y_{Pr} = \frac{1 - f}{1 - f_{stoic}}. \qquad (9.44c)$$

*At the flame ($f = f_{stoic}$)*

$$Y_F = 0, \qquad (9.45a)$$

$$Y_{Ox} = 0, \qquad (9.45b)$$

$$Y_{Pr} = 1. \qquad (9.45c)$$

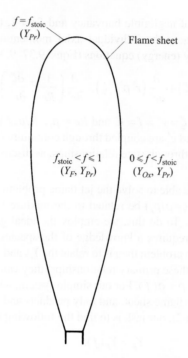

**Figure 9.7**    Simplified model of jet diffusion flame employing the flame-sheet approximation where inside the flame sheet only fuel and products exist, and outside the sheet only oxidizer and products exist.

*Outside the flame* $(0 \leq f < f_{\text{stoic}})$

$$Y_F = 0, \tag{9.46a}$$

$$Y_{Ox} = 1 - f/f_{\text{stoic}}, \tag{9.46b}$$

$$Y_{Pr} = f/f_{\text{stoic}}, \tag{9.46c}$$

where the stoichiometric mixture fraction relates to the stoichiometric (mass) coefficient $\nu$ as

$$f_{\text{stoic}} = \frac{1}{\nu + 1}. \tag{9.47}$$

Note that all of the mass fractions are related linearly to the mixture fraction (Fig. 9.8a).

To determine the mixture temperature as a function of mixture fraction requires a calorific equation of state (see Eqn. 2.4). As we have done in previous chapters, we follow Spalding [28] and make the following assumptions:

1. The specific heats of each species (fuel, oxidizer, and products) are constant and equal: $c_{p,F} = c_{p,Ox} = c_{p,Pr} \equiv c_p$.

2. The enthalpies of formation of the oxidizer and the products are zero: $h_{f,Ox}^o = h_{f,Pr}^o = 0$. This results in the enthalpy of formation of the fuel equaling its heat of combustion.

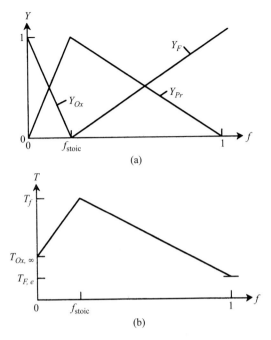

**Figure 9.8** (a) Simplified-state relationships for species mass fractions $Y_F(f)$, $Y_{Ox}(f)$ and $Y_{Pr}(f)$. (b) Simplified-state relationships for mixture temperature $T(f)$.

These assumptions are invoked only to provide a simple illustration of how to construct state relationships and are not essential to the basic concept. With these assumptions, our calorific equation of state is simply

$$h = \sum Y_i h_i = Y_F \, \Delta h_c + c_p \,(T - T_{ref}). \tag{9.48}$$

Substituting Eqn. 9.48 into the definition of the dimensionless enthalpy $h^*$ (Eqn. 9.35e) and recalling that, because of the similarity of the governing equations, $h^* = f$, we obtain

$$h^* = \frac{Y_F \, \Delta h_c + c_p \,(T - T_{Ox,\infty})}{\Delta h_c + c_p \,(T_{F,e} - T_{Ox,\infty})} \equiv f, \tag{9.49}$$

where the definitions $h_{Ox,\infty} \equiv c_p(T_{Ox,\infty} - T_{ref})$ and $h_{F,e} \equiv \Delta h_c + c_p(T_{F,e} - T_{ref})$ have also been substituted. Solving Eqn. 9.49 for $T$ provides the following general state relationship, $T = T(f)$, recognizing that $Y_F$ is also a function of the mixture fraction $f$:

$$T = (f - Y_F)\frac{\Delta h_c}{c_p} + f(T_{F,e} - T_{Ox,\infty}) + T_{Ox,\infty}. \tag{9.50}$$

Using Eqn. 9.50 with the appropriate expressions for $Y_F$ inside the flame sheet (Eqn. 9.44a), at the flame (Eqn. 9.45a), and outside the flame sheet (Eqn. 9.46a), yields the following:

*Inside the flame* ($f_{stoic} < f \leq 1$)

$$T = T(f) = f\left[(T_{F,e} - T_{Ox,\infty}) - \frac{f_{stoic}}{1 - f_{stoic}}\frac{\Delta h_c}{c_p}\right] + T_{Ox,\infty} + \frac{f_{stoic}}{(1 - f_{stoic})c_p}\Delta h_c. \quad (9.51a)$$

*At the flame* ($f = f_{stoic}$)

$$T \equiv T(f) = f_{stoic}\left(\frac{\Delta h_c}{c_p} + T_{F,e} - T_{Ox,\infty}\right) + T_{Ox,\infty}. \quad (9.51b)$$

*Outside the flame* ($0 \leq f < f_{stoic}$)

$$T = T(f) = f\left(\frac{\Delta h_c}{c_p} + T_{F,e} - T_{Ox,\infty}\right) + T_{Ox,\infty}. \quad (9.51c)$$

Note that with our simplified thermodynamics, the temperature depends linearly on $f$ in the regions inside and outside the flame, with a maximum at the flame as illustrated in Fig. 9.8b. It is also interesting to note that the flame temperature given by Eqn. 9.51b is identical to the constant-pressure adiabatic flame temperature calculated from the first law (Eqn. 2.40) for fuel and oxidizer with initial temperatures of $T_{F,e}$ and $T_{Ox,\infty}$, respectively. Our task is now complete in that, with the state relationships $Y_F(f)$, $Y_{Ox}(f)$, $Y_{Pr}(f)$, and $T(f)$, the mixture density can be determined solely as a function of mixture fraction $f$ using the ideal-gas equation of state (Eqn. 9.28). It is important to point out that our use of only three species (fuel, oxidizer, and products) and greatly simplified thermodynamics allows us to formulate simple, closed-form state relationships and illustrate the basic concepts and procedures involved in creating a conserved scalar model of diffusion flames. More sophisticated state relationships are frequently employed using equilibrium, partial equilibrium, or experimental state relationships for complex mixtures. Nevertheless, the basic concepts are essentially as presented here.

Table 9.1 summarizes our developments of conserved scalar models of laminar jet diffusion flames, and the following sections discuss various solutions to the problem.

## Various Solutions

**Burke–Schumann**    As mentioned previously, the earliest approximate solution to the laminar jet flame problem is that of Burke and Schumann [10], who analyzed circular and two-dimensional fuel jets emerging into coflowing oxidizer streams. For both the axisymmetric and two-dimensional problems, they used the flame-sheet approximation and assumed that a single velocity characterized the flow ($v_x = v$, $v_r = 0$). This latter assumption eliminates the need to solve the axial-momentum equation (Eqn. 9.24) and, by default, neglects buoyancy. Although the concept of a conserved scalar had not been formally developed at the time of Burke and Schumann's study (1928), their treatment of species conservation cast the problem in a form equivalent to that of a conserved scalar. With the assumption that $v_r = 0$, mass conservation (Eqn. 9.23) requires that $\rho v_x$ be a constant; thus, the variable-density species conservation equation (Eqn. 9.25) becomes

**Table 9.1**        Summary of conserved scalar models of laminar jet diffusion flame

| Assumptions | Solution Variables Required | Conservation Equations Required | State Relationships Required[a] |
|---|---|---|---|
| Primary assumptions only for conservation equations<br>+<br>Simplified thermodynamics for state relationships | $v_r^*(r^*, x^*),$<br>$v_x^*(r^*, x^*),$<br>$f(r^*, x^*)$ or $h^*(r^*, x^*)$ | Eqn. 9.36, 9.37, 9.38, or 9.39 | Eqn. 9.28 and Eqns. 9.44, 9.45, 9.46, and 9.51 (or equivalent) |
| Primary assumptions<br>+<br>No buoyancy and $Sc = 1$<br>+<br>Simplified thermodynamics for state relationships | $v_r^*(r^*, x^*),$<br>$\zeta(r^*, x^*),$ i.e., $v_x^*$ or $f$<br>or $h^*$ | Eqns. 9.36 and 9.42 | As above |

[a]Additional relations required for temperature-dependent transport properties $\mu$ and/or $\rho\mathcal{D}$.

$$\rho v_x \frac{\partial Y_i}{\partial x} - \frac{1}{r}\frac{\partial}{\partial r}\left(r\rho\mathcal{D}\frac{\partial Y_i}{\partial r}\right) = 0. \tag{9.52}$$

Since the above equation has no species production term, its solution requires *a priori* knowledge of the flame boundary. Burke and Schumann circumvented this problem by defining the flowfield in terms of a single fuel species whose mass fraction takes on a value of unity in the fuel stream, zero at the flame, and $-1/v$, or $-f_{\text{stoic}}/(1-f_{\text{stoic}})$, in the pure oxidizer. Thus, with this convention, "negative" fuel concentrations occur outside of the flame. In a modern context, their fuel mass fraction $Y_F$ is defined by the mixture fraction as

$$Y_F = \frac{f - f_{\text{stoic}}}{1 - f_{\text{stoic}}}. \tag{9.53}$$

Substitution of this definition into Eqn. 9.52 recovers the familiar conserved scalar equation (Eqn. 9.30). Although Burke and Schumann assumed constant properties and constant $v_x$, we can recover their governing equation by making the less restrictive assumption that the product of the density and diffusivity is constant, i.e., $\rho\mathcal{D} = $ constant $\equiv \rho_{\text{ref}}\mathcal{D}_{\text{ref}}$. In Chapter 3, we saw that the $\rho\mathcal{D}$ product varies approximately as $T^{1/2}$; thus, this assumption is clearly an approximation. Substituting $\rho_{\text{ref}}\mathcal{D}_{\text{ref}}$ for $\rho\mathcal{D}$ into Eqn. 9.52, removing this product from inside the radial derivative, and noting that $\rho v_x = $ constant $\equiv \rho_{\text{ref}}v_{x,\text{ref}}$, we see that the density $\rho_{\text{ref}}$ cancels, yielding the following final result:

$$v_{x,\text{ref}}\frac{\partial Y_F}{\partial x} = \mathcal{D}_{\text{ref}}\frac{1}{r}\frac{\partial}{\partial r}\left(r\frac{\partial Y_F}{\partial r}\right), \tag{9.54}$$

where $v_{x,\text{ref}}$ and $\mathcal{D}_{\text{ref}}$ are reference values of velocity and diffusivity, respectively, both evaluated at the same temperature.

The solution to the above partial differential equation, $Y_F(x, r)$, is a rather complicated expression involving **Bessel functions.** The flame length is not given explicitly but can be found by solving the following transcendental equation for the flame length, $L_f$:

$$\sum_{m=1}^{\infty} \frac{J_1(\lambda_m R)}{\lambda_m [J_0(\lambda_m R_o)]^2} \exp\left(-\frac{\lambda_m^2 \mathcal{D}}{v} L_f\right) - \frac{R_o^2}{2R}\left(1 + \frac{1}{S}\right) + \frac{R}{2} = 0. \qquad (9.55)$$

In the above, $J_0$ and $J_1$ are the zeroth- and first-order Bessel functions, which are described in mathematical reference books (e.g., Ref. [29]); the $\lambda_m$ are defined by all the positive roots of the equation $J_1(\lambda_m R_o) = 0$ [29]; $R$ and $R_o$ are the fuel tube and outer flow radii, respectively; and $S$ is the molar stoichiometric ratio of oxidizer (outer-flow fluid) to fuel (nozzle fluid). Flame lengths predicted by the Burke–Schumann theory are in reasonable agreement with theory for circular-port burners, primarily as a result of offsetting assumptions; the effect of buoyancy is to cause a narrowing of the flame, which, in turn, increases diffusion rates. Burke and Schumann recognized this possibility, foreshadowing the work of Roper [12] that showed this to be true. The numerical study of Kee and Miller [19] also explicitly showed this effect by running comparison cases with and without buoyancy.

**Roper**   Roper [12] proceeded in the spirit of the Burke–Schumann approach, but allowed the characteristic velocity to vary with axial distance as modified by buoyancy and in accordance with continuity. In addition to circular-port burners, Roper analyzed rectangular-slot and curved-slot [12, 14] burners. Roper's analytical solutions, and those as modified slightly by experiment, are presented in a separate section below.

**Constant-Density Solution**   If the density is assumed to be constant, the solutions to Eqns. 9.23, 9.24, and 9.30 are identical to those of the nonreacting jet. In this case, the flame length is given by Eqn. 9.22:

$$L_f \approx \frac{3}{8\pi} \frac{1}{\mathcal{D}} \frac{Q_F}{Y_{F,\text{stoic}}}. \qquad (9.56)$$

**Variable-Density Approximate Solution**   Fay [11] solved the variable-density, laminar, jet-flame problem. In his solution, buoyancy is neglected, thereby simplifying the axial momentum equation. With regard to properties, the Schmidt and Lewis numbers are assumed to be unity, consistent with our formulation of the governing equations and, furthermore, the absolute viscosity, $\mu$, is assumed to be directly proportional to temperature, i.e.,

$$\mu = \mu_{\text{ref}} \, T/T_{\text{ref}}.$$

Fay's solution for the flame length is

$$L_f \approx \frac{3}{8\pi} \frac{1}{Y_{F,\text{stoic}}} \frac{\dot{m}_F}{\mu_{\text{ref}}} \frac{\rho_\infty}{\rho_{\text{ref}}} \frac{1}{I\,(\rho_\infty/\rho_f)}, \qquad (9.57)$$

**Table 9.2**    Momentum integral estimates[a] for variable-density laminar jet flames

| $\rho_\infty/\rho_f$ | $\rho_\infty/\rho_{\mathrm{ref}}$ | $I(\rho_\infty/\rho_f)$ |
|:---:|:---:|:---:|
| 1 | 1 | 1 |
| 3 | 2 | 2.4 |
| 5 | 3 | 3.7 |
| 7 | 4 | 5.2 |
| 9 | 5 | 7.2 |

[a]Estimated from Fig. 3 of Ref. [11].

where $\dot{m}_F$ is the mass flow issuing from the nozzle, $\rho_\infty$ is the ambient fluid density far from the flame, and $I(\rho_\infty/\rho_f)$ is a function obtained by numerical integration as a part of Fay's solution. Tabulated values of $I(\rho_\infty/\rho_f)$ and $\rho_\infty/\rho_{\mathrm{ref}}$ are given in Table 9.2 for various ratios of ambient-to-flame densities, $\rho_\infty/\rho_f$.

Equation 9.57 can be recast in a form similar to the constant-density solution, Eqn. 9.56, by noting that $\dot{m}_F = \rho_F Q_F$ and $\mu_{\mathrm{ref}} = \rho_{\mathrm{ref}}\mathcal{D}(Sc = 1)$:

$$L_f \approx \frac{3}{8\pi} \frac{1}{\mathcal{D}_{\mathrm{ref}}} \frac{Q_F}{Y_{F,\,\mathrm{stoic}}} \left( \frac{\rho_F \rho_\infty}{\rho_{\mathrm{ref}}^2} \right) \frac{1}{I(\rho_\infty/\rho_f)}. \tag{9.58}$$

Thus, we see that the flame lengths predicted by the variable-density theory are longer than those of the constant-density theory by the factor

$$\frac{\rho_F \rho_\infty}{\rho_{\mathrm{ref}}^2} \frac{1}{I(\rho_\infty/\rho_f)}.$$

For $\rho_\infty/\rho_f = 5$ and $\rho_F = \rho_\infty$ (reasonable values for hydrocarbon–air flames) the predicted flame length for the variable-density theory is about 2.4 times that of the constant-density theory. Regardless of the ability of either theory to predict actual flame lengths, they both show that flame length is directly proportional to the nozzle-fluid volumetric flowrate and inversely proportional to the stoichiometric nozzle-fluid mass fraction, independent of the nozzle diameter.

**Numerical Solutions**    The use of digital computers and finite-difference techniques allows laminar jet flames to be modeled in much greater detail than in the analytical approaches discussed above. For example, the flame-sheet approximation with **frozen flow** assumed inside and outside of the flame can be replaced by a reacting mixture governed by chemical kinetics (see Chapter 4). Kee and Miller [18, 19] modeled an $H_2$–air flame using 16 reversible reactions involving 10 species. Smooke et al. [22] modeled a $CH_4$–air flame taking into account 79 reactions and 26 species and, more recently [6], modeled a $C_2H_4$–air flame using a chemical mechanism that contains 476 reactions with 66 species. Dynamical and chemical models for soot formation are also included in their recent work [6]. With chemical transformation effects no longer relegated to boundary conditions, the species conservation equations, Eqn. 9.25, would now have to include species production (source) and destruction (sink) terms (see Chapter 7). Numerical models also allow the relaxation of the assumption of simple binary diffusion. Detailed multicomponent diffusion is included in the computer models of Heys *et al.* [21] and Smooke *et al.* [6, 22]. Similarly, temperature-dependent properties can be incorporated easily into numerical models [6, 18–22]. In the Mitchell *et al.* [20] and Smooke *et al.* [6, 22] models, both radial and axial diffusion terms are retained in the conservation equation, thereby avoiding the boundary-layer approximations. Heys *et al.* [21] incorporate thermal radiation in their model of a $CH_4$–air flame and show that its inclusion results in peak temperature predictions of about 150 K lower than when no radiation losses are considered. Temperature differences of this order can greatly affect temperature-sensitive chemical reaction rates, such as those involved in nitric oxide formation (see Chapter 5). Smooke *et al.* [6] include both emission and reabsorption of radiation in their model. Davis *et al.* [26] studied the influence of buoyancy using a conserved scalar model for species and energy while including the gravitational body force in the axial momentum equation. (The dimensionless body force term appears on the right-hand side of Eqn. 9.37). Their analysis shows that the effects of gravity can be simulated by varying the pressure while maintaining fixed flowrates of fuel and coflowing air. Specifically, they found the dimensionless gravitational force to be proportional to the square of the pressure, i.e., $gR/v_e^2 \propto P^2$, when $\rho_e v_e = $ constant. Their numerical solutions show that the flickering and pulsing exhibited by flames at pressures greater than atmospheric result from the increased importance of buoyancy at higher pressures.

## FLAME LENGTHS FOR CIRCULAR-PORT AND SLOT BURNERS

### Roper's Correlations

Roper developed [12, 14], and verified by experiment [13, 14], expressions to predict laminar jet flame lengths for various burner geometries (circular, square, slot, and curved slot) and flow regimes (momentum-controlled, buoyancy-controlled, and transitional). Roper's results [12, 13] are summarized in Table 9.3 and detailed below.

For circular and square burner ports, the following expressions can be used to estimate flame lengths. These results apply regardless of whether or not buoyancy is important, and are applicable for fuel jets emerging into either a quiescent oxidizer or a coflowing stream, as long as the oxidizer is in excess, i.e., the flames are overventilated.

**Circular Port**

$$L_{f,\text{thy}} = \frac{Q_F(T_\infty/T_F)}{4\pi\mathcal{D}_\infty \ln(1+1/S)}\left(\frac{T_\infty}{T_f}\right)^{0.67}, \qquad (9.59)$$

$$L_{f,\text{expt}} = 1330\frac{Q_F(T_\infty/T_F)}{\ln(1+1/S)}, \qquad (9.60)$$

**Table 9.3**  Empirical and theoretical correlations for estimating lengths of vertical laminar jet flames

| Port Geometry | | | Conditions | Applicable Equation[a] |
|---|---|---|---|---|
| | $2R$ | Circular | Momentum- or buoyancy-controlled | Circular—Eqns. 9.59 and 9.60 |
| | $b$ | Square | Momentum- or buoyancy-controlled | Circular—Eqns. 9.61 and 9.62 |
| $h$ | $b$ | Slot | Momentum-controlled | Eqns. 9.63 and 9.64 |
| | | | Buoyancy-controlled | Eqns. 9.65 and 9.66 |
| | | | Mixed momentum-buoyancy-controlled | Eqn. 9.70 |

[a]For the circular and square geometries, the indicated equations apply for either a stagnant or coflowing oxidizer stream. For the slot geometry, the equations apply only to the stagnant oxidizer case.

where $S$ is the molar stoichiometric oxidizer–fuel ratio, $\mathcal{D}_\infty$ is a mean diffusion coefficient evaluated for the oxidizer at the oxidizer stream temperature, $T_\infty$, and $T_F$ and $T_f$ are the fuel stream and mean flame temperatures, respectively. In Eqn. 9.60, all quantities are evaluated in SI units (m, m³/s, etc.). Note that the burner diameter does not explicitly appear in either of these expressions.

**Square Port**

$$L_{f,\,\text{thy}} = \frac{Q_F (T_\infty / T_F)}{16\mathcal{D}_\infty \, [\text{inverf}((1+S)^{-0.5})]^2} \left( \frac{T_\infty}{T_f} \right)^{0.67}, \qquad (9.61)$$

$$L_{f,\,\text{expt}} = 1045 \, \frac{Q_F (T_\infty / T_F)}{[\text{inverf}((1+S)^{-0.5})]^2}, \qquad (9.62)$$

where **inverf** is the **inverse error function.** Values of the **error function, erf,** are tabulated in Table 9.4. Values for the inverse error function are generated from the error function tables in the same way that you would deal with inverse trigonometric functions, i.e., $\omega = \text{inverf}\,(\text{erf}\,\omega)$. Again, all quantities are evaluated in SI units.

**Slot Burner—Momentum-Controlled**

$$L_{f,\,\text{thy}} = \frac{b\beta^2 Q_F}{hI\mathcal{D}_\infty \, Y_{F,\,\text{stoic}}} \left( \frac{T_\infty}{T_F} \right)^2 \left( \frac{T_f}{T_\infty} \right)^{0.33}, \qquad (9.63)$$

$$L_{f,\,\text{expt}} = 8.6 \cdot 10^4 \, \frac{b\beta^2 Q_F}{hI Y_{f,\,\text{stoic}}} \left( \frac{T_\infty}{T_F} \right)^2, \qquad (9.64)$$

**Table 9.4**    Gaussian error function[a]

| $\omega$ | erf $\omega$ | $\omega$ | erf $\omega$ | $\omega$ | erf $\omega$ |
|---|---|---|---|---|---|
| 0.00 | 0.00000 | 0.36 | 0.38933 | 1.04 | 0.85865 |
| 0.02 | 0.02256 | 0.38 | 0.40901 | 1.08 | 0.87333 |
| 0.04 | 0.04511 | 0.40 | 0.42839 | 1.12 | 0.88679 |
| 0.06 | 0.06762 | 0.44 | 0.46622 | 1.16 | 0.89910 |
| 0.08 | 0.09008 | 0.48 | 0.50275 | 1.20 | 0.91031 |
| 0.10 | 0.11246 | 0.52 | 0.53790 | 1.30 | 0.93401 |
| 0.12 | 0.13476 | 0.56 | 0.57162 | 1.40 | 0.95228 |
| 0.14 | 0.15695 | 0.60 | 0.60386 | 1.50 | 0.96611 |
| 0.16 | 0.17901 | 0.64 | 0.63459 | 1.60 | 0.97635 |
| 0.18 | 0.20094 | 0.68 | 0.66378 | 1.70 | 0.98379 |
| 0.20 | 0.22270 | 0.72 | 0.69143 | 1.80 | 0.98909 |
| 0.22 | 0.24430 | 0.76 | 0.71754 | 1.90 | 0.99279 |
| 0.24 | 0.26570 | 0.80 | 0.74210 | 2.00 | 0.99532 |
| 0.26 | 0.28690 | 0.84 | 0.76514 | 2.20 | 0.99814 |
| 0.28 | 0.30788 | 0.88 | 0.78669 | 2.40 | 0.99931 |
| 0.30 | 0.32863 | 0.92 | 0.80677 | 2.60 | 0.99976 |
| 0.32 | 0.34913 | 0.96 | 0.82542 | 2.80 | 0.99992 |
| 0.34 | 0.36936 | 1.00 | 0.84270 | 3.00 | 0.99998 |

[a]The Gaussian error function is defined as erf $\omega \equiv \dfrac{2}{\sqrt{\pi}} \displaystyle\int_0^\omega e^{-v^2} dv$.

The complementary error function is defined as erfc $\omega \equiv 1 - \text{erf}\,\omega$.

where $b$ is the slot width, $h$ is the slot length (see Table 9.3), and the function $\beta$ is given by

$$\beta = \frac{1}{4\,\text{inverf}[1/(1+S)]},$$

and $I$ is the ratio of the actual initial momentum flow from the slot to that of uniform flow, i.e.,

$$I = \frac{J_{e,\text{act}}}{\dot{m}_F v_e}.$$

If the flow is uniform, $I = 1$, and for a fully developed, parabolic exit velocity profile (assuming $h \gg b$), $I = 1.5$. Equations 9.63 and 9.64 apply only if the oxidizer is stagnant. For a coflowing oxidizer stream, the reader is referred to Refs. [12] and [13].

### Slot Burner—Buoyancy-Controlled

$$L_{f,\text{thy}} = \left[ \frac{9\beta^4 Q_F^4 T_\infty^4}{8\mathcal{D}_\infty^2 a h^4 T_F^4} \right]^{1/3} \left[ \frac{T_f}{T_\infty} \right]^{2/9} \tag{9.65}$$

$$L_{f,\text{expt}} = 2 \cdot 10^3 \left[ \frac{\beta^4 Q_F^4 T_\infty^4}{a h^4 T_F^4} \right]^{1/3}. \tag{9.66}$$

where $a$ is the mean buoyant acceleration evaluated from

$$a \cong 0.6g \left( \frac{T_f}{T_\infty} - 1 \right), \tag{9.67}$$

where $g$ is the gravitational acceleration. Roper et al. [12] used a mean flame temperature, $T_f = 1500$ K, to evaluate the acceleration. As can be seen in Eqns. 9.65 and 9.66, the predicted flame length depends only weakly $(-\frac{1}{3}$ power) on $a$.

**Slot Burner—Transition Regime**   To determine whether a flame is momentum- or buoyancy-controlled, the flame **Froude number, $Fr_f$,** must be evaluated. The Froude number physically represents the ratio of the initial jet momentum flow to the buoyant force experienced by the flame. For the laminar jet flame issuing into a stagnant medium,

$$Fr_f \equiv \frac{(v_e I Y_{F,\,\text{stoic}})^2}{aL_f}, \tag{9.68}$$

and the flow regime can be established by the following criteria:

$$Fr_f \gg 1 \quad \text{Momentum-controlled,} \tag{9.69a}$$

$$Fr_f \approx 1 \quad \text{Mixed (transition),} \tag{9.69b}$$

$$Fr_f \ll 1 \quad \text{Buoyancy-controlled.} \tag{9.69c}$$

Note that in order to establish the appropriate flow regime, a value for $L_f$ is required; thus, an *a posteriori* check is required to see if the correct flow regime has been selected.

For the transitional region where both jet momentum and buoyancy are important, Roper [12, 13] recommends the following treatment:

$$L_{f,T} = \frac{4}{9} L_{f,M} \left( \frac{L_{f,B}}{L_{f,M}} \right)^3 \left\{ \left[ 1 + 3.38 \left( \frac{L_{f,M}}{L_{f,B}} \right)^3 \right]^{2/3} - 1 \right\}, \tag{9.70}$$

where the subscripts $M$, $B$, and $T$ refer to momentum-controlled, buoyancy-controlled, and transition (mixed), respectively.

---

It is desired to operate a square-port diffusion flame burner with a 50-mm-high flame in a laboratory. Determine the volumetric flowrate required if the fuel is propane. Also, determine the heat release ($\dot{m}\Delta h_c$) of the flame. What flowrate is required if methane is substituted for propane?

**Example 9.3**

**Solution**

We will apply Roper's correlation for square-port burners (Eqn. 9.62) to determine the volumetric flowrate:

$$L_f = \frac{1045 Q_F (T_\infty / T_F)}{[\text{inverf}((1+S)^{-0.5})]^2}.$$

If we assume that $T_\infty = T_F = 300$ K, the only parameter we need to calculate before we can find $Q_F$ is the molar stoichiometric air–fuel ratio, $S$. From Chapter 2, $S = (x + y/4) \, 4.76$; so, for propane ($C_3H_8$),

$$S = (3 + 8/4)4.76 = 23.8 \, \frac{\text{kmol}}{\text{kmol}}.$$

Thus,

$$\text{inverf}\,[(1 + 23.8)^{-0.5}] = \text{inverf}\,(0.2008) = 0.18,$$

where Table 9.4 was used to evaluate inverf (0.2008). Solving Eqn. 9.62 for $Q_F$, and evaluating, yields

$$Q_F = \frac{0.050(0.18)^2}{1045(300/300)} = 1.55 \cdot 10^{-6} \text{ m}^3/\text{s}$$

or

$$\boxed{Q_F = 1.55 \text{ cm}^3/\text{s}}$$

Using the ideal-gas law to estimate the propane density ($P = 1$ atm, $T = 300$ K), and taking the heat of combustion from Appendix B, the heat release rate is

$$\dot{m}\Delta h_c = \rho_F Q_F \, \Delta h_c$$

$$= 1.787(1.55 \cdot 10^{-6})46,357,000$$

$$\boxed{\dot{m}\Delta h_c = 128 \text{ W}}$$

Repeating the problem for methane, we find $S = 9.52$, $\rho_F = 0.65$, and $\Delta h_c = 50,016,000$ J/kg; thus,

$$\boxed{Q = 3.75 \text{ cm}^3/\text{s}}$$

and

$$\boxed{\dot{m}\Delta h_c = 122 \text{ W}}$$

**Comment**

Here we see that, although the volumetric flowrate required for $CH_4$ is about 2.4 times greater than the $C_3H_8$ requirement, both flames release approximately the same energy.

In the next two sections, we will explore the correlations above to see how various important engineering parameters affect flame lengths.

## Flowrate and Geometry Effects

Figure 9.9 compares flame lengths for a circular-port burner with slot burners having various exit aspect ratios, $h/b$, with all burners having the same port area. With the port area fixed, the mean exit velocity is the same for each configuration. From the figure, we see the linear dependence of flame length on flowrate for the circular port and the somewhat greater-than-linear dependence for the slot burners. For the conditions selected, flame Froude numbers are quite small, and, hence, the flames are dominated by buoyancy. As the slot-burner ports become

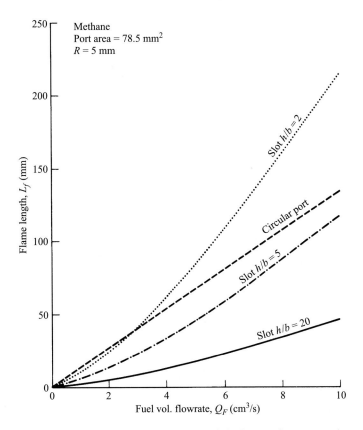

**Figure 9.9**   Predicted flame lengths for circular and slot burners having equal port areas.

more narrow ($h/d$ increasing), the flames become significantly shorter for the same flowrate.

## Factors Affecting Stoichiometry

The molar stoichiometric ratio, $S$, employed in the above expressions is defined in terms of the nozzle fluid and the surrounding reservoir fluid, i.e.,

$$S = \left( \frac{\text{moles ambient fluid}}{\text{moles nozzle fluid}} \right)_{\text{stoic}}. \tag{9.71}$$

Thus, $S$ depends on the chemical compositions of both the nozzle-fluid stream and the surrounding fluid. For example, the values of $S$ would be different for a pure fuel burning in air and a nitrogen-diluted fuel burning in air. Similarly, the mole fraction of oxygen in the ambient also affects $S$. In many applications, the following parameters, all of which affect $S$, are of interest.

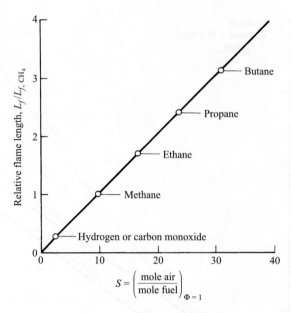

**Figure 9.10**    Dependence of flame length on fuel stoichiometry. Flame lengths for various fuels are shown relative to methane.

**Fuel Type**    The molar stoichiometric air–fuel ratio for a pure fuel can be calculated by applying simple atom balances (see Chapter 2). For a generic hydrocarbon, $C_xH_y$, the stoichiometric ratio can be expressed as

$$S = \frac{x + y/4}{\chi_{O_2}},$$    (9.72)

where $\chi_{O_2}$ is the mole fraction of oxygen in the air.

Figure 9.10 shows flame lengths, relative to methane, for hydrogen, carbon monoxide, and the $C_1$–$C_4$ alkanes, calculated using the circular-port expression, Eqn. 9.60, assuming equal flowrates for all fuels. Note that in using this expression, it is assumed that the same mean diffusivity applies to all of the mixtures, which is only approximately true. For hydrogen, this assumption may not be very good at all.

From Fig. 9.10, we see that flame lengths increase as the H/C ratio of the fuel decreases. For example, the propane flame is nearly $2\frac{1}{2}$ times as long as the methane flame. The flame lengths of the higher hydrocarbons, within a hydrocarbon family, do not differ greatly since the H/C ratios change much less with the addition of each C atom in comparison with the light hydrocarbons. We also see from Fig. 9.10 that carbon monoxide and hydrogen flames are much smaller than the hydrocarbon flames.

**Primary Aeration**    Gas-burning appliances premix some air with the fuel gas before it burns as a laminar jet diffusion flame. This **primary aeration,** which is typically 40–60 percent of the stoichiometric air requirement, makes the flames short and prevents soot from forming, resulting in the familiar blue flame. The maximum amount of air that can be added is limited by safety considerations. If too much air is added, the rich flammability limit may be exceeded, which implies that the mixture will support a premixed flame. Depending on flow conditions and burner geometry, this flame may propagate upstream, a condition referred to as **flashback.** If the flow velocity is too great for flashback to occur, an inner premixed flame will form inside the diffusion flame envelope, as in the Bunsen burner. Flashback limits are illustrated in Fig. 8.25.

Figure 9.11 shows the effect of primary aeration on the length of methane flames on circular-port burners. Note that in the range of 40–60 percent primary aeration, flame lengths are reduced approximately 85–90 percent from their original no-air-added lengths. The stoichiometric ratio $S$, defined by Eqn. 9.71, can be evaluated for the case of primary air addition by treating the "fuel", i.e., the nozzle fluid, as a mixture of the true fuel and primary air:

$$S = \frac{1 - \psi_{pri}}{\psi_{pri} + (1/S_{pure})}, \qquad (9.73)$$

where $\psi_{pri}$ is the fraction of the stoichiometric requirement met by the primary air, i.e., the primary aeration, and $S_{pure}$ is the molar stoichiometric ratio associated with the pure fuel.

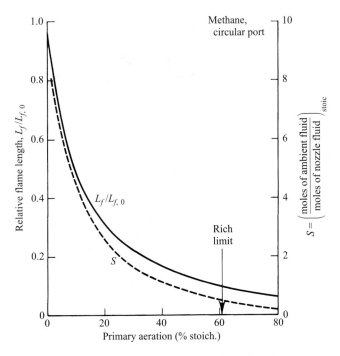

**Figure 9.11**     Effect of primary aeration on laminar jet flame lengths. For primary aerations greater than the rich limit, premixed burning (and flashback) is possible.

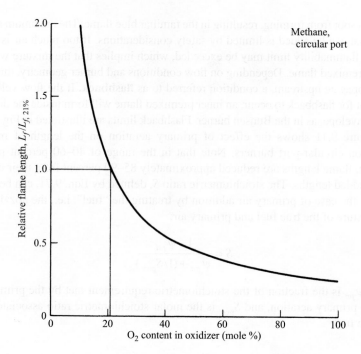

**Figure 9.12**    Effect of oxygen content in the oxidizing stream on flame length.

**Oxygen Content of Oxidizer**    The amount of oxygen in the oxidizer has a strong influence on flame lengths, as can be seen in Fig. 9.12. Small reductions in $O_2$ content from the nominal 21 percent value for air result in greatly lengthened flames. With a pure-$O_2$ oxidizer, flame lengths for methane are about one-quarter their value in air. To calculate the effect of $O_2$ content, Eqn. 9.72 can be used to evaluate the stoichiometric ratio for hydrocarbon fuels.

**Fuel Dilution with Inert Gas**    Diluting the fuel with an inert gas also has the effect of reducing flame length via its influence on the stoichiometric ratio. For hydrocarbon fuels,

$$S = \frac{x + y/4}{\left(\dfrac{1}{1 - \chi_{\text{dil}}}\right) \chi_{O_2}},$$

(9.74)

where $\chi_{\text{dil}}$ is the diluent mole fraction in the fuel stream.

---

**Example 9.4**

Design a natural-gas burner for a commercial cooking range that has a number of circular ports arranged in a circle. The circle diameter is constrained to be 160 mm (6.3 in.). The burner must deliver 2.2 kW at full load and operate with 40 percent primary aeration. For stable operation, the loading of an individual port should not exceed 10 W per mm$^2$ of port area. (See Fig. 8.25 for typical design constraints for natural-gas burners.) Also, the full-load flame height should not exceed 20 mm. Determine the number and the diameter of the ports.

**Solution**

We will assume that the fuel gas is methane, although for an accurate design, actual natural-gas properties should be used. Our overall strategy will be to relate the number of ports, $N$, and their diameter, $D$, to the port loading constraint; choose an $N$ and $D$ that satisfy this constraint; and then check to see if the flame-length constraint is violated. Having a design that meets these two constraints, we will check to see that the overall design makes physical sense.

*Step 1.* Apply port loading constraint. The total port area is

$$A_{\text{tot}} = N\pi D^2/4$$

and the constraint is

$$\frac{\dot{m}_F \Delta h_c}{A_{\text{tot}}} = \frac{2200 \text{ W}}{A_{\text{tot}} \text{ (mm}^2)} = 10 \text{ W/mm}^2;$$

thus,

$$ND^2 = \frac{4(2200)}{10\pi} = 280 \text{ mm}^2.$$

At this point, we can choose (more or less arbitrarily) a value for $N$ (or $D$) and calculate $D$ (or $N$) as a first trial for the design. Choosing $N = 36$ yields $D = 2.79$ mm.

*Step 2.* Determine flowrates. The design heat-release rate determines the fuel flowrate:

$$\dot{Q} = 2200 \text{ W} = \dot{m}_F \Delta h_c,$$

$$\dot{m}_F = \frac{2200 \text{ W}}{50,016,000 \text{ J/kg}} = 4.4 \cdot 10^{-5} \text{ kg/s}.$$

The primary aeration determines the flowrate of air premixed with the fuel:

$$\dot{m}_{A,\text{pri}} = 0.40(A/F)_{\text{stoic}} \dot{m}_F$$

$$= 0.40(17.1)4.4 \cdot 10^{-5} = 3.01 \cdot 10^{-4} \text{ kg/s}.$$

The total volumetric flowrate is

$$Q_{\text{TOT}} = (\dot{m}_{A,\text{pri}} + \dot{m}_F)/\bar{\rho}.$$

To determine $\bar{\rho}$, we apply the ideal-gas law where the mean molecular weight is calculated from the composition of the air–fuel mixture:

$$\chi_{A,\text{pri}} = \frac{N_A}{N_A + N_F} = \frac{Z}{Z+1},$$

where $Z$ is the primary molar air–fuel ratio:

$$Z = (x + y/4)4.76(\% \text{ aeration}/100)$$

$$= (1 + 4/4)(4.76)(40/100)$$

$$= 3.81.$$

Thus,

$$\chi_{A,\text{pri}} = \frac{3.81}{3.81+1} = 0.792,$$

$$\chi_{F,\text{pri}} = 1 - \chi_A = 0.208,$$

$$MW_{\text{mix}} = 0.792(28.85) + 0.208(16.04) = 26.19,$$

$$\bar{\rho} = \frac{P}{\left(\dfrac{R_u}{MW_{\text{mix}}}\right)T} = \frac{101,325}{\left(\dfrac{8315}{26.19}\right)300} = 1.064 \text{ kg/m}^3$$

and

$$Q_{TOT} = \frac{3.01 \cdot 10^{-4} + 4.4 \cdot 10^{-5}}{1.064} = 3.24 \cdot 10^{-4} \text{ m}^3/\text{s}.$$

*Step 3.* Check flame length constraint. The flowrate per port is

$$Q_{PORT} = Q_{TOT}/N = 3.24 \cdot 10^{-4}/36$$
$$= 9 \cdot 10^{-6} \text{ m}^3/\text{s}.$$

The molar ambient-air to nozzle-fluid stoichiometric ratio, $S$, is given by Eqn. 9.73 and evaluated as

$$S = \frac{1 - \psi_{pri}}{\psi_{pri} + (1/S_{pure})} = \frac{1 - 0.40}{0.40 + (1/9.52)} = 1.19.$$

We can calculate the flame length using Eqn. 9.60:

$$L_f = 1330 \frac{Q_F(T_\infty / T_F)}{\ln(1 + 1/S)}$$
$$= \frac{1330(9 \cdot 10^{-6})(300/300)}{\ln(1 + 1/1.19)} = 0.0196 \text{ m}$$
$$= 19.6 \text{ mm}.$$

A flame length of 19.6 mm meets our requirement of $L_f \leq 20$ mm.

*Step 4.* Check practicability of design. If we arrange 36 ports equally spaced on a 160-mm-diameter circle, the spacing between ports is

$$l = r\theta = \frac{160}{2} \text{ (mm)} \frac{2\pi}{36} \text{ (rad)}$$
$$= 14 \text{ mm}.$$

**Comment**

This spacing seems reasonable, although it is not clear whether or not the flames will form independently or merge. If the flames merged, our method of estimating flame height would not be valid. Since all of the constraints are satisfied with the 36-port design, iteration is not required.

## SOOT FORMATION AND DESTRUCTION

As mentioned in our overview of laminar flame structure at the beginning of this chapter, the formation and destruction of soot is an important feature of nonpremixed hydrocarbon–air flames. The incandescent soot within the flame is the primary source of a diffusion flame's luminosity, and the oil lamp is a prime example of a practical application dating back to ancient times. Soot also contributes to radiant heat losses from flames, with peak emission at wavelengths in the infrared region of the spectrum. Although soot formation in practical applications of laminar diffusion flames, e.g., in gas ranges, is to be avoided, the laminar diffusion flame is frequently used as a research tool in fundamental studies of soot formation in combustion systems and, hence, has a large literature associated with it. References [30–36] provide general reviews of soot formation in combustion.

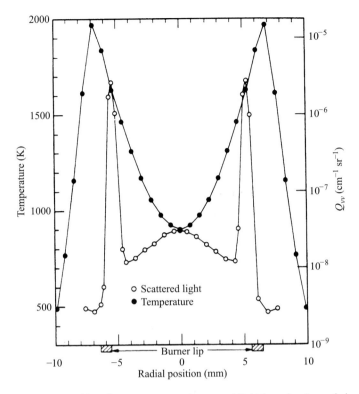

**Figure 9.13**    Radial profiles of temperature and scattered light for a laminar ethylene jet diffusion flame. Soot is contained in the region where the scattered light intensity is high.
SOURCE: Reprinted from Ref. [1] by permission of Gordon & Breach Science Publishers, © 1987.

It is generally agreed that soot is formed in diffusion flames over a limited range of temperatures, say, 1300 K < $T$ < 1600 K. Figure 9.13 illustrates this point for an ethylene jet flame. Here we see radial temperature profiles measured at an axial location between the burner and the flame tip. Also shown are measurements of light scattered by soot particles, where the two peaks correspond to regions containing significant amounts of soot. Note that the soot peaks are at radial locations interior to the temperature peaks and correspond to temperatures of about 1600 K. The soot region is very narrow and confined to a restricted range of temperatures. Although the chemistry and physics of soot formation in diffusion flames is exceedingly complex, the view has emerged that soot formation proceeds in a four-step sequence:

1.  Formation of precursor species.
2.  Particle inception.
3.  Surface growth and particle agglomeration.
4.  Particle oxidation.

In the first step, the formation of soot precursor species, polycyclic aromatic hydrocarbons (PAH) are thought to be important intermediates between the original fuel molecule and what can be considered a primary soot particle [31]. Chemical kinetics plays an important role in this first step. Although the detailed chemical mechanisms involved and the identity of the specific precursors are still subjects of research, the formation of ring structures and their growth via reactions with acetylene have been identified as important processes. The particle inception step involves the formation of small particles of a critical size (3,000–10,000 atomic mass units) from growth by both chemical means and coagulation. It is in this step that large molecules are transformed to, or become identified as, particles. When the small primary soot particles continue to be exposed to the bath of species from the pyrolizing fuel as they travel through the flame, they experience surface growth and agglomeration, the third step. At some point in their history, the soot particles must pass through an oxidizing region of the flame. For a jet flame, this region is invariably the flame tip, since the soot is always formed interior to the reaction zone lower in the flame, and the flow streamlines, which the particles largely follow, do not cross the reaction zone until near the flame tip [1]. If all of the soot particles are oxidized, the flame is termed nonsooting, while, conversely, incomplete oxidation yields a sooting flame. Figure 9.14 illustrates sooting and nonsooting conditions for nonpremixed laminar propylene and butane fuels. The nonzero value of the soot volume fraction beyond the flame tip ($x/x_{stoic} \gtrsim 1.1$) indicates a sooting flame.

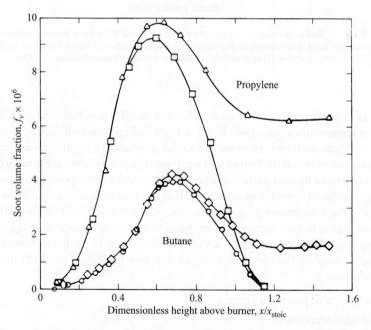

**Figure 9.14**    Measured soot volume fractions as functions of height above burner for propylene and butane at both sooting and nonsooting conditions.

SOURCE: Reprinted by permission of Elsevier Science, Inc., from Ref. [37]. © 1986, The Combustion Institute.

As suggested by Fig. 9.14, the amount of soot formed in a diffusion flame is strongly dependent on fuel type. An experimentally determined measure of a fuel's tendency to soot is the so-called **smoke point.** The smoke-point test was originally devised for liquid fuels and has also been used to characterize gaseous fuels. The basic concept is to increase the fuel flowrate until smoke is observed to escape from the flame tip. The greater the fuel flowrate at the incipient sooting condition, the lower the sooting propensity of the fuel. Sometimes smoke points are expressed as the height of the flame at the incipient sooting condition. As with flowrates, the larger the flame height, the lower the sooting propensity. Table 9.5, from Kent [37], lists smoke points for a large number of fuels. It is interesting to note that methane does not appear on the list since it is not possible to produce a stable sooting laminar methane flame.

If we group the fuels listed in Table 9.5 by families, we note that the fuel sooting propensities, from least to greatest, are in the order of alkanes, alkenes, alkynes, and aromatics. Table 9.6 shows these groupings. Quite obviously, the parent-fuel molecular structure is very important in determining the fuel's sooting propensity, and the groupings given above are consistent with the ideas that ring compounds and their growth via acetylene are important features of soot-formation chemistry.

**Table 9.5** Smoke points, $\dot{m}_{sp}$; maximum soot volume fractions, $f_{v,m}$; and maximum soot yields, $Y_s$; for selected fuels[a]

| Fuel | | $\dot{m}_{sp}$ (mg/s) | $f_{v,m} \times 10^6$ | $Y_s$ (%) |
|---|---|---|---|---|
| Acetylene | $C_2H_2$ | 0.51 | 15.3 | 23 |
| Ethylene | $C_2H_4$ | 3.84 | 5.9 | 12 |
| Propylene | $C_3H_6$ | 1.12 | 10.0 | 16 |
| Propane | $C_3H_8$ | 7.87 | 3.7 | 9 |
| Butane | $C_4H_{10}$ | 7.00 | 4.2 | 10 |
| Cyclohexane | $C_6H_{12}$ | 2.23 | 7.8 | 19 |
| $n$-Heptane | $C_7H_{16}$ | 5.13 | 4.6 | 12 |
| Cyclooctane | $C_8H_{16}$ | 2.07 | 10.1 | 20 |
| Isooctane | $C_8H_{18}$ | 1.57 | 9.9 | 27 |
| Decalin | $C_{10}H_{18}$ | 0.77 | 15.4 | 31 |
| 4-Methylcyclohexene | $C_7H_{12}$ | 1.00 | 13.3 | 22 |
| 1-Octene | $C_8H_{16}$ | 1.73 | 9.2 | 25 |
| 1-Decene | $C_{10}H_{20}$ | 1.77 | 9.9 | 27 |
| 1-Hexadecene | $C_{16}H_{32}$ | 1.93 | 9.2 | 22 |
| 1-Heptyne | $C_7H_{12}$ | 0.65 | 14.7 | 30 |
| 1-Decyne | $C_{10}H_{18}$ | 0.80 | 14.7 | 30 |
| Toluene | $C_7H_8$ | 0.27 | 19.1 | 38 |
| Styrene | $C_8H_8$ | 0.22 | 17.9 | 40 |
| $o$-Xylene | $C_8H_{10}$ | 0.28 | 20.0 | 37 |
| 1-Phenyl-1-propyne | $C_9H_8$ | 0.15 | 24.8 | 42 |
| Indene | $C_9H_8$ | 0.18 | 20.5 | 33 |
| $n$-Butylbenzene | $C_{10}H_{14}$ | 0.27 | 14.5 | 29 |
| 1-Methylnaphthalene | $C_{11}H_{10}$ | 0.17 | 22.1 | 41 |

[a]SOURCE: From Ref. [37].

**Table 9.6** Smoke points by hydrocarbon family[a]

| Alkanes | | Alkenes | | Alkynes | | Aliphatic aromatics | |
|---|---|---|---|---|---|---|---|
| Fuel | $\dot{m}_{sp}^{b}$ | Fuel | $\dot{m}_{sp}^{b}$ | Fuel | $\dot{m}_{sp}^{b}$ | Fuel | $\dot{m}_{sp}^{b}$ |
| Propane | 7.87 | Ethylene | 3.84 | Acetylene | 0.51 | Toluene | 0.27 |
| Butane | 7.00 | Propylene | 1.12 | 1-Heptyne | 0.65 | Styrene | 0.22 |
| n-Heptane | 5.13 | 1-Octene | 1.73 | 1-Decyne | 0.80 | o-Xylene | 0.28 |
| Isooctane | 1.57 | 1-Decene | 1.77 | | | n-Butylbenzene | 0.27 |
| | | 1-Hexadecene | 1.93 | | | | |

[a]SOURCE: Data from Ref. [37].
[b]Smoke point flowrate in mg/s.

In the design of practical burners, soot formation is generally avoided altogether. The effects of port loading and primary aeration on the "yellow-tip" zone conditions, where soot is formed within the flame, are shown for natural gas and for manufactured gas flames in Fig. 8.25.

# COUNTERFLOW FLAMES[1]

In the past few decades, many theoretical and experimental studies have been conducted on flames fed by opposing jets of fuel and oxidizer (Fig. 9.15). Such flames are of fundamental research interest because they approximate a one-dimensional character and because residence times within the flame zone can be easily varied. In the previous section, we saw the complexities associated with the 2-D (axisymmetric) jet flame; in contrast, the one-dimensionality of the counterflow flame makes both experiments and calculations much more practicable. For example, in

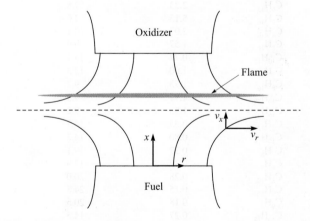

**Figure 9.15** Counterflow diffusion flame lies above stagnation plane (dashed line) created by opposing streams of fuel and oxidizer.

[1]This section may be skipped without any loss of continuity.

experiments, temperature and species conservation measurements need only be made along a single line; while in theoretical studies, only modest run times are needed for computations using extremely complex chemical kinetics (see Table 5.3). The counterflow flame provides fundamental understanding of the detailed structure of diffusion flames and their extinction characteristics. Furthermore, the laminar counterflow flame has been proposed as a fundamental element in the complex structure of turbulent nonpremixed flames [38] (Chapter 13). A rather large counterflow flame literature has developed (e.g., Refs. [39–42]), and continues to expand.

Before we present a mathematical description, it is important to understand the basic features of the counterflow flame. A typical experimental arrangement is illustrated in Fig. 9.15. Here we see opposing jets of fuel and oxidizer, which create a stagnation plane ($v_x = 0$) whose location depends on the relative magnitudes of the oxidizer and fuel jet initial momentum fluxes. For equal momentum fluxes ($\dot{m}_F v_F = \dot{m}_{Ox} v_{Ox}$), the stagnation plane lies at the midpoint between the nozzle exit planes; however, if the momentum flux of one stream is increased over the other, the stagnation plane moves closer to the low-momentum-flux stream outlet. Given appropriate conditions, a diffusion flame can be established between the two nozzles, the location of which is where the mixture fraction is nominally stoichiometric. For most hydrocarbon fuels burning in air, stoichiometric conditions require considerably more air than fuel ($f_{stoic} \approx 0.06$). In this case, then, the fuel must diffuse across the stagnation plane to the flame location, as shown in Fig. 9.15. Conversely, for a reactant pair in which more fuel than oxidizer is required for stoichiometric conditions ($f_{stoic} > 0.5$), the flame would lie on the fuel side of the stagnation plane. An important characteristic of the opposed flow is that the flame established between the nozzles is essentially flat (a disk for round nozzles) and one-dimensional, having dependencies only in the $x$-direction.

## Mathematical Description

Two different approaches to modeling counterflow flames can be found in the literature. The first approach couples a stagnation-point potential flow from a point source at infinity with a boundary-layer type analysis (see, e.g., Ref. [42]). In this analysis, the finite separation between the nozzles cannot be taken into account. A second approach [43, 44] was developed that explicitly accounts for the flows exiting from the nozzles rather than being generated by a far-removed point source. The initial formulation of this model was developed for premixed flames [43] and, subsequently, was extended to nonpremixed flames [44]. This second approach is summarized briefly below. For additional details, the reader is referred to the original references [43, 44]. After presenting the model, we will examine the detailed structure of a $CH_4$–air diffusion generated by a numerical solution.

The overall objective of the analysis is to transform the set of axisymmetric, governing, partial differential equations into a coupled system of ordinary differential equations, cast as a boundary-value problem. The continuity and conservation of momentum equations used as the starting point for the analysis are the axisymmetric forms given in Chapter 7: Eqn. 7.7 for mass conservation and Eqns. 7.43 and 7.44

for axial and radial momentum, respectively. To effect the desired transformation, the following stream function is employed:

$$\Psi \equiv r^2 F(x),$$  (9.75)

where

$$\frac{\partial \Psi}{\partial r} = r\rho v_x = 2rF$$  (9.76a)

and

$$-\frac{\partial \Psi}{\partial x} = r\rho v_r = -r^2 \frac{dF}{dx}.$$  (9.76b)

From the above, it is a simple matter to show that the stream function (Eqn. 9.75) satisfies continuity (Eqn. 7.7). To reduce the order of the radial momentum equation, discussed below, a new variable $G(\equiv dF/dx)$ is defined. This defining equation is the first ordinary differential equation in our set:

$$\frac{dF}{dx} = G.$$  (9.77)

Equations 9.76a, 9.76b, and 9.77 are now substituted into the momentum conservation relations, Eqns. 7.43 and 7.44, but buoyancy is ignored. The equations resulting from this substitution are of the following form:

$$\frac{\partial P}{\partial x} = f_1(x)$$  (9.78a)

$$\frac{1}{r}\frac{\partial P}{\partial r} = f_2(x).$$  (9.78b)

This result is used to generate an eigenvalue equation for the radial pressure gradient. From mathematical operations alone, we can relate the left-hand sides of Eqns. 9.78a and 9.78b:

$$\frac{\partial}{\partial x}\left(\frac{1}{r}\frac{\partial P}{\partial r}\right) = \frac{1}{r}\frac{\partial}{\partial x}\left(\frac{\partial P}{\partial r}\right) = \frac{1}{r}\frac{\partial}{\partial r}\left(\frac{\partial P}{\partial x}\right).$$

Furthermore, since both $\partial P/\partial x$ and $(1/r)(\partial P/\partial r)$ are functions of $x$ only, it follows that

$$\frac{\partial}{\partial x}\left(\frac{1}{r}\frac{\partial P}{\partial r}\right) = \frac{1}{r}\frac{\partial}{\partial r}\left(\frac{\partial P}{\partial x}\right) = 0$$  (9.79)

and

$$\frac{1}{r}\frac{\partial P}{\partial r} = \text{constant} \equiv H.$$  (9.80)

The radial-pressure-gradient eigenvalue, $H$, thus enters into the set of ordinary differential equations as

$$\frac{dH}{dx} = 0.$$  (9.81)

Since the pressure is uniform throughout the flowfield (low Mach-number approximation), we have no further use for the axial momentum equation (Eqn. 9.78a); the radial momentum equation, however, must be retained. Substituting Eqn. 9.80 into Eqn. 9.78b and fleshing out the right-hand side results in the following:

$$\frac{d}{dx}\left[\mu\frac{d}{dx}\left(\frac{G}{\rho}\right)\right] - 2\frac{d}{dx}\left(\frac{FG}{\rho}\right) + \frac{3}{\rho}G^2 + H = 0. \tag{9.82}$$

The corresponding energy and species conservation equations are, respectively,

$$2Fc_p\frac{dT}{dx} - \frac{d}{dx}\left(k\frac{dT}{dx}\right) + \sum_{i=1}^{N}\rho Y_i v_{i,\,\text{diff}}\, c_{p,i}\frac{dT}{dx} - \sum_{i=1}^{N}h_i\dot{\omega}_i MW_i = 0 \tag{9.83}$$

and

$$2F\frac{dY_i}{dx} + \frac{d}{dx}(\rho Y_i v_{i,\,\text{diff}}) - \dot{\omega}_i MW_i = 0 \qquad i = 1, 2, \ldots, N. \tag{9.84}$$

In summary, the counterflow diffusion flame model consists of the set of five ordinary differential equations (Eqns. 9.77, 9.81, 9.82, 9.83, and 9.84) for the four functions $F(x)$, $G(x)$, $T(x)$, and $Y_i(x)$ and the eigenvalue, $H$. In addition to these basic equations, the following ancillary relations or data are required:

- Ideal-gas equation of state (Eqn. 2.2).
- Constitutive relations for diffusion velocities (Eqns. 7.23 and 7.25 or Eqn. 7.31).
- Temperature-dependent species properties: $h_i(T)$, $c_{p,i}(T)$, $k_i(T)$, and $\mathcal{D}_{ij}(T)$.
- Mixture property relations to calculate $MW_{\text{mix}}$, $k$, $D_{ij}$, and $D_i^T$ from individual species properties and mole (or mass) fractions (e.g., for the $D_{ij}$s, Eqn. 7.26).
- A detailed chemical kinetic mechanism to obtain the $\dot{\omega}_i$s (e.g., Table 5.3).
- Interconversion relations for $\chi_i$s, $Y_i$s, and $[X_i]$s (Eqns. 6A.1–6A.10).

Boundary conditions are applied at the two nozzle exits (Fig. 9.15), defined as $x \equiv 0$ at the fuel-nozzle exit and $x \equiv L$ at the oxidizer-nozzle exit, to complete the formulation of the boundary-value problem. The necessary conditions are the exit velocities and their gradients, exit temperatures, and exit species mass fractions (or mass flux fractions), which we specify as follows:

At $x = 0$:  $\qquad\qquad\qquad$  At $x = L$:

$F = \rho_F v_{e,F}/2,$  $\qquad\qquad\qquad$  $F = \rho_{Ox} v_{e,Ox}/2,$

$G = 0,$  $\qquad\qquad\qquad\qquad\quad$  $G = 0,$  $\qquad\qquad$ (9.85)

$T = T_F,$  $\qquad\qquad\qquad\qquad$  $T = T_{Ox},$

$\rho v_x Y_i + \rho Y_i v_{i,\text{diff}} = (\rho v_x Y_i)_F;$  $\qquad$  $\rho v_x Y_i + \rho Y_i v_{i,\text{diff}} = (\rho v_x Y_i)_{Ox}.$

## Structure of CH$_4$–Air Flame

We now employ the counterflow flame model described above to analyze the structure of a CH$_4$–air diffusion flame. The OPPDIF software described in Ref. [44], together with the CHEMKIN library codes [45], were used with chemical kinetics taken from Miller and Bowman [46]. Figure 9.16 shows computed temperature and velocity profiles between the fuel (left) and air (right) nozzles, and Fig. 9.17 presents corresponding major species mole-fraction profiles. Also shown in Fig. 9.16 is the local equivalence ratio computed from carbon and oxygen balances.

Focusing on the velocity profile, we see in Fig. 9.16 that the stagnation plane ($v_x = 0$) lies to the left of the center plane between the two nozzle flows, as expected, since the higher density of air causes the momentum flux of the air stream to be greater than that of the fuel stream for equal outlet velocities (50 cm/s). The velocity profile exhibits interesting behavior in the heat-release region of the flame, where a minimum value ($v_x = -57.6$ cm/s) occurs slightly to the air side of the peak temperature. We note that this is the maximum absolute value of the velocity, as velocities directed to the left are negative. Simplistically, this result is understood as a consequence of continuity, with the gas speed increasing in response to the density decreasing. Frequently, a velocity gradient, $dv_x/dx$, is used to characterize the strain rate in counterflow flames; for the twin-nozzle geometry, the relatively long region of essentially constant slope before the velocity minimum is used as the characteristic gradient. For the particular case illustrated in Fig. 9.16, the value of the velocity gradient is approximately 360 $s^{-1}$.

An essential feature of nonpremixed flame is a continuous variation of the mixture fraction, $f$, or, alternatively, the equivalence ratio, $\Phi$ from pure fuel at the left nozzle ($f = 1$, $\Phi \rightarrow \infty$), to pure air at the right nozzle ($f = 0$, $\Phi = 0$). Figure 9.16 shows this variation of equivalence ratio over the range from 2 to zero. A close inspection of Fig. 9.16 reveals that the location of the maximum temperature occurs at a slightly rich stoichiometry ($\Phi = 1.148$), with temperatures about 40 K lower at $\Phi = 1$. From thermodynamic considerations alone, one expects peak temperatures to occur at slightly rich conditions (see Chapter 2). For our $CH_4$–air system, the peak adiabatic flame temperature occurs at $\Phi = 1.035$. This value of $\Phi$ is considerably less rich than for the diffusion flame ($\Phi = 1.148$), where convection, diffusion, and chemical kinetics all combine to determine the peak-temperature equivalence ratio.

Turning now to the species profiles (Fig. 9.17), we focus first on the reactants. Here we see that both $CH_4$ and $O_2$ mole fractions fall to near-zero values at an axial location of approximately 0.75 cm; this corresponds closely with the occurrence of the peak temperature, as can be seen from Fig. 9.16. We also note some overlap, or simultaneous presence of $CH_4$ and $O_2$, however small, in the region immediately preceding the maximum temperature. For the particular conditions of this simulation, the combustion kinetics are not sufficiently fast to obtain a true approximation to a flamesheet; thus, a distributed reaction zone results. Another interesting reactants-related feature of Fig. 9.17 is the presence of $N_2$ deep on the fuel side. Since all $N_2$ in the flame system has its origins with the air, the $N_2$ must diffuse across the stagnation plane to yield the relatively high concentrations we observe on the fuel side of the flame. Of course, the presence of fuel to the right of the stagnation plane ($x = 0.58$ cm) is a result of diffusion in the direction opposite to the $N_2$.

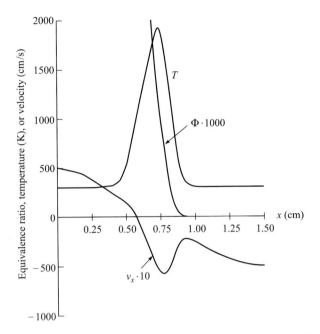

**Figure 9.16**    Equivalence ratio, temperature, and velocity profiles through $CH_4$–air counterflow diffusion flame. The $CH_4$ and air streams both exit at 50 cm/s; $L = 1.5$ cm.

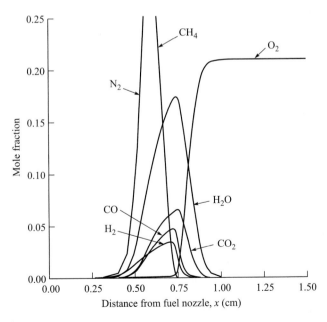

**Figure 9.17**    Major species mole-fraction profiles through $CH_4$–air counterflow flame. Same conditions as Fig. 9.16.

**Table 9.7**    Location of peak species mole fractions and temperature in simulated $CH_4$–air counterflow diffusion flame

| Species | Maximum Mole Fraction | Location of Maximum, $x$ (cm) | $\Phi$ | $T$ (K) |
|---|---|---|---|---|
| $H_2$ | 0.0345 | 0.7074 | 1.736 | 1786.5 |
| CO | 0.0467 | 0.7230 | 1.411 | 1862.6 |
| $H_2O$ | 0.1741 | 0.7455 | 1.165 | 1926.8 |
| | | $T_{max} = 1925.8$ K at $x = 0.7468$ cm, $\Phi = 1.148$ | | |
| $CO_2$ | 0.0652 | 0.7522 | 1.085 | 1913.8 |

**Table 9.8**    Comparison of flame composition with that for adiabatic equilibrium at locations of maximum temperature and stoichiometric mixture fraction ($\Phi = 1$)

| Condition | $O_2$ | CO | $H_2$ | $CO_2$ | $H_2O$ | $N_2$ |
|---|---|---|---|---|---|---|
| | | | $T = T_{max}$ (1925.8 K), $\Phi = 1.148$ | | | |
| Flame | 0.0062 | 0.0394 | 0.0212 | 0.0650 | 0.174 | 0.686 |
| Adiabatic equilibrium | $2.15 \cdot 10^{-6}$ | 0.0333 | 0.0207 | 0.0714 | 0.189 | 0.686 |
| | | | $\Phi = 1.000$, $T = 1887.5$ K | | | |
| Flame | 0.0148 | 0.0280 | 0.0132 | 0.0648 | 0.170 | 0.697 |
| Adiabatic equilibrium | 0.0009 | 0.0015 | 0.0007 | 0.0934 | 0.189 | 0.714 |

From Fig. 9.17, we see a progression of species mole-fraction maxima from left to right. This is illustrated more clearly by the data presented in Table 9.7, where we see that the $H_2$ peak occurs at the richest condition of all species shown, followed by the peaks of CO, $H_2O$, and $CO_2$. All of these maxima occur at rich equivalence ratios, as might be expected for all but the $CO_2$. Some effects of chemical kinetics can be illuminated by comparing the predicted mole fractions for the major species in the flame with equilibrium values based on identical temperatures and stoichiometry. Two such comparisons are shown in Table 9.8: the first, at the location of the maximum temperature, and the second, at the location of stoichiometric conditions. In both cases, we see substantially less $CO_2$ and $H_2O$ in the flame compared with the equilibrium conditions. With lower levels of fully oxidized products ($CO_2$ and $H_2O$) in the flame, one expects incomplete products of combustion to appear in greater abundance. For the $\Phi = 1.0$ condition, in particular, CO, $H_2$, and $O_2$ levels are approximately 15–20 times higher in the flame.

## SUMMARY

This chapter opened with a discussion of laminar constant-density jets, whose behavior has much in common with laminar jet flames, but is easier to describe mathematically. You should be able to describe the general characteristics of the velocity and nozzle-fluid concentration fields of laminar jets and understand the Reynolds-number dependence of the spreading characteristics. We also saw that the nozzle-fluid concentration distribution is identical for equal flowrates, which translates to flame lengths depending only on flowrate for a given fuel–oxidizer combination. You should be able to describe the general characteristics of the temperature, fuel and oxidizer mass fractions, and velocity fields of laminar jet flames, and have a firm grasp of how stoichiometry determines the flame contour. The conserved scalar formulation of the laminar jet diffusion flame problem was developed and emphasized. This approach results in a considerable mathematical simplification. For historical perspective, we reviewed the theoretical Burke–Schumann and Fay solutions of the laminar jet flame problem and then focused on Roper's simplified analyses, presenting flame length correlations for circular-, square-, and slot-port burners. You should be familiar with the use of these relationships. The Roper theory clearly showed the importance of the ambient oxidizer to nozzle-fluid stoichiometric ratio, which is determined by fuel type, $O_2$ content in the oxidizer, primary aeration, and fuel dilution by an inert gas. We also saw in this chapter how soot formation is an essential characteristic of diffusion flames, although soot formation can be avoided if residence times (flame lengths) are short enough. You should be familiar with the four-step sequence of soot formation and destruction occurring in flames and have an appreciation of the importance of fuel type (structure) in determining sooting propensities. The chapter concluded with a discussion of counterflow flames.

## NOMENCLATURE

| | |
|---|---|
| $a$ | Buoyant acceleration, Eqn. 9.67 $(m/s^2)$ |
| $A/F$ | Mass air–fuel ratio |
| $b$ | Port width, Table 9.3 (m) |
| $c_p$ | Specific heat (J/kg-K) |
| $D_{ij}$ | Multicomponent diffusion coefficient $(m^2/s)$ |
| $D_i^T$ | Thermal diffusion coefficient (kg/m-s) |
| $\mathcal{D}_{ij}$ | Binary diffusion coefficient $(m^2/s)$ |
| $f$ | Mixture fraction (kg/kg) |
| $f_v$ | Soot volume fraction |
| $F$ | Function defined in Eqn. 9.76 |
| $Fr$ | Froude number, Eqn. 9.68 |
| $g$ | Gravitational acceleration $(m/s^2)$ |
| $G$ | Function defined in Eqn. 9.77 |
| $h$ | Enthalpy (J/kg) or port length, Table 9.3 (m) |
| $h_f^o$ | Enthalpy of formation (J/kg) |
| $H$ | Radial momentum eigenvalue, Eqn. 9.86 |
| $I$ | Momentum flow ratio or momentum integral, Table 9.2 |
| $J$ | Momentum flow, Eqn. 9.1 $(kg\text{-}m/s^2)$ |
| $J_0, J_1$ | Bessel functions |
| $k$ | Thermal conductivity (W/m-K) |
| $L_f$ | Flame length (m) |
| $Le$ | Lewis number |
| $m$ | Mass (kg) |

| $\dot{m}$ | Mass flowrate (kg/s) |
|---|---|
| $MW$ | Molecular weight (kg/kmol) |
| $N$ | Number of moles (kmol) |
| $P$ | Pressure (Pa) |
| $Pr$ | Prandtl number |
| $Q$ | Volumetric flowrate (m³/s) |
| $r$ | Radial coordinate (m) |
| $r_{1/2}$ | Radius at half-height (m) |
| $R$ | Radius (m) |
| $R_o$ | Outer flow radius, Eqn. 9.55 (m) |
| $R_u$ | Universal gas constant (J/kmol-K) |
| $Re$ | Reynolds number |
| $S$ | Molar stoichiometric oxidizer to nozzle-fluid ratio (kmol/kmol) |
| $Sc$ | Schmidt number |
| $T$ | Temperature (K) |
| $v$ | Velocity (m/s) |
| $v_r, v_x$ | Radial and axial velocity components, respectively (m/s) |
| $x$ | Axial coordinate (m) or number of carbon atoms in fuel molecule |
| $y$ | Number of hydrogen atoms in fuel molecule |
| $Y$ | Mass fraction (kg/kg) |

### *Greek Symbols*

| $\alpha$ | Spreading angle (rad) or thermal diffusivity (m²/s) |
|---|---|
| $\beta$ | Defined following Eqn. 9.64 |
| $\varsigma$ | General conserved scalar variable, Eqn. 9.42 |
| $\mu$ | Absolute viscosity (N-s/m²) |
| $\nu$ | Kinematic viscosity (m²/s) or stoichiometric air–fuel ratio (kg/kg) |
| $\xi$ | Defined by Eqn. 9.11 |
| $\rho$ | Density (kg/m³) |
| $\Phi$ | Equivalence ratio |
| $\chi$ | Mole fraction (kmol/kmol) |
| $\psi$ | Primary aeration |
| $\Psi$ | Stream function |
| $\dot{\omega}$ | Species production rate (kmol/m³-s) |

### *Subscripts*

| act | Actual |
|---|---|
| $A$ | Air |
| $B$ | Buoyancy-controlled |
| $c$ | Core |
| diff | Diffusion |
| dil | Diluent |
| $e$ | Exit |
| expt | Experiment |
| $f$ | Flame |

| $F$ | Fuel |
|---|---|
| $i$ | $i$th species |
| $j$ | Jet |
| $m$ | Maximum |
| mix | Mixture |
| $M$ | Momentum-controlled |
| $Ox$ | Oxidizer |
| $Pr$ | Products |
| $pri$ | Primary |
| pure | Pure fuel |
| ref | Reference |
| $sp$ | Smoke point |
| stoic | Stoichiometric |
| thy | Theory |
| $T$ | Transition |
| 0 | Centerline |
| $\infty$ | Ambient |

**Superscripts**

| * | Nondimensional quantity |
|---|---|

**Other Notation**

| $[X]$ | Molar concentration of species $X$ ($kmol/m^3$) |
|---|---|

# REFERENCES

1. Santoro, R. J., Yeh, T. T., Horvath, J. J., and Semerjian, H. G., "The Transport and Growth of Soot Particles in Laminar Diffusion Flames," *Combustion Science and Technology,* 53: 89–115 (1987).

2. Santoro, R. J., and Semerjian, H. G., "Soot Formation in Diffusion Flames: Flow Rate, Fuel Species and Temperature Effects," *Twentieth Symposium (International) on Combustion,* The Combustion Institute, Pittsburgh, PA, p. 997, 1984.

3. Santoro, R. J., Semerjian, H. G., and Dobbins, R. A., "Soot Particle Measurements in Diffusion Flames," *Combustion and Flame,* 51: 203–218 (1983).

4. Puri, R., Richardson, T. F., Santoro, R. J., and Dobbins, R. A., "Aerosol Dynamic Processes of Soot Aggregates in a Laminar Ethene Diffusion Flame," *Combustion and Flame,* 92: 320–333 (1993).

5. Quay, B., Lee, T.-W., Ni, T., and Santoro, R. J., "Spatially Resolved Measurements of Soot Volume Fraction Using Laser-Induced Incandescence," *Combustion and Flame,* 97: 384–392 (1994).

6. Smooke, M. D., Long, M. B., Connelly, B. C., Colket, M. B., and Hall, R. J., "Soot Formation in Laminar Diffusion Flames," *Combustion and Flame,* 143: 613–628 (2005).

7. Thomson, K. A., Gülder, Ö. L., Weckman, E. J., Fraser, R. A., Smallwood, G. J., and Snelling, D. R., "Soot Concentration and Temperature Measurements in Co-Annular, Nonpremixed, $CH_4$/Air Laminar Flames at Pressures up to 4 MPa," *Combustion and Flame,* 140: 222–232 (2005).

8. Bento, D. S., Thomson, K. A., and Gülder, Ö. L., "Soot Formation and Temperature Field Structure in Laminar Propane–Air Diffusion Flames at Elevated Pressures," *Combustion and Flame,* 145: 765–778 (2006).

9. Williams, T. C., Shaddix, C. R., Jensen, K. A., and Suo-Antilla, J., M., "Measurement of the Dimensionless Extinction Coefficient of Soot within Laminar Diffusion Flames," *International Journal of Heat and Mass Transfer,* 50: 1616–1630 (2007).

10. Burke, S. P., and Schumann, T. E. W., "Diffusion Flames," *Industrial & Engineering Chemistry,* 20(10): 998–1004 (1928).

11. Fay, J. A., "The Distributions of Concentration and Temperature in a Laminar Jet Diffusion Flame," *Journal of Aeronautical Sciences,* 21: 681–689 (1954).

12. Roper, F. G., "The Prediction of Laminar Jet Diffusion Flame Sizes: Part I. Theoretical Model," *Combustion and Flame,* 29: 219–226 (1977).

13. Roper, F. G., Smith, C., and Cunningham, A. C., "The Prediction of Laminar Jet Diffusion Flame Sizes: Part II. Experimental Verification," *Combustion and Flame,* 29: 227–234 (1977).

14. Roper, F. G., "Laminar Diffusion Flame Sizes for Curved Slot Burners Giving Fan-Shaped Flames," *Combustion and Flame,* 31: 251–259 (1978).

15. Lin, K.-C., Faeth, G. M., Sunderland, P. B., Urban, D. L., and Yuan, Z.-G., "Shapes of Nonbuoyant Round Luminous Hydrocarbon/Air Laminar Jet Diffusion Flames," *Combustion and Flame,* 116: 415–431 (1999).

16. Aalburg, C., Diez, F. J., Faeth, G. M., Sunderland, P. B., Urban, D. L., and Yuan, Z.-G., "Shapes of Nonbuoyant Round Hydrocarbon-Fueled Laminar-Jet Diffusion Flames in Still Air," *Combustion and Flame,* 142: 1–16 (2005).

17. Lin, K.-C., and Faeth, G. M., "Shapes of Nonbuoyant Round Luminous Laminar-Jet Diffusion Flames in Coflowing Air," *AIAA Journal,* 37: 759–765 (1999).

18. Miller, J. A., and Kee, R. J., "Chemical Nonequilibrium Effects in Hydrogen–Air Laminar Jet Diffusion Flames," *Journal of Physical Chemistry,* 81(25): 2534–2542 (1977).

19. Kee, R. J., and Miller, J. A., "A Split-Operator, Finite-Difference Solution for Axisymmetric Laminar-Jet Diffusion Flames," *AIAA Journal,* 16(2): 169–176 (1978).

20. Mitchell, R. E., Sarofim, A. F., and Clomburg, L. A., "Experimental and Numerical Investigation of Confined Laminar Diffusion Flames," *Combustion and Flame,* 37: 227–244 (1980).

21. Heys, N. W., Roper, F. G., and Kayes, P. J., "A Mathematical Model of Laminar Axisymmetrical Natural Gas Flames," *Computers and Fluids,* 9: 85–103 (1981).

22. Smooke, M. D., Lin, P., Lam, J. K., and Long, M. B., "Computational and Experimental Study of a Laminar Axisymmetric Methane-Air Diffusion Flame," *Twenty-Third Symposium (International) on Combustion,* The Combustion Institute, Pittsburgh, PA, p. 575, 1990.

23. Anon., *Fundamentals of Design of Atmospheric Gas Burner Ports,* Research Bulletin No. 13, American Gas Association Testing Laboratories, Cleveland, OH, August 1942.

24. Weber, E. J., and Vandaveer, F. E., "Gas Burner Design," Chapter 12 in *Gas Engineers Handbook,* The Industrial Press, New York, pp. 12/193–12/210, 1965.

25. Schlichting, H., *Boundary-Layer Theory,* 6th Ed., McGraw-Hill, New York, 1968.

26. Davis, R. W., Moore, E. F., Santoro, R. J., and Ness, J. R., "Isolation of Buoyancy Effects in Jet Diffusion Flames," *Combustion Science and Technology,* 73: 625–635 (1990).

27. Kuo, K. K., *Principles of Combustion,* 2nd Ed., John Wiley & Sons, Hoboken, NJ, 2005.

28. Spalding, D. B., *Combustion and Mass Transfer,* Pergamon, New York, 1979.

29. Beyer, W. H. (ed.), *Standard Mathematical Tables,* 28th Ed., The Chemical Rubber Co., Cleveland, OH, 1987.

30. Kennedy, I. M., "Models of Soot Formation and Oxidation," *Progress in Energy and Combustion Science,* 23: 95–132 (1997).

31. Glassman, I., "Soot Formation in Combustion Processes," *Twenty-Second Symposium (International) on Combustion,* The Combustion Institute, Pittsburgh, PA, p. 295, 1988.

32. Wagner, H. G., "Soot Formation–An Overview," in *Particulate Carbon Formation during Combustion* (D. C. Siegla and G. W. Smith, eds.), Plenum Press, New York, p. 1, 1981.

33. Calcote, H. F., "Mechanisms of Soot Nucleation in Flames—A Critical Review," *Combustion and Flame,* 42: 215–242 (1981).

34. Haynes, B. S., and Wagner, H. G., "Soot Formation," *Progress in Energy and Combustion Science,* 7: 229–273 (1981).

35. Wagner, H. G., "Soot Formation in Combustion," *Seventeenth Symposium (International) on Combustion,* The Combustion Institute, Pittsburgh, PA, p. 3, 1979.

36. Palmer, H. B., and Cullis, C. F., "The Formation of Carbon in Gases," *The Chemistry and Physics of Carbon* (P. L. Walker, Jr., ed.), Marcel Dekker, New York, p. 265, 1965.

37. Kent, J. H., "A Quantitative Relationship between Soot Yield and Smoke Point Measurements," *Combustion and Flame,* 63: 349–358 (1986).

38. Marble, F. E., and Broadwell, J. E., "The Coherent Flames Model for Turbulent Chemical Reactions," *Project SQUID,* 29314-6001-RU-00, 1977.

39. Tsuji, H., and Yamaoka, I., "The Counterflow Diffusion Flame in the Forward Stagnation Region of a Porous Cylinder," *Eleventh Symposium (International) on Combustion,* The Combustion Institute, Pittsburgh, PA, p. 979, 1967.

40. Hahn, W. A., and Wendt, J. O. L., "The Flat Laminar Opposed Jet Diffusion Flame: A Novel Tool for Kinetic Studies of Trace Species Formation," *Chemical Engineering Communications,* 9: 121–136 (1981).

41. Tsuji, H., "Counterflow Diffusion Flames," *Progress in Energy and Combustion Science,* 8: 93–119 (1982).

42. Dixon-Lewis, G., *et al.*, "Calculation of the Structure and Extinction Limit of a Methane–Air Counterflow Diffusion Flame in the Forward Stagnation Region of a Porous Cylinder," *Twentieth Symposium (International) on Combustion,* The Combustion Institute, Pittsburgh, PA, p. 1893, 1984.

43. Kee, R. J., Miller, J. A., Evans, G. H., and Dixon-Lewis, G., "A Computational Model of the Structure and Extinction of Strained, Opposed-Flow, Premixed Methane–Air Flames," *Twenty-Second Symposium (International) on Combustion,* The Combustion Institute, Pittsburgh, PA, p. 1479, 1988.

44. Lutz, A. E., Kee, R. J., Grcar, J. F., and Rupley, F. M., "OPPDIF: A Fortran Program for Computing Opposed-Flow Diffusion Flames," Sandia National Laboratories Report SAND96-8243, 1997.

45. Kee, R. J., Rupley, F. M., and Miller, J. A., "Chemkin-II: A Fortran Chemical Kinetics Package for the Analysis of Gas-Phase Chemical Kinetics," Sandia National Laboratories Report SAND89-8009, March 1991.

46. Miller, J. A., and Bowman, C. T., "Mechanism and Modeling of Nitrogen Chemistry in Combustion," *Progress in Energy and Combustion Science,* 15: 287–338 (1989).

## REVIEW QUESTIONS

**1.** Make a list of all the boldfaced words in Chapter 9 and define each.

**2.** Describe the velocity and nozzle-fluid concentration fields of a constant-density laminar jet.

**3.** Describe the fields of temperature, and the fuel, oxidizer, and product mass fractions of a laminar jet flame.

**4.** Explain why the flame boundary is at the $\Phi = 1$ contour. *Hint:* Consider what would happen if the flame boundary were slightly inside of the $\Phi = 1$ contour ($Y_{Ox} = 0$, $Y_F > 0$) or outside of the $\Phi = 1$ contour ($Y_F = 0$, $Y_{Ox} > 0$).

**5.** How do the combined assumptions of unity Lewis number and unity Schmidt number simplify the governing conservation equations for a laminar flame?

**6.** Explain what it means for a jet flame to be buoyancy- or momentum-controlled. What dimensionless parameter determines the flow regime? What is the physical meaning of this parameter?

**7.** Light a flame on a butane lighter. Hold the lighter at an angle from the vertical, being careful not to burn your fingers. What happens to the flame shape? Explain.

**8.** List and discuss the four steps involved in soot formation and destruction in diffusion flames.

**9.** Explain how the use of conserved scalars simplifies the mathematical description of laminar jet flames. What assumptions are required to effect these simplifications?

## PROBLEMS

**9.1**  Starting with the more general axial momentum equation for a reacting axisymmetric flow (Eqn. 7.48), derive the constant-property, constant-density form given by Eqn. 9.4. *Hint:* You will need to apply continuity.

**9.2**  Repeat problem 9.1 for species conservation, starting with Eqn. 7.20 to obtain Eqn. 9.5.

**9.3**  Using the definition of the volumetric flowrate ($Q = v_e \pi R^2$), show that the centerline mass fraction $Y_{F,0}$ for a laminar jet (Eqn. 9.18) depends only on $Q$ and $v$.

**9.4\***  Calculate the velocity decay for a jet with an initial top-hat velocity profile (Eqn. 9.13) and for a jet where the profile is parabolic, specifically, $v(r) = 2v_e[1 - (r/R)^2]$, where $v_e$ is the mean velocity. Both jets have the same flowrates. Plot your results versus axial distance and discuss.

**9.5**  Two isothermal (300 K, 1 atm) air-in-air laminar jets with different diameters have the same volumetric flowrates.

    A.  Expressed in terms of $R$, what is the ratio of their Reynolds numbers?

    B.  For $Q = 5$ cm$^3$/s, and $R_1 = 3$ mm and $R_2 = 5$ mm for the two jets, respectively, calculate and compare $r_{1/2}/x$ and $\alpha$ for the two jets.

    C.  Determine the axial location where the centerline fuel velocity has decayed to $1/10$ of the nozzle-exit velocity for each jet.

**9.6**  The laminar jet spreading rate, $r_{1/2}/x$, defined in Eqn. 9.14 is based on $r_{1/2}$; i.e., the radial location where the axial velocity is half the centerline value.

    A.  Evaluate the spreading rate for $r_{1/10}/x$, where now the radial location is that where the axial velocity has decayed to one-tenth the value on the centerline. Compare $r_{1/10}/x$ with $r_{1/2}/x$.

    B.  Determine the average velocity, normalized by the centerline velocity, over this portion of the jet; i.e., determine $\bar{v}_x(0 \le r \le r_{1/10})/v_{x,o}$.

**9.7**  From the definition of the mixture fraction, $f$, show that Eqn. 9.44a is correct.

**9.8**  Using isothermal jet theory, estimate the length of an ethane–air diffusion flame for an initial velocity of 5 cm/s. The velocity profile issuing from the 10-mm-diameter port is uniform. Both the air and ethane are at 300 K and 1 atm. The viscosity of ethane is approximately $9.5 \cdot 10^{-6}$ N-s/m$^2$. Compare your estimate using an average viscosity, $(\mu_{air} + \mu_{ethane})/2$, with the predicted value from Roper's experimental correlations.

**9.9**  A circular-port and a square-port burner operate with the same mean velocities and produce the same flame lengths. What is the ratio of the circular-port diameter, $D$, to the length of a side $b$, of the square port? The fuel is methane.

*indicates optional use of computer.

**9.10**   Consider a slot burner with an aspect ratio $h/b = 10$ and slot width $b = 2$ mm. The slot has a contoured entrance that produces a uniform exit flow. Operating with methane, the burner has a heat release of 500 W. Determine the flame length.

**9.11**   Two circular-port burners have the same mean velocity $\bar{v}_e$, but the velocity profile of one burner is uniform, whereas the other has a parabolic distribution, $v(r) = 2\bar{v}_e[1 - (r/R)^2]$. Determine the ratio of the momentum flows of the two burners.

**9.12**   In a study of nitric oxide formation in laminar jet flames, the propane fuel is diluted with $N_2$ to suppress soot formation. The nozzle fluid is 60 percent $N_2$ by mass. The burner has a circular port. The fuel, nitrogen, and the air are all at 300 K and 1 atm. Compare the flame lengths for the following two cases with the undiluted base case ($\dot{m}_F = 5 \cdot 10^{-6}$ kg/s). What is the physical significance of your results? Discuss.

     A.   The total flowrate of the diluted flow ($C_3H_8 + N_2$) is $5 \cdot 10^{-6}$ kg/s.

     B.   The flowrate of the $C_3H_8$ in the diluted flow is $5 \cdot 10^{-6}$ kg/s.

**9.13**   In determining the constant associated with the experimental flame-length correlation (Eqn. 9.60), Roper used a flame temperature of 1500 K. What value of the mean diffusion coefficient $\mathcal{D}_\infty$ appearing in Eqn. 9.59 is consistent with the constant in Eqn. 9.60? How does this value compare with the value of the binary diffusivity for oxygen in air at 298 K ($\mathcal{D}_{O_2\text{–air}} = 2.1 \cdot 10^{-5}$ m$^2$/s)?

**9.14**   Estimate both the Lewis and Schmidt numbers for a dilute mixture of $O_2$ in air at 298 K and 1 atm. Assume $\mathcal{D}_{O_2\text{–air}} = 2.1 \cdot 10^{-5}$ m$^2$/s at these conditions. Discuss the implications of your results.

**9.15**   Evaluate the Prandtl number ($Pr$), Schmidt number ($Sc$), and Lewis number ($Le$) for air at the following conditions:

     A.   $P = 1$ atm,   $T = 298$ K.

     B.   $P = 1$ atm,   $T = 2000$ K.

     C.   $P = 10$ atm,   $T = 298$ K.

     Assume a binary diffusion coefficient $\mathcal{D}_{O_2\text{–air}}$ of $2.1 \cdot 10^{-5}$ m$^2$/s at 1 atm and 298 K.

**9.16**   The following fuels are burned in laminar jet flames, each with a mass flowrate of 3 mg/s: acetylene, ethylene, butane, and isooctane. Which flames emit soot from their tips? Discuss your answer.

**9.17**   A temperature sensor is to be placed above a $CH_4$–air jet flame along the extension of the flame's centerline. It is desired to locate the sensor as close as possible to the flame; however, the maximum-temperature operating limit of the sensor is 1200 K. The system operates at 1 atm and the air and fuel enter at 300 K.

A.  Using the simplified state relationships for a diffusion flame, determine the mixture fraction that would correspond to a sensor temperature limit of 1200 K. Assume a constant specific heat $c_p$ of 1087 J/kg-K.

B.  Using the constant-density laminar jet equations, determine the axial distance along the centerline from the burner nozzle exit to the sensor location corresponding to the conditions in part A for a nonreacting jet. Use a jet Reynolds number $(\rho_e v_e R/\mu)$ of 30 and a jet radius $R$ of 1 mm.

**9.18**  Design a multijet gas burner capable of heating a pot of water from room temperature (25°C) to the boiling point (100°C) in 5 min. Assume that 30 percent of the fuel energy enters the water. The design is constrained to be a number of equally spaced and equally sized individual diffusion flames arranged on a ring with a diameter not to exceed 160 mm. The fuel is methane (natural gas). A minimum flame height is desired. Primary aeration is to be used; however, the burner must operate in a pure diffusion mode. Your design should specify the following: fuel gas flowrate, number of flames, burner ring diameter, flame port diameter, primary aeration, and flame height. List all of the assumptions you invoke.

# Droplet Evaporation and Burning

## OVERVIEW

In this chapter, we consider our second nonpremixed combustion system: the evaporation and burning of spherical liquid droplets. With appropriate assumptions, this system is relatively simple to analyze and thus provides a good opportunity to see clearly how various physical phenomena interrelate. For both droplet evaporation and burning, closed-form analytic solutions to the simplified governing conservation equations are possible. These solutions allow us to explore the influence of droplet size and ambient conditions on droplet evaporation or burning times. Knowledge of droplet gasification rates and droplet lifetimes is important in the design and operation of practical devices. In addition to developing simple droplet gasification models, we show how the simple evaporation model can be incorporated into a one-dimensional analysis of a combustor. Before we begin our various analyses, however, it is worthwhile to review some of the practical applications that are related to or impacted by droplet evaporation and/or burning.

## SOME APPLICATIONS

Droplet burning has relevance to many practical combustion devices, including diesel, rocket, and gas-turbine engines, as well as oil-fired boilers, furnaces, and process heaters. To maintain perspective, it is important to point out that in these devices, spray combustion, rather than individual droplet burning, is the dominant feature; however, understanding isolated droplet burning is an essential prerequisite to dealing with more complex flames, as well as being important in its own right in some situations.

## Diesel Engines

There are two primary types of diesel engines, the indirect-injection type and the direct-injection type, shown in Figs. 10.1 and 10.2, respectively. With advances in fuel injection technology, direct-injection engines predominate in most applications. In the **indirect-injection** engine, fuel is injected under high pressure into a precombustion chamber. The fuel droplets begin to evaporate and the fuel vapor mixes with the air in the chamber. Portions of this fuel–air mixture autoignite (Example 6.1) and initiate nonpremixed combustion. Because of the heat release, the pressure rises inside the prechamber, forcing its contents through a throat or orifice into the main chamber. The partially reacted fuel–air mixture and any remaining fuel droplets mix with additional air in the main chamber and burn to completion. In the **direct-injection** engine, the fuel is introduced by a multiholed fuel injector. Fuel–air mixing is governed by both the injection process and the air motion within the combustion space. As suggested

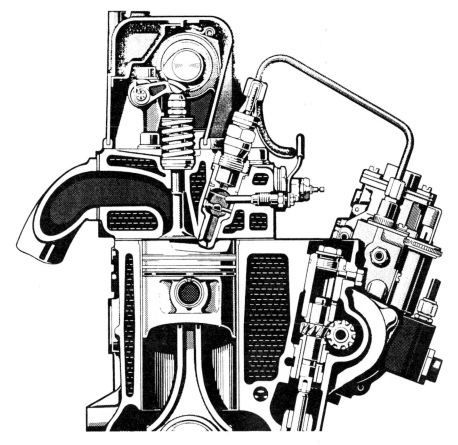

**Figure 10.1**   Light-duty indirect-injection diesel engine. Cutaway shows prechamber with glow plug, used for starting, and fuel injector.

SOURCE: Reprinted from Ref. [1] with permission, © 1971, Society of Automotive Engineers, Inc.

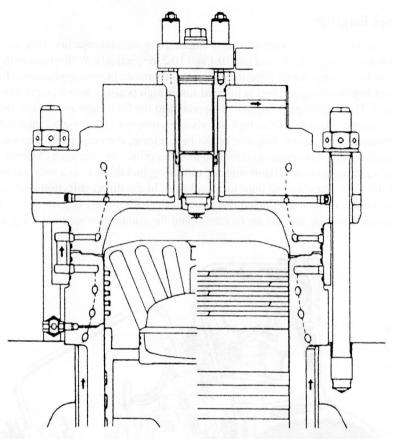

**Figure 10.2**    Combustion chamber and piston cutaway view of large direct-injection diesel engine.
SOURCE: Reprinted from Ref. [2] by permission of the American Society of Mechanical Engineers.

above, diesel-engine combustion takes place in both premixed and diffusion-controlled modes. Diesel fuels are much less volatile than spark-ignition fuels, but are more readily autoigniting (see Chapter 17). The rate at which the fuel vaporizes and mixes with the air is in competition with the rate of reaction leading to autoignition. Thus, the first fuel injected into the chamber may become premixed before it is subjected to an ignition source (a pocket of gas that has autoignited) and burns in a premixed flame; the fuel injected subsequently, however, burns in a diffusion mode, since an ignition source (the preexisting flame) is present as the fuel is injected. Clearly, droplet evaporation and burning are important in both indirect- and direct-injection engines. Appendix 10A presents Sir Harry Ricardo's imaginative account of the physical processes occurring within a direct-injection diesel engine combustion chamber. Ricardo (1885–1974), a pioneer engine researcher, contributed greatly to our understanding of combustion in reciprocating engines.

# Gas-Turbine Engines

Liquid-fueled gas-turbine engines propel most aircraft. Figure 10.3 shows a cutaway view of an aircraft turbine engine. In spite of the critical importance of the combustor to the engine system, it occupies a surprisingly small amount of space. Fuel is injected into and atomized within the annular combustor, where the flame is typically stabilized by swirling air that creates a recirculation zone. Aircraft gas-turbine combustor design is influenced by several factors: safety, combustion efficiency, combustion stability, relight-at-altitude capability, durability, and emissions, among others [7]. Prior to the advent of increasingly stringent emission requirements, aircraft engines employed nonpremixed combustion systems, with a near-stoichiometric primary flame zone integrated with secondary air streams to complete combustion and dilute the products to the proper temperature before entry into the turbine (Fig. 10.4a). Many modern designs and experimental systems utilize various degrees of premixing and/or staging to avoid high-temperature, $NO_x$-forming zones and to control soot formation [4–7]. Premixed combustion is achieved by vaporizing the fuel and mixing with a portion of the air before the mixture enters the hot combustion zone where it then mixes with additional air, ignites, and burns. Figure 10.4a illustrates the traditional **primary, secondary,** and **dilution zones** of an aircraft gas-turbine combustor. In Chapter 15, we will explore alternative state-of-the-art configurations.

Figure 10.4b shows the metal liner that separates the inner combustion space from the surrounding outer annular air-flow passage. A portion of the air is used to

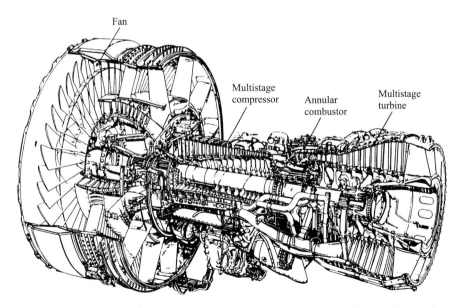

**Figure 10.3**     Aircraft gas-turbine engine. Note the combustor is a small part of the total volume of the engine.

SOURCE: Reprinted from Ref. [3] by permission of Gordon & Breach Science Publishers, SA.

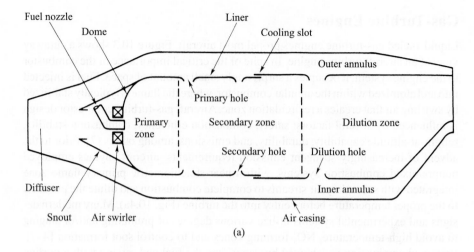

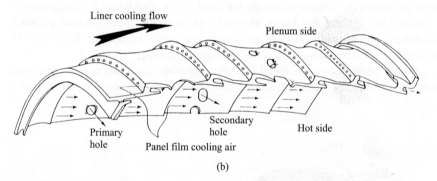

**Figure 10.4**      (a) Schematic of annular turbine combustor showing primary, secondary, and dilution zones. (b) Schematic of panel film cooling of combustor liner walls.

| SOURCE: (a) Reprinted from Ref. [8] by permission of Taylor & Francis. (b) Reprinted from Ref. [3] by permission of Gordon & Breach Science Publishers, SA.

cool the liner on the hot-gas side. This air flows through the ring of small holes and runs parallel to the liner, thus providing a cool boundary-layer flow. Combustion air is directed through the larger holes, forming high-velocity jets that penetrate into the core of the combustion space and rapidly mix with the hot gases. Both types of holes can be seen in the photographs of a real combustor in Fig. 10.5. A critical concern in the design of turbine combustors is the gas temperature distribution in the radial direction at the entrance to the turbine section. Air injection in the dilution zone is used to control this distribution. Hot spots that can damage turbine blades are avoided, and the gas temperature profile is tailored to increase from the blade root through a maximum and then decrease at the tip. At the root, the blade stresses are greatest; hence, the cooler gas at the root allows the blade to be cooler in this region. This is important since the blade material strength decreases with temperature. Optimizing

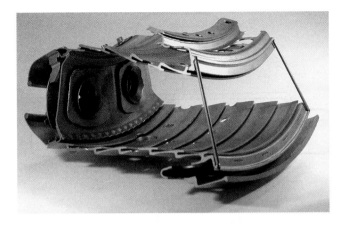

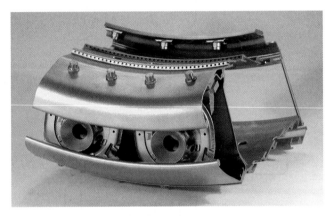

**Figure 10.5**     Segment of CFM56-7 turbo-fan jet engine produced by CFM International, a joint company of Snecma (France) and General Electric (U.S.A.). Fuel nozzles (not shown) fit into the large holes in the back (lower panel). Note the cooling air holes in the louvers of the combustion liner. The CFM56-7 engine is used in the Boeing 737 aircraft.

the temperature profile allows the maximum average turbine inlet temperature and high efficiencies as a result. The temperature distribution at the combustor outlet is frequently referred to as the **pattern factor.**

## Liquid-Rocket Engines

Of all the combustion devices considered here, the modern rocket engine produces the most intense combustion, i.e., the greatest energy release per volume of combustion space. There are two types of liquid rockets: **pressure-fed,** in which the fuel and oxidizer are pushed into the combustion chamber by a high-pressure gas; and **pump-fed,** where turbopumps deliver the propellants. These two arrangements are illustrated in Fig. 10.6. The pump-fed system delivers the higher performance of

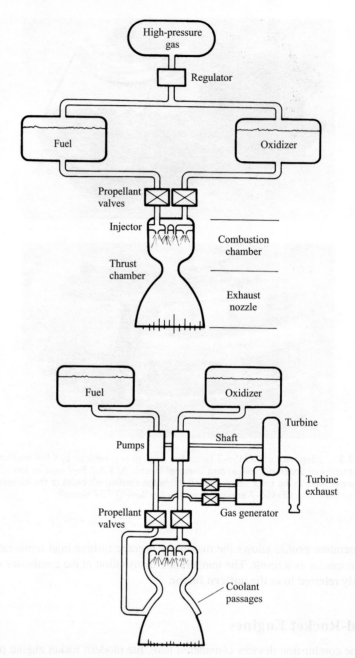

**Figure 10.6**    Schematic illustration of the two types of liquid-rocket systems: (top) pressure-fed system, (lower) pump-fed system.
SOURCE: From Ref. [9]. Reprinted with permission of the Jet Propulsion Laboratory.

(a)

(b)

**Figure 10.7**    The J-2 engine test firing (top) and *Saturn V* second stage with cluster of five J-2 engines.
| SOURCE: Courtesy of Rockwell Aerospace, Rocketdyne Division.

the two systems; however, it is much more complex. Figure 10.7 shows the liquid-hydrogen, liquid-oxygen pump-fed J-2 rocket engine used in both the second and third stages of the *Saturn* vehicle used in the *Apollo* program.

The thrust of the engine is produced by creating a hot, high-pressure gas from the combustion of the fuel and oxidizer in the combustion chamber that is subsequently accelerated through a supersonic, converging–diverging nozzle. Unlike any of the other

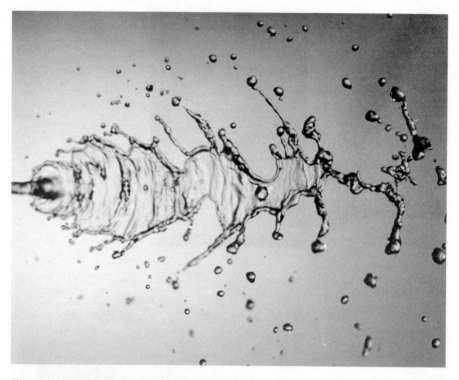

**Figure 10.8**     Impinging water jets form unstable liquid sheet, ligaments, and droplets.
The jet diameters and velocities are 1.4 mm and 3.4 m/s, respectively.
| SOURCE: Photograph courtesy of R. J. Santoro.

combustion devices discussed previously, the oxidizer for the rocket engine is supplied
as a liquid. To establish combustion thus requires the vaporization of both the fuel and
the oxidizer. A common injector arrangement is to have two liquid jets impinge to form
a liquid sheet (Fig. 10.8). This sheet is unstable and sheds strands, or ligaments, which,
in turn, break up to form droplets. Typically, a large number of individual injectors
are used to distribute propellant and oxidizer across the diameter of the combustion
chamber. Both premixed and diffusion burning are likely to be important in rocket-
engine combustion. Because of the extreme difficulty of instrumenting the interior of
the combustion chamber, relatively little is known about the details of the combustion
process. Work is in progress [10, 11] to learn more about the processes going on in a
rocket combustion chamber by using laser probes and other techniques.

## SIMPLE MODEL OF DROPLET EVAPORATION

To illustrate principles of mass transfer in Chapter 3, we developed a droplet evaporation model by transforming the Stefan problem to spherical coordinates. That
model involved only mass transfer, because it was assumed that the droplet surface
temperature was a known parameter. In the present analysis, we assume that the

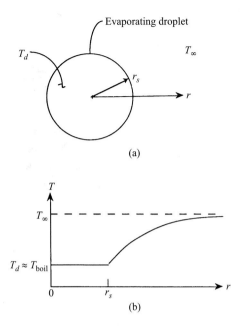

(a)

(b)

**Figure 10.9**     Evaporation of a liquid fuel droplet in a quiescent environment, where it is assumed that the droplet surface temperature is nearly at the liquid boiling point.

droplet surface temperature is near the droplet boiling point and, hence, the evaporation rate is controlled by the heat-transfer rate from the ambient to the droplet surface. This can be a good approximation in combustion environments where ambient temperatures are high. The mathematical description of this evaporation process is perhaps the simplest possible, and yet, is useful for engineering calculations. Later in this chapter, we develop a more general droplet-burning model (which can also be used to deal with pure evaporation) that couples both heat and mass transfer.

Figure 10.9 defines the spherically symmetric coordinate system. The radius $r$ is the only coordinate variable. It has its origin at the center of the droplet, and the droplet radius at the liquid–vapor interface is denoted $r_s$. Very far from the droplet surface ($r \rightarrow \infty$), the temperature is $T_\infty$.

Physically, heat from the ambient environment supplies the energy necessary to vaporize the liquid fuel, and the fuel vapor then diffuses from the droplet surface into the ambient gas. The mass loss causes the droplet radius to shrink with time until the droplet is completely evaporated ($r_s = 0$). The problem that we wish to solve is the determination of the mass flowrate of the fuel vapor from the surface at any instant in time. Knowledge of this will then enable us to calculate the droplet radius as a function of time and the droplet lifetime.

## Assumptions

The following assumptions for a droplet evaporating in a hot gas are commonly invoked because they lead to great simplification, principally by eliminating the need to deal with the mass-transfer aspects of the problem, yet still agree reasonably well with experimental results:

1. The droplet evaporates in a quiescent, infinite medium.

2. The evaporation process is quasi-steady. This means that at any instant in time the process can be described as if it were in steady state. This assumption eliminates the need to deal with partial differential equations.

3. The fuel is a single-component liquid with zero solubility for gases.

4. The droplet temperature is uniform, and, furthermore, the temperature is assumed to be the boiling point of the fuel, $T_d = T_{boil}$. In many problems, the transient heating of the liquid does not greatly affect the droplet life-time, and more rigorous calculations show that the droplet surface temperature is only slightly less than the liquid boiling point in combustion environments. This assumption eliminates the need to solve a liquid-phase (droplet) energy equation, and, more importantly, eliminates the need to solve the fuel vapor transport (species) equation in the gas phase! Implicit in this assumption is that $T_\infty > T_{boil}$. In our subsequent analysis, when we eliminate the assumption that the droplet is at the boiling point, you will see how much more complicated the analysis becomes.

5. We assume binary diffusion with a unity Lewis number ($\alpha = \mathcal{D}$). This permits us to use the simple Shvab–Zeldovich energy equation developed previously, in Chapter 7.

6. We also assume that all thermophysical properties, such as thermal conductivity, density, and specific heat, are constant. Although these properties may vary greatly as we move through the gas phase from the droplet surface to the faraway surroundings, constant properties allow a simple closed-form solution. In the final analysis, a judicious choice of mean values allows reasonably accurate predictions to be made.

## Gas-Phase Analysis

With the above assumptions, we can find the mass evaporation rate, $\dot{m}$, and the droplet radius history, $r_s(t)$, by writing a gas-phase mass conservation equation, a gas-phase energy equation, a droplet gas-phase interface energy balance, and a droplet liquid mass conservation equation. The gas-phase energy equation gives us the temperature distribution in the gas phase, which, in turn, allows us to evaluate the conduction heat transfer into the droplet at the surface. This is necessary to evaluate the surface energy balance that yields the evaporation rate $\dot{m}$. Knowing $\dot{m}(t)$, we can easily find the drop size as a function of time.

**Mass Conservation**   With the assumption of quasi-steady burning, the mass flow-rate, $\dot{m}(r)$, is a constant, independent of radius; thus,

$$\dot{m} = \dot{m}_F = \rho v_r 4\pi r^2 = \text{constant} \tag{10.1}$$

and

$$\frac{d\left(\rho v_r r^2\right)}{dr} = 0, \tag{10.2}$$

where $v_r$ is the bulk flow velocity.

**Energy Conservation**  As previously derived in Chapter 7, conservation of energy for the situation depicted in Fig. 10.10a is expressed by Eqn. 7.65. With the assumption of constant properties and unity Lewis number, this can be rearranged and written as:

$$\frac{d\left(r^2 \dfrac{dT}{dr}\right)}{dr} = \frac{\dot{m} c_{pg}}{4\pi k} \frac{dT}{dr}, \tag{10.3}$$

where the reaction rate term is zero, since no reactions occur for pure evaporation.

For convenience in the following development, we define $Z \equiv c_{pg}/4\pi k$; thus,

$$\frac{d\left(r^2 \dfrac{dT}{dr}\right)}{dr} = Z\dot{m}\frac{dT}{dr}. \tag{10.4}$$

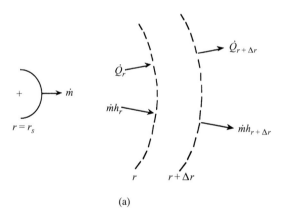

(a)

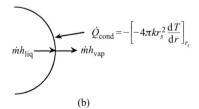

(b)

**Figure 10.10**  Evaporating-droplet energy balance for (a) gas phase and (b) droplet surface.

Solving Eqn. 10.4 gives us the temperature distribution, $T(r)$, in the gas phase. To effect this solution requires two boundary conditions:

$$\text{Boundary condition 1:} \qquad T(r \to \infty) = T_\infty \qquad (10.5a)$$

$$\text{Boundary condition 2:} \qquad T(r = r_s) = T_{\text{boil}}. \qquad (10.5b)$$

Equation 10.4 can be easily solved by twice separating variables and integrating. After the first integration, we obtain

$$r^2 \frac{dT}{dr} = Z\dot{m}T + C_1,$$

where $C_1$ is the integration constant. The second separation and integration yields the general solution,

$$\frac{1}{Z\dot{m}} \ln(Z\dot{m}T + C_1) = -\frac{1}{r} + C_2, \qquad (10.6)$$

where $C_2$ is the second integration constant. Applying Eqn. 10.5a to Eqn. 10.6 gives $C_2$ in terms of $C_1$:

$$C_2 = \frac{1}{Z\dot{m}} \ln(Z\dot{m}T_\infty + C_1).$$

Substituting $C_2$ back into Eqn. 10.6, applying the second boundary condition (Eqn. 10.5b), and using exponentiation to remove the logarithm, we can find $C_1$; i.e.,

$$C_1 = \frac{Z\dot{m}[T_\infty \exp(-Z\dot{m}/r_s) - T_{\text{boil}}]}{1 - \exp(-Z\dot{m}/r_s)}.$$

The second constant is now readily found by substituting $C_1$ into the expression for $C_2$ above. Thus,

$$C_2 = \frac{1}{Z\dot{m}} \ln\left[ \frac{Z\dot{m}(T_\infty - T_{\text{boil}})}{1 - \exp(-Z\dot{m}/r_s)} \right].$$

Finally, the temperature distribution can be found by substituting these expressions for $C_1$ and $C_2$ back into the general solution given by Eqn. 10.6. The final, and somewhat complicated, result is

$$T(r) = \frac{(T_\infty - T_{\text{boil}}) \exp(-Z\dot{m}/r) - T_\infty \exp(-Z\dot{m}/r_s) + T_{\text{boil}}}{1 - \exp(-Z\dot{m}/r_s)}. \qquad (10.7)$$

**Droplet-Gas-Phase Interface Energy Balance**    By itself, Eqn. 10.7 does not allow us to solve for the evaporation rate $\dot{m}$; however, it does allow us to evaluate the heat transferred to the droplet surface, which appears in the interface (surface) energy balance illustrated in Fig. 10.10b. Heat is conducted to the interface from the hot gas, and since we assume that the droplet temperature is uniform at $T_{boil}$, all of this heat must be used to vaporize the fuel, with no heat flowing into the droplet interior. It is relatively easy to allow transient droplet heating, as will be shown in our burning droplet analysis. The surface energy balance can be written as

$$\dot{Q}_{cond} = \dot{m}(h_{vap} - h_{liq}) = \dot{m}h_{fg}. \tag{10.8}$$

Substituting Fourier's law for $\dot{Q}_{cond}$, paying careful attention to sign conventions, we have

$$4\pi k_g r_s^2 \left. \frac{dT}{dr} \right|_{r_s} = \dot{m}h_{fg}. \tag{10.9}$$

Differentiating Eqn. 10.7, the temperature gradient in the gas phase at the droplet surface is

$$\left. \frac{dT}{dr} \right|_{r_s} = \frac{Z\dot{m}}{r_s^2} \left[ \frac{(T_\infty - T_{boil}) \exp(-Z\dot{m}/r_s)}{1 - \exp(-Z\dot{m}/r_s)} \right]. \tag{10.10}$$

Substituting this result into Eqn. 10.9 and solving for $\dot{m}$, we obtain

$$\dot{m} = \frac{4\pi k_g r_s}{c_{pg}} \ln \left[ \frac{c_{pg}(T_\infty - T_{boil})}{h_{fg}} + 1 \right]. \tag{10.11}$$

In the combustion literature, the first term in the brackets is defined as

$$B_q = \frac{c_{pg}(T_\infty - T_{boil})}{h_{fg}}, \tag{10.12}$$

so

$$\dot{m} = \frac{4\pi k_g r_s}{c_{pg}} \ln(B_q + 1). \tag{10.13}$$

The parameter $B$ is one of those dimensionless parameters, such as the Reynolds number, that has special significance in combustion and is commonly referred to by those knowledgeable in the field. It is sometimes referred to as the **Spalding number,** or simply as the **transfer number, $B$**. Recall that, in Chapter 3, we developed a similar expression for droplet evaporization controlled by mass transfer. The definition of $B$ given by Eqn. 10.12 applies only to the set of assumptions listed previously, with the subscript $q$ indicating that it is based on heat-transfer considerations alone. Other definitions exist, and their functional form depends on the assumptions made. For example, $B$ has a different definition if a spherically symmetric flame is assumed to surround the droplet. This will be elaborated upon later.

This completes our analysis of the gas phase. Knowing the instantaneous (quasi-steady) evaporation rate now allows us to calculate the droplet lifetime.

## Droplet Lifetimes

Following the same analysis as in Chapter 3 for mass-transfer-controlled evaporation, we obtain the droplet radius (or diameter) history by writing a mass balance that states that the rate at which the mass of the droplet decreases is equal to the rate at which the liquid is vaporized, i.e.,

$$\frac{dm_d}{dt} = -\dot{m}, \tag{10.14}$$

where the droplet mass, $m_d$, is given by

$$m_d = \rho_l V = \rho_l \pi D^3/6 \tag{10.15}$$

and $V$ and $D$ are the droplet volume and diameter, respectively.

Substituting Eqns. 10.15 and 10.13 into Eqn. 10.14, and performing the differentiation, yields

$$\frac{dD}{dt} = -\frac{4k_g}{\rho_l c_{pg} D} \ln(B_q + 1). \tag{10.16}$$

As discussed previously (see Chapter 3), Eqn. 10.16 is more commonly expressed in terms of $D^2$ rather than $D$, i.e.,

$$\frac{dD^2}{dt} = -\frac{8k_g}{\rho_l c_{pg}} \ln(B_q + 1). \tag{10.17}$$

Equation 10.17 shows that the time derivative of the square of the droplet diameter is constant; hence, $D^2$ varies linearly with $t$ with the slope $-(8k/\rho_l c_{pg})\ln(B_q + 1)$ as illustrated in Fig. 10.11. This slope is defined to be the **evaporation constant $K$**:

$$K = \frac{8k_g}{\rho_l c_{pg}} \ln(B_q + 1). \tag{10.18}$$

Note the similarity of this relation to Eqn. 3.58: the form of each equation is the same, and the equations become identical if the Lewis number is unity ($k_g / c_{pg} = \rho \mathcal{D}$),

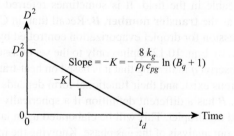

**Figure 10.11**     The $D^2$ law for droplet evaporation resulting from simplified analysis.

although different definitions apply to $B$. We can integrate Eqn. 10.17 to provide a general relationship expressing the variation of $D$ (or $D^2$) with time:

$$\int_{D_0^2}^{D^2} d\hat{D}^2 = -\int_0^t K \, d\hat{t},$$

which yields

$$D^2(t) = D_0^2 - Kt. \tag{10.19}$$

Equation 10.19 is the same **$D^2$ law** for droplet evaporation that we introduced in Chapter 3. Experiments (e.g., Ref. [12]) show that the $D^2$ law holds after an initial transient period associated with the heating of the droplet to near the boiling point (see Fig. 3.7b).

We can find the time it takes a droplet of given initial size to evaporate completely, i.e., the droplet lifetime, $t_d$, by letting $D^2(t_d) = 0$:

$$t_d = D_0^2 / K. \tag{10.20}$$

To use Eqns. 10.19 and 10.20 to predict droplet evaporation is straightforward; however, we have the problem of choosing appropriate mean values for $c_{pg}$ and $k_g$, the gas-phase specific heat and thermal conductivity, respectively, which appear in the evaporation rate constant. In our analysis, we assumed that $c_{pg}$ and $k_g$ were constant, whereas, in reality, they vary considerably in going from the droplet surface to the freestream conditions. Following the approach of Law and Williams [13] for burning droplets, the following approximations for $c_{pg}$ and $k_g$ are suggested:

$$c_{pg} = c_{pF}(\overline{T}) \tag{10.21}$$

$$k_g = 0.4 k_F(\overline{T}) + 0.6 k_\infty(\overline{T}), \tag{10.22}$$

where the subscript $F$ represents the fuel vapor and $\overline{T}$ is the average of the fuel boiling point temperature and that of the free stream,

$$\overline{T} = (T_{\text{boil}} + T_\infty)/2. \tag{10.23}$$

Other approximations for property evaluations exist that are more accurate [14]; however, the above technique is the easiest to apply.

---

**Example 10.1**

Consider a 500-$\mu$m-diameter liquid $n$-hexane ($C_6H_{14}$) droplet evaporating in hot, stagnant nitrogen at 1 atm. The $N_2$ temperature is 850 K. Determine the lifetime of the $n$-hexane droplet, assuming the droplet temperature is at its boiling point.

**Solution**

Find $t_d$. The droplet lifetime is evaluated from

$$t_d = D_0^2 / K, \tag{10.20}$$

where

$$K = \frac{8k_g}{\rho_l c_{pg}} \ln(B_q + 1) \tag{10.18}$$

and

$$B_q = \frac{c_{pg}(T_\infty - T_{boil})}{h_{fg}}. \tag{10.12}$$

The solution is straightforward, with, perhaps, the greatest complication being the property evaluations. The *n*-hexane properties are evaluated at

$$\overline{T} = \tfrac{1}{2}(T_{boil} + T_\infty) = \tfrac{1}{2}(342 + 850) = 596 \text{ K},$$

where the boiling point ($T_{boil} = 342$ K) is found in Appendix B, Table B.1, as are the liquid density and heat of vaporization. The *n*-hexane specific heat and thermal conductivity are evaluated from the curvefit coefficients provided in Appendix Tables B.2 and B.3.

$$c_{pg} = c_{pC_6 H_{14}}(\overline{T}) = 2872 \text{ J/kg-K} \qquad \text{(Appendix Table B.2)},$$

$$k_F = k_{C_6 H_{14}}(\overline{T}) = 0.0495 \text{ W/m-K} \qquad \text{(Appendix Table B.3)},$$

$$k_\infty = k_{N_2}(\overline{T}) = 0.0444 \text{ W/m-k}. \qquad \text{(Appendix Table C.2)}.$$

The appropriate mean thermal conductivity is (Eqn. 10.22)

$$k_g = 0.4(0.0495) + 0.6(0.0444) = 0.0464 \text{ W/m-K}$$

and the *n*-hexane heat of vaporization and liquid density are, respectively,

$$h_{fg} = 335,000 \text{ J/kg} \qquad \text{(Appendix Table B.1)},$$

$$\rho_l \approx 659 \text{ kg/m}^3 \qquad \text{(Appendix Table B.1)}.$$

The dimensionless transfer number $B$ can now be evaluated using the properties estimated above:

$$B_q = \frac{2872(850 - 342)}{335,000} = 4.36$$

and the evaporation constant $K$ is

$$K = \frac{8(0.0464)}{659(2872)} \ln(1 + 4.36)$$

$$= 1.961 \cdot 10^{-7}(1.679) = 3.29 \cdot 10^{-7} \text{ m}^2/\text{s}.$$

Thus, the droplet lifetime is

$$\boxed{t_d} = D_0^2/K = \frac{(500 \cdot 10^{-6})^2}{3.29 \cdot 10^{-7}} = \boxed{0.76 \text{ s}}$$

**Comments**

The example results can give us a physical feel for the timescales involved in droplet evaporation. Using the evaporation constant calculated above, but with $d \approx 50 \; \mu m$, $t_d$ is of the order of 10 ms. In many spray combustion systems, mean drop sizes are of the order of 50 $\mu m$ and smaller.

# SIMPLE MODEL OF DROPLET BURNING

Our approach here extends the development above to include a spherically symmetric diffusion flame that surrounds the droplet. We will retain the assumption of a quiescent environment and spherical symmetry in the initial development, but subsequently show how the spherically symmetric results can be adjusted to take into account the enhancement of burning caused by convection, either natural convection produced by the flame or a forced flow. We will also remove the restriction that the droplet is at the boiling point, which requires using species conservation in the gas phase.

## Assumptions

The following assumptions lead to a greatly simplified model of droplet combustion that preserves the essential physics and agrees reasonably well with experimental results:

1. The burning droplet, surrounded by a spherically symmetric flame, exists in a quiescent, infinite medium. There are no interactions with any other droplets, and the effects of convection are ignored.

2. As in our previous analysis, the burning process is quasi-steady.

3. The fuel is a single-component liquid with zero solubility for gases. Phase equilibrium prevails at the liquid–vapor interface.

4. The pressure is uniform and constant.

5. The gas phase consists of only three "species:" fuel vapor, oxidizer,[1] and combustion products. The gas-phase region is divided into two zones. The inner zone between the droplet surface and the flame contains only fuel vapor and products, and the outer zone consists of oxidizer and products. Thus, binary diffusion prevails in each region.

6. Fuel and oxidizer react in stoichiometric proportions at the flame. Chemical kinetics are assumed to be infinitely fast, resulting in the flame being represented as an infinitesimally thin sheet.

7. The Lewis number, $Le = \alpha/\mathcal{D} = k_g/\rho c_{pg}\mathcal{D}$, is unity.

8. Radiation heat transfer is negligible.

---

[1] The oxidizer may be a mixture of gases, some of which may be inert. For example, we here consider air to be an oxidizer, even though only approximately 21 percent of the air chemically reacts. At the flame, the nitrogen conceptually is transformed from oxidizer to products even though it does not actually react.

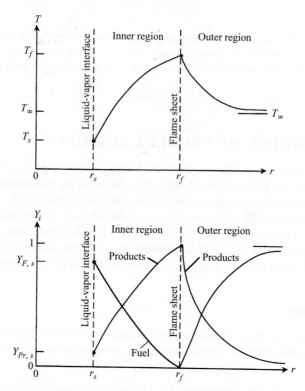

**Figure 10.12**    Temperature (top) and species profiles (bottom) for simple droplet-burning model.

9. The gas-phase thermal conductivity, $k_g$, specific heat, $c_{pg}$, and the product of the density and mass diffusivity, $\rho D$, are all constants.

10. The liquid fuel droplet is the only condensed phase; no soot or liquid water is present.

The basic model embodied by the above assumptions is illustrated in Fig. 10.12, which shows temperature and species profiles through the inner region between the droplet surface and the flame, $r_s \le r \le r_f$, and the outer region beyond the flame, $r_f \le r < \infty$. We see that there are three important temperatures: the droplet surface temperature, $T_s$; the flame temperature, $T_f$; and the temperature of the medium at infinity, $T_\infty$. The fuel vapor mass fraction, $Y_F$, is a maximum at the droplet surface and monotonically decreases to zero at the flame, where the fuel is totally consumed. The oxidizer mass fraction, $Y_{Ox}$, mirrors this, being a maximum (unity), far from the flame and decreasing to zero at the flame sheet. The combustion products have their maximum concentration at the flame sheet (unity), diffusing both inward toward the droplet and outward away from the flame. Since assumption 3 prevents the products from dissolving in the liquid, there can be no net flow of products inward from the flame to the droplet surface. Thus, the products in the inner region form a stagnant film through which fuel vapor moves. The species fluxes in this inner region behave in a manner reminiscent of the Stefan problem discussed in Chapter 3.

## Problem Statement

In the following analysis, our number one objective is to determine the droplet mass burning rate, $\dot{m}_F$, given the initial droplet size and the conditions far from the droplet, i.e., the temperature, $T_\infty$, and the oxidizer mass fraction, $Y_{Ox,\infty}$ $(=1)$. On the way to achieving this objective, we will also obtain expressions describing the temperature and species profiles in each region, together with relationships that allow us to calculate the flame radius, $r_f$; flame temperature, $T_f$; droplet surface temperature, $T_s$; and the fuel vapor mass fraction at the droplet surface, $Y_{F,s}$. In summary, we seek expressions that allow us to evaluate a total of five parameters: $\dot{m}_F$, $Y_{F,s}$, $T_s$, $T_f$, and $r_f$.

As an overview, the five relationships needed to solve for the five unknowns have their origins in the following: (1) an energy balance at the droplet surface, (2) an energy balance at the flame sheet, (3) the oxidizer distribution in the outer region, (4) the fuel vapor distribution in the inner region, and (5) phase equilibrium at the liquid–vapor interface expressed, for example, by the Clausius–Clapeyron relationship. Finally, knowing the instantaneous mass burning rate, droplet lifetimes will be calculated in the same manner as employed in our evaporation analysis. In the analysis, we will recover a $D^2$ law for burning droplets.

The problem of droplet combustion has been widely studied and has a very large literature associated with it, as evidenced by the reviews of the past few decades [15–21]. The physical model we describe here has its origins in the 1950s [22–23]. The solution approach taken here is not elegant. Our treatment, however, seeks to maintain a view of the physical processes at all times by retaining the important physical variables: temperature and species mass fractions. A more elegant approach, which is described in more advanced textbooks (e.g., Ref. [24]), combines the species and energy equations to create a conserved-scalar variable.

## Mass Conservation

Overall mass conservation in the gas phase is described as in our previous development (Eqns. 10.1 and 10.2), i.e.,

$$\dot{m}(r) = \dot{m}_F = \text{constant.} \tag{10.1}$$

Note that the total flowrate is everywhere identical to the fuel flowrate, that is, the burning rate. This will be important in applying species conservation.

## Species Conservation

**Inner Region**   In the *inner region*, the important diffusing species is the *fuel vapor*. We can apply Fick's law (Eqn. 3.5) to the inner region:

$$\dot{m}''_A = Y_A(\dot{m}''_A + \dot{m}''_B) - \rho \mathcal{D}_{AB} \nabla Y_A \tag{3.5}$$

where the subscripts $A$ and $B$ here denote fuel and products, respectively:

$$\dot{m}_A'' \equiv \dot{m}_F'' = \dot{m}_F / 4\pi r^2 \tag{10.24}$$

$$\dot{m}_B'' \equiv \dot{m}_{Pr}'' = 0, \tag{10.25}$$

and the $\nabla$-operator in spherical coordinates, where the only variations are in the $r$-direction, becomes $\nabla(\ ) = d(\ )/dr$. Thus, Fick's law is

$$\dot{m}_F = -4\pi r^2 \frac{\rho \mathcal{D}}{1-Y_F} \frac{dY_F}{dr}. \tag{10.26}$$

This first-order ordinary differential equation must satisfy two boundary conditions: at the droplet surface, liquid–vapor equilibrium prevails, so

$$Y_F(r_s) = Y_{F,s}(T_s), \tag{10.27a}$$

and at the flame, the fuel disappears, so

$$Y_F(r_f) = 0. \tag{10.27b}$$

The existence of two boundary conditions allows us to treat the burning rate, $\dot{m}_F$, as an eigenvalue, i.e., a parameter that can be calculated from the solution of Eqn. 10.26 subject to Eqns. 10.27a and b. Defining $Z_F \equiv 1/4\pi\rho\mathcal{D}$, the general solution to Eqn. 10.26 is

$$Y_F(r) = 1 + C_1 \exp(-Z_F \dot{m}_F / r). \tag{10.28}$$

Applying the droplet surface condition (Eqn. 10.27a) to evaluate $C_1$, we obtain

$$Y_F(r) = 1 - \frac{(1-Y_{F,s})\exp(-Z_F \dot{m}_F / r)}{\exp(-Z_F \dot{m}_F / r_s)}. \tag{10.29}$$

Applying the flame boundary condition (Eqn. 10.27b), we obtain a relationship involving the three unknowns $Y_{F,s}$, $\dot{m}_F$, and $r_f$:

$$Y_{F,s} = 1 - \frac{\exp(-Z_F \dot{m}_F / r_s)}{\exp(-Z_F \dot{m}_F / r_f)}. \tag{10.30}$$

Completing the species conservation solution in the inner region—although not necessary to accomplish our objective—the combustion product mass fraction can be expressed as

$$Y_{Pr}(r) = 1 - Y_F(r). \tag{10.31}$$

**Outer Region**   In the *outer region*, the important diffusing species is the *oxidizer*, which is transported radially inward to the flame. At the flame, the oxidizer and fuel

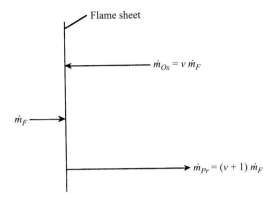

**Figure 10.13**    Mass flow relationships at flame sheet. Note that the net mass flow in both the inner and outer regions is equal to the fuel flowrate, $\dot{m}_F$.

combine to stoichiometric proportions according to

$$1 \, \text{kg fuel} + v \, \text{kg oxidizer} = (v+1) \, \text{kg products},  \tag{10.32}$$

where $v$ is the stoichiometric (mass) ratio and includes any nonreacting gases that may be a part of the oxidizer stream. This relationship is illustrated in Fig. 10.13. The vector mass fluxes in Fick's law are thus

$$\dot{\mathbf{m}}''_A \equiv \dot{\mathbf{m}}''_{Ox} = -v\dot{\mathbf{m}}''_F  \tag{10.33a}$$

$$\dot{\mathbf{m}}''_B \equiv \dot{\mathbf{m}}''_{Pr} = +(v+1)\dot{\mathbf{m}}''_F.  \tag{10.33b}$$

Fick's law in the outer region is thus

$$\dot{m}_F = +4\pi r^2 \frac{\rho \mathcal{D}}{v+Y_{Ox}} \frac{dY_{Ox}}{dr},  \tag{10.34}$$

with boundary conditions

$$Y_{Ox}(r_f) = 0  \tag{10.35a}$$

$$Y_{Ox}(r \rightarrow \infty) = Y_{Ox,\infty} \equiv 1.  \tag{10.35b}$$

Integrating Eqn. 10.34 yields

$$Y_{Ox}(r) = -v + C_1 \exp(-Z_F \dot{m}_F / r).  \tag{10.36}$$

Applying the flame condition (Eqn. 10.35a) to eliminate $C_1$ yields

$$Y_{Ox}(r) = v \left[ \frac{\exp(-Z_F \dot{m}_F / r)}{\exp(-Z_F \dot{m}_F / r_f)} - 1 \right].  \tag{10.37}$$

Applying the condition at $r \rightarrow \infty$ (Eqn. 10.35b), we obtain an algebraic relation between the burning rate, $\dot{m}_F$, and the flame radius, $r_f$, i.e.,

$$\exp(+Z_F \dot{m}_F / r_f) = (v+1)/v. \tag{10.38}$$

Complementing the oxidizer distribution (Eqn. 10.37), the product mass fraction distribution is

$$Y_{Pr}(r) = 1 - Y_{Ox}(r). \tag{10.39}$$

## Energy Conservation

Again, we use the Shvab–Zeldovich form of the energy equation. Since we confine chemical reactions to the boundary, i.e., the flame sheet, the reaction rate term is zero both inside the flame and outside the flame. Thus, the same energy equation that we developed for pure evaporation (Eqn. 10.3) also applies to the burning droplet:

$$\frac{d\left(r^2 \dfrac{dT}{dr}\right)}{dr} = \frac{\dot{m}_F c_{pg}}{4\pi k_g} \frac{dT}{dr}. \tag{10.3}$$

Again, for convenience, we define $Z_T = c_{pg}/4\pi k_g$; thus, our governing equation is

$$\frac{d\left(r^2 \dfrac{dT}{dr}\right)}{dr} = Z_T \dot{m}_F \frac{dT}{dr}. \tag{10.40}$$

As an aside, note that the parameter $Z_F$, defined in the species conservation analysis, equals $Z_T$ when the Lewis number is unity, since, then, $c_{pg}/k_g = \rho \mathcal{D}$. Since the form of Eqn. 10.40 is predicated on equal diffusivities for heat and mass, i.e., unity Lewis number (refer to the derivation of the Shvab–Zeldovich energy equation in Chapter 7), we require $Z_F = Z_T$.

The boundary conditions applied to Eqn. 10.40 are

$$\text{Inner region} \begin{cases} T(r_s) = T_s & (10.41a) \\ T(r_f) = T_f & (10.41b) \end{cases}$$

$$\text{Outer region} \begin{cases} T(r_f) = T_f & (10.41c) \\ T(r \rightarrow \infty) = T_\infty. & (10.41d) \end{cases}$$

Of the three temperatures, only $T_\infty$ is considered to be known; $T_s$ and $T_f$ are two of the five unknowns in our problem.

**Temperature Distributions**    The general solution to Eqn. 10.40 is

$$T(r) = \frac{C_1 \exp(-Z_T \dot{m}_F / r)}{Z_T \dot{m}_F} + C_2, \tag{10.42}$$

and in the *inner zone,* the temperature distribution resulting from application of Eqns. 10.41a and b is

$$T(r) = \frac{(T_s - T_f)\exp(-Z_T \dot{m}_F / r) + T_f \exp(-Z_T \dot{m}_F / r_s) - T_s \exp(-Z_T \dot{m}_F / r_f)}{\exp(-Z_T \dot{m}_F / r_s) - \exp(-Z_T \dot{m}_F / r_f)} \tag{10.43}$$

for $r_s \le r \le r_f$.

In the *outer region,* application of Eqns. 10.41c and d to Eqn. 10.42 yields

$$T(r) = \frac{(T_f - T_\infty)\exp(-Z_T \dot{m}_F / r) + \exp(-Z_T \dot{m}_F / r_f) T_\infty - T_f}{\exp(-Z_T \dot{m}_F / r_f) - 1} \tag{10.44}$$

for $r_f < r < \infty$.

**Energy Balance at Droplet Surface**    Figure 10.14 shows the conduction heat-transfer rates and the enthalpy fluxes at the surface of the evaporating droplet. Heat is conducted from the flame through the gas phase to the droplet surface. Some of this energy is used to vaporize fuel, with the remainder conducted into the droplet interior. Mathematically, this is expressed as

$$\dot{Q}_{g-i} = \dot{m}_F (h_{vap} - h_{liq}) + \dot{Q}_{i-l} \tag{10.45a}$$

or

$$\dot{Q}_{g-i} = \dot{m}_F h_{fg} + \dot{Q}_{i-l}. \tag{10.45b}$$

The heat conducted into the droplet interior, $\dot{Q}_{i-l}$, can be handled in several ways. One common approach is to model the droplet as consisting of two zones: an interior region existing uniformly at its initial temperature, $T_0$; and a thin surface layer at the surface temperature, $T_s$. For this so-called "onion-skin" model,

$$\dot{Q}_{i-l} = \dot{m}_F c_{pl} (T_s - T_0) \tag{10.46}$$

is the energy required to heat the fuel that is vaporized from $T_0$ to $T_s$. For convenience, we define

$$q_{i-l} \equiv \dot{Q}_{i-l} / \dot{m}_F, \tag{10.47}$$

so for the *onion-skin model,*

$$q_{i-l} = c_{pl} (T_s - T_0). \tag{10.48}$$

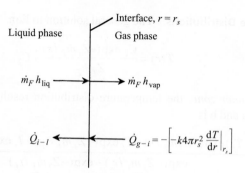

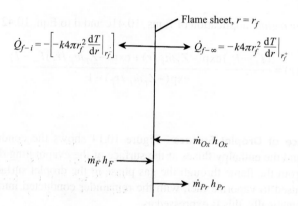

**Figure 10.14** Surface energy balance at droplet liquid–vapor interface (top) and surface energy balance at flame sheet (bottom).

Another common treatment of $\dot{Q}_{i-l}$ is to assume that the droplet behaves as a lumped parameter, i.e., it has a uniform temperature, with a transient heat-up period. For the *lumped parameter,*

$$\dot{Q}_{i-l} = m_d c_{pl} \frac{dT_s}{dt} \qquad (10.49)$$

and

$$q_{i-l} = \frac{m_d c_{pl}}{\dot{m}_F} \frac{dT_s}{dt}, \qquad (10.50)$$

where $m_d$ is the droplet mass. Implementation of the lumped-parameter model requires solving energy and mass conservation equations for the droplet as a whole in order to obtain $dT_s / dt$.

A third approach, and the simplest, is to assume that the droplet rapidly heats up to a steady temperature, $T_s$. This, in effect, says the thermal inertia of the droplet is negligible. With this assumption of *negligible thermal inertia,*

$$q_{i-l} = 0. \qquad (10.51)$$

Returning to the surface energy balance expressed by Eqn. 10.45b, the conduction heat transfer from the gas phase, $\dot{Q}_{g-i}$, can be evaluated by applying Fourier's law and using the temperature distribution in the inner region (Eqn. 10.43) to obtain the temperature gradient, i.e.,

$$-\left[-k_g 4\pi r^2 \frac{dT}{dr}\right]_{r_s} = \dot{m}_F (h_{fg} + q_{i-l}), \tag{10.52}$$

where

$$\frac{dT}{dr} = \frac{(T_s - T_f)Z_T \dot{m}_F \exp(-Z_T \dot{m}_F / r)}{r^2 [\exp(-Z_T \dot{m}_F / r_s) - \exp(-Z_T \dot{m}_F / r_f)]} \tag{10.53}$$

for $r_s \leq r \leq r_f$.

Evaluating the heat-transfer rate at $r = r_s$, Eqn. 10.52 becomes, after rearranging and substituting the definition of $Z_T$,

$$\frac{c_{pg}(T_f - T_s)}{(q_{i-l} + h_{fg})} \frac{\exp(-Z_T \dot{m}_F / r_s)}{[\exp(-Z_T \dot{m}_F / r_s) - \exp(-Z_T \dot{m}_F / r_f)]} + 1 = 0. \tag{10.54}$$

Equation 10.54 contains four unknowns: $\dot{m}_F$, $T_f$, $T_s$, and $r_f$.

**Energy Balance at Flame Sheet**  Referring to Fig. 10.14, we see how the various energy fluxes at the flame sheet relate. Since the flame temperature is the highest temperature in the system, heat is conducted both toward the droplet, $\dot{Q}_{f-i}$, and away to infinity, $\dot{Q}_{f-\infty}$. The chemical energy released at the flame is taken into account by using standardized enthalpy fluxes for the fuel, oxidizer, and products. A surface energy balance at the flame sheet can be written as

$$\dot{m}_F h_F + \dot{m}_{Ox} h_{Ox} - \dot{m}_{Pr} h_{Pr} = \dot{Q}_{f-i} + \dot{Q}_{f-\infty}. \tag{10.55}$$

The enthalpies are defined as

$$h_F \equiv h_{f,F}^o + c_{pg}(T - T_{ref}), \tag{10.56a}$$

$$h_{Ox} \equiv h_{f,Ox}^o + c_{pg}(T - T_{ref}), \tag{10.56b}$$

$$h_{Pr} \equiv h_{f,Pr}^o + c_{pg}(T - T_{ref}), \tag{10.56c}$$

and the heat of combustion, $\Delta h_c$, per unit mass of fuel is given by

$$\Delta h_c(T_{ref}) = (1)h_{f,F}^o + (\nu)h_{f,Ox}^o - (1+\nu)h_{f,Pr}^o. \tag{10.57}$$

The mass flowrates of fuel, oxidizer, and products are related by stoichiometry (see Eqns. 10.32 and 10.33a and b). Note that although products exist in the inner region, there is no net flow of products between the droplet surface and the flame;

thus, all of the products flow radially outward away from the flame. Knowing this, Eqn. 10.55 becomes

$$\dot{m}_F[h_F + \nu h_{Ox} - (\nu+1)h_{Pr}] = \dot{Q}_{f-i} + \dot{Q}_{f-\infty}. \qquad (10.58)$$

Substituting Eqns. 10.56 and 10.57 into the above yields

$$\dot{m}_F \Delta h_c + \dot{m}_F c_{pg} [(T_f - T_{ref}) + \nu(T_f - T_{ref}) - (\nu+1)(T_f - T_{ref})] = \dot{Q}_{f-i} + \dot{Q}_{f-\infty}. \qquad (10.59)$$

Since we assume that $c_{pg}$ is constant, then $\Delta h_c$ is independent of temperature; thus, we can choose the flame temperature as a reference state to simplify Eqn. 10.59:

$$\underset{\substack{\text{Rate at which chemical} \\ \text{energy is converted to} \\ \text{thermal energy at the flame}}}{\dot{m}_F \Delta h_c} \quad = \quad \underset{\substack{\text{Rate at which heat} \\ \text{is conducted away} \\ \text{from the flame}}}{\dot{Q}_{f-i} + \dot{Q}_{f-\infty}.} \qquad (10.60)$$

Once again, we rely on Fourier's law and the previously derived temperature distributions to evaluate the conduction heat-transfer terms, $\dot{Q}_{f-i}$ and $\dot{Q}_{f-\infty}$; i.e.,

$$\dot{m}_F \Delta h_c = k_g 4\pi r_f^2 \left.\frac{dT}{dr}\right|_{r_f^-} - k_g 4\pi r_f^2 \left.\frac{dT}{dr}\right|_{r_f^+}. \qquad (10.61)$$

To evaluate the temperature gradient at $r = r_f^-$, we can employ Eqn. 10.53; for the gradient at $r = r_f^+$, the temperature distribution in the outer region is differentiated to yield

$$\frac{dT}{dr} = \frac{Z_T \dot{m}_F (T_\infty - T_f) \exp(-Z_T \dot{m}_F / r)}{r^2 [1 - \exp(-Z_T \dot{m}_F / r_f)]} \qquad (10.62)$$

and evaluated at $r_f^+$. Performing these substitutions and rearranging, the flame-sheet energy balance is finally expressed as

$$\frac{c_{pg}}{\Delta h_c} \left[ \frac{(T_s - T_f) \exp(-Z_T \dot{m}_F / r_f)}{\exp(-Z_T \dot{m}_F / r_s) - \exp(-Z_T \dot{m}_F / r_f)} - \frac{(T_\infty - T_f) \exp(-Z_T \dot{m}_F / r_f)}{[1 - \exp(-Z_T \dot{m}_F / r_f)]} \right] - 1 = 0. \qquad (10.63)$$

Equation 10.63 is a nonlinear algebraic relation involving the same four unknowns ($\dot{m}_F$, $T_f$, $T_s$, and $r_f$) that appear in Eqn. 10.54.

**Liquid-Vapor Equilibrium**    Thus far in our development, we have four equations and five unknowns. Assuming equilibrium between the liquid and vapor phases of the fuel at the surface, and applying the Clausius–Clapeyron equation, we obtain a fifth equation to provide closure to the problem. Of course, other, more accurate expressions could be used to express this equilibrium; however, the Clausius–Clapeyron approach is quite simple to apply. At the liquid–vapor interface, the partial pressure of the fuel vapor is given by

$$P_{F,s} = A\exp(-B/T_s),\qquad(10.64)$$

where $A$ and $B$ are constants obtained from the Clausius–Clapeyron equation and take on different values for different fuels. The fuel partial pressure can be related to the fuel mole fraction and mass fraction as follows:

$$\chi_{F,s} = P_{F,s}/P\qquad(10.65)$$

and

$$Y_{F,s} = \chi_{F,s}\frac{MW_F}{\chi_{F,s}MW_F + (1-\chi_{F,s})MW_{Pr}}.\qquad(10.66)$$

Substituting Eqns. 10.64 and 10.65 into Eqn. 10.66 yields an explicit relation between $Y_{F,s}$ and $T_s$:

$$Y_{F,s} = \frac{A\exp(-B/T_s)MW_F}{A\exp(-B/T_s)MW_F + [P - A\exp(-B/T_s)]MW_{Pr}}.\qquad(10.67)$$

This completes the mathematical description of our simplified droplet-burning model. It is important to note that if we allow $T_f \rightarrow T_\infty$ and $r_f \rightarrow \infty$, we recover a pure evaporation model, but with coupled heat- and mass-transfer effects, unlike the previous simple models that assume either heat- or mass-transfer effects dominate.

---

**Example 10.2**

Determine the Clausius–Clapeyron constants, $A$ and $B$, which appear in Eqn. 10.64, for $n$-hexane. The 1-atm boiling point for $n$-hexane is 342 K, its enthalpy of vaporization is 334,922 J/kg, and its molecular weight is 86.178.

**Solution**

For an ideal gas, the Clausius–Clapeyron relationship between vapor pressure and temperature is

$$\frac{dP_v}{dT} = \frac{P_v h_{fg}}{RT^2},$$

which can be separated and then integrated as follows:

$$\frac{dP_v}{P_v} = \frac{h_{fg}}{R}\frac{dT}{T^2}$$

and

$$\ln P_v = -\frac{h_{fg}}{R}\frac{1}{T} + C$$

or

$$P_v = \exp(C)\exp\left[\frac{-h_{fg}}{RT}\right].$$

Letting $P_v = 1$ atm and $T = T_{boil}$,

$$\exp(C) = \exp\left[\frac{h_{fg}}{RT_{boil}}\right],$$

which we identify as the constant $A$. By inspection,

$$B = h_{fg}/R.$$

We can now numerically evaluate $A$ and $B$:

$$A = \exp\left[\frac{334,922}{\left(\frac{8315}{86.178}\right)342}\right]$$

$$\boxed{A = 25,580 \text{ atm}}$$

and

$$B = \frac{334,922}{\left(\frac{8315}{86.178}\right)}$$

$$\boxed{B = 3471.2 \text{ K}}$$

**Comments**

For conditions not too far from the normal boiling point, the vapor pressure equation above, with the calculated values of $A$ and $B$, should be a useful approximation.

## Summary and Solution

Table 10.1 summarizes the five equations that must be solved for the five unknowns: $\dot{m}_F$, $r_f$, $T_f$, $T_s$, and $Y_{F,s}$. The system of nonlinear equations can be reduced to a useful level by simultaneously solving Eqns. II, III, IV for $\dot{m}_F$, $r_f$, and $T_f$, treating $T_s$ as a

**Table 10.1**     Summary of droplet burning model

| Equation No. | Unknowns Involved | Underlying Physical Principles |
|---|---|---|
| I  (10.30) | $\dot{m}_F, r_f, Y_{F,s}$ | Fuel species conservation in the inner region |
| II  (10.38) | $\dot{m}_F, r_f$ | Oxidizer species conservation in the outer region |
| III (10.54) | $\dot{m}_F, r_f, T_f, T_s$ | Energy balance at droplet liquid–vapor interface |
| IV (10.63) | $\dot{m}_F, r_f, T_f, T_s$ | Energy balance at flame sheet |
| V (10.67) | $T_s, Y_{F,s}$ | Liquid–vapor phase equilibrium at interface with Clausius–Clapeyron applied |

known parameter for the time being. Following this procedure, the burning rate is

$$\dot{m}_F = \frac{4\pi k_g r_s}{c_{pg}} \ln\left[1 + \frac{\Delta h_c / v + c_{pg}(T_\infty - T_s)}{q_{i-l} + h_{fg}}\right], \qquad (10.68a)$$

or, in terms of the **transfer number, $B_{o,q}$,** defined as

$$B_{o,q} = \frac{\Delta h_c / v + c_{pg}(T_\infty - T_s)}{q_{i-l} + h_{fg}}, \qquad (10.68b)$$

$$\dot{m}_F = \frac{4\pi k_g r_s}{c_{pg}} \ln(1 + B_{o,q}). \qquad (10.68c)$$

The flame temperature is

$$T_f = \frac{q_{i-l} + h_{fg}}{c_{pg}(1 + v)}[vB_{o,q} - 1] + T_s \qquad (10.69)$$

and the flame radius is

$$r_f = r_s \frac{\ln[1 + B_{o,q}]}{\ln[(v+1)/v]}. \qquad (10.70)$$

The fuel mass fraction at the droplet surface is

$$Y_{F,s} = \frac{B_{o,q} - 1/v}{B_{o,q} + 1}. \qquad (10.71)$$

Equations 10.68–10.71 can all be evaluated for an assumed value of $T_s$. Equation V (10.67) can be used to provide an improved value for $T_s$ (shown below as Eqn. 10.72, where $T_s$ has been isolated on the left-hand side) and then Eqns. 10.68–10.71 can be re-evaluated and the process repeated until convergence is obtained.

$$T_s = \frac{-B}{\ln\left[\dfrac{-Y_{F,s}PMW_{Pr}}{A(Y_{F,s}MW_F - Y_{F,s}MW_{Pr} - MW_F)}\right]}. \qquad (10.72)$$

As in our pure evaporation analysis, if we assume the fuel is at the boiling point, the problem is greatly simplified. With this assumption, Eqns. 10.68–10.70 are used to find $\dot{m}_F$, $T_f$, and $r_f$ without iteration, and Eqn. 10.71 is irrelevant since $Y_{F,s}$ is unity when $T_s = T_{boil}$. This is a reasonable assumption when the droplet is burning vigorously after its initial heat-up transient.

## Burning Rate Constant and Droplet Lifetimes

The droplet mass burning rate in terms of the transfer number, $B_{o,q}$, expressed by Eqn. 10.68c, is of identical form to the expression derived for the evaporation rate (see Eqn. 10.13). Thus, without further development we can immediately define the **burning rate constant, $K$,** as

$$K = \frac{8k_g}{\rho_l c_{pg}} \ln(1 + B_{o,q}). \tag{10.73}$$

The burning rate constant is truly a constant only after a steady-state surface temperature is reached, since then $B_{o,q}$ is a constant.

Assuming the transient heat-up period is small in comparison with the droplet lifetime, we recover once again a **$D^2$ law** for droplet burning:

$$D^2(t) = D_0^2 - Kt, \tag{10.74}$$

where the droplet lifetime is found by letting $D^2(t_d) = 0$; i.e.,

$$t_d = D_0^2 / K. \tag{10.75}$$

Figure 10.15 shows that the $D^2$ law is a good representation of experimental results [25] after the heat-up transient.

As with the pure evaporation problem, we need to define appropriate values of the properties $c_{pg}$, $k_g$, and $\rho_l$ that appear in Eqn. 10.73. Law and Williams [13] suggest the following empiricism:

$$c_{pg} = c_{pF}(\bar{T}), \tag{10.76a}$$

$$k_g = 0.4k_F(\bar{T}) + 0.6k_{Ox}(\bar{T}), \tag{10.76b}$$

$$\rho_l = \rho_l(T_s), \tag{10.76c}$$

where

$$\bar{T} = 0.5(T_s + T_f). \tag{10.76d}$$

---

**Example 10.3**

Consider the combustion of an *n*-heptane ($C_7H_{16}$) droplet when its diameter is 100 $\mu$m. Determine (A) the mass burning rate, (B) the flame temperature, and (C) the ratio of the flame radius to the droplet radius for $P = 1$ atm and $T_\infty = 300$ K. Assume quiescent surroundings and that the droplet is at its boiling point.

**Solution**

We will employ Eqns. 10.68, 10.69, and 10.70 to find $\dot{m}_F$, $T_f$, and $r_f/r_s$, respectively. Our first task will be to determine values for the average properties required (Eqn. 10.76). From our knowledge of Chapter 2, we guess that the flame temperature will be about 2200 K; thus,

$$\bar{T} = 0.5(T_s + T_f) = 0.5(371.5 + 2200) \cong 1285 \text{ K},$$

where $T_{boil}(=T_s)$ was obtained from Appendix Table B.1.

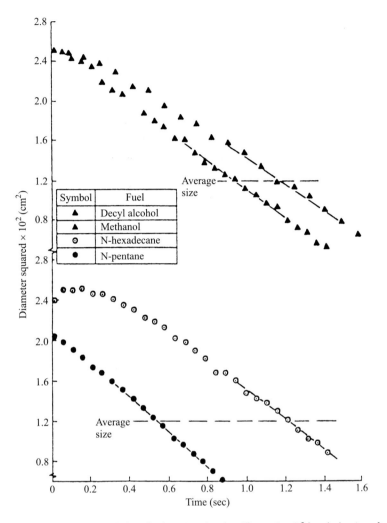

**Figure 10.15**    Experimental data for burning droplets illustrating $D^2$ law behavior after initial transient.

SOURCE: Reprinted from Ref. [25] Reprinted with permission of the American Institute of Aeronautics and Astronautics.

From Appendices B and C, we obtain

$$k_{Ox}(\overline{T}) = 0.081 \ \text{W/m-K} \qquad (\text{Appendix Table C.1})$$

$$k_F(\overline{T}) = k_F(1000 \ \text{K})\left(\frac{\overline{T}}{1000 \ \text{K}}\right)^{1/2}$$

$$= 0.0971\left(\frac{1285}{1000}\right)^{1/2} = 0.110 \ \text{W/m-K} \qquad (\text{Appendix Table B.3}),$$

where we use the $T^{1/2}$-dependence (see Eqn. 3.27) to extrapolate from 1000 K to 1285 K. Thus,

$$k_g = 0.4(0.110) + 0.6(0.081) = 0.0926 \text{ W/m-K}$$

and

$$c_{p,F}(\overline{T}) = 4.22 \text{ kJ/kg-K} \qquad \text{(Appendix Table B.3)},$$
$$h_{fg}(T_{\text{boil}}) = 316 \text{ kJ/kg} \qquad \text{(Appendix Table B.1)},$$
$$\Delta h_c = 44{,}926 \text{ kJ/kg}. \qquad \text{(Appendix Table B.1)}.$$

The stoichiometric air–fuel ratio, $\nu$, is (Eqns. 2.31 and 2.32)

$$\nu = (x + y/4)4.76 \frac{MW_{Ox}}{MW_F}$$

$$= (7 + 16/4)(4.76)\frac{28.85}{100.20} = 15.08.$$

We can now evaluate the transfer number, $B_{o,q}$,

$$B_{o,q} = \frac{\Delta h_c / \nu + c_{pg}(T_\infty - T_s)}{q_{i-l} + h_{fg}}$$

$$= \frac{(44{,}926/15.08) + 4.22(300 - 371.5)}{0 + 316} = 8.473,$$

where droplet heating is ignored ($q_{i-l} = 0$).

A. The mass burning rate is found as follows (Eqn. 10.68):

$$\dot{m}_F = \frac{4\pi k_g r_s}{c_{pg}}\ln(1 + B_{o,q})$$

$$= \frac{4\pi 0.0926\left(\dfrac{100 \cdot 10^{-6}}{2}\right)}{4220}\ln(1 + 8.473)$$

$$\boxed{\dot{m}_F = 3.10 \cdot 10^{-8} \text{ kg/s}}$$

B. To find the flame temperature, we evaluate as follows (Eqn. 10.69):

$$T_f = \frac{q_{i-l} + h_{fg}}{c_{pg}(1 + \nu)}(\nu B_{o,q} - 1) + T_s$$

$$= \frac{0 + 316}{4.22(1 + 15.08)}[15.08(8.473) - 1] + 371.5$$

$$= 590.4 + 371.5$$

$$\boxed{T_f = 961.9 \text{ K}}$$

This value is quite low. See comments below.

C. The dimensionless flame radius can be calculated from (Eqn. 10.70):

$$\frac{r_f}{r_s} = \frac{\ln[1+B_{o,q}]}{\ln[(v+1)/v]} = \frac{\ln[1+8.473]}{\ln(16.08/15.08)}$$

$$\boxed{\frac{r_f}{r_s} = 35}$$

## Comments

We note that the calculated temperature is much less than our guess of 2200 K! Unfortunately, the problem is not with our guess, but, rather, with the simplified theory. The choice of $c_{pg} = c_{pF}(\overline{T})$ does a good job at predicting $\dot{m}_F$; however, the large value ($c_{pF} = 4.22$ kJ/kg-K) is inappropriate for calculating $T_f$. A more reasonable choice would be to use a value that is typical of air (or products). For example, using $c_{pg}(\overline{T}) = c_{p,\,air} = 1.187$ kJ/kg-K, yields $T_f = 2470$ K, a much more reasonable temperature.

Experimental values of dimensionless flame radii (~10) are considerably smaller than that calculated above. Law [17] indicates that fuel-vapor accumulation effects account for the difference. In spite of the shortcomings of the theory, useful estimates of burning rates and droplet lifetimes are obtained.

---

**Example 10.4**

Determine the lifetime of the 100-$\mu$m-diameter $n$-heptane droplet from Example 10.3. How does the result compare with that for pure vaporization (no flame) with $T_\infty = 2200$ K?

## Solution

We first evaluate the burning rate constant using a liquid density of 684 kg/m³ (Appendix Table B.1):

$$K = \frac{8k_g}{\rho_l c_{pg}} \ln(1 + B_{o,q})$$

$$= \frac{8(0.0926)}{684(4220)} \ln(1 + 8.473) = 5.77 \cdot 10^{-7}\ \text{m}^2/\text{s}.$$

The lifetime of the burning droplet is then (Eqn. 10.75)

$$t_d = D_0^2/K = (100 \cdot 10^{-6})^2/5.77 \cdot 10^{-7}$$

$$\boxed{t_d = 0.0173\ \text{s}}$$

For the pure evaporation problem, we use the same expression to evaluate $K$ and $t_d$; however, the transfer number is now expressed (Eqn. 10.12):

$$B = \frac{c_{pg}(T_\infty - T_{\text{boil}})}{h_{fg}}$$

$$= \frac{4.220(2200 - 371.5)}{316} = 24.42.$$

The evaporation rate constant is then $K = 8.30 \cdot 10^{-7}$ m²/s and the droplet lifetime is

$$\boxed{t_d = 0.0120\ \text{s}}$$

**Comments**

We expect that the droplet lifetime should be longer for the pure evaporation case when $T_f = T_\infty$; however, the "theoretical" flame temperature of 961.9 K (Example 10.2) is much less than the 2200 K ambient. Using the pure evaporation model with $T_\infty = 961.9$ K, yields a predicted droplet lifetime of 0.0178 s, which is slightly longer than that for the burning droplet, as expected.

## Extension to Convective Environments

To achieve the conditions of spherically symmetric burning in a stagnant medium, as assumed in our analysis above, requires no relative velocity between the droplet and the free stream, and no buoyancy. This latter condition can only be achieved in the absence of gravity or in a weightless freefall. Interest in buoyancy-free combustion has a long history (e.g., Refs. [26–28]), while a renewed interest has been shown with the advent of the space shuttle and a permanent space station in low-Earth orbit [21, 29–32].

There are several approaches to incorporating convection in the droplet-burning problem [15–19]. The approach taken here is that of chemical engineering "film theory." This approach is straightforward and in keeping with our desire for simplicity.

The essence of film theory is the replacement of the heat- and mass-transfer boundary conditions at infinity with the same conditions moved inward to the so-called film radius, $\delta_M$ for species and $\delta_T$ for energy. Figure 10.16 illustrates how the film radius steepens concentration and temperature gradients, and, hence, increases

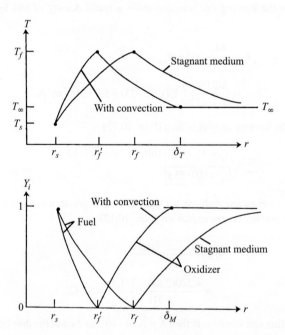

**Figure 10.16**   Comparison of temperature and species profiles with and without convection. Theoretical film thicknesses are indicated by $\delta_T$ (temperature) and $\delta_M$ (species).

mass- and heat-transfer rates at the droplet surface. This, of course, means that convection enhances droplet burning rates and, as a result, burning times decrease.

The film radii are defined in terms of the **Nusselt number, $Nu$,** for heat transfer, and the **Sherwood number, $Sh$,** for mass transfer. Physically, the Nusselt number is the dimensionless temperature gradient at the droplet surface, and the Sherwood number is the dimensionless concentration (mass-fraction) gradient at the surface. Formally, the film radii are given by

$$\frac{\delta_T}{r_s} = \frac{Nu}{Nu-2} \qquad (10.77a)$$

$$\frac{\delta_M}{r_s} = \frac{Sh}{Sh-2}. \qquad (10.77b)$$

Note that for a stagnant medium, $Nu = 2$; thus, we recover $\delta_T \rightarrow \infty$ in the absence of convection. Consistent with the unity Lewis number assumption, we assume $Sh = Nu$. For droplet burning with forced convection, Faeth [16] recommends the following correlation to evaluate $Nu$:

$$Nu = 2 + \frac{0.555\,Re^{1/2}Pr^{1/3}}{[1+1.232/(RePr^{4/3})]^{1/2}}, \qquad (10.78)$$

where the Reynolds number, $Re$, is based on the droplet diameter and the relative velocity. For simplicity, thermophysical properties can be evaluated at the mean temperature (Eqn. 10.76d).

In terms of the basic conservation principles, convection affects the species conservation relations in the outer region (oxidizer distribution, Eqns. 10.37 and 10.38) and the energy conservation relations that involve the outer region ($T$-distribution in the outer region, Eqn. 10.44, and the energy balance at the flame sheet, Eqn. 10.63).

Application of the film-theory boundary condition for species conservation,

$$Y_{Ox}(\delta_M) = 1, \qquad (10.79)$$

to Eqn. 10.37 yields

$$\frac{\exp[-Z_M \dot{m}_F/[r_s Nu/(Nu-2)]]}{\exp(-Z_M \dot{m}_F/r_f)} - \frac{v+1}{v} = 0. \qquad (10.80)$$

Equation 10.80 (with convection) is equivalent to Eqn. 10.38 (without convection).

Application of the film-theory boundary condition for energy conservation,

$$T(\delta_T) = T_\infty, \qquad (10.81)$$

to Eqn. 10.40 yields the outer-zone temperature distribution:

$$\begin{aligned} T(r) = &[(T_\infty - T_f)\exp(-Z_T\dot{m}_F/r) + T_\infty\exp(-Z_T\dot{m}_F/r_f) \\ &- T_f\exp(-Z_T\dot{m}_F(Nu-2)/r_s Nu)]/[\exp(-Z_T\dot{m}_F/r_f) \\ &- \exp(-Z_T\dot{m}_F(Nu-2)/r_s Nu)]. \end{aligned} \qquad (10.82)$$

Using Eqn. 10.82 to evaluate the energy balance at the flame (Eqn. 10.61) yields

$$\frac{c_{pg}}{\Delta h_c}\left[\frac{(T_f - T_s)\exp(-Z_T \dot{m}_F / r_f)}{\exp(-Z_T \dot{m}_F / r_f) - \exp(-Z_T \dot{m}_F / r_s)}\right.$$ 
(10.83)

$$\left. - \frac{(T_f - T_\infty)\exp(-Z_T \dot{m}_F / r_f)}{\exp(-Z_T \dot{m}_F / r_f) - \exp(-Z_T \dot{m}_F (Nu-2)/r_s Nu)}\right] - 1 = 0,$$

which is the film-theory equivalent to Eqn. 10.54 for the stagnant case.

We now, once again, have a system of five nonlinear algebraic equations involving the five unknowns $\dot{m}_F$, $T_f$, $r_f$, $Y_{F,s}$, and $T_s$ (see Table 10.1). Simultaneously solving for three of these—$\dot{m}_F$, $T_f$, and $r_f$—gives the burning rate:

$$\dot{m}_F = \frac{2\pi k_g r_s Nu}{c_{pg}}\ln\left[1 + \frac{\Delta h_c / \nu + c_{pg}(T_\infty - T_s)}{q_{i-l} + h_{fg}}\right]$$
(10.84a)

or

$$\dot{m}_F = \frac{2\pi k_g r_s Nu}{c_{pg}}\ln(1 + B_{o,q}),$$
(10.84b)

where the transfer number has been introduced (see Eqn. 10.68b). Note that this expression (Eqn. 10.84b) differs from that for the stagnant case (Eqn. 10.68c) only by the explicit appearance of the Nusselt number. In the stagnant flow case, $Nu = 2$, and the two expressions are identical, as expected.

## ADDITIONAL FACTORS

There are many complicating features of real burning droplets that are ignored in the simplified theory presented above. It is beyond the scope of this book to discuss any of the topics in detail; instead, we merely list some of them and cite a few references to provide starting points for further study.

In the simplified model, all properties are treated as constants, and a judicious selection of mean values is used to provide agreement with experiment. In reality, many of the properties possess strong temperature and/or composition dependencies. Various approaches have been taken to include **variable properties,** with, perhaps, the most comprehensive treatment found in Refs. [33] and [34]. Reference [14] compares various simplified approaches with comprehensive numerical results. Also related to the issue of variable properties is the correct formulation of the conservation equations when the ambient temperature and/or pressure exceeds the thermodynamic critical point of the evaporating liquid. This is an important issue for modeling burning droplets in both diesel and rocket engines. More information on **supercritical droplet combustion and evaporation** can be found in Refs. [35] and [36].

The unrealistically large flame stand-off distances predicted by the $D^2$ law (see Example 10.3) are shown by Law *et al.* [37] to be a result of neglecting the unsteady effect of **fuel-vapor accumulation** between the droplet surface and the flame. Inclusion of this effect in a modified $D^2$-law model results in capturing the flame movement observed in experiments [17, 37].

More sophisticated models of **droplet heating** have also been developed that take into account the time-varying temperature field within the droplet [38]. Proper treatment of the liquid phase is important in the evaporation and combustion of **multicomponent fuel** droplets [39, 40]. Related to this issue is the shear-driven fluid motion within droplets in a convective environment, i.e., **internal recirculation** [19, 40].

More recent approaches to single droplet gasification include numerical models that treat the problem in its appropriate axisymmetric coordinate system. This approach incorporates the effects of convection directly, rather than as an ad-hoc modification to assumed spherical symmetry [40]. Recent models [41] also take into account the detailed chemical kinetics associated with envelope flames, abandoning the simplified flame-sheet or one-step chemistry approximations.

An interesting, and somewhat ironic, result of experiments conducted at near-zero gravity is the observation of the formation of a **soot shell** between the droplet surface and the flame [31]. In normal gravity, buoyancy-induced flow convects the soot away from the droplet. In zero gravity, however, there is no buoyant flow to remove the soot and, hence, a soot shell forms. This shell greatly alters the combustion process and causes radiation effects to play important roles in burning and extinction [30, 32].

Finally, we come to the issue of **interactions among multiple droplets,** as would be present in a fuel spray. Spray evaporation and combustion is of great practical importance and has a huge literature associated with it. Useful starting points for further study of sprays are Refs. [18, 24, 42]. A primary issue in spray combustion is the coupling of the variable ambient conditions, to which a droplet is exposed during its lifetime, and the droplet evaporation rate.

## ONE-DIMENSIONAL VAPORIZATION-CONTROLLED COMBUSTION

In this section, we will apply the concepts developed above to the analysis of a simple, one-dimensional, steady-flow, liquid-fuel combustor. This section stands alone in that it is a large-scale example of how droplet evaporation and equilibrium concepts (Chapter 2) can be combined to model spray combustors, such as are used in gas-turbine and rocket engines (see Figs. 10.4a and 10.6), and brings together the previously developed theory and applications, albeit in a simplified fashion. Since the primary objective of this section is to provide a framework from which various design problems or projects can be developed, no new fundamental combustion concepts are introduced here; therefore, depending on the reader's objectives, this section can be skipped without jeopardizing understanding of subsequent chapters.

## Physical Model

Figure 10.17a schematically illustrates a simple combustor with a constant cross-section area. Fuel droplets, uniformly distributed across the combustor, evaporate as they move downstream in an oxidizer stream. The fuel vapor is assumed to mix with the gas phase and burn instantaneously. This causes the gas temperature to rise, speeding the vaporization of the droplets. The gas velocity increases because droplet vaporization, and, possibly, secondary oxidizer addition, add mass to the gas phase, and because combustion decreases the gas density.

Obviously, the model discussed above neglects many detailed phenomena occurring in a real combustor; however, for flows without recirculation and back-mixing, or downstream of such regions, the model is a useful first approximation when mixing

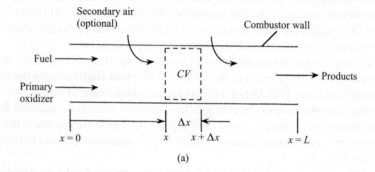

(a)

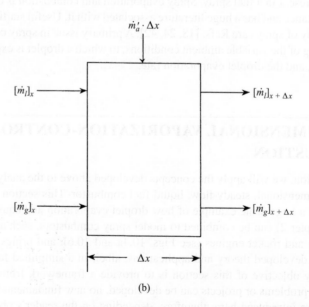

(b)

**Figure 10.17**     (a) Overall schematic of one-dimensional combustor with (b) details of control-volume analysis for overall mass conservation.

and burning rates are substantially faster than fuel vaporization rates. Priem and Heidmann [43] and Dipprey [9] have applied one-dimensional, vaporization-controlled models to liquid-rocket engine combustion, and Turns and Faeth [44] have used the approach to model slurry-fuel combustion in a gas-turbine combustor. In the following sections, we develop the analysis of a one-dimensional combustor that can be easily implemented in a computer code.

## Assumptions

The following assumptions will be embodied in our simple analysis of a liquid-fueled combustor:

1. The system comprises only two phases: a gas phase, consisting of combustion products; and a liquid, single-component fuel phase.

2. The properties of the gas phase and liquid phase depend only on the coordinate in the flow direction, i.e., the flow is one-dimensional. This implies that the gas phase has uniform properties in the radial direction perpendicular to the flow at any axial position, $x$. Furthermore, we neglect diffusion, even though axial concentration gradients exist.

3. The effects of friction and velocity changes on pressure are neglected. This implies the pressure is constant, i.e., $dP/dx = 0$, and simplifies the vaporization problem.

4. The fuel is introduced as a stream of monodisperse droplets, i.e., all droplets have the same diameter and velocity at any axial location.

5. The droplets in the stream all obey the droplet evaporation theory presented earlier in Chapter 10, with the droplet temperature assumed fixed at the boiling point.

6. The gas-phase properties are determined by equilibrium, or alternatively, by the even simpler no-dissociation model with water-gas equilibrium (see Chapter 2).

## Mathematical Problem Statement

Given a set of initial conditions for the gas phase and the liquid phase, we wish to determine the following functions:

| | |
|---|---|
| *Gas phase:* | $T_g(x)$, temperature; |
| | $\dot{m}_g(x)$, mass flowrate; |
| | $\Phi_g(x)$, equivalence ratio; |
| | $v_g(x)$, velocity; |
| *Liquid phase:* | $D(x)$, droplet diameter; |
| | $\dot{m}_l(x)$, fuel evaporation rate; |
| | $v_d(x)$, droplet velocity. |

In the section below, we apply the basic conservation principles (mass, momentum, and energy conservation) to determine the desired functions listed above.

## Analysis

For this problem, we choose a steady-state, steady-flow, control-volume analysis. The selected control volume extends across the combustor and is $\Delta x$ in length (Fig. 10.17a).

**Mass Conservation**    Figure 10.17b shows the mass flows of both the liquid and gas phases into and out of the control volume. Since there is no change in mass within the control volume, the total mass flows entering and exiting the control volume are equal:

$$[\dot{m}_l]_x + [\dot{m}_g]_x + \dot{m}_a' \cdot \Delta x = [\dot{m}_l]_{x+\Delta x} + [\dot{m}_g]_{x+\Delta x},$$

where $\dot{m}_l$ and $\dot{m}_g$ are the mass flowrates (kg/s) of the liquid and gas phases, respectively, and $\dot{m}_a'$ is the mass flow of secondary oxidizer per unit length (kg/s-m) entering the control volume. We assume that $\dot{m}_a'$ is a given function of $x$.

Rearranging the above, we obtain

$$\frac{[\dot{m}_l]_{x+\Delta x} - [\dot{m}_l]_x}{\Delta x} + \frac{[\dot{m}_g]_{x+\Delta x} - [\dot{m}_g]_x}{\Delta x} = \dot{m}_a'.$$

Taking the limit of $\Delta x \to 0$ and recognizing the definition of a derivative, we obtain the following governing equation for conservation of total mass:

$$\frac{d\dot{m}_l}{dx} + \frac{d\dot{m}_g}{dx} = \dot{m}_a'. \tag{10.85}$$

Equation 10.85 can be integrated to provide an expression for $\dot{m}_g(x)$ :

$$\dot{m}_g(x) = \dot{m}_g(0) + \dot{m}_l(0) - \dot{m}_l(x) + \int_0^x \dot{m}_a' dx. \tag{10.86}$$

We next focus solely on the liquid phase; thus, the control volume shown in Fig. 10.18a contains only liquid. The $\dot{m}_{lg}'$ term is the mass flowrate of fuel per unit length going from the liquid phase into the gas phase, i.e., the fuel vaporization rate. Thus, the rate at which liquid fuel exits the control volume at $x + \Delta x$ is less than that entering at $x$ by the amount $\dot{m}_{lg}' \Delta x$, i.e.,

$$[\dot{m}_l]_x - [\dot{m}_l]_{x+\Delta x} = \dot{m}_{lg}' \cdot \Delta x.$$

Rearranging, and taking the limit $\Delta x \to 0$, yields

$$\frac{d\dot{m}_l}{dx} = -\dot{m}_{lg}'.$$

The flowrate of liquid through the combustor can be related to the number of fuel droplets entering the chamber per unit time, $\dot{N}$, and the mass of an individual droplet, $m_d$. Thus,

$$\dot{m}_l = \dot{N} m_d \tag{10.87}$$

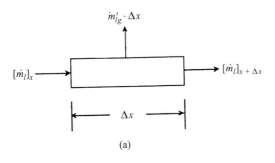

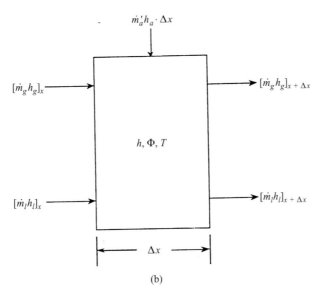

**Figure 10.18** Control-volume analysis for (a) fuel mass conservation and (b) overall energy conservation.

or

$$\dot{m}_l = \dot{N}\rho_l \pi D^3 / 6, \tag{10.88}$$

where $\rho_l$ and $D$ are the droplet density and diameter, respectively.

Differentiating Eqn. 10.88, we obtain

$$\frac{d\dot{m}_l}{dx} = (\pi/4)\dot{N}\rho_l D \frac{dD^2}{dx}. \tag{10.89}$$

The derivative of $D^2$ with respect to $x$ can be related to the time derivative through the droplet velocity $v_d$, i.e., $dx = v_d dt$; thus,

$$\frac{dD^2}{dx} = \frac{1}{v_d} \frac{dD^2}{dt}. \tag{10.90}$$

In an earlier section, we derived an expression for $dD^2/dt$. Substituting Eqn. 10.17 into Eqn. 10.90 yields

$$\frac{dD^2}{dx} = -K/v_d, \tag{10.91}$$

where $K$ is the evaporation coefficient (see Eqn. 10.18).

The number of droplets entering the combustor per unit time is easily related to the initial fuel flowrate and the assumed initial droplet size, $D_0$, i.e.,

$$\dot{m}_{l,0} = \dot{N}\rho_l \pi D_0^3 /6 \tag{10.92a}$$

or

$$\dot{N} = 6\dot{m}_{l,0}/\left(\pi\rho_l D_0^3\right). \tag{10.92b}$$

The above mass conservation analysis provides a single ordinary differential equation, Eqn. 10.91, which, when solved, yields $D^2(x)$ or $D(x)$. When the droplet diameter is known, $\dot{m}_g$ can be calculated using Eqn. 10.86 and appropriate ancillary relations.

Finding the droplet velocity as a function of axial distance, $v_d(x)$, requires application of conservation of momentum to a droplet, which will be discussed subsequently. The velocity of the gas phase can be written as

$$v_g = \dot{m}_g/\rho_g A, \tag{10.93}$$

where $\rho_g$ is the gas-phase density, and $A$ is the cross-sectional area of the combustor. The density can be related to the gas temperature and pressure through the ideal-gas law:

$$\rho_g = P/R_g T_g, \tag{10.94}$$

where

$$R_g = R_u/MW_g. \tag{10.95}$$

Gas-phase energy conservation is required to find $T_g$, and since the composition of the gas phase is continually varying as we move downstream, the molecular weight, $MW_g$, is a variable as well. Substituting Eqns. 10.94 and 10.95 into Eqn. 10.93 yields

$$v_g = \dot{m}_g R_u T_g /(MW_g PA). \tag{10.96}$$

**Gas-Phase Energy Conservation**   With reference to Fig. 10.18b, and following the same procedures used to derive the mass conservation relationships, conservation of energy for the control volume is expressed as

$$\frac{d(\dot{m}_g h_g)}{dx} + \frac{d(\dot{m}_l h_l)}{dx} = \dot{m}_a' h_a, \tag{10.97}$$

where we assume no heat or work interactions. Expanding the derivatives of the products and recognizing that the enthalpy of the liquid is a constant, Eqn. 10.97 becomes, after rearrangement,

$$\frac{dh_g}{dx} = \left[ \dot{m}_a' h_a - h_g \frac{d\dot{m}_g}{dx} - h_l \frac{d\dot{m}_l}{dx} \right] \Bigg/ \dot{m}_g. \tag{10.98}$$

Since we are interested in finding the temperature distribution, $T(x)$, the enthalpy can be related to the temperature and the other thermodynamic variables:

$$h_g = f(T_g, P_g, \Phi_g), \tag{10.99}$$

where the relationships among these variables are determined by equilibrium. Applying the chain rule, recognizing that $P_g$ is constant, $dh_g/dx$ can be expressed as

$$\frac{dh_g}{dx} = \frac{\partial h_g}{\partial T} \frac{dT}{dx} + \frac{\partial h_g}{\partial \Phi} \frac{d\Phi}{dx}, \tag{10.100}$$

where the subscript $g$ has been dropped from $T$ and $\Phi$.

Equating the right-hand sides of Eqns. 10.98 and 10.100, and solving for $dT/dx$, yields

$$\frac{dT}{dx} = \left[ \left( \dot{m}_a' h_a - h_g \frac{d\dot{m}_g}{dx} - h_l \frac{d\dot{m}_l}{dx} \right) \Bigg/ \dot{m}_g - \frac{\partial h_g}{\partial \Phi} \frac{d\Phi}{dx} \right] \Bigg/ \frac{\partial h_g}{\partial T}. \tag{10.101}$$

The gas enthalpy, $h_g$, and its partial derivatives, $\partial h_g/\partial T$ and $\partial h_g/\partial \Phi$, can all be calculated knowing the equilibrium state at $T$, $P$, and $\Phi$. The Olikara and Borman [45] equilibrium code (see Appendix F) can be employed to perform these calculations. Equation 10.101 can be simplified further by using Eqn. 10.85 to eliminate $d\dot{m}_g/dx$ i.e.,

$$\frac{dT}{dx} = \left\{ \left[ (h_a - h_g) \dot{m}_a' + (h_g - h_l) \frac{d\dot{m}_l}{dx} \right] \Bigg/ \dot{m}_g - \frac{\partial h_g}{\partial \Phi} \frac{d\Phi}{dx} \right\} \Bigg/ \frac{\partial h_g}{\partial T}. \tag{10.102}$$

To maintain perspective on where we have been and where we are going, a brief summary is helpful at this point in our analysis. So far, we have derived first-order ordinary differential equations giving expressions for $dD^2/dx$ and $dT_g/dx$. These equations can be integrated (numerically) to yield $D(x)$ and $T_g(x)$ for given initial conditions. It remains for us to determine expressions for $\Phi$ and $d\Phi/dx$, and to derive one more differential equation that can be integrated to find the droplet velocity, $v_d(x)$.

**Gas-Phase Composition**     Our goal here is to determine the axial distribution of the equivalence ratio, $\Phi(x)$. This calculation is just a subset of mass conservation, since all that is required is a knowledge of the mass of the fuel and oxidizer at any axial location.

Assume that at $x = 0$, the chamber entrance, there is an initial gas-phase flow. This flow may be a burned or unburned mixture of oxidizer and fuel (or pure oxidizer or pure fuel), and can be expressed as

$$\dot{m}_g(0) = \dot{m}_F(0) + \dot{m}_a(0). \tag{10.103}$$

The fuel–oxidizer ratio at an arbitrary position, $x$, downstream, is just the ratio of the mass flow of the gas phase that had its origin as fuel to the mass flow of the gas phase that had its origin as oxidizer:

$$(F/O)_x = \frac{\dot{m}_{F,x}}{\dot{m}_{a,x}} = \frac{\dot{m}_g(x) - \dot{m}_{a,0} - \int_0^x \dot{m}_a' \, dx}{\dot{m}_{a,0} + \int_0^x \dot{m}_a' \, dx}. \tag{10.104a}$$

Equation 10.104a can be rearranged to yield

$$(F/O)_x = \dot{m}_g \left[ \dot{m}_{a,0} + \int_0^x \dot{m}_a' dx \right]^{-1} - 1, \tag{10.104b}$$

which, when differentiated with respect to $x$, becomes

$$\frac{d(F/O)_x}{dx} = \frac{d\dot{m}_g}{dx} \left[ \dot{m}_{a,0} + \int_0^x \dot{m}_a' dx \right]^{-1} - \dot{m}_g \dot{m}_a' \left[ \dot{m}_{a,0} + \int_0^x \dot{m}_a' dx \right]^{-2}. \tag{10.105}$$

The equivalence ratio and its derivative are then easily obtained by definition:

$$\Phi(x) \equiv (F/O)_x / (F/O)_{\Phi=1} \tag{10.106}$$

and

$$\frac{d\Phi(x)}{dx} = \frac{1}{(F/O)_{\Phi=1}} \frac{d(F/O)_x}{dx}, \tag{10.107}$$

where we assume that the fuel entering the chamber in the gas phase ($\dot{m}_F(0)$) has the same hydrogen-to-carbon ratio as the injected fuel ($\dot{m}_l(0)$). If this is not true, then $(F/O)_{\Phi=1}$ varies through the chamber, and this variation would have to be taken into account.

**Droplet Momentum Conservation**    Fuel droplets initially injected at high velocities into the lower-velocity gas stream will slow down because of drag. As the fuel evaporates and burns, the velocity of the gas increases. This may cause the droplets to decelerate at a lesser rate, or accelerate, depending on the relative velocity between the gas and the droplet. The relative velocity also affects the evaporation rate. We assume that drag is the only force acting on the particle. This force will act in

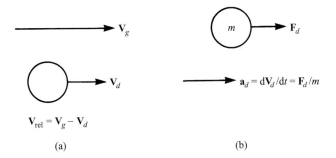

**Figure 10.19**     (a) Definition of relative velocity for a droplet in a gas stream and (b) Newton's second law applied to a droplet.

the same direction as $\mathbf{V}_{rel}$. Figure 10.19a defines the relative velocity, $\mathbf{V}_{rel}$. Applying Newton's second law ($\mathbf{F} = m\mathbf{a}$) to the droplet yields

$$F_d = m_d \frac{dv_d}{dt}. \tag{10.108}$$

Transforming from the time to space domain via $dx = v_d\,dt$, we obtain

$$F_d = m_d v_d \frac{dv_d}{dx}. \tag{10.109}$$

The drag force can be obtained by using an appropriate drag coefficient correlation [46], i.e.,

$$C_D = f(Re_{D,\,rel}) \approx \frac{24}{Re_{D,rel}} + \frac{6}{1 + \sqrt{Re_{D,\,rel}}} + 0.4 \tag{10.110}$$

for $0 \le Re_{D,\,rel} \le 2 \cdot 10^5$, where

$$C_D = \frac{F_d/(\pi D^2/4)}{\rho_g v_{rel}^2/2}. \tag{10.111}$$

Substituting Eqn. 10.111 into 10.109 and rearranging gives

$$\frac{dv_d}{dx} = \frac{3C_D \rho_g v_{rel}^2}{4\rho_l D v_d}, \tag{10.112}$$

where the sign of $dv_d/dx$ is the same as $v_{rel}$, or

$$\frac{dv_d}{dx} = \frac{3C_D \rho_g (v_g - v_d)|v_g - v_d|}{4\rho_l D v_d}. \tag{10.113}$$

## Model Summary

In the preceding, we developed a system of ordinary differential equations and ancillary algebraic relations that describe the operation of a vaporization-controlled, one-dimensional combustor. Mathematically, the problem is an initial-value problem, with the temperature and flowrates specified at the inlet ($x = 0$). To make the solution procedure clear, the governing equations and their initial conditions are summarized below.

*Liquid mass (droplet diameter)*

$$\frac{dD^2}{dx} = f_1, \tag{10.91}$$

with

$$D(0) = D_0.$$

*Gas-phase mass*

$$\dot{m}_g(x) = \dot{m}_g(0) + \dot{m}_l(0) - \dot{m}_l(x) + \int_0^x \dot{m}_a dx, \tag{10.86}$$

where

$$\dot{m}_g(0) = \dot{m}_{F,0} + \dot{m}_{a,0}.$$

*Gas-phase energy*

$$\frac{dT_g}{dx} = f_2, \tag{10.102}$$

with

$$T_g(0) = T_{g,0}.$$

*Droplet momentum (velocity)*

$$\frac{dv_d}{dx} = f_3, \tag{10.113}$$

with

$$v_d(0) = v_{d,0}.$$

In addition to the above, there are several other purely algebraic relationships necessary to provide closure, together with the relationships expressing the complex equilibrium for the C–H–O–N system.

To solve the problem, the above governing equations can be integrated numerically using one of the many readily available subroutines [47, 48]. To calculate the equilibrium properties of the gas phase, the methods presented in Chapter 2 can be employed to solve the problem, or an existing program or subroutine, e.g. Ref. [45], can be adopted for this purpose. The Olikara and Borman code [45] is convenient to use since it routinely calculates the various useful partial derivatives $((\partial h/\partial T),$ $(\partial h/\partial \Phi),$ etc.) in its determination of the equilibrium composition.

---

**Example 10.5**

Consider the liquid-fueled rocket engine illustrated in Fig. 10.20, where a cylindrical combustion chamber is attached to a converging–diverging supersonic nozzle. A portion of the fuel is used to cool the nozzle and, hence, is vaporized; thus, both gaseous and liquid fuel enter the combustion chamber along with the gaseous oxidizer. The liquid fuel is atomized into small droplets at the injector plate (head end of combustion chamber at $x = 0$). Using the parameters shown below, apply the one-dimensional analysis developed above to explore the effects of initial droplet diameter, $D_0$, on the temperature distribution, $T(x)$, and the gas-phase equivalence ratio distribution, $\Phi(x)$, along the axis of the combustion chamber. Also, show the droplet diameter history and the droplet- and gas-phase velocities.

| | |
|---|---|
| Combustor cross-sectional area | 0.157 m², |
| Combustor length | 0.725 m, |
| Total fuel injection area | 0.0157 m², |
| | |
| Fuel | $n$-heptane ($C_7H_{16}$), |
| Overall equivalence ratio, $\Phi_{OA}$ | 2.3, |
| Premixed equivalence ratio, $\Phi(0)$ | 0.45, |
| | |
| Initial temperature, $T(0)$ | 801 K, |
| Combustion chamber pressure | 3.4474 MPa, |
| Initial droplet velocity, $v_d(0)$ | 10 m/s. |

**Solution**

The mathematical one-dimensional combustor model, as summarized above (Eqns. 10.91, 10.86, 10.102, and 10.113), was coded in Fortran. A main program calculated the initial conditions from the given input parameters and carried out the integration of the governing ordinary differential equations by calling the IMSL routine DGEAR [47]. The equilibrium properties and the thermodynamic partial derivatives were obtained using a version of the Olikara and Borman routine [45], modified to deal with a nitrogen-free oxidizer. Curvefit properties

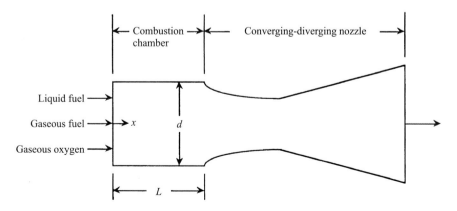

**Figure 10.20** Schematic diagram of liquid-fueled rocket engine with cylindrical combustion chamber.

($k$, $\mu$, and $c_p$) for $n$-heptane vapor were taken from Appendix B, and the heptane heat of vaporization was estimated from vapor-pressure data [50] using the Clausius–Clapeyron relation. For simplicity, the gas-phase transport properties were assumed to be those of pure oxygen and obtained from curvefits [49]. Convective enhancement of evaporation was ignored.

Results from exercising this model for five different initial droplet diameters, ranging from 30 $\mu$m to 200 $\mu$m, are shown in Figs. 10.21 and 10.22. Here we see that, for the two

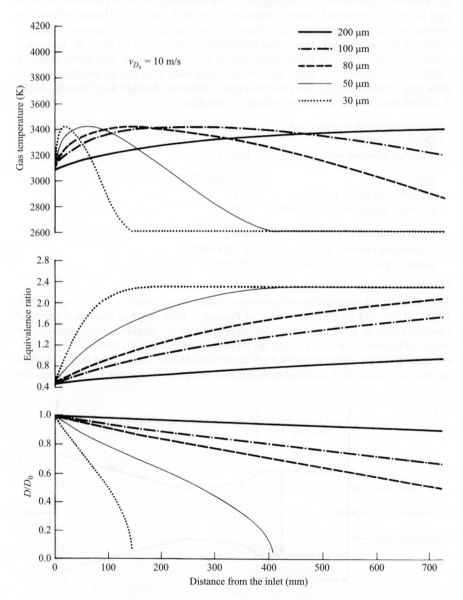

**Figure 10.21**   The effects of initial droplet size on gas temperature, equivalence ratio, and droplet size. The initial droplet velocity is 10 m/s.

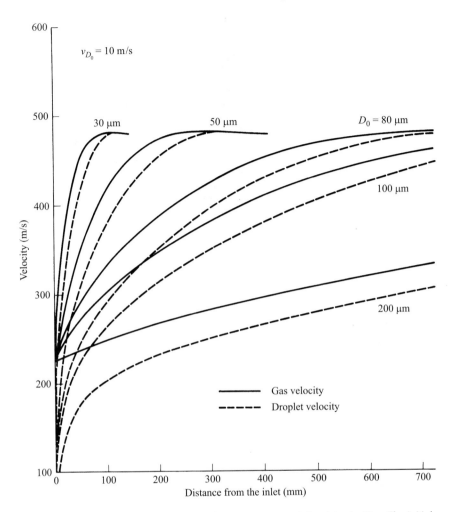

**Figure 10.22**    The effects of initial droplet size on gas and droplet velocities. The initial droplet velocity is 10 m/s.

smallest initial droplet sizes ($D_0 = 30$ and $50$ $\mu$m), combustion is complete within the length of the combustion chamber, i.e., $D/D_0$ becomes zero at a distance less than $L$ (=725 mm). The larger droplets, however, exit the chamber after being partially vaporized. To achieve complete combustion within the combustion chamber, we see from Fig. 10.21 that the initial drop size must be somewhat less than 80 $\mu$m. Since the combustor operates rich overall, the gas-phase equivalence ratio, $\Phi_g$, monotonically increases from the initial value based on the entering fuel vapor. Peak temperatures of approximately 3400 K occur near $\Phi_g = 1$ (see Chapter 2). From Fig. 10.21, we see that for the smallest droplets (30 $\mu$m), temperatures are highest near the injector face, whereas for the largest droplets (200 $\mu$m), the peak temperature is not quite reached before the exit of the combustor. In practice, excessive heat transfer from the combustion gases can damage or destroy the injector assembly. We see then, that an optimum initial droplet size exists such that combustion can be completed within the combustion chamber, while maintaining the highest-temperature zone as far downstream from the injector face as possible. Figure 10.22 shows the tremendous increase in gas velocity associated with the vaporization and combustion of the fuel droplets. With the assumption that the initial droplet velocity is 10 m/s, we see that the droplet is always moving more slowly than the gas and, hence, is continually being accelerated. We also see that, with such relatively large slip velocities, neglecting convective enhancement of vaporization is not justified.

**Comments**

This example clearly shows that even a simple model can be used to gain insight into complex situations. Since parametric investigations are easily conducted, such models can be used to provide guidance in both engine design and development efforts. One must be cautious, however, in that a model is only as good as its incorporated physics, which is always incomplete. Nevertheless, computer codes, much more sophisticated than those employed here, are routinely used in the design and development of combustion systems.

## SUMMARY

In this chapter, we first saw how the operation of many practical devices depends on the evaporation and/or combustion of fuel droplets. To illuminate these processes, several mathematical models were developed. The first of these was a simple heat-transfer-controlled model of droplet vaporization in a hot gas. This model is analogous to the mass-transfer-controlled evaporation model introduced in Chapter 3. With the assumption that the droplet surface is at the boiling point, the problem is uncoupled from mass-transfer considerations and very simple and easy-to-use expressions result for calculating evaporation rates and droplet lifetimes. The second, more general, analysis models droplet gasification with a spherically symmetric flame surrounding the droplet. The flame is assumed to be an infinitesimally thin sheet where the fuel vapor from the evaporating droplet reacts instantaneously with the inward-diffusing oxidizer. This simple model allows prediction of the droplet burning rate, flame radius, droplet surface and flame temperatures, and fuel-vapor concentration at the droplet surface. You should be thoroughly familiar with the conservation principles involved and how they apply to this particular problem. In particular, you should have a physical understanding of the various mass-flux relationships. In the burning-droplet model, we have coupled the heat-transfer and mass-transfer problems, unlike the isolated mass-transfer analysis (Stefan problem) of Chapter 3 and the heat-transfer (evaporating-droplet) analyses presented in this chapter. Of course, if one assumes that the droplet surface is at the boiling point, we return to the heat-transfer-dominated problem. You should be able to apply the Clausius–Clapeyron relationship relating vapor pressure and temperature to determine the fuel-vapor concentration at the droplet surface. Applying the results of the simple gas-phase theory to estimate droplet lifetimes, we recovered the $D^2$ law for droplet combustion. Lumped-parameter and onion-skin models treating transient droplet heat-up were also developed. Using film theory, we extended the analysis to a droplet burning in a convective environment and saw that convection enhances burning rates and, thus, shortens droplet lifetimes. The chapter also contains a brief discussion of the many complicating factors, of which you should be aware, that are ignored in the simple model.

The chapter concludes with the development of a simple, but general, one-dimensional model of a combustion chamber in which the combustion rate is governed by the vaporization rate of the fuel droplets injected into the chamber. Applications of this model include liquid-rocket engine and gas-turbine combustors. You saw how a steady-flow, control-volume approach could be coupled with droplet-evaporation theory and chemical equilibrium (Chapter 2). The control-volume analyses (mass, energy, and droplet momentum) are sufficiently simple that you should be able to develop similar analyses on your own. The 1-D model was applied to the problem of a liquid-hydrocarbon-fueled rocket engine, and the effect of initial droplet size on important chamber design parameters were explored. With appropriate computer codes, you should be able to conduct similar design analyses.

# NOMENCLATURE

| | |
|---|---|
| $A$ | Clausius–Clapeyron constant, Eqn. 10.64 (Pa or atm), or combustor flow area ($m^2$) |
| $B$ | Clausius–Clapeyron constant, Eqn. 10.64 (K) |
| $B_q, B_{o,q}$ | Transfer, or Spalding, number |
| $c_p$ | Specific heat (J/kg-K) |
| $C_D$ | Drag coefficient |
| $C_1, C_2$ | Integration constants |
| $D$ | Diameter (m) |
| $\mathcal{D}$ | Mass diffusivity ($m^2/s$) |
| $F$ | Force (N) |
| $F/O$ | Mass fuel–oxidizer ratio (kg/kg) |
| $h$ | Enthalpy (J/kg) |
| $h_f^o$ | Enthalpy of formation (J/kg) |
| $h_{fg}$ | Latent heat of vaporization (J/kg) |
| $k$ | Thermal conductivity (W/m-K) |
| $K$ | Evaporation rate constant ($m^2/s$) |
| $L$ | Combustor length (m) |
| $Le$ | Lewis number, $\alpha/\mathcal{D}$ |
| $m$ | Mass (kg) |
| $\dot{m}$ | Mass flowrate (kg/s) |
| $\dot{m}'$ | Mass flowrate per unit length of combustor (kg/s-m) |
| $\dot{m}''$ | Mass flux (kg/s-$m^2$) |
| $MW$ | Molecular weight (kg/kmol) |
| $\dot{N}$ | Number of droplets per unit time (1/s) |
| $Nu$ | Nusselt number |
| $P$ | Pressure (Pa or atm) |

| $Pr$ | Prandtl number |
| $q$ | Heat per unit mass (J/kg) |
| $\dot{Q}$ | Heat-transfer rate (W) |
| $r$ | Radial coordinate or radius (m) |
| $R$ | Specific gas constant (J/kg-K) |
| $R_u$ | Universal gas constant (J/mol-K) |
| $Re$ | Reynolds number |
| $Sh$ | Sherwood number |
| $t$ | Time (s) |
| $T$ | Temperature (K) |
| $v$ | Velocity (m/s) |
| $v_r$ | Radial velocity (m/s) |
| $V$ | Volume (m$^3$) |
| $\mathbf{V}$ | Velocity vector (m/s) |
| $x$ | Number of carbon atoms in fuel molecule or axial coordinate (m) |
| $y$ | Number of hydrogen atoms in fuel molecule |
| $Y$ | Mass fraction (kg/kg) |
| $Z, Z_T$ | $c_p g/(4\pi k_g)$ (m-s/kg) |
| $Z_F$ | $1/(4\pi\rho\mathcal{D})$ (m-s/kg) |

### Greek Symbols

| $\alpha$ | Thermal diffusivity (m$^2$/s) |
| $\delta_T, \delta_M$ | Film thickness based on heat or mass transfer, respectively (m) |
| $\Delta h_c$ | Enthalpy of combustion (J/kg) |
| $\nu$ | Oxidizer-to-fuel stoichiometric ratio (kg/kg) |
| $\rho$ | Density (kg/m$^3$) |
| $\Phi$ | Equivalence ratio |
| $\chi$ | Mole fraction (kmol/kmol) |

### Subscripts

| $a$ | Oxidizer (or air) |
| boil | Boiling point |
| cond | Conduction |
| $d$ | Droplet |
| $f$ | Flame |
| $F$ | Fuel |
| $g$ | Gas |
| $i$ | Interface |
| $l$, liq | Liquid |
| $lg$ | Liquid-to-gas |
| $Ox$ | Oxidizer |
| $Pr$ | Product |
| ref | Reference state |
| rel | Relative |
| $s$ | Surface |

| sat | Saturation |
| vap | Vapor |
| ∞ | Freestream or far-removed from surface |
| 0 | Initial condition |

# REFERENCES

1. Hoffman, H., "Development Work on the Mercedes-Benz Commercial Diesel Engine, Model Series 400," SAE Paper 710558, 1971.

2. Eberle, M. L., "The Marine Diesel Engine—The Answer to Low-Grade Fuels," ASME Paper 80-DGP-16, 1980.

3. Correa, S. M., "A Review of $NO_x$ Formation Under Gas-Turbine Combustion Conditions," *Combustion Science and Technology,* 87: 329–362 (1992).

4. Davis, L. B., and Washam, R. M., "Development of a Dry Low $NO_x$ Combustor," ASME 89-GT-255, Gas Turbine and Aeroengine Congress and Exposition, Toronto, ON, 4–8 June 1988.

5. Shaw, R. J., "Engine Technology Challenges for a 21st Century High Speed Civil Transport," NASA Technical Memorandum 104363, 1991.

6. Mongia, H. C., "TAPS—A 4th Generation Propulsion Technology for Low Emissions," AIAA Paper 2003-2657, AIAA/ICAS International Air and Space Symposium and Exposition, Dayton, OH, 14–17 July 2003.

7. McDonell, V., "Lean Combustion in Gas Turbines," Chapter 5 in *Lean Combustion: Technology and Control,* (D. Dunn-Rankin, ed.), Academic Press, London, pp. 121–160, 2007.

8. Lefebvre, A. H., *Gas Turbine Combustion,* 2nd Ed., Taylor & Francis, Philadelphia, PA, 1999.

9. Dipprey, D. F., "Liquid Rocket Engines," in *Chemistry in Space Research* (R. F. Landel and A. Rembaum, eds.), Elsevier, New York, pp. 464–597, 1972.

10. Santoro, R. J., "Applications of Laser-Based Diagnostics to High Pressure Rocket and Gas Turbine Combustor Studies," AIAA Paper 98-2698, 1998.

11. Chehroudi, B., Talley, D., Mayer, W., Branam, R., Smith, J. J., Schik, A., and Oschwald, M., "Understanding Injection into High Pressure Supercritical Environments," Fifth International Conference on Liquid Space Propulsion, Chattanooga, TN, 28 [1/n] October 2003.

12. Nishiwaki, N., "Kinetics of Liquid Combustion Processes: Evaporation and Ignition Lag of Fuel Droplets," *Fifth Symposium (International) on Combustion,* Reinhold, New York, pp. 148–158, 1955.

13. Law, C. K., and Williams, F. A., "Kinetics and Convection in the Combustion of Alkane Droplets," *Combustion and Flame,* 19(3): 393–406 (1972).

14. Hubbard, G. L., Denny, V. E., and Mills, A. F., "Droplet Evaporation: Effects of Transients and Variable Properties," *International Journal of Heat and Mass Transfer,* 18: 1003–1008 (1975).

15. Williams, A., "Combustion of Droplets of Liquid Fuels: A Review," *Combustion and Flame,* 21: 1–21 (1973).

16. Faeth, G. M., "Current Status of Droplet and Liquid Combustion," *Progress in Energy and Combustion Science,* 3: 191–224 (1977).

17. Law, C. K., "Recent Advances in Droplet Vaporization and Combustion," *Progress in Energy and Combustion Science,* 8: 171–201 (1982).

18. Faeth, G. M., "Evaporation and Combustion of Sprays," *Progress in Energy and Combustion Science,* 9: 1–76 (1983).

19. Sirignano, W. A., "Fuel Droplet Vaporization and Spray Combustion Theory," *Progress in Energy and Combustion Science,* 9: 291–322 (1983).

20. Chiu, H. H., "Advances and Challenges in Droplet and Spray Combustion. I. Toward a Unified Theory of Droplet Aerothermochemistry," *Progress in Energy and Combustion Science,* 16: 381–416 (2000).

21. Choi, M. Y., and Dryer, F. L., "Microgravity Droplet Combustion," Chapter 4 in *Microgravity Combustion: Fire in Free Fall,* (H. D. Ross, ed.), Academic Press, London, pp. 183–298, 2001.

22. Godsave, G. A. E., "Studies of the Combustion of Drops in a Fuel Spray: The Burning of Single Drops of Fuel," *Fourth Symposium (International) on Combustion,* Williams & Wilkins, Baltimore, MD, pp. 818–830, 1953.

23. Spalding, D. B., "The Combustion of Liquid Fuels," *Fourth Symposium (International) on Combustion,* Williams & Wilkins, Baltimore, MD, pp. 847–864, 1953.

24. Kuo, K. K., *Principles of Combustion,* 2nd Ed., John Wiley & Sons, Hoboken, NJ, 2005.

25. Faeth, G. M., and Lazar, R. S., "Fuel Droplet Burning Rates in a Combustion Gas Environment," *AIAA Journal,* 9: 2165–2171 (1971).

26. Kumagai, S., and Isoda, H., "Combustion of Fuel Droplets in a Falling Chamber," *Sixth Symposium (International) on Combustion,* Reinhold, New York, pp. 726–731, 1957.

27. Faeth, G. M., "The Kinetics of Droplet Ignition in a Quiescent Air Environment," Ph.D. Thesis, The Pennsylvania State University, University Park, PA, 1964.

28. Kumagai, S., Sakai, T., and Okajima, S., "Combustion of Free Fuel Droplets in a Freely Falling Chamber," *Thirteenth Symposium (International) on Combustion,* The Combustion Institute, Pittsburgh, PA, pp. 779–785, 1971.

29. Hara, H., and Kumagai, S., "Experimental Investigation of Free Droplet Combustion Under Microgravity," *Twenty-Third Symposium (International) on Combustion.* The Combustion Institute, Pittsburgh, PA, pp. 1605–1610, 1990.

30. Dietrick, D. L., Haggard, J. B., Dryer, F. L., Nayagam, V., Shaw, B. D., and Williams, F. A., "Droplet Combustion Experiments in Spacelab," *Twenty-Sixth Symposium (International) on Combustion,* The Combustion Institute, Pittsburgh, PA, pp. 1201–1207, 1996.

31. Avedisian, C. T., "Soot Formation in Spherically Symmetric Droplet Combustion," in *Physical and Chemical Aspects of Combustion: A Tribute to Irvin Glassman* (F. L. Dryer and R. F. Sawyer, eds.), Gordon & Breach, Amsterdam, pp. 135–160, 1997.

32. Jackson, G. S., and Avedisian, C. T., "Experiments of the Effect of Initial Diameter in Spherically Symmetric Droplet Combustion of Sooting Fuel," *Proceedings of the Royal Society of London,* A466: 257–278 (1994).

33. Law, C. K., and Law, H. K., "Theory of Quasi-Steady One-Dimensional Diffusional Combustion with Variable Properties Including Distinct Binary Diffusion Coefficients," *Combustion and Flame,* 29: 269–275 (1977).

34. Law, C. K., and Law, H. K., "Quasi-Steady Diffusion Flame Theory with Variable Specific Heats and Transport Coefficients," *Combustion Science and Technology,* 12: 207–216 (1977).

35. Shuen, J. S., Yang, V., and Hsiao, C. C., "Combustion of Liquid-Fuel Droplets in Super-critical Conditions," *Combustion and Flame,* 89: 299–319 (1992).

36. Canada, G. S., and Faeth, G. M., "Combustion of Liquid Fuels in a Flowing Combustion Gas Environment at High Pressures," *Fifteenth Symposium (International) on Combustion,* The Combustion Institute, Pittsburgh, PA, pp. 419–428, 1975.

37. Law, C. K., Chung, W. H., and Srinivasan, N., "Gas-Phase Quasi-Steadiness and Fuel Vapor Accumulation Effects in Droplet Burning," *Combustion and Flame,* 38: 173–198 (1980).

38. Law, C. K., and Sirignano, W. A., "Unsteady Droplet Combustion and Droplet Heating II: Conduction Limit," *Combustion and Flame,* 28: 175–186 (1977).

39. Law, C. K., "Multicomponent Droplet Combustion with Rapid Internal Mixing," *Combustion and Flame,* 26: 219–233 (1976).

40. Megaridis, C. M., and Sirignano, W. A., "Numerical Modeling of a Vaporizing Multi-component Droplet," *Twenty-Third Symposium (International) on Combustion,* The Combustion Institute, Pittsburgh, PA, pp. 1413–1421, 1990.

41. Cho, S. Y., Yetter, R. A., and Dryer, F. L., "Computer Model for Chemically Reactive Flow with Coupled Chemistry / Multi-component Molecular Diffusion / Heterogeneous Processes," *Journal of Computational Physics,* 102: 160–179 (1992).

42. Sirignano, W. A., "Fluid Dynamics of Sprays—1992 Freeman Scholar Lecture," *Journal of Fluids Engineering,* 115: 345–378 (1993).

43. Priem, R. J., and Heidmann, M. F., "Propellant Vaporization as a Design Criterion for Rocket-Engine Combustion Chambers," NASA Technical Report R-67, 1960.

44. Turns, S. R., and Faeth, G. M., "A One-Dimensional Model of a Carbon-Black Slurry-Fueled Combustor," *Journal of Propulsion and Power,* 1(1): 5–10 (1985).

45. Olikara, C., and Borman, G. L., "A Computer Program for Calculating Properties of Equilibrium Combustion Products with Some Application to I. C. Engines," SAE Paper 750468, 1975.

46. White, F. M., *Viscous Fluid Flow,* McGraw-Hill, New York, p. 209, 1974.

47. Visual Numerics, Inc., "DGEAR," IMSL Numerical Library, Houston, TX.

48. Press, W. H., Teukolsky, S. A., Vetterling, W. T., and Flannery, B. P., *Numerical Recipes 3rd Edition—The Art of Scientific Computing,* Cambridge University Press, New York, 2007.

49. Andrews, J. R., and Biblarz, O., "Temperature Dependence of Gas Properties in Polynomial Form," Naval Postgraduate School, NPS67-81-00l, January 1981.

50. Weast, R. C. (ed.), *CRC Handbook of Chemistry and Physics,* 56th Ed., CRC Press, Cleveland, OH, 1975.

51. Evans, A. F., *The History of the Oil Engine: A Review in Detail of the Development of the Oil Engine from the Year 1680 to the Beginning of the Year 1930,* Sampson Low Marston & Co., London, pp. vii–x (Foreword by Sir Dugald Clerk), undated (1932).

## PROBLEMS

**10.1**  Calculate the evaporation rate constant for a 1-mm-diameter water droplet evaporating into dry, hot air at 500 K and 1 atm.

**10.2**  Calculate the lifetimes for *n*-hexane droplets evaporating in a quiescent ambient environment of air at 800 K and 1 atm for initial diameters of 1000, 100, and 10 $\mu$m. Also, calculate the mean evaporation rate defined as $m_0/t_d$, where $m_0$ is the initial droplet mass. Assume the liquid density is 664 kg/m$^3$.

**10.3**  Determine the influence of ambient temperature on droplet lifetimes. Use as a base case *n*-hexane with $D_0 = 100$ $\mu$m and $T = 800$ K, as determined in problem 10.2. Use temperatures of 600, 800, and 1000 K. Separate the effects of temperature proper and temperature-dependent properties by first calculating lifetimes assuming properties fixed at the same values as the base case, and then repeat your calculations with appropriate properties. Plot your results. Discuss.

**10.4**  Determine the influence of pressure on the lifetimes of 500-$\mu$m water droplets. The ambient environment is dry air at 1000 K. Use pressures of 0.1, 0.5, and 1.0 MPa. Plot your results. Discuss.

**10.5**  Estimate the mass burning rate of a 1-mm-diameter *n*-hexane droplet burning in air at atmospheric pressure. Assume no heat is conducted into the interior of the liquid droplet and that the droplet temperature is equal to the boiling point. The ambient air is at 298 K.

**10.6**  Calculate the ratio of the flame radius to droplet radius and flame temperature corresponding to the conditions given in problem 10.5.

**10.7\***  For the conditions associated with problems 10.5 and 10.6, plot the temperature distribution for the region $r_s \leq r \leq 2r_f$. Based on your plot, determine which is greater: the heat conduction from the flame toward the droplet, or the heat conduction from the flame to the surroundings. Discuss the physical significance of your result.

**10.8**  Repeat problems 10.5 and 10.6, removing the assumption that the droplet temperature is at the boiling point. The Clausius–Clapeyron constants for *n*-hexane are $A = 25{,}591$ atm and $B = 3471.2$ K. Compare these results with those of problems 10.5 and 10.6 and discuss.

*Indicates required use of computer.

**10.9**   A rogue fuel droplet escapes from the primary zone of a gas-turbine combustor. If both the droplet and the combustion gases are traveling at 50 m/s
through the combustor, estimate the combustor length required to burn out
the rogue droplet. Discuss the implications of your results.

   To simplify property evaluation, assume the combustion products
have the same properties as air. For the fuel droplet, use $n$-hexane properties. Assume the liquid density is 664 kg/m$^3$. Also assume $T_s = T_{boil}$, using
the Clausius–Clapeyron constants to evaluate $T_{boil}$ at the given pressure.
Other pertinent data are $P = 10$ atm, $T_\infty = 1400$ K, and drop diameter
$D = 200$ $\mu$m.

**10.10**   What is the influence of introducing a 10-m/s slip velocity between the
droplet and gas stream in problem 10.9? Answer this by comparing the
initial burning rates with and without slip.

**10.11**   Derive a "power-law" expression for droplet burning in a convective environment, i.e.,

$$D^n(t) = D_0^n(t) - K't,$$

where the value for $n$ results from your analysis. You will also have to
define an appropriate burning rate constant $K'$. To keep your analysis simple, assume

$$Nu = CRe^{0.8}Pr^{1/3},$$

where $C$ is a known constant, rather than trying to employ Eqn. 10.78.

**10.12**   Determine the influence of droplet heat-up on droplet lifetimes. Use the
results from problem 10.5 as the base case (heat-up ignored) and compare
with results obtained using the "onion-skin" model to account for droplet
heat-up. For the heat-up model, assume the bulk droplet temperature is
300 K and the liquid specific heat is 2265 J/kg-K.

**10.13**   Starting with Eqn. 10.69, and assuming that the heat transfer into the interior of the droplet, $q_{i-l}$, is zero, show that

$$\Delta h_c = c_{pg}(T_f - T_s) + vc_{pg}(T_f - T_\infty) + h_{fg}.$$

Discuss the physical significance of each term in the above expression.

---

## PROJECTS

*Note:* The detailed implementations of the following suggested projects
depend upon available software. It is assumed that subroutines to calculate
equilibrium properties and their partial derivatives (Appendix F) and a
numerical integration routine are available.

1. Model the primary zone of a gas-turbine combustor can assuming one-dimensionality and the other Chapter 10 assumptions. The following conditions are known or given:

$$\Phi_{supplied} = 1.0,$$
$$\dot{m}_{a,0} = 0.66 \text{ kg/s},$$
$$A = 0.0314 \text{ m}^2,$$
$$L = 0.30 \text{ m},$$
$$P = 10 \text{ atm},$$
$$T_{inlet, air} = 600 \text{ K}.$$

A. Determine the maximum allowed drop size if 95 percent of the fuel is to be burned in the primary zone. Simplify the vaporization problem by using the following approximate properties and treating them as independent of temperature:

$$C_{12}H_{26}(n\text{-dodecane}),$$
$$\rho_l = 749 \text{ kg/m}^3,$$
$$h_{fg} \approx 263 \text{ kJ/kg},$$
$$k_g = 0.05 \text{ W/m-K},$$
$$c_{pg} = 1200 \text{ J/kg-K}.$$
$$T_b = 447 \text{ K}.$$

Use properties of air at 1600 K to evaluate $Re$, $Pr$, and $Nu$.

B. Plot $D(x)$, $T_g(x)$, $v_g(x)$, and $\Phi_g(x)$ for the conditions determined in part A. Discuss.

2. Consider a gas-turbine combustor can that has a diameter of 0.2 m. Assume the fuel can be approximated as $n$-dodecane with the properties listed in Project 1. Gas-phase conditions at the exit of the primary zone are

$$\Phi_g = 1.0,$$
$$P = 10 \text{ atm},$$
$$T_g = 2500 \text{ K},$$
$$\dot{m}_g = 0.6997 \text{ kg/s}.$$

Not all of the fuel injected into the combustor is burned in the primary zone, with 0.00441 kg/s of fuel entering the secondary zone as liquid drops of 25-$\mu$m diameter.

Determine the length of the secondary zone required to completely burn the remaining fuel, assuming secondary air is added at a constant rate per unit length, $\dot{m}_a'$, and that the total amount of secondary air added is 1.32 kg/s.

How much time is required to burn out the droplets? (*Hint:* $dt = dx/v_d$.) Also, determine the temperature and equivalence ratio at the end of the secondary zone.

Discuss your results and support your discussion with various graphs showing variation of key parameters with axial distance, $x$.

# APPENDIX 10A
# SIR HARRY R. RICARDO'S DESCRIPTION
# OF COMBUSTION IN DIESEL ENGINES [51]

"Before concluding, I am going to take a rather unconventional course, in a technical lecture, of asking you to accompany me, in imagination, inside the cylinder of a Diesel engine. Let us imagine ourselves seated comfortably on the top of the piston, at or about the end of the compression stroke. We are in complete darkness, the atmosphere is a trifle oppressive, for the shade temperature is well over 500 °C—almost a dull red heat—and the density of the air is such that the contents of an average sitting-room would weigh about a ton: also it is very draughty, in fact, the draught is such that in reality we should be blown off our perch and hurled about like autumn leaves in a gale. Suddenly, above our heads a valve is opened and a rainstorm of fuel begins to descend. I have called it a rainstorm, but the velocity of droplets approaches much more nearly that of rifle bullets than of raindrops. For a while nothing startling happens, the rain continues to fall, the darkness remains intense. Then suddenly, away to our right perhaps, a brilliant gleam of light appears moving swiftly and purposefully; in an instant this is followed by a myriad others all around us, some large and some small, until on all sides of us the space is filled with a merry blaze of moving lights; from time to time the smaller lights wink and go out, while the larger ones develop fiery tails like comets; occasionally these strike the walls, but being surrounded with an envelope of burning vapour they merely bounce off like drops of water spilt on a red-hot plate. Right overhead all is darkness still, the rainstorm continues, and the heat is becoming intense; and now we shall notice that a change is taking place. Many of the smaller lights around us have gone out, but new ones are beginning to appear, more overhead, and to form themselves into definite streams shooting rapidly downwards or outwards from the direction of the injector nozzles. Looking round again we see that the lights around are growing yellower; they no longer move in definite directions, but appear to be drifting listlessly hither and thither; here and there they are crowding together in dense nebulae, and these are burning now with a sickly smoky flame, half suffocated for want of oxygen. Now we are attracted by a dazzle overhead, and, looking up, we see that what at first was cold rain falling through utter darkness, has given place to a cascade of fire, as from a rocket. For a little while this continues, then ceases abruptly as the fuel valve closes. Above and all around us are still some lingering fireballs, now trailing long tails of sparks and smoke and wandering aimlessly in search of the last dregs of oxygen which will consume them finally and set their souls at rest. If so, well and good; if not, some unromantic engineer outside will merely grumble that the exhaust is dirty and will set the fuel valve to close a trifle earlier. So ends the scene, or rather my conception of the scene, and I will ask you to realize that what has taken me nearly five minutes to describe may all be enacted in one five-hundredth of a second or even less."

# 11

# Introduction to Turbulent Flows

## OVERVIEW

In practical devices that involve flowing fluids, turbulent flows are more frequently encountered than are laminar flows. This rule of thumb is especially true for combustion devices such as reciprocating internal combustion engines, gas turbines, furnaces, boilers, rocket engines, etc. Unfortunately, the mathematical description of a turbulent flow and the solution of the resulting governing conservation equations is much more difficult than for laminar flows. Analytical and numerical solutions to turbulent flows are engineering approximations, even for the most simple geometries, and can be subject to large errors. In contrast, exact solutions can be found for many laminar flows, especially simple geometries, and very accurate solutions are obtainable for even complicated situations using numerical methods. The essential dilemma of solving a turbulent flow problem is that if all the information necessary for describing a flow is tracked, there is no computer in the world large enough to handle the job; on the other hand, if the problem is simplified to be tractable using available computers, large errors may result, especially for flow conditions that have not been studied experimentally. Considerable progress, however, is being made on this front. Because of the tremendous importance of turbulent flows in engineering and other applications, considerable effort has been expended over many decades to understand turbulence and to develop descriptive and predictive methods that are useful in the design of practical devices.

Our objectives in this chapter are by necessity quite limited. First, we will review a few key concepts related to turbulent flow that you may have seen in an introductory fluid mechanics course. These include the general behavior of the velocity and scalars (e.g., temperature and species) in turbulent flows, some physical notions of turbulence, definitions of length scales, and the most simple mathematical description of a turbulent flow. Our second objective is to discuss the essential features of free (unconfined) turbulent jets. This archetypal flow occurs frequently in many practical combustion devices, and, hence, is a useful starting point to understand more complex flows. For example, during the intake stroke of a spark-ignition engine, a jetlike flow is created as the incoming charge passes into the cylinder through the relatively narrow passage created by the opening intake valve. Inside the cylinder, the intake jet and other flows created by piston motion interact with the cylinder walls and head. In gas-turbine engines, the secondary and dilution zones comprise a combination of jet and wall flows (see Fig. 10.4). Industrial burners and furnaces also are examples of practical devices where both turbulent jet and wall flows are important. Of course, in all of these devices, the actual flows are much more complicated than the simple situation that we will study in this chapter; however, an understanding of simpler flows is an essential prerequisite to dealing with complex flows.

In this chapter, we treat only nonreacting, incompressible flows to illuminate the essential features of turbulent flows uncomplicated by combustion. Subsequent chapters explore premixed turbulent flames, with the specific application of spark-ignition engines in mind (Chapter 12), and nonpremixed turbulent jet flames (Chapter 13).

## DEFINITION OF TURBULENCE

Turbulent flow results when instabilities in a flow are not sufficiently damped by viscous action and the fluid velocity at each point in the flow exhibits random fluctuations. All students of fluid mechanics should be familiar with the famous experiment of Osborne Reynolds [1] in which the transition from a laminar to a turbulent flow in a tube was visualized with a dye streak and related to the dimensionless parameter that now bears the name of this important early fluid mechanician, i.e., the **Reynolds number.** The random unsteadiness associated with various flow properties is the hallmark of a turbulent flow and is illustrated for the axial velocity component in Fig. 11.1. One particularly useful way to characterize a turbulent flowfield is to define **mean** and **fluctuating** quantities. Mean properties are defined by taking a time-average of the flow property over a sufficiently large time interval, $\Delta t = t_2 - t_1$; i.e.,

$$\bar{p} \equiv \frac{1}{\Delta t} \int_{t_i}^{t_2} p(t)\,dt, \tag{11.1}$$

where $p$ is any flow property, e.g., velocity, temperature, pressure, etc. The fluctuation, $p'(t)$, is the difference between the instantaneous value of the property, $p(t)$, and the mean value, $\bar{p}$, or

$$p(t) = \bar{p} + p'(t). \tag{11.2}$$

Figure 11.1 illustrates the fluctuating component of $v_x$ at a specific time, $t_0$; i.e., $v_x(t_0) = \bar{v}_x + v'_x(t_0)$. In turbulent flows with combustion, there are also frequently large random fluctuations in temperature, density, and species, as defined by

$$T(t) = \bar{T} + T'(t),$$

$$\rho(t) = \bar{\rho} + \rho'(t),$$

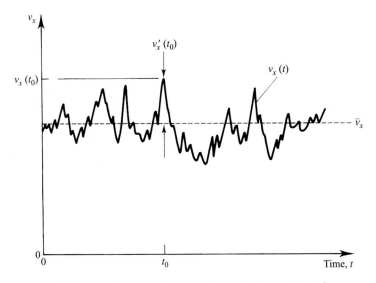

**Figure 11.1**    Velocity as a function of time at a fixed point in a turbulent flow.

and

$$Y_i(t) = \overline{Y}_i + Y_i'(t).$$

This manner of expressing variables as a mean and a fluctuating component is referred to as the **Reynolds decomposition.**

It is common practice to define the **intensity** of the turbulent fluctuations in terms of root-mean-square quantities; that is,

$$p'_{rms} \equiv \sqrt{\overline{p'^2}}. \qquad (11.3a)$$

The **relative intensity** is defined as

$$p'_{rms}/\overline{p}, \qquad (11.3b)$$

and is usually expressed as a percentage.

What is the physical nature of a turbulent flow? Figure 11.2 gives a partial answer to this question. In this figure, we see fluid blobs and filaments of fluid intertwining. A common notion in fluid mechanics is the idea of a fluid **eddy.** An eddy is considered to be a macroscopic fluid element in which the microscopic elements composing the eddy behave in some ways as a unit. For example, a **vortex** imbedded in a flow would be considered an eddy. A turbulent flow comprises many eddies with a multitude of sizes and **vorticities,** a measure of angular velocities. A number of smaller eddies may be imbedded in a larger eddy. A characteristic of a fully turbulent flow is the existence of a wide range of length scales, i.e., eddy sizes. For a turbulent

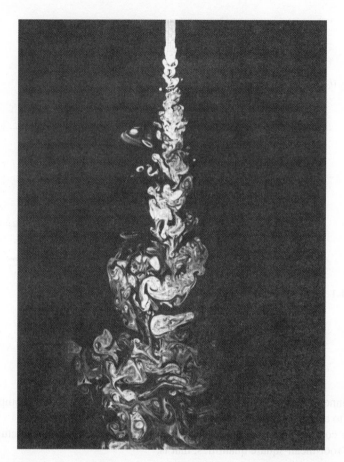

**Figure 11.2**     Turbulent jet issuing into a quiescent reservoir. The flow visualization technique makes visible the jet fluid in a plane containing the jet centerline.
SOURCE: From Ref. [2], reprinted by permission of PWN-Polish Scientific Publishers, Warsaw. Also appears in Ref. [3], p. 97.

flow, the Reynolds number is a measure of the range of scales present; the greater the Reynolds number, the greater the range of sizes from the smallest eddy to the largest. It is this large range of length scales that makes calculating turbulent flows from first principles intractable. We will discuss length scales in more detail in the next section.

The rapid intertwining of fluid elements is a characteristic that distinguishes turbulent flow from laminar flow. The turbulent motion of fluid elements allows momentum, species, and energy to be transported in the cross-stream direction much more rapidly than is possible by the molecular diffusion processes controlling transport in laminar flows. Because of this, most practical combustion devices employ turbulent flows to enable rapid mixing and heat release in relatively small volumes.

# LENGTH SCALES IN TURBULENT FLOWS

We can improve our qualitative understanding of the structure of turbulent flows by exploring some of the length scales that are used to characterize such flows. Furthermore, some understanding of the important length scales will be essential to our discussion of premixed turbulent combustion in Chapter 12, since the relationships among length scales determine the intrinsic character of turbulent combustion.

## Four Length Scales

In the turbulence literature, many length scales have been defined; however, the following four scales are of general relevance to our discussion and, in general, are frequently cited. In decreasing order of size, these scales are as follows:

| | |
|---|---|
| $L$ | Characteristic width of flow or macroscale |
| $\ell_0$ | Integral scale or turbulence macroscale |
| $\ell_\lambda$ | Taylor microscale |
| $\ell_K$ | Kolmogorov microscale |

In the following sections, we examine each of these length scales, providing some physical interpretation and expanding our view of turbulence structure.

**Characteristic Width of Flow or Macroscale, $L$**     This is the largest length scale in the system and is the upper bound for the largest possible eddies. For example, in a pipe flow, the largest eddy would be equal to the pipe diameter; and for a jet flow, $L$ would be the local width of the jet at any axial location. In a reciprocating internal combustion engine, $L$ might be taken as the time-varying clearance between the piston top and the head, or perhaps the cylinder bore. In general, this length scale is defined by the actual hardware or device being considered. This length scale is frequently used to define a Reynolds number based on the mean flow velocity, but is not used to define a turbulence Reynolds number, as are the other three length scales. Of particular interest to a study of combustion is the ability of the largest structures in the flow to stir a fluid. For example, the largest eddies in a fuel jet **engulf,** or **stir,** air well into the central region of the jet. This large-scale engulfment or stirring process is clearly evident in the jet flow pictured in Fig. 11.2. In certain turbulent flows, persistent organized motion can coexist with random motions. The most familiar example of this is the two-dimensional mixing layer, where coherent spanwise vortical structures dominate the large-scale motion [4, 5].

**Integral Scale or Turbulence Macroscale, $\ell_0$**     The **integral scale** physically represents the mean size of the large eddies in a turbulent flow: those eddies with low frequency and large wavelength. The integral scale is always smaller than $L$, but is of the same order of magnitude. Operationally, the integral scale can be measured by integrating the correlation coefficient for the fluctuating velocities obtained as a function of the distance between two points, i.e.,

$$\ell_0 = \int_0^\infty R_x(r)\,\mathrm{d}r, \qquad (11.4a)$$

where

$$R_x(r) \equiv \frac{\overline{v'_x(0)v'_x(r)}}{v'_{x,rms}(0)v'_{x,rms}(r)}. \tag{11.4b}$$

In less precise terms, $\ell_0$ represents the distance between two points in a flow where there ceases to be a correlation between the fluctuating velocities at the two locations. Reference [6] presents physical pictures of $\ell_0$ in terms of possible structural models of turbulence: one where $\ell_0$ represents the spacing between narrow vortex tubes that make up the fine structure of the flow [7], and another where $\ell_0$ is the spacing between thin vortex sheets [8]. Figure 11.3 affords a visual appreciation of a length scale based on the spacing between vortex tubes.

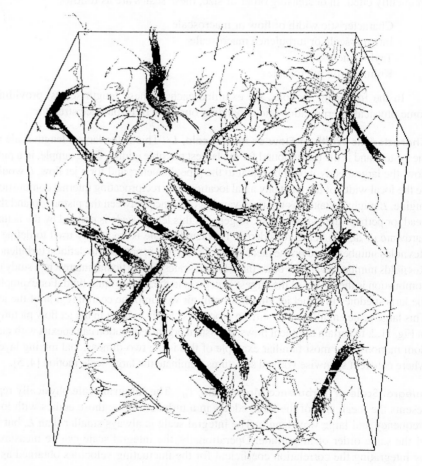

**Figure 11.3**     Direct numerical simulation of isotropic turbulence reveals tubelike structure of intense vorticity (dark lines), while regions of moderate vorticity (gray lines) lack structure. $Re_{\ell_\lambda} \approx 77$.

| SOURCE: From Ref. [13], reprinted by permission.

**Taylor Microscale, $\ell_\lambda$**  The **Taylor microscale** is an intermediate length scale between $\ell_0$ and $\ell_K$, but is weighted more towards the smaller scales. This scale is related to the mean rate of strain and can be formally expressed [6] as

$$\ell_\lambda = \frac{v'_{x,rms}}{\left[\overline{\left(\frac{\partial v_x}{\partial x}\right)^2}\right]^{1/2}},$$
(11.5)

where the denominator represents the mean strain rate. Physically, the Taylor microscale is the length scale at which viscous dissipation begins to affect the eddies [9].

**Kolmogorov Microscale, $\ell_K$**  The **Kolmogorov microscale** is the smallest length scale associated with a turbulent flow and, as such, is representative of the dimension at which the dissipation of turbulent kinetic energy to fluid internal energy occurs. Thus, the Kolmogorov scale is the scale at which molecular effects (kinematic viscosity) are significant. Dimensional arguments [10, 11] show that $\ell_K$ can be related to the rate of dissipation, $\epsilon_0$ as

$$\ell_K \approx (\nu^3/\epsilon_0)^{1/4},$$
(11.6)

where $\nu$ is the molecular kinematic viscosity, and the dissipation rate is approximately expressed as

$$\epsilon_0 \equiv \frac{\delta(ke_{turb})}{\delta t} \approx \frac{3v'^2_{rms}/2}{\ell_0/v'_{rms}}.$$
(11.7)

Note that the integral length scale, $\ell_0$, appears in the approximation for the dissipation rate, thereby linking the two scales. We will shortly show the relationships among the various length scales. As an aside, it is helpful to new students of turbulence to point out that dimensional, and sometimes heuristic, arguments are invoked in discussions of turbulence, the above (Eqns. 11.6 and 11.7) being examples of these.

The final point we wish to make concerning $\ell_K$ is possible physical interpretations. In Tennekes' [7] model of a turbulent flow, $\ell_K$ represents the thickness of the smallest vortex tubes or filaments that permeate a turbulent flow, whereas Ref. [8] suggests that $\ell_K$ represents the thickness of vortex sheets imbedded in the flow. Results from direct numerical simulation of turbulence (Fig. 11.3) show that the most intense vorticity is contained in vortex tubes that have diameters on the order of the Kolmogorov scale [12, 13].

## Turbulence Reynolds Numbers

Three of the four length scales discussed above are used to define three turbulence Reynolds numbers. In all of the Reynolds numbers, the characteristic velocity is the root-mean-square fluctuating velocity, $v'_{rms}$. Thus, we define

$$Re_{\ell_0} \equiv v'_{rms} \ell_0/\nu,$$
(11.8a)

$$Re_{\ell_\lambda} \equiv v'_{rms} \ell_\lambda/\nu,$$
(11.8b)

and

$$Re_{\ell_K} \equiv v'_{rms}\ell_K/\nu. \tag{11.8c}$$

Equations 11.6 and 11.7 defining $\ell_K$ and the dissipation rate, $\epsilon_0$, can be used to relate the largest (the integral) and the smallest (the Kolmogorov) turbulence length scales:

$$\ell_0/\ell_K = Re_{\ell_0}^{3/4}. \tag{11.9}$$

The Taylor microscale, $\ell_\lambda$, also can be related to $Re_{\ell_0}$ as follows [11]:

$$\ell_0/\ell_\lambda = Re_{\ell_0}^{1/2}. \tag{11.10}$$

Equation 11.9 expresses, in a semiquantitative way, the wide separation of length scales in high-$Re$ flows discussed earlier in this chapter. For example, with $Re_{\ell_0} = 1000$, the ratio $\ell_0/\ell_K$ is about 178:1; but when $Re_{\ell_0}$ is increased to 10,000, by increasing the mean flow velocity, say, the ratio becomes 1000:1. Figure 11.4 schematically illustrates this development of increasingly finer small-scale turbulence with increasing $Re$, while the largest scales in the flow remain unchanged. As we will

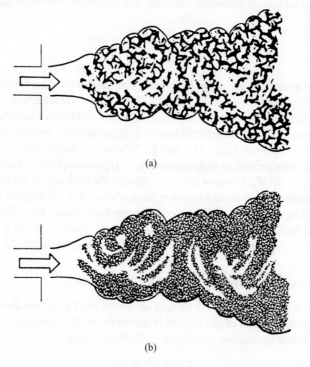

(a)

(b)

**Figure 11.4**    Turbulent jets at (a) low and (b) high Reynolds numbers. The shading pattern illustrates the small-scale turbulence structure.
SOURCE: From Ref. [10], reprinted by permission of MIT Press.

see in Chapter 12, the size of the scales of turbulence, in relation to the laminar flame thickness, determines the character of the turbulent flame.

---

A natural-gas-fired industrial gas turbine rated at 3950 kW has the following specifications:          **Example 11.1**

> Air flowrate = 15.9 kg/s,
> $F/A = 0.017$,
> Primary/secondary air split = 45/55,
> Combustor pressure = 10.2 atm,
> Combustor inlet temperature = 600 K,
> Primary zone temperature = 1900 K,
> Dilution zone temperature = 1300 K.

As shown in Fig. 11.5, gas-turbine combustors can be configured in several ways. The combustor for the turbine specified above is arranged in a can-annular (cannular) configuration with eight cans each of 0.20 m diameter. A single combustor can be seen in Fig. 2.4. The can length-to-diameter ratio is 1.5.

   Assuming that the relative turbulence intensity is approximately 10 percent and that the integral length scale is about 1/10 the can diameter, estimate the Kolmogorov length scales at (i) the combustor inlet, (ii) within the primary zone, and (iii) at the end of the dilution zone.

**Solution**

We can utilize Eqn. 11.9 to calculate the Kolmogorov length scale $\ell_K$. This requires that we first estimate the turbulence Reynolds number based on the integral scale $\ell_0$. To get the mean velocity through each can at the desired locations, overall continuity is applied:

$$v_j = \frac{\dot{m}_j}{\rho_j A}, \qquad (I)$$

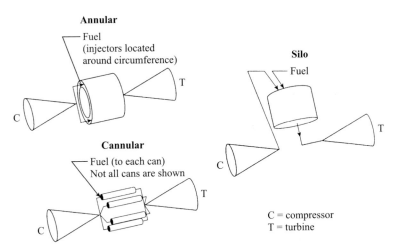

**Figure 11.5**   Schematic representation of the three types of combustor configurations used in large stationary gas turbines.

SOURCE: From Ref. [14], Reprinted by permission of Taylor & Francis Group, http://www.informaworld.com.

where the $j$ subscript denotes the location of interest. The flowrates at the three locations can be calculated:

$$\dot{m}_j = \frac{\left(\begin{array}{c}\text{fraction of total} \\ \text{air at location } j\end{array}\right)\dot{m}_A + (F/A)\dot{m}_A}{(\text{number of cans})}.$$

Thus, at the combustor inlet and within the primary zone,

$$\dot{m}_i = \dot{m}_{ii} = \frac{0.45(15.9) + 0.017(15.9)}{8} = 0.928 \text{ kg/s}.$$

At the end of the dilution zone, i.e., at the turbine inlet,

$$\dot{m}_{iii} = \frac{1.0(15.9) + 0.017(15.9)}{8} = 2.02 \text{ kg/s}.$$

The flow area is

$$A = \frac{\pi D^2}{4} = \frac{\pi(0.20)^2}{4} = 0.0314 \text{ m}^2$$

and the densities at each location are estimated using the ideal-gas law with properties of air:

$$\rho_j = \frac{P}{(R_u/MW)T_j}.$$

Thus,

$$\rho_i = \frac{10.2(101,325)}{(8315/28.85)600} = 5.97 \text{ kg/m}^3,$$

$$\rho_{ii} = \rho_i \frac{T_i}{T_{ii}} = 5.97\frac{600}{1900} = 1.89 \text{ kg/m}^3,$$

$$\rho_{iii} = \rho_i \frac{T_i}{T_{iii}} = 5.97\frac{600}{1300} = 2.76 \text{ kg/m}^3.$$

From Eqn. I, the velocities are determined to be

$$v_i = \frac{\dot{m}_i}{\rho_i A} = \frac{0.928}{5.97(0.0314)} = 4.95 \text{ m/s},$$

$$v_{ii} = \frac{0.928}{1.89(0.0314)} = 15.6 \text{ m/s},$$

$$v_{iii} = \frac{2.02}{2.76(0.0314)} = 23.3 \text{ m/s}.$$

We now calculate Reynolds numbers based on mean flow and turbulence quantities:

$$Re_{L,i} = \frac{\rho_i v_i D}{\mu_i} = \frac{5.97(4.95)(0.20)}{305.8 \cdot 10^{-7}} = 1.93 \cdot 10^5,$$

$$Re_{L,\text{ii}} = \frac{1.89(15.6)(0.20)}{663 \cdot 10^{-7}} = 8.89 \cdot 10^4,$$

$$Re_{L,\text{iii}} = \frac{2.76(23.3)(0.20)}{496 \cdot 10^{-7}} = 2.59 \cdot 10^5,$$

$$Re_{\ell_0,\text{i}} = \frac{\rho_\text{i} v'_{rms} \ell_0}{\mu_\text{i}} = \frac{5.97(0.1)(4.95)(0.20/10)}{305.8 \cdot 10^{-7}} = 1930,$$

$$Re_{\ell_0,\text{ii}} = \frac{1.89(0.1)(15.6)(0.20/10)}{663 \cdot 10^{-7}} = 889,$$

$$Re_{\ell_0,\text{iii}} = \frac{2.76(0.1)(23.3)(0.20/10)}{496 \cdot 10^{-7}} = 2590.$$

In the above, viscosity values are approximated as those of air, $v'_{rms} = 0.1v$, and $\ell_0 = L/10$. The Kolmogorov scales are now estimated from Eqn. 11.9 as follows:

$$\ell_K = \ell_0 Re_{\ell_0}^{-3/4},$$

$$\ell_{K,\text{i}} = \frac{20 \text{ mm}}{10}(1930)^{-0.75} = 0.069 \text{ mm}$$

$$\ell_{K,\text{ii}} = \frac{20 \text{ mm}}{10}(889)^{-0.75} = 0.123 \text{ mm}$$

$$\ell_{K,\text{iii}} = \frac{20 \text{ mm}}{10}(2590)^{-0.75} = 0.055 \text{ mm}$$

## Comment

We see that in this application the Kolmogorov scales are quite small, on the order of 0.1 mm, with the smallest at the turbine inlet.

## ANALYZING TURBULENT FLOWS

The basic conservation equations describing fluid motion that we have applied to laminar flows apply equally well to turbulent flows, provided, of course, that the unsteady terms are retained. For example, we can write the continuity (mass conservation) and Navier–Stokes (momentum conservation) equations for unsteady, constant-density, Newtonian fluid as follows (see Chapter 7):

### Mass Conservation

$$\frac{\partial v_x}{\partial x} + \frac{\partial v_y}{\partial y} + \frac{\partial v_z}{\partial z} = 0. \tag{11.11}$$

### x-Direction Momentum Conservation

$$\frac{\partial}{\partial t}\rho v_x + \frac{\partial}{\partial x}\rho v_x v_x + \frac{\partial}{\partial y}\rho v_y v_x + \frac{\partial}{\partial z}\rho v_z v_x$$

$$= \mu\left(\frac{\partial^2 v_x}{\partial x^2} + \frac{\partial^2 v_x}{\partial y^2} + \frac{\partial^2 v_x}{\partial z^2}\right) - \frac{\partial P}{\partial x} + \rho g_x. \tag{11.12}$$

The above equations and their companions, the additional two components of momentum, conservation of energy and species, conceptually could be solved for the functions $v_x(t)$, $v_y(t)$, $v_z(t)$, $T(t)$, and $Y_i(t)$ at discrete points in the flow using numerical techniques; however, the number of grid points required to resolve the details of the flow (recall the large range of length scales inherent in turbulent flows) would be huge. For example, Ref. [15] indicates that approximately 2 billion grid points would be necessary to deal with a flow having an integral-scale Reynolds number of 200. Using a computer operating at 1 gigaflop, i.e., $10^9$ flops, where one flop is one floating-point operation per second, would require 20 months of computer time. Currently, the fastest computers operate at teraflop ($10^{12}$) rates.

## Reynolds Averaging and Turbulent Stresses

Rather than relying entirely on experimental methods, techniques have been developed that allow turbulent flows to be analyzed to obtain useful information and to allow predictions. One fruitful method of analyzing turbulent flows is to write out the partial differential equations that embody the basic conservation principles—mass, momentum, energy, and species—perform a Reynolds decomposition, and then average the equations over time. The resulting governing equations are called the **Reynolds-averaged** equations. The averaging process has two major consequences: First, it eliminates the fine details of the flow; for example, the complex time-dependent velocity at a point, as illustrated in Fig. 11.1, could not be predicted using time-averaged equations of motion. The second consequence of averaging is the appearance of new terms in the time-averaged governing equations that have no counterpart in the original time-dependent equations. Finding a way to calculate or approximate these new terms is frequently referred to as the **closure problem** of turbulence.

**Two-Dimensional Boundary Layer**    To develop some feel for the time-averaged approach, we will develop the time-averaged x-momentum equation for a two-dimensional boundary layer flow over a flat plate. For this flow, with $x$ in the flow direction and $y$ perpendicular to the plate, Eqn. 11.12 reduces to

$$\frac{\partial}{\partial t}\rho v_x + \frac{\partial}{\partial x}\rho v_x v_x + \frac{\partial}{\partial y}\rho v_y v_x = \mu\frac{\partial^2 v_x}{\partial y^2}. \tag{11.13}$$

      ①        ②      ③      ④

The first step in the process is to substitute for each fluctuating quantity its representation as a mean and a fluctuating component, i.e., the Reynolds decomposition $v_x = \bar{v}_x + v'_x$ and $v_y = \bar{v}_y + v'_y$. We will then time-average Eqn. 11.13, term by term, using the overbar to signify a time-averaged quantity. Thus, the first term can be averaged:

$$\overline{\frac{\partial}{\partial t}\rho(\bar{v}_x + v'_x)} = \frac{\partial}{\partial t}\rho\bar{v}_x + \frac{\partial}{\partial t}\rho v'_x = 0, \tag{11.14}$$

where each term on the right-hand side of Eqn. 11.14 is identically zero; the first because we assume the flow is steady in an overall sense, i.e., $\bar{v}_x$ is a constant, and the second, because the time derivative of a random function with a zero mean is also a random function with a zero mean.

The third term in Eqn. 11.13 is the most interesting because it yields the dominant additional momentum flux arising from the velocity fluctuations:

$$\overline{\frac{\partial}{\partial y}\rho(\bar{v}_y + v'_y)(\bar{v}_x + v'_x)} = \overline{\frac{\partial}{\partial y}\rho(\bar{v}_x\bar{v}_y + \bar{v}_x v'_y + v'_x v'_y + \bar{v}_y v'_x)}. \tag{11.15}$$

Splitting out each term under the large overbar on the right-hand side and evaluating their individual averages yields

$$\overline{\frac{\partial}{\partial y}\rho\bar{v}_x\bar{v}_y} = \frac{\partial}{\partial y}\rho\bar{v}_x\bar{v}_y, \tag{11.16a}$$

$$\overline{\frac{\partial}{\partial y}\rho\bar{v}_x v'_y} = 0, \tag{11.16b}$$

$$\overline{\frac{\partial}{\partial y}\rho v'_x v'_y} = \frac{\partial}{\partial y}\rho\overline{v'_x v'_y}, \tag{11.16c}$$

and

$$\overline{\frac{\partial}{\partial y}\rho\bar{v}_y v'_x} = 0. \tag{11.16d}$$

Similar operations can be performed on the second and fourth terms of Eqn. 11.13 and are left as an exercise for the reader. Reassembling, the now time-averaged Eqn. 11.13 yields

$$\frac{\partial}{\partial x}\rho\bar{v}_x\bar{v}_x + \frac{\partial}{\partial y}\rho\bar{v}_y\bar{v}_x + \boxed{\frac{\partial}{\partial x}\rho\overline{v'_x v'_x}}$$
$$+ \boxed{\frac{\partial}{\partial y}\rho\overline{v'_x v'_y}} = \mu\frac{\partial^2\bar{v}_x}{\partial y^2}. \tag{11.17}$$

The terms in the dashed boxes are new terms arising from the turbulent nature of the flow. In a laminar flow, $v'_x$ and $v'_y$ are zero; thus, Eqn. 11.17 becomes identical with Eqn. 11.13 for a steady flow.

It is common practice to define

$$\tau_{xx}^{\text{turb}} \equiv -\rho\overline{v_x'v_x'} \tag{11.18a}$$

$$\tau_{xy}^{\text{turb}} \equiv -\rho\overline{v_x'v_y'}, \tag{11.18b}$$

which represent additional momentum fluxes resulting from the turbulent fluctuations. These terms have several designations and are referred to as components of the **turbulent momentum flux** (momentum flow per unit area), **turbulent stresses,** or **Reynolds stresses.** In a general formulation of the Reynolds-averaged momentum equation, a total of nine Reynolds stresses appear, analogous to the nine components of the laminar viscous shear stresses, i.e., $\tau_{ij}^{\text{turb}} \equiv -\rho\overline{v_i'v_j'}$ where $i$ and $j$ represent the coordinate directions.

Rearranging Eqn. 11.17, and applying continuity to the evaluation of the derivatives of the velocity products, yields our final result for a two-dimensional turbulent boundary-layer flow:

$$\rho\left(\overline{v}_x\frac{\partial\overline{v}_x}{\partial x} + \overline{v}_y\frac{\partial\overline{v}_x}{\partial y}\right) = \mu\frac{\partial^2\overline{v}_x}{\partial y^2} - \frac{\partial}{\partial y}\rho\overline{v_x'v_y'}, \tag{11.19}$$

where we assume that $(\partial/\partial x)\tau_{xx}^{\text{turb}}$ can be neglected.

**Axisymmetric Jet**   The turbulent axisymmetric jet is a particularly important flow in our study of combustion, so in this section we introduce the Reynolds-averaged form of the momentum equation analogous to the two-dimensional cartesian boundary layer presented in the previous section. For a constant-density jet, the continuity equation and axial momentum equation describing the steady flowfield were given in Chapter 9 (Eqns. 9.3 and 9.4). Performing the same operations of velocity decomposition and time-averaging to the axial momentum equation for an axisymmetric flow, yields, after simplification,

$$\rho\left(\overline{v}_x\frac{\partial\overline{v}_x}{\partial x} + \overline{v}_r\frac{\partial\overline{v}_x}{\partial r}\right) = \mu\frac{1}{r}\frac{\partial}{\partial r}\left(r\frac{\partial\overline{v}_x}{\partial r}\right) - \frac{1}{r}\frac{\partial}{\partial r}\left(r\rho\overline{v_r'v_x'}\right). \tag{11.20}$$

In Eqn. 11.20, we now see the turbulent shear to be related to the fluctuations of the axial and radial velocities, $\tau_{rx}^{\text{turb}} = -\rho\overline{v_r'v_x'}$.

In a subsequent section, we will see how Eqn. 11.20 can be employed to find the velocity distributions in a free (unconfined) turbulent jet.

## The Closure Problem

The Reynolds-averaging discussed above introduces new unknowns into the equations of motion, viz., the turbulent stresses, $-\rho\overline{v_i'v_j'}$. Finding a way to evaluate these stresses and solving for any other additional unknowns that may be introduced in the process is termed the **closure problem** of turbulence. Presently, there are many techniques that are used to "close" the system of governing equations, ranging from the very straightforward to the quite sophisticated [16–19]. We now investigate the simplest ideas that can be employed to achieve closure, and apply these to solving our archetypal flow, the free jet.

**Eddy Viscosity**   The first concept that we wish to explore is that of the **eddy viscosity.** The eddy viscosity is a fiction arising out of our treatment of the turbulent momentum fluxes as turbulent stresses. For example, we can rewrite Eqn. 11.20 explicitly in terms of the laminar (Newtonian) and turbulent (Reynolds) stresses as follows:

$$\rho\left(\bar{v}_x\frac{\partial \bar{v}_x}{\partial x}+\bar{v}_r\frac{\partial \bar{v}_x}{\partial r}\right)=\frac{1}{r}\frac{\partial}{\partial r}[r(\tau_{\text{lam}}+\tau_{\text{turb}})]. \tag{11.21}$$

Furthermore, we can define the stresses to be proportional to the mean velocity gradient, i.e.,

$$\tau_{\text{lam}}=\mu\frac{\partial \bar{v}_x}{\partial r} \tag{11.22a}$$

and

$$\tau_{\text{turb}}=\rho\varepsilon\frac{\partial \bar{v}_x}{\partial r}, \tag{11.22b}$$

where $\mu$ is the molecular viscosity and $\varepsilon$ is the kinematic eddy viscosity ($\varepsilon = \mu_{\text{turb}}/\rho$, where $\mu_{\text{turb}}$ is the apparent turbulent viscosity). The relationship expressed by Eqn. 11.22a is just the familiar expression derived for a Newtonian fluid for the flow of interest; the relationship expressed by Eqn. 11.22b, however, is the definition of the so-called eddy viscosity. The idea of an eddy viscosity as defined in Eqn. 11.22b was first proposed by Boussinesq [20] in 1877. For our cartesian-coordinate problem (see Eqn. 11.19), the stresses are given by

$$\tau_{\text{lam}}=\mu\frac{\partial \bar{v}_x}{\partial y}, \tag{11.23a}$$

$$\tau_{\text{turb}}=\rho\varepsilon\frac{\partial \bar{v}_x}{\partial y}. \tag{11.23b}$$

An **effective viscosity, $\mu_{\text{eff}}$,** is defined as

$$\mu_{\text{eff}}=\mu+\mu_{\text{turb}}=\mu+\rho\varepsilon, \tag{11.24}$$

so that

$$\tau_{\text{tot}}=(\mu+\rho\varepsilon)\frac{\partial \bar{v}_x}{\partial y}. \tag{11.25}$$

For turbulent flows far from a wall, $\rho\varepsilon \gg \mu$, so $\mu_{\text{eff}} \cong \rho\varepsilon$; however, near a wall, both $\mu$ and $\rho\varepsilon$ contribute to the total stress in the fluid.

Note that the introduction of the eddy viscosity *per se* does not achieve closure; the problem is now transformed into how to determine a value or the functional form of $\varepsilon$. Unlike the molecular viscosity $\mu$, which is a thermophysical property of the fluid itself, the eddy viscosity $\varepsilon$ depends on the flow itself. One would not necessarily expect the same value of $\varepsilon$ for two distinctly different flows; for example, a swirling confined flow with recirculation, and a free jet, are not likely to have the same values for $\varepsilon$. Moreover, since $\varepsilon$ depends on the local flow properties, it will take on different values at different locations within the flow. Therefore, one must be careful not to take too literally the analogy between laminar and turbulent flows implied by Eqns. 11.22a and b and 11.23a and b. Also, in some flows, the turbulent stresses are not proportional to the mean velocity gradients, as is required by Eqns. 11.22b and 11.23b [17].

**Mixing-Length Hypothesis**   The simplest closure scheme would be to assume the eddy viscosity is a constant throughout the flowfield; unfortunately, experimental evidence shows this not to be a generally useful assumption, as expected. More sophisticated hypotheses, therefore, need to be applied. One of the most useful, and yet simple, hypotheses is that proposed by Prandtl [21]. Prandtl's hypothesis, by analogy with the kinetic theory of gases, states that the eddy viscosity is proportional to the product of the fluid density, a length scale called the **mixing length,** and a characteristic turbulent velocity, i.e.,

$$\mu_{\text{turb}} = \rho\varepsilon = \rho\ell_m v_{\text{turb}}. \tag{11.26}$$

Furthermore, Prandtl [21] assumed that the turbulent velocity, $v_{\text{turb}}$, is proportional to the product of the mixing length, $\ell_m$, and the magnitude of the mean velocity gradient, $|\partial\bar{v}_x/\partial y|$. Thus,

$$\mu_{\text{turb}} = \rho\varepsilon = \rho\ell_m^2 \left|\frac{\partial\bar{v}_x}{\partial y}\right|. \tag{11.27}$$

Equation 11.27 is useful in dealing with flows near a wall. For free (unconfined) turbulent flows, Prandtl [22] proposed an alternative hypothesis for the characteristic turbulent velocity in Eqn. 11.26, i.e., $v_{\text{turb}} \propto \bar{v}_{x,\text{max}} - \bar{v}_{x,\text{min}}$. The corresponding turbulent viscosity is then

$$\mu_{\text{turb}} = \rho\varepsilon = 0.1365\rho\ell_m(\bar{v}_{x,\text{max}} - \bar{v}_{x,\text{min}}), \tag{11.28}$$

where the numerical constant is chosen to agree with experimental results. Equation 11.28 is a key relation that we will invoke in our subsequent solution to the jet problem. Note, however, that we have yet to obtain closure since we have introduced yet another unknown! All that has been done so far is to replace the unknown velocity correlation $\overline{v_x'v_y'}$ with an expression involving an unknown eddy viscosity, and, in turn, to relate that eddy viscosity to an unknown mixing length. Our next step finally attains closure by specifying the mixing length. Since the mixing length depends on the flow itself, different specifications generally are required for each kind of flow. Mixing-length functions are given for a wide variety of flows in Ref. [17]; we will focus, however, only on the mixing lengths required to solve the jet and wall-flow problems.

For a free axisymmetric jet,

$$\ell_m = 0.075\delta_{99\%},$$  (11.29)

where $\delta_{99\%}$ is the half-width of the jet measured from the jet centerline to the radial location where $\bar{v}_x$ has decayed to be only 1 percent of its centerline value. Note that the mixing length increases with distance from the jet origin since $\delta_{99\%}$ grows with axial distance. Note also that there is no radial dependence specified for the mixing length; hence, Eqn. 11.29 implies that $\ell_m$ is constant over the jet width at any axial station.

A different $\ell_m$ obtains for flows near a wall, where a cross-stream dependence is inherent in the mixing length. For a wall boundary layer, the flow is conveniently divided into three zones: the **viscous sublayer** adjacent to the wall, a **buffer layer,** and a **fully developed** turbulent region far from the wall. The mixing length for each of these three regions is given by the following equations [17]:

*The laminar sublayer,*

$$\ell_m = 0.41y\left[1 - \exp\left(-\frac{y\sqrt{\rho\tau_w}}{26\mu}\right)\right],$$  (11.30a)

*The buffer layer,*

$$\ell_m = 0.41y \quad \text{for} \quad y \le 0.2195\delta_{99\%},$$  (11.30b)

*The fully developed region,*

$$\ell_m = 0.09\delta_{99\%}.$$  (11.30c)

In the above expressions, $\tau_w$ is the local wall shear stress, and $\delta_{99\%}$ is the local boundary-layer thickness, defined as the $y$-location at which the velocity equals 99 percent of the freestream value. Note that in Eqn. 11.30a, proposed by van Driest [23], the mixing length vanishes as the distance from the wall $y$ goes to zero; hence, $\mu_{eff} = \mu$. For sufficiently large $y$-values, Eqn. 11.30a degenerates to Eqn. 11.30b.

For turbulent flows in a circular pipe or tube, the mixing-length distribution given by Nikuradse [24] is frequently invoked:

$$\ell_m/R_0 = 0.14 - 0.08(r/R_0)^2 - 0.06(r/R_0)^4,$$  (11.31)

where $R_0$ is the pipe radius. With the closure problem solved, i.e., the specification of mixing lengths, we are now in a position to solve the governing equations to yield the velocity fields for our chosen example, the jet. This solution is discussed in the next section.

---

**Example 11.2**

Determine a numerical value for the viscosity, $\mu_{turb}$, for a free jet of air at an axial location where the mean centerline velocity has decayed to 60 percent of the initial velocity, i.e., $\bar{v}_{x,0}/v_e = 0.6$.

The jet width, $\delta_{99\%}$, at this location is 15 cm. The initial jet velocity is 70 m/s. The pressure is 1 atm and the temperature is 300 K. Also, compare the turbulent viscosity with the molecular (laminar) viscosity.

**Solution**

To evaluate $\mu_{turb}$, we will employ the defining relationship, Eqn. 11.28, together with the expression defining the jet mixing length, Eqn. 11.29. We start by finding the mixing length:

$$\ell_m = 0.075\delta_{99\%}$$
$$= 0.075(0.15) = 0.01125 \text{ m}.$$

We obtain the density via the ideal-gas law:

$$\rho = \frac{P}{(R_u/MW)T} = \frac{101{,}325}{(8315/28.85)300} = 1.17 \text{ kg/m}^3.$$

The turbulent viscosity is (Eqn. 11.28)

$$\mu_{turb} = 0.1365\rho\ell_m(\bar{v}_{x,\max} - \bar{v}_{x,\min})$$
$$= 0.1365(1.17)(0.01125)[0.6(70) - 0] = 0.0755.$$

We then check units:

$$\mu_{turb}\ [=] \frac{\text{kg}}{\text{m}^3}(\text{m})\left(\frac{\text{m}}{\text{s}}\right) = \left(\frac{\dfrac{\text{N·s}^2}{\text{m}}}{\text{m}^3}\right)(\text{m})\left(\frac{\text{m}}{\text{s}}\right) = \frac{\text{N·s}}{\text{m}^2}$$

$$\boxed{\mu_{turb} = 0.0755 \text{ N·s/m}^2}$$

From Appendix Table C.1, the molecular viscosity of air at 300 K is $184.6 \cdot 10^{-7}$ N-s/m$^2$. Thus,

$$\boxed{\frac{\mu_{turb}}{\mu_{lam}} = \frac{0.0755}{184 \cdot 10^{-7}} = 4090}$$

**Comment**

These calculations show indeed that the turbulent viscosity dominates over the molecular viscosity, i.e., $\mu_{eff} \cong \mu_{turb}$. We also note that the width of the jet is approximately $13(\delta_{99\%}/\ell_m = 1/0.075)$ mixing lengths.

# AXISYMMETRIC TURBULENT JET

The fundamental conservation equations (mass and axial momentum) that describe the turbulent jet are essentially the same as those used in Chapter 9 for the laminar jet; however, mean velocities replace instantaneous ones, and the molecular viscosity is replaced by the effective viscosity. For the turbulent jet, the molecular viscosity is negligible in comparison with the eddy viscosity (see Example 11.2). Thus, the axial momentum equation, Eqn. 9.4, becomes

$$\bar{v}_x\frac{\partial \bar{v}_x}{\partial x} + \bar{v}_r\frac{\partial \bar{v}_x}{\partial r} = \frac{1}{r}\frac{\partial}{\partial r}\left(r\varepsilon\frac{\partial \bar{v}_x}{\partial r}\right), \tag{11.32}$$

and that for mass conservation (see Eqn. 9.3),

$$\frac{\partial(\overline{v}_x r)}{\partial x} + \frac{\partial(\overline{v}_r r)}{\partial r} = 0. \tag{11.33}$$

The boundary conditions for the mean velocities are identical to those used to solve the laminar jet problem, Eqns. 9.7a, b, and d.

To determine the eddy viscosity, $\varepsilon$, we combine the mixing-length relationships of Eqns. 11.28 and 11.29 to yield

$$\varepsilon = 0.0102\delta_{99\%}(x)\overline{v}_{x,\max}(x), \tag{11.34}$$

where both the jet width, $\delta_{99\%}(x)$, and maximum axial velocity, $\overline{v}_{x,\max}(x)$, are functions of axial position. The maximum axial velocity occurs on the jet centerline, so $\overline{v}_{x,\max}(x) = \overline{v}_{x,0}(x)$, and, in arriving at Eqn. 11.34, we assume that $\overline{v}_{x,\min}$ is zero, since the jet is issuing into a quiescent medium. At this juncture, we will employ some empirical information [25]. First, measurements show that for turbulent jets, $\delta_{99\%} \approx 2.5r_{1/2}$, where $r_{1/2}$ is the jet radius at which the axial velocity has fallen to half its centerline value. Figure 11.6 illustrates measurements of $r_{1/2}$ and $\delta_{99\%}$. The other experimental evidence we wish to employ is that $r_{1/2}$ grows in direct proportion to $x$, and $\overline{v}_{x,0}$ declines in an inverse manner with $x$, i.e.,

$$r_{1/2} \propto x^1 \tag{11.35a}$$

$$\overline{v}_{x,0} \propto x^{-1}. \tag{11.35b}$$

Thus, the $x$-dependencies in Eqn. 11.34 cancel, yielding

$$\varepsilon = 0.0256r_{1/2}(x)\overline{v}_{x,0}(x) = \text{constant}. \tag{11.36}$$

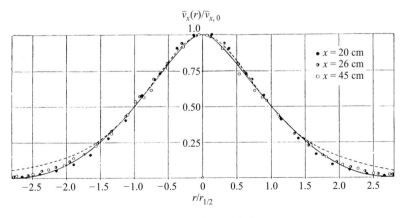

**Figure 11.6**    Radial profile of axial velocity for a turbulent jet.
SOURCE: From Ref. [25], reprinted by permission, © 1968, McGraw Hill.

This is a rather fortunate result in that we can now use the laminar jet solutions (Eqns. 9.8 and 9.9) by simply replacing the constant molecular viscosity with the just-revealed constant eddy viscosity (times density), $\rho\varepsilon$. Thus, the mean velocity components are

$$\bar{v}_x = \frac{3}{8\pi}\frac{J_e}{\rho\varepsilon x}\left[1+\frac{\xi^2}{4}\right]^{-2} \tag{11.37}$$

$$\bar{v}_r = \left[\frac{3J_e}{16\pi\rho_e}\right]^{1/2}\frac{1}{x}\frac{\xi-\dfrac{\xi^3}{4}}{\left[1+\dfrac{\xi^2}{4}\right]^2}, \tag{11.38}$$

where $J_e$ is the initial jet momentum flow, which for a uniform exit velocity, $v_e$, is expressed

$$J_e = \rho_e v_e^2 \pi R^2, \tag{11.39}$$

and

$$\xi = \left[\frac{3J_e}{16\rho_e\pi}\right]^{1/2}\frac{1}{\varepsilon}\frac{r}{x}. \tag{11.40}$$

For the time being, let us retain the unknown constant $\varepsilon$, first finding relationships for $r_{1/2}(x)$ and $\bar{v}_{x,0}(x)$. With these relationships and Eqn. 11.37, we can relate $\varepsilon$, $r_{1/2}(x)$, and $\bar{v}_{x,0}(x)$ solely to known jet parameters, $v_e$ and $R$, and, thereby, conclude our theoretical development.

The axial velocity, normalized by the assumed uniform exit velocity $v_e$, is obtained by substituting Eqn. 11.39 into Eqn. 11.37 and rearranging:

$$\bar{v}_x/v_e = 0.375(v_e R/\varepsilon)(x/R)^{-1}[1+\xi^2/4]^{-2}, \tag{11.41}$$

and the dimensionless centerline velocity decay is obtained by setting $r=0$ ($\xi=0$):

$$\bar{v}_{x,0}(x)/v_e = 0.375(v_e R/\varepsilon)(x/R)^{-1}. \tag{11.42}$$

To find an expression for $r_{1/2}$, we divide Eqn. 11.41 by 11.42, set the result equal to $1/2$, i.e., $\bar{v}_x/\bar{v}_{x,0} = 1/2$, and solve for $r(=r_{1/2})$. Thus,

$$r_{1/2} = 2.97\left(\frac{v_e R}{\varepsilon x}\right)^{-1}. \tag{11.43}$$

Simultaneously solving Eqns. 11.36, 11.42, and 11.43 yields our desired final results:

The jet velocity decay is

$$\bar{v}_{x,0}/v_e = 13.15(x/R)^{-1}, \tag{11.44}$$

the jet spreading rate is

$$r_{1/2}/x = 0.08468, \tag{11.45}$$

and the eddy viscosity is

$$\varepsilon = 0.0285 v_e R. \tag{11.46}$$

It is interesting to compare the velocity decays and spreading rates of the turbulent jet with those of the laminar axisymmetric jet (Eqns. 9.13 and 9.14). For the turbulent jet, we have the interesting behavior that neither the velocity decay nor the spreading rate depend on the jet Reynolds number, whereas for the laminar jet, the velocity decay is directly proportional to $Re_j$ (Eqn. 9.13) and the spreading rate is inversely proportional to $Re_j$ (Eqn. 9.14). Thus, the character of a turbulent jet is independent of exit conditions, provided the Reynolds number is sufficiently high to assure fully turbulent flow. This result has interesting implications for turbulent jet flames, which we will discuss in Chapter 13.

## BEYOND THE SIMPLEST MODEL

We previously indicated the existence of closure methods more sophisticated than the simple prescribed mixing-length models presented earlier in this chapter. The usual step in the hierarchy of turbulence models beyond the mixing-length model is the so-called **two-equation models** for turbulence. Two-equation models comprise the Reynolds-averaged equations of motion for the mean flow variables, $\bar{v}_x$ and $\bar{v}_r$, along with coupled partial differential equations for two turbulence variables. The overwhelming choice for the first variable is $k$, the turbulence kinetic energy, whereas the second variable is less generally agreed upon. Perhaps the most well-known choice for the second variable, but not necessarily the best [19], is the turbulence kinetic energy dissipation rate, $\epsilon$ (see Eqn. 11.7). This choice results in the **$k$-$\epsilon$ model.** Numerical solutions for the mean and turbulence quantities are generated most frequently using grid-based approximations to the coupled set of governing equations. In effect, two-equation models calculate local mixing lengths throughout the flowfield, rather than requiring that the mixing length be known *a priori*. As in the mixing-length model, the Reynolds stresses appearing in the time-averaged momentum equations are the modeled quantities: The Boussinesq approximation is invoked, i.e., $-\overline{v'_x v'_y} = \mu_{\text{turb}} \, \partial \bar{v}_x/\partial y$ (see Eqn. 11.22b), and the local turbulent viscosity is calculated from $k$ and $\epsilon$: $\mu_{\text{turb}} = C_\mu k^2/\epsilon$, where $C_\mu$ is a constant. Spalding [26] presents some interesting historical insights in the development of two-equation models, while a very readable discussion of the formulation and implementation of these models can be found in Wilcox [19].

   **Higher-order models** use additional partial differential equations to calculate the Reynolds stresses directly, rather than to model them, as is done in the two-equation models. In these so-called **Reynolds-stress models,** the closure problem is relegated to modeling the higher-order terms that now appear in the Reynolds-stress transport equations, for example, terms like $\rho \overline{v'_x v'_y v'_z}$ and $\overline{p' v'_x}$. The philosophy here is that delaying the invocation of modeling assumptions to higher-order terms provides a closer representation of the true physics of the flow. Discussions of Reynolds-stress models are found in Refs. [19, 27, 28].

Both two-equation and higher-order approaches have their roots in the solution of time-averaged mathematical representations of the flow. Workers in the field admit the shortcomings of the statistical approach and associated heuristic closure assumptions in representing the true physics of the flow [19, 27]; however, an extensive database has been built upon such methods that has been found to be quite useful in engineering practice.

An alternative to time-averaged, or statistical, approaches is **direct numerical simulation (DNS).** We alluded to this approach in the section immediately preceding our discussion of Reynolds averaging. With DNS, the complete time history of the velocity is resolved down to the Kolmogorov spatial and temporal scales throughout the flowfield. Computer requirements preclude the application of DNS to practical flows of engineering interest [15]; however, DNS calculations are used to test models of turbulence and to provide insights into the correct physics of turbulent flows. The DNS results shown in Fig. 11.3 are an example of this. Good overviews of DNS and how this area of research has developed are found in the reviews of Moin and Mahesh [29] and Rogallo and Moin [30].

Probability density function (pdf) methods [15] have also been developed to close the Reynolds stress terms in the momentum equation; moreover, these methods are particularly useful in dealing with closure issues in reacting flows. Haworth [31] provides a good review of pdf methods.

Large-eddy simulation is an approach that combines attributes of both direct numerical simulation and statistical turbulence modeling methods. **Large-eddy simulation (LES)** temporally and spatially resolves the flow for scales much larger than Kolmogorov scales, but models the turbulence at scales smaller than those resolved. The attractiveness of LES is that all of the large-scale, energy-containing motions of the flow are calculated, not modeled, thus preserving the true physics of the flow. As previously mentioned, this is particularly important for mixing and reaction. Although more economical than direct numerical simulation, large-eddy simulation still requires substantial computational effort, presently preventing it from being practical for solution of routine engineering problems. With continued advances in computing capability, LES may become a common engineering tool. Reviews of LES methods are presented in Refs. [15, 30, 32–34].

# SUMMARY

In this chapter, the groundwork is laid for subsequent discussions of premixed (Chapter 12) and nonpremixed (Chapter 13) turbulent combustion. Turbulence was defined and the concepts of mean and fluctuating properties and turbulence intensity were introduced. We discussed, albeit superficially, the structure of turbulent flows and defined four length scales that help to characterize turbulent flow structure.

You should be familiar with the length scales and how they interrelate through various turbulent Reynolds numbers. In particular, you should be aware of how the range of length scales (the difference between the largest and smallest eddies) increases with Reynolds numbers. In introducing the mathematical concept of Reynolds averaging, we saw how turbulent, or Reynolds, stresses arise in the time-averaged conservation equations, which, in turn, brought to light the problem of closure. You should have some appreciation for this issue. An example of closure, Prandtl's mixing-length hypothesis, was presented, with formulae given for free jets and wall-bounded flows. The mixing-length theory was applied to a free jet, and we saw, as a consequence, that the normalized velocity field ($v(r, x)/v_e$) scales with the initial jet diameter, indicating a universal (i.e., $Re$-independent) structure of turbulent jets. Thus, the spreading angle is a constant, a finding distinctly different from the laminar case; this difference results in turbulent jet flames (Chapter 13) having properties that are considerably different than those of laminar flames.

## NOMENCLATURE

| | |
|---|---|
| $A$ | Area (m$^2$) |
| $C_\mu$ | $k$-$\epsilon$ model constant |
| $D$ | Diameter (m) |
| $g$ | Gravitational acceleration (m/s$^2$) |
| $J$ | Momentum flow (kg-m/s$^2$) |
| $k, ke$ | Kinetic energy per unit mass (m$^2$/s$^2$) |
| $\ell_0$ | Integral scale or turbulence macroscale (m) |
| $\ell_\lambda$ | Taylor microscale (m) |
| $\ell_K$ | Kolmogorov microscale (m) |
| $\ell_m$ | Mixing length (m) |
| $\dot{m}$ | Flowrate |
| $MW$ | Molecular weight |
| $L$ | Characteristic width of flow or macroscale (m) |
| $p$ | Arbitrary property |
| $P$ | Pressure (Pa) |
| $r$ | Radial coordinate (m) |
| $r_{1/2}$ | Jet half-width at half-height (m) |
| $R$ | Initial jet radius (m) |
| $R_u$ | Universal gas constant |
| $R_x$ | Correlation coefficient, Eqn. 11.4 |
| $R_0$ | Pipe radius (m) |
| $Re$ | Reynolds number (dimensionless) |
| $t$ | Time (s) |
| $T$ | Temperature (K) |
| $v$ | Velocity (m/s) |
| $v_r, v_x$ | Radial and axial velocity components, respectively (m/s) |
| $v_x, v_y, v_z$ | Velocity components in cartesian coordinates (m/s) |

| $x, y, z$ | Cartesian coordinates (m) |
| $Y$ | Mass fraction (kg/kg) |

**Greek Symbols**

| $\delta_{99\%}$ | Jet width (m) |
| $\varepsilon$ | Eddy viscosity (m$^2$/s) |
| $\epsilon_0$ | Dissipation rate, Eqn. 11.7 (m$^2$/s$^3$) |
| $\mu$ | Absolute (molecular) viscosity (N-s/m$^2$) |
| $\mu_{\text{eff}}$ | Effective viscosity, Eqn. 11.24 (N-s/m$^2$) |
| $\nu$ | Kinematic viscosity (m$^2$/s) |
| $\rho$ | Density (kg/m$^3$) |
| $\tau$ | Shear stress (N/m$^2$) |

**Subscripts**

| $e$ | Exit |
| lam | Laminar |
| max | Maximum |
| min | Minimum |
| *rms* | Root-mean-square |
| turb | Turbulent |
| $w$ | Wall |

**Other Notation**

| $\overline{(\ )}$ | Time-averaged quantity |
| $(\ )'$ | Fluctuating quantity |

# REFERENCES

1. Reynolds, O., "An Experimental Investigation of the Circumstances which Determine Whether the Motion of Water shall be Direct or Sinuous, and of the Law of Resistance in Parallel Channels," *Phil. Trans. Royal Society of London,* 174: 935–982, 1883.

2. Dimotakis, P. E., Lye, R. C., and Papantoniou, D. Z., "Structure and Dynamics of Round Turbulent Jets," *Fluid Dynamics Transactions,* 11: 47–76 (1982).

3. Van Dyke, M., *An Album of Fluid Motion,* Parabolic Press, Stanford, CA, p. 95, 1982.

4. Brown, G. L., and Roshko, A., "On Density Effects and Large Structure in Turbulent Mixing Layers," *Journal of Fluid Mechanics,* 64: 775–816 (1974).

5. Wygnanski, I., Oster, D., Fiedler, H., and Dziomba, B., "On the Perseverance of a Quasi-Two-Dimensional Eddy-Structure in a Turbulent Mixing Layer," *Journal of Fluid Mechanics,* 93: 325–335 (1979).

6. Andrews, G. E., Bradley, D., and Lwakabamba, S. B., "Turbulence and Turbulent Flame Propagation—A Critical Appraisal," *Combustion and Flame,* 24: 285–304 (1975).

7. Tennekes, H., "Simple Model for the Small-Scale Structure of Turbulence," *Physics of Fluids,* 11: 669–671 (1968).

8. Townsend, A. A., "On the Fine-Scale Structure of Turbulence," *Proceedings of the Royal Society of London, Series A,* 208: 534–542 (1951).

9. Glickman, T. S., *Glossary of Meteorology,* 2nd Ed., American Meteorological Society, Boston, MA, 2000.

10. Tennekes, H., and Lumley, J. L., *A First Course in Turbulence,* MIT Press, Cambridge, MA, 1972.

11. Libby, P. A., and Williams F. A., "Fundamental Aspects," in *Turbulent Reacting Flows* (P. A. Libby and F. A. Williams, eds.), Springer-Verlag, New York, 1980.

12. She, Z.-S., Jackson, E., and Orsag, S. A., "Intermittent Vortex Structures in Homogeneous Isotropic Turbulence," *Nature,* 344: 226–228 (1990).

13. She, Z.-S., Jackson, E., and Orsag, S. A., "Structure and Dynamics of Homogeneous Turbulence: Models and Simulations," *Proceedings of the Royal Society of London, Series A,* 434: 101–124 (1991).

14. Correa, S. M., "A Review of $NO_x$ Formation Under Gas-Turbine Combustion Conditions," *Combustion Science and Technology,* 87: 329–362 (1992).

15. Pope, S. B., *Turbulent Flows,* Cambridge University Press, New York, 2000.

16. Patankar, S. V., and Spalding, D. B., *Heat and Mass Transfer in Boundary Layers,* 2nd Ed., International Textbook, London, 1970.

17. Launder, B. E., and Spalding, D. B., *Lectures in Mathematical Models of Turbulence,* Academic Press, New York, 1972.

18. Schetz, J. A., *Injection and Mixing in Turbulent Flow,* Progress in Astronautics and Aeronautics, Vol. 68, American Institute of Aeronautics and Astronautics, New York, 1980.

19. Wilcox, D. C., *Turbulence Modeling for CFD,* 3rd Ed., DCW Industries, Inc., La Cañada, CA, 2006.

20. Boussinesq, T. V., "Théorie de l'écoulement Tourbillant," *Mém. prés. Acad. Sci.,* Paris, XXIII, 46 (1877).

21. Prandtl, L., "Über die ausgebildete Turbulenze," *Z.A.M.M,* 5: 136–139 (1925).

22. Prandtl, L., "Bemerkungen zur Theorie der Freien Turbulenz," *Z.A.M.M.,* 22: 241–243 (1942).

23. van Driest, E. R., "On Turbulent Flow Near a Wall," *Journal of the Aeronautical Sciences,* 23: 1007 (1956).

24. Nikuradse, J., "Laws of Flow in Rough Pipes," English Translation in NACA Technical Memorandum 1292, November 1950. (Original published in German, 1933.)

25. Schlichting, H., *Boundary-Layer Theory,* 6th Ed., McGraw-Hill, New York, 1968.

26. Spalding, D. B., "Kolmogorov's Two-Equation Model of Turbulence," *Proceedings of the Royal Society of London, Series A,* 434: 211–216 (1991).

27. Libby, P. A., *Introduction to Turbulence,* Taylor & Francis, Washington, DC, 1996.

28. Hanjalic, K., "Advanced Turbulence Closure Models: A View of Current Status and Future Prospects," *International Journal of Heat and Fluid Flow,* 15: 178–203 (1994).

29. Moin, P., and Mahesh, K., "Direct Numerical Simulation: A Tool in Turbulence Research," *Annual Review of Fluid Mechanics,* 30: 539–578 (1998).

30. Rogallo, R. S., and Moin, P., "Numerical Simulation of Turbulent Flows," *Annual Review of Fluid Mechanics,* 16: 99–137 (1984).

31. Haworth, D. C., "Progress in Probability Density Function Methods for Turbulent Reacting Flows," *Progress in Energy and Combustion Science,* 36:168–259 (2010).

32. Lesieur, M., and Métais, O., "New Trends in Large-Eddy Simulations of Turbulence," *Annual Review of Fluid Mechanics,* 28: 45–82 (1996).

33. Sagaut, P., *Large Eddy Simulation for Incompressible Flows: An Introduction,* 3rd Ed., Springer, New York, 2005.

34. Lesieur, M., Métais, O., and Comte, P., *Large-Eddy Simulations of Turbulence,* Cambridge University Press, Cambridge, 2005.

## QUESTIONS AND PROBLEMS

**11.1** Make a list of all of the boldfaced words appearing in Chapter 11 and define each.

**11.2** The axial velocity at a point in a flow is given by

$$v_x(t) = A\sin(\omega_1 t + \phi_1) + B\sin(\omega_2 t + \phi_2) + C.$$

A. Develop an expression for the time-mean velocity, $\bar{v}_x$.

B. Develop an expression for $v_x'$.

C. Develop an expression for $v_{x,rms}'$.

**11.3** Perform the Reynolds-averaging process on the second and fourth terms of Eqn. 11.13. Compare your results with Eqn. 11.17.

**11.4** In the realm of turbulence theory, what is meant by closure? Discuss. Illustrate closure concretely using the free jet as an example, citing appropriate equations as necessary.

**11.5** In a spark-ignition engine, the integral scale, $\ell_0$, is found to be approximately equal to one-third the clearance height. At top-dead-center (TDC), the residual inlet jet has a velocity of 30 m/s and a relative turbulence intensity of 30 percent. Estimate the Taylor and Kolmogorov microscales at TDC, in millimeters, for a motoring (nonfiring) engine. Assume isentropic compression of air from $P = 1$ atm and $T = 300$ K with a volumetric compression ratio of 7:1. The TDC clearance height is 10 mm.

**11.6** In play, a child blows through a 6-mm-diameter straw aimed at a sibling. If a velocity of 3 m/s is needed for the sibling to feel the jet, and the child can maintain an initial velocity of 35 m/s sufficiently long, how far away can the child stand from the sibling and the jet still be felt? Also, verify that the jet is turbulent.

# 12

# Turbulent Premixed Flames

## OVERVIEW

Paradoxically, turbulent premixed flames are of tremendous practical importance, being encountered in many useful devices, as discussed below, while their theoretical description is still a matter of uncertainty, or at least controversy. Because there is, as, yet, no comprehensive, generally accepted theory of turbulent premixed flames, and because many of the descriptions are highly mathematical, we will avoid developing or discussing any particular theory in detail, departing from the approach of previous chapters. Rather, our approach will be more phenomenological and empirical, discussing some of the important issues that complicate a rigorous understanding of premixed combustion. Several reviews of this subject have been presented in the past few decades and are recommended as a starting point for those wishing to explore this complex subject in greater detail [1–12]. The most recent of these [9–11] clearly frame the many issues involved in understanding turbulent premixed combustion.

## SOME APPLICATIONS

### Spark-Ignition Engines

The spark-ignition engine is a prime example of the use of premixed combustion. Figure 12.1 shows several spark-ignition, automobile-engine combustion chamber configurations. The fuel–air mixture is produced either by a carburetor system or, now almost universally in the United States, by a fuel-injection system. Even though the fuel may be introduced as a liquid, spark-ignition engine fuels are highly volatile, and the liquid has time to vaporize and thoroughly mix with the air before the

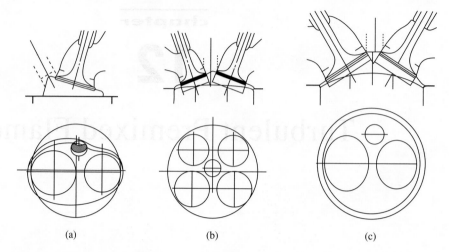

**Figure 12.1**     Various configurations of spark-ignition engine combustion chambers:
(a) wedge, (b) four-valve pent roof and (c) hemispherical.
SOURCE: Adapted from Ref. [13] and used with permission.

mixture is ignited with a spark. The combustion duration is an important parameter in the operation of spark-ignition engines and is controlled by the turbulent flame speed and the distribution of the combustion volume. Compact combustion chambers, such as shown in Fig. 12.1a, produce short combustion durations. Combustion duration governs, to a large degree, the lean-limit of stable operation, tolerance to exhaust-gas recirculation, thermal efficiency, and the production of oxides of nitrogen ($NO_x$) emissions. More information directly related to spark-ignition engine combustion can be found in internal combustion engine textbooks, e.g., Refs. [13] and [14].

## Gas-Turbine Engines

Although gas-turbine engines are commonly used to power aircraft, these engines are being employed more and more in stationary power systems. Current turbine combustor design is greatly influenced by the need to control simultaneously soot, carbon monoxide, and oxides of nitrogen emissions. Older engines employed purely nonpremixed combustion systems, with a near-stoichiometric flame zone integrated with secondary air streams to complete combustion and dilute the products to the proper temperature before entry into the turbine. Current designs and experimental systems utilize various degrees of premixing to avoid high-temperature, $NO_x$-forming zones [15–17]. Figure 12.2 illustrates a silo-type combuster (see Fig. 11.5) that employs natural-gas burners that can operate in a premixed mode. A low-$NO_x$ gas-turbine combustor can from a premixed cannular-type combustor is shown in Fig. 2.4. With the benefit of $NO_x$ control from premixed combustion comes a number

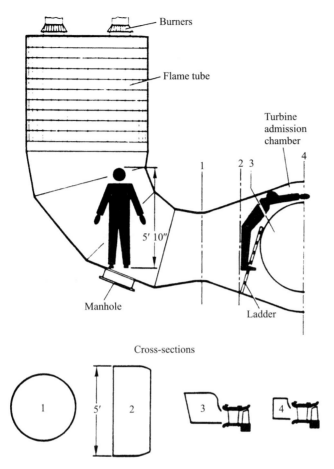

**Figure 12.2**    Schematic of silo-type, gas-turbine combustor with ceramic-tile-lined chamber and access for inspection of the combustion chamber and the turbine inlet.
SOURCE: From Ref. [16]. Reprinted by permission of the American Society of Mechanical Engineers.

of concerns; primary among them are turndown ratio (ratio of maximum to minimum flowrates), flame stability, and carbon monoxide emissions.

## Industrial Gas Burners

Premixed flames are used in a large number of industrial applications. Premixing of the fuel gas and air may be accomplished upstream of the burner nozzle using a mixer (see Fig. 12.3), or the fuel gas and air may be mixed within the burner nozzle. Figure 12.4 shows two examples of nozzle-mixing burners. Depending upon the degree of premixing achieved in the burner's nozzle, the flame produced by this type of burner may exhibit some nonpremixed or diffusion flame characteristics.

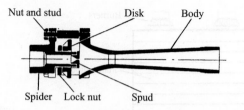

**Figure 12.3**    Inspirator mixer uses high-pressure gas flow to entrain air. Mixing occurs downstream of the venturi throat as the fuel–air mixture passes to the burner (not shown).
SOURCE: From Ref. [18]. Reprinted by permission of Fives North American Combustion, Inc.

Fully premixed burners are illustrated in Fig. 12.5. Small-port or ported-manifold burners (Fig. 12.5, top) are used in domestic appliances and industrial applications such as make-up air heating, drying ovens, baking ovens, food roasters, and deep-fat vats [19]. The large-port or pressure-type burners (Fig. 12.5, bottom), as well as nozzle-mixing burners, are used in many industrial applications: for example, kilns used in the manufacturing of bricks, porcelain, and tile, and furnaces for heat-treating, forging, and melting [19].

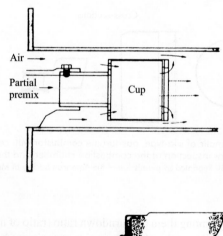

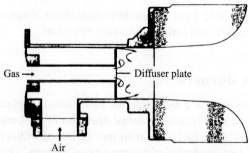

**Figure 12.4**    Nozzle-mixing gas burners for industrial applications.
SOURCE: From Ref. [19]. Reprinted by permission of Fives North American Combustion, Inc.

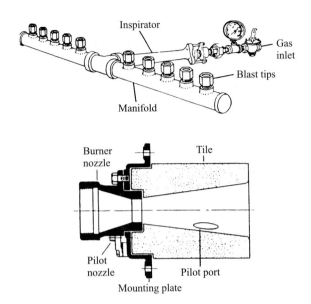

**Figure 12.5**    Examples of premixed burners. Top example includes inspirator mixer illustrated in Fig. 12.3.

SOURCE: From Ref. [18]. Reprinted by permission of Fives North American Combustion, Inc.

## DEFINITION OF TURBULENT FLAME SPEED

Unlike a laminar flame, which has a propagation velocity that depends uniquely on the thermal and chemical properties of the mixture, a turbulent flame has a propagation velocity that depends on the character of the flow, as well as on the mixture properties. This is consistent with the basic nature of turbulent flows as discussed in Chapter 11. For an observer traveling with the flame, we can define a **turbulent flame speed,** $S_t$, as the velocity at which unburned mixture enters the flame zone in a direction normal to the flame. In this definition, we assume that the flame surface is represented as some time-mean quantity, recognizing that the instantaneous position of the high-temperature reaction zone may be fluctuating wildly. Because the direct measurement of unburned gas velocities at a point near a turbulent flame is exceedingly difficult, flame speeds frequently are determined from measurements of reactant flowrates. Thus, the turbulent flame speed can be expressed as

$$S_t = \frac{\dot{m}}{\overline{A}\rho_u},\qquad(12.1)$$

where $\dot{m}$ is the reactant flow rate, $\rho_u$ is the unburned gas density, and $\overline{A}$ is the time-smoothed flame area. Driscoll [11] also refers to this expression as the **global consumption speed.** Experimental determinations of turbulent flame speeds are complicated by determining a suitable flame area, $\overline{A}$, for thick, and frequently curved, flames. The ambiguity associated with determining this flame area can result in considerable uncertainty (or ambiguity) in measurements of turbulent burning velocities. Although the definition of turbulent burning velocity expressed by Eqn. 12.1 is commonly used, other definitions are also employed [11].

**Example 12.1**

Consider the measurement of turbulent flame speeds. An air–fuel mixture passes through a 40-mm × 40-mm flow channel with a flame anchored at the channel exit along the top and bottom walls, as shown in the sketch. Quartz sidewalls contain the flame beyond the exit, whereas the top and bottom are open to the laboratory. For a mean flow velocity of 68 m/s, the resulting wedge-shaped flame has an included angle of 13.5°, estimated from time-exposure photographs. Estimate the turbulent burning velocity at this condition. The properties of the unburned mixture are $T = 293$ K, $P = 1$ atm, and $MW = 29$ kg/kmol.

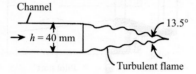

**Solution**

Equation 12.1 can be applied directly to find the turbulent burning velocity. The reactants flowrate is

$$\dot{m} = \rho_u A_{\text{duct}} \overline{V}_{\text{duct}}$$

$$= 1.206(0.04)^2 68 = 0.131 \text{ kg/s},$$

where the reactants density is estimated using the ideal-gas law, i.e.,

$$\rho_u = \frac{P}{RT} = \frac{101,325}{(8315/29)293} = 1.206 \text{ kg/m}^3.$$

From the flame geometry (wedge), we estimate the apparent flame area, $\overline{A}$, first by finding the length of the flame sheet, $L$:

$$\frac{h/2}{L} = \sin\left(\frac{13.5°}{2}\right)$$

or

$$L = \frac{h/2}{\sin 6.75°} = \frac{(0.04)/2}{\sin 6.75°} = 0.17 \text{ m}.$$

Thus,

$$\overline{A} = 2 \cdot \text{width} \cdot \text{length} = 2(0.04)0.17 = 0.0136 \text{ m}^2.$$

The turbulent burning velocity is then

$$\boxed{S_t = \frac{\dot{m}}{\overline{A}\rho_u} = \frac{0.131}{0.0136(1.206)} = 8.0 \text{ m/s}}$$

**Comment**

Note that we have calculated a mean turbulent burning velocity for the entire flame. The implied assumptions that the flame is perfectly wedge-shaped and that the unburned mixture enters the reaction zone at a uniform velocity are only approximations; thus, local turbulent flame speeds may differ from the calculated mean.

# STRUCTURE OF TURBULENT PREMIXED FLAMES

## Experimental Observations

One particular view of the structure of turbulent flames is illustrated in Fig. 12.6. Here we see superimposed instantaneous contours of convoluted thin reaction zones (Fig. 12.6a), obtained by visualizing the large temperature gradients in the flame using schlieren photography at different instants. The flow is from bottom to top with the flame stabilized above a tube through which the reactants flow into the open atmosphere. The instantaneous flame front is highly convoluted, with the largest folds near the top of the flame. The positions of the reaction zones move rapidly in space, producing a time-averaged view that gives the appearance of a thick reaction zone (Fig. 12.6b). This apparently thick reaction zone is frequently referred to as a **turbulent flame brush.** The instantaneous view, however, clearly shows the actual reaction front to be relatively thin, as in a laminar premixed flame. These reaction fronts are sometimes referred to as **laminar flamelets.**

As mentioned above, spark-ignition engines operate with turbulent premixed flames. Recent developments in laser-based instrumentation have allowed researchers to explore, in much more detail than previously possible, the hostile environment of the internal combustion engine combustion chamber. Figure 12.7 shows a time sequence of two-dimensional flame visualizations in a spark-ignition engine from a study

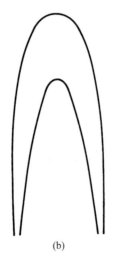

(a)                                    (b)

**Figure 12.6**     (a) Superposition of instantaneous reaction fronts obtained at different times. (b) Turbulent flame "brush" associated with a time-averaged view of the same flame.
SOURCE: (a) After Ref. [20] from Ref. [21]. Copyright 1962 by Royal Society Publishing.

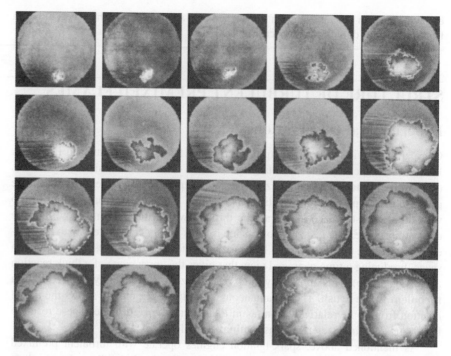

**Figure 12.7**    Visualization of turbulent flame propagation in a spark-ignition engine operating at 1200 rpm. Images represent a planar slice through the combustion chamber with the sequence starting shortly after ignition (upper left) and proceeding until the flame nears the cylinder walls (lower right).

SOURCE: Reprinted, with permission, from Ref. [22], © 1987, Society of Automotive Engineers, Inc.

by zur Loye and Bracco [22]. The flame begins to propagate outward from the spark plug, as shown in the first frame, and moves across the chamber until nearly all the gas is burned. In these flame visualizations, we see that the division between the unburned and burned gases occurs over a very short distance and the flame front is distorted by both relatively large- and small-scale wrinkles. For the specific conditions associated with Fig. 12.7, the burned gas and unburned gas are generally simply connected, i.e., there are very few islands or pockets of unburned gas within the burned gas, and vice versa. At higher engine speeds (2400 rpm), however, islands and pockets are seen frequently.

## Three Flame Regimes

The visualizations of turbulent flames presented above (Figs. 12.6 and 12.7) suggest that the effect of turbulence is to wrinkle and distort an essentially laminar flame front (Chapter 8). Turbulent flames of this type are referred to as being in the **wrinkled laminar-flame (or flamelet) regime.** This is one pole in our classification of turbulent premixed flames. At the other pole is the **distributed-reaction regime.** Falling between these two regimes is a region sometimes referred to as the **flamelets-in-eddies regime.**

**Regime Criteria** Before we discuss each regime in more detail, it is helpful to develop a basic understanding of the factors that differentiate these three regimes. To do this, we appeal to some of the basic concepts of turbulent flow presented in Chapter 11, in particular, the idea that various length scales exist simultaneously in a turbulent flow. Recall that the smallest scale, the Kolmogorov microscale, $\ell_K$, represents the smallest eddies in the flow. These eddies rotate rapidly and have high vorticity, resulting in the dissipation of the fluid kinetic energy into internal energy, i.e., fluid friction results in a temperature rise of the fluid. At the other extreme of the length-scale spectrum is the integral scale, $\ell_0$, which characterizes the largest eddy sizes. The basic structure of a turbulent flame is governed by the relationships of $\ell_K$ and $\ell_0$ to the laminar flame thickness, $\delta_L$ (see Chapter 8). The laminar flame thickness characterizes the thickness of a reaction zone controlled by molecular, not turbulent, transport of heat and mass. More explicitly, the three regimes are defined by

$$\text{Wrinkled laminar flames:} \quad \delta_L \leq \ell_K, \tag{12.2a}$$

$$\text{Flamelets in eddies:} \quad \ell_0 > \delta_L > \ell_K, \tag{12.2b}$$

$$\text{Distributed reaction:} \quad \delta_L > \ell_0. \tag{12.2c}$$

Equations 12.2a and 12.2c have clear physical interpretations; when the flame thickness, $\delta_L$, is much thinner than the smallest scale of turbulence, $\ell_K$, the turbulent motion can only wrinkle or distort the thin laminar flame zone (Eqn. 12.2a). The criterion for the existence of a wrinkled laminar flame (Eqn. 12.2a) is sometimes referred to as the **Williams–Klimov criterion** [5]. At the other extreme, if all scales of turbulent motion are smaller than the reaction zone thickness ($\delta_L$), then transport within the reaction zone is no longer governed solely by molecular processes, but is controlled, or at least influenced, by the turbulence. This criterion for the existence of a distributed-reaction zone is sometimes referred to as the **Damköhler criterion** [23].

It is convenient to discuss flame structure in terms of dimensionless parameters. The turbulence length scales and laminar flame thickness can be expressed as two dimensionless parameters: $\ell_K/\delta_L$ and $\ell_0/\delta_L$. Two additional parameters are important to our discussion: the turbulence Reynolds number, $Re_{\ell_0}$, defined in Chapter 11, and the Damköhler number, which is introduced below.

**Damköhler Number** An important dimensionless parameter in combustion is the **Damköhler number, $Da$**. This parameter appears in the description of many combustion problems and is quite important in understanding turbulent premixed flames. The fundamental meaning of the Damköhler number, $Da$, used here is that it represents the ratio of a characteristic flow or mixing time to a characteristic chemical time; thus,

$$Da \equiv \frac{\text{characteristic flow time}}{\text{characteristic chemical time}} = \frac{\tau_{\text{flow}}}{\tau_{\text{chem}}}. \tag{12.3}$$

The evaluation of $Da$ depends on the situation under study, in the same sense that the Reynolds number has many particular definitions derived from its fundamental meaning as the ratio of inertia to viscous forces. For our study of premixed flames, particularly useful characteristic times are the lifetimes of large eddies in the flow ($\tau_{flow} \equiv \ell_0/v'_{rms}$) and a chemical time based on a laminar flame ($\tau_{chem} \equiv \delta_L/S_L$). Using these characteristic times, we define the Damköhler number as

$$Da = \frac{\ell_0/v'_{rms}}{\delta_L/S_L} = \left(\frac{\ell_0}{\delta_L}\right)\left(\frac{S_L}{v'_{rms}}\right). \tag{12.4}$$

When chemical reaction rates are fast in comparison with fluid mixing rates, then $Da \gg 1$, and a **fast-chemistry regime** is defined. Conversely, when reaction rates are slow in comparison with mixing rates, then $Da \ll 1$. Note that the characteristic rates are inversely proportional to their corresponding characteristic times. The definition of $Da$ in Eqn. 12.4 is also instructive in that it represents the product of the length-scale ratio, $\ell_0/\delta_L$, and the reciprocal of a relative turbulence intensity, $v'_{rms}/S_L$. Thus, if we fix the length-scale ratio, the Damköhler number falls as turbulence intensity goes up.

Thus far in our discussion, we have identified five dimensionless parameters: $\ell_K/\delta_L$, $\ell_0/\delta_L$, $Re_{\ell_0}$, $Da$, and $v'_{rms}/S_L$. From their fundamental definitions, these five groups can be interrelated, and one such way of viewing these interrelationships is shown in Fig. 12.8 [23]. With such a representation (Fig. 12.8), we can estimate the flame regime that might occur in practical combustion devices, provided we have sufficient information characterizing the turbulent flowfield. The two bold lines in Fig. 12.8 define three separate regions on the graph of $Da$ versus $Re_{\ell_0}$ corresponding to the three regions defined by Eqns. 12.2a–c. Above the bold line denoted $\ell_K/\delta_L = 1$, reactions can take place in thin sheets, i.e., the wrinkled laminar-flame regime; below the other bold line, denoted $\ell_0/\delta_L = 1$, reactions will take place over a distributed region in space. The region between the two bold lines is our so-called flamelets-in-eddies regime. The boxed region containing the data symbols represents an estimate of the domain associated with spark-ignition engine combustion [23]. Here we see that combustion is predicted to occur with either wrinkled laminar flames or flamelets in eddies, depending on the specific operating conditions. Interestingly, experimental evidence is mounting that the laminar flamelet structure is preserved even under conditions of high turbulence intensity (small $Da$) [11], so caution must be exercised in using graphs such as Fig. 12.8 to predict real flame behavior. With modern advances in both laser-based flame diagnostics and predictions of flame behavior from first principles, i.e., use of direct numerical simulation (DNS), the ideas presented in this chapter are likely to be superseded by a much more rigorous understanding of turbulent-flame physics in the not-too-distant future.

| **Example 12.2** | Estimate the Damköhler number and the ratio of the Kolmogorov length scale to the laminar flame thickness for conditions prevailing in the combustor of a utility-class gas-turbine engine. What flame regime does this suggest? Assume the unburned gas temperature is 650 K, |
| --- | --- |

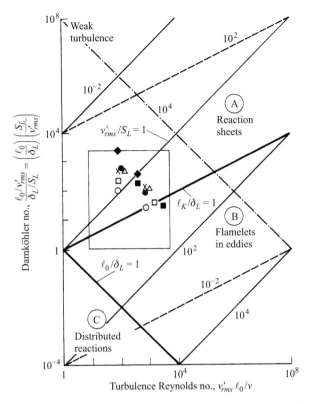

**Figure 12.8**    Important parameters characterizing turbulent premixed combustion. Conditions satisfying the Williams–Klimov criterion for the existence of wrinkled flames lie above the solid line ($\ell_K = \delta_L$), and conditions satisfying the Damköhler criterion for distributed reactions fall below the solid line ($\ell_0 = \delta_L$).

SOURCE: After Ref. [23], used with permission from SAE paper 850345 © 1985 SAE International.

the burned gas temperature is 2000 K, the pressure is 15 atm, the mean velocity is 100 m/s, equivalence ratio is unity, fuel properties are those of isooctane, and the combustor can diameter is 0.3 m. Assume a relative turbulence intensity of 10 percent and an integral scale of 1/10 the can diameter. Treat the fuel and air as essentially premixed.

**Solution**

The Damköhler number is estimated from Eqn. 12.4, where the characteristic flow time is

$$\tau_{\text{flow}} = \frac{\ell_0}{v'_{rms}} = \frac{D/10}{0.10\overline{v}} = \frac{0.30/10}{0.10/(100)}$$

$$= 0.003\,\text{s}\quad\text{or}\quad 3\,\text{ms},$$

and the characteristic chemical time is $\delta_L/S_L$. To estimate the chemical time, we will use the laminar flame speed correlations of Chapter 8 to find $S_L$ (Eqn. 8.33), and the simple theory result (Eqn. 8.21b) to estimate the laminar flame thickness $\delta_L$. Following the procedures of Example 8.3, we obtain the laminar flame speed, assuming the fuel burns similarly to isooctane:

$$S_L = S_{L,\text{ref}} \left( \frac{T_u}{T_{u,\text{ref}}} \right)^{\gamma} \left( \frac{P}{P_{\text{ref}}} \right)^{\beta}$$

$$= 24.9 \left( \frac{650}{298} \right)^{2.18} \left( \frac{15}{1} \right)^{-0.16} = 88.4 \text{ cm/s}.$$

The laminar flame thickness (Eqn. 8.21b) is

$$\delta \cong 2\alpha/S_L.$$

The thermal diffusivity is estimated at the average temperature, $0.5(T_b + T_u) = 1325$ K, using air properties (Appendix Table C.1) and corrected for the elevated pressure:

$$\alpha = 254 \cdot 10^{-6} \left( \frac{1 \text{ atm}}{15 \text{ atm}} \right) = 1.7 \cdot 10^{-5} \text{ m}^2/\text{s}.$$

Thus,

$$\delta_L \cong 2(1.7 \cdot 10^{-5})/0.884 = 3.85 \cdot 10^{-5} \text{ m}$$

$$\delta_L \cong 0.039 \text{ mm}.$$

The characteristic chemical time is then

$$\tau_{\text{chem}} = \frac{\delta_L}{S_L} = \frac{3.85 \cdot 10^{-5}}{0.884} = 4.4 \cdot 10^{-5} \text{ s}$$

$$= 0.0435 \text{ ms}.$$

The Damköhler number can now be estimated as

$$\boxed{Da = \frac{\tau_{\text{flow}}}{\tau_{\text{chem}}} = \frac{3 \text{ ms}}{0.0435 \text{ ms}} = 69}$$

We can locate a point on Fig. 12.8 after calculating $Re_{\ell_0}$:

$$Re_{\ell_0} = \frac{\rho v'_{rms} \ell_0}{\mu} = \frac{(P/RT_b)(0.1\bar{v})(D/10)}{\mu_b}$$

$$= \frac{15(101,325)/[(288.3)(2000)](0.1)100(0.30/10)}{689 \cdot 10^{-7}} = 11,477.$$

This corresponds to $\ell_K/\delta_L \approx 1$ from Fig. 12.8 and falls on the boundary between the wrinkled-flame and the flamelets-in-eddies regimes. More accurately, we can estimate $\ell_K/\ell_L$ by calculating $\ell_K$ from Eqn. 11.9, as in Example 11.1:

$$\ell_K = \ell_0 Re_{\ell_0}^{-3/4} = \left( \frac{0.30}{10} \right) (11,477)^{-0.75} = 2.7 \cdot 10^{-5} \text{ m}$$

$$\ell_K/\delta_L = 2.7 \cdot 10^{-5}/3.85 \cdot 10^{-5} = 0.70$$

$$\boxed{\ell_K/\delta_L = 0.7 \approx 1}$$

This confirms our estimate obtained from the graph.

**Comment**

In Fig. 12.8, our turbine example conditions fall just to the right of the box representing reciprocating engine conditions. Thus, we see that the turbulent combustion regimes of the two devices are not that far apart. We also note that our estimate of the mean thermal diffusivity was based on a density evaluated at the mean temperature, rather than the unburned gas density as employed previously in the theoretical developments in Chapter 8. Using $\rho_u$ instead of $\rho(\bar{T})$ to evaluate $\alpha(\bar{T})$ yields $\alpha = 8.8 \cdot 10^{-6}$ m$^2$/s, which is about half as large as that calculated above. Regardless of which value is used for $\alpha(\bar{T})$, we still estimate that the conditions are near the boundary between the wrinkled-flame and flamelets-in-eddies regimes.

## WRINKLED LAMINAR-FLAME REGIME

In this regime, chemical reactions occur in thin sheets. Referring again to Fig. 12.8, we see that reaction sheets are predicted only for Damköhler numbers greater than unity, depending on the turbulence Reynolds numbers, clearly indicating that the reaction-sheet regime is characterized by fast chemistry (in comparison with fluid mechanical mixing). For example, typical values for the engine-data Damköhler numbers in the reaction-sheet regime are on the order of 500 for $Re_{\ell_0} \approx 100$. For these conditions, the turbulence intensity $v'_{rms}$ is of the same order of magnitude as the laminar flame speed $S_L$.

One of the simplest ways of dealing with the wrinkled laminar-flame regime of turbulent combustion is to assume that the flamelets propagate with a velocity consistent with a plane (one-dimensional) laminar flame. In this view, the only effect of turbulence is the wrinkling of the flame, which results in an increase in flame area; therefore, the ratio of the turbulent flame speed to the laminar flame speed is simply a ratio of the wrinkled flamelet area to the time-mean flame area $\bar{A}$ defined in Eqn. 12.1. This can be seen by expressing the turbulent burning rate, $\dot{m}$, in terms of flamelet area and speed, i.e.,

$$\dot{m} = \rho_u \bar{A} S_t = \rho_u A_{\text{flamelets}} S_L. \tag{12.5}$$

Thus,

$$S_t / S_L = A_{\text{flamelets}} / \bar{A}, \tag{12.6}$$

which is illustrated in Fig. 12.9. In reality, $S_L$ will not be a single value but will depend on local flow properties. In particular, flame curvature, velocity gradients in the flow, molecular diffusion effects, and recirculation of burned products into the flame, can all alter the local laminar flame velocity [1].

Many theories have been developed to relate turbulent flame speeds to flow properties, based on the underlying concept of wrinkled laminar flamelets; for example, Andrews *et al.* [1], in 1975, cite 13 different models and more have been developed since [8–12]. We will not attempt to review the many theories, but, rather, present four models as examples. The first model is by Damköhler [24], and is expressed as

$$S_t / S_L = 1 + v'_{rms} / S_L; \tag{12.7}$$

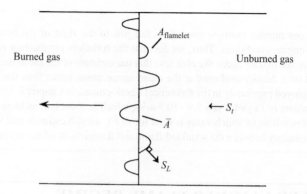

**Figure 12.9**     Sketch of wrinkled laminar flame illustrating instantaneous flamelet area together with the mean area, $\bar{A}$, used to define the turbulent flame speed, $S_t$.

the second, by Clavin and Williams [25], is given by

$$S_t/S_L = \left\{ 0.5 \left[ 1 + \left( 1 + 8C v_{rms}'^2 / S_L^2 \right)^{1/2} \right] \right\}^{1/2}, \qquad (12.8a)$$

where $C$ is a constant with a value of approximately 1. For small values of $v_{rms}'/S_L$, Eqn. 12.8a reduces to

$$S_t/S_L = 1 + C v_{rms}'^2 / S_L^2. \qquad (12.8b)$$

A third relationship, proposed by Klimov [26], is

$$S_t/S_L = 3.5 (v_{rms}'/S_L)^{0.7}, \qquad (12.9)$$

for $v_{rms}'/S_L \gg 1$.

The Damköhler model (Eqn. 12.7) is based on the fact that for a constant laminar flame speed in a purely laminar flow, the flame area, $A_{\text{flame}}$, is directly proportional to the flow velocity, $v_u$, i.e., $\dot{m} = \rho_u v_u A = \rho_u S_L A_{\text{flame}}$. Hence, the area ratio $A_{\text{flame}}/A$ is equal to $v_u/S_L$, where $A$ is the flow area, in accordance with the laminar flame concepts developed in Chapter 8. Analogously, this idea is then extended to the turbulent flow case by assuming that

$$A_{\text{wrinkles}}/\bar{A} = v_{rms}'/S_L, \qquad (12.10)$$

where the area of the wrinkles is defined as the area in excess of the time-mean flame area, i.e., $A_{\text{wrinkles}} \equiv A_{\text{flamelets}} - \bar{A}$. So, from our definition of turbulent flame speed (Eqn. 12.5), we recover the Damköhler model (Eqn. 12.7):

$$S_t = \frac{A_{\text{flamelets}}}{\bar{A}} S_L = \frac{\bar{A} + A_{\text{wrinkles}}}{\bar{A}} S_L = \left( 1 + \frac{v_{rms}'}{S_L} \right) S_L. \qquad (12.11)$$

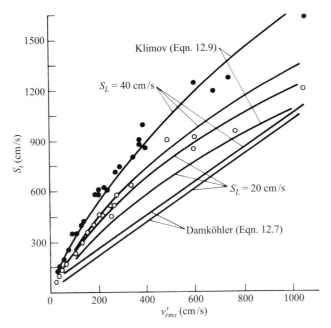

**Figure 12.10**    Experimental data for $S_t$ versus $v'_{rms}$ compared with wrinkled laminar-flame theories of turbulent flame propagation.
SOURCE: Data from Ref. [27].

The Clavin and Williams expressions (Eqns. 12.8a and b) are derived from a more rigorous treatment of the dynamics of wrinkled laminar flames for small values of $v'_{rms}$. Further discussion is beyond the scope of this book, and more information can be found in Refs. [6, 7, 25]. The functional form of the Klimov relationship (Eqn. 12.9) is based on theoretical grounds; however, the proportionality constant and exponent are based on experimental data in the Russian literature (Ref. [27] and others) cited by Klimov [26].

Figure 12.10 shows experimental data for turbulent flame speeds [27] and the predictions from the wrinkled laminar-flame models of Damköhler (Eqn. 12.7) and Klimov (Eqn. 12.9). Since the Klimov model [26] has constants partially derived from this experimental dataset, we expect good agreement. Note, however, that the straight-line form of the Damköhler model does not capture the experimental trend. Since the Clavin and Williams model (Eqn. 12.8) is based on low levels of $v'_{rms}$, the predictions of this model are not shown.

All of the correlations presented thus far depend only on the ratio of turbulent velocity fluctuations and the laminar flame speed, i.e., $v'_{rms}/S_L$. Such correlations do not necessarily capture the following influences on flamelet wrinkling or flamelet stability: (1) the turbulence integral scale, (2) the stretching of the flame surface by velocity gradients in the turbulent flow field, and (3) the molecular transport (thermo-diffusive) properties of the fuel–air mixture. The stretching of a flame by a velocity gradient affects the local flame speed, and, if sufficiently strong, can extinguish a

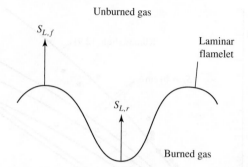

Unburned gas

$S_{L,f}$

Laminar flamelet

$S_{L,r}$

Burned gas

**Figure 12.11**    The local flame speed determines whether or not a wrinkle caused by turbulence will lead to further wrinkling or to smoothing of the flame surface. For the case shown, we assume that stretch and thermo-diffusive effects are such that the positive curvature at the leading point of the wrinkle results in an enhanced flame speed denoted as $S_{L,f}$. Conversely, these same effects result in a lesser flame speed $S_{L,r}$ at the negatively curved rearward point of a wrinkle. For this situation ($S_{L,f} > S_{L,r}$), the flame will become more wrinkled because the distance between the hill and valley of the wrinkle increases with time. On the other hand, the thermo-diffusive properties of the mixture may be such that $S_{L,r} > S_{L,f}$. For this case, the distance between the hill and valley of the wrinkle will decrease with time, causing the flamelet to become more planar. In the absence of any curvature, each element of the flamelet will propagate at the same speed, causing the flame to remain planar.

flamelet. The thermo-diffusive properties of the reactants play a large role in determining whether or not a wrinkle caused by turbulence will lead to further wrinkling as a result of the local laminar flame speed being enhanced at the forward point of a wrinkle, or oppositely, the smoothing of the flame surface as a result of the local laminar flame speed being enhanced at the rearward point of a wrinkle. Figure 12.11 illustrates the effects of local flame speed on flamelet wrinkling.

There is a large body of information that indicates all three of these influences are important. The connections, however, are still a subject of research [11]. The correlation of Bradley [12], Eqn. 12.12, attempts to account for two of these factors explicitly:

$$\frac{S_t}{S_L} = 1 + 0.95 Le^{-1} \left( \frac{v'_{rms}}{S_L} \frac{\ell_0}{\delta_L} \right)^{1/2} . \qquad (12.12)$$

In this correlation, the effect of integral scale $\ell_0$ appears as a square-root dependence; increasing the integral scale results in an increase in the turbulent flame speed, but at a decreasing rate. The effects of the thermo-diffusive properties of the mixture can be expressed in several ways; Bradley [12] chooses to use a Lewis number, which expresses the ratio of heat conduction to mass transfer (see Chapter 7), whereas the Markstein number is more frequently used.[1] In Eqn. 12.12,

---

[1]The dimensionless parameters most frequently used to characterize the effects of flame stretching and molecular transport are the Karlovitz and Markstein numbers. We leave discussion of these to more advanced textbooks [28, 29].

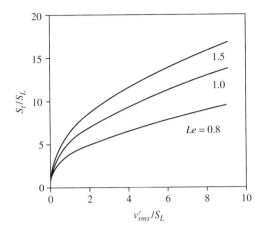

**Figure 12.12**     Plot of Eqn. 12.12 showing the strong influence of the Lewis number. This correlation (Eqn. 12.12) is based on a variety of flame types and, hence, represents a curvefit to scattered data. Adapted from Ref. [11].

the reciprocal Lewis-number term $Le^{-1}$ determines the response of the local laminar flame speed to positive and negative curvature of the local flame front. The review of Lipatnikov and Chomiak [10] focuses on the role that molecular transport processes play in turbulent flames. A detailed look at these effects is beyond the scope of this book; we refer the interested reader to Refs. [10–12] for more information.

Figure 12.12 shows the Bradley correlation (Eqn. 12.12) for fixed values of the Lewis number. The cause of the bending over of these curves with increasing $v'_{rms}/S_L$ is currently a subject of debate. Reasons proposed [11] include the merging and extinguishment of flamelets, the diverging nature of the local velocity field caused by the gas expansion in going from cold reactants to hot products, and specific geometric factors related to a particular flame type.

---

Laser Doppler anemometry is used to measure the mean and fluctuating velocities in a specially instrumented spark-ignition engine. Estimate the turbulent flame speed for the following conditions: $v'_{rms} = 3$ m/s, $P = 5$ atm, $T_u = 500°C$, and $\Phi = 1.0$ (propane–air). The mass fraction of the residual burned gases mixed with the fresh charge is 0.09.

**Example 12.3**

**Solution**

We will employ the theoretical relations of Klimov (Eqn. 12.9) and Damköhler (Eqn. 12.7) to provide estimates of the turbulent burning velocity. Both relations require a value for the laminar flame speed $S_L$ to evaluate $S_t$; thus, we first employ the correlations of Chapter 8 to determine $S_L$ (Eqn. 8.33) as follows:

$$S_{L,\text{ref}} = B_M + B_2(\Phi - 1.08)^2$$
$$= 33.33 - 138.65(1 - 1.08)^2 = 32.44 \text{ cm/s}$$

$$S_L = 32.44 \left( \frac{T_u}{T_{u,\text{ref}}} \right)^\gamma \left( \frac{P}{P_{\text{ref}}} \right)^\beta (1 - 2.1 Y_{\text{dil}})$$

$$= 32.44 \left( \frac{773}{298} \right)^{2.18} \left( \frac{5}{1} \right)^{-0.16} [1 - 2.1(0.09)] = 162.4 \text{ cm/s}.$$

Thus,

$$\frac{v'_{rms}}{S_L} = \frac{3}{1.62} = 1.85.$$

From Eqn. 12.9,

$$S_t = 3.5 S_L \left( \frac{v'_{rms}}{S_L} \right)^{0.7}$$

$$= 3.5(1.67)(1.8)^{0.7}$$

$$\boxed{S_t = 8.8 \text{ m/s}}$$

Or, from Eqn. 12.7,

$$S_t = S_L + v'_{rms}$$

$$= 1.67 + 3$$

$$\boxed{S_t = 4.7 \text{ m/s}}$$

**Comment**

We note that the values from the two theories differ by nearly a factor of two. Also, since $v'_{rms}/S_L$ is only 1.8, it is not clear that the condition of $v'_{rms}/S_L \gg 1$ for Eqn. 12.9 has been met.

# DISTRIBUTED-REACTION REGIME

One way to enter this regime is to require small integral length scales, $\ell_0/\delta_L < 1$, and small Damköhler numbers ($Da < 1$). This is difficult to achieve in a practical device, since these requirements imply that, simultaneously, $\ell_0$ must be small and $v'_{rms}$ must be large, i.e., small flow passages and very high velocities. (These conditions can be inferred easily from Fig. 12.8.) Pressure losses in such devices surely would be huge and, hence, render them impractical. Also, it is not clear that a flame can be sustained under such conditions [26]. Nonetheless, it is instructive to look at how chemical reactions and turbulence might interact in this regime, even if it is generally inaccessible for flame reactions, because many important reactions are slow and, hence, occur in distributed regions. Nitric oxide formation is an example of this.

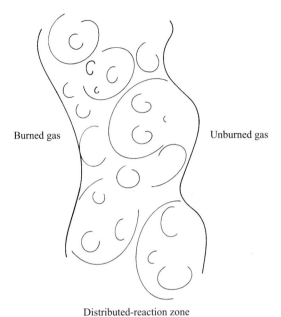

Burned gas

Unburned gas

Distributed-reaction zone

**Figure 12.13**    Conceptual view of turbulent flame propagation in the distributed-reaction regime showing various length scales of turbulence within the reaction zone proper.

Figure 12.13 heuristically illustrates a distributed-reaction zone in which all turbulence length scales are within the reaction zone. Since reaction times are longer than eddy lifetimes ($Da < 1$), fluctuations in velocity, $v'_{rms}$, temperature, $T'_{rms}$, and species mass fractions, $Y'_{i,rms}$, all occur simultaneously. Thus, the instantaneous chemical reaction rates depend on the instantaneous values of $T\,(=\overline{T}+T')$ and $Y_i\,(=\overline{Y}_i+Y'_i)$, and fluctuate as well. Moreover, the time-average reaction rate is not simply evaluated using mean values, but involves correlations among the fluctuating quantities. (Recall from Chapter 11 that the correlation of velocity components results in the Reynolds stresses.) In the following, we develop a mean reaction rate for a single bimolecular reaction step,

$$A + B \xrightarrow{k(T)} AB.$$

Recall from Chapter 4 that the reaction rate for species A, $\dot{\omega}_A$ (kmol/m$^3$-s) can be expressed

$$\dot{\omega}_A = -k[X_A][X_B], \tag{12.13}$$

where the brackets denote molar concentrations (kmol/m$^3$) and $k$ is the temperature-dependent rate coefficient. Expressing Eqn. 12.13 in terms of mass fractions yields

$$\frac{dY_A}{dt} = -kY_A\,Y_B\,\frac{\rho}{MW_B}, \tag{12.14}$$

where $\rho$ is the mixture density. Defining the instantaneous values for each variable,

$$k = \bar{k} + k',$$

$$Y_A = \bar{Y}_A + Y_A',$$

$$Y_B = \bar{Y}_B + Y_B',$$

$$\rho = \bar{\rho} + \rho',$$

substituting into Eqn. 12.14, and taking the time average, as denoted by the overbar, we arrive at the average reaction rate, $\overline{dY_A/dt}$,

$$\overline{dY_A/dt} = -\overline{k\rho Y_A Y_B} = -\overline{(\bar{k} + k')(\bar{\rho} + \rho')(\bar{Y}_A + Y_A')(\bar{Y}_B + Y_B')}, \qquad (12.15)$$

where $MW_B$ has been absorbed in the rate coefficient $k$. The right-hand side of Eqn. 12.15 can be expanded to yield

$$-\frac{\overline{dY_A}}{dt} = \bar{\rho}\bar{k}\bar{Y}_A\bar{Y}_B$$

$$+ \bar{\rho}\bar{k}\overline{Y_A'Y_B'} + \text{five additional two-variable correlation terms}$$

$$+ \bar{k}\overline{\rho'Y_A'Y_B'} + \text{three additional three-variable correlation terms}$$

$$+ \overline{\rho'k'Y_A'Y_B'}. \qquad (12.16)$$

From Eqn. 12.16, the complications caused by turbulent fluctuations are immediately obvious, and multitudinous. The leading term on the right-hand side of Eqn. 12.16 is the rate of reaction that would be obtained simply by substituting the mean values of $\rho$, $k$, $Y_A$, and $Y_B$ into Eqn. 12.14. This term is then augmented by six two-variable correlations, four three-variable correlations, and a single four-variable correlation! Obviously, many terms could be eliminated by assuming isothermal ($k = \text{constant}$) and constant-density flow, leaving only the correlations involving $Y_A'$ and $Y_B'$; however, the problem would still be formidable.

## FLAMELETS-IN-EDDIES REGIME

This regime lies in the wedge-shaped region between the wrinkled laminar-flame and distributed-reaction regimes as shown in Fig. 12.8. This region is typified by moderate Damköhler numbers and high turbulence intensities ($v_{rms}'/S_L \gg 1$). This region is of particular interest in that it is likely that some practical combustion devices operate in this regime. For example, we note that a portion of the estimated region of spark-ignition engine combustion [23] lies in the flamelets-in-eddies regime.

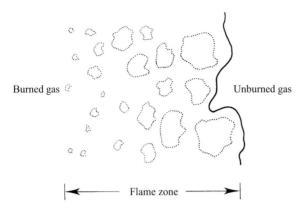

Burned gas                                    Unburned gas

$\longleftarrow$ Flame zone $\longrightarrow$

**Figure 12.14**    Conceptual view of turbulent flame propagation in the flamelets-in-eddies regime.

The experiments of Ballal and Lefebvre [30, 31], utilizing premixed propane flames in a confined flow, provide much of the experimental data related to this regime.

Figure 12.14 illustrates, conceptually, how combustion might proceed in this regime and follows the ideas that support the **eddy-breakup model** [4, 32]. This concept has met with success in predicting combustion rates in some devices. As illustrated in Fig. 12.14, the burning zone consists of parcels of burned gas and almost fully burned gas. The intrinsic idea behind the eddy-breakup model is that the rate of combustion is determined by the rate at which parcels of unburned gas are broken down into smaller ones, such that there is sufficient interfacial area between the unburned mixture and hot gases to permit reaction [4]. The implication of this is that chemical reaction rates play no role in determining the burning rates, but, rather, turbulent mixing rates completely control combustion. Implementing these ideas in mathematical form, the fuel mass burning rate per unit volume, $\bar{m}_F'''$, can be expressed [7]:

$$\bar{m}_F''' = -\rho C_F Y_{F,rms}' \epsilon_0 / \left( 3 v_{rms}'^2 / 2 \right), \tag{12.17}$$

where $C_F$ is a constant ($0.1 < C_F < 100$, but is typically of the order of unity), $Y_{F,rms}'$ is the root-mean-square fuel mass fraction fluctuation, $\epsilon_0$ is the turbulence dissipation rate (defined in Chapter 11 by Eqn. 11.7) and $3 v_{rms}'^2 / 2$ is the turbulence kinetic energy (per unit mass) assuming $v_{rms}' = v_{x,rms}' = v_{y,rms}' = v_{z,rms}'$, i.e., isotropic turbulence. Utilizing the definition of $\epsilon_0$ and pulling all of the constants into $C_F$, we obtain

$$\bar{m}_F''' = -\rho C_F Y_{F,rms}' v_{rms}' / \ell_0. \tag{12.18}$$

In Eqn. 12.18, we see that the volumetric mass burning rate depends directly on the characteristic fluctuation of $Y$, $Y_{F,rms}'$, and the characteristic turnaround time of an eddy, $v'_{rms}/\ell_0$. This model predicts that the turbulence length scale is quite important in determining turbulent burning rates.

## FLAME STABILIZATION

As mentioned earlier in this chapter, there are several important practical applications or specific devices that employ essentially turbulent premixed flames: premixed and nozzle-mixed gas burners for industrial or commercial applications, premixing/prevaporizing gas-turbine combustors, turbojet after-burners, and spark-ignition engines. In spark-ignition engines, no provision for flame stabilization is required since the flame freely propagates in a sealed chamber, and the concepts of flashback, liftoff, and blowoff (see Chapter 8) have no meaning here. Instability in spark-ignition (SI) engines is characterized, rather, by failure to achieve consistent ignition, or early flame growth, and incomplete propagation of the flame to all reaches of the combustion chamber. Therefore, our attention in this section will be directed toward applications other than SI engines.

In a practical device, a stable flame is one that is anchored at a desired location and is resistant to flashback, liftoff, and blowoff over the device's operating range. Several methods are employed to hold and stabilize flames:

- Low-velocity bypass ports.
- Refractory burner tiles.
- Bluff-body flameholders.
- Swirl or jet-induced recirculating flows.
- Rapid increase in flow area creating recirculating separated flow.

In some applications, several methods may be used simultaneously, and, in general, these same methods are employed in stabilizing turbulent non-premixed flames, as discussed in Chapter 13.

The essential principle involved in anchoring a flame in a turbulent flow, at conditions sufficiently removed from blowout, is that the local turbulent flame speed matches the local mean flow velocity. This is the same principle that we applied to the stabilization of laminar flames, except that now the turbulent flame speed replaces the laminar flame speed. Criteria establishing blowout, however, are more similar to those associated with ignition, as discussed below.

### Bypass Ports

Figure 12.15 illustrates the application of low-velocity bypass ports to stabilize an industrial burner flame. The same principle is used in handheld propane torches and modern laboratory Bunsen burners, so it might be possible for you to get a firsthand look at such a device.

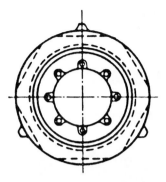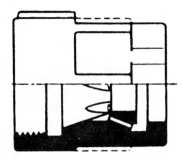

**Figure 12.15**    Flame-retention-type industrial burner. The lower half of this burner nozzle is shown in section to illustrate the bypass ports which serve to relight the main flame in the event that it is blown out. A gas mixer must be provided upstream.
SOURCE: From Ref. [18]. Reprinted by permission of Fives North American Combustion, Inc.

## Burner Tiles

Flames are frequently stabilized within a refractory passageway, referred to as a **burner tile,** in industrial burners. Figure 12.5 (bottom) illustrates a typical burner tile for a premixed burner. After startup, the refractory tile represents a nearly adiabatic boundary, reradiating back to the flame and thereby maintaining near-adiabatic flame temperatures. This helps to keep turbulent burning velocities high through the strong temperature dependence of the laminar flame speed (see Eqns. 12.7 and 8.29). From Fig. 12.5, we also observe that the specific burner tile shown has been designed with a relatively large divergence angle; therefore, it is possible that the boundary layer separates and creates a recirculation zone within the tile. Recirculation of hot combustion products back upstream promotes ignition of the unburned mixture. Swirling flows, discussed below, are also frequently employed in conjunction with burner tiles. Presser *et al.* [33] present results of a numerical study that explored the effects of burner tile geometry and recirculation on flame stability.

## Bluff Bodies

Turbulent flames can also be stabilized in the wake of a bluff, i.e., an unstreamlined, body, as shown in Fig. 12.16. A considerable amount of research has been conducted on bluff-body stabilization, and a good, but somewhat dated, review can be found in Ref. [35]. Applications utilizing bluff-body stabilizers include ramjets, turbojet afterburners, and nozzle-mixing burners, as shown in Fig. 12.4. With the trend to increasingly lean mixtures for oxides of nitrogen control, understanding bluff-body flame dynamics near the blowout limit is a topic of current interest. See Ref. [36], for example.

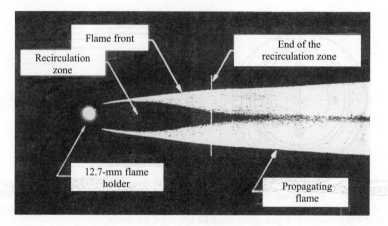

**Figure 12.16**    Turbulent flame stabilized in the wake of a bluff body.
SOURCE: After Ref. [34], reprinted by permission of the authors and Butterworth-Heinemann.

The principal characteristic of bluff-body stabilized flames is the existence of a strong recirculation zone behind the flameholding device, which is frequently in the form of a "vee gutter" with the point of the vee facing upstream. The recirculation zone consists of burned products of nearly uniform and adiabatic temperature. The flame stabilization point lies close to the edge of the flameholder, as suggested by Fig. 12.16. Blowout is governed by the processes occurring near the recirculation zone [35].

Several theories have been proposed to explain blowout from bluff-body stabilized flames; however, we will confine our discussion to the ideas of Spalding [37] and Zukoski and Marble [34], summarized and expanded by Williams [7, 35]. This model requires that a critical amount of energy must be released during the time that the initially unburned gases are in contact with the hot gases of the recirculation zone. Applying this model results in the following formula for the **blowoff velocity,** i.e., the value of the upstream velocity that causes the flame to blow off the flame holder:

$$v_{\text{blowout}} = 2\rho_0 L \left( \frac{S_L^2}{\rho \alpha_T} \right), \tag{12.19}$$

where $\rho_0$ and $\rho$ are the unburned and burned gas densities, respectively, $L$ is a characteristic length of the recirculation zone, and $\alpha_T$ denotes a so-called **turbulent thermal diffusivity.** If we assume a turbulent Prandtl number of unity, then $\alpha_T = \varepsilon$, where $\varepsilon$ is the turbulent diffusivity, or eddy viscosity, as defined in Chapter 11.

Stabilization of flames by a rapid increase in flow area, the most drastic being a step, has much in common with bluff-body stabilization. Again, the presence of a strong recirculation zone of hot products ignites the unburned gases and provides a region where the local turbulent flame speed can match the local flow velocity. Steps for stabilization are used in premixed gas-turbine combustors and premixed industrial burners.

## Swirl or Jet-Induced Recirculating Flows

As discussed above, the creation of a recirculation zone using a solid obstruction or a rapid area change can anchor a flame. Recirculation zones also can be created by introducing a swirl component to the incoming gases or by directing jets in an appropriate manner into the combustion space. Figure 12.17 illustrates the recirculation zone created by inlet swirl, whereas Fig. 12.18 shows several configurations of can combustors with jet-induced recirculation zones [38]. Huang and Yang [39] provide an excellent modern review of swirl-stabilized combustion.

   Swirl stabilization is frequently used in industrial burners and gas-turbine combustors for both premixed and nonpremixed modes of combustion. Of course, with nonpremixed flames, the swirl-induced flow patterns affect not only mixing of products and reactants, but also the mixing of fuel and air. Further discussion of this aspect of swirling flows is presented in Chapter 13.

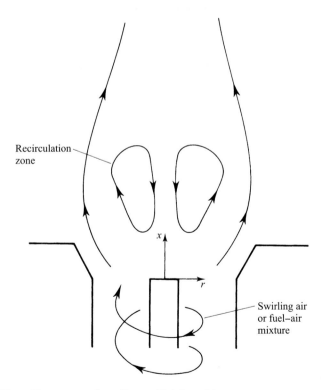

**Figure 12.17**     Flow patterns for a flame with inlet swirl.

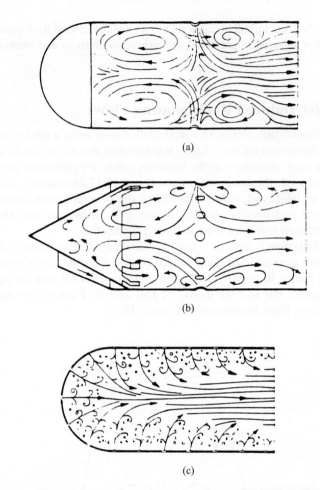

**Figure 12.18**    Flow patterns for can combustors with (a) single row of holes with enclosed end, (b) shrouded cone, and (c) multiple rows of holes.
| SOURCE: From Ref. [38], reprinted by permission of The Combustion Institute.

## SUMMARY

In this chapter, you were introduced to the many applications of premixed turbulent combustion. The concept of a turbulent flame speed was introduced and defined, and the nature of turbulent premixed flames was described. We defined three regimes of turbulent premixed combustion: the wrinkled laminar-flame regime, the flamelets-in-eddies regime, and the distributed-reaction regime. Theoretical criteria for combustion occurring in each of these regimes were defined. These criteria are based on the ratio of the laminar flame thickness to various turbulence length scales. The Damköhler number was defined as the ratio of a characteristic flow time to a characteristic chemical time, and the idea of a fast-chemistry regime was presented.

We briefly explored wrinkled laminar-flame theories and showed that there is no single theory that is generally agreed upon. We also explored the nature of the distributed-reaction regime and saw the tremendous complexity introduced into the species conservation equations as a result of the various turbulent correlations. In dealing with the intermediate flamelet-in-eddies regime, we discussed the eddy-breakup model, which assumes that the combustion rate is dominated by fluid mechanical, rather than chemical, phenomena. We also noted that recent laser-based flame visualization experiments suggest that even under conditions of high turbulence (small *Da*) wrinkled flamelets exist, casting doubt on the existence of distributed-reaction regimes for fuel combustion. The chapter ends with a discussion of the several ways turbulent flames are stabilized in practice. These included low-velocity bypass ports, burner tiles, bluff bodies, swirl or jet-induced recirculation zones, and recirculation zones created by area-change-induced separation. All told, the understanding of turbulent premixed flames is in a tremendous state of flux. Therefore, our goal in this chapter has been to highlight some applications and to present some of the key theoretical concepts, supplemented by some experimental findings, to introduce this subject, cautioning the reader that, unlike some other mature engineering sciences, definitive answers to many questions are not yet available.

# NOMENCLATURE

| | |
|---|---|
| $A$ | Area ($m^2$) |
| $Da$ | Damköhler number |
| $k$ | Reaction rate coefficient (various units) |
| $\ell_K$ | Kolmogorov microscale (m) |
| $\ell_0$ | Integral scale or turbulence macroscale (m) |
| $\dot{m}$ | Mass flowrate (kg/s) |
| $\bar{\dot{m}}'''$ | Volumetric mass production rate (kg/s-$m^3$) |
| $MW$ | Molecular weight (kg/kmol) |
| $Re$ | Reynolds number |
| $S_L$ | Laminar flame speed (m/s) |
| $S_t$ | Turbulent flame speed (m/s) |
| $t$ | Time (s) |
| $T$ | Temperature (K) |
| $v$ | Velocity (m/s) |
| $Y$ | Mass fraction (kg/kg) |

### Greek Symbols

| | |
|---|---|
| $\alpha$ | Thermal diffusivity ($m^2$/s) |
| $\alpha_T$ | Turbulent thermal diffusivity ($m^2$/s) |
| $\delta_L$ | Laminar flame thickness (m) |
| $\varepsilon$ | Eddy viscosity ($m^2$/s) |

| $\epsilon_0$ | Dissipation rate, Eqn. 11.7 ($m^2/s^3$) |
|---|---|
| $\nu$ | Kinematic viscosity ($m^2/s$) |
| $\rho$ | Density ($kg/m^3$) |
| $\tau$ | Time (s) |
| $\dot{\omega}$ | Reaction rate ($kmol/m^3$-s) |

**Subscripts**

| | |
|---|---|
| $b$ | Burned |
| chem | Chemical |
| dil | Diluent |
| $f$ | Forward point of wrinkled flamelet |
| $F$ | Fuel |
| $r$ | Rearward point of wrinkled flamelet |
| ref | Reference state |
| $rms$ | Root-mean-square |
| $u$ | Unburned |

**Other Notation**

| | |
|---|---|
| $\overline{(\ )}$ | Time-averaged |
| $(\ )'$ | Fluctuating component |
| $[X]$ | Molar concentration of species $X$ ($kmol/m^3$) |

# REFERENCES

1. Andrews, G. E., Bradley, D., and Lwakabamba, S. B., "Turbulence and Turbulent Flame Propagation—A Critical Appraisal," *Combustion and Flame,* 24: 285–304 (1975).

2. Abdel-Gayed, R. G., and Bradley, D., "Dependence of Turbulent Burning Velocity on Turbulent Reynolds Number and Ratio of Laminar Burning Velocity to R.M.S. Turbulent Velocity," *Sixteenth Symposium (International) on Combustion,* The Combustion Institute, Pittsburgh, PA, pp. 1725–1735, 1976.

3. Libby, P. A., and Williams, F. A., "Turbulent Flows Involving Chemical Reactions," *Annual Review of Fluid Mechanics,* Vol. 8, Annual Reviews, Inc., Palo Alto, CA, pp. 351–376, 1976.

4. Bray, K. N. C., "Turbulent Flows with Premixed Reactants," in *Topics in Applied Physics,* Vol. 44, *Turbulent Reacting Flows* (P. A. Libby and F. A. Williams, eds.), Springer-Verlag, New York, pp. 115–183, 1980.

5. Williams F. A., "Asymptotic Methods in Turbulent Combustion," *AIAA Journal,* 24: 867–875 (1986).

6. Clavin, P., "Dynamic Behavior of Premixed Flame Fronts in Laminar and Turbulent Flows," *Progress in Energy and Combustion Science,* 11: 1–59 (1985).

7. Williams, F. A., *Combustion Theory,* 2nd Ed., Addison-Wesley, Redwood City, CA, 1985.

8. Chomiak, J., *Combustion: A Study in Theory, Fact and Application,* Gordon & Breach, New York, 1990.

9. Lipatnikov, A. N., and Chomiak, J., "Turbulent Flame Speed and Thickness: Phenomenology, Evaluation, and Application in Multi-Dimensional Simulations," *Progress in Energy and Combustion Science,* 28: 1–74 (2002).

10. Lipatnikov, A. N., and Chomiak, J., "Molecular Transport Effects on Turbulent Flame Propagation and Structure," *Progress in Energy and Combustion Science,* 31: 1–73 (2005).

11. Driscoll, J. F., "Turbulent Premixed Combustion: Flamelet Structure and Its Effect on Turbulent Burning Velocities," *Progress in Energy and Combustion Science,* 34: 91–134 (2008).

12. Bradley, D., "How Fast Can We Burn?" *Twenty-Fourth Symposium (International) on Combustion,* The Combustion Institute, Pittsburgh, PA, pp. 247–262, 1992.

13. Heywood, J. B., *Internal Combustion Engine Fundamentals,* McGraw-Hill, New York, 1988.

14. Obert, E. F., *Internal Combustion Engines and Air Pollution,* Harper & Row, New York, 1973.

15. David, L. B., and Washam, R. M., "Development of a Dry Low $NO_x$ Combustor," ASME 89-GT-255, Gas Turbine and Aeroengine Congress and Exposition, Toronto, ON, 4–8 June, 1988.

16. Maghon, H., Berenbrink, P., Termuehlen, H., and Gartner, G., "Progress in $NO_x$ and CO Emission Reduction of Gas Turbines," ASME 90-JPGC/GT-4, ASME/IEEE Power Generation Conference, Boston, 21–25 October, 1990.

17. Lovett, J. A., and Mick, W. J., "Development of a Swirl and Bluff-Body Stabilized Burner for Low-$NO_x$, Lean-Premixed Combustion," ASME paper, 95-GT-166, 1995.

18. North American Manufacturing Co., *North American Combustion Handbook,* The North American Manufacturing Co., Cleveland, OH, 1952.

19. North American Manufacturing Co., *North American Combustion Handbook,* 2nd Ed., North American Manufacturing Co., Cleveland, OH, 1978.

20. Fox, M. D., and Weinberg, F. J., "An Experimental Study of Burner Stabilized Turbulent Flames in Premixed Reactants," *Proceedings of the Royal Society of London, Series A,* 268: 222–239 (1962).

21. Glassman, I., *Combustion,* 2nd Ed., Academic Press, Orlando, FL, 1987.

22. zur Loye, A. O., and Bracco, F. V., "Two-Dimensional Visualization of Premixed Charge Flame Structure in an IC Engine," Paper 870454, SAE SP-715, Society of Automotive Engineers, Warrendale, PA, 1987.

23. Abraham, J., Williams, F. A., and Bracco, F. V., "A Discussion of Turbulent Flame Structure in Premixed Charges," Paper 850345, SAE P-156, Society of Automotive Engineers, Warrendale, PA, 1985.

24. Damköhler, G., "The Effect of Turbulence on the Flame Velocity in Gas Mixtures," *Zeitschrift Electrochem,* 46: 601–626 (1940) (English translation, NACA TM 1112, 1947).

25. Clavin, P., and Williams, F. A., "Effects of Molecular Diffusion and of Thermal Expansion on the Structure and Dynamics of Premixed Flames in Turbulent Flows of Large Scale and Low Intensity," *Journal of Fluid Mechanics,* 116: 251–282 (1982).

26. Klimov, A. M., "Premixed Turbulent Flames—Interplay of Hydrodynamic and Chemical Phenomena," in *Flames, Lasers, and Reactive Systems* (J. R. Bowen, N. Manson, A. K. Oppenheim, and R. I. Soloukhin, eds.), Progress in Astronautics and Aeronautics, Vol. 88, American Institute of Aeronautics and Astronautics, New York, pp. 133–146, 1983.

27. Ill'yashenko, S. M., and Talantov, A. V., *Theory and Analysis of Straight-Through-Flow Combustion Chambers,* Edited Machine Translation, FTD-MT-65-143, Wright-Patterson AFB, Dayton, OH, 7 April 1966.

28. Kuo, K. K., *Principles of Combustion,* 2nd Ed., John Wiley & Sons, Hoboken, NJ, 2005.

29. Law, C. K., *Combustion Physics,* Cambridge University Press, New York, 2006.

30. Ballal, D. R., and Lefebvre, A. H., "The Structure and Propagation of Turbulent Flames," *Proceedings of the Royal Society of London, Series A,* 344: 217–234 (1975).

31. Ballal, D. R., and Lefebvre, A. H., "The Structure of a Premixed Turbulent Flame," *Proceedings of the Royal Society of London, Series A,* 367: 353–380 (1979).

32. Mason, H. B., and Spalding, D. B., "Prediction of Reaction Rates in Turbulent Premixed Boundary-Layer Flows," *Combustion Institute European Symposium* (F. J. Weinberg, ed.,), Academic Press, New York, pp. 601–606, 1973.

33. Presser, C., Greenberg, J. B., Goldman, Y., and Timnat, Y. M., "A Numerical Study of Furnace Flame Root Stabilization Using Conical Burner Tunnels," *Nineteenth Symposium (International) on Combustion,* The Combustion Institute, Pittsburgh, PA, pp. 519–527, 1982.

34. Zukoski, E. E., and Marble, F. E., "The Role of Wake Transition in the Process of Flame Stabilization on Bluff Bodies," in *Combustion Researches and Reviews, 1955,* AGARD, Butterworth, London, pp. 167–180, 1955.

35. Williams, F. A., "Flame Stabilization of Premixed Turbulent Gases," in *Applied Mechanics Surveys* (N. N. Abramson, H. Liebowitz, J. M. Crowley, and S. Juhasz, eds.), Spartan Books, Washington, DC, pp. 1157–1170, 1966.

36. Nair, S., and Lieuwen, T., "Near-Blowoff Dynamics of a Bluff-Body Stabilized Flame," *Journal of Propulsion and Power,* 23: 421–427 (2007).

37. Spalding, D. B., "Theoretical Aspects of Flame Stabilization," *Aircraft Engineering,* 25: 264–268 (1953).

38. Jeffs, R. A., "The Flame Stability and Heat Release Rates of Some Can-Type Combustion Chambers," *Eighth Symposium (International) on Combustion,* Williams & Wilkins, Baltimore, MD, pp. 1014–1027, 1962.

39. Huang, Y., and Yang, V., "Dynamics and Stability of Lean-Premixed Swirl-Stabilized Combustion, *Progress in Energy and Combustion Science,* 35: 293–364 (2009).

# PROBLEMS

**12.1** Make a list of all the boldfaced words in Chapter 12 and discuss.

**12.2** A turbulent flame is stabilized in a rectangular duct in the wake of a circular rod as shown in the sketch. The included angle of the V-shaped flame is 11.4°. Assuming the approach velocity of the unburned mixture is a uniform 50 m/s, estimate the turbulent flame speed.

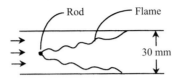

**12.3** A turbulent flame is stabilized by a pilot flame at the exit of a 2-mm-diameter pipe coannular with a 25-mm-diameter duct. The flame is in the shape of a truncated cone. Assume that the flame goes all the way to the duct wall and that the unburned mixture velocity is uniform. The flowrate through the duct is 0.03 kg/s. The unburned mixture is at 310 K at 1 atm with a molecular weight of 29.6 kg/kmol. Estimate the length of the flame zone, $L$, if the turbulent burning velocity is 5 m/s.

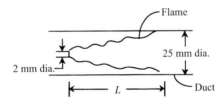

**12.4** Estimate the turbulence intensity $v'_{rms}$ and the relative intensity $v'_{rms}/\bar{v}$ for the experiment described in Example 12.1.

**12.5** Show that Eqn. 12.8b follows from 12.8a for $v'_{rms}/S_L \ll 1$.

**12.6** If $v'_{rms}$ is doubled in the calculations given in Example 12.2, does the estimated combustion regime change? If so, in what way?

**12.7** Consider turbulent premixed combustion given the following conditions and parameters:

Propane–air mixture,

$\Phi = 0.6$,

$T_u = 350$ K,

$P = 2$ atm,

$v'_{rms} = 4$ m/s,

$\ell_0 = 5$ mm.

Calculate the characteristic chemical and flow times and determine the value of the Damköhler number. Are these conditions representative of "fast" or "slow" chemistry?

**12.8*** Reference [23] suggests that turbulence intensities in spark-ignition engines can be estimated as $v'_{rms}$ (at time of spark) $\approx v_p/2$, where $v_p$ is the piston velocity. Also, the integral scale $\ell_0 \approx h/2$, where $h$ is the instantaneous clearance between the top of the piston and the cylinder head in disk-shaped combustion chambers. The instantaneous piston speed (in m/s) is related to the rotational speed, $N$ (rev/s), the crank angle after top-dead-center (TDC), $\theta$, and the ratio of the connecting rod length to crank radius, $R^*$, by the following:

$$v_P = 2LN\frac{\pi}{2}\sin\theta\left[1+\frac{\cos\theta}{(R^{*2}-\sin^2\theta)^{1/2}}\right],$$

where $L$ is the engine stroke in meters. The instantaneous clearance height is expressed as

$$\frac{h}{h_{TDC}} = 1+\tfrac{1}{2}(r_c-1)[R^*+1-\cos\theta-(R^{*2}-\sin^2\theta)^{1/2}],$$

where $r_c$ is the geometric compression ratio.

For the following conditions, estimate the turbulent Reynolds number based on the integral scale, $Re_{\ell_0}$, at the time of spark. Also, estimate the combustion regime implied by these parameters using Fig. 12.8. Use $R^* = 3.5$. For simplicity, use air properties to evaluate the thermal diffusivity at a mean temperature of $\overline{T} = (T + T_{flame})/2$. Although the flame temperature is different in each case, assume, again for simplicity, a constant value of ~2200 K.

| Parameter | Case i | Case ii | Case iii | Case iv |
|---|---|---|---|---|
| $r_c$ | 8.7 | 4.8 | 7.86 | 8.5 |
| $L$ (mm) | 83 | 114.3 | 89 | 95 |
| $h_{TDC}$ (mm) | 10.8 | 30 | 13 | 12.5 |
| $N$ (rev/min) | 1500 | 1380 | 1220 | 5000 |
| $\Phi$ | 0.6 | 1.15 | 1.13 | 1.0 |
| $P$ at spark (atm) | 4.5 | 1.75 | 7.2 | 6 |
| $T$ at spark (K) | 580 | 450 | 570 | 650 |
| Mass fraction of residual burned gases | 0.10 | 0.20 | 0.20 | 0.10 |
| Spark timing (° before TDC) | 40 | 55 | 30 | 25 |
| Fuel | Propane | Propane | Isooctane | Isooctane |

**12.9**   A turbulent methane–air flame is stabilized in the wake of a circular rod ($d = 6.4$ mm) held perpendicular to the flow. For inlet conditions of 1 atm and 298 K, the blowout velocity is 61 m/s when the equivalence ratio is unity. The recirculation zone behind the rod is six diameters long. Estimate the turbulent thermal diffusivity, $\alpha_T$. How does this value compare with the molecular thermal diffusivity, $\alpha$? (Use air properties for $\alpha$.)

**12.10**   A turbulent premixed flame from a jet torch is used to heat-treat a metal plate. The plate is located 0.05 m from the torch exit. Estimate the turbulence intensity required to close the flame cone before the flame strikes the surface. The torch has an exit diameter of 10 mm and fires a preheated (600 K) mixture of propane and air ($\Phi = 0.8$) at 1 atm. The firing rate is 20 kW.

# 13

# Turbulent Nonpremixed Flames

## OVERVIEW

Historically, turbulent nonpremixed flames have been employed in the majority of practical combustion systems, principally because of the ease with which such flames can be controlled [1]. With current concern for pollutant emissions, however, this advantage becomes something of a liability in that there is also less ability to control, or tailor, the combustion process for low emissions. For example, current practice for low-$NO_x$ gas-turbine combustors now employs premixed primary zones, rather than the nonpremixed systems used almost exclusively in the past.

Because of the many applications of nonpremixed combustion, there are many types of nonpremixed flames. For example, simple jet flames are employed in glass melting and cement clinker operations, among others; unsteady, liquid-fuel sprays of many shapes are burned in diesel engines (see Chapter 10); flames stabilized by strong recirculation zones created by swirling flows or diverging walls are used in many systems, such as utility and industrial boilers; flames stabilized behind bluff bodies are used in turbojet afterburners; and partially premixed flames, created by nozzle-mix burners (see Fig. 12.4), are employed in many industrial furnaces. Figure 13.1 shows a radiant-tube burner that employs a confined nonpremixed jet flame with some swirl imparted. In this application, a long flame length is desirable to provide uniform heating of the tube walls. The load is heated indirectly by radiation from the outer tube and, thus, avoids contact with the combustion products. A nonpremixed, or partially premixed, flame employing pure oxygen is illustrated in Fig. 13.2. Here we see that the mixing among several fluid streams is used to tailor the flame characteristics [2]. Other examples include flaring operations in refineries and oil fields, and pool and other natural fires. As you can see, the list, although incomplete, is long.

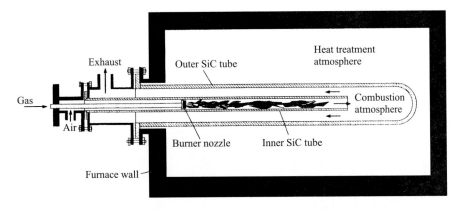

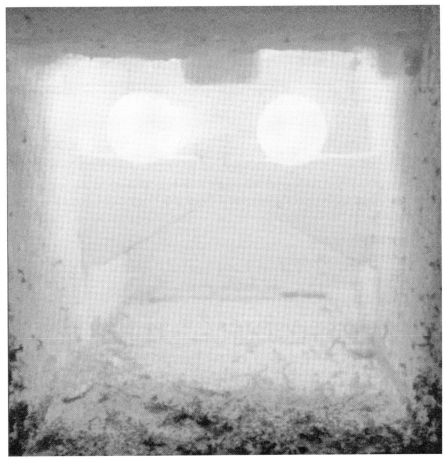

**Figure 13.1** Radiant-tube burner schematic (top). Photograph of two radiant-tube burners in an aluminum holding furnace (bottom).

SOURCE: (top) Reprinted with permission, the Center for Advanced Materials, Penn State University. (bottom) Courtesy of Eclipse Combustion, Rockford, IL.

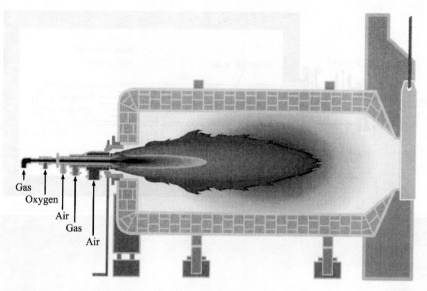

**Figure 13.2**    Both air and pure oxygen are employed as oxidizers in some applications. This oxy-fuel burner is used in aluminum recycling furnaces.
| SOURCE: Courtesy of Air Products.

The large variety of flame types reflects the wide range of needs in practical devices. For any particular application, the designer is faced with many issues. Foremost among them are the following:

- Flame shape and size
- Flame holding and stability
- Heat transfer
- Pollutant emissions.

The first three of these issues will guide our discussion in this chapter. To achieve a firm grasp of the nature of turbulent nonpremixed combustion, we will use the simple jet flame as our primary example. The jet flame is relatively uncomplicated, and, because of this, it has been the subject of many theoretical and experimental investigations. We will rely on this wealth of information.[1] Also, the limitations of the jet flame in dealing with the design issues listed above provide motivation for looking at other flame systems frequently employed in practice.

---

[1]A particularly rich source of information on turbulent nonpremixed combustion results from the International Workshop on Measurement and Computation of Turbulent Nonpremixed Flames (or TNF Workshop). The proceedings of past workshops and related information can be found at the TNF Workshop website [3].

# JET FLAMES

## General Observations

Turbulent nonpremixed jet flames visually have brushy or fuzzy edges, similar to premixed flames; however, the nonpremixed hydrocarbon flames generally are more luminous than their premixed counterparts, since some soot is usually present within the flame. Figure 13.3 illustrates these observations for an ethylene flame burning in air. The last flame in the sequence of photographs shown here is a 4-s time exposure. This image closely resembles the visual perception of a turbulent jet flame. The other photographs shown in Fig. 13.3 are instantaneous views of the flame. From these, we see that the instantaneous visible length of the flame is highly variable, as is the overall flame shape. Mungal and co-workers [4] investigated the far-field global structure of jet flames using computer-based volume rendering of motion picture records. Figure 13.4 is an example of the application of this technique. Figure 13.4a shows the burnout of a large eddy at the tip of an ethylene jet flame in a stack of images. The volume rendering (Fig. 13.4b) is a view of a large 3-D $(x, y, t)$ stack of images—many more than presented in Fig. 13.4a—and shows the trajectories of the large eddies. The character of the flame luminosity can be inferred from Fig. 13.3. The luminosity at the base of the flame is quite weak, is blue in color, and is characteristic of a soot-free region. At higher levels in the flame, considerable quantities of soot exist, and the flame takes on a bright-yellow appearance. An instantaneous image of the soot field for the flames of Fig. 13.3 is shown in the left panel of Fig. 13.5 [5]. Maximum soot volume fractions in this flame are approximately 6 ppm. The right panel of Fig. 13.5 shows that soot precursor molecules, polycyclic aromatic hydrocarbons or PAH, fill the core of the flame along the centerline up to about one-third of the flame length; beyond that, the precursors form small islands and then disappear altogether. For fuels with less sooting propensity, e.g., methane, the length of the blue-flame region is much longer and the luminosity in the soot-containing region is much less. Spectral radiation measurements by Gore [6] from laboratory methane–air flames showed that, although a flame may have some visible luminosity from soot, the contribution of the soot radiation to the total radiant loss may be insignificant. For more highly sooting fuels, however, soot radiation can be a major contribution to the total radiant heat loss [6]. Later in this chapter, we will discuss flame radiation in greater detail.

The central panel of Fig. 13.5 shows an instantaneous planar image of OH radicals in a $C_2H_4$–air jet flame [5]. This image clearly shows the convoluted nature of the high-temperature region in which OH radicals are in abundance. We see that the width of the OH zones increase with downstream distance and the merging of the large high-temperature zones at the flame tip. A similar structure of the OH field has been observed in $H_2$–air flames [7, 8]. The reaction zone in which a hydrocarbon fuel decomposes can be defined by imaging of CH radicals. Donbar *et al.* [9] used this technique to analyze the structure of a non-sooting $CH_4$–$N_2$–$O_2$ jet flame. They found that the fuel-decomposition zone remains thin throughout the length of the turbulent flame, with thicknesses comparable to those in laminar flames, indicating that turbulence does not broaden this reaction zone. Everest *et al.* [10] have made planar

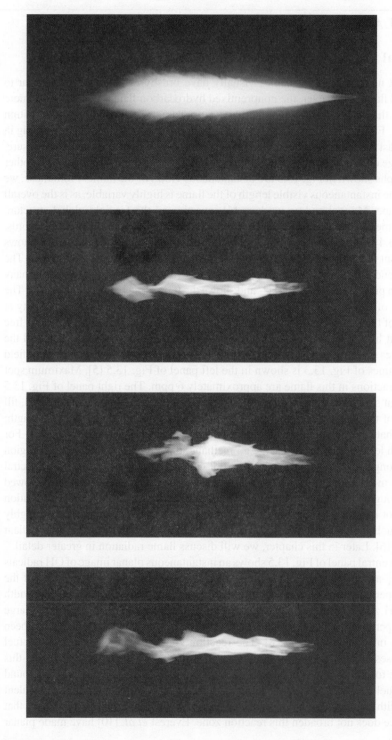

**Figure 13.3**   Instantaneous and time-averaged photographs of turbulent jet diffusion flames ($C_2H_4$–air, $d_j = 2.18$ mm). Because of the low luminosity at the base of the flame, this region is not visible on the instantaneous images (the first three photographs).

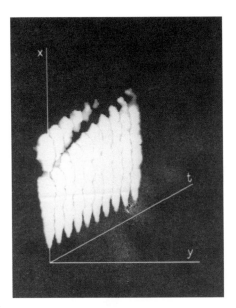 

**Figure 13.4** (a) This shows a stack of individual images of an ethylene–air jet flame from cine records ($v_e$ = 250 m/s, $d_j$ = 3 mm). (b) This shows volume renderings (computer image of computer-shaded stack) viewed 70° from the time axis.
SOURCE: Reprinted from Ref. [4] with permission of Gordon & Breach Science Publishers.

temperature measurements in jet flames. These temperature-field images show regions where thermal layers are thin and temperature gradients are large, as well as broad regions where temperatures are nearly uniform. The structure in the central core of a jet flame interior to the fuel combustion zone can be quite complicated. Figure 13.6 shows the rollup of the shear layer from the nozzle in the lower portion of the flame. As we will see in Chapter 15, the flame structure implied by these various images [4, 5, 7–11] has important consequences for pollutant emissions.

The second general observation we wish to discuss is the influence of initial jet diameter and fuel flowrate on the size of the flame. Figure 13.7 shows the results from the classical experiments of Wohl *et al.* [12]. Several important features of nonpremixed flames are apparent in this figure. First, at low flowrates, where the flames are laminar, the flame height is independent of the initial jet diameter, and depends only on the flowrate. This characteristic of laminar jet diffusion flames was discussed in Chapter 9. As the flowrate increases, turbulence begins to influence the flame height and a transitional regime can be seen. Over this transitional region, the increasing turbulence levels with flowrate result in the fully turbulent flames being shorter than their laminar counterparts, indicated by the local minimum in each curve. As the flowrate is increased further, flame lengths either remain essentially constant (tube diameters less than 0.133 in.), or increase at decreasing rates (tube diameters of 0.152 in. and greater). This is a consequence of the air entrainment and mixing rates being proportional, more or less, to the fuel flow. Furthermore,

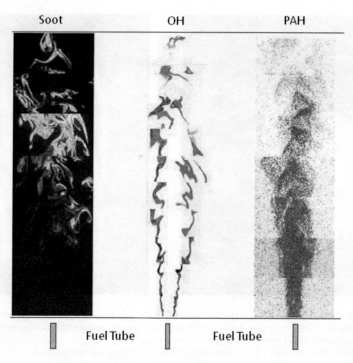

**Figure 13.5**     The panel on the left shows a quantitative instantaneous planar image of the soot field in a turbulent jet diffusion flame obtained using laser-induced incandescence. The field of view in this figure ranges from 20 to 230 nozzle diameters downstream of the nozzle exit. This flame was produced using the same fuel and burner used in Fig. 13.3, i.e., $C_2H_4$–air, $d_j = 2.18$ mm. The brightest part of the soot image corresponds to soot volume fractions of approximately 6 ppm. Note the absence of soot near the base of the flame and the appearance of narrow, and sometimes highly convoluted, soot zones, as well as broad soot-containing regions. The central panel shows OH radicals, which occupy narrow zones on the periphery of the flame in the lower third of the flame. These zones broaden and merge toward the flame tip. The panel on the right shows a planar image of polycyclic aromatic hydrocarbons (PAH), known precursors to soot formation. The darkest regions, which fill the center region of the flame, correspond to high concentrations of PAH. The lighter gray regions result from background noise. Figure adapted from Lee *et al.* [5] and used with permission.

we observe a significant dependence of flame length on initial jet diameter. These interesting characteristics of turbulent jet flames will be discussed more fully in subsequent sections of this chapter.

Our last general observation concerns flame stability. At sufficiently low flowrates, the base of a jet flame lies quite close to the burner tube outlet (say, within a few millimeters) and is said to be **attached.** As the fuel flowrate is increased, holes begin to form in the flame sheet at the base of the flame, and with further increase in the flowrate, more and more holes form until there is no continuous flame close to the burner port. This condition is called **liftoff.** A lifted jet flame is shown in the photograph in Fig. 13.8. With yet further increases in flowrate, the **liftoff distance**—the distance from the burner port to the flame base—increases. At a sufficiently large flowrate, the

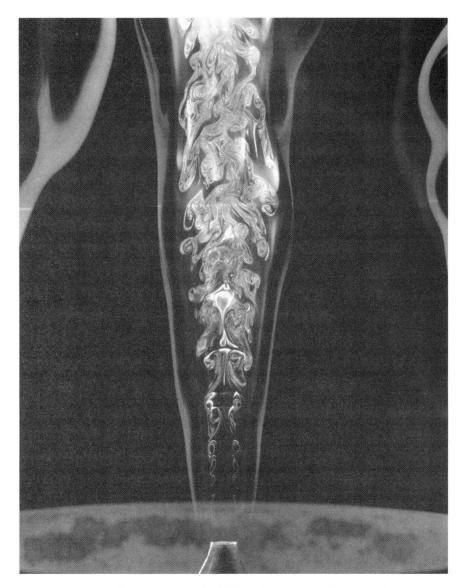

**Figure 13.6**    Visualization of nozzle-fluid flow patterns using reactive Mie scattering in methane–air jet flame ($v_e$ = 10 m/s, $d_j$ = 10 mm). The total length of the image is approximately 20 nozzle-diameters. Near the nozzle, the flame zone resides just outside of the rollup of the nozzle shear layer. Near the top of the photograph, the flame zone interacts with the turbulent core. The visualization of an outer layer results from water vapor from the flame reacting with $TiCl_4$ vapor in the surrounding air flow.

SOURCE: Photograph courtesy of W. M. Roquemore. See also Ref. [11].

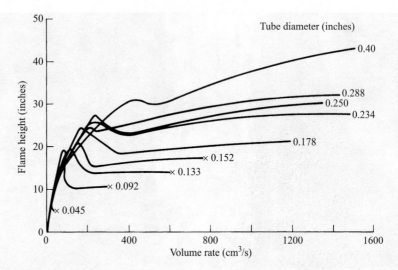

**Figure 13.7**     Effects of flowrate on flame height illustrating behavior in laminar, transitional, and fully turbulent regimes.

SOURCE: Reprinted from Ref. [12] by permission of The Combustion Institute.

flame blows out. Thus, there are two critical flow conditions related to flame stability: liftoff and **blowout.** In Chapter 8, we studied these phenomena as they apply to laminar premixed flames.

Flame stability has many implications for practical applications. For example, liftoff should be avoided so that the flame is close to the burner and its position is independent of the flowrate. This allows for positive ignition by a spark or pilot flame at the burner and assures that the flame position is controlled. In some applications, a certain amount of liftoff may be desirable to prevent overheating of critical burner components. Obviously, for reasons of safety, operation near the blowout limit should be avoided. A large furnace filled with a raw fuel–air mixture would be an immense hazard.

## Simplified Analysis

Mathematical modeling of turbulent flames is a formidable task. As we saw in Chapter 11, analyzing an isothermal turbulent flow is in itself challenging; moreover, with combustion, variable density and chemical reactions need to be taken into account in some manner. In keeping with the spirit of other chapters in this book, we will develop a very simplified mathematical model, this time for a jet flame, to illuminate some of the essential underlying physics. The modeling of turbulent combustion is a field where the state of the art is changing rapidly. Our intent here is to provide a basic foundation from which a greater understanding can be developed by study at a more advanced level. Recently published reviews [13–16] and books [17–21] provide entry points for more in-depth study.

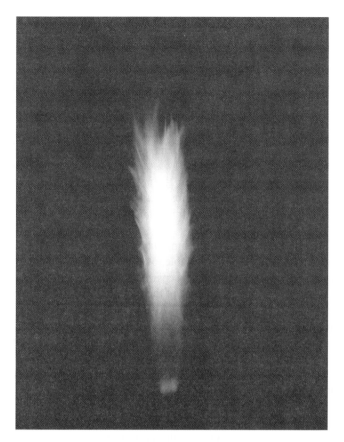

**Figure 13.8**    Photograph of lifted ethylene–air jet flame ($d_j$ = 2.18 mm). The base of the lifted flame is wide and appears blue. The burner exit is at the bottom of the frame.

**Comparison with Nonreacting Jets**    In Chapter 11, we developed a simple mixing-length model of a constant-density turbulent jet and discovered the following three important characteristics: first, the velocity field is a "universal" function when all velocities are scaled by the exit velocity and the spatial coordinates $x$ and $r$ are scaled by the jet nozzle radius $R(d_j \equiv 2R)$; second, the jet spreading angle is a constant, independent of both jet exit velocity and diameter; and third, the so-called eddy viscosity, $\varepsilon$, is independent of location in the flowfield and is directly proportional to both the nozzle exit velocity, $v_e$, and diameter, $d_j$. Also, these simple theoretical results are in reasonable agreement with experimental findings.

It seems reasonable, at least to a first approximation, to propose that jet flames share some of the characteristics of isothermal jets. If we make this assumption, what implications follow? The primary implications concern flame shape and length. If we make the further assumption that the turbulent diffusivity for mass transfer is the same as that for momentum transfer, then the resulting fuel mass-fraction distribution should be equivalent to the dimensionless velocity distribution, i.e.,

$Y_F(x, r) = v(r, x)/v_e = f(x/d_j$ only). However, this can only be true in the absence of combustion since $Y_F$ is zero outside of the flame boundary, while, obviously, the velocity has not yet decayed to zero. As we saw in Chapter 9, the mixture fraction can be used instead of the fuel mass fraction and has behavior analogous to the dimensionless velocity field. Thus, we expect that, for a given fuel type, the flame length should be independent of jet velocity $v_e$ and proportional to the nozzle diameter $d_j$, and the spreading angle should be independent of both $v_e$ and $d_j$. The experimental results in Fig. 13.7 show that, for small-diameter tubes, flame lengths are indeed independent of $v_e$ and that the flame lengths are roughly proportional to $d_j$. As we will see later, buoyancy causes a breakdown of this simple analogy of jet flames to isothermal jets and can be used to explain the non-constancy of flame lengths for the larger diameter tubes. Nonetheless, it is instructive to see how the underlying physics of simple jets has much in common with jet flames.

**Conserved Scalars Revisited**   In our simplified analysis of the jet flame, we wish to use the mixture fraction as a single variable to describe the composition of the flame at any location, rather than employing separate species conservation relations for fuel, oxidizer, and products. The concept of the mixture fraction and its conservation throughout the flowfield was discussed in Chapter 7 and developed for the laminar jet diffusion flame in Chapter 9. Although key definitions and relationships will be repeated here, it is recommended that the sections of Chapter 7 and 9 dealing with the conserved scalar concept be reviewed before proceeding.

From Chapter 7, the definition of the mixture fraction, $f$, which applies locally for a sufficiently small control volume, is

$$f \equiv \frac{\text{Mass of material having its origin in the fuel stream}}{\text{Mass of mixture}}. \tag{7.68}$$

This quantity has two particularly important characteristics: first, it can be used to define the flame boundaries, since it is uniquely related to the equivalence ratio as

$$f \equiv \frac{\Phi}{(A/F)_{\text{stoic}} + \Phi}, \tag{13.1}$$

where $(A/F)_{\text{stoic}}$ is the stoichiometric air–fuel ratio for the fuel issuing from the nozzle. Thus, as we did for the laminar diffusion flame in Chapter 9, a flame boundary is defined as the location where $\Phi = 1$; hence, the flame boundary is also defined by the location where the mixture fraction takes on the value $f_s$, defined as

$$f_s \equiv \frac{1}{(A/F)_{\text{stoic}} + 1}. \tag{13.2}$$

The second important characteristic of the mixture fraction is that it is conserved throughout the flowfield as defined by a "sourceless" governing equation. It is this property that allows us to use $f$ in place of the individual species conservation equations and, as a result, greatly simplifies the mathematical formulation of the problem.

**Assumptions**   The following assumptions are used to construct a simple mathematical model of a nonpremixed turbulent jet flame:

1. The time-averaged flow is steady and axisymmetric and is produced by a jet of fuel emerging from a circular nozzle of radius $R$, which burns in a quiescent, infinite reservoir of air.

2. Molecular transport of momentum, species, and thermal energy is unimportant in comparison with turbulent transport.

3. The turbulent momentum diffusivity, or eddy viscosity, $\varepsilon\,(=\mu_{\mathrm{turb}}/\rho)$, is constant throughout the flowfield and is equal to $0.0285v_e R$ (see Eqn. 11.46). This is an extension of the mixing-length hypothesis for the constant-density jet (Chapter 11) to the variable-density reacting jet by neglecting density fluctuations.

4. All correlations involving density fluctuations are neglected.

5. The turbulent transport of momentum, species, and thermal energy are all equal; i.e., the turbulent Schmidt, Prandtl, and Lewis numbers are all unity, so $Sc_T = Pr_T = Le_T$. The consequence of this assumption is that the turbulent momentum diffusivity (eddy viscosity, $\varepsilon$) can be substituted for the turbulent mass diffusivity and thermal diffusivity, $\varepsilon = \mathcal{D}_T = \alpha_T$.

6. Buoyancy is neglected.

7. Radiation heat transfer is negligible.

8. Only radial turbulent diffusion of momentum, species, and thermal energy is important; axial turbulent diffusion is neglected.

9. The fuel jet velocity is uniform at the nozzle exit, i.e., a top-hat profile.

10. Mixture properties are defined by three species: fuel, oxidizer, and products. All have equal molecular weights ($MW = 29$ kg/kmol) and specific heats ($c_p = 1200$ J/kg-K), and the fuel heat-of-combustion is $4 \cdot 10^7$ J/kg$_F$. The stoichiometric mass air–fuel ratio is 15:1 ($f_s = 1/16 = 0.0625$).

11. The fluctuations in mixture fraction are ignored in the state relations used to determine the mean density. In a more rigorous analysis, the mean density would be calculated from some assumed mixture-fraction probability distribution function characterized by a mean value and a variance.

**Application of Conservation Laws**   With the above assumptions, the basic conservation equations are identical to those we derived in Chapter 7 and applied to the laminar jet flame in Chapter 9, except that time-averaged variables replace the instantaneous ones, and the turbulent transport properties (momentum, species, and thermal energy diffusivities) replace their molecular counterparts. Defining dimensionless variables using a characteristic length, $R$, and a characteristic velocity, $v_e$, overall mass conservation is written (Eqn. 9.36) as

$$\frac{\partial}{\partial x^*}(\overline{\rho}^*\overline{v}_x^*) + \frac{1}{r^*}\frac{\partial}{\partial r^*}(r^*\overline{\rho}^*\overline{v}_r^*) = 0. \tag{13.3}$$

Axial momentum conservation (Eqn. 9.37) becomes

$$\frac{\partial}{\partial x^*}(r^*\overline{\rho}^*\overline{v}_x^*\overline{v}_x^*) + \frac{\partial}{\partial r^*}(r^*\overline{\rho}^*\overline{v}_r^*\overline{v}_x^*) = \frac{\partial}{\partial r^*}\left[\left(\frac{1}{Re_T}\right)r^*\frac{\partial\overline{v}_x^*}{\partial r^*}\right] \tag{13.4}$$

and the mixture-fraction equation (Eqn. 9.38) is given by

$$\frac{\partial}{\partial x^*}(r^*\overline{\rho}^*\overline{v}_x^*\overline{f}) + \frac{\partial}{\partial r^*}(r^*\overline{\rho}^*\overline{v}_r^*\overline{f}) = \frac{\partial}{\partial r^*}\left[\left(\frac{1}{Re_T Sc_T}\right)r^*\frac{\partial\overline{f}}{\partial r^*}\right], \tag{13.5}$$

where the turbulent Reynolds number is defined as

$$Re_T \equiv \frac{v_e R}{\varepsilon}. \tag{13.6}$$

With our third assumption, $Re_T = 35$. Note that with a constant Reynolds number, the dimensionless solution to the problem, $\overline{v}_x^*(r^*, x^*)$, $\overline{v}_r^*(r^*, x^*)$, and $\overline{f}(r^*, x^*)$, does not depend on either the initial jet velocity or nozzle diameter.

The boundary conditions are

$$\overline{v}_r^*(0, x^*) = 0, \tag{13.7a}$$

$$\frac{\partial\overline{v}_x^*}{\partial r^*}(0, x^*) = \frac{\partial\overline{f}}{\partial r^*}(0, x^*) = 0 \quad \text{(symmetry)}, \tag{13.7b}$$

$$\overline{v}_x^*(\infty, x^*) = \overline{f}(\infty, x^*) = 0, \tag{13.7c}$$

$$\overline{v}_x^*(r^* \leq 1, 0) = \overline{f}(r^* \leq 1, 0) = 1,$$
$$\qquad\qquad\qquad\qquad\qquad\qquad \text{(top-hat exit profiles)} \tag{13.7d}$$
$$\overline{v}_x^*(r^* > 1, 0) = \overline{f}(r^* > 1, 0) = 0.$$

Note that the boundary conditions on $\overline{v}_x^*$ and $\overline{f}$ are identical, and that the governing equations for $\overline{v}_x^*$ and $\overline{f}$ become identical for $Sc_T = 1$ (assumption 5). With assumptions 10 and 11, the state relationship for temperature is a simple, piecewise-linear function of $\overline{f}$ (see Chapter 9) as shown in Fig. 13.9. The mean density is then related simply to the mean temperature through the ideal-gas equation of state, $\overline{\rho} = PMW/(R_u\overline{T})$.

**Solution**   A finite-difference numerical scheme was used to solve the problem posed above. The results of these calculations are shown in Fig. 13.10 where mixture-fraction contours are shown on the left, and mixture fractions are plotted versus radial distance for fixed axial distance on the right. Using the definition of flame length as the axial

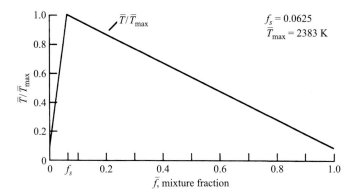

**Figure 13.9**     State relationship for temperature for simplified chemistry.

location where the mixture fraction takes on the stoichiometric value ($\overline{f}/f_s = 1$), we see that the predicted flame length is about 45 nozzle diameters ($L_f/d_j = 45$). Again, using the stoichiometric contour to define the flame boundary, the aspect ratio (ratio of flame length to width) is found to be about 11:1. This value is somewhat greater than experimental values of about 7:1 found for hydrocarbon jet flames. In spite of the great simplicity of our turbulent jet flame model, the general features of the

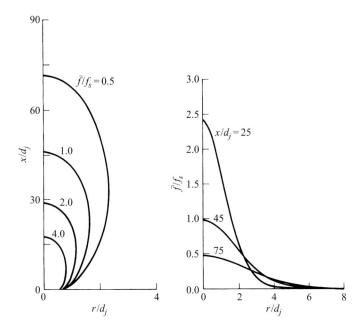

**Figure 13.10**     Calculated mixture-fraction distribution using Prandtl's mixing-length model applied to a jet flame.

flame are well predicted, although accuracy certainly cannot be claimed. Neglecting density fluctuations is probably the most serious oversimplification; for more realistic mathematical models of turbulent jet flames, which explicitly account for density fluctuations, the reader is referred to the literature [22–24].

## Flame Length

Interest in being able to explain and predict turbulent jet flame lengths has a long history. The earliest, and now classical, studies are those of Hottel [25], Hawthorne *et al.* [26], and Wohl *et al.* [12] reported in the late 1940s and early 1950s. Major reviews of this subject were performed by Becker and Liang [27] and, more recently, by Delichatsios [28] and Blake and McDonald [29].

**Flame Length Definitions**   Many definitions and techniques for measuring flame lengths are found in the literature, and no single definition is accepted as preferred. Therefore, care must be exercised in comparing results of different investigators and in the application of correlation formulae. Common definitions of flame length include visual determinations by a trained observer, averaging a number of individual instantaneous visible flame lengths from photographic records, measuring the axial location of the average peak center-line temperature using thermocouples, and measuring the axial location where the mean mixture fraction on the flame axis is the stoichiometric value using gas sampling. In general, visible flame lengths tend to be larger than those based on temperature or concentration measurements. For example, Ref. [30] reports temperature-based flame lengths to range between approximately 65 percent and 80 percent of time-averaged visible flame lengths, depending on fuel type.

**Factors Affecting Flame Length**   For vertical flames created by a fuel jet issuing into a quiescent environment, four primary factors determine flame length:

- Relative importance of initial jet momentum flux and buoyant forces acting on the flame, $Fr_f$.
- Stoichiometry, $f_s$.
- Ratio of nozzle fluid to ambient gas density, $\rho_e / \rho_\infty$.
- Initial jet diameter, $d_j$.

The first of these factors, the relative importance of initial momentum and buoyancy, can be characterized by a flame Froude number, $Fr_f$. Recall that in Chapter 9, a flame Froude number was defined to establish momentum-controlled and buoyancy-controlled regimes for laminar jet flames (see Eqns. 9.68 and 9.69). For turbulent jet flames, the following definition of the flame Froude number is useful [28]:

$$Fr_f = \frac{v_e f_s^{3/2}}{\left(\dfrac{\rho_e}{\rho_\infty}\right)^{1/4}\left[\dfrac{\Delta T_f}{T_\infty}gd_j\right]^{1/2}},$$ 

$$(13.8)$$

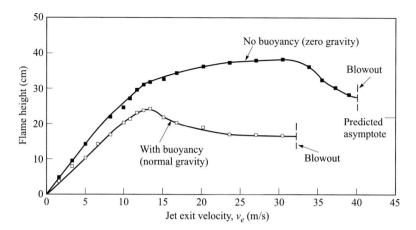

**Figure 13.11**    Comparison of jet flame heights with and without buoyancy ($C_3H_8$–air, $d_j$ = 0.8 mm). Nonbuoyant conditions result from tests being conducted with near-zero (< $10^{-5}$ g) gravitational acceleration. Flames are nominally laminar at exit velocities up to approximately 10 m/s.

SOURCE: Data from Bahadori et al. [32].

where $\Delta T_f$ is the characteristic temperature rise resulting from combustion. Other similar definitions are defined in the literature [27, 31]. For very small values of $Fr_f$, flames are dominated by buoyancy, whereas for very large values, the initial jet momentum controls the mixing and, hence, the velocity field within the flame. For example, the simplified analysis presented above ignores buoyancy and, thus, applies only in the limit of large $Fr_f$. Direct measurements of the effects of buoyancy on jet flames have been made under microgravity conditions by Bahadori *et al.* [32]. Figure 13.11 shows how increased mixing, produced from buoyancy-induced motion, results in relatively shorter flames in comparison with the nonbuoyant case. Presumably, in the absence of the blowout phenomenon (discussed later), the flame heights for the two cases should approach the same asymptotic limit as the jet exit velocity increases. Using the correlation of Fig. 13.12, the asymptotic flame height is estimated to be 21.8 cm and is indicated in Fig. 13.11. Thus, we expect the nonbuoyant flame height to decrease, and the normal gravity flame height to increase. Experiments utilizing larger jet diameters would be required to verify this predicted behavior.

   The remaining three factors are important regardless of whether or not buoyancy exerts an influence. For example, fuels with small values of stoichiometric mixture fraction, $f_s$, require greater quantities of air to be entrained per unit mass of fuel to achieve stoichiometric proportions in comparison with fuels with large values of $f_s$. This implies longer flames for smaller stoichiometric mixture fractions. As an illustration, the stoichiometric air requirement of propane is about six times that of carbon monoxide, and a propane flame is approximately seven times longer than a carbon monoxide flame [26].

   The density ratio, $\rho_e/\rho_\infty$, and initial jet diameter, $d_j$, can conveniently be combined as a single parameter, frequently referred to as the **momentum diameter, $d_j^*$,** and defined as

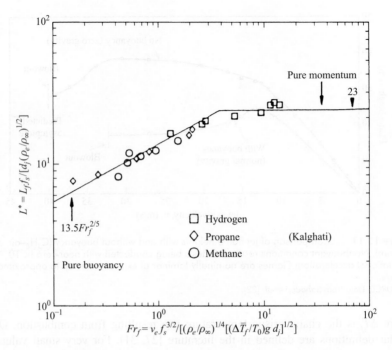

**Figure 13.12**    Flame lengths for jet flames correlated with flame Froude number.
| SOURCE: Reprinted by permission of Elsevier Science, Inc., from Ref. [28]. © 1993, The Combustion Institute.

$$d_j^* = d_j \, (\rho_e / \rho_\infty)^{1/2}, \tag{13.9}$$

where a uniform exit velocity profile has been assumed. The basic idea embodied in this definition is that jets with identical initial jet momentum fluxes should have identical velocity fields. Thus, increasing the density of the nozzle fluid produces the same effect as increasing the nozzle diameter in accordance with Eqn. 13.9. Such a result is predicted by jet theory [33], and experimental evidence [34] shows this to be true. Tacina and Dahm [35], more recently, have developed scaling laws that apply equally well to isothermal and reacting jets by defining an extended momentum diameter. This parameter captures the effect of heat release on the density of the jet fluid.

**Correlations**    Experimental results showing the combined influence of the four primary factors affecting flame length, discussed above, are shown in Fig. 13.12. The flame Froude number, $Fr_f$, appears as the abscissa, and the remaining three factors are used to define a dimensionless flame length, which is plotted on the ordinate, and given by

$$L^* \equiv \frac{L_f f_s}{d_j (\rho_e / \rho_\infty)^{1/2}} \tag{13.10}$$

or

$$L^* = \frac{L_f f_s}{d_j^*}.$$  (13.11)

Two regimes are identified in Fig. 13.12: (1) a buoyancy-dominated regime at small flame Froude numbers, and (2) a momentum-dominated regime at flame Froude numbers greater than about 5. Delichatsios [28] developed the following correlation for the dimensionless flame length $L^*$ that spans both regimes:

$$L^* = \frac{13.5 Fr_f^{2/5}}{\left(1 + 0.07 Fr_f^2\right)^{1/5}}.$$  (13.12)

At small values of $Fr_f$, Eqn. 13.12 simplifies to $L^* = 13.5 Fr_f^{2/5}$, the buoyancy-dominated limit. As the value of $Fr_f$ increases, dimensionless flame lengths predicted by this correlation asymptotically approach the momentum-dominated dimensionless flame length value $L^* = 23$. These two limiting expressions are indicated in Fig. 13.12.

Other correlations have been proposed [27]; however, the above is easy to use and probably as accurate as others. As previously discussed, various investigators use different definitions and experimental techniques to determine flame lengths; hence, the accuracy of any correlation depends on the consistency of the database used to establish it. The datapoints shown in Fig. 13.12 represent averages of measurements taken from individual video frames.

---

Estimate the flame length for a propane jet flame in air at ambient conditions ($P = 1$ atm, $T_\infty = 300$ K). The propane mass flowrate is $3.66 \cdot 10^{-3}$ kg/s and the nozzle exit diameter is 6.17 mm. Assume the propane density at the nozzle exit is 1.854 kg/m³.          **Example 13.1**

### Solution

We will use the Delichatsios correlation (Eqn. 13.12) to find the propane flame length. To do so first requires us to determine the flame Froude number, $Fr_f$. The properties required are the following:

$$\rho_\infty = \rho_{air} = 1.1614 \text{ kg/m}^3 \quad \text{(Appendix Table C.1)},$$
$$T_f \cong T_{ad} = 2267 \text{ K} \quad \text{(Appendix Table B.1)},$$
$$f_s = \frac{1}{(A/F)_{stoic} + 1} = \frac{1}{15.57 + 1} = 0.06035.$$

The stoichiometric air–fuel ratio in the above is calculated from Eqn. 2.32. The nozzle exit velocity is calculated from the mass flowrate:

$$v_e = \frac{\dot{m}_e}{\rho_e \pi d_j^2 / 4}$$

$$= \frac{3.66 \cdot 10^{-3}}{1.854 \pi (0.00617)^2 / 4} = 66.0 \text{ m/s}.$$

We now have all the information necessary to evaluate $Fr_f$ (Eqn. 13.8):

$$Fr_f = \frac{v_e f_s^{3/2}}{\left(\dfrac{\rho_e}{\rho_\infty}\right)^{1/4}\left(\dfrac{T_f - T_\infty}{T_\infty} gd_j\right)^{1/2}}$$

$$= \frac{66.0(0.06035)^{1.5}}{\left(\dfrac{1.854}{1.1614}\right)^{0.25}\left[\left(\dfrac{2267-300}{300}\right)9.81(0.00617)\right]^{0.5}}$$

$$= 1.386.$$

We now employ Eqn. 13.12 to find the dimensionless flame length, $L^*$:

$$L^* = \frac{13.5 Fr_f^{2/5}}{\left(1+0.07 Fr_f^2\right)^{1.5}}$$

$$= \frac{13.5(1.386)^{0.4}}{\left[1+0.07(1.386)^2\right]^{0.2}}$$

$$= 15.0.$$

From the definition of $L^*$ (Eqn. 13.11) and $d_j^*$ (Eqn. 13.9), we determine the actual flame length in meters:

$$d_j^* = d_j \left(\frac{\rho_e}{\rho_\infty}\right)^{1/2} = 0.00617\left(\frac{1.854}{1.1614}\right)^{0.5}$$

$$= 0.0078 \text{ m}$$

and

$$L_f = \frac{L^* d_j^*}{f_s} = \frac{15.0(0.0078)}{0.06035}$$

$$\boxed{L_f = 1.94 \text{ m}}$$

or

$$\boxed{L_f / d_j = 314}$$

**Comment**

From Fig. 13.12, we see that this flame is in the mixed regime where both the initial momentum flow and flame-generated buoyancy are important. It is also interesting to note that the flame length predicted above is somewhat less than, but quite close to, the visible flame length measured ($L_f/d_j = 341$) in Ref. [30]. That the flame length estimated from Eqn. 13.12 is less than that measured is consistent with the different experimental techniques used to determine $L_f$.

---

**Example 13.2**

For the same heat release rate and nozzle exit diameter used in Example 13.1, determine the flame length when the fuel is methane and compare with the propane flame length. The density of methane is 0.6565 kg/m³.

## Solution

We can use the same procedures to find $L_f$ as in Example 13.1, but first we must determine the methane flowrate. Since both flames release the same chemical energy,

$$\dot{m}_{CH_4} LHV_{CH_4} = \dot{m}_{C_3H_8} LHV_{C_3H_8}.$$

Using the lower heating values from Appendix Table B.1, we can find the $CH_4$ flowrate:

$$\dot{m}_{CH_4} = \dot{m}_{C_3H_8} \frac{LHV_{C_3H_8}}{LHV_{CH_4}} = 3.66 \cdot 10^{-3} \frac{46,357}{50,016}$$

$$= 3.39 \cdot 10^{-3} \text{ kg/s}.$$

Following the approach in Example 13.1, we obtain the following:

$$\rho_\infty = 1.1614 \text{ kg/m}^3,$$
$$T_f = 2226 \text{ K},$$
$$f_s = 0.0552,$$
$$v_e = 172.7 \text{ m/s}.$$

Using the above with Eqns. 13.8–13.12, we obtain the following:

$$Fr_f = 4.154,$$
$$L^* = 20.36,$$
$$d_j^* = 0.0046 \text{ m}.$$

Thus,

$$\boxed{L_f = 1.71 \text{ m}}$$

or

$$\boxed{L_f/d_j = 277}$$

Comparing the methane flame length with that of the propane flame,

$$\frac{L_{f,CH_4}}{L_{f,C_3H_8}} = \frac{1.71}{1.94} = 0.88,$$

we see that the $CH_4$ flame is only about 12 percent shorter.

## Comment

What factors contribute to the methane flame being shorter? First, we see that the $CH_4$ flame is more nearly momentum controlled ($Fr_{f,CH_4} > Fr_{f,C_3H_8}$), which implies a greater dimensionless flame length ($L^*_{CH_4} = 20.36$ vs. $L^*_{C_3H_8} = 15.0$). However, the lower density of the methane results in a significantly smaller momentum diameter ($d^*_{j,CH_4} = 0.0046$ m vs. $d^*_{j,C_3H_8} = 0.0078$ m). This smaller $d_j^*$ is the controlling factor in producing the shorter $CH_4$ flame and outweighs the small opposing difference in stoichiometries.

## Flame Radiation

Turbulent nonpremixed flames can be highly radiating. In certain practical applications, this radiation is a desired attribute, contributing to heating a load. In other applications, however, radiant losses can contribute to loss of efficiency (e.g., diesel engines) or safety hazards (e.g., flaring operations as illustrated in Fig.13.13 [36]). In gas-turbine engines, the radiant heat load is a major factor in the durability of the combustor liner. In this section, we will present a brief overview of the radiation characteristics of jet flames to acquaint the reader with this important subject. The review articles of Refs. [37] and [38] are recommended for readers seeking more information on radiation from combustion systems and applications other than jet flames.

A frequently used term in the combustion literature is the **radiant fraction, $\chi_R$.** The radiant fraction represents the ratio of the radiant heat-transfer rate from the flame to the surroundings, $\dot{Q}_{rad}$, to the total heat released by the flame, $\dot{m}_F \Delta h_c$, i.e.,

$$\chi_R \equiv \frac{\dot{Q}_{rad}}{\dot{m}_F \Delta h_c}, \tag{13.13}$$

where $\dot{m}_F$ is the mass flowrate of fuel supplied to the flame and $\Delta h_c$ is the fuel heat of combustion. Depending on fuel type and flow conditions, radiant fractions for jet flames range from a few percent to more than 50 percent. Figure 13.14 shows the variation of $\chi_R$ with heat release rate, $\dot{Q}(= \dot{m}\Delta h_c)$, for three fuels and various sized burners [39].

Several interesting observations can be made from Fig. 13.14. First, we see that the radiant fractions for the three fuels arrange in the same order as the fuel's sooting propensity. From Table 9.5, the smoke points are $\dot{m}_{sp}$ (mg/s) $= 0.51$ and $1.12$ for $C_2H_2$ and $C_3H_6$, respectively, whereas methane is nonsooting; the corresponding maximum radiant fractions are approximately 0.6, 0.45, and 0.15, respectively. This clearly suggests that the in-flame soot is important in determining a flame's radiant fraction. More will be said about this below. The second observation is that $\chi_R$ depends on both flame size ($d_j$) and heat release rate. The general trend is for $\chi_R$ to decrease both as flame size decreases at a fixed firing rate, and as the firing rate increases for a fixed flame size. Although the actual processes involved are quite complicated, especially since soot kinetics is involved, these trends can be interpreted by a simple scaling analysis [31]. If we consider the entire flame to be a uniform source of both heat release and radiation, the rate at which energy is lost from the flame can be approximated as

$$\dot{Q}_{rad} \approx a_p V_f \sigma T_f^4, \tag{13.14}$$

where $a_p$ is an appropriate absorption coefficient for the flame, and $V_f$ and $T_f$ are the flame volume and temperature, respectively. The radiant fraction, from Eqn. 13.13, becomes

$$\chi_R = \frac{a_p V_f \sigma T_f^4}{\dot{m}_F \Delta h_c}. \tag{13.15}$$

**Figure 13.13**     Flame radiation and soot are greatly reduced by injecting air during a flaring operation. Flaring of an unsaturated hydrocarbon shows extensive soot and visible radiation (top panel) before the air is injected. As the air blower comes up to speed, the visible soot diminishes (middle panel). At steady state, visible soot is eliminated and the radiation from the flame is greatly reduced by the injection of the air (bottom panel).
SOURCE: Photographs from Ref. [36] and used with permission of John Zink Company and CRC Press.

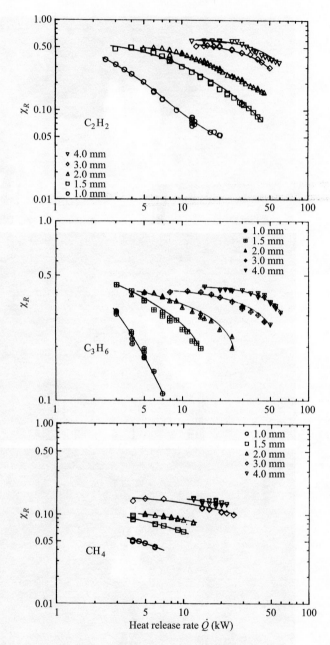

**Figure 13.14**   Jet flame radiant fractions as functions of heat release rate and burner size.
SOURCE: After Ref. [39]. Reprinted by permission of the Gas Research Institute.

Now, from our previous discussion of flame lengths, we know that, in the momentum-dominated regime, the flame length is directly proportional to the burner diameter, $d_j$; hence, we expect the following for the flame volume:

$$V_f \propto d_j^3.$$

Furthermore,

$$\dot{m}_F = \rho_F v_e \pi d_j^2 / 4.$$

Combining the $V_f$ and $\dot{m}_F$ scalings above into Eqn. 13.15 yields

$$\chi_R \propto a_p T_f^4 d_j / v_e, \tag{13.16}$$

which is consistent with the diameter and heat release rate influences on $\chi_R$ shown in Fig. 13.14 for the larger $\dot{Q}$ values. Note that $\dot{Q} \propto v_e$ for fixed $d_j$. The basic idea expressed by Eqn. 13.16 is that $\chi_R$ depends directly on the time available for radiant energy to be lost from the flame, where the characteristic time is proportional to $d_j / v_e$. It should be kept in mind that this simple analysis ignores buoyant effects and the interrelationships among $a_p$, $T_f$, and $\chi_R$. More sophisticated approaches to radiation scaling in jet flames are available in the literature [40].

As mentioned above, the sooting propensity of the fuel is a major factor in determining the radiant loss from a flame. There are two sources of radiation in flames: molecular radiation, primarily from $CO_2$ and $H_2O$; and essentially blackbody radiation from in-flame soot. The molecular radiation is concentrated in broadened bands of the infrared spectrum. Figures 13.15 and 13.16 show spectral radiation intensities for a methane flame, which contains very little in-flame soot (Fig. 13.15), and an ethylene flame (Fig. 13.16), which produces substantial in-flame soot [38, 41]. Comparing Figs. 13.15 and 13.16, we see that both flames exhibit significant intensity peaks in the bands at 2.5–3 $\mu$m and 4–5 $\mu$m. These correspond to the molecular radiation. In contrast, the ethylene flame (Fig. 13.16) exhibits strong continuum radiation at the short wavelengths, peaking near 1.5 $\mu$m. This continuum radiation is absent in the methane flame (Fig. 13.15). We can use Wien's displacement law [42] to estimate the wavelength associated with the maximum intensity of the soot blackbody radiation in the ethylene flame as follows:

$$\lambda_{max} = \frac{2897.8 \; \mu\text{m-K}}{T_f(\text{K})} \approx 1.22 \; \mu\text{m},$$

where the adiabatic flame temperature for a stoichiometric ethylene–air mixture (2369 K) has been used for $T_f$. Since the flame is nonadiabatic, the true flame temperature must be less than 2369 K and $\lambda_{max}$ larger than 1.22 $\mu$m. Using a $\lambda_{max}$ of 1.5 $\mu$m, which is consistent with the experimental data of Fig. 13.16, a characteristic flame temperature of about 1930 K is predicted.

Accurate modeling of radiation is important in simulations of many combustion systems. For example, to predict pollutant emissions (e.g., nitric oxide [43] and soot) requires accurate predictions of local temperatures, which, in turn, are greatly

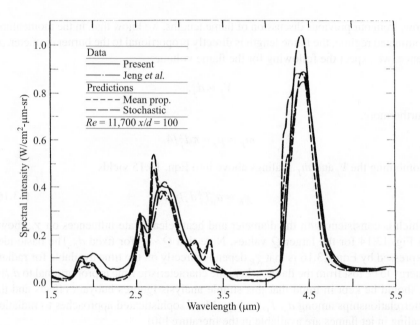

**Figure 13.15**     Spectral measurements of thermal radiation from a nonpremixed methane–air jet flame showing strong peaks associated with molecular band radiation.
| SOURCE: From Ref. [38]. Reproduced with permission of Taylor & Francis Group, http://www.informaworld.com.

influenced by radiation. The computational effort to deal with the detailed spectral behavior of participating species (particularly, CO, $CO_2$, and $H_2O$) can be huge [44]; however, with the advances in computing power in the past decade, considerable progress has been made in modeling the radiation from turbulent flames. Understanding the interactions between the turbulence and radiation, which result from the nonlinear couplings between the local fluctuating radiation intensities and the local fluctuating absorption coefficients, is a research topic of current interest [45–47].

## Liftoff and Blowout

As mentioned above and illustrated in Fig. 13.8, a jet flame will lift from an attached position at the burner exit if the exit velocity is sufficiently high. The **liftoff height,** the distance between the burner port and the base of the flame, will increase with additional increases in velocity until the flame blows out. The phenomena of lift-off and blowout of jet flames have been the subject of much research, and a good overview of this work can be found in Pitts [48, 49]. Pitts delineated three different theories that have been proposed to explain liftoff. The criteria for establishing the liftoff height are different for each theory and can be given as follows:

Theory I      The local flow velocity at the position where the laminar flame speed is a maximum matches the turbulent burning velocity of a premixed flame, i.e., $\overline{v}(S_{L,\max}) = S_T$.

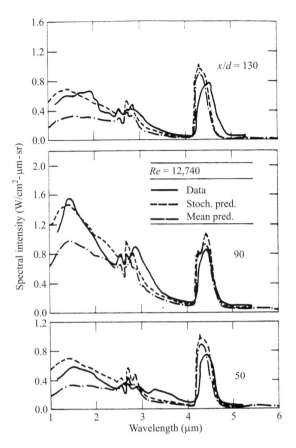

**Figure 13.16**    Spectral measurements of thermal radiation from a nonpremixed ethylene–air jet flame showing both continuum radiation from soot and molecular band radiation.
SOURCE: Reprinted from Ref. [41] with permission of The Combustion Institute.

Theory II    The local strain rates in the fluid exceed the extinction strain rate for a laminar diffusion flamelet, i.e., $\epsilon > \epsilon_{crit}$.

Theory III    The time available for backmixing by large-scale flow structures of hot products with fresh mixture is less than a critical chemical time required for ignition, i.e., $\tau_{local\ mixmg} < \tau_{chem,crit}$.

The first of these theories has its origins with Wohl *et al.* [12] and was articulated by Vanquickenborne and Van Tiggelen [50], whereas the other two theories are of more recent origins [51–53]. Lyons [54] in his recent review of liftoff adds two more theories to these three. One is a variant of Theory I. The other postulates that edge, or triple,[2] flames [55, 56] play an important role in stabilizing a lifted

[2]Triple, or tribrachial, flames have been observed in the mixing layer between a fuel and air stream. Such flames consist of a rich premixed branch, a lean premixed branch, and a diffusion flame trailing the premixed branches where they meet.

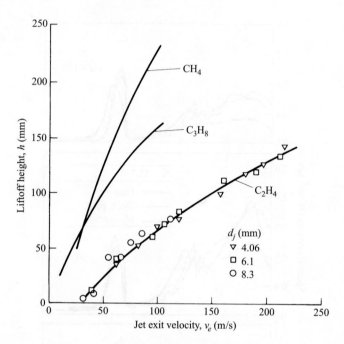

**Figure 13.17**     Liftoff height versus jet exit velocity for methane, propane, and ethylene jet flames.

SOURCE: After Kalghatgi [58].

turbulent flame. Lyons [54] also notes that the five theories are not necessarily completely independent. Takahashi and Schmoll [57] also discuss the liftoff characteristics associated with thick-lipped, straight-tube burners, where additional phenomena are important.

Figure 13.17 shows liftoff heights, $h$, for methane, propane, and ethylene jet flames as functions of initial velocity, $v_e$. It is interesting to note the lack of a burner-diameter dependence and the ordering of the curves with respect to their laminar flame speeds ($S_{L,CH_4} < S_{L,C_3H_8} < S_{L,C_2H_4}$). Kalghatgi [58], interpreting his data in terms of Theory I, developed the following correlation that can be used to describe the liftoff behavior of hydrocarbon–air flames:

$$\frac{\rho_e S_{L,\max} h}{\mu_e} = 50 \left( \frac{v_e}{S_{L,\max}} \right) \left( \frac{\rho_e}{\rho_\infty} \right)^{1.5}, \tag{13.17}$$

where $S_{L,\max}$ is the maximum laminar flame speed, which occurs near stoichiometric ratios for hydrocarbons.

The phenomenon of blowout can also be interpreted in terms of premixed turbulent flame concepts, i.e., Theory I, where it is assumed that the base of the lifted flame is a premixed flame. In this view, blowout occurs at a flowrate where the turbulent burning velocity falls more rapidly with distance downstream than does the local velocity at the position of $S_{L,\,\mathrm{max}}$. This explains the sudden occurrence of blowout just beyond a critical liftoff height, even though the mixture is still within the flammability limits at the flame base.

Kalghatgi [59] proposes the following correlation to estimate blowout flowrates for jet flames:

$$\frac{v_e}{S_{L,\,\mathrm{max}}}\left(\frac{\rho_e}{\rho_\infty}\right)^{1.5} = 0.017\,Re_H\left(1 - 3.5 \cdot 10^{-6}\,Re_H\right), \qquad (13.18\mathrm{a})$$

where the Reynolds number, $Re_H$, is given by

$$Re_H = \frac{\rho_e S_{L,\,\mathrm{max}} H}{\mu_e}. \qquad (13.18\mathrm{b})$$

The characteristic length, $H$, is the distance along the burner axis where the mean fuel concentration has fallen to its stoichiometric value and can be estimated by the following [60]:

$$H = \left[\frac{4Y_{F,e}}{Y_{F,\,\mathrm{stoic}}}\left(\frac{\rho_e}{\rho_\infty}\right)^{1/2} + 5.8\right]d_j. \qquad (13.18\mathrm{c})$$

Figure 13.18 shows the applicability of Eqn. 13.18 to a wide range of fuels. Note that for a fixed fuel type, blowout velocities increase with jet diameter. This helps to explain the difficulty associated with blowing out oil-well fires.

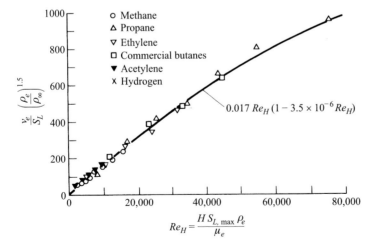

**Figure 13.18**    Universal blowout stability curve from Kalghatgi [59].
SOURCE: Reprinted by permission of Taylor & Francis Group, http://www.informaworld.com.

| Example 13.3 | Determine the blowoff velocity for a propane jet flame in air for a nozzle diameter of 6.17 mm. Also, estimate the liftoff height at the incipient blowoff condition. The ambient conditions are $P = 1$ atm and $T_\infty = 300$ K. The propane temperature at the nozzle exit is also 300 K. The propane density is $1.854 \text{ kg/m}^3$. |

**Solution**

To determine the blowout velocity, we will employ Kalghatgi's correlation, Eqn. 13.18a. We must first determine the characteristic length $H$ (Eqn. 13.18c) to find the Reynolds number $Re_H$ (Eqn. 13.18b) required by the correlation.

To evaluate $H$, we note that

$$Y_{F,\text{stoic}} = f_s = 0.06035 \qquad \text{(see Example 13.1)}$$

$$Y_{F,e} = 1 \qquad \text{(pure fuel at nozzle exit)}.$$

Thus,

$$H = d_j \left[ \frac{4Y_{F,e}}{f_s} \left( \frac{\rho_e}{\rho_\infty} \right)^{1/2} + 5.8 \right]$$

$$= (0.00617) \left[ \frac{4(1)}{0.06035} \left( \frac{1.854}{1.1614} \right)^{0.5} + 5.8 \right]$$

$$= 0.449 \text{ m}.$$

To evaluate $Re_H$, we also need to find the maximum laminar flame speed for a propane–air mixture, $S_{L,\text{max}}$, and the propane viscosity, $\mu$, at 300 K:

$$\mu_e = 8.26 \cdot 10^{-6} \text{ N-s/m}^2 \qquad \text{(Appendix Table B.3 curvefits)}.$$

Using the Metghalchi and Keck flame-speed correlations (Table 8.3):

$$S_{L,\text{max}} = S_{L,\text{ref}} = 0.3422 \text{ m/s},$$

noting that $S_{L,\text{max}}$ occurs when $\Phi = \Phi_M = 1.08$.

We can now calculate $Re_H$:

$$Re_H = \frac{\rho_e S_{L,\text{max}} H}{\mu_e} = \frac{1.854(0.3422)0.449}{8.26 \cdot 10^{-6}}$$

$$= 34,500.$$

Using the blowout velocity correlation (Eqn. 13.18a), we evaluate

$$\frac{v_e}{S_{L,\text{max}}} \left( \frac{\rho_e}{\rho_\infty} \right)^{1.5} = 0.017 Re_H \left( 1 - 3.5 \cdot 10^{-6} Re_H \right)$$

$$= 0.017(34,500) \left[ 1 - 3.5 \cdot 10^{-6}(34,500) \right]$$

$$= 516.$$

Finally, we can use the above to solve for the blowout velocity, $v_e$:

$$v_e = 516(0.3422) \left( \frac{1.1614}{1.854} \right)^{1.5}$$

$$\boxed{v_e = 87.6 \text{ m/s}}$$

We can use this velocity to determine the liftoff height at incipient blowout. From Fig. 13.17, we obtain

$$\boxed{h_{\text{liftoff}} \cong 150 \text{ mm}}$$

**Comment**

Comparing the above results with the conditions in Example 13.1, we see that the Example 13.1 flame is close to the blowout condition and requires stabilization of some sort to prevent lifting. Experimental results corresponding to the Example 13.1 flame [30] show that a hydrogen pilot flame with a flowrate of 1.4 percent that of the propane was required to prevent lifting.

## OTHER CONFIGURATIONS

In many practical devices employing nonpremixed flames, the combustion air is introduced in some coflowing arrangement with the fuel, unlike the simple jet flames discussed above where the fuel jet flowed into a reservoir of air. Figure 13.19 illustrates a simple gas burner designed to produce long, luminous flames [64]. In this particular design, the combustion air flows through an annulus surrounding the fuel tube. A burner tile (see Chapter 12) is used to anchor the flame to the burner outlet. By minimizing the relative velocity between the fuel and air streams, mixing rates are slow, resulting in long flames. Furthermore, the long residence times within the fuel jet core are favorable for the production of soot (see Chapter 9); thus, a highly luminous flame is produced. The burner type illustrated in Fig. 13.19 is useful for applications desiring a relatively uniform radiative heating over a large distance. Similarly, slow-mixing designs are used in radiant-tube burners (see Fig. 13.1) where uniform heating over the length of the tube is required.

Nozzle-mix burners (shown previously in Fig. 12.4), partially premix the air and fuel within the nozzle, causing the flame to have some attributes of both premixed

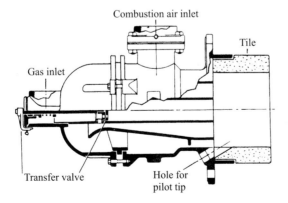

**Figure 13.19**    Nonpremixed-flame burner employing annular flow of air around central fuel jet.

Source: From Ref. [64], reprinted by permission of North American Manufacturing Co.

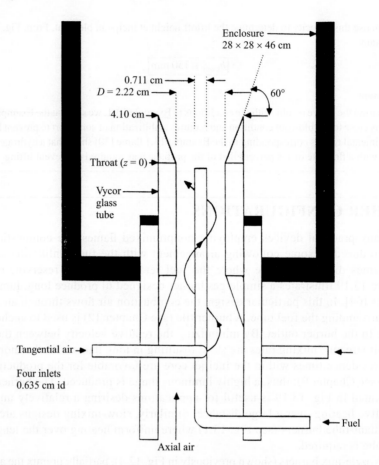

**Figure 13.20**    Schematic diagram of experimental swirl-flame burner having both tangential and axial air entries.

| SOURCE: Reprinted from Ref. [61] with permission. © 1987, Gordon & Breach Science Publishers.

and nonpremixed combustion, the relative proportions depending upon the degree of mixing achieved at the nozzle.

Swirl is also frequently used in practical nonpremixed burners, particularly those employing sprays of liquid fuel or pulverized coal. Figure 13.20 illustrates an experimental burner arrangement where the degree of swirl is controlled by the relative amounts of air introduced tangentially and axially. In practical applications, vanes are typically used to impart swirl to the air and/or fuel streams. Figure 13.21 shows an adjustable swirl-vane arrangement for a furnace burner.

Swirl is useful for two reasons: first, swirl can stabilize a flame by creating a recirculation zone, if the amount of swirl is sufficiently great; and second, the length of the flame can be controlled by the amount of swirl. These two aspects of swirl combustion are illustrated in Figs. 13.22 and 13.23, respectively. In Fig. 13.22, we

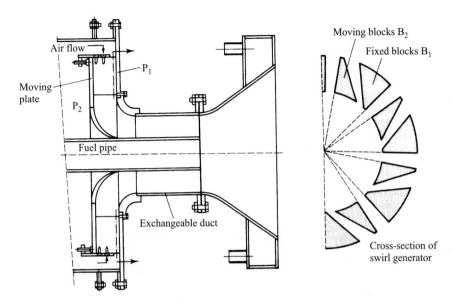

**Figure 13.21**    Movable-block swirl-vane arrangement for furnace burner.
SOURCE: From Ref. [62] (Figure 5.3). Reprinted by permission of Krieger Publishing Co.

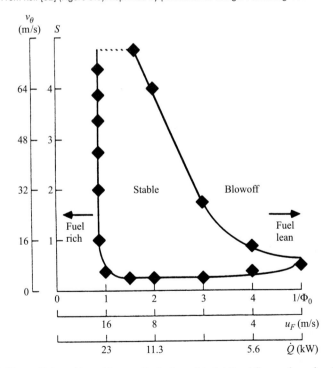

**Figure 13.22**    Rich and lean blowout limits for swirl-stabilized flames for a fixed velocity of air of 19 m/s. Burner geometry corresponds to that of Fig. 13.20.
SOURCE: Reprinted from Ref. [61] with permission. © 1987, Gordon & Breach Science Publishers.

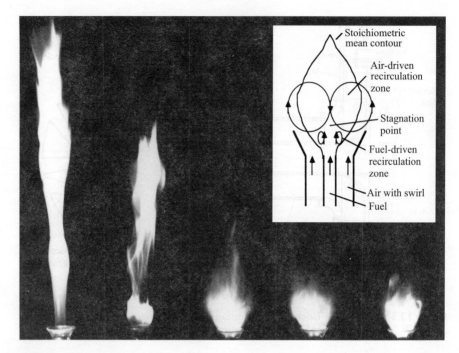

**Figure 13.23**     Effect of swirl on flame length. Photographic sequence showing no swirl (left) progressing to a swirl number of $S = 1.1$.
| SOURCE: Reprinted from Ref. [63] with permission of The Combustion Institute.

see the dimensionless swirl number, $S$, and the tangential velocity component, $v_\theta$, plotted on the vertical axis, and the reciprocal of the overall equivalence ratio, $1/\Phi$, on the horizontal axis. Also shown on the horizontal axis are the corresponding fuel velocity, $u_F$, and the flame heat-release rate $\dot{Q}$. To understand the influence of swirl on blowout stability, let us consider first the lower boundary of the region labeled stable. Below this line, no flame can be stabilized; however, as swirl is added to the inlet air, a condition is reached where a stable flame is possible. For example, at $\Phi = 1$, a stable flame results when the swirl number reaches approximately 0.4. The length of a horizontal line lying within the stable region gives a measure of the overall stability, expressed as a range of equivalence ratios over which stable flames occur. Hence, the maximum overall stability occurs at about $S = 0.6$, with smaller ranges of stability for swirl numbers both larger and smaller than this value.

The dramatic effect of swirl on flame length is illustrated in the sequence of photographs shown in Fig. 13.23. Here we see that the addition of swirl ($S = 1.1$) reduces the flame length by a factor of about five in comparison to the flame with no swirl. Swirl greatly enhances the mixing of air and fuel streams, thereby causing the shorter flame. Chen and Driscoll [63] provide scaling relations that relate flame length to both jet and swirl mixing parameters.

# SUMMARY

In this chapter, we explored turbulent nonpremixed combustion, focusing primarily on the simple turbulent jet flame. A simple mathematical analysis was developed that built upon our previous study of laminar nonpremixed jet flames (Chapter 9) and nonreacting turbulent jets (Chapter 11). The usefulness of conserved scalars was again demonstrated. We saw that when buoyancy is unimportant, the turbulent flame length is directly proportional to the nozzle diameter and independent of the initial jet velocity. Empirical correlations for turbulent jet flame lengths, covering a wide range of flow conditions, were presented and applications of their use were shown. We also saw that radiation heat transfer is an important aspect of nonpremixed combustion. The concepts of liftoff and blowout were explored and, again, empirical correlations for these phenomena were presented for simple jet flames. Swirling flames, which are often used in practical devices, were introduced, emphasizing the ability of swirl to create a recirculation zone that acts to both stabilize and shorten the flame.

# NOMENCLATURE

| | |
|---|---|
| $a_p$ | Mean absorption coefficient ($m^{-1}$) |
| $A/F$ | Air–fuel ratio (kg/kg) |
| $d$ | Diameter (m) |
| $\mathcal{D}$ | Mass diffusivity ($m^2/s$) |
| $f$ | Mixture fraction (kg/kg) |
| $Fr$ | Froude number (dimensionless) |
| $g$ | Gravitational acceleration ($m/s^2$) |
| $h$ | Liftoff height (m) |
| $H$ | Characteristic length (m) |
| $L$ | Length (m) |
| $Le$ | Lewis number (dimensionless) |
| LHV | Lower heating value (J/kg) |
| $\dot{m}$ | Mass flowrate (kg/s) |
| $P$ | Pressure (Pa) |
| $Pr$ | Prandtl number (dimensionless) |
| $\dot{Q}$ | Heat transfer or heat release rate (W) |
| $r$ | Radial coordinate (m) |
| $R$ | Nozzle radius (m) |
| $Re$ | Reynolds number (dimensionless) |
| $S$ | Swirl number (dimensionless) |
| $S_L$ | Laminar flame speed (m/s) |
| $Sc$ | Schmidt number (dimensionless) |
| $T$ | Temperature (K) |
| $v$ | Velocity (m/s) |

| $V$ | Volume ($m^3$) |
| $x$ | Axial coordinate (m) |
| $Y$ | Mass fraction (kg/kg) |

**Greek Symbols**

| $\alpha$ | Thermal diffusivity ($m^2$/s) |
| $\Delta h_c$ | Heat of combustion (J/kg) |
| $\varepsilon$ | Eddy viscosity ($m^2$/s) |
| $\epsilon$ | Strain rate ($s^{-1}$) |
| $\lambda$ | Wavelength |
| $\mu$ | Absolute viscosity (N-s/$m^2$) |
| $\nu$ | Kinematic viscosity ($m^2$/s) |
| $\rho$ | Density (kg/$m^3$) |
| $\Phi$ | Equivalence ratio |
| $\chi_R$ | Radiant fraction |

**Subscripts**

| crit | Critical |
| $e$ | Exit |
| $f$ | Flame |
| $F$ | Fuel |
| $j$ | Jet |
| max | Maximum |
| $r$ | Radial |
| rad | Radiation |
| $s$, stoic | Stoichiometric |
| $sp$ | Smoke point |
| turb, $T$ | Turbulent |
| $\infty$ | Ambient |

**Other Notation**

| $(\ )^*$ | Dimensionless variable |
| $\overline{(\ )}$ | Time-averaged quantity |

# REFERENCES

1. Weinberg, F. J., "The First Half-Million Years of Combustion Research and Today's Burning Problems," *Progress in Energy and Combustion Science,* 1: 1731 (1975).

2. Baukal, C. E., Jr., (ed.), *Oxygen-Enhanced Combustion,* CRC Press, Boca Raton, FL, 1998.

3. International Workshop on Measurement and Computation of Nonpremixed Flames, Sandia National Laboratories, U.S. Department of Energy, http://public.ca.sandia.gov/TNF/abstract.html.

4. Mungal, M. G., Korasso, P. S., and Lozano, A., "The Visible Structure of Turbulent Jet Diffusion Flames—Large-Scale Organization and Flame Tip Oscillation," *Combustion Science and Technology,* 76: 165–185 (1991).

5. Lee, S.-Y., Turns, S. R., and Santoro, R. J., "Measurements of Soot, OH, and PAH Concentrations in a Turbulent Ethylene/Air Jet Flames," *Combustion and Flame,* 156: 2264–2275 (2009).

6. Gore, J. P., *A Theoretical and Experimental Study of Turbulent Flame Radiation,* Ph.D. Thesis, The Pennyslvania State University, University Park, PA, p. 119, 1986.

7. Seitzman, J. M., "Quantitative Applications of Fluorescence Imaging in Combustion," Ph.D. Thesis, Stanford University, Stanford, CA, June 1991.

8. Seitzman, J. M., Üngüt, A., Paul, P. H., and Hanson, R. K., "Imaging and Characterization of OH Structures in a Turbulent Nonpremixed Flame," *Twenty-Third Symposium (International) on Combustion,* The Combustion Institute, Pittsburgh, PA, pp. 637–644, 1990.

9. Donbar, J. M., Driscoll, J. F., and Carter, C. D., "Reaction Zone Structure in Turbulent Nonpremixed Jet Flames—From CH-OH PLIF Images," *Combustion and Flame,* 122: 1–19 (2000).

10. Everest, D. A., Driscoll, J. F., Dahm, W. J. A., and Feikema, D. S., "Images of the Two-Dimensional Field and Temperature Gradients to Quantify Mixing Rates within a Non-Premixed Turbulent Jet Flame," *Combustion and Flame,* 101: 58–68 (1995).

11. Chen, L.-D., Roquemore, W. M., Goss, L. P., and Vilimpoc, V., "Vorticity Generation in Jet Diffusion Flames," *Combustion Science and Technology,* 77: 41–57 (1991).

12. Wohl, K., Gazley, C., and Kapp, N., "Diffusion Flames," *Third Symposium on Combustion and Flame and Explosion Phenomena,* Williams & Wilkins, Baltimore, MD, p. 288, 1949.

13. Haworth, D. C., "Progress in Probability Density Function Methods for Turbulent Reacting Flows," *Progress in Energy and Combustion Science,* 36: 168–259 (2009).

14. Bilger, R. W., Pope, S. B., Bray, K. N. C., and Driscoll, J. F., "Paradigms in Turbulent Combustion Research," *Proceeding of the Combustion Institute,* 30: 21–42 (2005).

15. Westbrook, C. K., Mizobuchi, Y., Poinsot, T. J., Smith, P. J., and Warnatz, J., "Computational Combustion," *Proceedings of the Combustion Institute,* 30: 125–157 (2005).

16. Veyante, D., and Vervisch, L., "Turbulent Combustion Modeling," *Progress in Energy and Combustion Science,* 28: 193–266 (2002).

17. Cant, R. S., and Mastorakos, E., *An Introduction to Turbulent Reacting Flows,* Imperial College Press, London, 2008.

18. Poinsot, T., and Veynante, D., *Theoretical and Numerical Combustion,* 2nd Ed., R. T. Edwards, Philadelphia, PA, 2005.

19. Kee, R. J., Coltrin, M. E., and Glarborg, P., *Chemically Reacting Flow: Theory and Practice,* John Wiley & Sons, Hoboken, NJ, 2003.

20. Fox, R. O., *Computational Models for Turbulent Reacting Flows,* Cambridge University Press, Cambridge, UK, 2003.

21. Peters, N., *Turbulent Combustion,* Cambridge University Press, Cambridge, UK, 2000.

22. Chen, J.-Y., Kollman, W., and Dibble, R. W., "Pdf Modeling of Turbulent Non-premixed Methane Jet Flames," *Combustion Science and Technology,* 64: 315–346 (1989).

23. Bilger, R. W., "Turbulent Flows with Nonpremixed Reactants," in *Turbulent Reacting Flows* (P. A. Libby, and F. A. Williams, eds.), Springer-Verlag, New York, pp. 65–113, 1980.

24. Bilger, R. W., "Turbulent Jet Diffusion Flames," *Progress in Energy and Combustion Science,* 1: 87–109 (1976).

25. Hottel, H. C., "Burning in Laminar and Turbulent Fuel Jets," *Fourth Symposium (International) on Combustion,* Williams & Wilkins, Baltimore, MD, p. 97, 1953.

26. Hawthorne, W. R., Weddell, D. S., and Hottel, H. C., "Mixing and Combustion in Turbulent Gas Jets," *Third Symposium on Combustion and Flame and Explosion Phenomena,* Williams & Wilkins, Baltimore, MD, p. 266, 1949.

27. Becker, H. A., and Liang, P., "Visible Length of Vertical Free Turbulent Diffusion Flames," *Combustion and Flame,* 32: 115–137 (1978).

28. Delichatsios, M. A., "Transition from Momentum to Buoyancy-Controlled Turbulent Jet Diffusion Flames and Flame Height Relationships," *Combustion and Flame,* 92: 349–364 (1993).

29. Blake, T. R., and McDonald, M., "An Examination of Flame Length Data from Vertical Turbulent Diffusion Flames," *Combustion and Flame,* 94: 426–432 (1993).

30. Turns, S. R., and Bandaru, R. B., "Oxides of Nitrogen Emissions from Turbulent Hydrocarbon/Air Jet Diffusion Flames," Final Report-Phase II, GRI 92/0470, Gas Research Institute, September 1992.

31. Turns, S. R., and Myhr, F. H., "Oxides of Nitrogen Emissions from Turbulent Jet Flames: Part I—Fuel Effects and Flame Radiation," *Combustion and Flame,* 87: 319–335 (1991).

32. Bahadori, M. Y., Small, J. F., Jr., Hegde, U. G., Zhou, L., and Stocker, D. P., "Characteristics of Transitional and Turbulent Jet Diffusion Flames in Microgravity," NASA Conference Publication 10174, Third International Microgravity Combustion Workshop, Cleveland, OH, pp. 327–332, 11–13 April, 1995.

33. Thring, M. W., and Newby, M. P., "Combustion Length of Enclosed Turbulent Jet Flames," *Fourth Symposium (International) on Combustion,* Williams & Wilkins, Baltimore, MD, p. 789, 1953.

34. Ricou, F. P., and Spalding, D. B., "Measurements of Entrainment by Axisymmetrical Turbulent Jets," *Journal of Fluid Mechanics,* 11: 21–32 (1963).

35. Tacina, K. M., and Dahm, W. J. A., "Effects of Heat Release on Turbulent Shear Flows. Part 1. A General Equivalence Principle for Non-Buoyant Flows and Its Application to Turbulent Jet Flames," *Journal of Fluid Mechanics,* 415: 23–44 (2000).

36. Baukal, C. E., Jr., ed., *The John Zink Combustion Handbook,* CRC Press, Boca Raton, FL, 2000.

37. Viskanta, R., and Mengüç, M. P., "Radiation Heat Transfer in Combustion Systems," *Progress in Energy and Combustion Science,* 13: 97–160 (1987).

38. Faeth, G. M., Gore, J. P., Chuech, S. G., and Jeng, S.-M., "Radiation from Turbulent Diffusion Flames," in *Annual Review of Numerical Fluid Mechanics and Heat Transfer,* Vol. 2, Hemisphere, Washington, DC, pp. 1–38, 1989.

39. Delichatsios, M. A., Markstein, G. H., Orloff, L., and deRis, J., "Turbulent Flow Characterization and Radiation from Gaseous Fuel Jets," Final Report, GRI 88/0100, Gas Research Institute, 1988.

40. Orloff, L., deRis, J., and Delichatsios, M. A., "Radiation from Buoyant Turbulent Diffusion Flames," *Combustion Science and Technology,* 84: 177–186 (1992).

41. Gore, J. P., and Faeth, G. M., "Structure and Spectral Radiation Properties of Turbulent Ethylene/Air Diffusion Flames," *Twenty-First Symposium (International) on Combustion,* The Combustion Institute, Pittsburgh, PA, p. 1521, 1986.

42. Siegel, R., and Howell, J. R., *Thermal Radiation Heat Transfer,* 2nd Ed., McGraw-Hill, New York, 1981.

43. Frank, J. H., Barlow, R. S., and Lundquist, C., "Radiation and Nitric Oxide Formation in Turbulent Non-Premixed Jet Flames," *Proceedings of the Combustion Institute,* 28: 447–454 (2000).

44. Modest, M. F., Radiative Heat Transfer, 2nd Ed., Academic Press, New York, 2003.

45. Coelho, P. J., Teerling, O. J., and Roekaerts, D., "Spectral Radiative Effects and Turbulence/Radiation Interaction in a Non-Luminous Turbulent Jet Diffusion Flame," *Combustion and Flame,* 133: 75–91 (2003).

46. Tessé, L., Dupoirieux, F., and Taine, J., "Monte Carlo Modeling of Radiative Transfer in a Turbulent Sooty Flame," *International Journal of Heat and Mass Transfer,* 47: 555–572 (2004).

47. Wang, A., Modest, M. F., Haworth, D. C., and Wang, L., "Monte Carlo Simulation of Radiative Heat Transfer and Turbulence Interactions in Methane/Air Jet Flames," *Journal of Quantitative Spectroscopy & Radiative Transfer,* 109: 269–279 (2008).

48. Pitts, W. M., "Assessment of Theories for the Behavior and Blowout of Lifted Turbulent Jet Diffusion Flames," *Twenty-Second Symposium (International) on Combustion,* The Combustion Institute, Pittsburgh, PA, p. 809, 1988.

49. Pitts, W. M., "Importance of Isothermal Mixing Processes to the Understanding of Lift-Off and Blowout of Turbulent Jet Diffusion Flames," *Combustion and Flame,* 76: 197–212 (1989).

50. Vanquickenborne, L., and Van Tiggelen, A., "The Stabilization Mechanism of Lifted Diffusion Flames," *Combustion and Flame,* 10: 59–69 (1966).

51. Peters, N., "Local Quenching Due to Flame Stretch and Non-Premixed Turbulent Combustion," *Combustion Science and Technology,* 30: 1–17 (1983).

52. Janicka, J., and Peters, N. "Prediction of Turbulent Jet Diffusion Flame Lift-Off Using a PDF Transport Equation," *Nineteenth Symposium (International) on Combustion,* The Combustion Institute, Pittsburgh, PA, p. 367, 1982.

53. Broadwell, J. E., Dahm, W. J. A., and Mungal, M. G., "Blowout of Turbulent Diffusion Flames," *Twentieth Symposium (International) on Combustion,* The Combustion Institute, Pittsburgh, PA, p. 303, 1984.

54. Lyons, K. M., "Toward an Understanding of the Stabilization Mechanisms of Lifted Turbulent Jet Flames: Experiments," *Progress in Energy and Combustion Science,* 33: 211–231 (2007).

55. Buckmaster, J., "Edge-Flames," *Progress in Energy and Combustion Science,* 28: 435–475 (2002).

56. Joedicke, A., Peters, N., and Mansour, M., "The Stabilization Mechanism and Structure of Turbulent Hydrocarbon Lifted Flames," *Proceedings of the Combustion Institute,* 30: 901–909 (2005).

57. Takahashi, F., and Schmoll, W. J., "Lifting Criteria of Jet Diffusion Flames," *Twenty-Third Symposium (International) on Combustion,* The Combustion Institute, Pittsburgh, PA, p. 677, 1990.

58. Kalghatgi, G. T., "Lift-Off Heights and Visible Lengths of Vertical Turbulent Jet Diffusion Flames in Still Air," *Combustion Science and Technology,* 41: 17–29 (1984).

59. Kalghatgi, G. T., "Blow-Out Stability of Gaseous Jet Diffusion Flames. Part I: In Still Air," *Combustion Science and Technology,* 26: 233–239 (1981).

60. Birch, A. D., Brown, D. R., Dodson, M. G., and Thomas, J. R., "The Turbulent Concentration Field of a Methane Jet," *Journal of Fluid Mechanics,* 88: 431–449 (1978).

61. Tangirala, V., Chen, R. H., and Driscoll, J. F., "Effect of Heat Release and Swirl on the Recirculation within Swirl-Stabilized Flames," *Combustion Science and Technology,* 51: 75–95 (1987).

62. Beér, J. M., and Chigier, N. A., *Combustion Aerodynamics,* Krieger, Malabar, FL, 1983.

63. Chen, R.-H., and Driscoll, J. F., "The Role of the Recirculation Vortex in Improving Fuel-Air Mixing within Swirling Flames," *Twenty-Second Symposium (International) on Combustion,* The Combustion Institute, Pittsburgh, PA, pp. 531–540, 1988.

64. North American Manufacturing Co., *North American Combustion Handbook,* North American Manufacturing Co., Cleveland, OH, 1952.

---

# REVIEW QUESTIONS

**1.** Make a list of all of the boldfaced words in Chapter 13 and discuss.

**2.** Discuss the factors that cause different fuels to produce different jet-flame lengths for identical initial conditions, i.e., equal $d_j$ values and $v_e$ values.

**3.** Discuss why flame lengths for momentum-dominated jet flames do not depend on initial velocity, but, rather, are directly proportional to the initial jet diameter. Use the Prandtl mixing-length model applied to jets to support your arguments.

**4.** What happens to the length of a jet flame when an inert diluent is added to the fuel? Discuss.

**5.** Discuss the concept of flame stability and the various methods employed to stabilize nonpremixed turbulent flames.

**6.** Make a list of applications that employ turbulent nonpremixed combustion.

**7.** Discuss the factors that affect the radiant heat transfer from turbulent, nonpremixed hydrocarbon–air flames.

## PROBLEMS

**13.1** What nozzle exit velocities are required to achieve momentum-dominated propane and carbon monoxide jet flames for a nozzle diameter of 5 mm? At nozzle exit conditions, the density of propane is 1.854 kg/m$^3$ and for carbon monoxide 1.444 kg/m$^3$. The adiabatic flame temperature for a CO–air flame is 2400 K. Discuss your results.

**13.2** Calculate and compare the flame lengths for momentum-dominated propane–air and hydrogen–air jet flames for a nozzle exit diameter of 5 mm. Assume ambient and fuel temperatures of 300 K. Discuss your results.

**13.3** Determine if the flames in problem 13.2 will require some form of stabilization to prevent them from being blown out. Useful hydrogen properties are $T_{ad} = 2383$ K, $\mu_{300\,K} = 8.96 \cdot 10^{-6}$ N-s/m$^2$, and $S_{L,\,max} = 3.25$ m/s. Assume that the velocity is the minimum required for a momentum-dominated flame.

**13.4** A propane–air jet flame has a nozzle exit velocity of 200 m/s. Determine the minimum nozzle exit diameter required to prevent the flame from being blown out.

**13.5** Determine the length of the flame in problem 13.4.

**13.6** A propene (C$_3$H$_6$)–air jet flame is stabilized on a 3-mm-diameter nozzle. The jet exit velocity is 52.7 m/s. Determine the radiant heat-transfer rate (kW) to the cold surroundings. Assume the propene density is approximately 1.76 kg/m$^3$ at the jet exit conditions.

**13.7** Consider a methane–air jet flame. The nozzle exit diameter is 4 mm and the heat release rate of the flame is 25 kW. The density of the methane at the exit conditions is 0.6565 kg/m$^3$.

　　A. Estimate the liftoff height of the flame.

　　B. Compare the liftoff length calculated in part A above with the total flame length.

**13.8** Estimate the change in flame length in a momentum-dominated ethane–air jet flame when the pure fuel stream is diluted with nitrogen to 50 percent N$_2$ by volume. Assume ideal-gas behavior.

**13.9**  A defense contractor has developed an exotic fuel whose composition is thought to be $CO_2H_4$. A combustion test is to be conducted in which this fuel is injected axially at 367 K as a turbulent jet into a 1.83-m-long combustor. Air is provided at 811 K and 20 atm. Premixed combustion tests at the same temperatures and pressure indicate that the maximum laminar flame speed is approximately 0.61 m/s. The viscosity of the fuel is estimated to be $5 \cdot 10^{-5}$ N-s/m$^2$ at 367 K. Determine the minimum number of jet flames and the corresponding nozzle size required to burn 1.82 kg/s of fuel within the length of the combustor. Assume that the length of the turbulent jet flames is momentum controlled, that there are no interactions among the flames, and that the coflowing air has negligible effect on the flame length. To achieve a practicable design, be sure to consider the possibility of blowout.

# 14

# Burning of Solids

## OVERVIEW

Thus far in our study of combustion, we have considered primarily the burning of gaseous fuels (Chapters 8, 9, 12 and 13), with some attention to liquid fuels (Chapters 3 and 10) where the combustion reactions ultimately involve gaseous species, i.e., the liquid must first vaporize before it can burn. In this chapter, we introduce the subject of burning solids. One of the most important solid fuels is coal (see Chapter 17), which is typically burned as a pulverized powder in utility boilers, as discussed in this chapter. Other applications involving solids combustion include refuse burning, metals combustion, hybrid rocket engines, wood burning, and carbon combustion (coal char or coke), among others.

As suggested by the diversity of the applications and types of fuels listed above, the problem of solids combustion is very complex, with the details depending upon both the nature of the fuel and the specific application. For example, an entire book (or several) would be required to treat coal combustion in detail. Therefore, our approach will be, first, to introduce some of the fundamental concepts that are important to solids combustion and, second, to apply these concepts to the development of simple models of the burning of a spherical solid carbon particle. The carbon combustion models provide some insight into the general nature of solids combustion and also have relevance to the problem of coal combustion. The chapter concludes with a brief overview of the combustion of coal and a few other solids.

## COAL-FIRED BOILERS

Figure 14.1 illustrates a typical pulverized-coal-fired boiler. The boiler produces steam for steam-turbine generation of electricity. Utility boilers are huge. The combustion space may be as large as 15 m × 20 m, with total heights reaching over 50 m.

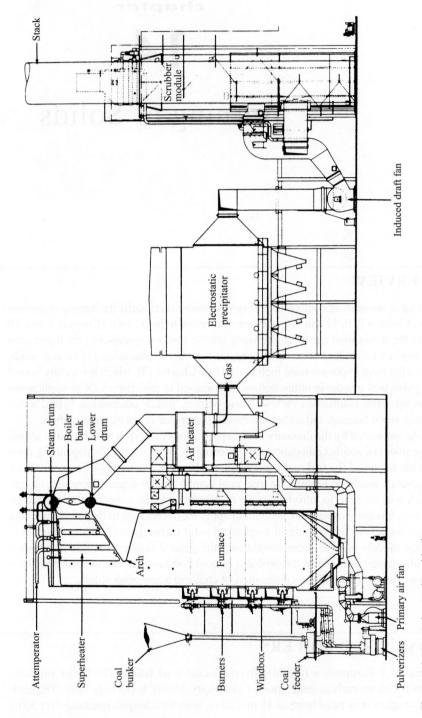

**Figure 14.1** Pulverized-coal boiler.

SOURCE: Reprinted from Ref. [1] with permission of Babcock & Wilcox Co.

Crushed coal is pulverized so that most of the coal particles are smaller than about 75 $\mu$m. The coal is blown into the primary zone (lower furnace) by the primary air. The primary air is on the order of 20 percent of the total air supplied, and the oxygen in the primary air is consumed in the combustion of the volatiles. Secondary air enters through the overfire port at high velocities, mixing with the char and the combustion products from the lower furnace. Heat from the flame gases is transferred through the tubes that line the combustion space to the superheated steam. Further downstream where the combustion gases are cooler, steam from the turbines is reheated, feedwater heated, and combustion air preheated. The combustion gases are then treated to remove particulate matter and to reduce sulfur oxides and, in some cases, nitrogen oxides. A large portion of the boiler-system volume is occupied by the pollution control system(s).

Other configurations are also employed in coal burning. For example, larger coal particles can be burned on fixed or traveling grates. Cyclone combustors, with air and/or fuel streams directed to create a swirling flow, are used in some boilers. Because of environmental constraints, the use of fluidized-bed combustors is becoming attractive. The use of fluidized bed boilers in the United States has increased dramatically in the last few decades [2]. Expanded application of this technology is likely in the future [2].

## HETEROGENEOUS REACTIONS

Solids combustion differs in a fundamental way from our previous topics in that, now, heterogeneous chemical reactions assume importance. A **heterogeneous reaction** is a reaction involving species that exist in different physical states, i.e., gas–liquid or gas–solid reactions. We introduced heterogeneous reactions in Chapter 4; however in all of our previous discussions (Chapters 5–13), we assumed chemical reactions occur as a result of colliding gas-phase molecules, i.e., **homogeneous reactions.** A detailed study of heterogeneous reactions would require several volumes; therefore, our treatment here is but a very brief introduction. For further study beyond our discussion in Chapter 4, basic textbooks on chemical kinetics are suggested, e.g., Ref. [3].

Gardiner [3] subdivides the overall process of gas–solid reactions into the following constituent processes:

1. Transport of the reactant molecule to the surface by convection and/or diffusion.
2. Adsorption of the reactant molecule on the surface.
3. Elementary reaction steps involving various combinations of adsorbed molecules, the surface itself, and gas-phase molecules.
4. Desorption of product molecules from the surface.
5. Transport of the product molecules away from the surface by convection and/or diffusion.

The first and fifth steps are familiar to us and can be treated using the mass-transfer concepts developed in Chapter 3. The intervening steps are more complex, especially step 3 (see Chapter 4), and are more properly in the realm of physical chemistry. Rather than elaborate on these steps, we will cite three "rate laws" [3] that arise as a result of how strongly the reactant and/or product molecules are adsorbed on the surface. First, if the reactant molecule A is weakly adsorbed, then the reaction rate, $\Re$, is proportional to the gas-phase concentration of A adjacent to the surface:

$$\Re = k(T)[A], \tag{14.1}$$

where $k(T)$ is the rate coefficient. Second, if A is strongly adsorbed, then the reaction rate is independent of the gas-phase concentration of A, i.e.,

$$\Re = k(T). \tag{14.2}$$

The last case we consider is when the reactant molecule A is weakly adsorbed, while the product molecule B is strongly adsorbed. For this case,

$$\Re = k(T)\frac{[A]}{[B]}, \tag{14.3}$$

where [A] and [B] are the gas-phase concentrations of A and B immediately adjacent to the surface. The point in presenting Eqns. 14.1–14.3 is to illustrate how gas–solid reaction-rate expressions differ from those for elementary homogeneous reactions with which you are familiar. Furthermore, we will employ expressions of the form of Eqn. 14.1 in the development of our models of carbon combustion.

## BURNING OF CARBON

In addition to providing a nice example to illustrate the general nature of solids combustion, carbon combustion is interesting for practical reasons. For example, in the combustion of pulverized coal, a carbon char is produced after the volatile matter has been driven from the coal particle and burned. The subsequent burnout of the carbon char is the limiting process establishing the necessary residence time requirements, and hence volume, of the combustion space. Moreover, a substantial amount of heat is radiated to the load from the burning char particles. Even though the real processes involved in coal-char combustion are much more complicated than implied by the models developed here, these models provide insight into the real processes and can serve as order-of-magnitude approximations.

Because of its importance to coal combustion, carbon combustion has a large literature associated with it, for example, Refs. [4–6]; and many models have been developed. Caram and Amundson [7] review and compare 12 models developed between 1924 and 1977. Before beginning our simplified analysis, we first present an overview of the various processes involved.

## Overview

Figure 14.2 schematically shows a burning carbon surface within a reacting boundary layer. At the surface, the carbon can be attacked by either $O_2$, $CO_2$, or $H_2O$, depending primarily upon the surface temperature, via the following global reactions:

$$C + O_2 \xrightarrow{k_1} CO_2, \qquad\qquad\qquad (R14.4)$$

$$2C + O_2 \xrightarrow{k_2} 2CO, \qquad\qquad\qquad (R14.5)$$

$$C + CO_2 \xrightarrow{k_3} 2CO, \qquad\qquad\qquad (R14.6)$$

$$C + H_2O \xrightarrow{k_4} CO + H_2. \qquad\qquad (R14.7)$$

The principal product at the carbon surface is CO. The CO diffuses away from the surface through the boundary layer, where it combines with the inward-diffusing $O_2$ according to the following global homogeneous reaction:

$$CO + \tfrac{1}{2}O_2 \rightarrow CO_2. \qquad\qquad\qquad (R14.8)$$

Of course, many elementary reaction steps are involved in R14.8, with one of the most important being $CO + OH \rightarrow CO_2 + H$, as you may recall from Chapter 5.

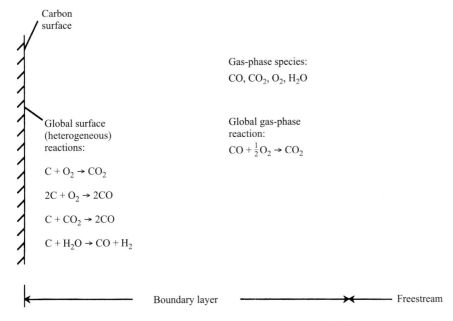

**Figure 14.2**    General scheme for carbon combustion showing global heterogeneous and homogeneous reactions.

In principle, the problem of carbon oxidation could be solved by writing the appropriate conservation equations for species, energy, and mass, defining all of the elementary reaction steps, and then solving these equations subject to appropriate boundary conditions at the surface and free stream. A major complication to this scenario, however, is that the carbon surface is porous and the detailed nature of the surface changes as the carbon oxidation proceeds. Thus, the process of **intraparticle diffusion** plays a major role in carbon combustion under certain conditions. A review of this subject can be found in Simons [8].

Simplified models of carbon combustion rely on the global reactions given in Fig. 14.2 and usually assume that the surface is impervious to diffusion. Depending upon the assumptions made for both the surface and gas-phase chemistry, different scenarios emerge which are generally classified as **one-film, two-film,** or **continuous-film** models. In the one-film models, there is no flame in the gas phase and the maximum temperature occurs at the carbon surface. In the two-film models, a flame sheet lies at some distance from the surface, where the CO produced at the surface reacts with incoming $O_2$. In the continuous-film models, a flame zone is distributed within the boundary layer, rather than occurring in a sheet.

In the following analyses, we will develop both one-film and two-film models. The one-film model is quite simple and illustrates conveniently and clearly the combined effects of heterogeneous kinetics and gas-phase diffusion. The two-film model, although also still quite simplified, is more realistic in that it shows the sequential production and oxidation of CO. We then use these models to obtain estimates of carbon-char burning times.

## One-Film Model

The basic approach to the problem of carbon combustion is quite similar to our previous treatment of droplet evaporation in Chapter 3, except that now chemical reaction at the surface replaces evaporation. We consider the burning of a single spherical carbon particle subject to the following assumptions. Because of the similarity of the present analysis to previous ones, we will not elaborate on those assumptions that have been employed and explained previously. If the implications of certain assumptions are not clear, a review of the appropriate sections of Chapters 3 and 10 is recommended.

### Assumptions

1. The burning process is quasi-steady.

2. The spherical carbon particle burns in a quiescent, infinite ambient medium that contains only oxygen and an inert gas, such as nitrogen. There are no interactions with other particles, and the effects of convection are ignored.

3. At the particle surface, the carbon reacts kinetically with oxygen to produce carbon *dioxide,* i.e., reaction R14.4, $C + O_2 \rightarrow CO_2$, prevails. In general, this reaction choice is not particularly good since carbon *monoxide* is the preferred product at combustion temperatures. Nonetheless, this assumption eliminates the problem of how and where the CO oxidizes, and we will do better in our two-film model.

4. The gas phase consists of only $O_2$, $CO_2$, and inert gas. The $O_2$ diffuses inward, reacts with the surface to form $CO_2$, which then diffuses outward. The inert gas forms a stagnant layer as in the Stefan problem (see Chapter 3).

5. The gas-phase thermal conductivity, $k$, specific heat, $c_p$, and the product of the density and mass diffusivity, $\rho D$, are all constants. Furthermore, we assume that the Lewis number is unity, i.e., $Le = k/(\rho c_p D) = 1$.

6. The carbon particle is impervious to gas-phase species; i.e., intraparticle diffusion is ignored.

7. The particle is of uniform temperature and radiates as a gray body to the surroundings without participation of the intervening medium.

Figure 14.3 illustrates the basic model embodied by these assumptions, showing how the species mass fraction and temperature profiles vary with the radial coordinate. Here we see that the $CO_2$ mass fraction is a maximum at the surface and is zero far from the particle surface. Conversely, the $O_2$ mass fraction takes on its smallest value at the surface. Later, we will see that if the chemical kinetic rate of $O_2$ consumption is very fast, the oxygen concentration at the surface, $Y_{O_2,s}$, approaches zero. If the kinetics are slow, there will be an appreciable concentration of $O_2$ at the surface. Since we assume that there are no reactions occurring in the gas phase and all of the heat release occurs at the surface, the temperature monotonically falls from a maximum at the surface, $T_s$, to its value far from the surface, $T_\infty$.

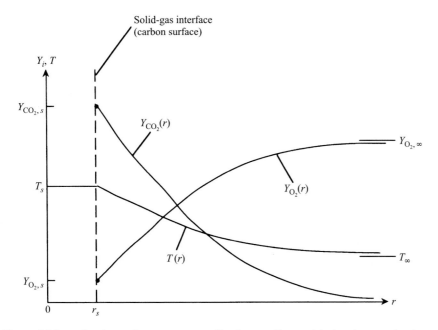

**Figure 14.3**    Species and temperature profiles for one-film model of carbon combustion assuming that $CO_2$ is the only product of combustion at the carbon surface.

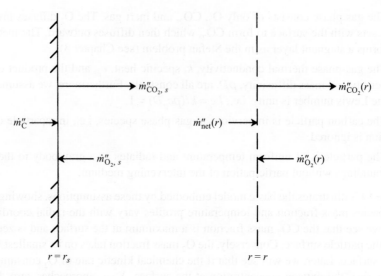

**Figure 14.4**     Species mass fluxes at the carbon surface and at an arbitrary radial location.

**Problem Statement**   Our primary objective in the following analysis is to determine expressions that allow evaluation of the mass burning rate of the carbon, $\dot{m}_C$, and the surface temperature, $T_s$. Intermediate variables of interest are the mass fractions of $O_2$ and $CO_2$ at the carbon surface. The problem is straightforward and requires dealing with only species and energy conservation.

**Overall Mass and Species Conservation**   The relationship among the three species mass fluxes, $\dot{m}''_C$, $\dot{m}''_{O_2}$, and $\dot{m}''_{CO_2}$, is illustrated in Fig. 14.4. At the surface, the mass flow of carbon must equal the difference between the outgoing flow of $CO_2$ and incoming flow of $O_2$, i.e.,

$$\dot{m}''_C = \dot{m}''_{CO_2} - \dot{m}''_{O_2}. \tag{14.9}$$

Similarly, at any arbitrary radial position, $r$, the net mass flux is the difference between the $CO_2$ and $O_2$ fluxes:

$$\dot{m}''_{net} = \dot{m}''_{CO_2} - \dot{m}''_{O_2}. \tag{14.10}$$

Since the mass flowrates of each species are constant with respect to both radial position (no gas-phase reactions) and time (steady state), we have

$$\dot{m}''_C 4\pi r_s^2 = \dot{m}''_{net} 4\pi r^2 \tag{14.11}$$

or

$$\dot{m}''_C = \dot{m}''_{net} = \dot{m}''_{CO_2} - \dot{m}''_{O_2}. \tag{14.12}$$

Thus, we see that the outward flowrate is just the carbon combustion rate, as expected. The $CO_2$ and $O_2$ flowrates can be related by the stoichiometry associated with the reaction at the surface:

$$12.01 \text{ kg C} + 31.999 \text{ kg O}_2 \rightarrow 44.01 \text{ kg CO}_2. \quad (14.13a)$$

On a per-kilogram-of-carbon basis, we have

$$1 \text{ kg C} + v_I \text{ kg O}_2 \rightarrow (v_I + 1) \text{ kg CO}_2, \quad (14.13b)$$

where the mass stoichiometric coefficient is

$$v_I = \frac{31.999 \text{ kg O}_2}{12.01 \text{ kg C}} = 2.664. \quad (14.14)$$

The subscript I is used to denote that this coefficient applies to the one-film model. A different value of the stoichiometric coefficient results for the two-film model.

We can now relate the gas-phase species flowrates to the carbon burning rate:

$$\dot{m}_{O_2} = v_I \dot{m}_C \quad (14.15a)$$

and

$$\dot{m}_{CO_2} = (v_I + 1)\dot{m}_C. \quad (14.15b)$$

Thus, the problem now is to find any one of the species flowrates. To do this, we apply Fick's law (Eqn. 3.1) to express the conservation of $O_2$:

$$\dot{m}''_{O_2} = Y_{O_2}\left(\dot{m}''_{O_2} + \dot{m}''_{CO_2}\right) - \rho D \frac{dY_{O_2}}{dr} i_r. \quad (14.16)$$

Recognizing that the mass fluxes are simply related to the mass flows as $\dot{m}_i = 4\pi r^2 \dot{m}''_i$ and substituting Eqns. 14.15a and 14.15b, taking care to account for the direction of the flows (inward flow is negative, outward flow is positive), Eqn. 14.16 becomes, with some additional manipulation,

$$\dot{m}_C = \frac{4\pi r^2 \rho D}{\left(1 + Y_{O_2}/v_I\right)} \frac{d\left(Y_{O_2}/v_I\right)}{dr}. \quad (14.17)$$

The boundary conditions that apply to the equation are

$$Y_{O_2}(r_s) = Y_{O_2,s} \quad (14.18a)$$

and

$$Y_{O_2}(r \rightarrow \infty) = Y_{O_2,\infty}. \quad (14.18b)$$

Having two boundary conditions for our first-order ordinary differential equation allows us to determine an expression for $\dot{m}_C$, the eigenvalue of the problem. Separating Eqn. 14.17 and integrating between the two limits given by Eqns. 14.18a and b yields

$$\dot{m}_C = 4\pi r_s \rho D \ln\left[\frac{1 + Y_{O_2,\infty}/v_I}{1 + Y_{O_2,s}/v_I}\right]. \quad (14.19)$$

Since $Y_{O_2, \infty}$ is treated as a given quantity, our problem would be solved if we knew the value of $Y_{O_2, s}$, the oxygen mass fraction at the carbon surface. To find this value, we apply our model of the chemical kinetics at the surface.

**Surface Kinetics**    We assume that the reaction $C + O_2 \rightarrow CO_2$ is first-order with respect to $O_2$ and follows the form of Eqn. 14.1. Adopting the conventions of Ref. [7], the carbon reaction rate is expressed as

$$\mathfrak{R}_C (\text{kg/s-m}^2) = \dot{m}''_{C,s} = k_c MW_C [O_{2,s}], \tag{14.20}$$

where $[O_{2,s}]$ is the molar concentration ($\text{kmol}/\text{m}^3$) of $O_2$ at the surface and $k_c$ is the rate coefficient, which is usually expressed in Arrhenius form, i.e., $k_c = A \exp[-E_A / R_u T_s]$. Converting the molar concentration to a mass fraction using

$$[O_{2,s}] = \frac{MW_{\text{mix}}}{MW_{O_2}} \frac{P}{R_u T_s} Y_{O_2, s}$$

and relating the burning rate to the carbon mass flux at the surface ($r = r_s$), Eqn. 14.20 becomes

$$\dot{m}_C = 4 \pi r_s^2 k_c \frac{MW_C MW_{\text{mix}}}{MW_{O_2}} \frac{P}{R_u T_s} Y_{O_2, s}, \tag{14.21}$$

or, more compactly,

$$\dot{m}_C = K_{\text{kin}} Y_{O_2, s}, \tag{14.22}$$

where all of the kinetic parameters except $Y_{O_2, s}$ have been absorbed into the factor $K_{\text{kin}}$. Note that $K_{\text{kin}}$ depends on the pressure, the surface temperature, and the carbon particle radius.

Solving Eqn. 14.22 for $Y_{O_2, s}$ and substituting into Eqn. 14.19 yields a single transcendental (and awkward) equation for the burning rate $\dot{m}_C$. Rather than working with such a result, we will adopt an electrical circuit analogy, similar to those that you may be familiar with for heat transfer, to more conveniently express our solution. Moreover, such an approach yields considerable physical insight.

**Circuit Analog**    To develop our circuit analogy, we need to transform our two expressions for $\dot{m}_C$ (Eqns. 14.19 and 14.22) into forms involving a so-called potential difference, or driving force, and a resistance. For Eqn. 14.22, this is trivial:

$$\dot{m}_C = \frac{\left( Y_{O_2, s} - 0 \right)}{(1/K_{\text{kin}})} \equiv \frac{\Delta Y}{R_{\text{kin}}}, \tag{14.23}$$

where the zero has been added to indicate a "potential difference," and the "resistance" is the reciprocal of our kinetic factor $K_{\text{kin}}$. Equation 14.23 is thus analogous to Ohm's law ($i = \Delta V / R$), where $\dot{m}_C$ is the "flow variable" or current analog.

Dealing with Eqn. 14.19 requires a bit of manipulation. First, we rearrange the logarithm term to yield

$$\dot{m}_C = 4\pi r_s \rho D \ln\left[1 + \frac{Y_{O_2,\infty} - Y_{O_2,s}}{\nu_I + Y_{O_2,s}}\right]. \tag{14.24}$$

As an aside, if we define a transfer number $B_{O,m}$ to be

$$B_{O,m} \equiv \frac{Y_{O_2,\infty} - Y_{O_2,s}}{\nu_I + Y_{O_2,s}}, \tag{14.25}$$

Eqn. 12.24 becomes

$$\dot{m}_C = 4\pi r_s \rho D \ln(1 + B_{O,m}), \tag{14.26}$$

which is of the same form as the expression developed for liquid droplet evaporation (Eqn. 3.53) and combustion (Eqn. 10.68c). Returning to the problem at hand, Eqn. 14.26 can be linearized, if the numerical value of $B_{O,m}$ is small, by truncating the series

$$\ln(1 + B_{O,m}) = B_{O,m} - \tfrac{1}{2}B_{O,m}^2 + \tfrac{1}{3}B_{O,m}^3 - \cdots \tag{14.27a}$$

after the first term, i.e.,

$$\ln(1 + B_{O,m}) \approx B_{O,m}. \tag{14.27b}$$

Since $\nu_I = 2.664$, and $Y_{O_2,s}$ must range between zero and $Y_{O_2,\infty}$ ($= 0.233$ for air), it is easy to show that our approximation (Eqn. 14.27b) is reasonable. The desired linearized version of Eqn. 14.19 is written

$$\dot{m}_C = 4\pi r_s \rho D \left[\frac{Y_{O_2,\infty} - Y_{O_2,s}}{\nu_I + Y_{O_2,s}}\right], \tag{14.28}$$

where, in turn, $\dot{m}_C$ can be expressed as the ratio of a "potential difference" and a "resistance," i.e.,

$$\dot{m}_C = \frac{\left(Y_{O_2,\infty} - Y_{O_2,s}\right)}{\left(\dfrac{\nu_I + Y_{O_2,s}}{4\pi r_s \rho D}\right)} \equiv \frac{\Delta Y}{R_{\text{diff}}}. \tag{14.29}$$

Note that the appearance of $Y_{O_2,s}$ in $R_{\text{diff}}$ is nonconstant, which creates a nonlinear relationship between $\dot{m}_C$ and $\Delta Y$.

Since the burning rate expressed by Eqn. 14.23, derived from chemical kinetics, must be the same as that expressed by Eqn. 14.29, which was derived from mass-transfer considerations alone, a two-resistor series circuit results. Figure 14.5 illustrates the resulting circuit analog. Note that, since we chose our potentials to be

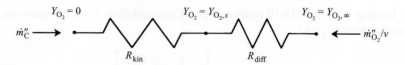

**Figure 14.5**     Electrical circuit analog for a burning carbon particle showing chemical-kinetic and diffusional resistances in series.

$O_2$ mass fractions, the carbon flows from a low potential to a high potential, just the opposite of the electrical analog. Our analogy is perfectly consistent, however, when the flow variable is $\dot{m}_{O_2}/v_I(=-\dot{m}_C)$, which is also indicated in Fig. 14.5.

We can now determine the burning rate $\dot{m}_C$ with the aid of the circuit analog. Referring to Fig. 14.5, we write

$$\dot{m}_C = \frac{\left(Y_{O_2,\infty} - 0\right)}{R_{kin} + R_{diff}},\tag{14.30}$$

where

$$R_{kin} \equiv 1/K_{kin} = \frac{v_I R_u T_s}{4\pi r_s^2 MW_{mix} k_c P}\tag{14.31}$$

and

$$R_{diff} \equiv \frac{v_I + Y_{O_2,s}}{\rho D 4\pi r_s}.\tag{14.32}$$

Since $R_{diff}$ involves the unknown value of $Y_{O_2,s}$, some iteration is still required in this approach. Alternatively, a quadratic equation can be formed and solved directly. In the following section, we further exploit the circuit analog (Eqns. 14.30–14.32).

**Limiting Cases**     Depending primarily on the particle temperature and size, one of the resistors may be much larger than the other, thus causing $\dot{m}_C$ to depend essentially only on that resistor. For example, let us assume that $R_{kin}/R_{diff} \ll 1$. In this case, the burning rate is said to be **diffusionally controlled.** When does this occur? What does this imply? Using the definitions of $R_{kin}$ and $R_{diff}$ (Eqns. 14.31 and 14.32) and taking their ratio, we obtain

$$\frac{R_{kin}}{R_{diff}} = \left(\frac{v_I}{v_I + Y_{O_2,s}}\right)\left(\frac{R_u T_s}{MW_{mix} P}\right)\left(\frac{\rho D}{k_c}\right)\left(\frac{1}{r_s}\right),\tag{14.33}$$

and can now see how individual parameters affect this quantity. This ratio can be made small in several ways. First, $k_c$ can be very large; this implies a fast surface reaction. We also see that a large particle size, $r_s$, or high pressure, $P$, has the same effect. Although the surface temperature appears explicitly in the numerator of Eqn. 14.33, its effect is primarily through the temperature-dependence of $k_c$, where $k_c$ typically increases rapidly with temperature since $k_c = A \exp(-E_A/R_u T)$. As a result of

**Table 14.1**   Summary of carbon combustion regimes

| Regime | $R_{kin}/R_{diff}$ | Burning Rate Law | Conditions of Occurrence |
|---|---|---|---|
| Diffusionally controlled | $\ll 1$ | $\dot{m}_C = Y_{O_2,\infty}/R_{diff}$ | $r_s$ large, $T_s$ high, $P$ high |
| Intermediate | ~1 | $\dot{m}_C = Y_{O_2,\infty}/(R_{diff} + R_{kin})$ | — |
| Kinetically controlled | $\gg 1$ | $\dot{m}_C = Y_{O_2,\infty}/R_{kin}$ | $r_s$ small, $T_s$ low, $P$ low |

the burning being diffusionally controlled, we see that none of the chemical-kinetic parameters influence the burning rate and that the $O_2$ concentration at the surface approaches zero.

The other limiting case, **kinetically controlled** combustion, occurs when $R_{kin}/R_{diff} \gg 1$. In this case, the $R_{diff}$ is small and the nodes $Y_{O_2,s}$ and $Y_{O_2,\infty}$ are essentially at the same value, i.e., the concentration of $O_2$ at the surface is large. Now the chemical-kinetic parameters control the burning rate and the mass-transfer parameters are unimportant. Kinetically controlled combustion occurs when particle sizes are small, pressures low, and temperatures low (a low temperature causes $k_c$ to be small).

Table 14.1 summarizes our discussion of the limiting regimes of carbon combustion.

---

**Example 14.1**

Estimate the burning rate of a 250-μm-diameter carbon particle burning in still air ($Y_{O_2,\infty} = 0.233$) at 1 atm. The particle temperature is 1800 K, and the kinetic rate constant $k_c$ is 13.9 m/s. Assume the mean molecular weight of the gases at the surface is 30 kg/kmol. Also, what combustion regime prevails?

**Solution**

We will employ the circuit analogy to find $\dot{m}_C$. The diffusional resistance is calculated from Eqn. 14.32, where the density is estimated from the ideal-gas law at the surface temperature:

$$\rho = \frac{P}{\left(\dfrac{R_u}{MW_{mix}}\right)T_s} = \frac{101{,}325}{\left(\dfrac{8315}{30}\right)1800} = 0.20 \text{ kg/m}^3,$$

and the mass diffusivity is estimated using a value for $CO_2$ in $N_2$ from Appendix Table D.1, corrected to 1800 K:

$$\mathcal{D} = \left(\frac{1800 \text{ K}}{393 \text{ K}}\right)^{1.5} 1.6 \cdot 10^{-5}\frac{\text{m}^2}{\text{s}} = 1.57 \cdot 10^{-4} \text{ m}^2/\text{s}.$$

Thus, assuming $Y_{O_2,s} \approx 0$ for the time being,

$$R_{diff} = \frac{\nu_I + Y_{O_2,s}}{\rho \mathcal{D} 4\pi r_s} = \frac{2.664 + 0}{0.2(1.57 \cdot 10^{-4})4\pi(125 \cdot 10^{-6})}$$

$$= 5.41 \cdot 10^7 \text{ s/kg}.$$

The chemical kinetic resistance is calculated from Eqn. 14.31:

$$R_{kin} = \frac{\nu_1 R_u T_s}{4\pi r_s^2 MW_{mix} k_c P}$$

$$= \frac{2.664(8315)1800}{4\pi(125 \cdot 10^{-6})^2 30(13.9)101,325}$$

$$= 4.81 \cdot 10^6 \text{ s/kg}.$$

From these calculations, we see that $R_{diff}$ is slightly more than 10 times the value of $R_{kin}$; thus, the combustion is **nearly diffusion controlled.** We can now estimate $\dot{m}_C$ using Eqn. 14.30, and then find $Y_{O_2,s}$ to get an improved value for $R_{diff}$, and iterate if necessary to get an improved value for $\dot{m}_C$:

$$\dot{m}_C = \frac{Y_{O_2,\infty}}{R_{kin} + R_{diff}} = \frac{0.233}{4.81 \cdot 10^6 + 5.41 \cdot 10^7}$$

$$\dot{m}_C = 3.96 \cdot 10^{-9} \text{ kg/s} \qquad \text{1st iteration.}$$

From the circuit diagram (Fig. 14.5),

$$Y_{O_2,s} - 0 = \dot{m}_C R_{kin}$$

$$= 3.96 \cdot 10^{-9}(4.81 \cdot 10^6)$$

$$= 0.019 \text{ or } 1.9\%.$$

Thus,

$$R_{diff} = \frac{2.664 + 0.019}{2.664}(R_{diff})_{1st\ iter}$$

$$= 1.007(5.41 \cdot 10^7) = 5.45 \cdot 10^7 \text{ s/kg}.$$

Since $R_{diff}$ changes by less than 1 percent, no further iteration is required.

**Comment**

This example shows how the circuit analog provides a simple calculation procedure with easy iteration. We also see that an appreciable $O_2$ concentration exists at the surface because of the non-negligible kinetic resistance. It should be emphasized that the one-film model, as developed, is not an accurate representation of the actual chemical processes occurring, but, rather, serves as a pedagogical tool to illuminate key concepts with a minimum of complexity.

---

**Energy Conservation**    So far in our analysis we have treated the surface temperature $T_s$ as a known parameter; however, this temperature cannot be any arbitrary value but is a unique value that depends upon energy conservation at the particle surface. As we will see, the controlling surface energy balance depends strongly on the burning rate, i.e., the energy and mass-transfer processes are coupled.

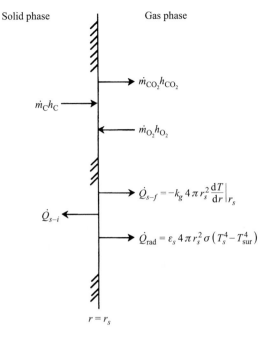

**Figure 14.6**    Energy flows at surface of a spherical carbon particle burning in air.

Figure 14.6 illustrates the various energy fluxes associated with the burning carbon surface. Writing the surface energy balance yields

$$\dot{m}_C h_C + \dot{m}_{O_2} h_{O_2} - \dot{m}_{CO_2} h_{CO_2} = \dot{Q}_{s-i} + \dot{Q}_{s-f} + \dot{Q}_{rad}. \tag{14.34}$$

Since we assume combustion occurs in a steady state, there is no heat conducted into the particle interior; thus, $\dot{Q}_{s-i} = 0$. Following the development in Chapter 10 (Eqns. 10.55–10.60), it is easy to show that the left-hand side of Eqn. 14.34 is simply $\dot{m}_C \Delta h_c$, where $\Delta h_c$ is the carbon–oxygen reaction heat of combustion (J/kg$_C$). Thus, Eqn. 14.34 becomes

$$\dot{m}_C \Delta h_c = -k_g 4\pi r_s^2 \frac{dT}{dr}\bigg|_{r_s} + \varepsilon_s 4\pi r_s^2 \sigma \left(T_s^4 - T_{sur}^4\right). \tag{14.35}$$

To obtain an expression for the gas-phase temperature gradient at the surface requires that we write an energy balance within the gas phase and solve for the temperature distribution. Since we have already done this for our droplet evaporation model, we simply use the result from Chapter 10 (Eqn. 10.10) substituting $T_s$ for $T_{boil}$:

$$\frac{dT}{dr}\bigg|_{r_s} = \frac{Z\dot{m}_C}{r_s^2}\left[\frac{(T_\infty - T_s)\exp(-Z\dot{m}_C/r_s)}{1 - \exp(-Z\dot{m}_C/r_s)}\right], \tag{14.36}$$

where $Z \equiv c_{pg}/(4\pi k_g)$. Substituting Eqn. 14.36 into Eqn. 14.35 yields, upon rearrangement, our final result,

$$\dot{m}_C \Delta h_c = \dot{m}_C c_{pg} \left[ \frac{\exp\left(\dfrac{-\dot{m}_C c_{pg}}{4\pi k_g r_s}\right)}{1 - \exp\left(\dfrac{-\dot{m}_C c_{pg}}{4\pi k_g r_s}\right)} \right] (T_s - T_\infty) + \varepsilon_s 4\pi r_s^2 \sigma \left(T_s^4 - T_{sur}^4\right). \quad (14.37)$$

Note that Eqn. 14.37 contains two unknowns, $\dot{m}_C$ and $T_s$. To effect the complete solution of the carbon burning problem requires the simultaneous solution of Eqns. 14.37 and 14.30. Since both of these equations are nonlinear, an iterative method is probably the best approach to their solution. Note too that, in the intermediate regime between diffusional and kinetic control, $Y_{O_2,s}$ also becomes an unknown. Thus, Eqn. 14.21 will have to be added to the equation set.

---

**Example 14.2**

In the combustion of solid fuels, radiation usually plays a key role. Estimate the gas temperature required to keep a 250-$\mu$m-diameter burning carbon particle at 1800 K for (i) when there is no radiation ($T_s = T_{sur}$), and (ii) when the particle radiates as a blackbody to surroundings at 300 K. Conditions are identical to those presented in Example 14.1.

**Solution**

The surface energy balance (Eqn. 14.37) can be used to determine $T_\infty$ for the two cases. We will estimate the gas-phase properties using those of air at 1800 K:

$$c_{pg}(1800\ \text{K}) = 1286\ \text{J/kg-K} \quad (\text{Appendix Table C.1})$$

$$k_g(1800\ \text{K}) = 0.12\ \text{W/m-K} \quad (\text{Appendix Table C.1}).$$

Since the particle is a blackbody, the emissivity, $\varepsilon_s$, is unity, and the heat of combustion for carbon $\bar{h}_{f,CO_2}^o / MW_C$ is

$$\Delta h_c = 3.2765 \cdot 10^7\ \text{J/kg} \quad (\text{Appendix Table A.2}).$$

i. In the absence of radiation, Eqn. 14.37 can be rearranged to solve for $T_\infty$, as

$$T_\infty = T_s - \frac{\Delta h_c}{c_{pg}} \frac{\left[1 - \exp\left(\dfrac{-\dot{m}_C c_{pg}}{4\pi k_g r_s}\right)\right]}{\exp\left(\dfrac{-\dot{m}_C c_{pg}}{4\pi k_g r_s}\right)}$$

and evaluated using the burning rate ($\dot{m}_C = 3.96 \cdot 10^{-9}$ kg/s) from Example 14.1:

$$T_\infty = 1800 - \frac{3.2794 \cdot 10^7}{1286} \frac{\left[1 - \exp\left(\dfrac{-3.96 \cdot 10^{-4}(1286)}{4\pi(0.12)125 \cdot 10^{-6}}\right)\right]}{\exp\left(\dfrac{-3.96 \cdot 10^{-4}(1286)}{4\pi(0.12)125 \cdot 10^{-6}}\right)}$$

$$= 1800 - 698$$

$$T_\infty = 1102 \text{ K} \qquad \text{without radiation.}$$

ii. With surroundings at 300 K, the radiation loss is

$$\dot{Q}_{\text{rad}} = \varepsilon_s 4\pi r_s^2 \sigma\left(T_s^4 - T_{\text{sur}}^4\right)$$

$$= (1.0)4\pi(125 \cdot 10^{-6})^2 5.67 \cdot 10^{-8}(1800^4 - 300^4)$$

$$= 0.1168 \text{ W.}$$

The chemical heat release is

$$\dot{m}_C \Delta h_c = 3.96 \cdot 10^{-4}(3.2765 \cdot 10^7)$$

$$= 0.1299 \text{ W.}$$

The energy conducted from the particle surface is then

$$\dot{Q}_{\text{cond}} = \dot{m}_C \Delta h_c - \dot{Q}_{\text{rad}} = \dot{m}_C c_{pg} \frac{\exp\left(\dfrac{-\dot{m}_C c_{pg}}{4\pi k_g r_s}\right)}{1 - \exp\left(\dfrac{-\dot{m}_C c_{pg}}{4\pi k_g r_s}\right)}(T_s - T_\infty).$$

Solving for $T_\infty$ using these numerical values yields

$$T_\infty = 1730 \text{ K} \qquad \text{with radiation.}$$

**Comment**

We see that the gas temperature needs to be quite high to maintain the 1800 K surface temperature in the presence of radiation.

## Two-Film Model

As the one-film model was intended to be more instructive than realistic, we now present a model that somewhat more realistically captures the chemical and physical processes involved in carbon combustion, at least for some conditions. In particular, the two-film model has the carbon oxidize to carbon monoxide, rather than carbon dioxide. Since the basic methods employed to develop the two-film model are the same as those for the one-film model, we will be more brief in this section, leaving the gaps to be filled in by the interested reader, or completed as homework assignments.

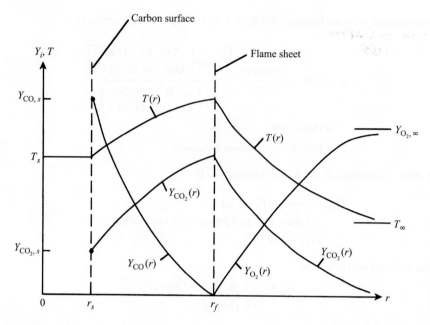

**Figure 14.7**    Species mass fractions and temperature profiles for a two-film model of a burning spherical carbon particle.

Figure 14.7 illustrates the species concentration and temperature profiles through the two gas films: one interior to the flame sheet and one exterior. In the two-film model, the carbon surface is attacked by $CO_2$ according to the global reaction R.14.6, $C + CO_2 \rightarrow 2CO$. The CO produced at the surface diffuses outward and is consumed at a flame sheet where it meets an inward-diffusing flow of $O_2$ in stoichiometric proportions. The global reaction $CO + \frac{1}{2}O_2 \rightarrow CO_2$ is assumed to be infinitely fast; thus, both CO and $O_2$ are identically zero at the flame sheet. The temperature peaks at the flame sheet. Except for the carbon surface reaction, the scenario illustrated in Fig. 14.7 is identical to our previous treatment of droplet combustion in Chapter 10 (see Fig. 10.11). The basic assumptions of the one-film model still hold, except as modified by the preceding discussion. Our specific objective now is to find the burning rate $\dot{m}_C$.

**Stoichiometry**    The mass flowrates of the various species can be related by simple mass balances at the particle surface and at the flame, as illustrated in Fig. 14.8. These are

*At the surface*

$$\dot{m}_C = \dot{m}_{CO} - \dot{m}_{CO_2,i} \qquad (14.38a)$$

*At the flame*

$$\dot{m}_{CO} = \dot{m}_{CO_2,i} - \dot{m}_{CO_2,o} - \dot{m}_{O_2} \qquad (14.38b)$$

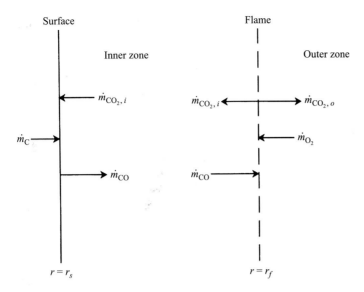

**Figure 14.8**    Species mass flowrates at the carbon surface and the flame sheet.

or

$$\dot{m}_C = \dot{m}_{CO_2,o} - \dot{m}_{O_2}. \tag{14.38c}$$

Using the stoichiometric relations at the surface and at the flame, all of the flowrates can be related to the burning rate, $\dot{m}_C$:

*At the surface*

$$1\,kg\,C + \nu_s\;kg\,CO_2 \rightarrow (\nu_s + 1)\,kg\,CO \tag{14.39a}$$

*At the flame*

$$1\,kg\,C + \nu_f\;kg\,O_2 \rightarrow (\nu_f + 1)\,kg\,CO_2, \tag{14.39b}$$

where

$$\nu_s = \frac{44.01}{12.01} = 3.664 \tag{14.40a}$$

and

$$\nu_f = \nu_s - 1. \tag{14.40b}$$

Thus, the magnitudes of the flowrates are

$$\dot{m}_{CO_2,i} = \nu_s \dot{m}_C, \tag{14.41a}$$

$$\dot{m}_{O_2} = \nu_f \dot{m}_C = (\nu_s - 1)\dot{m}_C, \tag{14.41b}$$

$$\dot{m}_{CO_2,o} = (\nu_f + 1)\dot{m}_C = \nu_s \dot{m}_C. \tag{14.41c}$$

**Species Conservation** We proceed by applying Fick's law to obtain differential equations describing the distribution of $CO_2$ in both the inner and outer zones and, similarly, an equation for the inert species ($N_2$). These are

*Inner zone $CO_2$*

$$\dot{m}_C = \frac{4\pi r^2 \rho D}{\left(1 + Y_{CO_2}/\nu_s\right)} \frac{d\left(Y_{CO_2}/\nu_s\right)}{dr},$$  (14.42a)

with boundary conditions

$$Y_{CO_2}(r_s) = Y_{CO_2,s}$$  (14.42b)

$$Y_{CO_2}(r_f) = Y_{CO_2,f}.$$  (14.42c)

*Outer zone $CO_2$*

$$\dot{m}_C = \frac{-4\pi r^2 \rho D}{\left(1 - Y_{CO_2}/\nu_s\right)} \frac{d\left(Y_{CO_2}/\nu_s\right)}{dr},$$  (14.43a)

with boundary conditions

$$Y_{CO_2}(r_f) = Y_{CO_2,f}$$  (14.43b)

$$Y_{CO_2}(r \rightarrow \infty) = 0.$$  (14.43c)

*Inert ($N_2$)*

$$\dot{m}_C = \frac{4\pi r^2 \rho D}{Y_I} \frac{dY_I}{dr},$$  (14.44a)

with boundary conditions

$$Y_I(r_f) = Y_{I,f}$$  (14.44b)

$$Y_I(r \rightarrow \infty) = Y_{I,\infty}.$$  (14.44c)

Integration of these three equations (Eqns. 14.42–14.44) between the limits indicated by the boundary conditions yields the following three algebraic equations involving five unknowns—$\dot{m}_C$, $Y_{CO_2,s}$, $Y_{CO_2,f}$, $Y_{I,f}$, and $r_f$:

$$\dot{m}_C = 4\pi \left(\frac{r_s r_f}{r_f - r_s}\right) \rho D \ln \left[\frac{1 + Y_{CO_2,f}/\nu_s}{1 + Y_{CO_2,s}/\nu_s}\right],$$  (14.45)

$$\dot{m}_C = -4\pi r_f \rho D \ln \left(1 - Y_{CO_2,f}/\nu_s\right),$$  (14.46)

$$Y_{I,f} = Y_{I,\infty} \exp(-\dot{m}_C/(4\pi r_f \rho D)).$$  (14.47)

From $\sum Y_i = 1$, we also know that

$$Y_{CO_2, f} = 1 - Y_{I, f}. \tag{14.48}$$

The remaining equation for closure results from writing a chemical kinetic expression involving $\dot{m}_C$ and $Y_{CO_2, s}$, which is done in the next section.

**Surface Kinetics**   The reaction $C + CO_2 \rightarrow 2CO$ is first-order in $CO_2$ concentration [9] and thus the rate is expressed in a form identical to that developed for the one-film model reaction:

$$\dot{m}_C = 4\pi r_s^2 k_c \frac{MW_C MW_{mix}}{MW_{CO_2}} \frac{P}{R_u T_s} Y_{CO_2, s}, \tag{14.49}$$

where

$$k_c (m/s) = 4.016 \cdot 10^8 \exp\left[\frac{-29,790}{T_s (K)}\right], \tag{14.50}$$

as given by Mon and Amundson [9]. Equation 14.49 can be written more compactly as

$$\dot{m}_C = K_{kin} Y_{CO_2, s}, \tag{14.51}$$

where

$$K_{kin} = 4\pi r_s^2 k_c \frac{MW_C MW_{mix}}{MW_{CO_2}} \frac{P}{R_u T_s}. \tag{14.52}$$

**Closure**   To present a tractable solution to our problem, Eqns. 14.45–14.48 can be manipulated to eliminate all variables but $\dot{m}_C$ and $Y_{CO_2, s}$; i.e.,

$$\dot{m}_C = 4\pi r_s \rho \mathcal{D} \ln\left(1 + B_{CO_2, m}\right), \tag{14.53}$$

where

$$B_{CO_2, m} = \frac{2Y_{O_2, \infty} - [(v_s - 1)/v_s] Y_{CO_2, s}}{v_s - 1 + [(v_s - 1)/v_s] Y_{CO_2, s}}. \tag{14.54}$$

Equation 14.53 can be solved iteratively with Eqn. 14.51 to find $\dot{m}_C$. For diffusion-controlled burning, $Y_{O_2, s}$ is zero and $\dot{m}_C$ can be evaluated directly from Eqn. 14.53.

   To obtain the surface temperature, it is necessary to write and solve energy balances at the surface and at the flame sheet. The procedures are the same as those employed earlier in this chapter and previously in Chapter 10. Determining the surface temperature is left as an exercise.

| **Example 14.3** | Assuming diffusion-controlled combustion and identical conditions ($Y_{O_2, \infty} = 0.233$), compare the burning rates predicted by the one-film and two-film models. |

**Solution**

The burning rates for both the one-film and two-film models are of the form

$$\dot{m}_C = 4\pi r_s \rho D \ln(1 + B_m).$$

For identical conditions, the $B_m$s are the only parameters having different values; thus,

$$\frac{\dot{m}_C(\text{two-film})}{\dot{m}_C(\text{one-film})} = \frac{\ln\left(1 + B_{CO_2, m}\right)}{\ln(1 + B_{O, m})},$$

where the transfer numbers $B_{CO_2, m}$ and $B_{O, m}$ are evaluated from Eqns. 14.54 and 14.25, respectively. For diffusion control, the surface concentrations of $CO_2$ (two-film model) and $O_2$ (one-film model) are zero. The transfer numbers are thus evaluated:

$$B_{CO_2, m} = \frac{2Y_{O_2, \infty} - [(v_s - 1)/v_s]Y_{CO_2, s}}{v_s - 1 + [(v_s - 1)/v_s]Y_{CO_2, s}}$$

$$= \frac{2(0.233) - 0}{3.664 - 1 + 0} = 0.175$$

and

$$B_{O, m} = \frac{Y_{O_2, \infty} - Y_{O_2, s}}{v_I + Y_{O_2, s}} = \frac{0.233 - 0}{2.664 + 0} = 0.0875.$$

The ratio of burning rates is then

$$\frac{\dot{m}_C(\text{two-film})}{\dot{m}_C(\text{one-film})} = \frac{\ln(1 + 0.175)}{\ln(1 + 0.087)} = 1.92.$$

**Comment**

It is interesting to note that this difference in burning rates is not a result of the type of model *per se*, but rather is a result of the reaction assumed to occur at the gasifying carbon surface. We can show that this is true by assuming that CO, rather than $CO_2$, is the product formed at the surface in a one-film model. In this case, $v_I = 31.999/24.01 = 1.333$ and $B_m$ then becomes 0.175, which is identical to the two-film model. In this case then, the one-film and two-film models predict identical burning rates independent of the fate of the CO produced at the surface and independent of the surface-attacking species ($O_2$ or $CO_2$), as long as the product is CO.

| **Example 14.4** | Use the two-film model to estimate the burning rate of a 70-$\mu$m-diameter carbon particle burning in air ($Y_{O_2, \infty} = 0.233$). The surface temperature is 1800 K, and the pressure is 1 atm. Assume the molecular weight of the gaseous mixture at the particle surface is 30 kg/kmol. |

**Solution**

The conditions of this problem are identical to those of Example 14.1, so the gas-phase properties are the same, i.e., $\rho = 0.2$ kg/m³ and $D = 1.57 \cdot 10^{-4}$ m²/s.

From Eqn. 14.50, the kinetic rate constant for the C–$CO_2$ reaction is

$$k_c = 4.016 \cdot 10^8 \exp\left[\frac{-29,790}{T_s}\right]$$

$$= 4.016 \cdot 10^8 \exp\left[\frac{-29,790}{1800}\right] = 26.07 \, \text{m/s}.$$

The combustion rate can be expressed in terms of the surface $CO_2$ concentration (Eqns. 14.49 and 14.51):

$$\dot{m}_C = 4\pi r_s^2 k_c \frac{MW_C MW_{mix}}{MW_{CO_2}} \frac{P}{R_u T_s} Y_{CO_2, s}$$

$$= \frac{4\pi(35 \cdot 10^{-6})^2(26.07)12.01(30)101,325}{44.01(8315)1800} Y_{CO_2, s}$$

$$= 2.22 \cdot 10^{-8} Y_{CO_2, s} \quad \text{(kg/s)}. \tag{I}$$

Equations 14.53 and 14.54 also provide an expression for $\dot{m}_C$ in terms of $Y_{CO_2, s}$:

$$\dot{m}_C = 4\pi r_s \rho \mathcal{D} \ln(1 + B)$$

$$= 4\pi(35 \cdot 10^{-6})0.20(1.57 \cdot 10^{-4}) \ln(1 + B)$$

$$= 1.381 \cdot 10^{-8} \ln(1 + B) \quad \text{(kg/s)} \tag{II}$$

and

$$B = \frac{2Y_{O_2, \infty} - [(\nu_s - 1)/\nu_s]Y_{CO_2, s}}{\nu_s - 1 + [(\nu_s - 1)/\nu_s]Y_{CO_2, s}}$$

$$= \frac{2(0.233) - [(3.664 - 1)/3.664]Y_{CO_2, s}}{3.664 - 1 + [(3.664 - 1)/3.664]Y_{CO_2, s}}$$

$$= \frac{0.466 - 0.727 Y_{CO_2, s}}{2.664 + 0.727 Y_{CO_2, s}}. \tag{III}$$

We now iteratively solve Eqns. I, II, and III for $\dot{m}_C$, $B$, and $Y_{CO_2, s}$. We start the iteration assuming $Y_{CO_2, s}$ is zero, the diffusion-controlled limit:

| Iteration | $Y_{CO_2, s}$ | B | $\dot{m}_C$ (kg/s) |
|---|---|---|---|
| 1 | 0 | 0.1749 | $2.225 \cdot 10^{-9}$ |
| 2 | 0.1003 | 0.1436 | $1.853 \cdot 10^{-9}$ |
| 3 | 0.0835 | 0.1488 | $1.915 \cdot 10^{-9}$ |
| 4 | 0.0863 | 0.1479 | $1.905 \cdot 10^{-9}$ |

Given two-significant-figure accuracy, the solution converges to

$$\dot{m}_C = 1.9 \cdot 10^{-9} \, \text{kg/s}.$$

**Comment**

Ignoring chemical kinetics at the surface (see Example 14.1) results in overestimating the burning rate by about 16.8 percent ($=100\%(2.22 - 1.9)/1.9$). A lower surface temperature (or lower pressure) would make kinetics even more important. Kinetics also becomes increasingly important as the particle diameter decreases as it burns out.

## Particle Burning Times

For diffusion-controlled burning, it is a simple matter to find particle burning times. The procedure follows directly from those used in Chapters 3 and 10 to establish the $D^2$ law and will not be repeated here. The particle diameter can be expressed as a function of time as follows:

$$D^2(t) = D_0^2 - K_B t, \tag{14.55}$$

where the burning rate constant, $K_B$, is given by

$$K_B = \frac{8\rho\mathcal{D}}{\rho_C} \ln(1+B). \tag{14.56}$$

Setting $D = 0$ in Eqn. 14.55 gives the particle lifetime,

$$t_C = D_0^2/K_B. \tag{14.57}$$

Depending upon whether a one-film or two-film analysis is employed, the transfer number $B$ is either $B_{O,m}$ (Eqn. 14.25) or $B_{CO_2,m}$ (Eqn. 14.54), with the surface mass fractions set to zero. Note the appearance of two densities in Eqn. 14.56, where $\rho$ applies to the gas phase and $\rho_C$ to the solid carbon.

Thus far, our analyses have assumed a quiescent gaseous medium. To take into account the effects of a convective flow over a burning carbon particle, the *film-theory* analysis of Chapter 10 can be applied. For diffusion-controlled conditions with convection, the mass burning rates are augmented by the factor $Sh/2$, where $Sh$ is the Sherwood number and plays the same role for mass transfer as the Nusselt number does for heat transfer. For unity Lewis number, $Sh = Nu$; thus,

$$(\dot{m}_{C,\text{diff}})_{\substack{\text{with} \\ \text{convection}}} = \frac{Nu}{2}(\dot{m}_{C,\text{diff}})_{\substack{\text{without} \\ \text{convection}}}. \tag{14.58}$$

The Nusselt number can be evaluated using Eqn. 10.78 as a reasonable approximation.

---

**Example 14.5**

Estimate the lifetime of a 70-$\mu$m-diameter carbon particle assuming diffusionally controlled combustion at the conditions given in Example 14.4. Assume the carbon density is 1900 kg/m$^3$.

**Solution**

The particle lifetime is straightforwardly calculated from Eqn. 14.57. The burning rate constant (Eqn. 14.56) is evaluated as

$$K_B = \frac{8\rho\mathcal{D}}{\rho_C} \ln\left(1 + B_{CO_2,m}\right)$$

$$= \frac{8(0.20)1.57 \cdot 10^{-4}}{1900} \ln(1+0.1749)$$

$$= 2.13 \cdot 10^{-8} \text{ m}^2/\text{s},$$

where the value for the transfer number $B_{CO_2, m}$ is that for the first iteration in Example 14.4. The lifetime is evaluated as

$$t_C = D_0^2 / K_B = (70 \cdot 10^{-6})^2 / 2.13 \cdot 10^{-8}$$

$$t_C = 0.23 \text{ s.}$$

### Comment

This value seems reasonable in that residence times in coal-fired boilers are on the order of a few seconds and the upper size limit of pulverized coal is about 70 $\mu$m. It is important to note that in a real boiler $Y_{O_2, \infty}$ does not remain constant, but decreases as the combustion process runs its course, which tends to increase particle combustion times. Opposing this effect, however, is the augmentation of burning rates by convection, which is present in real combustors. Note also that heterogeneous chemical kinetics controls the latter stages of burnout as the particle diameter decreases. Under these conditions, the particle surface temperature becomes quite important, being greatly influenced by the radiation field within the combustor.

## COAL COMBUSTION

As mentioned at the outset of this chapter, coal combustion has a huge literature associated with it. Our objective here is merely to outline the various physical and chemical processes involved in coal combustion. In Chapter 17, we discuss some of the important physical and chemical properties of coal.

Coal is a heterogeneous substance consisting of moisture, volatiles, mineral matter, and carbon-based char, in varying proportions (see Chapter 17). Because of this, the combustion of coal is quite complex, with the detailed behavior depending on the specific properties and composition of any particular coal. The general scheme, however, is for the moisture to be driven off, followed by the evolution of the volatiles. The volatiles may burn homogeneously in the gas phase or at the coal particle surface. The coal may swell and become porous as the volatiles are given off. The material left after devolatilization is the char and its associated mineral matter. The char then burns and the mineral matter is transformed into ash, slag, and fine particulate fume in various proportions. The combustion of the char proceeds in a manner suggested by our previous discussion of carbon combustion, although the pore structure of the char can be quite important in the actual process.

## OTHER SOLIDS

Wood burning has much in common with coal burning in that the evolution of volatiles and char combustion are both important. A review of wood burning can be found in Tillman *et al.* [10]. Metals that burn heterogeneously include boron, silicon, titanium, and zirconium [11]. Boron has been investigated as a fuel additive for military applications because of its high energy density. The combustion of boron is complicated by the fact that the liquid product of combustion, $B_2O_3$, can form an inhibiting layer on the surface of the burning boron. A review of boron combustion can be found in King [12].

## SUMMARY

In this chapter, the concept of heterogeneous chemical reactions, i.e., those occurring on or at a surface, was revisited, and the importance of these reactions to solids combustion was discussed. Simple models of carbon combustion were developed to illustrate certain fundamental aspects common to the combustion of most solids. In particular, the idea of chemical kinetic processes occurring in series with diffusional processes was illustrated by developing an electrical circuit analog. Diffusionally controlled and kinetically controlled combustion are key concepts of this chapter and should be clearly understood. Means of estimating carbon-char particle combustion times were presented for diffusionally controlled combustion. The similarities and differences between solid combustion and droplet combustion should be understood by the reader. Finally, ever-so-brief discussions of coal and other solids combustion were presented.

## NOMENCLATURE

| | |
|---|---|
| $B_m$ | Spalding transfer number, Eqns. 14.25 and 14.54 |
| $c_p$ | Specific heat (J/kg-K) |
| $D$ | Diameter (m) |
| $\mathcal{D}$ | Mass diffusivity (m²/s) |
| $h$ | Enthalpy (J/kg) |
| $i_r$ | Unit vector in radial direction |
| $k, k_c$ | Thermal conductivity (W/m-K) or kinetic rate constant (various units) |
| $K_B$ | Burning rate constant (m²/s) |
| $K_{kin}$ | Parameter defined in Eqns. 14.22 and 14.52 |
| $Le$ | Lewis number |
| $\dot{m}$ | Mass flowrate (kg/s) |
| $\dot{m}''$ | Mass flux (kg/s-m²) |
| $MW$ | Molecular weight (kg/kmol) |
| $Nu$ | Nusselt number |
| $P$ | Pressure (Pa) |
| $\dot{Q}$ | Heat-transfer rate (W) |
| $r$ | Radius or radial coordinate (m) |
| $R$ | Mass-transfer resistance (s/kg) |
| $\mathfrak{R}$ | Reaction rate (kg/s-m²) |
| $R_u$ | Universal gas constant (J/kmol-K) |
| $Sh$ | Sherwood number |
| $t$ | Time (s) |
| $t_C$ | Carbon particle lifetime (s) |
| $T$ | Temperature (K) |
| $Y$ | Mass fraction (kg/kg) |
| $Z$ | Parameter in Eqn. 14.36, $c_{pg}/(4\pi k_g)$ |

## Greek Symbols

| | |
|---|---|
| $\Delta h_c$ | Heat of combustion (J/kg) |
| $\varepsilon$ | Emissivity |
| $\nu$ | Mass stoichiometric coefficient |
| $\rho$ | Density |
| $\sigma$ | Stefan–Boltzmann constant (W/m$^2$-K$^4$) |

## Subscripts

| | |
|---|---|
| cond | Conduction |
| diff | Diffusion |
| $f$ | Flame |
| $g$ | Gas |
| $i$ | Interior |
| I | One-film model or inert |
| mix | Mixture |
| $o$ | Outward or exterior |
| rad | Radiation |
| $s$ | Surface |
| sur | Surroundings |
| $\infty$ | Freestream |

## Other Notation

| | |
|---|---|
| $[X]$ | Molar concentration of $X$ (kmol/m$^3$) |

## REFERENCES

1. Stultz, S. C., and Kitto, J. B. (eds.), *Steam: Its Generation and Use,* 40th Ed., Babcock & Wilcox, Barberdon, OH, 1992.

2. Miller, B. G., *Coal Energy Systems,* Elsevier Academic Press, Burlington, MA, 2005.

3. Gardiner, W. C., Jr., *Rates and Mechanisms of Chemical Reactions,* Benjamin, Menlo Park, CA, 1972.

4. Smith, I. W., "The Combustion Rates of Coal Chars: A Review," *Nineteenth Symposium (International) on Combustion,* The Combustion Institute, Pittsburgh, PA, p. 1045, 1983.

5. Laurendeau, N. M., "Heterogeneous Kinetics of Coal Char Gasification and Combustion," *Progress in Energy and Combustion Science,* 4: 221–270 (1978).

6. Mulcahy, M. F. R., and Smith, I. W., "Kinetics of Combustion of Pulverized Fuel: Review of Theory and Experiment," *Reviews of Pure and Applied Chemistry,* 19: 81–108 (1969).

7. Caram, H. S., and Amundson, N. R., "Diffusion and Reaction in a Stagnant Boundary Layer about a Carbon Particle," *Industrial Engineering Chemistry Fundamentals,* 16(2): 171–181 (1977).

8. Simons, G. A., "The Role of Pore Structure in Coal Pyrolysis and Gasification," *Progress in Energy and Combustion Science,* 9: 269–290 (1983).

9. Mon, E., and Amundson, N. R., "Diffusion and Reaction in a Stagnant Boundary Layer about a Carbon Particle. 2. An Extension," *Industrial Engineering Chemistry Fundamentals,* 17(4): 313–321 (1978).

10. Tillman, D. A., Amadeo, J. R., and Kitto, W. D., *Wood Combustion, Principles, Processes, and Economics,* Academic Press, New York, 1981.

11. Glassman, I., *Combustion,* 2nd Ed., Academic Press, Orlando, FL, 1987.

12. King, M. K., "Ignition and Combustion of Boron Particles and Clouds," *Journal of Spacecraft,* 19(4): 294–306 (1982).

## QUESTIONS AND PROBLEMS

**14.1**   List and define all of the boldfaced words in Chapter 14.

**14.2**   A.   Determine the reaction rates in $kg_C/m^2$-s for the following reactions for surface temperatures of 500, 1000, 1500, and 2000 K:

$$2C + O_2 \xrightarrow{k_1} 2CO \tag{R.1}$$

$$C + CO_2 \xrightarrow{k_2} 2CO, \tag{R.2}$$

where

$$k_1 = 3.007 \cdot 10^5 \exp(-17{,}966/T_s)[=] \text{ m/s}$$
$$k_2 = 4.016 \cdot 10^8 \exp(-29{,}790/T_s)[=] \text{ m/s}.$$

Assume unity mass fractions for $Y_{O_2,s}$ and $Y_{CO_2,s}$.

B.   Determine the ratio $\Re_1/\Re_2$ for the same temperatures. Discuss.

**14.3***   Use the one-film model to determine burning fluxes $(=\dot{m}_C/4\pi r_s^2)$ for carbon particles with radii of 500, 50 and 5 $\mu$m burning in air at 1 atm. Assume a surface temperature of 1500 K and that the kinetic rate constant, $k_c$, can be approximated as $3 \cdot 10^5 \exp(-17{,}966/T_s)$ with units of m/s. Assume the mixture molecular weight at the surface is 29 kg/kmol. In which combustion regime, diffusionally controlled, kinetically controlled, or intermediate, does the burning occur for each particle size?

**14.4***   Use the one-film model to determine the surface temperature of a 1-mm-diameter carbon particle burning in air ($Y_{O_2,\infty} = 0.233$). Both the air and the surroundings are at 300 K. Assume diffusional control and $\varepsilon_s = 1$. Is diffusional control a good assumption? Discuss.

---

*Indicates required use of computer.

**14.5**   Estimate the lifetime of the carbon particle in problem 14.4.

**14.6**   Explain why Eqns. 14.55 and 14.56 apply only for diffusionally controlled combustion. How would you find the droplet lifetime if diffusionally controlled combustion does not prevail?

**14.7***   Use the two-film model to determine the burning rate of a 10-$\mu$m-diameter carbon particle burning in air ($Y_{O_2,\infty} = 0.233$). Assume a surface temperature of 2000 K. Is the burning kinetically or diffusionally controlled?

**14.8**   Determine the influence of the freestream oxygen mole fraction on the burning times of 1-mm-diameter carbon particles. Assume diffusional control. Let $Y_{O_2,\infty} = 0.1165, 0.233$, and 0.466. Assume a surface temperature of 2000 K.

**14.9**   Fill in the missing steps in the development of the two-film model.

**14.10**   Rederive the two-film solution treating $Y_{CO_2,\infty}$ as a parameter, rather than setting it equal to zero as was done in the Chapter 14 development.

**14.11**   Derive an expression to allow evaluation of the surface temperature for the two-film problem. Note that the reaction at the surface is endothermic and that heat will be conducted from the flame to the surface. Use the Chapter 10 development as a guide.

**14.12**   Consider the conditions of problem 14.3.

   A.   Estimate the lifetime of the 500-$\mu$m-radius particle. Assume the particle density is 2100 kg/m$^3$.

   B.   Calculate the ambient gas and surroundings temperature. Assume $T_\infty = T_{sur}$.

**14.13**   Estimate the burning rate (kg/s) of a carbon particle (diameter = 50 $\mu$m) in air ($Y_{O_2,\infty} = 0.233$) at $P = 1$ atm. Assume the particle temperature is $T_s = 1500$ K, the kinetic rate constant is $k_c = 1.9$ m/s, and mean molecular weight of the gas at the particle surface is 30 kg/kmol. Estimate the diffusion coefficient using the value for $CO_2$ in $N_2$ given in Appendix Table D.1, being sure to correct for temperature. Which combustion regime prevails? If the burning rate were constant, independent of particle diameter, how long would it take to completely burn this particle? Assume the density of the particle is 2300 kg/m$^3$.

# chapter

# 15

# Emissions

## OVERVIEW

Control of pollutant emissions is a major factor in the design of modern combustion systems. Pollutants of concern include particulate matter, such as soot, fly ash, metal fumes, various aerosols, etc.; the sulfur oxides, $SO_2$ and $SO_3$; unburned and partially burned hydrocarbons, such as aldehydes; oxides of nitrogen, $NO_x$, which consist of NO and $NO_2$; and carbon monoxide, among others. Although not traditional pollutants, greenhouse gases associated with combustion, most notably $CO_2$, $CH_4$, and $N_2O$, are receiving considerable attention because of their role in global climate change. Control of greenhouse gases is a contentious topic, and progress on this front has been slow because of many competing interests. Table 15.1 summarizes general areas of current concern and the various combustion-generated pollutants associated with each. Early concerns with air pollution focused on the visible particulate emissions from industrial processes and stationary power generation. As was shown in Chapter 1, dramatic reductions in particulate emissions in the United States were achieved, starting in 1950 and leveling out in about 1980. In the 1950s, it became clear that the photochemical smog in the Los Angeles basin was primarily the result of automobile emissions of unburned hydrocarbons and $NO_x$ [6]. Emission controls on vehicles were instituted in California in the 1960s, and national air quality standards were set in the federal Clean Air Act of 1963 for a variety of pollutants. Amendments to the Clean Air Act in 1970, 1977, and 1990 have generally imposed stricter standards and have brought more and more sources under scrutiny. In general, California emission standards have been more stringent than federal standards, and California frequently leads the nation in subjecting new sources to control. Other nations also have adopted stringent emission standards and implemented emission controls.

This chapter takes a somewhat different approach than previous ones, being more descriptive and less theoretical. All of the theoretical concepts necessary to understanding this chapter have been previously developed.

**Table 15.1**     Combustion-generated or related air pollution concerns

| Concern (International Treaty/U.S. Regulation) | Combustion-Generated or Related Species |
|---|---|
| Local/Regional Air Quality (National Ambient Air Quality Standards [1, 2]) | Criteria pollutants: particles ($PM_{10}$ and $PM_{2.5}$),[a] $O_3$, $NO_2$, $SO_2$, CO, lead |
| Air Toxics/Hazardous Air Pollutants [3, 4] (1990 Clean Air Act Amendments) | 187 Substances[b]: selected aliphatic, aromatic, and polycyclic aromatic hydrocarbons; selected halogenated hydrocarbons; various oxygenated organics; metals; and other compounds |
| Greenhouse Effect/Global Warming [5] (Kyoto Protocol, 1997) | $CO_2$, $CH_4$, $N_2O$, stratospheric $H_2O$, tropospheric and stratospheric $O_3$, C (soot), sulfates[c] |
| Stratospheric Ozone Destruction [5] (Montreal Protocol, 1987) | $CH_4$, $N_2O$, $CH_3Cl$, $CH_3Br$, stratospheric $H_2O$, stratospheric $O_3$ |

[a]The subscripts 10 and 2.5 designate particles having aerodynamic diameters equal to or less than 10 $\mu m$ and 2.5 $\mu m$, respectively.
[b]Number on list as modified in 2005.
[c]Sulfates ($SO_2/SO_4^{2-}$) act as anti-greenhouse gases.

Also, much of Chapter 15 is specific to particular devices. We begin our discussion with a brief review of some of the consequences of combustion-generated pollutants, followed by a discussion of the many ways emissions are quantified. A large portion of the chapter is concerned with the origins of pollutants from premixed and nonpremixed systems. We conclude the chapter with a discussion of greenhouse gases.

## EFFECTS OF POLLUTANTS

Primary air pollutants (those emitted directly from the source) and secondary pollutants (those formed via reactions involving primary pollutants in the atmosphere) affect our environment and human health in many ways. Seinfeld [7] indicates four principal effects of air pollutants in the troposphere:

1. Altered properties of the atmosphere and precipitation.
2. Harm to vegetation.
3. Soiling and deterioration of materials.
4. Potential increase of morbidity (sickness) and mortality in humans.

Each of these effects is briefly addressed in the following discussion.

Altered properties of the atmosphere affecting local areas include reduced visibility, resulting from the presence of carbon-based particulate matter, sulfates, nitrates, organic compounds, and nitrogen dioxide; increased fog formation and precipitation, resulting from high concentrations of $SO_2$ that form sulfuric acid droplets, which then serve as condensation nuclei; reduced solar radiation; and altered temperature and wind distributions. On a larger scale, greenhouse gases (see Table 15.1) alter the global climate. References [5, 8–11] provide an introduction to this fascinating, technically complex, and politically charged subject. Also, acid rain, produced from $SO_x$ and $NO_x$ emissions, affects lakes and susceptible soils [12–17].

Vegetation is harmed by the phytotoxicants $SO_2$, peroxyacetyl nitrate (PAN), $C_2H_4$, and others. Phytotoxicants destroy chlorophyll and disrupt photosynthesis.

Particulate matter soils clothing, buildings, and other structures, creating not only a reduced aesthetic quality, but also additional cleaning costs over pollution-free environments. Acid and alkaline particles—in particular, those containing sulfur—corrode paint, masonry, electrical contacts, and textiles, while ozone severely deteriorates rubber.

Because of the difficulty of conducting research on human subjects, and the large number of uncontrolled variables, assessing the effects of pollution on human health is difficult. Epidemiological studies, however, have shown statistically significant correlations between pollutant levels and health effects [18–20]. It is well known that pollutants can aggravate preexisting respiratory ailments. The occurrence of both acute and chronic bronchitis, as well as emphysema, can be correlated with $SO_2$ and particulate matter. The famous air pollution episodes in Donora, Pennsylvania (1948); London (1952); and New York (1966) resulted in many excess deaths and other effects. These episodes were all consequences of simultaneously high levels of $SO_2$ and particles. In a recent reevaluation of the lethal 1952 London fog, Bell and Davis [21] estimate that 12,000 excess deaths resulted from the episode. Carbon-based particles also may contain adsorbed carcinogens. Only recently have researchers begun to understand the physical and biological interactions of particulate matter and other pollutants with the human body [22]. For example, Oberdörster *et al*. [23] report that sufficiently small particles can pass between cell walls in the lung and enter the blood and lymph streams. These particles then can be deposited in bone marrow, lymph nodes, spleen, and heart where they can facilitate the production of oxidants. Such behavior thus may provide a causal link between fine particles and cardiovascular disease [22, 23]. Secondary pollutants in photochemical smog cause eye irritation. These pollutants—ozone, organic nitrates, oxygenated hydrocarbons, and photochemical aerosol—are formed primarily by the reactions among nitric oxide and various hydrocarbons. The health effects of carbon monoxide are well documented. Figure 15.1 shows the effects on humans of exposure to various levels of CO. As indicated in Table 15.1, National Ambient Air Quality Standards [1, 2] exist for the six criteria pollutants (particulate matter, $PM_{10}$ and $PM_{2.5}$; $O_3$; $NO_2$; $SO_2$; CO; and Pb); the designation *criteria* refers to emissions that meet the criteria specified in the Clean Air Act, i.e., (1) emissions that cause or contribute to air pollution that may reasonably be anticipated to endanger public health or welfare, and (2) the presence of which in the ambient air results from numerous or diverse mobile or stationary sources. The 1990 Clean Air Act Amendments also identified 189 **hazardous air pollutants (HAPs)** for control [3, 4]. The number of HAPs was reduced to 187 in 2005.

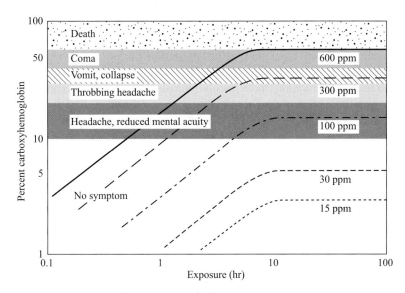

**Figure 15.1**    Effects of CO exposure on humans.
SOURCE: From Ref. [7]. © 1986, Reproduced with permission of John Wiley & Sons, Inc.

## QUANTIFICATION OF EMISSIONS

Emission levels are expressed in many different ways, which can make comparisons difficult and, sometimes, ambiguous. These differences arise from the needs of different technologies; for example, automobile emissions are expressed in grams per mile, utility boiler emissions in pounds per million BTUs, and many measurements are reported as parts per million (by volume) for a stated $O_2$ concentration. Although the interconversion among various ways of quantifying emissions is not necessarily difficult, we develop some of them here because they are so frequently useful. Reference [24] provides information additional to that given here.

### Emission Indices

The **emission index** for species $i$ is the ratio of the mass of species $i$ to the mass of fuel burned by the combustion process:

$$EI_i = \frac{m_{i,\,\text{emitted}}}{m_{F,\,\text{burned}}}. \tag{15.1}$$

In principle, the emission index is a dimensionless quantity, like the Reynolds number and other dimensionless groups; however, units such as g/kg, g/lb, etc., are used to avoid working with very small numbers, so caution must be exercised. The emission index is particularly useful because it unambiguously expresses the amount of pollutant formed per mass of fuel, independent of any dilution of the product stream or efficiency of the combustion process. Thus, the emission index can be thought of as a measure of the efficiency of a particular combustion process in producing a particular pollutant, uncoupled from the specific application.

For the combustion of a hydrocarbon fuel in air, the emission index can be determined from concentration (mole fraction) measurements of the species of interest, together with those of all of the C-containing species. Assuming all of the fuel carbon appears as either $CO_2$ or CO, the emission index is expressed

$$EI_i = \left( \frac{\chi_i}{\chi_{CO} + \chi_{CO_2}} \right) \left( \frac{x\, MW_i}{MW_F} \right), \tag{15.2}$$

where the $\chi$s are mole fractions; $x$ is the number of moles of carbon in a mole of fuel, $C_xH_y$; and $MW_i$ and $MW_F$ are the molecular weights of species $i$ and the fuel, respectively. Physically, the first bracketed term in Eqn. 15.2 represents the number of moles of $i$ per mole of carbon originating in the fuel, whereas the other term provides the necessary conversion of C moles to fuel moles and their respective conversion to mass units. From Eqn. 15.2, it is obvious that the measurement of an emission index is independent of any dilution of the sample stream by air, for example, because all of the measured concentrations appear as a ratio, with the effect of the diluent canceling.

---

**Example 15.1**

A spark-ignition engine is running on a dynamometer test stand and the following measurements of the exhaust products are made:

$$CO_2 = 12.47\%,$$
$$CO = 0.12\%,$$
$$O_2 = 2.3\%,$$
$$C_6H_{14}\,(\text{equivalent}) = 367\ \text{ppm},$$
$$NO = 76\ \text{ppm}.$$

All concentrations are by volume on a dry basis. The engine is fueled by isooctane. Determine the emission index of the unburned hydrocarbons expressed as equivalent hexane.

**Solution**

If we ignore the fact that not all of the fuel appears as CO and $CO_2$, i.e., unburned hydrocarbons are measured, we can use Eqn. 15.2 directly. The molecular weights of hexane and isooctane are 86.2 and 114.2 kg/kmol, respectively. Thus,

$$EI_{C_6H_{14}} = \left( \frac{\chi_{C_6H_{14}}}{\chi_{CO} + \chi_{CO_2}} \right) \left( \frac{x\, MW_{C_6H_{14}}}{MW_{C_8H_{18}}} \right)$$

$$= \left( \frac{367 \cdot 10^{-6}}{0.0012 + 0.1247} \right) \frac{8(86.2)}{114.2}$$

$$EI_{C_6H_{14}} = 0.0176\ \text{kg/kg} \quad \text{or} \quad 17.6\ \text{g/kg}$$

We can redo this calculation, taking into account the unburned hydrocarbons, by adding $6\chi_{C_6H_{14}}$ to $\chi_{CO} + \chi_{CO_2}$ in the denominator of Eqn. 15.2:

$$EI_{C_6H_{14}} = \left( \frac{367 \cdot 10^{-6}}{0.0012 + 0.1247 + 6(367 \cdot 10^{-6})} \right) \frac{8(86.2)}{114.2}$$

$$EI_{C_6H_{14}} = 17.3 \text{ g/kg}$$

This value is lower than the previous value by about 1.7 percent; thus, the amount of fuel carbon appearing as unburned hydrocarbons has only a small effect on the calculated emission indices.

---

## Corrected Concentrations

Concentrations, corrected to a particular level of $O_2$ in the product stream, are frequently reported in the literature and used in practice. The purpose of correcting to a specific $O_2$ level is to remove the effect of various degrees of dilution so that true comparisons of emission levels can be made, while still retaining a familiar mole-fraction-like variable. Before discussing corrected concentrations, we need to define "wet" and "dry" concentrations (mole fractions) of an arbitrary species in a combustion product stream, since corrected concentrations may be expressed on either a **wet** or **dry** basis. Assuming stoichiometric or lean combustion with only trace amounts of CO, $H_2$, and pollutants, the combustion of one mole of fuel with air (21 percent $O_2$, 79 percent $N_2$ by volume) can be expressed as

$$C_xH_y + aO_2 + 3.76aN_2 \rightarrow$$
$$xCO_2 + (y/2)H_2O + bO_2 + 3.76aN_2 + \text{trace species.} \qquad (15.3)$$

In many applications, moisture is removed from the exhaust sample prior to analysis, yielding so-called dry concentrations, and sometimes the sample is heated and the moisture retained. Assuming all the moisture is removed, the dry mole fraction for species $i$ is then defined as

$$\chi_{i,\text{dry}} = \frac{N_i}{N_{\text{mix,dry}}} = \frac{N_i}{x + b + 3.76a}, \qquad (15.4a)$$

and the corresponding wet mole fraction is

$$\chi_{i,\text{wet}} = \frac{N_i}{N_{\text{mix,wet}}} = \frac{N_i}{x + y/2 + b + 3.76a}. \qquad (15.4b)$$

Using these definitions (Eqn. 15.4) and an atom balance for O atoms, we can find the ratio of the total number of moles in a wet mixture to the total number of moles in a dry mixture:

$$\frac{N_{\text{mix,wet}}}{N_{\text{mix,dry}}} = 1 + \frac{y}{2(4.76a - y/4)}, \qquad (15.5)$$

where the oxygen coefficient $a$ is defined by the measured $O_2$ mole fraction as

$$a = \frac{x + \left(1 + \chi_{O_2, \text{wet}}\right) y/4}{1 - 4.76 \chi_{O_2, \text{wet}}} \tag{15.6a}$$

or

$$a = \frac{x + \left(1 - \chi_{O_2, \text{dry}}\right) y/4}{1 - 4.76 \chi_{O_2, \text{dry}}}. \tag{15.6b}$$

Equation 15.5 can then be used to interconvert wet and dry concentrations, i.e.,

$$\chi_{i, \text{dry}} = \chi_{i, \text{wet}} \frac{N_{\text{mix, wet}}}{N_{\text{mix, dry}}}. \tag{15.7}$$

Remember that the above relations were derived assuming stoichiometric or lean mixtures. For rich mixtures, the situation is more complex since CO and $H_2$ need to be considered (see Chapter 2 and Ref. [24]).

We return now to the concept of corrected concentrations, in which a "raw" measured mole fraction (wet or dry) is expressed as a mole fraction (wet or dry) corrected to a specific $O_2$ mole fraction. An example of this usage is "200 ppm NO corrected to 3 percent $O_2$." To correct a measured concentration or convert from one $O_2$ level to another, one simply applies

$$\chi_i \left( \begin{array}{c} \text{corrected to} \\ O_2\text{-level 2} \end{array} \right) = \chi_i \left( \begin{array}{c} \text{raw, or corrected, at} \\ O_2\text{-level 1} \end{array} \right) \frac{N_{\text{mix}, O_2\text{-level 1}}}{N_{\text{mix}, O_2\text{-level 2}}}, \tag{15.8}$$

where, for wet concentrations,

$$N_{\text{mix, wet}} = 4.76 \left[ \frac{x + \left(1 + \chi_{O_2, \text{wet}}\right) y/4}{1 - 4.76 \chi_{O_2, \text{wet}}} \right] + y/4 \tag{15.9a}$$

and, for dry concentrations,

$$N_{\text{mix, dry}} = 4.76 \left[ \frac{x + \left(1 - \chi_{O_2, \text{dry}}\right) y/4}{1 - 4.76 \chi_{O_2, \text{dry}}} \right] - y/4. \tag{15.9b}$$

---

**Example 15.2**

Using the data provided in Example 15.1, convert the given NO concentration to a wet basis.

**Solution**

The conversion from wet- to dry-basis concentrations is accomplished using Eqn. 15.7. We first calculate the ratio of the total number of moles in a wet mixture to the number of moles in a dry mixture using Eqns. 15.5 and 15.6b:

$$a = \frac{x + \left(1 - \chi_{O_2, dry}\right) y/4}{1 - 4.76\chi_{O_2, dry}}$$

$$= \frac{8 + (1 - 0.023)18/4}{1 - 4.76(0.023)} = 13.92$$

$$\frac{N_{mix, wet}}{N_{mix, dry}} = 1 + \frac{y}{2(4.76a - y/4)}$$

$$= 1 + \frac{18}{2[(4.76)13.92 - 18/4]} = 1.146.$$

The wet NO concentration can now be found as

$$\chi_{NO, wet} = \chi_{NO, dry} \frac{N_{mix, dry}}{N_{mix, wet}}$$

$$= 76 \text{ ppm} \frac{1}{1.146}$$

$$\chi_{NO, wet} = 66.3 \text{ ppm}.$$

Thus, the wet-basis concentration is about 12.7 percent less than the dry-basis value.

### Comment

From the calculation of the wet-to-dry total moles ratio, we see that the product stream origi-nally contained 12.7 percent $H_2O$. Presumably, nearly all of this moisture was removed prior to the gases being analyzed.

---

In Example 15.1, 76 ppm (dry) NO was measured in an exhaust stream containing 2.3 percent $O_2$. What is the NO concentration corrected to 5 percent $O_2$?          **Example 15.3**

### Solution

To correct the NO concentration from 2.3 percent to 5 percent $O_2$, we first calculate the total moles associated with each condition (Eqn. 15.9b):

$$N_{mix} @ \chi_{O_2} = 4.76 \left[ \frac{x + \left(1 - \chi_{O_2}\right) y/4}{1 - 4.76\chi_{O_2}} \right] - y/4,$$

$$N_{mix} @ 2.3\% \ O_2 = 4.76 \left[ \frac{8 + (1 - 0.023)18/4}{1 - 4.76(0.023)} \right] - \frac{18}{4} = 61.76,$$

$$N_{mix} @ 5\% \ O_2 = 4.76 \left[ \frac{8 + (1 - 0.05)18/4}{1 - 4.76(0.05)} \right] - \frac{18}{4} = 72.18.$$

The corrected concentration is then (Eqn. 15.8):

$$\chi_{NO} @ 5\% \ O_2 = \chi_{NO} @ 2.3\% \ O_2 \frac{N_{mix} @ 2.3\% \ O_2}{N_{mix} @ 5\% \ O_2}$$

$$= 76 \ \text{ppm} \frac{61.76}{72.18}$$

$$\chi_{NO} @ 5\% \ O_2 = 65 \ \text{ppm}.$$

**Comment**

Correcting to the 5 percent level reduces the given concentration by about 15 percent. When emissions are reported as concentrations, it is essential that the effect of dilution be taken into account when comparisons are made.

## Various Specific Emission Measures

In the dynamometer testing of spark-ignition and diesel engines, emissions are frequently expressed as

$$\text{Mass specific emission} = \frac{\text{Mass flow of pollutant}}{\text{Brake power produced}}, \tag{15.10}$$

where the units are typically g/kW-hr, or the mixed units of g/hp-hr. Mass specific emissions (MSE) are conveniently related to the emission index as

$$(\text{MSE})_i = \dot{m}_F \ EI_i / \dot{W}, \tag{15.11}$$

where $\dot{m}_F$ is the fuel mass flowrate and $\dot{W}$ is the power delivered.

Another frequently employed specific emission measure is the mass of pollutant emitted per amount of fuel energy supplied, which is expressed as

$$\frac{\text{Mass of pollutant } i}{\text{Fuel energy supplied}} = \frac{EI_i}{\Delta h_c}, \tag{15.12}$$

where $\Delta h_c$ is the fuel heat of combustion. Usual units are g/MJ or lb/MMBtu, where MMBtu refers to $10^6$ Btu. (Note that confusion can arise with the British units where MBtu refers to $10^3$ Btu.)

Other measures of emissions may depend on obtaining a particular weighted average over a specified test cycle. Examples include the use of a driving cycle to determine emissions in grams per mile from automobiles [25], and test cycles for aircraft engines to establish emission levels in grams per kilonewtons thrust [26]. Such procedures are used to define legislated emission standards. Other emission measures are defined for various industrial processes.

Although all of the quantities previously discussed make good sense and are quite useful as measures in specific applications, the emission index is a particularly useful parameter from a combustion point of view.

# EMISSIONS FROM PREMIXED COMBUSTION

The primary pollutants that we wish to deal with are oxides of nitrogen, carbon monoxide, unburned and partially burned hydrocarbons, and soot. Sulfur oxides are emitted quantitatively based on the sulfur content of the fuel. Since nearly all premixed combustion deals with very low sulfur-content fuels, $SO_x$ ($SO_2$ and $SO_3$) emissions from such systems are usually not of concern. Natural gas contains only trace amounts of sulfur as $H_2S$ and other compounds, and gasoline specifications require no more than 80 ppm S by weight (see Chapter 17). For nonpremixed systems burning coal or low-quality oils, $SO_x$ is a major concern. Our discussion of $SO_x$ emissions, therefore, will be taken up later.

## Oxides of Nitrogen

**Review of Chemical Mechanisms**    The chemical kinetics of NO and $NO_2$ formation and destruction were discussed in Chapter 5, and a review of that material now would be helpful. In Chapter 5, we saw that NO is formed through several mechanisms and variations thereof. Bowman [27] classifies these into the following three categories:

1. The extended Zeldovich (or thermal) mechanism in which O, OH, and $N_2$ species are at their equilibrium values and N atoms are in steady state (Reactions N.1–N.3 in Chapter 5).

2. Mechanisms whereby NO is formed more rapidly than predicted by the thermal mechanism above, either by (i) the Fenimore CN and HCN pathways (Reactions N.7 and N.8), (ii) the $N_2O$-intermediate route (Reactions N.4–N.6), or (iii) as a result of superequilibrium concentrations of O and OH radicals in conjunction with the extended Zeldovich scheme. With the discovery of the role of NNH, we can add a fourth mechanism here, (iv) the NNH route.

3. Fuel nitrogen mechanism, in which fuel-bound nitrogen is converted to NO.

The primary nitrogen oxide from combustion systems is NO; although, in some systems, appreciable $NO_2$ is produced, usually as a result of $NO \rightarrow NO_2$ conversion in low-temperature mixing regions of nonpremixed systems (see Reactions N.14–N.17 in Chapter 5). Therefore, in this section, our discussion focuses on NO.

Calculation of NO formed via the thermal mechanism with equilibrium O and OH radicals is relatively straightforward, following the principles of Chapter 4. In fact, several of the Chapter 4 examples and homework problems are based on this mechanism. Neglecting reverse reactions, we found that

$$\frac{d[NO]}{dt} = 2k_{1f}[O]_e[N_2]_e \qquad (15.13)$$

where $k_{1f}$ is the forward rate coefficient for the rate-limiting reaction $O + N_2 \rightarrow NO + N$. Of course, reverse reactions need to be included when NO levels become appreciable. The basic premise behind the use of Eqn. 15.13 is that the NO chemistry is much slower than the combustion chemistry; thus, the O- and OH-atom concentrations have had time to equilibrate. This assumption of uncoupled combustion and NO-formation chemistry breaks down, however, when O and OH atoms are formed in quantities well over equilibrium (up to a factor of $10^3$!), which can occur in flame zones. In this case, NO is formed via the Zeldovich reactions much more rapidly than if O atoms were in equilibrium. Calculation of superequilibrium radical concentrations is quite complex and must be coupled to the fuel oxidation kinetics. Also, within the flame zone, the Fenimore prompt-NO pathways can be important, and the possibility exists for NO formation through NNH as well. Although workers have had some success in using only the thermal-equilibrium-O mechanism to predict NO emissions in the past, our current understanding of NO production shows that the situation is indeed more complex [28–30]. Of course, there are applications where the thermal mechanism dominates, and the simple mechanism is thus useful.

Table 15.2 presents calculated relative contributions of the various NO production pathways in premixed combustion [27]. The effect of pressure on NO from laminar flames is given by the first data grouping. Here we see that at low pressures, the NO yield is dominated by the Fenimore (HC–$N_2$) and superequilibrium O, OH routes. At 10 atm, the simple thermal mechanism produces a little more than half of the $NO_x$, with the other three routes making substantial contributions. These data have significance for spark-ignition engines where stoichiometric mixtures are employed and pressures can go up to more than 20 atm.

The second data grouping in Table 15.2 shows the effect of equivalence ratio for rich conditions. The principal observation here is that as the mixture is made increasingly rich, the Fenimore mechanism dominates, yielding 95 percent of the total NO at $\Phi = 1.32$. At sufficiently rich mixtures, however, this mechanism no longer dominates [34].

The well-stirred-reactor data in Table 15.2 show that under conditions of strong backmixing of reactants and products, the superequilibrium O, OH route dominates for lean mixtures, whereas the Fenimore mechanism controls for stoichiometric and rich mixtures.

Also of interest is the question of what mechanisms are important at conditions employed in lean premixed gas-turbine combustors. Early work by Correa and Smooke [35] showed the importance of the $N_2O$ route for NO formation in lean ($\Phi = 0.6$) methane flames. More recent modeling studies [36] indicate that, for a variety of fuels, the NNH route dominates NO formation for lean ($\Phi = 0.61$) combustion in a perfectly stirred reactor at 1 atm. These results are shown in Fig. 15.2.

The third NO-formation category, the fuel-N route, is not generally important in premixed combustion applications because most fuels used in premixed combustion (natural gas and gasolines) contain little or no bound nitrogen. Pulverized coal and heavy distillate fuels, however, can contain significant quantities of fuel nitrogen; thus, we defer our discussion of the fuel-N mechanism until the section on nonpremixed combustion.

**$NO_x$ Control Strategies**    For processes dominated by thermal NO formation, time, temperature, and oxygen availability are the primary variables affecting $NO_x$ yields.

**Table 15.2** Relative contributions of various mechanisms to $NO_x$ formation in premixed combustion[a]

| Flame | Φ | P(atm) | Total $NO_x$ (ppm) | Fraction of Total NO Formation | | | |
|---|---|---|---|---|---|---|---|
| | | | | Equilibrium Thermal | Superequilibrium | HC–$N_2$ | $N_2O$ |
| Premixed, laminar, $CH_4$–air [31] | 1 | 0.1 | 9 @ 5 ms | 0.04 | 0.22 | 0.73 | 0.01 |
| | 1 | 1.0 | 111 | 0.50 | 0.35 | 0.10 | 0.05 |
| | 1 | 10.0 | 315 | 0.54 | 0.15 | 0.21 | 0.10 |
| Premixed, laminar, $CH_4$–air [32] | 1.05 | 1 | 29 @ 5 mm | 0.53 | 0.30 | 0.17 | — |
| | 1.16 | 1 | 20 | 0.30 | 0.20 | 0.50 | — |
| | 1.27 | 1 | 20 | 0.05 | 0.05 | 0.90 | — |
| | 1.32 | 1 | 23 | 0.02 | 0.03 | 0.95 | — |
| Well-stirred reactor, $CH_4$–air [32, 33] | 0.7 | 1 | 12 @ 3 ms | ≈0 | 0.65 | 0.05 | 0.30 |
| | 0.8 | 1 | 20 | — | 0.85 | 0.10 | 0.05 |
| | 1.0 | 1 | 70 | — | 0.30 | 0.70 | — |
| | 1.2 | 1 | 110 | — | 0.10 | 0.90 | — |
| | 1.4 | 1 | 55 | — | — | 1.00 | — |

[a] SOURCE: From Ref. [27]. Sources of original data are Refs. [31–33].

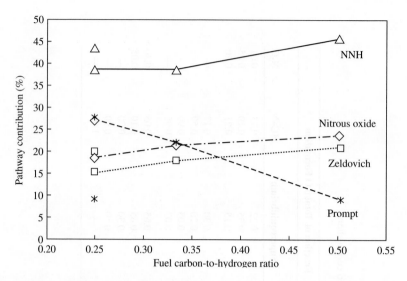

**Figure 15.2**    The results from the modeling study of Rutar *et al.* [36] show the contributions of various NO-formation pathways to the total NO formed for lean premixed combustion for $\Phi = 0.61$, 1 atm, and 1790 K. A sequence of perfectly stirred reactors was modeled to simulate the experiments used for comparison.

The rate coefficient for the $O + N_2 \rightarrow NO + N$ reaction has a very large activation temperature ($E_A/R_u = 38{,}370$ K) and thus increases rapidly at temperatures above about 1800 K. From Fig. 2.14, we see that for adiabatic, constant-pressure combustion, the maximum equilibrium O-atom mole fraction lies near $\Phi = 0.9$. This is also approximately the same equivalence ratio where the maximum kinetically formed NO is found for spark-ignition engines, as can be seen in Fig. 15.3. Unfortunately, from the viewpoint of emission control, maximum efficiency also is achieved near this equivalence ratio for many practical devices.

Reducing peak temperatures can significantly reduce $NO_x$ emissions. In industrial burners and spark-ignition engines, this can be achieved by mixing flue gases or exhaust gases with the fresh air or fuel. Figure 15.4 shows experimental results for exhaust-gas recirculation, frequently denoted EGR, for a spark-ignition engine. The effect of EGR or FGR (flue-gas recirculation) is to increase the heat capacity of the burned gases for a given quantity of heat release, thus lowering the combustion temperature (see Example 2.9). Figure 15.5 shows the correlation of NO reduction with diluent heat capacity for an SI engine. The effect of ultra lean operation also is to increase the heat capacity of the products and, consequently, reduce temperatures. Lean combustion is a subject of much current interest, as evidenced by the recent publication of a book devoted to this topic [40].

The **homogeneous charge compression ignition (HCCI) engine** is a concept that has been the subject of much recent research [41, 42]. In the HCCI engine, an ultra lean, or highly dilute, premixed fuel–air mixture autoignites, rather than

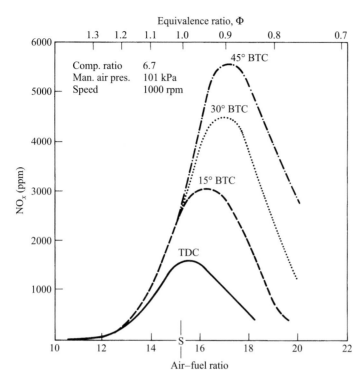

**Figure 15.3** $NO_x$ concentrations as functions of air–fuel and equivalence ratios for various spark timings.

SOURCE: After Ref. [37]. Reprinted with permission of the Air and Waste Management Association.

being ignited by a spark. The potential advantage of this mode of operation over a standard spark-ignition engine is twofold: First, combustion temperatures are very low, which results in low, $NO_x$ emissions; and second, considerably less throttling is required for part-load operation, and consequently, pumping losses are small and the thermal efficiency of the HCCI engine approaches that of a diesel engine. Moreover, particulate matter emissions are low. However, numerous problems need to be overcome for the HCCI engine to be a viable alternative to more conventional engines. For example, concerns include load control from idle to full power, control of combustion timing, and high emissions of carbon monoxide and unburned hydrocarbons. For more information on HCCI engines, Refs. [41] and [42] are highly recommended.

---

Consider nitric oxide formation in the postflame gases of a stoichiometric propane–air mixture at atmospheric pressure. Assuming adiabatic conditions, how does the initial rate of NO formation (ppm/s) from the Zeldovich mechanism compare for no dilution and 25 percent dilution by $N_2$ (moles $N_2$ added equals 0.25 the number of moles of air)? The reactants and $N_2$ diluent are initially at 298 K.

**Example 15.4**

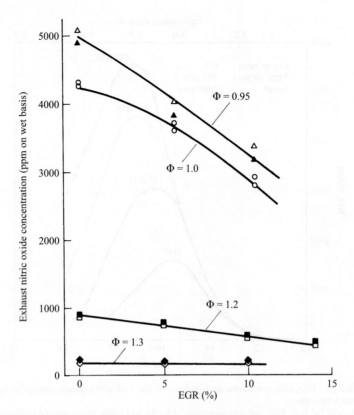

**Figure 15.4** Effect of exhaust-gas recirculation (EGR) on $NO_x$ emissions from a spark-ignition engine.
SOURCE: Reprinted with permission from Ref. [38], © 1973, Society of Automotive Engineers, Inc.

**Solution**

We first calculate the equilibrium adiabatic flame temperature and equilibrium composition for the two cases using an equilibrium routine (see Chapter 2 and Appendix F). Results of this computation are

|  | **No Dilution** | **25% Dilution $N_2$** |
|---|---|---|
| $T_{ad}$ (K) | 2267.9 | 2033.2 |
| $\chi_{O,e}$ | $3.12 \cdot 10^{-4}$ | $3.59 \cdot 10^{-5}$ |
| $\chi_{N_2,e}$ | 0.721 | 0.777 |

We can use Eqn. 15.13 to evaluate $d[NO]/dt$:

$$\frac{d[NO]}{dt} = 2k_{1f}[O]_e[N_2]_e,$$

where the rate coefficient $k_{1f}$ from Chapter 5 is

$$k_{1f} = 1.82 \cdot 10^{11} \exp[-38,370/T(K)] \ m^3/kmol\text{-}s.$$

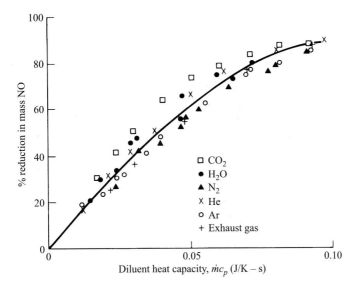

% reduction in mass NO

Diluent heat capacity, $\dot{m}c_p$ (J/K – s)

□ $CO_2$
● $H_2O$
▲ $N_2$
✕ He
○ Ar
+ Exhaust gas

**Figure 15.5**    Correlation of NO reduction with diluent heat capacity, $\dot{m}c_p$ for a spark-ignition engine.
SOURCE: Reprinted from Ref. [39], © 1971, Society of Automotive Engineers, Inc.

For the case of no dilution,

$$k_{1f} = 1.82 \cdot 10^{11} \exp[-38,370/2267.9] = 8173 \text{ m}^3/\text{kmol-s}$$

and, converting molar concentrations to mole fractions,

$$\frac{d\chi_{NO}}{dt} = 2k_{1f}\chi_{O,e}\,\chi_{N_2,e}\frac{P}{R_u T_{ad}}$$

$$= 2(8173)3.12 \cdot 10^{-4}(0.721)\frac{101,325}{8315(2267.9)}$$

$$\left(\frac{d\chi_{NO}}{dt}\right)_{no\ dil} = 1.98 \cdot 10^{-2} \text{ 1/s  or  19,750 ppm/s.}$$

With 25 percent $N_2$ dilution,

$$k_{1f} = 1.82 \cdot 10^{11} \exp[-38,370/2033.2] = 1159 \text{ m}^3/\text{kmol-s}$$

and

$$\left(\frac{d\chi_{NO}}{dt}\right)_{25\%\ N_2} = 2(1159)3.59 \cdot 10^{-5}(0.777)\frac{101,325}{8315(2033.2)}$$

$$= 3.875 \cdot 10^{-4} \text{ 1/s  or  338 ppm/s.}$$

The ratio of the formation rates with and without dilution is

$$\frac{\left(\dfrac{d\chi_{NO}}{dt}\right)_{25\%\ N_2}}{\left(\dfrac{d\chi_{NO}}{dt}\right)_{no\ dil}} = \frac{388}{19,750} = 0.0196$$

Thus, without dilution the initial NO formation rate is about 50 times greater!

**Comment**

From this example, we see that thermal $NO_x$ can be greatly reduced by dilution with cold inerts. The two major factors in reducing the formation rates are the decreased value of the rate coefficient (a factor of ~7) and the decreased equilibrium O-atom concentration (a factor of ~9), both of which are controlled by the temperature.

Another means to lower combustion temperatures in spark-ignition engines is to retard the spark timing. Late spark timing shifts the combustion event so that peak pressures occur when the piston is well beyond top-dead-center (minimum volume), resulting in lower pressures and temperatures. This effect can easily be seen in Fig. 15.3 where each curve represents a different spark timing. Significant fuel-economy penalties result from retarded spark timings.

The amount of thermal $NO_x$ produced in a device is strongly linked to the time that combustion products spend at high temperatures. For conditions where NO levels are well below their equilibrium values and reverse reactions are unimportant, the NO yield is directly proportional to time (see Example 4.4). In the design of a combustion system, therefore, the temperature-versus-time relationship is key to the control of NO emissions; however, drastic alteration of the time–temperature relationship for the gas flow may compromise the useful operation of the device. Obviously, a furnace that has low $NO_x$ emissions, but fails to properly heat the load, is not a useful device.

Staged combustion, in which a rich–lean or lean–rich combustion sequence takes place, is also an $NO_x$ control strategy. The basic concept is illustrated in Fig. 15.6 for a rich–lean sequence. The idea here is first to take advantage of both the good stability and low $NO_x$ emissions associated with rich combustion and, subsequently, to complete the combustion of the unburned CO and $H_2$ in a lean stage where additional $NO_x$ production is also low. For staging to be effective, the mixing of rich products and air must be very rapid, or a substantial amount of heat must be removed between stages. Consider the ideal staged-combustion process in Fig. 15.6 represented by the path 0–1–2′–2, where the bell-shaped curve represents the $NO_x$ yield for a fixed residence time, $\Delta t \, (= \Delta t_{rich})$. In the rich stage, the amount of $NO_x$ formed in the time $\Delta t_{rich}$ is represented by the segment 0–1. Secondary air is then instantaneously mixed ($\Delta t_{mix} = 0$) with the rich products (segment 1–2′) with no additional $NO_x$ formed. In the lean stage, the CO and $H_2$ are oxidized and an additional amount of $NO_x$ is formed (segment 2′–2) in the time associated with the lean stage ($\Delta t_{lean}$). If the mixing is not instantaneous, as it must be in any real process, additional $NO_x$ is formed during the mixing process as the stoichiometry passes through the region of high $NO_x$ formation rates (path 1–3). Obviously, the success of staging depends on how well the mixing process can be controlled in practice. Although the ideal staged combustion (0–1–2′–2) is represented as a sequence of two premixed combustion processes, most real processes become non-premixed because of the *in-situ* mixing of rich products and secondary air. Siewert and Turns [43] describe the implementation of an ideal staged sequence in which two different cylinders of a spark-ignition engine are employed. The rich products

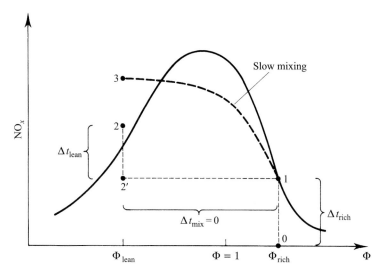

**Figure 15.6**    Schematic representation of staged combustion on $NO_x$-equivalence ratio coordinates. The path 0–1–2′–2 represents the ideal case of instantaneous secondary air mixing, while 0–1–3 indicates the path for slow secondary mixing.

are cooled and mixed with air before combustion is again initiated and completed in the second cylinder. Very low $NO_x$ levels were demonstrated, although CO and unburned hydrocarbon emissions were found to be unacceptably high without aftertreatment.

In automotive applications, combustion system modifications alone have not yet been able to reduce $NO_x$ levels below legislated standards, and, thus, catalytic converters are also employed to reduce $NO_x$ in the exhaust stream. We will discuss catalytic converters later, in conjunction with CO and unburned hydrocarbon control strategies.

## Carbon Monoxide

As we saw in Chapter 2, CO is a major species in rich-combustion products; thus, substantial CO will be produced whenever rich mixtures are used. In normal operation of most devices, rich conditions are generally avoided; however, spark-ignition engines employ rich mixtures during startup to prevent stalling, and at wide-open-throttle conditions to provide maximum power. Modern computer-controlled electronic fuel injection systems enable precise air–fuel ratio control over a wide range of conditions. For stoichiometric and slightly lean mixtures, CO is found in substantial quantities at typical combustion temperatures as a result of the dissociation of $CO_2$. The upper curve in Fig. 15.7 shows CO concentrations ranging from 1.2 percent (by vol.) at $\Phi = 1$ to 830 ppm at $\Phi = 0.8$ in the product gases of adiabatic, atmospheric-pressure, propane–air flames. Carbon monoxide

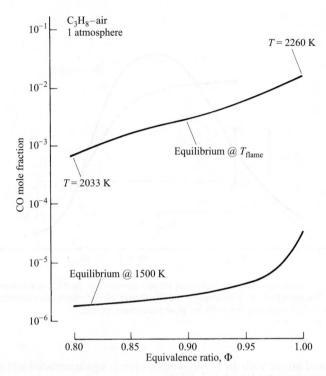

**Figure 15.7**     Equilibrium carbon monoxide mole fractions in propane–air combustion products at adiabatic flame temperatures and at 1500 K.

concentrations rapidly fall with temperature, as illustrated by the values shown in Fig. 15.7 for equilibrium at a temperature of 1500 K. Thus, if CO were to remain equilibrated as useful energy is extracted from the combustion products, very low levels of CO would be found in the exhaust system. In furnaces, where residence times are on the order of a second, equilibrium is likely to prevail; however, in spark-ignition engines, temperatures fall very rapidly during the expansion and exhaust processes. As a result, CO is not equilibrated and it passes into the exhaust stream frozen at levels somewhere between equilibrium at peak temperatures and pressures, and exhaust temperatures and pressures [44, 45]. Calculations indicate that the important reaction, $CO + OH \Leftrightarrow CO_2 + H$ (Reaction CO.3 from Chapter 5) is equilibrated during the expansion and blowdown processes; however, the three-body recombination processes, e.g., $H + OH + M \rightarrow H_2O + M$, are not fast enough to maintain equilibrium among all the radicals. This causes the partial-equilibrium CO concentrations to be well above those for full equilibrium [45, 46].

Other CO production mechanisms include quenching by cold surfaces, following the ideas presented above, and partial oxidation of unburned fuel. The latter is the subject of the next section.

## Unburned Hydrocarbons

In most devices employing premixed reactants, unburned hydrocarbons are usually negligible. An exception to this is the spark-ignition engine. The problem of SI-engine hydrocarbon emissions has a large literature, and Heywood's textbook [47] provides an excellent review of this subject. In this section, we will briefly discuss some of the most salient aspects of this problem.

In Chapter 8, we discussed the process of flame quenching, whereby a flame is extinguished a short distance from a cold surface. The quenching process leaves a thin layer of unburned fuel–air mixture adjacent to the wall. Whether or not this **quench layer** contributes to unburned hydrocarbon emissions depends upon subsequent diffusion, convection, and oxidation processes. In an SI engine, most of the hydrocarbons from wall quenching ultimately mix with hot gases and are oxidized; however, unburned hydrocarbons can result from flame quenching within and at the entrance to **crevices,** such as those formed by the piston topland and ring pack [48]. The helical spark-plug thread crevice can also be a source of unburned hydrocarbons. Figure 15.8 illustrates this crevice-volume mechanism for unburned hydrocarbons emissions in engines. Other known contributors to unburned hydrocarbon emissions in engines are absorption and subsequent desorption of fuel into and out of oil layers on the cylinder walls (Fig. 15.8). A similar process can occur for wall deposits, which for unleaded-fuel operations are carbonaceous. Unburned hydrocarbon emissions can also result from incomplete flame propagation in the bulk of the charge. This occurs for lean and/or dilute mixtures approaching the flammability limits [49, 50]. Excessive exhaust gas recirculation can result in misfire or incomplete flame propagation.

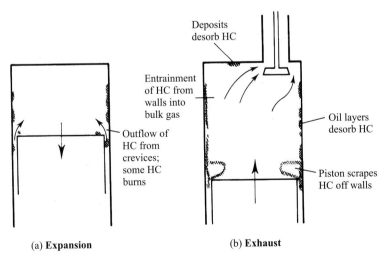

(a) **Expansion**                    (b) **Exhaust**

**Figure 15.8**     Schematic representation of unburned hydrocarbon emission mechanism for spark-ignition engines.
SOURCE: After Ref. [47].

**Table 15.3**     Typical composition of spark-ignition engine unburned hydrocarbons without aftertreatment[a]

| Species | Fraction of Total Unburned Constituents |
|---|---|
| Ethylene | 19.0 |
| Methane | 13.8 |
| Propylene | 9.1 |
| Toluene | 7.9 |
| Acetylene | 7.8 |
| 1-Butane, *i*-Butane, 1,3-Butadiene | 6.0 |
| *p*-Xylene, *m*-Xylene, *o*-Xylene | 2.5 |
| *i*-Pentane | 2.4 |
| *n*-Butane | 2.3 |
| Ethane | 2.3 |
| Total | 73.1 |

[a] SOURCE: From Ref. [51].

Only about a third of the unburned hydrocarbons found in the untreated exhaust are fuel molecules [51]. The remainder are fuel pyrolysis and partial oxidation products, as shown in Table 15.3. The partial oxidation of hydrocarbons results in the production of CO, aldehydes, and other particularly undesirable compounds. For lean operation, this partial oxidation of crevice, oil, and deposit-layer hydrocarbons is the major source of CO, since the previously discussed CO-mechanisms cannot account for the level of CO found at such conditions.

## Catalytic Aftertreatment

Catalytic aftertreatment is the primary technique applied to control, simultaneously, nitric oxide, carbon monoxide, and unburned hydrocarbon emissions from spark-ignition engines. Figure 15.9 illustrates a honeycomb ceramic monolith catalytic converter. In this design, a thin washcoat containing small catalyst particles is deposited on the ceramic substrate. Noble-metal catalysts, e.g., platinum, palladium, and rhodium, provide active sites for reactions that oxidize carbon monoxide and unburned hydrocarbons while simultaneously reducing nitric oxide. To achieve high conversion efficiencies, i.e., pollutant destruction, requires that the composition of the stream through the converter be maintained in a narrow range near the stoichiometric ratio ($\Phi = 1$). Typical three-way catalyst conversion efficiencies are illustrated in Fig. 15.10. Reference [53] contains a wealth of information on three-way catalyst systems.

## Particulate Matter

Emissions of particulate matter from premixed combustion are generally considered to be much less of a problem than those associated with nonpremixed combustion.

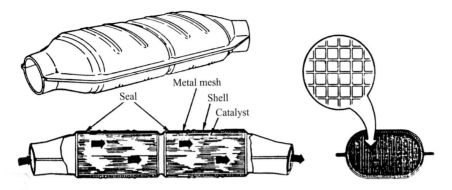

**Figure 15.9**    Monolith catalytic converter.
| SOURCE: Reprinted from Ref. [52] with permission of the American Society of Mechanical Engineers.

With the removal of tetraethyl lead from gasoline, this source of particulate matter from spark-ignition engines has been eliminated. Similarly, the reduction of the allowed sulfur content in gasoline to very low levels (see Chapter 17) minimizes the potential to form sulfate particles or aerosols within the combustion system or in the atmosphere. Fuel–air mixtures sufficiently rich to produce soot in premixed combustion are usually the result of some malfunction, rather than typical operation. Table 15.4 shows limiting equivalence ratios ($\Phi_c$) for the formation of soot in atmospheric-pressure premixed flames for selected fuels with air. Soot forms at equivalence ratios equal to, or greater than, $\Phi_c$. For premixed combustion, the difference in sooting tendencies of various fuel types is related not only to fuel structure but also to differences in flame temperature, as discussed by Glassman [55].

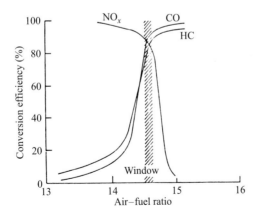

**Figure 15.10**    Conversion efficiencies for a typical three-way automotive catalyst, showing narrow air–fuel ratio window for simultaneous destruction of unburned hydrocarbons, CO, and NO.
| SOURCE: Reprinted from Ref. [52] with permission of the American Society of Mechanical Engineers.

**Table 15.4**     Limiting equivalence ratios for the formation of soot in premixed flames[a]

| Fuel | Limiting Equivalence Ratio, $\Phi_c$ |
|---|---|
| Ethane | 1.67 |
| Propane | 1.56 |
| $n$-Hexane | 1.45 |
| $n$-Octane | 1.39 |
| Isooctane | 1.45 |
| Isodecane | 1.41 |
| Acetylene | 2.08 |
| Ethylene | 1.82 |
| Propylene | 1.67 |
| Ethanol | 1.52 |
| Benzene | 1.43 |
| Toluene | 1.33 |

[a] SOURCE: From Ref. [54].

Although soot is not usually a source of particulate matter from spark-ignition engines, these engines do emit fine particles composed primarily of organic carbon, presumably from the condensation of hydrocarbon vapors. The mass emission rates of these fine particles are typically small, being approximately 10 to 100 times less than those from diesel engines [56]. In spite of the low emission rates, the collective particulate matter emitted by spark-ignition-engine–powered vehicles can be significant in highly polluted areas. Recent studies [57, 58] show that spark-ignition-engine–powered vehicles contribute 12–22 percent of the fine particulate matter ($PM_{2.5}$) found in the ambient air in non-attainment areas of the mid-Atlantic United States. Furthermore, the number of ultra fine particles ($PM_{0.1}$, i.e., particulate matter with particle sized 0.1 $\mu$m and smaller) emitted by spark-ignition engines can be relatively large [56]. At some operating conditions, the total number of particles emitted can be comparable to that emitted by diesel engines [56]. The proposed 2014 European emission standards for motor vehicles consider both the number of particles and the mass emitted [59]. Such standards will apply to both gasoline-engine and diesel-powered vehicles. The development of particulate matter controls and understanding the fundamental processes involved are currently hot topics.

## EMISSIONS FROM NONPREMIXED COMBUSTION

Although the chemical processes are the same in premixed and nonpremixed combustion, the additional physical processes associated with nonpremixed combustion, e.g., evaporation and mixing, can produce a range of local compositions spanning a wide range of stoichiometries. For example, the overall mixture may be stoichiometric, but within the combustion space there may be regions that are quite rich, while others may be quite lean. This aspect of nonpremixed combustion adds considerable complexity to the problem of pollutant formation in such systems. In some situations, however, combustion can occur essentially in a premixed mode when fuel evaporation and subsequent mixing are sufficiently rapid, even though the fuel and air are introduced separately into the combustion space. For such systems, pollutant production should be interpreted within the premixed framework discussed earlier. Because of the great complexity involved in pollutant formation in non-premixed systems, and because emissions in such systems frequently depend on specific details of the system (e.g., the droplet size distribution of the fuel spray), we present here only a brief introduction to the subject and point to the literature for further information.

## Oxides of Nitrogen

In this section, we first look at the $NO_x$ emission characteristics of a simple non-premixed system, a vertical jet flame in a quiescent environment. This continues our study of this flame type begun in Chapter 13 and serves as a good introduction to the discussions of the more complex nonpremixed systems that follow.

**Simple Turbulent Jet Flames**    The $NO_x$ emissions from turbulent jet flames have been extensively studied, and overviews of the issues involved can be found in Refs. [60, 61]. The structure of jet flames, indicated by the OH imaging shown previously in Fig. 13.5, suggests that NO is produced in thin laminarlike flamelet regions in the lower-to-mid regions of the flame and in relatively large and broadened reaction zones in the upper regions of the flame. The simple thermal, superequilibrium-O, and Fenimore mechanisms for NO formation are all likely to be active in hydrocarbon jet flames, although the determination of the relative contribution of each mechanism to the total $NO_x$ yield is yet a subject of research. In applications where flame temperatures are quite high, such as flames in furnaces with reradiating walls or flames using oxygen-enriched air, $NO_x$ emissions are likely to be controlled by Zeldovich kinetics.

As indicated in our discussion of Zeldovich kinetics in premixed systems, temperature, composition, and time are the important variables determining $NO_x$ emissions, and these variables are also controlling in nonpremixed flames. In nonpremixed jet flames, however, the composition varies from point to point in the flow and is governed by fluid mechanical mixing. Likewise, the temperature distribution can be coupled to the composition distribution, as we saw in Chapter 13, through the use of the conserved scalar. Although simple analyses, such as those employed in Chapter 13, are useful for predicting global flame properties, such as flame length, more complicated approaches are required to capture kinetic effects associated with pollutant formation. State-of-the-art computer modeling seeks to couple all of the important physical and chemical phenomena to predict the $NO_x$ emissions of nonpremixed jet flames [62–64]. A discussion of these approaches is beyond the scope of this book, and we will have to be content with the generality that thermal NO is produced primarily in flame regions that have simultaneously high temperatures and high concentrations of O and OH atoms, i.e., conditions near or slightly lean of stoichiometric. These regions may be the thin laminarlike flame regions low in the jet flame or the broad regions near the flame tip. The detailed temperature and composition distributions within these regions are determined by fluid mechanical, chemical-kinetic, and thermal effects. The larger regions with longer residence times are particularly susceptible to radiation heat losses.

Figure 15.11 shows fuel-energy specific $NO_x$ emissions for simple propane–air and ethylene–air jet flames. We observe that the characteristic trend with heat release rate varies with both fuel type and initial jet diameter. These trends are explained by offsetting effects of residence time and temperature [65]. Increasing the fuel flowrate (heat release rate) reduces global and local residence times, which tends to reduce NO formation. Reduced residence times also allow the flames to be more adiabatic because less time is available for radiant losses to occur. Smaller flames (smaller $d_j$) also yield shorter residence times. Thus, we see that for the more luminous flames and larger flames, temperature effects prevail, causing the increasing trend of $NO_x$ with heat release.

The importance of flame radiation is more clearly delineated in the results shown in Fig. 15.12, where four fuels with different luminosities, and hence radiation characteristics, were employed. As shown at the top of the figure, the ethylene flames had the greatest radiant fractions followed by propane, methane, and a mixture of carbon monoxide and hydrogen. This is consistent with the discussion of sooting tendencies of nonpremixed flames (Chapter 9). The addition of $N_2$ to the fuel has two effects: first, dilution causes a decrease in adiabatic flame temperature; and, second, $N_2$ decreases the amount of soot formed in the flame. For the most-luminous flames ($C_2H_4$), the decrease in radiant heat losses more than compensates for the adiabatic flame temperature effect, and characteristic flame temperatures increase (Fig. 15.12, center). For the nonluminous (nonsooting) $CO$–$H_2$ flames, the adiabatic flame temperature effect controls. The $C_3H_8$ and $CH_4$ flames show intermediate characteristics. At the bottom of Fig. 15.12, we see that the $NO_x$-versus-$N_2$-dilution trends all mirror the flame-temperature-versus-dilution trends, clearly showing the importance of heat losses on $NO_x$ emissions [65, 66]. Finite-rate chemistry effects associated with prompt-NO, however, may also play a role [60, 62, 67].

For simple hydrocarbon–air turbulent jet flames radiating to room-temperature surroundings, Turns *et al.* [66] correlated $NO_x$ emissions with characteristic flame temperatures $T_f$ (K), and global residence times $\tau_G$ (s), as follows:

$$\ln\left[[NO_x]/\tau_G\right] = A + B \ln \tau_G + C/T_f, \qquad (15.14)$$

where

$A = 1.1146,$
$B = -0.7410,$
$C = -16,347,$

and $[NO_x]$ is defined as gmol of $NO_x$ per cm$^3$ at stoichiometric conditions at the nonadiabatic flame temperature, $T_f$. The global residence time is defined by

$$\tau_G \equiv \frac{\rho_f W_f^2 L_f f_{\text{stoic}}}{3\rho_{F,0} d_j^2 v_e} \qquad (15.15)$$

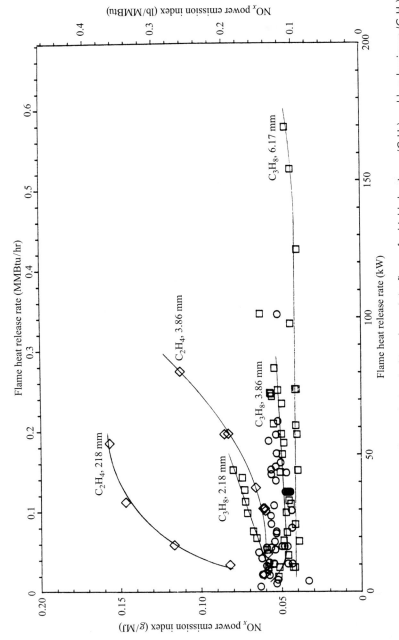

**Figure 15.11** Oxides of nitrogen emissions from turbulent hydrocarbon–air jet flames for highly luminous ($C_2H_4$) and less-luminous ($C_3H_8$) flames of various sizes ($d_j = 2.18, 3.86,$ and $6.17$ mm).

SOURCE: Reprinted by permission of Elsevier Science, Inc. From Ref. [65]. © 1991, The Combustion Institute.

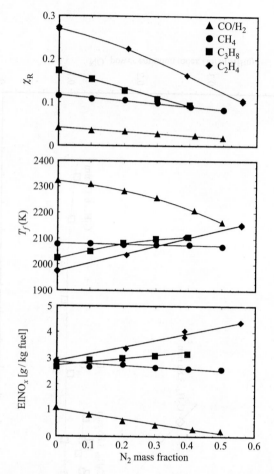

**Figure 15.12**     The influence of $N_2$ dilution on radiant fractions, characteristic flame temperatures, and $NO_x$ emission indices for jet flames having various luminosities (sooting propensities).

| SOURCE: Reprinted by permission from Elsevier. From Ref. [66].

where $\rho_f$, $W_f$, and $L_f$ are flame density, maximum visible width, and visible length, respectively; $\rho_{F,0}$ is the cold fuel density; and $f_{\text{stoic}}$ is the mass fraction of fuel in a stoichiometric mixture.

**Industrial Combustion Equipment**     In addition to boilers, this class of devices includes process heaters, furnaces, and ovens, all burning, primarily, natural gas. Discussion of oil- and coal-fired devices is limited to the section on utility boilers, which follows. The U.S. Environmental Protection Agency maintains a website that provides information on pollutants emitted from a wide array of sources [68]. Included here are tabulated emission factors [69]; for example, see Eqns. 15.1 and 15.10–15.12.

**Table 15.5** Emission factors for NO$_x$ from natural gas–fired boilers and residential furnaces [69]

| Combustor Type | NO$_x$ Emission Factor (lb/MMBtu) |
|---|---|
| Large wall-fired boilers (>100 MMBtu/hr) | |
|    Uncontrolled | 0.186 |
|    Controlled—Low-NO$_x$ burners | 0.137 |
|    Controlled—Flue gas recirculation (FGR) | 0.098 |
| Small boilers (<100 MMBtu/hr) | |
|    Uncontrolled | 0.098 |
|    Controlled—Low-NO$_x$ burners | 0.049 |
|    Controlled—Low-NO$_x$ burners/FGR | 0.137 |
| Tangential-fired boilers (all sizes) | |
|    Uncontrolled | 0.167 |
|    Controlled—Flue gas recirculation (FGR) | 0.075 |
| Residential furnaces (<0.3 MMBtu/hr) | |
|    Uncontrolled | 0.092 |

Table 15.5 shows emission factors for NO$_x$ for natural gas–burning boilers and residential furnaces, with and without emission controls, taken from the EPA compilation of emission factors for external combustion sources. The emission factors here are expressed as pounds of NO$_x$ (as NO$_2$) per million Btu of input fuel energy (see Eqn. 15.2). The larger devices have the larger NO$_x$ emission factors. Table 15.6 shows current California South Coast Air Quality Management District (SCAQMD) standards for a variety of gas-fired combustion systems [70]. The SCAQMD rules specify much tighter regulation for small boilers (2–5 MMBtu/hr) beginning in 2011. The federal Clean Air Act Amendments of 1990 also require reductions from industrial sources.

Figure 15.13 shows various strategies employed to reduce NO$_x$ emissions from gas–fired equipment [71]. Certain of these techniques also apply to oil-fired devices. The NO$_x$ reduction techniques are divided into those involving combustion modifications and those involving post-combustion controls. Within each broad classification there are several specific techniques (Fig. 15.13). Each of these will be briefly discussed here, and more detailed information can be found in Refs. [71–75].

**Table 15.6** The NO$_x$ emission regulations for industrial sources (California SCAQMD) [70]

| Process | Limit | Rule No. |
|---|---|---|
| Gas-fired industrial boilers | 30 ppm (3% O$_2$) | 1146, 1146.1 |
| Refining heaters | 0.03 lb/MMBtu | 1109 |
| Glass-melting furnaces | 4 lb/ton of glass | 1117 |
| Gas turbines (no SCR) | 12 ppm (15% O$_2$) | 1134 |
| Gas turbines (SCR) | 9 ppm (15% O$_2$) | 1134 |
| Others | Best available current technology | |

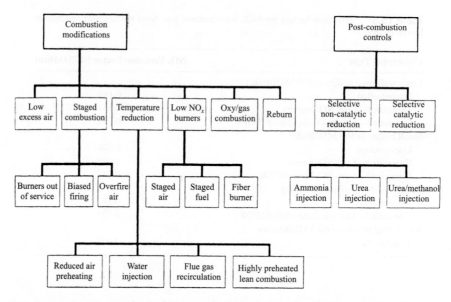

**Figure 15.13**    The NO$_x$ control technologies for gas–fired industrial combustion equipment.
SOURCE: Adapted from Ref. [71] with permission of the Gas Research Institute.

***Low Excess Air***    Thermal NO$_x$ emissions peak at leaner than stoichiometric equivalence ratios (see Fig. 15.3). This NO$_x$ reduction technique involves reducing the air supplied to move down the NO$_x$–$\Phi$ curve from the peak toward stoichiometric. Only limited NO$_x$ reductions are possible with this method since CO emissions rise as the amount of excess air is decreased.

***Staged Combustion***    In this method of NO$_x$ control, operation of existing burners in a multiburner device is modified to create, typically, a rich–lean staging of combustion. This is achieved by having upstream burners operate rich, while adjusting downstream burners to supply air only; or by adjusting some burners to run rich and some lean; or by adjusting all burners to operate rich, with additional air supplied through downstream ports. The NO$_x$ reductions attainable with these techniques range from 10 to 40 percent [71].

***Temperature Reduction***    In many combustion devices, the combustion air is preheated by the hot exhaust gases to improve thermal efficiency. Reducing the amount of air preheat reduces combustion temperatures and, consequently, NO$_x$ formation. Injecting water reduces flame temperatures because combustion energy is used to vaporize and superheat the water to combustion temperatures. In concept, water injection is the same as flue-gas recirculation (FGR) in that both act as diluents. The effectiveness of FGR depends on both the quantity and temperature of the recirculated gas. Figure 15.14 shows the effects of FGR on NO$_x$ for burners operating with ambient and 500 °F (533 K) combustion air [71]. NO$_x$ reductions from approximately 50 to 85 percent are possible with FGR in gas–fired industrial boilers [71].

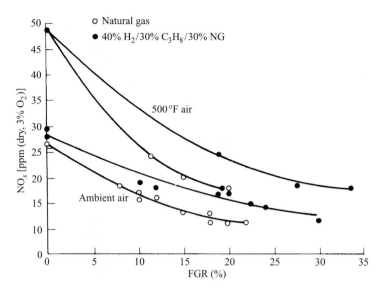

**Figure 15.14**    Effects of fuel-gas recirculation (FGR) on NO$_x$ emissions from staged-fuel burners using ambient or preheated air.
SOURCE: Courtesy of John Zink Co., Tulsa, OK.

Burners have been developed that recirculate product gases directly from within the furnace into the flame or reactant streams, as opposed to externally recirculating a portion of the flue gases [73, 74, 76]. When very large quantities of products are directly recycled, so-called flameless operation can be achieved [73, 74]. A related method to achieve low-temperature combustion is to operate the burner lean and highly diluted with products. The heat feedback from the hot gases allows combustion reactions to proceed at stoichiometries that would not normally be flammable [76]. Several variations of this concept have been identified [78].

*Low-NO$_x$ Burners*    Burners designed for low NO$_x$ emissions employ fuel or air staging. Fuel staging creates a sequential lean–rich (actually, less lean) combustion process (Fig. 15.15a), whereas air staging creates a rich–lean process (Fig. 15.15b). Figure 15.16 illustrates the relatively complex geometry of a commercial low-NO$_x$ burner. Flow paths here are much more complex than suggested by the simple schematics of Fig. 15.15. Low-NO$_x$ burners are a mature technology and, with the aid of advances in computational fluid dynamics, have undergone several generations of development in the past 10 to 15 years [79]. Another class of low-NO$_x$ burners are the fiber-matrix burners. These burners employ premixed combustion above or within a metal or ceramic fiber matrix. Because of radiation and convection heat transfer from the matrix, combustion temperatures are quite low, as are NO$_x$ emissions. A fiber burner is illustrated in Fig. 15.17.

*Oxy/Gas Combustion*    The concentration of nitrogen in the combustion system can be reduced by supplying additional oxygen to the combustion air. With sufficiently

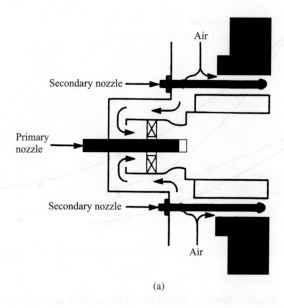

(a)

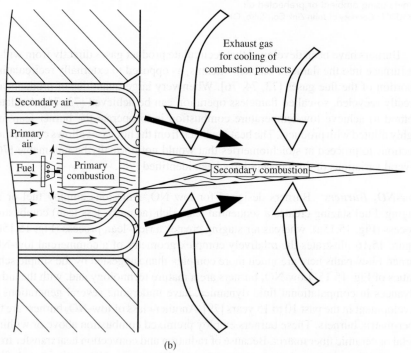

(b)

**Figure 15.15**     (a) Low-NOₓ burner employing fuel staging to achieve sequentially lean and rich combustion zones. From Ref. [72]. (b) Low-NOₓ burner employing air staging to achieve sequentially rich and lean combustion zones. The schematic also illustrates the use of the secondary air to entrain burned gases into the secondary combustion zone. Figure from Ref. [74]. Reprinted with permission from Elsevier.

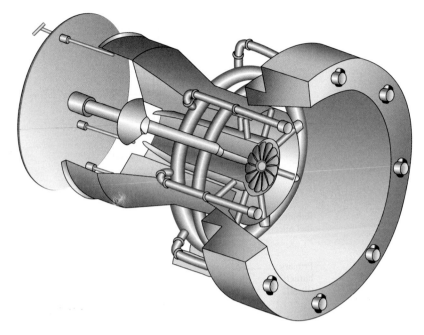

**Figure 15.16**      Commercial low-NO$_x$ burner. Drawing courtesy of Coen Company, Inc.

large O$_2$ additions, the decreased N$_2$ concentration outweighs the increased combustion temperatures and NO$_x$ levels can be reduced. If air leaks into the combustion chamber can be prevented, operation with pure O$_2$ ideally eliminates all NO$_x$ production, assuming that no nitrogen is contained in the fuel. See Refs. [73] and [75] for more information.

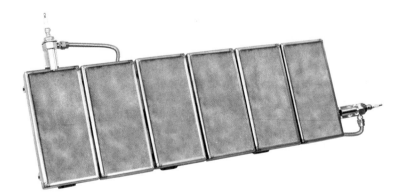

**Figure 15.17**      Low-NO$_x$ radiant burner used for moisture removal in textile and paper ovens, paint drying, and powder coating, as well as many pre-heat, plastic forming, heat treating and annealing operations.
SOURCE: Courtesy of Maxon Corporation.

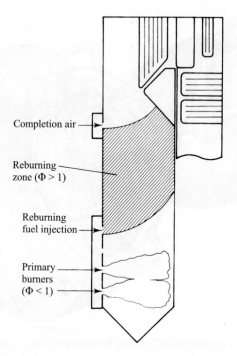

**Figure 15.18**    Industrial boiler equipped with reburn $NO_x$ control.
| SOURCE: After Ref. [81].

***Reburn***    In this method of $NO_x$ control [80], about 15 percent of the total fuel is introduced downstream of the main, fuel-lean combustion zone. Within the reburning zone
($\Phi > 1$), NO is reduced via reactions with hydrocarbons and hydrocarbon intermediates, such HCN, similar to those involved in the Fenimore mechanism. Additional
air is then supplied to provide the final burnout of the reburn fuel. Reductions of
$NO_x$ of about 60 percent are typical for boilers employing reburn technology [81].
The reburn process is schematically shown in Fig. 15.18. Additional information on
reburn chemistry and physical processes can be found in Refs. [27, 82–85].

***Selective Non-Catalytic Reduction (SNCR)***    In this post-combustion control technique, a nitrogen-containing additive, either ammonia, urea, or cyanuric acid, is injected and mixed with flue gases to effect chemical reduction of NO to $N_2$ without
the aid of a catalyst. The method that utilizes ammonia, discovered by Lyon [86], is
frequently referred to as the thermal de-$NO_x$ process. Temperature is a critical variable, and operation within a relatively narrow range of temperatures is required to
achieve large $NO_x$ reductions. Figure 15.19 illustrates this point [27]. With imperfect
mixing and nonuniformity of temperature in an actual exhaust stream, $NO_x$ reductions in practice are somewhat less than the maximum shown in Fig. 15.19 for a
laboratory-scale reactor. Additional information on SNCR can be found in Refs. [27,
32, 87–93].

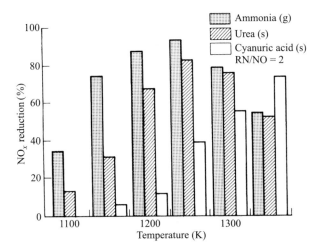

**Figure 15.19**    The $NO_x$ reduction effectiveness for selective non-catalytic reduction (SNCR) techniques.
SOURCE: Reprinted from Ref. [27] by permission of The Combustion Institute.

***Selective Catalytic Reduction (SCR)***    In this technique, a catalyst, typically $V_2O_5 - TiO_2$ [93] is used in conjunction with ammonia injection to reduce NO to $N_2$. The temperature window for effective reduction depends upon the catalyst used, but is contained within the range of about 480 K (400 °F) to 780 K (950 °F) [71]. The advantage of SCR over SNCR is that greater $NO_x$ reductions are possible, and the operating window is at lower temperatures. Costs of $NO_x$ removal ($/ton) with SCR are generally the highest of all $NO_x$ control techniques because of both the initial high cost and the operating costs associated with catalyst replacement. Because of stringent regulations, SCR technology is used extensively in Japan and Germany, and recently has seen more widespread use in the United States [93]. Sources of additional information on selective catalytic reduction are Refs. [27, 93–96].

**Utility Boilers**    As a class, electric utility boilers emitted approximately 22 percent of the $NO_x$ produced by combustion sources in the United States in 2002 [97], and they are subject to some of the most stringent emission standards worldwide [27]. In the United States in 2006, coal-fired boilers produced 69 percent of the total electricity generated, followed by natural gas–fired boilers with 28.2 percent, and only 2.2 percent generated from oil-fired boilers [98] (see Table 1.1 in Chapter 1).

Burning coal and heavy oils provides an additional source of $NO_x$ over devices burning natural gas and light distillate fuels because of the bound nitrogen in the fuel. Table 15.7 compares the nitrogen content of coal and liquid fuels. Light distillate fuels have small amounts of nitrogen in comparison with both coal and heavy distillate fuels, which can contain up to a few percent nitrogen by weight.

For gas-fired units, all of the combustion modification techniques and aftertreatment methods discussed in the previous section can be employed for $NO_x$ reduction. For oil- and coal-fired boilers, combustion modifications which tend to limit $O_2$ availability,

**Table 15.7**    Typical bound nitrogen content of coals and distillate fuels

| Fuel | Average Nitrogen, wt. % | Range, wt. % |
|---|---|---|
| Coal and coal-derived fuels[a] | 1.3 | 0.5–2.0 |
| Crude oil[b] | 0.65 | 0.2–3.0 |
| Heavy distillates[b] | 1.40 | 0.60–2.15 |
| Light distillates[b] | 0.07 | 0.002–0.60 |
| Natural gas | Nil | — |

[a] SOURCE: From Ref. [27].
[b] SOURCE: From Ref. [99].

e.g., low excess air and staged combustion, reduce both thermal and fuel $NO_x$, whereas temperature-reduction techniques (e.g., FGR and water injection) have their effect primarily on thermal $NO_x$ [102]. In oil- and coal-fired units, about 20–40 percent of the fuel-N is converted to $NO_x$ that appears in the flue gases, resulting in as much as half of the total $NO_x$ emitted [99]. In coal-burning systems that operate at relatively low temperatures (<1725 K), such as fluidized bed combustors, $NO_x$ emissions can be dominated by fuel-N conversion with contributions ranging from 75 to 95 percent [100, 101]. Also, the greenhouse gas nitrous oxide ($N_2O$) can be formed in low-temperature coal combustion, with the magnitude of its emission increasing with increasing coal rank (see Chapter 17) [100, 101]. Application of post-combustion $NO_x$ reduction techniques (SNCR and SCR) is complicated by the sulfur in coal (see Chapter 17) and in residual oils, and by particulate matter [93, 96, 103]. Ammonia used in either SNCR or SCR reacts with $SO_3$ to form ammonium bisulfate ($NH_4HSO_4$), an extremely corrosive substance [93, 95, 96, 103]. Poisoning of catalysts by $SO_3$ and plugging of the catalyst surface by particulate matter make application of SCR much more difficult for oil and coal than for natural gas [93, 95, 101]. Srivastava *et al.* [93] report that prior to about 2005, SNCR had been installed at 36 and SCR at more than 150 coal-fired utility boilers in the United States. Techniques are also being developed that simultaneously remove both $NO_x$ and $SO_2$, and in some cases, Hg as well [93].

**Diesel Engines**    Thus far in our discussion of $NO_x$ emissions from nonpremixed combustion systems, we have focused on stationary combustion devices that operate essentially at atmospheric pressure. Diesel engines, in contrast, may be mobile or stationary, and combustion occurs at high pressures. For example, advanced turbocharged, heavy-duty, on-road diesel engines have peak combustion pressures in the range of 220 to 250 atm [104]. Compression of air from atmospheric conditions to these high pressures results in a considerable temperature rise. For example, adiabatic compression of 300 K air from 1 atm to 200 atm results in a temperature of approximately 1360 K. Thus, peak combustion temperatures are high, and consequently, thermal (Zeldovich) NO formation rates are very rapid. In diesel engines, NO formation is complex because of the unsteady mixing of fuel and air and subsequent mixing of combustion products with air and fuel. A simplified view is that NO is formed in the high-temperature, near-stoichiometric regions of the burning mixture [105, 106]. Therefore, combustion modifications employed to reduce $NO_x$ are those that lower temperatures. Exhaust gas recirculation systems are continually being refined to get increasingly lower engine-out $NO_x$ levels [107–109]. State-of-the-art systems employ cooling loops to reduce the temperature of the exhaust gases before they are mixed with the air [109].

Other technologies employed to control $NO_x$ and other emissions from diesel engines include advanced fuel-injection systems capable of multiple injections, variable valve actuation, and closed-loop combustion control [107, 108].

To meet stringent current and proposed emission standards in Europe, the United States, and Japan, post-combustion controls are necessary. Three post-combustion systems have been developed [107, 108]: (1) urea-based selective catalytic reduction (SCR), (2) lean $NO_x$ traps, and (3) lean $NO_x$ catalysts.

Selective catalytic reduction has been applied to commercial vehicles for the past several years [104]. Issues associated with SCR applied to diesel engines are the relatively low $NO_x$ reductions achieved in light-duty test cycles [104]; the potential for unreacted ammonia emissions (ammonia slip); and the need to have a tank for the liquid reagent, the associated plumbing and hardware, and the control system. Avoiding freezing of the reagent is also a concern.

Lean $NO_x$ traps, also variously known as $NO_x$ adsorber catalysts, $NO_x$ storage catalysts, and $NO_x$ sorbate catalysts, avoid the use of a reagent. The basic design is similar to three-way catalysts used in spark-ignition engine applications; however, the ceramic honeycomb matrix is now coated with a washcoat containing both precious metal catalysts particles (Pt and Ph) and alkalis (e.g., BaO and $K_2CO_3$). During normal (lean) operation of the diesel engine, NO in the exhaust stream is oxidized to $NO_2$ by the precious metal catalyst, and the $NO_2$ in turn is adsorbed on the alkali [110]. Details of the chemistry are poorly understood [110], although progress is being made [110, 111]. After some hours of operation, the lean $NO_x$ trap must be regenerated. During regeneration, the alkali compounds release the adsorbed $NO_x$, which is then converted to $N_2$ at the precious metal sites, similar to the operation of a three-way catalyst. Regeneration occurs when rich conditions are momentarily (~2 seconds) obtained, for example, by a separate late injection of fuel in the engine [112]. An oxidation catalyst ahead of the lean $NO_x$ trap can be used to eliminate any oxygen in the incoming rich mixture [112]. After regeneration, the trap resumes operation as a $NO_x$ adsorber. Adding complexity to this system is the need to deal with sulfur poisoning of the catalyst [110, 112–114]. The latest requirements for diesel fuel specify a maximum sulfur content of 15 ppm (see Chapter 17), and techniques are being developed to prevent sulfur from reaching the trap [113] or to periodically drive off any accumulated sulfur within lean $NO_x$ traps [112, 113].

Lean $NO_x$ catalysts are often considered another application of selective catalytic reduction (HC-SCR), where the reagent now is a hydrocarbon (fuel) rather than urea ($[NH_2]_2CO$) or cyanuric acid (HNCO); however, HC-SCR is not truly selective, as many multiples of the stoichiometric quantity of the hydrocarbon must be added to affect the $NO_x$ conversion [115]. Although older catalyst formulations had low effectiveness, recent developments show $NO_x$ conversion efficiencies greater than 80 percent [104, 116]. Lean $NO_x$ catalysts offer potential lower catalyst costs as zeolites, silver, and other relatively inexpensive catalysts can be used; however, the use of fuel as a reagent generates a fuel economy penalty [107].

**Gas-Turbine Engines**   Like diesel engines, gas-turbine engines are used in both stationary and mobile applications. Similarly, they operate at high pressures. For stationary gas turbines employed for power generation, typical combustion pressures are in the range of 10–15 atm, whereas aircraft-engine combustors operate in a range of 20–40 atm. Residence times range from 10–20 ms in stationary applications and range over a few milliseconds in aircraft combustors [117]. To control $NO_x$ from stationary engines, both combustion modifications and post-combustion controls are used. The use of selective catalytic reduction (SCR) is a mature technology for stationary engines [118]; however, for aircraft engines post-combustion control is not practicable. To achieve combustion control of $NO_x$ in both types of applications, various degrees of premixing and combustion staging are used to reduce high-temperature regions in the combustor. Figure 15.20 illustrates the range of $NO_x$ emissions that might be expected in the operation of lean-premixed stationary engines [119, 120]. Development of a lean-premixed combustion system for power generation gas-turbine engines (see Fig. 15.21) is described in Refs. [121–123]. Engineering issues associated with lean-premixed are avoiding flashback and autoignition, maintaining combustion stability, and being able to operate over a sufficiently wide range of loads [119]. Understanding combustion stability is an active area of combustion research [124]. Because of safety considerations, low emissions are secondary considerations in aviation engines; however, standards have been established by the International Civil Aviation Organization (ICAO) [26],

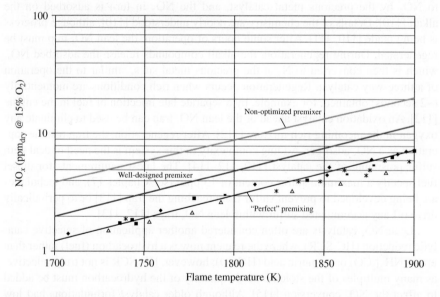

**Figure 15.20**   For a given mean flame temperature, $NO_x$ emissions decrease as the quality of the premixing improves in lean-premixed gas-turbine combustors. Adapted from Refs. [119] and [120] and used with permission of Elsevier.

**Figure 15.21**     Natural gas–fired General Electric 9H gas-turbine engine for stationary power generation has an annular array of combustors. The 9H engine has a power output of 480 MW. *Photograph courtesy of General Electric.*

which are typically mirrored in the regulations of various countries, e.g., the U.S. Environmental Protection Agency. To meet these standards, considerable combustor development has taken place. Techniques for $NO_x$ control [119, 125, 126] include rich-burn, quick-mix, lean-burn (RQL) combustion (see Fig. 15.22); variations on lean-premixed combustion (LPP); lean direct injection (LDI); and various forms of staged combustion. Figure 15.23 schematically illustrates the various mixing and combustion zones in a twin annular premixing swirler (TAPS) combustor [126, 127].

## Unburned Hydrocarbons and Carbon Monoxide

In nonpremixed combustion systems, there are two sources of unburned hydrocarbons and carbon monoxide that result directly from the nature of nonpremixed combustion.

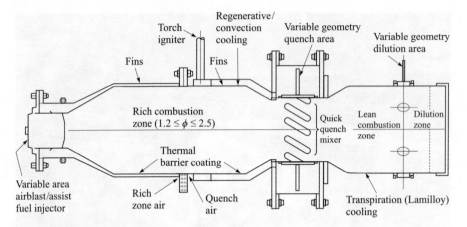

**Figure 15.22**    Sketch of rich / quench / lean (RQL) staged gas-turbine combustor. Very rapid mixing of the rich combustion products with air is required to keep peak temperatures low (cf. Fig. 15.6).
SOURCE: From Ref. [125]. Reprinted with permission of the authors.

First, overly lean regions are created within the combustion chamber that do not support rapid combustion. Fuel injector characteristics and fuel–air mixing patterns are important parameters in this mechanism. Since a normal flame does not propagate through overlean regions, fuel pyrolysis and partial oxidation products are formed. Among these are oxygenated species, such as aldehydes, and carbon monoxide. The overlean mechanism is particularly important at light loads when the amount of excess air is great.

The second source of incomplete combustion products directly related to non-premixed combustion is the creation of overly rich regions that subsequently do not mix with sufficient additional air, or, if they do, there is insufficient time for oxidation reactions to go to completion. Overrich regions are more likely to be a source of unburned and partially burned species at heavy loads, where excess air levels are low.

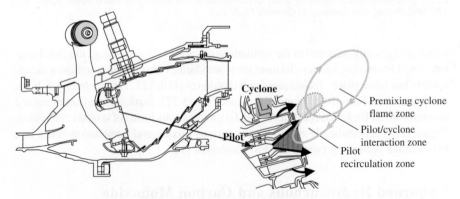

**Figure 15.23**    Multiple fuel-injection zones provide staged combustion and the ability to maintain a wide turndown ratio in the twin annular premixing swirler (TAPS) combustor. Adapted from Ref. [126] and used with permission.

Additional mechanisms that can result in unburned and partially burned species are the following:

- Wall quenching (diesel engines).

- Quenching by secondary or dilution air jets (gas-turbine engines).

- Fuel dribble from the volume between the injector valve seat and the hole exposed to the combustion space, i.e., the nozzle sac volume (diesel engines).

- Occasional formation of very large rogue fuel droplets (gas-turbine engines).

Whether or not these mechanisms contribute to emissions depends to a large degree on specific details of hardware design, particularly the fuel-injection system. For example, fuel issuing from the nozzle sac volume can be a major contributor to diesel engine hydrocarbons and is affected only by changes in nozzle design [128].

In addition to unburned hydrocarbons, partially burned species are emitted from nonpremixed systems without emission controls. These species (aldehydes, ketones, etc.) typically are quite reactive in photochemical smog, act as eye irritants, and are odorous. The wide variety of diesel exhaust odors result from the different compositions of oxygenates produced under various operating conditions [129].

## Particulate Matter

With the exception of the mineral-matter ash and unreacted char produced by coal burning, which is removed from the effluent stream by electrostatic precipitators, baghouse filters, or other means, the primary particulate matter produced in nonpremixed combustion is soot. Formation of soot can be considered an intrinsic property of most diffusion flames. In Chapter 9, we saw that soot is formed in the rich regions of diffusion flames, and whether or not soot is emitted from a flame depends upon competition between soot formation and soot oxidation processes. Combustion system modifications to minimize soot thus can act to reduce the amount of soot produced prior to oxidation and/or to increase oxidation rates.

The particulate matter issued from diesel engines is predominantly carbonaceous (soot), combined with other substances. Figure 15.24 illustrates the typical composition of particulate matter emitted by a heavy-duty diesel engine. Air quality studies have used the large carbon fraction as a signature to distinguish particulate matter emitted by diesel engines from that emitted by spark-ignition engines [57, 58]. For spark-ignition engines, particulate matter predominately comprises condensed hydrocarbons, as discussed previously. For diesel engines, simultaneous in-cylinder control of $NO_x$ and particulate emissions has been a longstanding challenge. Combustion modifications to reduce $NO_x$ frequently result in increased particulate matter (PM) emissions, and vice versa. This situation is referred to as the $NO_x$–PM trade-off and is graphically shown by plotting PM emissions versus $NO_x$ emissions to obtain a trade-off curve [104]. Regulations are sufficiently stringent that exhaust aftertreatment devices are needed to meet legislated levels for particulate matter, and $NO_x$, as discussed above. To minimize the load on aftertreatment devices, engine designers seek to provide the lowest practicable engine-out emissions. Combustion

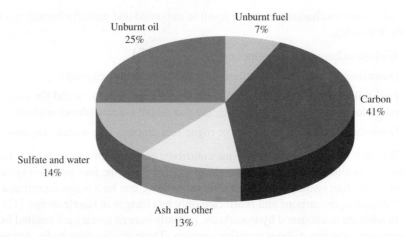

**Figure 15.24**    Typical composition of particulate matter emitted from heavy-duty diesel engine operating on a heavy-duty test cycle. From Ref. [56] and used with permission of Elsevier.

chamber configuration, injector design, and injection system controls are all critical to minimizing particulate emissions. These design factors affect $NO_x$ emissions as well, hence, the $NO_x$–PM trade-off. Diesel particulate filters (see Fig. 15.25) have been developed to remove the particulate matter from the exhaust stream [104, 107, 108]. After particles have built up in the filter, they must be burned off to prevent

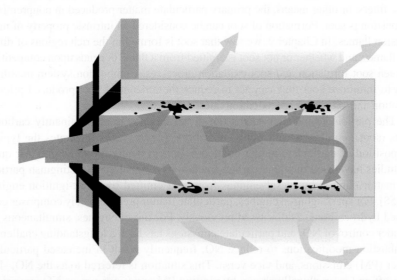

**Figure 15.25**    In a typical ceramic honeycomb diesel particulate filter, particulate matter–laden exhaust enters porous-walled ceramic tubes closed at their opposite ends. The particles accumulate on the inner surface of the tubes, allowing the clean exhaust to pass through the walls. The particulate matter is then burned off by one of several means. A catalyst within the wall is one method to promote oxidation of the accumulating particulate matter. Drawing adapted from U.S. Environmental Protection Agency.

further increases in pressure drop across the filter. Methods to accomplish this regeneration include (1) external means, such as fuel burners, electric heating, injection of fuel into the exhaust, etc., (2) engine operation, such as post-injection of fuel, injection timing retards, etc., and (3) catalytic processes, such as fuel-borne catalysts, catalytic filter coatings, and the generation of reactive species ($NO_2$) [108, 130]. For further information on diesel engine emission controls, we refer the interested reader to Ref. [131].

## Oxides of Sulfur

In combustion processes, all of the sulfur that is present in the fuel appears as $SO_2$ or $SO_3$ in the combustion products, the combination of the two being denoted $SO_x$. Because of this quantitative conversion of fuel sulfur, there are only two possible ways to control $SO_x$ emissions: remove the sulfur from the fuel, or remove the $SO_x$ from the product gases. Both of these techniques are used in varying degrees in practice. Table 15.8 provides estimates of the sulfur content of various fuels. Coal and residual oils are particularly high in sulfur, whereas there is very little sulfur in unleaded gasoline. Regulations in the United States and elsewhere limit the amount of sulfur in transportation fuels (see Chapter 17).

The amount of $SO_3$ produced is typically only a few percent of the amount of $SO_2$, although the $SO_3$ is usually found in greater than equilibrium concentrations. Sulfur trioxide readily reacts with water to form sulfuric acid ($SO_3 + H_2O \rightarrow H_2SO_4$); thus, sulfuric acid is formed in exhaust streams because of the simultaneous presence of $SO_3$ and $H_2O$. In addition to the obvious deleterious effects of producing sulfuric acid, $SO_3$ also poisons automotive three-way catalysts; thus, sulfur levels are required to be low in gasolines (see Table 15.8).

The fate of $SO_2$ in the atmosphere is for it to be oxidized by, primarily, OH radicals in gas-phase reactions, or after it has been absorbed on a particle or droplet [5]. Subsequent reaction with water then produces sulfuric acid.

The most commonly used method of removing $SO_x$ from flue gases involves reacting $SO_2$ with limestone ($CaCO_3$) or lime ($CaO$) [136, 137]. In this control method,

**Table 15.8**     Sulfur content of various fuels

| Fuel wt. % | Range % |
| --- | --- |
| Coal | $\leq 10$ |
| Heavy residual oil[a] | 0.5–4 |
| Blended residuals and crudes[a] | 0.2–3 |
| Diesel fuel (No. 2)[b] | 0.0015–0.50 |
| Unleaded gasoline[c] | 0.008 |

[a] Ref. [132].
[b] Ref. [133].
[c] Ref. [134].

an aqueous slurry of limestone or lime is sprayed in a tower through which the flue gases pass. The overall reactions for the process are, for limestone,

$$CaCO_3 + SO_2 + 2H_2O \rightarrow CaSO_3 \cdot 2H_2O + CO_2, \qquad (15.16)$$

and, for lime,

$$CaO + SO_2 + 2H_2O \rightarrow CaSO_3 \cdot 2H_2O. \qquad (15.17)$$

The process can be either **wet** or **dry** depending upon whether the reactions occur in solution (wet) or with particles from which the water has evaporated (dry). In a wet process, the final product, calcium sulfite dihydrate ($CaSO_3 \cdot 2H_2O$), precipitates from a retention tank and is disposed of in a pond or landfill. In a dry process, electrostatic precipitators are used to remove the particulate product. Other methods of $SO_x$ control are discussed in Refs. [100, 135, 136]. In 2005, 248 power plants in the United States employed $SO_x$ scrubbers. These units supplied approximately 25 percent of the total fossil-fuel generating capacity.

## Greenhouse Gases

Climate change associated with human activities presents a formidable challenge for the 21st century. A major contributor to climate change is the emission of so-called greenhouse gases. The principal greenhouse gases are carbon dioxide ($CO_2$), methane ($CH_4$), nitrous oxide ($N_2O$), and various fluorinated species. The first three of these are associated with the combustion of fossil fuels. Table 15.9 shows the lifetimes of greenhouse gases in the atmosphere and their global warming potentials (GWP) for a 20-year time horizon [137]. HCFC-22 and HFC-134a are commonly used refrigerants. The GWP of $CO_2$ is assigned a value of unity [137]. On a per-molecule basis, methane and nitrous oxide have GWPs much greater than $CO_2$, and the fluorinated species have GWPs thousands of times greater than $CO_2$. The relatively long lifetimes of the greenhouse gases indicates that any reductions in emissions will take years to be felt. Figure 15.26 shows the contributions of the various greenhouse gases in the United States from 1990 to 2006 [138]. Here we see the dominant contribution of $CO_2$.

**Table 15.9**     Atmospheric lifetimes and global warming potentials of greenhouse gases[a]

| Gas | Chemical Formula | Lifetime (years) | Global Warming Potential for 20-year Time Horizon |
|---|---|---|---|
| Carbon dioxide | $CO_2$ | 95[b] | 1 |
| Methane | $CH_4$ | 12 | 72 |
| Nitrous oxide | $N_2O$ | 114 | 289 |
| HCFC-22 | $CHClFH_2$ | 12 | 5,160 |
| HFC-134a | $CH_2FCF_3$ | 14 | 3,830 |

[a] Values from Ref. [137] except as noted.
[b] Lifetime calculated as pseudo $1/e$-decay time from formula presented in Ref. [137].

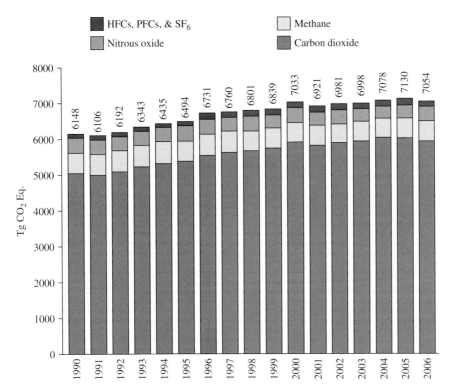

**Figure 15.26**   Contributions of the various greenhouse gases expressed at teragrams of $CO_2$ equivalent (see Table 15.9). Adapted from U.S. EPA [138].

A simplistic view of the role of greenhouse gases in climate change is that they trap outgoing infrared radiation from the earth and, hence, result in heating of the earth's surface. The detailed physics, however, is quite complex. Complicating factors include the complex nature of the spectral absorption and emission of each of the participating species, the radiative interaction of the greenhouse gases with the naturally occurring water vapor in the atmosphere, and the temperature distribution through the atmosphere. For readers desiring more information, Refs. [139] and [140] are recommended.

The source of $CO_2$ from combustion is obvious: the efficient combustion of a carbon-containing fuel converts essentially all of the fuel carbon to $CO_2$. Table 15.10 shows the $CO_2$ emission indices and fuel energy-based emission factors for several fuels. The fuels are ranked on the basis of their energy-based $CO_2$ emission factors, with coal having the highest value ($0.0948$ $kg_{CO_2}/MJ$), which is about 50% larger than the value for natural gas ($0.0494$ $kg_{CO_2}/MJ$). A coal-fired power plant switching to natural gas would result in a substantial reduction in the $CO_2$ emitted, assuming the same overall efficiencies. The rankings in Table 15.9 reflect the carbon-to-hydrogen ratio of the fuels—as this ratio goes down, so does the $CO_2$ emission factor. Because hydrogen contains no carbon, its combustion produces no $CO_2$. Sources of $CO_2$ in the United States by fuel type and end use are shown in Fig. 15.27 [141].

**Table 15.10**     Carbon dioxide emission indices and emission factors for various fuels

| Fuel | Actual or Equivalent Composition | Molecular Weight (kg/kmol) | Higher Heating Value (MJ/kg) | $CO_2$ Emission Index ($kg_{CO_2}/kg_{fuel}$) | $CO_2$ Emission Factor ($kg_{CO_2}/MJ$) |
|---|---|---|---|---|---|
| Coal (Pittsburgh #8) | $C_{65}H_{52}NSO_3$ | 927.18 | 32.55 | 3.08 | 0.0948 |
| Diesel | $C_{12.3}H_{22.2}$ | 170.1 | 44.8 | 3.18 | 0.0710 |
| Gasoline | $C_{7.9}H_{14.8}$ | 109.8 | 47.3 | 3.17 | 0.0669 |
| Ethanol | $C_2H_5OH$ | 46.06 | 29.7 | 1.91 | 0.0643 |
| Natural gas | $C_{1.16}H_{4.32}N_{0.11}$ | 19.83 | 50.0 | 2.57 | 0.0515 |
| Methane | $CH_4$ | 16.04 | 55.53 | 2.74 | 0.0494 |
| Hydrogen | $H_2$ | 2.016 | 142.0 | 0 | 0 |

The transportation and electricity sectors dominate, and the total $CO_2$ emissions from coal and petroleum are approximately the same.

Primary combustion-related sources of methane emissions in the United States include natural gas and petroleum systems (133.5 tg $CO_2$-equivalent per year), stationary combustion (6.6 tg $CO_2$-equivalent per year), and mobile combustion (2.3 tg $CO_2$-equivalent per year) [142]. The $CH_4$ emissions from these three sources represent 24.3 percent of the total U.S. emission in 2006 [142]. The first of these sources include the production, processing, storage, transmission, and distribution of natural

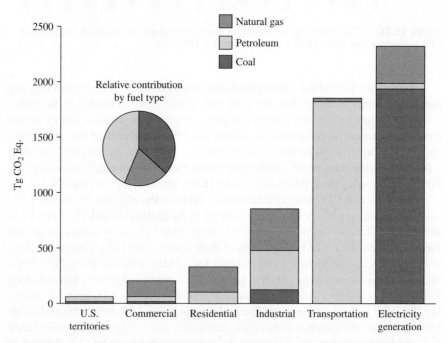

**Figure 15.27**     U.S. $CO_2$ emissions from fossil fuel combustion by sector and fuel type in 2006. Units are teragrams. Adapted from U.S. EPA [141].

gas and petroleum. Simple leaks are a major factor here. The largest human-related sources are enteric fermentation, which results in the emission of $CH_4$ in the breath of ruminants (139.0 tg $CO_2$-equivalent per year), and the escape of $CH_4$ from landfills (132.9 tg $CO_2$-equivalent per year) [142].

Nitrous oxide is emitted by a variety of natural and anthropogenic sources. The single largest source in the United States is agricultural soil management, with the emission of 207.9 tg $CO_2$-equivalent in 2006, contributing 66.7 percent of the total $N_2O$ emissions [143]. Principal combustion-related sources of $N_2O$ are mobile combustion (30.1 tg $CO_2$-equivalent per year), stationary combustion (14.7 tg $CO_2$-equivalent per year), and incineration of waste (0.4 tg $CO_2$-equivalent per year) [143]. These combined represent approximately 14.5 percent of the U.S. total $N_2O$ emission [143]. For mobile combustion, the $N_2O$ results from trace emission from three-way catalytic converters [144, 145], with emissions increasing with the age of the catalyst [145]. For stationary combustion, $N_2O$ emissions from pulverized coal-fired power plants are typically less than 1% of the $NO_x$ emissions [146]. Higher levels of $N_2O$, however, are associated with both fluidized bed combustors and refuse incineration systems [147].

There are no sources for halocarbon emission directly related to combustion. Sources of these gases include the production and use of refrigerants, the production and / or the processing of aluminum and magnesium, the manufacturing of semiconductors, and electrical power transmission ($SF_6$) [148].

Controlling the emission of greenhouse gases into the atmosphere is a monumental task. Pacala and Socolow [149] offer many measures that, when combined, have the potential to greatly reduce greenhouse gas emissions. Because stationary combustion is a large emitter of carbon dioxide, carbon capture and storage is likely to be an important technology in the near future. The United Nations Intergovernmental Panel on Climate Change has published a book devoted to this topic [150].

## SUMMARY

In this relatively long chapter, we discussed first the implications of pollutant emissions from combustion systems. You should have a knowledge of the major effects of the primary pollutants, $NO_x$, CO, unburned and partially burned hydrocarbons, particulate matter, and $SO_x$; that is, what is it that makes these species pollutants? Next we investigated the various ways used to quantify pollutants. The bases for all of the measures are simple species or atom balances and mass conservation. You should understand why some measures are more useful than others depending on the application, and be able to convert from one measure to another. The next major topics dealt with the formation and control of emissions; first, from premixed systems, and second, from nonpremixed systems. The chapter concluded with a discussion of the role of combustion in the emission of greenhouse gases. Much of the material concerning pollutant formation relies on previous chapters, particularly sections of Chapters 4, 5, 8, and 9. Review of that material would be appropriate now to consolidate your understanding. Although not comprehensive, many practical applications were also discussed in this chapter. You should be able to discuss the most important pollutant formation and control mechanisms and how they apply to spark-ignition engines, gas–fired industrial combustion equipment, utility boilers, and diesel and gas-turbine engines.

# NOMENCLATURE

| | |
|---|---|
| $c_p$ | Specific heat (J/kg-K) |
| $d_j$ | Jet diameter (m) |
| $EI$ | Emission index (kg/kg or related) |
| $f$ | Mixture fraction (kg/kg) |
| $L$ | Length (m) |
| $m$ | Mass (kg) |
| $\dot{m}$ | Mass flowrate (kg/s) |
| $MW$ | Molecular weight (kg/kmol) |
| $N$ | Number of moles (kmol) |
| $P$ | Pressure (Pa) |
| $R_u$ | Universal gas constant (J/kmol-K) |
| $t$ | Time (s) |
| $T$ | Temperature (K) |
| $v_e$ | Exit velocity (m/s) |
| $W$ | Width (m) |
| $\dot{W}$ | Power (W) |
| $x$ | Number of carbon atoms in fuel molecule |
| $y$ | Number of hydrogen atoms in fuel molecule |

*Greek Symbols*

| | |
|---|---|
| $\Delta h_c$ | Heat of combustion (J/kg) |
| $\rho$ | Density (kg/m$^3$) |
| $\tau_G$ | Global residence time (s) |
| $\Phi$ | Equivalence ratio |
| $\chi$ | Mole fraction (kmol/kmol) |
| $\chi_R$ | Radiant fraction |

*Subscripts*

| | |
|---|---|
| $ad$ | Adiabatic |
| $c$ | Critical for soot production |
| $e$ | Equilibrium |
| $f$ | Flame |
| $F$ | Fuel |
| $i$ | Species $i$ |
| mix | Mixture |
| stoic | Stoichiometric |

*Other Symbols*

| | |
|---|---|
| $[X]$ | Molar concentration of species $X$ (kmol/m$^3$) |

# REFERENCES

1. Code of Federal Regulations, "National Primary and Secondary Ambient Air Quality Standards," Title 40, Vol. 2, Part 50, U.S. Government Printing Office, July 1997.

2. U.S. Environmental Protection Agency, National Ambient Air Quality Standards (NAAQS), http://www.epa.gov/air/criteria.html, updated 14 July 2009.

3. Koshland, C. P., "Impacts and Control of Air Toxics from Combustion," *Twenty-Sixth Symposium (International) on Combustion,* The Combustion Institute, Pittsburgh, PA, pp. 2049–2065, 1996.

4. U.S. Environmental Protection Agency, Modifications to the 112(b)1 Hazardous Air Pollutants, http://www.epa.gov/ttn/atw/pollutants/atwsmod.html, updated 24 July 2007.

5. Prather, M. J., and Logan, J. A., "Combustion's Impact on the Global Atmosphere," *Twenty-Fifth Symposium (International) on Combustion,* The Combustion Institute, Pittsburgh, PA, pp. 1513–1527, 1994.

6. Haagen-Smit, A. J., "Chemistry and Physiology of Los Angeles Smog," *Industrial Engineering Chemistry,* 44: 1342–1346 (1952).

7. Seinfeld, J. H., *Atmospheric Chemistry and Physics of Air Pollution,* John Wiley & Sons, New York, 1986.

8. Solomon, S., *et al.* (eds.), *Climate Change 2007—The Physical Science Basis,* Contribution of Working Group I to the Fourth Assessment Report of the Intergovernmental Panel on Climate Change, Cambridge University Press, New York, 2007.

9. Hansen, J., *et al.,* "Global Temperature Change," *Proceedings of the National Academy of Science,* 103: 14,288–14,293 (2006).

10. Hansen, J., *et al.,* "Climate Change and Trace Gases," *Philosophical Transactions of the Royal Society A,* 365: 1925–1954 (2007).

11. Hansen, J., *et al.,* "Dangerous Human-Made Interference with Climate: A GISS ModelE Study," *Atmospheric Chemistry and Physics,* 7: 2287–2312 (2007).

12. Whelpdale, D. M., Summers, P. W., and Sanhueza, E., "Global Overview of Atmospheric Acid Deposition Fluxes," *Environmental Monitoring and Assessment,* 48: 217–247 (1977).

13. Irwin, J. G., and Williams, M. L., "Acid Rain: Chemistry and Transport," *Environmental Pollution,* 50: 29–59 (1988).

14. Rua, A., Gimeno, L., and Hernandez, E., "Relationships between Air Pollutants Emission Patterns and Rainwater Acidity," *Toxicological and Environmental Chemistry,* 59: 199–207 (1997).

15. Driscoll, C. T., *et al.,* "Acidic Deposition in the Northeastern United States: Sources and Inputs, Ecosystem Effects, and Management Strategies," *BioScience,* 51: 180–198 (2001).

16. Driscoll, C. T., Driscoll, K. M., Roy, K. M., and Mitchell, M. J., "Chemical Response of Lakes in the Adirondack Region of New York to Declines in Acid Deposition," *Environmental Science & Technology,* 37: 2036–2042 (2003).

17. Chestnut, L. G., and Mills, D. M., "A Fresh Look at the Benefits and Costs of the U.S. Acid Rain Program," *Journal of Environmental Management,* 77: 252–266 (2005).

18. Dockery, D. W., Schwartz, J., and Spengler, J. D., "Air Pollution and Daily Mortality: Associations with Particulates and Acid Aerosols," *Environmental Research,* 59: 362–373 (1992).

19. Dockery, D. W., *et al.,* "An Association between Air Pollution and Mortality in Six U.S. Cities," *New England Journal of Medicine,* 329: 1753–1759 (1993).

20. Samet, J. M., Dominici, F., Curriero, F. C., Coursac, I., and Zeger, S. L., "Fine Particulate Air Pollution and Mortality in 20 U.S. Cities, 1987–1994, *New England Journal of Medicine,* 343: 1742–1749 (2000).

21. Bell, M. L., and Davis, D. L., "Reassessment of the Lethal London Fog of 1952: Novel Indicators of Acute and Chronic Consequences of Acute Exposure to Air Pollution," *Environmental Health Perspectives,* 109 (supplement 3): 389–394 (2001).

22. Kennedy, I. M., "The Health Effects of Combustion-Generated Aerosols," *Proceedings of the Combustion Institute,* 31: 2757–2770 (2007).

23. Oberdörster, G., Oberdörster, E., and Oberdörster, J., "Nanotoxicology: An Emerging Discipline Evolving from Studies of Ultrafine Particles," *Environmental Health Perspectives,* 113: 823–839 (2005).

24. Stivender, D. L., "Development of a Fuel-Based Mass Emission Measurement Procedure." SAE Paper 710604, 1971.

25. Federal Register, "Final Regulations for Revisions to the Federal Test Procedure for Emissions from Motor Vehicles," Volume 61, No. 205, 22 October 1996. See http://www.epa.gov/EPA-AIR/1996/October/Day-22/pr-23769.txt.html.

26. International Civil Aviation Organization, *Environmental Protection, Annex 16 to the Convention on International Civil Aviation, Volume II, Aircraft Engine Emissions, International Standards and Recommended Practices,* 3rd Ed., International Civil Aviation Organization, Montreal, July 2008.

27. Bowman, C. T., "Control of Combustion-Generated Nitrogen Oxide Emissions: Technology Driven by Regulations," *Twenty-Fourth Symposium (International) on Combustion,* The Combustion Institute, Pittsburgh, PA, pp. 859–878, 1992.

28. Dagaut, P., Glarborg, P., and Alzueta, M. U., "The Oxidation of Hydrogen Cyanide and Related Chemistry," *Progress in Energy and Combustion Science,* 34: 1–46 (2008).

29. Glarborg, P., "Hidden Interactions—Trace Species Governing Combustion and Emissions," *Proceedings of the Combustion Institute,* 31: 77–98 (2007).

30. Dean, A., and Bozzelli, J., "Combustion Chemistry of Nitrogen," in *Gas-Phase Combustion Chemistry* (Gardiner, W. C., Jr., ed.), Springer, New York, pp. 125–341, 2000.

31. Drake, M. C., and Blint, R. J., "Calculations of $NO_x$ Formation Pathways in Propagating Laminar, High Pressure Premixed $CH_4$/Air Flames," *Combustion Science and Technology,* 75: 261–285 (1991).

32. Miller, J. A., and Bowman, C. T., "Mechanism and Modeling of Nitrogen Chemistry in Combustion," *Progress in Energy and Combustion Science,* 15: 287–338 (1989).

33. Glarborg, P., Miller, J. A., and Kee, R. J., "Kinetic Modeling and Sensitivity Analysis of Nitrogen Oxide Formation in Well-Stirred Reactors," *Combustion and Flame,* 65: 177–202 (1986).

34. Bachmeir, F., Eberius, K. H., and Just, Th., "The Formation of Nitric Oxide and the Detection of HCN in Premixed Hydrocarbon-Air Flames at 1 Atmosphere," *Combustion Science and Technology,* 7: 77–84 (1973).

35. Correa, S. M., and Smooke, M. D., "$NO_x$ in Parametrically Varied Methane Flames," *Twenty-Third Symposium (International) on Combustion,* The Combustion Institute, Pittsburgh, PA, pp. 289–295, 1990.

36. Rutar, T., Lee, J. C. Y., Dagaut, P., Malte, P. C., and Byrne, A. A., "$NO_x$ Formation Pathways in Lean-Premixed-Prevapourized Combustion of Fuels with Carbon-to-Hydrogen Ratios between 0.25 and 0.88," *Proceedings of the Institution of Mechanical Engineers, Part A: Journal of Power and Energy,* 221: 387–398 (2007).

37. Nebel, G. J., and Jackson, M. W., "Some Factors Affecting the Concentration of Oxides of Nitrogen in Exhaust Gases from Spark-Ignition Engines," *Journal of the Air Pollution Control Association,* 8: 213–219 (1958).

38. Komiyama, K., and Heywood, J. B., "Predicting $NO_x$ Emissions and Effects of Exhaust Gas Recirculation in Spark-Ignition Engines," SAE Paper 730475, 1973.

39. Quader, A. A., "Why Intake Charge Dilution Decreases NO Emissions from S. I. Engines," Paper 710009, *SAE Transactions,* 80: 20–30 (1971).

40. Dunn-Rankin, D. (ed.), *Lean Combustion: Technology and Control,* Academic Press, Burlington, MA, 2008.

41. Zhao, F., Asmus, T. W., Assanis, D. N., Dec, J. E., Eng, J. E., and Najt, P. M. (eds.), *Homogeneous Charge Compression Ignition (HCCI) Engines: Key Research and Development Issues,* PT-94, Society of Automotive Engineers, Inc., Warrendale, PA, 2003.

42. Dec, J. E., "Advanced Compression-Ignition Engines—Understanding In-Cylinder Processes, *Proceedings of the Combustion Institute,* 32: 2727–2742 (2009).

43. Siewert, R. M., and Turns, S. R., "The Staged Combustion Compound Engine (SCCE): Exhaust Emissions and Fuel Economy Potential," Paper 750889, *SAE Transactions,* 84: 2391–2420 (1975).

44. Newhall, H. K., "Kinetics of Engine-Generated Nitrogen Oxides and Carbon Monoxide," *Twelfth Symposium (International) on Combustion,* The Combustion Institute, Pittsburgh, PA, pp. 603–613, 1968.

45. Delichatsios, M. M., "The Kinetics of CO Emissions from an Internal Combustion Engine," S. M. Thesis, Massachusetts Institute of Technology, Cambridge, MA, June 1972.

46. Keck, J. C., and Gillespie, D., "Rate-Controlled Partial-Equilibrium Method for Treating Reacting Gas Mixtures," *Combustion and Flame,* 17: 237–241 (1971).

47. Heywood, J. B., *Internal Combustion Engine Fundamentals,* McGraw-Hill, New York, 1988.

48. Ishizawa, S., "An Experimental Study of Quenching Crevice Widths in the Combustion Chamber of a Spark-Ignition Engine," *Twenty-Sixth Symposium (International) on Combustion,* The Combustion Institute, Pittsburgh, PA, pp. 2605–2611, 1996.

49. Quader, A. A., "Lean Combustion and the Misfire Limit in Spark Ignition Engines," SAE Paper 741055, 1974.

50. Hadjiconstantinou, N., Min, K., and Heywood, J. B., "Relationship between Flame Propagation Characteristics and Hydrocarbon Emissions under Lean Operating Conditions in Spark-Ignition Engines," *Twenty-Sixth Symposium (International) on Combustion,* The Combustion Institute, Pittsburgh, PA, pp. 2637–2644, 1996.

51. Jackson, M. W., "Effects of Some Engine Variables and Control Systems on Composition and Reactivity of Exhaust Hydrocarbons," SAE Paper 660404, 1966.

52. Mondt, J. R., "An Historical Overview of Emission-Control Techniques for Spark Ignition Engines: Part B—Using Catalytic Converters," in *History of the Internal Combustion Engine* (E. F. C. Sommerscales and A. A. Zagotta, eds.), ICE-Vol. 8, American Society of Mechanical Engineers, New York, 1989.

53. Kubsh, J. (ed.), *Advanced Three-Way Catalysts,* PT-123, Society of Automotive Engineers, Inc., Warrendale, PA, 2006.

54. Street, J. C., and Thomas, A., "Carbon Formation in Pre-mixed Flames," *Fuel,* 34: 4–36 (1955).

55. Glassman, I., *Combustion,* 3rd Ed., Academic Press, San Diego, CA, 1996.

56. Kittleson, D. B., "Engines and Nanoparticles: A Review," *Journal of Aerosol Science,* 29: 575–588 (1998).

57. Kim, E., and Hopke, P. K., "Identification of Fine Particle Sources in Mid-Atlantic U.S. Area," *Water, Air, and Soil Pollution,* 168: 391–421 (2005).

58. Kim, E., and Hopke, P. K., "Improving Source Apportionment of Fine Particles in the Eastern United States Utilizing Temperature-Resolved Carbon Fractions," *Journal of the Air & Waste Management Association,* 55: 1456–1463 (2005).

59. Regulation (EC) No. 715/2007 of the European Parliament and of the Council of 20 June 2007 on type approval of motor vehicles with respect to emissions from light duty and commercial vehicles (Euro 5 and Euro 6) and on access to vehicle repair and maintenance information, *Official Journal of the European Union,* L 171/1–16, 29.6.2007.

60. Turns, S. R., "Understanding $NO_x$ Formation in Nonpremixed Flames: Experiments and Modeling," *Progress in Energy and Combustion Science,* 21: 361–385 (1995).

61. Driscoll, J. F., Chen, R.-H., and Yoon, Y., "Nitric Oxide Levels of Turbulent Jet Diffusion Flames: Effects of Residence Time and Damköhler Number," *Combustion and Flame,* 88: 37–49 (1992).

62. Frank, J. H., Barlow, R. S., and Lundquist, C., "Radiation and Nitric Oxide Formation in Turbulent Non-Premixed Jet Flames," *Proceedings of the Combustion Institute,* 28: 447–454 (2000).

63. Wang, L., Haworth, D. C., Turns, S. R., and Modest, M. F., "Interactions among Soot, Thermal Radiation, and $NO_x$ Emissions in Oxygen-Enriched Turbulent Nonpremixed Flames: A Computational Fluid Dynamics Modeling Study," *Combustion and Flame,* 141: 170–179 (2005).

64. Wang, L., Modest, M. F., Haworth, D. C., and Turns, S. R., "Modelling Nongrey Gas-Phase and Soot Radiation in Luminous Turbulent Nonpremixed Jet Flames," *Combustion Theory and Modelling,* 9: 479–498 (2005).

65. Turns, S. R., and Myhr, F. H., "Oxides of Nitrogen Emissions from Turbulent Jet Flames: Part I—Fuel Effects and Flame Radiation," *Combustion and Flame,* 87: 319–335 (1991).

66. Turns, S. R., Myhr, F. H., Bandaru, R. Y., and Maund, E. R., "Oxides of Nitrogen Emissions from Turbulent Jet Flames: Part II—Fuel Dilution and Partial Premixing Effects," *Combustion and Flame,* 43: 255–269 (1993).

67. Røkke, N. A., Hustad, J. E., Sønju, O. K., and Williams, F. A., "Scaling of Nitric Oxide Emissions from Buoyancy-Dominated Hydrocarbon Turbulent-Jet Diffusion Flames," *Twenty-Fourth Symposium (International) on Combustion,* The Combustion Institute, Pittsburgh, PA, pp. 385–393, 1992.

68. U.S. Environmental Protection Agency, Technology Transfer Network, Clearinghouse for Inventories and Emission Factors, http://www.epa.gov/ttn/chief/index.html, updated 5 May 2008.

69. U.S. Environmental Protection Agency, AP 42, Fifth Edition, Compilation of Air Pollutant Emission Factors, Volume 1: Stationary and Area Sources, http://www.epa.gov/ttn/chief/ap42, updated 7 October 2009.

70. South Coast Air Quality Management District, Regulation XI, Source Specific Standards, http://www.aqmd.gov/rules/reg/reg11_tofc.html. Accessed 9 October 2009.

71. Bluestein, J., "$NO_x$ Controls for Gas–Fired Industrial Boilers and Combustion Equipment: A Survey of Current Practices," GRI-92/0374, Gas Research Institute Report, October 1992.

72. U.S. Environmental Protection Agency, "Nitrogen Oxide Control of Stationary Combustion Sources," EPA-625/5-86/020, July 1986.

73. Baukal, C. E., Jr. (ed.), *Industrial Burners Handbook,* CRC Press, Boca Raton, FL, 2004.

74. Wünning, J. A., and Wünning, J. G., "Flameless Oxidation to Reduce Thermal NO-Formation," *Progress in Energy and Combustion Science,* 23: 81–94 (1997).

75. Baukal, C. E., Jr. (ed.), *Oxygen-Enhanced Combustion,* CRC Press, Boca Raton, FL, 1998.

76. Cavaliere, A., Joannon, M., and Ragucci, R., "Chapter 3: Highly Preheated Lean Combustion," Dunn-Rankin, D. (ed.), *Lean Combustion: Technology and Control,* Academic Press, Burlington, MA, 2008.

77. Waibel, R. T., Price, D. N., Tish, P. S., and Halprin, M. L., "Advanced Burner Technology for Stringent $NO_x$ Regulations," API Midyear Refining Meeting, Orlando, FL, 8 May 1990.

78. Cavaliere, A., and de Joannon, M., "Mild Combustion," *Progress in Energy and Combustion Science,* 30: 329–366 (2004).

79. Lani, B. W., Feeley, T. J. III, Miller, C. E., Carney, B. A., and Murphy, J. T., "DOE/NETL's $NO_x$ Emissions Control R&D Program—Bringing Advanced Technology to the Marketplace," *DOE/NETL $NO_x$ R&D Overview,* April 2008.

80. Wendt, J. O. L., Sternling, C. Y., and Matovich, M. A., "Reduction of Sulfur Trioxide and Nitrogen Oxides by Secondary Fuel Injection," *Fourteenth Symposium (International) on Combustion,* The Combustion Institute, Pittsburgh, PA, pp. 897–904, 1973.

81. U.S. Environmental Protection Agency, "Sourcebook: $NO_x$ Control Technology Data," EPA-600/2-91/029, Control Technology Center, July 1991.

82. Glarborg, P., Alzueta, M. U., Dam-Johansen, K., and Miller, J. A., "Kinetic Modeling of Hydrocarbon/Nitric Oxide Interactions in a Flow Reactor," *Combustion and Flame,* 115: 1–27 (1998).

83. Zamansky, Y. M., Ho, L., Maly, P. M., and Seeker, W. R., "Reburning Promoted by Nitrogen- and Sodium-Containing Compounds," *Twenty-Sixth Symposium (International) on Combustion,* The Combustion Institute, Pittsburgh, PA, pp. 2075–2082, 1996.

84. Lanier, W. S., Mulholland, J. A., and Beard, J. T., "Reburning Thermal and Chemical Processes in a Two-Dimensional Pilot-Scale System," *Twenty-First Symposium (International) on Combustion,* The Combustion Institute, Pittsburgh, PA, pp. 1171–1179, 1986.

85. Chen, S. L., *et al.,* "Bench and Pilot Scale Process Evaluation of Reburning for In-Furnace $NO_x$ Reduction," *Twenty-First Symposium (International) on Combustion,* The Combustion Institute, Pittsburgh, PA, pp. 1159–1169, 1986.

86. Lyon, R. K., "The $NH_3$-NO-$O_2$ Reaction," *International Journal of Chemical Kinetics,* 8: 315–318 (1976). (See also U.S. Patent 3,900,554, 1975.)

87. Perry, R. A., and Siebers, D. L., "Rapid Reduction of Nitrogen Oxides in Exhaust Gas Streams," *Nature,* 324: 657–658 (1986).

88. Muzio, L. J., Quartucy, G. C., and Chichanowiczy, J. E., "Overview and Status of Post-Combustion $NO_x$ Control: SNCR, SCR and Hybrid Technologies," *International Journal of Environment and Pollution,* 17: 4–30 (2002).

89. Chen, S. L., *et al.,* "Advanced $NO_x$ Reduction Processes Using -NH and -CN Compounds in Conjunction with Staged Air Addition," *Twenty-Second Symposium (International) on Combustion,* The Combustion Institute, Pittsburgh, PA, pp. 1135–1145, 1988.

90. Heap, M. P., Chen, S. L., Kramlick, J. C., McCarthy, J. M., and Pershing, D. W., "Advanced Selective Reduction Processes for $NO_x$ Control," *Nature,* 335: 620–622 (1988).

91. Muzio, L. J., Montgomery, T. A., Quartucy, G. C., Cole, J. A., and Kramlick, J. C., "$N_2O$ Formation in Selective Non-Catalytic $NO_x$ Rejection Processes," *Proceedings: 1991 Joint Symposium on Stationary Combustion $NO_x$ Control,* Vol. 2, EPRI GS-7447, Electric Power Research Institute, Palo Alto, CA, pp. 5A73–5A96, November 1991.

92. Kjærgaard, K., Glarborg, P., Dam-Johansen, K., and Miller, J. A., "Pressure Effects on the Thermal De-$NO_x$ Process," *Twenty-Sixth Symposium (International) on Combustion,* The Combustion Institute, Pittsburgh, PA, pp. 2067–2074, 1996.

93. Srivastava, R. K., Hall, R. E., Khan, S., Culligan, K., and Lani, B. W., "Nitrogen Oxides Emission Control Options for Coal-Fired Electric Utility Boilers," *Journal of the Air & Waste Management Association,* 55: 1367–1388 (2005).

94. May, P. A., Campbell, L. M., and Johnson, K. L., "Environmental and Economic Evaluation of Gas Turbine SCR NO, Control," *Proceedings: 1991 Joint Symposium on Stationary Combustion $NO_x$, Control,* Vol. 2, EPRI GS-7447, Electric Power Research Institute, Palo Alto, CA, pp. 5BI9–5B36, November 1991.

95. Behrens, E. S., Ikeda, S., Teruo, Y., Mittelbach, G., and Makato, Y., "SCR Operating Experience on Coal-Fired Boilers and Recent Progress," *Proceedings: 1991 Joint Symposium on Stationary Combustion NO$_x$ Control,* Vol. I, EPRI GS7447, Electric Power Research Institute, Palo Alto, CA, pp. 4B59–4B77, November 1991.

96. Robie, C. P., Ireland, P. A., and Cichanowicz, J. E., "Technical Feasibility and Cost of SCR for U.S. Utility Application," *Proceedings: 1991 Joint Symposium on Stationary Combustion NO$_x$ Control,* Vol. 1, EPRI GS-7447, Electric Power Research Institute, Palo Alto, CA, pp. 4B81–4Bl00, November 1991.

97. U.S. Environmental Protection Agency, "Nitrogen Oxides," http://www.epa.gov/air/emissions/nox.htm, updated on 21 October 2008.

98. U. S. Energy Information Agency, "Electricity," http://www.eia.doe.gov/fuelelectric.html. Accessed 30 July 2008.

99. Bowman, C. T., "Kinetics of Pollutant Formation and Destruction in Combustion," *Progress in Energy and Combustion Science,* 1: 33–45 (1975).

100. Miller, B. G., *Coal Energy Systems,* Elsevier Academic Press, Burlington, MA, 2005.

101. Miller, B. G., and Tillman, D. A., *Combustion Engineering Issues for Solid Fuel Systems,* Elsevier Academic Press, Burlington, MA, 2008.

102. Sarofim, A. F., and Flagan, R. C., "NO$_x$ Control for Stationary Combustion Sources," *Progress in Energy and Combustion Sciences,* 2: 1–25 (1976).

103. Rosenberg, H. S., Curran, L. M., Slack, A. Y., Ando, J., and Oxley, J. H., "Post Combustion Methods for Control of NO$_x$ Emissions," *Progress in Energy and Combustion Science,* 6: 287–302 (1980).

104. Johnson, T. V., "Diesel Emission Control in Review," SAE Technical Paper Series, 2008-01-0069, SAE International, Warrendale, PA (2008).

105. Plee, S. L., Ahmad, T., Myers, J. P., and Faeth, G. M., "Diesel NO$_x$ Emissions—A Simple Correlation Technique for Intake Air Effects," *Nineteenth Symposium (International) on Combustion,* The Combustion Institute, Pittsburgh, PA, pp. 1495–1502, 1983.

106. Ahmad, T., and Plee, S. L., "Application of Flame Temperature Correlations to Emissions from a Direct-Injection Diesel Engine," Paper 831734, *SAE Transactions,* 92: 4.910–4.921 (1983).

107. Johnson, T. V., "Diesel Emission Control in Review," SAE Technical Paper Series, 2006-01-0233, SAE International, Warrendale, PA (2006).

108. Charlton, S. J., "Developing Diesel Engines to Meet Ultra-Low Emission Standards," SAE paper 2005-01-3628, in *Diesel Exhaust Aftertreatment 2000–2007,* PT-126 (M. Khair and F. Millo, eds.) SAE International, Warrendale, PA, pp. 41–77, 2008.

109. Zheng, M., Reader, G. T., and Hawley, J. G., "Diesel Engine Exhaust Gas Recirculation—A Review on Advanced and Novel Concepts," *Energy Conversion and Management,* 45: 883–900 (2004).

110. Schmitz, P. J., and Baird, R. J., "NO and NO$_2$ Adsorption on Barium Oxide: Model Study of the Trapping Stage of NO$_x$ Conversion via Lean NO$_x$ Traps," *Journal of Physical Chemistry B,* 106: 4172–4180 (2002).

111. Epling, W. S., Campbell, L. E., Yezerets, A., Currier, N. W., and Parks, J. E. III, "Overview of the Fundamental Reactions and Degradation Mechanisms of $NO_x$ Storage/Reduction Catalysts," *Catalysis Reviews,* 46: 163–245 (2004).

112. Geckler, S., *et al.,* "Development of a Desulfurization Strategy for a $NO_x$ Adsorber Catalyst System," SAE paper 2001-01-0510, in *Diesel Exhaust Aftertreatment 2000–2007,* PT-126 (M. Khair and F. Millo, eds.), SAE International, Warrendale, PA, pp. 427–435, 2008.

113. Parks, J., *et al.,* "Sulfur Control for $NO_x$ Sorbate Catalysts: Sulfur Sorbate Catalysts and Desulfation," SAE Technical Paper Series, 2001-01-2001, SAE International, Warrendale, PA (2001).

114. Amberntsson, A., Skoglundh, M., Ljungström, S., and Fridell, E., "Sulfur Deactivation of $NO_x$ Storage Catalysts: Influence of Exposure Conditions and Noble Metal," *Journal of Catalysis,* 217: 253–263 (2003).

115. Brandenberger, S., Kröchner, O., Tissler, A., and Althoff, R., "The State of the Art in Selective Catalytic Reduction of $NO_x$ by Ammonia Using Metal-Exchanged Zeolite Catalysts," *Catalysis Reviews,* 50: 492–531 (2008).

116. Blint, R. C., Koermer, G., and Fitzgerald, G., "Discovery of New $NO_x$ Reduction Catalysts for CIDI Engines Using Combinatorial Techniques," Ultra Clean Transportation Fuels Program, Annual Technical Progress Report (FY05), February 2006. (See http://www.osti.gov/bridge/servlets/purl/907773-jXQ5I2/907773.pdf.)

117. Correa, S. M., "A Review of NO Formation Under Gas-Turbine Combustion Conditions," *Combustion Science and Technology,* 87: 329–362 (1992).

118. Forzatti, P., "Present Status and Perspectives in De-$NO_x$ SCR Catalysis," *Applied Catalysis A: General,* 222: 221–236 (2001).

119. McDonell, V., "Chapter 5: Lean Combustion in Gas Turbines," Dunn-Rankin, D. (ed.), *Lean Combustion: Technology and Control,* Academic Press, Burlington, MA, 2008.

120. Leonard, G., and Stegmaier, J., "Development of an Aeroderivative Gas Turbine Dry Low Emissions Combustion System," *Transactions of the ASME, Journal of Engineering for Gas Turbines and Power,* 116: 542–546 (1994).

121. Myers, G., *et al.,* "Dry, Low Emissions for the 'H' Heavy-Duty Industrial Gas Turbine: Full-Scale Combustion System Rig Test Results," Paper GT2003-38193, *Proceedings of 2003 ASME Turbo Expo: Power for Land, Sea, and Air,* 16–19 June 2003, Atlanta, GA.

122. Pritchard, J., "H-System™ Technology Update" Paper GT2003-38711, *Proceedings of 2003 ASME Turbo Expo: Power for Land, Sea, and Air,* 16–19 June 2003, Atlanta, GA.

123. Feigl, M., Setzer, F., Feigl-Varela, R., Myers, G. D., and Sweet, B, "Field Test Validation of the DLN2.5H Combustion System on the 9H Gas Turbine at Baglan Bay Power Station," Paper GT2005-68843, *Proceedings of GT2005 ASME Turbo Expo 2005: Power for Land, Sea, and Air,* 6–9 June 2005, Reno-Tahoe, NV.

124. Huang, Y., and Yang, V., "Dynamics and Stability of Lean-Premixed Swirl-Stabilized Combustion, *Progress in Energy and Combustion Science,* 35: 293–364 (2009).

125. Rizk, N. K., and Mongia, H. C., "Three-Dimensional $NO_x$ Model for Rich/Lean Combustor," AIAA Paper AIAA-93-0251, 1993.

126. Stouffer, S. D., Ballal, D. R., Zelina, J., Shouse, D. T., Hancock, R. D., and Mongia, H. C., "Development and Combustion Performance of a High-Pressure WSR and TAPS Combustor," AIAA paper AIAA 2005-1416, 43rd AIAA Aerospace Sciences Meeting and Exhibit, 10–13 January 2005, Reno, NV.

127. Mongia, H. C., "TAPS—A 4th Generation Propulsion Combustor Technology for Low Emissions," AIAA paper AIAA 2003-2657, AIAA/ICAS International Air and Space Symposium and Exposition, 14–17 July 2003, Dayton, OH.

128. Greeves, G., Khan, I. M., Wang, C. H. T., and Fenne, I., "Origins of Hydrocarbon Emissions from Diesel Engines," SAE Paper 770259, 1977.

129. Levins, P. C., Kendall, D. A., Caragay, A. B., Leonardos, G., and Oberholter, J. E., "Chemical Analysis of Diesel Exhaust Odor Species," SAE Paper 740216, 1974.

130. Kostandopoulos, A. G., *et al.*, "Fundamentals Studies of Diesel Particulate Filters: Transient Loading, Regeneration and Aging," SAE Paper 2000-01-1016, in *Diesel Exhaust Aftertreatment 2000–2007,* PT-126 (M. Khair and F. Millo, eds.), SAE International, Warrendale, PA, pp. 119–141, 2008.

131. Khair, M., and Millo, F. (eds.), *Diesel Exhaust Aftertreatment 2000–2007,* PT-126, SAE International, Warrendale, PA, 2008.

132. Lefebvre, A. H., *Gas Turbine Combustion,* Hemisphere, Washington, DC, 1983.

133. Anon., "Standard Specification for Automotive Spark-Ignition Engine Fuel, Designation: D 4814 – 08b," ASTM International, West Conshohocken, PA, 2008.

134. Anon., "Standard Test Method for Cetane Number of Diesel Fuel Oil, Designation: D 613 – 08b," ASTM International, West Conshohocken, PA, 2008.

135. Flagan, R. C, and Seinfeld, J. H., *Fundamentals of Air Pollution Engineering,* Prentice Hall, Englewood Cliffs, NJ, 1988.

136. Heinsohn, R. J., and Kabel, R. L., *Sources and Control of Air Pollution,* Prentice Hall, Upper Saddle River, NJ, 1999.

137. Solomon, S., *et al.,* (eds.), *Climate Change 2007—The Physical Science Basis,* Contribution of Working Group I to the Fourth Assessment Report of the Intergovernmental Panel on Climate Change, Cambridge University Press, 2007.

138. U.S. Environmental Protection Agency, "U.S. Greenhouse Gas Inventory," http://www.epa.gov/climatechange/emissions/usgginventory.html, updated on 8 September 2009.

139. Clough, S. A., and Iacono, M. J., "Line-by-Line Calculation of Atmospheric Fluxes and Cooling Rates. 2. Application to Carbon Dioxide, Ozone, Methane, Nitrous Oxide and the Halocarbons," *Journal of Geophysical Research,* 100: 16,519–16,535 (1995).

140. Bohren, C. F., and Clothiaux, E., *Fundamentals of Atmospheric Radiation,* Wiley –VCH, Weinheim, Germany, 2006.

141. U.S. Environmental Protection Agency, "Human-Related Sources and Sinks of Carbon Dioxide," http://www.epa.gov/climatechange/emissions/co2_human.html, updated on 8 October 2009.

142. U.S. Environmental Protection Agency, "Methane," http://www.epa.gov/methane/sources .html, updated on 20 July 2009.

143. U.S. Environmental Protection Agency, "Nitrous Oxide," http://www.epa.gov/nitrous oxide/sources.html, updated on 20 July 2009.

144. Becker, K. H., *et al.,* "Nitrous Oxide ($N_2O$) Emissions from Vehicles," *Environmental Science & Technology,* 33: 4,134–4,139 (1999).

145. Odaka, M., Koike, N., and Suzuki, H., "Influence of Catalyst Deactivation on $N_2O$ Emissions from Automobiles," *Chemosphere—Global Change Science,* 2: 413–423 (2000).

146. Yokoyama, T., Nishinomiya, S., and Matsuda, H., "$N_2O$ Emissions from Fossil Fuel Power Plants," *Environmental Science & Technology,* 25: 347–348 (1991).

147. Blok, K., and De Jager, D., "Effectiveness of Non-$CO_2$ Greenhouse Gas Emission Reduction Technologies," *Environmental Monitoring and Assessment,* 31: 17–40 (1994).

148. U.S. Environmental Protection Agency, "High Global Warming Potential (GWP) Gases," http://www.epa.gov/highgwp/sources.html, updated on 16 July 2009.

149. Pacala, S., and Socolow, R., "Stabilization Wedges: Solving the Climate Problem for the Next 50 Years with Current Technologies," *Science,* 305: 968–972 (2004).

150. Metz, B., Davidson, O., Coninck, H., Loos, M., and Meyer, L. (eds.), *IPCC Special Report on Carbon Dioxide Capture and Storage,* Cambridge University Press, New York, 2005.

## QUESTIONS AND PROBLEMS

**15.1**   Make a list of the boldfaced words in Chapter 15 and discuss their meaning and significance.

**15.2**   Discuss the implications of each of the following for environmental and human health effects: nitric oxide, sulfur trioxide, unburned hydrocarbons, carbon monoxide, and diesel exhaust particulate matter.

**15.3**   A value of 375 ppm was obtained for a "wet" measurement of nitric oxide in a combustion product stream containing 5 percent $O_2$. Does correcting the nitric oxide measurement to 3 percent $O_2$ result in a lower, higher, or unchanged value? Does converting the NO measurement to a "dry basis" result in a lower, higher, or unchanged value?

**15.4**   Calculate the nitric oxide emission index for a combustor burning ethane ($C_2H_6$) with air, given the following mole fraction measurements in the exhaust duct: $\chi_{CO_2} = 0.110$, $\chi_{O_2} = 0.005$, $\chi_{H_2O} = 0.160$, and $\chi_{NO} = 185 \cdot 10^{-6}$. Assume that the CO and unburned hydrocarbon concentrations are negligible. Do your results depend on whether the concentrations are "wet" or "dry"?

**15.5**   A natural gas–fired, stationary gas-turbine engine is used for electrical power generation. The exhaust emission level of nitric oxide is 20 ppm (by vol.), measured in the exhaust with an oxygen concentration of 13 percent (by vol.). No aftertreatment (SNCR or SCR) is used.

A.  What is the nitric oxide concentration corrected to 3 percent oxygen?

B.  Determine the $NO_x$ emission index in grams of $NO_x$ ($NO_2$ equivalent) per kilogram of fuel burned. Assume that the natural gas is essentially all methane.

C.  Would this engine meet the California South Coast Air Quality Management District—Rule 1134 emission standards?

**15.6**  A heavy-duty naturally aspirated diesel engine is being evaluated on a dynamometer test stand. Operating at an air–fuel ratio of 21:1 with a fuel flowrate of $4.89 \cdot 10^{-3}$ kg/s, the engine produces 80 kW of brake power. The multicomponent fuel has the equivalent formula $C_{12}H_{22}$. The unburned hydrocarbon concentration measured in the exhaust stream is 120 ppm $C_1$ (wet basis).

A.  Determine the unburned hydrocarbon concentration of a dry basis.

B.  Determine the unburned hydrocarbon emission index (g/kg) for the engine. Assume that the hydrogen–carbon ratio of the unburned $C_1$ equivalent is the same as the original fuel molecule.

C.  Find the brake-specific unburned hydrocarbon emission in g/kW-hr for the engine.

**15.7**  An engineer runs a test on an experimental aircraft engine burning isooctane ($C_8H_{18}$) and measures the burned gas composition using the sampling system sketched below:

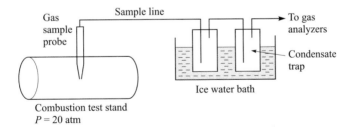

A water-cooled gas sample probe withdraws a sample of the hot burned gas. The sample gas passes through two condensate traps held in an ice bath to remove water, so all of the analyzers receive essentially dry gas.

The gas analyzers record the following gas composition: 8.44 percent $CO_2$, 8.79 percent $O_2$, 76 ppm $NO_x$, 44 ppm CO, and 15 ppm unburned hydrocarbons.

A.  What are the $NO_x$, CO, and unburned hydrocarbon concentrations on a dry 15-percent-$O_2$ basis?

B.   What are the $NO_x$, CO, and unburned hydrocarbon concentrations on a dry 3-percent-$O_2$ basis?

C.   What is the $NO_x$ emission index in g/kg-fuel?

**15.8**   A utility company operates a 250-MW combined-cycle power plant, fired with natural gas ($CH_4$). A government inspector has arrived to verify that the power plant is in compliance with all applicable emissions regulations. If the power plant is emitting more than 25 ppm $NO_x$ (on a dry-15-percent-$O_2$ basis), the plant will not be in compliance.

The inspector withdraws a gas sample from the stack and measures the following dry gas composition using portable gas analyzers: 4.4 percent $CO_2$, 14.5 percent $O_2$, 29 ppm $NO_x$, and 42 ppm CO.

A.   Based upon the inspector's measurements, what are the $NO_x$ emissions from the power plant (on a dry 15-percent-$O_2$ basis)? Is the plant in compliance?

B.   The plant supervisor has asked the chief engineer to review the inspector's gas measurements. The chief engineer knows that the plant runs at an overall equivalence ratio of 0.4. Based on this observation, and assuming complete combustion, calculate the dry $O_2$ concentration that the inspector's $O_2$ analyzer **should** have read. Using this new $O_2$ concentration, recalculate the $NO_x$ emissions from the power plant. Is the power plant in compliance now or not? Should the inspector check the calibration of his $O_2$ analyzer and visit the plant again?

**15.9**   Consider the idealized combustion of a hydrocarbon fuel ($CH_4$) and air in stoichiometric proportions. Assume no dissociation and constant and equal specific heats for reactants and products of 1200 J/kg-K. Assume a fuel heating value of $4 \cdot 10^7$ J/kg.

A.   Calculate the adiabatic constant-pressure flame temperatures for zero, 10 and 20 percent of the undiluted reactants (by mass) exhaust-gas recycle. Assume that both the reactants and recycled gas are initially at 300 K.

B.   Repeat the calculations of part A, but assume the recycled gas is at 1200 K and the air and fuel are at 300 K.

C.   Discuss the implications of your calculations for $NO_x$ production.

**15.10**   Explain the differences among the nitric oxide formation mechanisms denoted simple thermal (Zeldovich) mechanism, superequilibrium mechanism, and the prompt Fenimore mechanism. Use information from Chapter 5 as necessary.

**15.11**   Derive an expression for d[NO]/dt for the thermal NO mechanism assuming equilibrium [O] concentrations and steady-state [N]. Neglect reverse reactions. Eliminate $[O]_e$ as a variable, using $[O_2]_e$ instead from the equilibrium $\frac{1}{2}O_2 \Leftrightarrow O$. *Hint:* Your final result should involve only the forward rate constant $k_1$ for the reaction $O + N_2 \rightarrow NO + N$, the equilibrium constant $K_p$, $[O_2]_e$, $[N_2]_e$, and the temperature.

**15.12*** Starting with the result from problem 15.11,

    A. Convert your expression involving molar concentrations $(kmol/m^3)$ to one involving mole fractions.

    B. Calculate the initial rate of NO formation $(ppm/s)$ for $k_1 = 7.6 \cdot 10^{10}$ $\exp(-38{,}000/T(K))\, m^3/kmol\text{-}s$ and $K_p = 3.6 \cdot 10^3 \exp(-31{,}090/T(K))$ for the following conditions:

        i. $\chi_{O_2,e} = 0.20$, $\chi_{N_2,e} = 0.67$, $T = 2000$, $P = 1$ atm.

        ii. $\chi_{O_2,e} = 0.10$, with other conditions as part i.

        iii. $\chi_{O_2,e} = 0.05$, with other conditions as part i.

        iv. $\chi_{O_2,e} = 0.20$, $\chi_{N_2,e} = 0.67$, $T = 1800$, $P = 1$ atm.

        v. $T = 2200$, with other conditions as part iv.

    C. Plot your results from part B to show the effects of $\chi_{O_2,e}$ and $T$. Use separate graphs. Discuss.

**15.13** Discuss the fundamental differences between the origins of unburned hydrocarbons in spark-ignition engines and gas-turbine engines. What source of unburned hydrocarbons occurs with diesel engines that is not present in either SI or gas-turbine engines?

**15.14** The CO emissions measured in the tailpipe of a spark-ignition engine are 2000 ppm. Calculations show that the equilibrium concentration of CO for the tailpipe conditions is 2 ppm. Explain the origin of the 2000 ppm CO level in the tailpipe.

**15.15*** Use an equilibrium code to determine the equilibrium composition of the products of $CH_4$–air combustion over a range of equivalence ratios from 1.0 to 1.4 for the following conditions: (i) $T = 2000$ K and $P = 10$ atm, (ii) $T = 925$ K and $P = 1$ atm.

    A. Plot the CO mole fractions versus $\Phi$ for both cases on the same graph.

    B. Assuming conditions i approximate a point early in the SI engine expansion process, and conditions ii approximate conditions in the exhaust pipe, what are the implications of the plot you made in part A?

**15.16** Discuss why the formation of $SO_3$ in a combustion system is more undesirable than the formation of $SO_2$.

**15.17** An oil-fired industrial boiler uses No. 2 fuel oil. Estimate the likely range of concentrations of $SO_2$ in the flue gases for an $O_2$ level of 3 percent.

# chapter

# 16

# Detonations

## OVERVIEW

In this chapter, we explore the differences between detonations and flames (deflagrations) and then proceed to develop a simple, one-dimensional analysis of a detonation wave in the same spirit as previous analyses of flames. From this analysis, approximate expressions are developed to estimate detonation velocities. The chapter concludes with a brief discussion of both the idealized, one-dimensional structure of a detonation wave and the complex, three-dimensional structure of real detonations.

## PHYSICAL DESCRIPTION

### Definition

In Chapter 8, we simply defined a detonation as a combustion wave propagating at supersonic velocities. We now expand that definition to relate a detonation to a shock wave, a concept that you may have seen in an introductory fluid mechanics course: a **detonation** is a **shock wave** sustained by the energy released by combustion. The combustion process, in turn, is initiated by the shock-wave compression and the resulting high temperatures. Thus, we see that a detonation comprises the interaction between a hydrodynamic process, the shock wave, and a thermochemical process, the combustion.[1]

---

[1] In some literature, the term detonation refers to a rapid homogeneous energy release in a combustible mixture, i.e., an explosion. An example of this is the use of the word detonation to refer to engine knock (see Example 6.1).

616

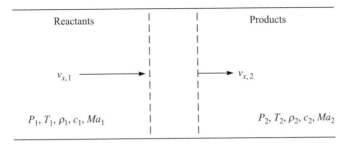

**Figure 16.1**    One-dimensional detonation wave in a constant-area duct. The coordinate system is fixed to the wave.

## Principal Characteristics

A detonation can be initiated by igniting a combustible mixture at the closed end of a long tube that is open at the opposite end. In this situation, the flame initiated at the closed end is accelerated as it propagates through the mixture as a result of the expansion of the burned gas confined between the flame and the closed end. This acceleration leads to the formation of a shock wave preceding the combustion zone and propagation at supersonic velocities.

Figure 16.1 illustrates a control volume containing the detonation wave. In the laboratory frame, the detonation wave moves downstream, i.e., from right to left in the figure; however, by fixing the coordinate system to the detonation wave, the wave is stationary, with the reactants entering at a velocity $v_{x,1}$, and combustion products exiting at a velocity $v_{x,2}$. The qualitative differences between the upstream and downstream properties across the detonation wave for the most part are quite similar to the property differences across an ordinary **normal shock.** The principal difference is that, while for the normal shock the downstream velocity is always subsonic, for the detonation this velocity is always the local speed of sound.

To obtain somewhat quantitative comparisons of normal shocks (no combustion), detonations, and deflagrations, Table 16.1 shows upstream and downstream Mach

**Table 16.1**    Typical properties of normal shocks, detonations, and deflagrations

| Property | Normal Shock[a] | Detonation[b] | Deflagration[c] |
|---|---|---|---|
| $Ma_1$ | 5.0 | 5–10 | 0.001 |
| $Ma_2$ | 0.42 | 1.0 | 0.003 |
| $v_{x,2}/v_{x,1}$ | 0.20 | 0.4–0.7 | 7.5 |
| $P_2/P_1$ | 29 | 13–55 | $\approx 1$ |
| $T_2/T_1$ | 5.8 | 8–21 | 7.5 |
| $\rho_2/\rho_1$ | 5.0 | 1.7–2.6 | 0.13 |

[a]Properties based on air with $\gamma = 1.4$ and the upstream Mach number arbitrarily assumed to be 5.0.

[b]Typical values from Friedman [1].

[c]Estimates from laminar, adiabatic combustion of a stoichiometric methane–air mixture at 1 atm with an initial temperature of 298 K.

numbers ($Ma \equiv v/c$) and the ratios of downstream-to-upstream properties. Here we see that the normal-shock property ratios are qualitatively similar to those of detonations and of the same order of magnitude, with the caveat that for the detonation, $v_{x,2}$ is the sonic velocity, as previously indicated. In contrast, when comparing deflagrations with detonations we see that the Mach number increases across the flame,[2] the velocity increases greatly, and the density drops significantly. All these effects are opposite in direction to those corresponding in either shocks or detonations.

Another striking difference is that the pressure is essentially constant across a deflagration (actually there is a very slight decrease), whereas a major feature of a detonation is the high pressure downstream of the propagating wave. The only characteristic shared by detonations, deflagration, and normal shock waves is the large increase in temperature across the wavefront.

## ONE-DIMENSIONAL ANALYSIS

Although the structure of real detonations is highly three-dimensional, considerable insight is provided by a one-dimensional analysis. The first attempts to explain detonations (Chapman in 1899 [2]) relied on a one-dimensional approach, and this approach is still useful today as it provides a foundation from which to build a more detailed understanding.

### Assumptions

If our control volume (Fig. 16.1) is chosen such that the upstream and downstream boundaries lie in regions where there are no temperature or species concentration gradients, we can perform a fairly rigorous analysis with only the following few assumptions:

1. One-dimensional, steady flow.
2. Constant area.
3. Ideal-gas behavior.
4. Constant and equal specific heats.
5. Negligible body forces.
6. Adiabatic conditions (no heat losses to surroundings).

These assumptions are the same as those typically invoked in one-dimensional analyses of normal shocks that you may have seen in your previous studies of thermodynamics or fluid mechanics.

---

[2]Note, however, that the Mach number is very small and thus is not a very useful parameter to characterize a deflagration.

## Conservation Laws

Because the flow is one-dimensional and steady, we can easily write the integral conservation laws for the finite control volume bounded by the dashed lines shown in Fig. 16.1.

**Mass Conservation**  For steady flow, the mass flowrate $\dot{m}$ is constant; moreover, if the area is fixed, the mass flux ($\dot{m}'' = \dot{m}/A$) is also a constant. Thus,

$$\dot{m}'' = \rho_1 v_{x,1} = \rho_2 v_{x,2}. \tag{16.1}$$

**Momentum Conservation**  Since there are neither shear nor body forces acting on the control volume, the sole force acting is pressure. Axial momentum conservation is thus simply expressed:

$$P_1 + \rho_1 v_{x,1}^2 = P_2 + \rho_2 v_{x,2}^2. \tag{16.2}$$

**Energy Conservation**  Energy conservation is expressed as

$$h_1 + v_{x,1}^2/2 = h_2 + v_{x,2}^2/2, \tag{16.3}$$

where the enthalpy is the standardized enthalpy. Splitting out the sensible and heat-of-formation contributions to the total enthalpy is helpful to our analysis in that it explicitly allows us to define a "heat release" or "heat addition" resulting from reaction.

We can write a calorific equation of state,

$$h(T) = \sum_i Y_i h_{f,i}^o + \sum_i Y_i \int_{T_{\text{ref}}}^{T} c_{p,i}\, dT, \tag{16.4}$$

where the $h_{f,i}^o$ values are found at the reference temperature. With our assumption of constant specific heats, Eqn. 16.4 simplifies to the following:

$$h(T) = \sum_i Y_i h_{f,i}^o + c_p (T - T_{\text{ref}}). \tag{16.5}$$

Substituting Eqn. 16.5 into our conservation of energy expression (Eqn. 16.3) yields

$$c_p T_1 + v_{x,1}^2/2 + \left[ \underbrace{\sum_i Y_i h_{f,i}^o}_{\text{state 1}} - \underbrace{\sum_i Y_i h_{f,i}^o}_{\text{state 2}} \right] = c_p T_2 + v_{x,2}^2/2. \tag{16.6}$$

The bracketed term in Eqn. 16.6 is just the heat of combustion per mass of mixture. Adopting the convention frequently found in the literature, we define the "heat addition" $q$ to be

$$q \equiv \underbrace{\sum_i Y_i h_{f,i}^o}_{\text{state 1}} - \underbrace{\sum_i Y_i h_{f,i}^o}_{\text{state 2}}. \tag{16.7}$$

This notation results in an energy equation that looks like one you may have encountered in your study of compressible gas dynamics:

$$c_p T_1 + v_{x,1}^2/2 + q = c_p T_2 + v_{x,2}^2/2. \tag{16.8}$$

It is important to emphasize that the heat release $q$ appearing in the above equation is a property of the mixture, the magnitude of which depends on the specific fuel–oxidizer combination and the mixture strength, i.e., the equivalence ratio, $\Phi$, as can be seen from its definition (Eqn. 16.7).

**State Relationships**  With the assumption of ideal-gas behavior, we can write

$$P_1 = \rho_1 R_1 T_1 \tag{16.9}$$

$$P_2 = \rho_2 R_2 T_2, \tag{16.10}$$

where $R_1$ and $R_2$ are the specific gas constants ($R_i = R_u/MW_i$). These state relationships, together with the calorific equation of state (Eqn. 16.5) presented above in our discussion of energy conservation, provide closure to our equation set.

## Combined Relations

In this section, we combine the above conservation and state relationships to obtain a qualitative understanding of a detonation wave.

**The Rayleigh Line**  Simultaneous solution of the continuity and momentum conservation equations, Eqns. 16.1 and 16.2, yields the following relationships:

$$\frac{P_2 - P_1}{1/\rho_2 - 1/\rho_1} = -\dot{m}''^2 \tag{16.11a}$$

or, using the specific volume, $v$,

$$\frac{P_2 - P_1}{v_2 - v_1} = -\dot{m}''^2. \tag{16.11b}$$

Using Eqn. 16.11 to plot $P$ versus $1/\rho (= v)$ for a fixed flowrate results in the **Rayleigh line**. For example, fixing $P_1$ and $v_1$ and rearranging Eqn. 16.11b yields the generic linear relationship

$$P = av_2 + b, \tag{16.12a}$$

where the slope $a$ is given by

$$a = -\dot{m}''^2 \tag{16.12b}$$

and the intercept $b$ is given by

$$b = P_1 + \dot{m}''^2 v_1. \tag{16.12c}$$

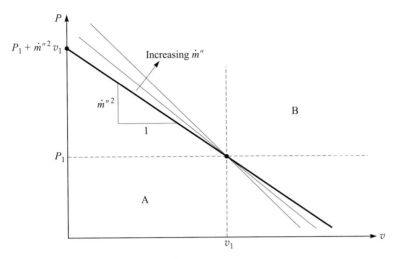

**Figure 16.2**  Bold line with slope $-\dot{m}''^2$ is the Rayleigh line for flow defined by $\dot{m}''$, $P_1$, and $v_1$. Increasing the mass flux $\dot{m}''$ while maintaining the initial state ($P_1$, $v_1$) causes the slope of the line to become steeper (more negative).

Figure 16.2 shows a plot of a Rayleigh line for state 1 fixed by $P_1$ and $v_1$. Increasing the mass flux $\dot{m}''$ causes the line to steepen, pivoting through the point ($P_1$, $v_1$). In the limit of an infinite mass flux, the Rayleigh line would be vertical; whereas in the opposing limit of zero mass flux, the line becomes horizontal. Since these two extremes contain all possible mass fluxes, no solutions to Eqn. 16.11 are possible in the two quadrants labeled A and B formed by the dashed lines going through point ($P_1$, $v_1$). Regions A and B are physically inaccessible. We will use this fact later to help decide what final states are possible for a detonation wave.

**The Rankine–Hugoniot Curve**  The **Rankine–Hugoniot curve**[3] results when we require that the energy equation (Eqn. 16.8) be satisfied in addition to the continuity and momentum conservation equations (Eqns. 16.1 and 16.2). Combining Eqns. 16.1, 16.2, and 16.8, and employing the ideal-gas equation of state and other ideal-gas relations (e.g., $\gamma \equiv c_p/c_v$), yields the following:

$$\frac{\gamma}{\gamma-1}(P_2/\rho_2 - P_1/\rho_1) - \tfrac{1}{2}(P_2 - P_1)(1/\rho_1 + 1/\rho_2) - q = 0 \qquad (16.13a)$$

or

$$\frac{\gamma}{\gamma-1}(P_2 v_2 - P_1 v_1) - \tfrac{1}{2}(P_2 - P_1)(v_1 + v_2) - q = 0. \qquad (16.13b)$$

Detailed derivation of the above is left as an exercise for the reader.

[3]This is sometimes referred to as simply the Hugoniot curve.

Let us assume that the heat release $q$ is a known parameter. We also fix the values of $P_1$ and $v_1$ as we did before in our analysis of the Rayleigh line. Equation 16.13b then becomes a transcendental relation between $P_2$ and $v_2$; i.e.,

$$f(P_2, v_2) = 0, \tag{16.14a}$$

or, more generically,

$$f(P, v) = 0. \tag{16.14b}$$

We can now plot the pressure $P$ as a function of specific volume $v$ for $P_1$, $v_1$, and $q$ fixed at particular known values. The point $(P_1, v_1)$ is sometimes referred to as the origin of the Rankine–Hugoniot curve. Figure 16.3 illustrates the particular Rankine–Hugoniot curve associated with the fixed values $P_1$, $v_1$, and $q_1$. In this figure, we indicate the origin as point A, and define the **upper branch** of the Hugoniot curve as those points on the Hugoniot lying above point B and the **lower branch** as those points lying below point C.

We now proceed to determine what points on the Hugoniot correspond to realizable physical states. Since any real process going from state 1 to state 2 must satisfy both the Rayleigh relation (Eqn. 16.11) **and** the Hugoniot relation (Eqn. 16.13), we immediately see that the points between B and C on the Hugoniot curve are unrealizable because no valid Rayleigh line can be drawn between point A and any point between B and C. We also see that, for the upper branch of the Hugoniot, there is a limiting Rayleigh line that is just tangent to the Hugoniot at point D. This point of tangency is designated the **upper Chapman–Jouguet point.** Similarly, there is a limiting Rayleigh line, just tangent to the lower branch of the Hugoniot. This point of tangency is called the **lower Chapman–Jouguet point.**

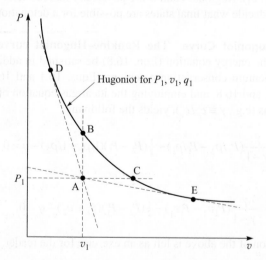

**Figure 16.3**     Rankine–Hugoniot curve for $q = q_1$ with origin A at $(P_1, v_1)$. Note that the curve does not actually pass through the so-called origin. Dashed lines passing through A are Rayleigh lines.

**Table 16.2**    Physical phenomena associated with various segments of the Hugoniot curve

| Region or Segment of Hugoniot Curve | Characteristic | Burned Gas Velocity,[a] $v_{x,2}$ |
|---|---|---|
| Above D | Strong detonation | Subsonic |
| D–B | Weak detonation | Supersonic |
| B–C | Inaccessible | — |
| C–E | Weak deflagration | Subsonic |
| Below E | Strong deflagration | Supersonic |

[a]Relative to wavefront.

The four limiting Rayleigh lines (A–D, A–B, A–C, and A–E) divide the Hugoniot into five segments. The physical characteristics associated with each Hugoniot segment are presented in Table 16.2. Above the upper Chapman–Jouguet point (D) are states associated with **strong detonations.** Although these states are mathematically realizable for our one-dimensional analysis, strong detonations are difficult to produce [3]. Between D and B is the region of **weak detonations.** Weak detonations, too, require special conditions (very rapid reaction rate) for them to occur. Although real detonations are not one-dimensional, conditions at the upper Chapman–Jouguet point reasonably approximate those associated with actual detonations. It can be shown that, at the upper Chapman–Jouguet point, $v_{x,2}$, the velocity of the burned gas relative to the traveling detonation wave is the sonic velocity.

Before we focus our attention on the structure of detonations, it is useful to see how we might fit together what we learned in Chapter 8 about laminar premixed flames, i.e., deflagrations, and the information contained in the Hugoniot curve of Fig. 16.3. Consider a point on the Hugoniot just below C to be representative of the conditions in the burned gases behind a one-dimensional flame. We note, first, that the pressure is just slightly less than that of the unburned gases. This is consistent with our assumption that the pressure is uniform throughout a flame and adds the information that there is actually a small pressure drop associated with the flame. The second point we wish to make is that the burned gas state must lie below C, as the Rayleigh line, A–C, associated with point C is a horizontal line. Since the slope of the Rayleigh line is $-\dot{m}''^2$, the flow rate associated with A–C is zero; hence, the physically inaccessible region includes the point C.

In Chapter 8, we learned that typical flame speeds for hydrocarbon–air mixtures are typically less than 1 m/s; thus, the mass fluxes ($\dot{m}'' = \rho_u S_L$) associated with such flames are quite small, especially when compared with the mass flux of a detonation traveling at supersonic speeds. For example, the mass flux for a stoichiometric methane–air mixture with reactants at 1 atm and 298 K is 0.45 kg/s-m²; whereas for an equivalent detonation, the estimated mass flux is 2000 kg/s-m². We conclude then that the slope of the Rayleigh line associated with a real flame in Fig. 16.3 must be quite small. This is our justification for choosing a point below C to be representative of a real flame.

| **Example 16.1** | A combustion wave propagates with a mass flux of 3500 kg/s-m² through a mixture initially at 298 K and 1 atm. The molecular weight and specific-heat ratio of the mixture (burned and unburned) are 29.0 kg/kmol and 1.30, respectively, and the heat release is $3.40 \cdot 10^6$ J/kg. Determine the state $(P_2, v_2)$ of the burned gas and determine in which region this state lies on the Rankine–Hugoniot curve. Also, determine the Mach number of the burned gases. |
| --- | --- |

### Solution

Before performing any calculations, we can obtain some insight into the problem solution by referring to Fig. 16.3. First, we note that the final state must lie above point B or below point C on the Hugoniot, as the region B–C is inaccessible. We also know that, if the burned gas state lies below C, the resulting combustion wave will be a deflagration. The given mass flux of 3500 kg/s-m² is quite large compared with deflagrations associated with ordinary fuels for the given unburned state properties; i.e., typical mixture densities and flame (deflagration) speeds are both of the order of unity, so $\dot{m}'' = \rho_u S_L$ is also of the order of unity, a value much smaller than 3500. We thus conclude that the final state most likely lies above B. Furthermore, unless by coincidence the final state happens to be at the upper Chapman–Jouguet point, two possible final states will exist since a Rayleigh line passing through the origin at A crosses the Hugoniot twice: first, at a location between B and D and, again, at a location above D. This preliminary exercise also shows that to obtain the solution requires the simultaneous solution of Eqn. 16.11b (the Rayleigh line) and Eqn. 16.13b (the Hugoniot curve).

To effect this simultaneous solution, we solve Eqn. 16.11b for $P_2$; that is,

$$P_2 = P_1 + \dot{m}''^2(v_1 - v_2),$$

and substitute this result into Eqn. 16.13b to yield

$$\frac{\gamma}{\gamma - 1}\left[\left[P_1 + \dot{m}''^2(v_1 - v_2)\right]v_2 - P_1 v_2\right]$$
$$-\frac{1}{2}\left[\left[P_1 + \dot{m}''^2(v_1 - v_2)\right] - P_1\right](v_1 + v_2) - q = 0.$$

Expanding the above and collecting terms, we obtain the following quadratic equation in $v_2$:

$$av_2^2 + bv_2 + c = 0,$$

where

$$a = \frac{1 + \gamma}{2(1 - \gamma)}\dot{m}''^2,$$

$$b = \frac{\gamma}{\gamma - 1}\left(P_1 + \dot{m}''^2 v_1\right),$$

$$c = \frac{\gamma}{1 - \gamma}P_1 v_1 - \frac{1}{2}\dot{m}''^2 v_1^2 - q.$$

To evaluate $a$, $b$, and $c$ numerically requires a value for $v_1$. This we easily obtain through the ideal-gas equation of state; i.e.,

$$v_1 = \frac{R_1 T_1}{P_1} = \frac{(8315/29)298}{101,325} = 0.843 \text{ m}^3/\text{kg}.$$

Thus,

$$a = \frac{1+1.3}{2(1-1.3)}(3500)^2 = -4.696 \cdot 10^7,$$

$$b = \frac{1.3}{1.3-1}(101,325+(3500)^2 0.843) = 4.519 \cdot 10^7,$$

$$c = \frac{1.3}{1-1.3}101,325(0.843) - 0.5(3500)^2(0.843)^2 - 3.4 \cdot 10^6 = -8.122 \cdot 10^6.$$

Solving for $v_2$,

$$v_2 = \frac{-b \pm (b^2 - 4ac)^{1/2}}{2a}$$

$$= \frac{-4.519 \cdot 10^7 \pm [(-4.519 \cdot 10^7)^2 - 4(-4.696 \cdot 10^7)(-8.122 \cdot 10^6)]^{1/2}}{2(-4.696 \cdot 10^7)}$$

$$= \frac{0.9623 \pm 0.4839}{2}$$

$$v_2 = 0.723 \text{ or } 0.239 \text{ m}^3/\text{kg}.$$

We can now apply Eqn. 16.11b to determine the pressures associated with each of the specific volumes:

$$P_2 = P_1 + \dot{m}''^2(v_1 - v_2),$$

which for $v_2 = 0.723$ yields

$$P_2 = 101,325 + (3500)^2(0.843 - 0.723)$$

$$P_2 = 1.57 \cdot 10^6 \text{ Pa}$$

and for $v_2 = 0.239$, yields

$$P_2 = 7.50 \cdot 10^6 \text{ Pa}$$

We see that, as expected, two burned gas states result. The first solution ($P_2 = 1.57 \cdot 10^6$ Pa, $v_2 = 0.723$ m$^3$/kg) lies on the Hugoniot somewhere between points B and D (see Fig. 16.3); while the second solution ($P_2 = 7.50 \cdot 10^6$ Pa, $v_2 = 0.239$ m$^3$/kg) lies above D.

To calculate the Mach numbers requires that we know both the velocities and local sound speeds. The velocities are calculated from the mass flux, $\dot{m}''(= \rho_2 v_{x,2} = v_{x,2}/v_2)$:

$$v_{x,2} = \dot{m}'' v_2$$

$$= 3500 (0.723) \text{ or } 3500 (0.239)$$

$$= 2530 \text{ or } 837 \text{ m/s}.$$

To calculate the sound speed, we first determine the state-2 temperature from the ideal-gas equation of state:

$$T_2 = P_2 v_2 / R_2$$

$$= \frac{1.57 \cdot 10^6 (0.723)}{286.7} = 3960 \text{ K}$$

or

$$T_2 = \frac{7.50 \cdot 10^6 (0.239)}{286.7} = 6250 \text{ K},$$

where $R_2 = 8315/29 = 286.7$ J/kg-K. Thus,

$$c_2 = (\gamma_2 R_2 T_2)^{1/2}$$
$$= [1.3(286.7)3960]^{1/2} = 1210 \text{ m/s}$$

or

$$c_2 = [1.3(286.7)6250]^{1/2} = 1530 \text{ m/s}.$$

Applying the definition $Ma = v_x/c$, the two corresponding Mach numbers are now readily determined:

$$Ma_2 = v_{x,2}/c_2$$

$$Ma_2 = 2530/1210 = 2.09$$

or

$$Ma_2 = 837/1530 = 0.55$$

**Comment**

This example clearly illustrates how the burned gas states in a detonation are determined by the simultaneous solution of the Rayleigh line and the Hugoniot curve. The first of two solutions ($Ma_2 = 2.09 > 1$) lies in the weak detonation region in which velocities are supersonic (see Table 16.2); whereas the second ($Ma_2 = 0.55 < 1$) represents a strong detonation, where velocities are subsonic.

## DETONATION VELOCITIES

In Chapter 8, the concept of flame speed was a major focus of our study of laminar premixed flames. In the same spirit, we now use the analysis of the previous section to determine detonation velocities. To accomplish this goal, we need add only one assumption to our previous list:

The pressure of the burned gases is much greater than that of the unburned mixture, i.e., $P_2 \gg P_1$.

We see from Table 16.1 that this is a reasonable approximation since $P_2$ is typically more than an order of magnitude greater than $P_1$.

To start our analysis, we define the **detonation velocity, $v_D$**, to be equal to the velocity at which the unburned mixture enters the detonation wave for an observer riding with the detonation. This definition is analogous to, and consistent with, our definition of the laminar flame speed in Chapter 8. From Fig. 16.1, it is obvious that

$$v_D \equiv v_{x,1}. \tag{16.15}$$

Recognizing that state 2 for the detonation is the upper Chapman–Jouguet point, where the gas velocity is sonic, we rewrite mass conservation (Eqn. 16.1) as

$$\rho_1 v_{x,1} = \rho_2 c_2.$$ (16.16)

Solving for $v_{x,1}$ and utilizing the fact that $c_2 = (\gamma R_2 T_2)^{1/2}$, we obtain

$$v_{x,1} = \frac{\rho_2}{\rho_1}(\gamma R_2 T_2)^{1/2}.$$ (16.17)

Our problem now is to relate the density ratio, $\rho_2/\rho_1$, and $T_2$ to upstream (state 1) or other known quantities. To find an appropriate expression for $\rho_2/\rho_1$, we start with momentum conservation Eqn. 16.2, divide through by $\rho_2 v_{x,2}^2$, and neglect $P_1$ compared with $P_2$, since we have assumed that $P_2 \gg P_1$. This yields the following:

$$\frac{\rho_1 v_{x,1}^2}{\rho_2 v_{x,2}^2} - \frac{P_2}{\rho_2 v_{x,2}^2} = 1.$$ (16.18)

We now apply continuity (Eqn. 16.1) to eliminate $v_{x,1} (= (\rho_2/\rho_1)v_{x,2})$ and solve the resulting expression for the density ratio $\rho_2/\rho_1$; i.e.,

$$\frac{\rho_2}{\rho_1} = 1 + \frac{P_2}{\rho_2 v_{x,2}^2}.$$ (16.19)

The next step is to replace $v_{x,2}^2$ with $c_2^2 (= \gamma R_2 T_2)$; thus,

$$\frac{\rho_2}{\rho_1} = 1 + \frac{P_2}{\rho_2 \gamma R_2 T_2}.$$ (16.20)

The final step is to recognize that, for an ideal gas, $P_2 = \rho_2 R_2 T_2$. With this, Eqn. 16.20 simplifies to the following:

$$\frac{\rho_2}{\rho_1} = 1 + \frac{1}{\gamma} = \frac{\gamma+1}{\gamma}.$$ (16.21)

We now focus on relating $T_2$ to known quantities. Solving the energy conservation equation (Eqn. 16.8) for $T_2$ yields

$$T_2 = T_1 + \frac{v_{1,x}^2 - v_{2,x}^2}{2c_p} + \frac{q}{c_p}.$$ (16.22)

As before, we can eliminate $v_{x,1}$ through continuity (Eqn. 16.1), and, in turn, substitute for $v_{x,2}^2$ the square of the sound speed, $c_2^2 = \gamma R_2 T_2$. We eliminate the density ratio, which appears from our use of continuity, using Eqn. 16.21. Performing these operations yields the following:

$$T_2 = T_1 + \frac{q}{c_p} + \frac{\gamma R_2 T_2}{2c_p}\left[\left(\frac{\gamma+1}{\gamma}\right)^2 - 1\right].$$ (16.23)

Solving Eqn. 16.23 for $T_2$ results in

$$T_2 = \frac{2\gamma^2}{\gamma+1}\left(T_1 + \frac{q}{c_p}\right), \tag{16.24}$$

where the substitution of $\gamma - 1$ for $\gamma R_2/c_p$ has been employed. This last substitution is easily derived from the property relation $c_p - c_v = R_2$.

We are now in a position to obtain closure on our derivation of an approximate expression for the detonation velocity. To accomplish this, Eqns. 16.21 and 16.24 are substituted into Eqn. 16.17 to yield

$$v_D = v_{x,1} = [2(\gamma+1)\gamma R_2(T_1 + q/c_p)]^{1/2}. \tag{16.25}$$

We caution that Eqn. 16.25 is approximate, not only because of the approximations employed in specifying the physics, but also because of the numerical approximation that $P_2 \gg P_1$.

If we relax the assumption of constant and equal specific heats, the following more accurate, but still approximate, expressions for the state-2 temperature and the detonation velocity can be derived:

$$T_2 = \frac{2\gamma_2^2}{\gamma_2+1}\left(\frac{c_{p,1}}{c_{p,2}}T_1 + \frac{q}{c_{p,2}}\right) \tag{16.26}$$

$$v_D = \left[2(\gamma_2+1)\gamma_2 R_2\left(\frac{c_{p,1}}{c_{p,2}}T_1 + \frac{q}{c_{p,2}}\right)\right]^{1/2}, \tag{16.27}$$

where $c_{p,1}$ and $c_{p,2}$ are mixture specific heats at states 1 and 2, respectively. Similarly, Eqn. 16.21 becomes

$$\rho_2/\rho_1 = (\gamma_2+1)/\gamma_2. \tag{16.28}$$

A more exact mathematical description of the one-dimensional detonation wave is presented by Kuo [3] and others [4]. A numerical solution method for such a formulation is described by Gordon and McBride [4] and implemented in the NASA Chemical Equilibrium Code (CEC).

---

**Example 16.2**

Estimate the detonation velocity for a stoichiometric $C_2H_2$–air mixture initially at 298 K and 1 atm. Neglect dissociation in the products. The molar specific heat of $C_2H_2$ at 298 K is 43.96 kJ/kmol-K.

**Solution**

To estimate the detonation velocity, we will apply Eqn. 16.27. To do this requires an estimation of $q$ and the burned gas properties $c_{p,2}$, $\gamma_2$, and $R_2$. Also needed is the unreacted mixture specific heat, $c_{p,1}$. To obtain these properties, we first determine the compositions of the unreacted and reacted mixtures. Applying Eqns. 2.30 and 2.31 yields

$$C_2H_2 + 2.5(O_2 + 3.76N_2) \rightarrow 2CO_2 + H_2O + 9.40N_2.$$

With this, we can determine the species mole and mass fractions:

| | $MW_i$ | $N_i$ | $\chi_i = N_i / N_{tot}$ | $Y_i = \chi_i MW_i / MW_{mix}$ |
|---|---|---|---|---|
| *Reactants (State 1)* | | | | |
| $C_2H_2$ | 26.038 | 1 | 0.0775 | 0.0705 |
| $O_2$ | 31.999 | 2.5 | 0.1938 | 0.2166 |
| $N_2$ | 28.013 | 9.4 | 0.7287 | 0.7129 |
| | | $MW_1 = \Sigma \chi_i MW_i = \mathbf{28.63}$ | | |
| *Products (State 2)* | | | | |
| $CO_2$ | 44.011 | 2 | 0.1613 | 0.2383 |
| $H_2O$ | 18.016 | 1 | 0.0806 | 0.0487 |
| $N_2$ | 28.013 | 9.4 | 0.7581 | 0.7129 |
| | | $MW_2 = \mathbf{29.79}$ | | |

Using the above and tabulated values of $\bar{c}_{p, O_2}$ and $\bar{c}_{p, N_2}$ from Appendix A, we calculate $c_{p,1}$:

$$c_{p,1} = \sum_{\text{state 1}} \chi_i \bar{c}_{p,i} / MW_1$$

$$= [0.0775\,(43.96) + 0.1938\,(29.315) + 0.7287\,(29.071)]/28.63$$

$$= 30.272/28.63 = 1.057 \text{ kJ/kg-K}.$$

To determine the properties at state 2 requires us to guess the temperature $T_2$ and, subsequently, iterate. We guess $T_2 = 3500$ K. Using $\bar{c}_{p,i}$s at 3500 K from Appendix A,

$$c_{p,2} = \sum_{\text{state 2}} \chi_i \bar{c}_{p,i} / MW_2$$

$$= [0.1613\,(62.718) + 0.0806\,(57.076) + 0.7581\,(37.302)]/29.79$$

$$= 42.995/29.79 = 1.443 \text{ kJ/kg-K}.$$

We now determine $R_2$ and $\gamma_2$:

$$R_2 = R_u / MW_2$$

$$= 8.315/29.79 = 0.2791 \text{ kJ/kg-K}$$

$$\gamma_2 = \frac{c_{p,2}}{c_{v,2}} = \frac{c_{p,2}}{c_{p,2} - R_2}$$

$$= 1.443/(1.443 - 0.2791) = 1.240.$$

We calculate $q$ from Eqn. 16.7 using enthalpies-of-formation from Appendices A and B converted to a mass basis:

$$q = \sum_{\text{state 1}} Y_i h_{f,i}^o - \sum_{\text{state 2}} Y_i h_{f,i}^o$$

$$= 0.0705(8708) + 0.2166(0) + 0.7129(0)$$

$$- 0.2383(-8942) - 0.0487(-13,424) - 0.7129(0)$$

$$= 3398.5 \text{ kJ/kg}.$$

We now have all of the information to calculate the detonation velocity from Eqn. 16.27:

$$v_D = \left[ 2(\gamma_2 + 1)\gamma_2 R_2 \left( \frac{c_{p,1}}{c_{p,2}} T_1 + \frac{q}{c_{p,2}} \right) \right]^{1/2}$$

$$= \left[ 2(2.240)(1.240)279.1 \left( \frac{1.057}{1.443} 298 + \frac{3398.5}{1.443} \right) \right]^{1/2}$$

$$v_D = 1998 \text{ m/s}.$$

Note that the above requires the use of $R_2$ with dimensions of J/kg-K, not kJ/kg-K. We now check the reasonableness of our initial guess of $T_2 = 3500$ K using Eqn. 16.26:

$$T_2 = \frac{2\gamma_2^2}{\gamma_2 + 1} \left( \frac{c_{p,1}}{c_{p,2}} T_1 + \frac{q}{c_{p,2}} \right)$$

$$= \frac{2(1.240)^2}{2.240} \left( \frac{1.057}{1.443} 298 + \frac{3398.5}{1.443} \right)$$

$$= 3533 \text{ K}.$$

We could now repeat the calculation with $c_{p,2}$ and $\gamma_2$ evaluated at 3533 K; however, we conclude that, for our purposes here, the estimate of $v_D$ is sufficiently accurate.

**Comment**

With a final temperature of 3533 K, dissociation—an effect we have neglected—is likely to be quite important. Knowing the density ratio (Eqn. 16.28) and the temperature ratio also allows us to estimate how well the approximation $P_2 \gg P_1$ is met: with $P_2/P_1 = (\rho_2/\rho_1)(MW_1/MW_2)(T_2/T_1) = (1.806)(28.63/29.79)(3533/298) = 20.6$, we conclude that the approximation holds to first order.

## STRUCTURE OF DETONATION WAVES

The structure of an actual detonation wave is quite complex. It is useful, however, to think conceptually of a detonation wave as consisting of a leading shock wave followed by a reaction zone. Since the thickness of a shock is of the order of a few molecular mean-free-paths, little or no chemical reaction occurs in this zone. In our study of chemical kinetics in Chapter 4, we saw that the probability of reaction occurring during a collision of reactive molecules is typically much less than unity; therefore, in the shock wave, where there are only a few collisions, reactive collisions become rare events. We thus conclude that the reaction zone follows the shock wave and is considerably greater in thickness. This general picture, idealized as one-dimensional, was conceived independently by Zeldovich [5], von Neumann [6], and Döring [7], and has come to be known as the ZND model of detonation structure. Figure 16.4 schematically illustrates this structure. For the ZND model, it is a relatively easy task to determine the state of the gas downstream of the shock zone

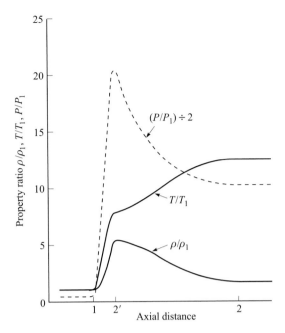

**Figure 16.4**  Schematic representation of ZND detonation structure. State points 1, 2′, and 2 denote upstream conditions, conditions at the end of the leading shock, and the Chapman–Jouguet point, respectively.

before any reaction occurs using ideal-gas shock relationships, and the final state downstream of the reaction zone is determined by the analysis shown in the previous section. The following example illustrates these ideas.

---

**Example 16.3**

Consider the detonation wave defined in Example 16.2. Using the ZND model for the detonation structure, estimate the gas-mixture properties ($T$, $P$, $\rho$, and $Ma$) immediately following the shock front (state 2′) and compare these with the properties at the end of the combustion zone (state 2).

**Solution**

To determine the state-2′ properties, we employ the following ideal-gas normal-shock equations, which are derived and presented in many basic fluid mechanics and thermodynamics textbooks [8–10]:

$$\frac{P_{2'}}{P_1} = \frac{1}{\gamma+1}\left[2\gamma\, Ma_1^2 - (\gamma-1)\right], \tag{I}$$

$$\frac{T_{2'}}{T_1} = \left[2+(\gamma-1)\,Ma_1^2\right]\frac{2\gamma\, Ma_1^2 - (\gamma-1)}{(\gamma+1)^2\, Ma_1^2}, \tag{II}$$

$$\frac{\rho_{2'}}{\rho_1} = \frac{(\gamma+1)\,Ma_1^2}{(\gamma-1)\,Ma_1^2+2}. \tag{III}$$

To evaluate Eqns. I, II, and III requires only a knowledge of the mixture specific-heat ratio, $\gamma$, and the Mach number at state 1. In Example 16.2, $c_{p,1}$ was calculated to be 1057 J/kg-K; thus,

$$c_{v,1} = c_{p,1} - R_u/MW_1$$
$$= 1057 - 8315/28.63$$
$$= 766.6 \text{ J/kg-K}$$

and

$$Ma_1 = v_{x,1}/c_1 = v_{x,1}/(\gamma R_1 T_1)^{1/2}$$
$$= \frac{1998}{[1.379(8315/28.63)\,298]^{1/2}} = \frac{1998}{345.5}$$
$$= 5.78.$$

The state-2′ properties are now straightforwardly calculated from Eqns. I and II:

$$\frac{P_{2'}}{P_1} = \frac{1}{1.379+1}[(2)1.379(5.78)^2 - (1.379-1)]$$
$$= 38.57,$$

so,

$$P_{2'} = 38.57 P_1 = 38.57(101,325)$$

$$P_{2'} = 3.908 \cdot 10^6 \text{ Pa}$$

$$\frac{T_{2'}}{T_1} = [2+(1.379-1)(5.78)^2]\frac{(2)1.379(5.78)^2 - (1.379-1)}{(1.379+1)^2(5.78)^2}$$

$$= \frac{14.66(91.76)}{189.1} = 7.11,$$

so

$$T_{2'} = 7.11 T_1 = 7.11(298)$$

$$T_{2'} = 2119 \text{ K}$$

Rather than applying Eqn. III to find $\rho_{2'}/\rho_1$, it is easier to use the ideal-gas equation of state, $\rho = P/RT$:

$$\rho_{2'} = \frac{3.908 \cdot 10^6}{(8315/28.63)\,2119}$$

$$\rho_{2'} = 6.35 \text{ kg/m}^3$$

For completeness, we calculate the density ratio,

$$\frac{\rho_{2'}}{\rho_1} = \frac{P_{2'}}{P_1}\frac{T_1}{T_{2'}} = \frac{38.57}{7.11} = 5.42.$$

To determine the state-2' Mach number, we first obtain $v_{x,\,2'}$ from mass conservation, $\dot{m}'' = \rho_{2'}v_{x,\,2'} = \rho_1 v_{x,\,1}$, using the value of $v_{x,\,1}$ $(= 1998$ m/s) obtained in Example 16.2:

$$v_{x,\,2'} = \frac{\rho_1}{\rho_{2'}}v_{x,\,1} = \frac{1998}{5.42} = 369 \text{ m/s}.$$

Thus,

$$Ma_{2'} = v_{x,\,2'}/c_{2'} = v_{x,\,2'}/(\gamma_{2'}R_{2'}T_{2'})^{1/2}$$

$$= \frac{369}{[1.379(8315/28.63)\,2119]^{1/2}} = \frac{369}{921}$$

$$= 0.40.$$

In the above calculation, we assume that $\gamma_{2'} \approx \gamma_1$ for simplicity and consistency with the constant-property assumption used to determine the state-2' properties from Eqns. I, II, and III. This completes our determination of the state-2' properties.

In Example 16.2, the temperature of the burned gases following the combustion zone of the detonation wave was determined; i.e., $T_2 = 3533$ K. Remaining for us is to calculate the other state-2 properties. We estimate the density using Eqn. 16.28, recognizing that this relation is approximate and holds only for $P_2 \gg P_1$:

$$\rho_2/\rho_1 = (\gamma_2 + 1)/\gamma_2 = (1.240 + 1)/1.240$$

$$= 1.806,$$

so,

$$\rho_2 = 1.806\rho_1 = 1.806\frac{P_1}{R_1 T_1}$$

$$= \frac{1.806(101{,}325)}{(8315/28.63)\,298} = 2.114 \text{ kg/m}^3.$$

The pressure is found from the ideal-gas equation of state:

$$P_2 = \rho_2 R_2 T_2$$

$$= 2.114\,(8315/29.79)\,3533$$

$$= 2.085 \cdot 10^6 \text{ Pa}.$$

The state-2 Mach number should be unity, the value associated with the upper Chapman–Jouguet point. We verify this as a check on our previous calculations. Once again, using the value of $v_{x,\,1}$ obtained in Example 16.2, the state-2 velocity is obtained through mass conservation $(\dot{m}'' = \rho_2 v_{x,\,2} = \rho_1 v_{x,\,1})$:

$$v_{x,\,2} = \frac{\rho_1}{\rho_2}v_{x,\,1} = \frac{1998}{1.806} = 1106 \text{ m/s}.$$

Thus

$$Ma_2 = v_{x,\,2}/c_2 = v_{x,\,2}/(\gamma_2 R_2 T_2)^{1/2}$$

$$= \frac{1106}{[1.240(8315/29.79)\,3533]^{1/2}} = \frac{1106}{1106}$$

$$= 1,$$

as expected!

The following table of results facilitates comparison of the properties at states 1, 2′, and 2:

| Property | State 1 | State 2′ | State 2 |
|---|---|---|---|
| $\rho / \rho_1$ | 1 | 5.42 | 1.806 |
| $P / P_1$ | 1 | 38.6 | 20.6 |
| $T / T_1$ | 1 | 7.11 | 11.85 |
| $Ma$ | 5.78 | 0.40 | 1.00 |

### Comment

From the table above, we see that the large density and pressure ratios across the initial shock front, 5.42 and 38.6, respectively, are ameliorated somewhat through the combustion zone as the temperature continues to increase as a result of the heat release. It is also interesting to point out that the flow is subsonic after the shock front (state 2′) and then accelerates to the sonic point at the trailing edge of the detonation wave (state 2).

Unfortunately, real detonations do not globally conform to the ZND structure, although the general idea of a thin, reactionless shock wave followed by a much thicker reaction zone is a useful substructure. Various researchers [11–14] have investigated detonations in the laboratory and found that there are several shock fronts interacting in the traveling detonation wave. A characteristic structure within a detonation wave is the triple-shock interaction, schematically illustrated in Fig. 16.5. In this figure, we see that the nominally normal shock front is composed of convex

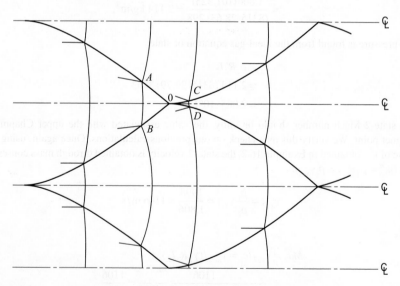

**Figure 16.5**    Shock fronts in an idealized two-dimensional detonation wave. Points labeled A, B, C, and D indicate triple-shock interactions. The wave is moving from left to right, thus the flow direction is opposite to that shown in Fig. 16.1.

SOURCE: From Strehlow [11]. Reprinted with permission of Krieger Publishing Co.

segments facing upstream (e.g., A–B) and that there is a transverse wave structure formed by the oblique shock fronts emanating from the triple-shock interactions. The detailed nature of this transverse wave structure is strongly dependent on the geometry of any confining tube, whereas the transverse wave structure is random for unconfined spherical detonations [11]. In tube-confined flows, the transverse detonation waves couple with transverse acoustic modes of the tube [11]. For a detailed description of detonation wave structure, the reader is referred to Williams [15].

## SUMMARY

We began this chapter with comparisons of detonations, shocks, and deflagrations. You should now be familiar with the commonalities and differences among these. Using a simplified one-dimensional analysis, we developed the concept of the Rankine–Hugoniot curve. Five separate segments of the Hugoniot were distinguished, with each of these corresponding to various physical phenomena. We discovered that the upper Chapman–Jouguet point defines the typical state of the gases after passage of a detonation wave. The one-dimensional analysis was further extended to obtain an approximate expression for calculating the detonation velocity. We examined the one-dimensional Zeldovich–von Neumann–Döring (ZND) model of shock structure, which consists of a thin normal shock with no reaction followed by a somewhat thicker reaction zone. You should be able to calculate the properties after the shock zone and at the end of the reaction zone using idealized one-dimensional approaches. The chapter concluded with a brief discussion of real detonation waves. You should be aware that these are highly three-dimensional and that the detailed structure of the detonation depends on the geometry of the confining environment, or the lack of such confinement.

## NOMENCLATURE

| | |
|---|---|
| $A$ | Area (m$^2$) |
| $c$ | Sound speed (m/s) |
| $c_p$ | Constant-pressure specific heat (J/kg-K) |
| $c_v$ | Constant-volume specific heat (J/kg-K) |
| $h$ | Enthalpy (J/kg) |
| $h_f^o$ | Enthalpy of formation (J/kg) |
| $\dot{m}$ | Mass flowrate (kg/s) |
| $\dot{m}''$ | Mass flowrate (kg/s) |
| $Ma$ | Mach number |
| $MW$ | Molecular weight (kg/kmol) |
| $P$ | Pressure (Pa) |
| $q$ | Heat addition, Eqn. 16.7 (J/kg) |
| $R$ | Specific gas constant (J/kg-K) |
| $R_u$ | Universal gas constant (J/kmol-K) |

| $T$ | Temperature (K) |
|---|---|
| $v_D$ | Detonation velocity (m/s) |
| $v_x$ | Axial velocity (m/s) |
| $v$ | Specific volume (m$^3$/kg) |
| $Y$ | Mass fraction |

**Greek Symbols**

| $\gamma$ | Specific-heat ratio, $c_p/c_v$ |
|---|---|
| $\rho$ | Density (kg/m$^3$) |
| $\Phi$ | Equivalence ratio |

**Subscripts**

| mix | Mixture |
|---|---|
| 1 | Upstream condition |
| 2 | Downstream condition |
| 2′ | Shock-wave trailing edge |

# REFERENCES

1. Friedman, R., "Kinetics of the Combustion Wave," *American Rocket Society Journal,* 23: 349–354 (1953).

2. Chapman, D. L., "On the Rate of Explosion of Gases," *Philosophical Magazine,* 47: 90–103 (1899).

3. Kuo, K. K., *Principles of Combustion,* 2nd Ed., Wiley, Hoboken, NJ, 2005.

4. Gordon, S., and McBride, B. J., "Computer Program for Calculation of Complex Chemical Equilibrium Compositions, Rocket Performance, Incident and Reflected Shocks, and Chapman–Jouguet Detonations," NASA SP-273, 1976.

5. Zeldovich, Y. B., "The Theory of the Propagation of Detonation in Gaseous Systems," *Experimental and Theoretical Physics, S.S.S.R.,* 10: 542 (1940). (English translation, NACA TM 1261, 1950.)

6. von Neumann, J., "Theory of Detonation Waves," Progress Report No. 238 (April 1942). OSRD Report No. 549, 1949.

7. Döring, W., "Über den Detonationvorgang in Gasen," *Ann. Phys. Leipzig,* 43: 421–436 (1943).

8. White, F. M., *Fluid Mechanics,* 3rd Ed., McGraw-Hill, New York, p. 530, 1994.

9. Shames, I. H., *Mechanics of Fluids,* 3rd Ed., McGraw-Hill, New York, p. 506, 1992.

10. Moran, M. J., and Shapiro, H. N., *Fundamentals of Engineering Thermodynamics,* 3rd Ed., Wiley, New York, p. 435, 1995.

11. Strehlow, R. A., *Fundamentals of Combustion,* Krieger Publishing Co., Malabar, FL, 1979.

12. Strehlow, R. A., "Multi-Dimensional Detonation Wave Structure," *Astronautica Acta,* 15: 345–357 (1970).

13. Oppenheim, A. K., and Soloukin, R. I., "Experiments in Gas Dynamics of Explosions," *Annual Review of Fluid Mechanics,* 5: 31–58 (1973).

14. Lee, J. H. S., "Dynamic Parameters of Gaseous Detonations," *Annual Review of Fluid Mechanics,* 16: 311–336 (1984).

15. Williams, F. A., *Combustion Theory,* 2nd Ed., Addison-Wesley, Redwood City, CA, 1985.

16. Lewis, B., and von Elbe, G., *Combustion, Flames and Explosions of Gases,* 3rd Ed., Academic Press, Orlando, FL, p. 545, 1987.

## PROBLEMS

**16.1** Derive the Rankine–Hugoniot relation, Eqn. 16.13.

**16.2** Determine the velocity of a detonation wave propagating through an ideal-gas mixture having a constant specific heat, $c_p$, of 1200 J/kg-K. The mixture is initially at 2 atm and 500 K. The heat release per unit mass of mixture is $3 \cdot 10^6$ J/kg and the mixture molecular weight is 29 kg/kmol.

**16.3** Assuming a ZND structure, determine the key state properties for the detonation wave described in problem 16.2; i.e., calculate $\rho$, $P$, $T$, and $Ma$ at states 1, 2′, and 2.

**16.4** Using the detonation wave of problem 16.2 and the results of problem 16.3 as a base case, determine the effect of the heat release parameter, $q$, on all of the properties of the detonation wave: $\rho$, $P$, $T$, and $Ma$ at states 1, 2′ and 2. To accomplish this, use a value of $q$ twice as large as that given in problem 16.2. Does the detonation velocity increase or decrease? Does the maximum pressure increase or decrease? Discuss.

**16.5** Consider a detonation wave traveling in a $H_2$–$O_2$ mixture having a molar oxidizer-to-fuel ratio of 3.0. Estimate the detonation velocity for an initial state $P_1 = 1$ atm and $T_1 = 291$ K. Neglect dissociation. How does your result compare to the experimentally determined value of 1700 m/s [16]? Discuss reasons why these values differ.

# chapter

# 17

# Fuels

## OVERVIEW

Concerns over global warming, environmental degradation, and national energy independence, among others, have resulted in a renewed interest in fuels. This interest is not limited to the scientific and engineering communities; articles frequently appear in newspapers and magazines about the many alternative fuels being considered for transportation and other applications. Examples include biodiesel, ethanol (either corn-based or cellulosic), Fischer–Tropsch liquids from coal or biomass, hydrogen, and others. In this chapter, we present the naming conventions and molecular structure of hydrocarbons, and other fuel molecules, found in fuel mixtures. We also discuss conventional and alternative fuels, highlighting interesting aspects of their production and/or use.

## NAMING CONVENTIONS AND MOLECULAR STRUCTURES

The purpose of this section is to acquaint the reader with some of the common hydrocarbons and alcohols used as fuels or in fuel blends. Here, we present naming conventions, fuel molecular structures, and related information on several fuel classes.

### Hydrocarbons

The various hydrocarbon families (Table 17.1) are differentiated by whether the fuel molecules consist entirely of single carbon-carbon bonds (C—C) or contain one double (C=C) or one triple (C≡C) bond, and whether the molecules are open chains (all chain ends unconnected) or form rings. The **alkanes, alkenes,** and **alkynes** are all open-chain structures, whereas the **cyclanes** and **aromatics** exhibit ring structures.

**Table 17.1**    Basic hydrocarbon families

| Family Name | Other Designations | Molecular Formula | Carbon-Carbon Bonding | Primary Molecular Structure |
|---|---|---|---|---|
| Alkanes | Paraffins | $C_nH_{2n+2}$ | Single bonds only | Straight or branched open chains |
| Alkenes | Olefins | $C_nH_{2n}$ | One double bond, remainder single | Straight or branched open chains |
| Alkynes | Acetylenes | $C_nH_{2n-2}$ | One triple bond, remainder single | Straight or branched open chains |
| Cyclanes | Cycloalkanes, Cycloparaffins, Naphthenes | $C_nH_{2n}$ or $(CH_2)_n$ | Single bonds only | Closed rings |
| Aromatics | Benzene family | $C_nH_{2n-6}$ | Resonance hybrid bonds (Aromatic bonds) | Closed rings |

For the open-chain families (alkanes, alkenes, and alkynes), the following nomenclature is used to denote the number of carbon atoms contained in a particular family member:

| | | |
|---|---|---|
| 1—meth | 5—pent | 9—non |
| 2—eth | 6—hex | 10—dec |
| 3—prop | 7—hept | 11—undec |
| 4—but | 8—oct | 12—dodec |

Using this nomenclature, and knowing that the word endings **–ane, –ene,** and **–yne** indicate how the carbon atoms are bound in the molecule, alkane, alkene, and alkyne family members containing three carbon atoms, for example, would be given as

Note that the older conventions can sometimes complicate fuel names for alkene and alkyne family members that have two, three, or four carbon atoms:

| | $C_1$ | $C_2$ | $C_3$ | $C_4$ |
|---|---|---|---|---|
| **Alkanes** | Methane | Ethane | Propane | Butane |
| **Alkenes** | — | Ethene | Propene | Butene |
| | — | Ethylene | Propylene | Butylene |
| **Alkynes** | — | Ethyne | Propyne | Butyne |
| | — | Acetylene | Methyl acetylene | Ethyl actylene |

The adjective **saturated** is used to denote hydrocarbon molecules in which the maximum number of hydrogen atoms is associated with the carbon atoms, i.e., the molecule has no double or triple bonds. Alkanes are saturated in that it is not possible to add another H atom without breaking a chain; on the other hand, both the alkenes and alkynes are **unsaturated** in that one or two additional H atoms could be added, respectively, by converting the double and triple carbon-carbon bonds to single bonds and using the valence electrons thus freed to form carbon-hydrogen bonds. Using the preceding $C_3$ families as examples, we see that propene can be saturated by the addition of two H atoms: $C_3H_6 + H_2 \rightarrow C_3H_8$; and propyne is saturated with the addition of four H atoms: $C_3H_4 + 2H_2 \rightarrow C_3H_8$.

Many of the higher hydrocarbons (carbon atom number $n \geq 3$) appear as **branched chains** rather than simple **straight chains.** Compounds with straight chains are denoted as **normal,** or with the prefix $n$-. For example, normal-pentane, or $n$-pentane, is represented as

$$n\text{-Pentane} \qquad
\begin{array}{ccccc}
\text{H} & \text{H} & \text{H} & \text{H} & \text{H} \\
| & | & | & | & | \\
\text{H}-\text{C}-&\text{C}-&\text{C}-&\text{C}-&\text{C}-\text{H} \\
| & | & | & | & | \\
\text{H} & \text{H} & \text{H} & \text{H} & \text{H}
\end{array}
\qquad C_5H_{12}$$

Alternatively, compounds having the same number of carbon and hydrogen atoms can be formed using branched chains. Such compounds are referred to as **structural isomers.** For example,

$$2\text{-Methyl butane} \qquad
\begin{array}{c}
\text{H} \\
| \\
\text{H}-\text{C}-\text{H} \\
\\
\text{H} \qquad | \qquad \text{H} \quad \text{H} \\
| \qquad | \qquad | \quad | \\
\text{H}-\text{C}-\text{C}-\text{C}-\text{C}-\text{H} \\
| \qquad | \qquad | \quad | \\
\text{H} \quad \text{H} \quad \text{H} \quad \text{H}
\end{array}
\qquad C_5H_{12}$$

is the branched-chain isomer of pentane: 2-methyl butane. The generic formula for both $n$-pentane and 2-methyl butane is $C_5H_{12}$. The "2" in 2-methyl butane refers to the carbon atom position (numbering from left to right) where the chain branch is attached. In this example, note that, by symmetry, the same molecular structure obtains with the addition of the chain at either the second or the third carbon atom;

thus, it makes no sense to refer to 3-methyl butane. One alkane isomer of particular engineering importance is **2,2,4-trimethyl pentane,** which is often denoted simply as **isooctane:**

2,2,4-Trimethyl pentane

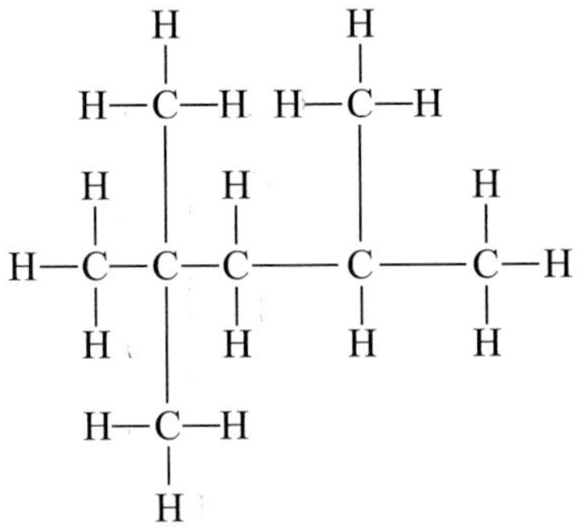

$C_8H_{18}$

This compound is used as a reference fuel in the knock rating of spark-ignition engines and is assigned an **octane rating** of 100. Fuels more prone to knock than isooctane thus have octane numbers less than 100, whereas fuels with greater knock resistance than isooctane have numerical ratings larger than 100. For aviation gasoline, ratings greater than 100 are called **performance numbers.** Octane ratings are discussed in a subsequent section.

The basic molecular structure of the **cyclanes** is a closed ring, with all carbon atoms singly bonded; for example, cyclopropane and cyclohexane have the following structure (hydrogen atoms not shown):

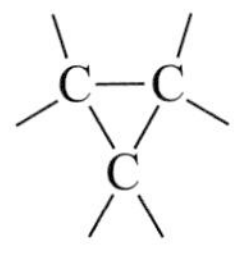

Cyclopropane, $C_3H_6$ 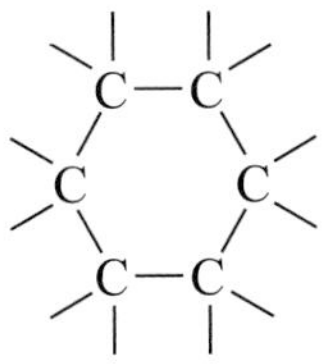        Cyclohexane, $C_6H_{12}$

More complex cyclanes are formed by the substitution of paraffinic groups for H atoms.

The **aromatic,** or **benzene derivative,** family is based on a ring of six carbon atoms, but with only one hydrogen atom associated with each carbon atom. The resulting six free valence electrons form so-called resonance hybrid bonds among the six carbon atoms in the ring. The six carbon-carbon bonds are all equivalent, with the bonding electrons delocalized over several atoms. The following diagrams are frequently used to indicate this special kind of bonding:

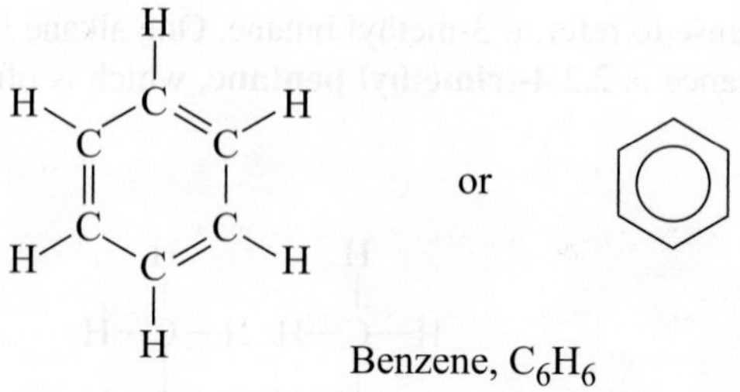

Benzene, $C_6H_6$

Benzene rings can combine to form polycyclic aromatics, and side chains may be substituted for hydrogen atoms.

## Alcohols

Common **alcohols** are formed by the substitution of a hydroxyl group (OH) for an H atom in an alkane molecule. For example, the one-, two-, and three-carbon alcohols are given as

Methanol
(Methyl alcohol)

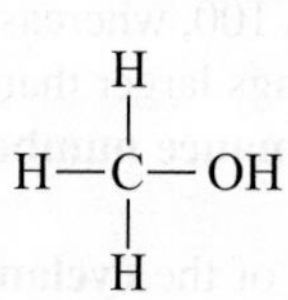

$CH_3OH$

Ethanol
(Ethyl alcohol)

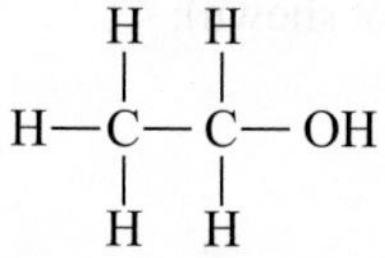

$C_2H_5OH$

Propanol
(Propyl alcohol)

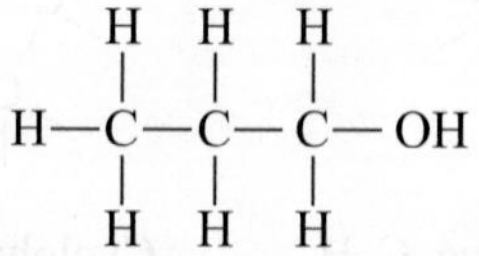

$C_3H_7OH$

Alcohols are also generically designated as ROH, where R is the parent hydrocarbon radical. The abbreviation EtOH is commonly used for ethanol.

## Other Organic Compounds

Many other important organic compounds are constituents in fuel blends or form as partial oxidation products of hydrocarbon combustion. Here we highlight several of these: ethers, aldehydes, ketones, and esters.

The general structure of an **ether** is R—O—R. Dimethyl ether (DME) is a common ether and is represented as

Dimethyl ether          $CH_3$—O—$CH_3$          $C_2H_6O$

The solvent and anesthetic, simply known as ether, is diethyl ether:

Diethyl ether          $CH_3$—$CH_2$—O—$CH_2$—$CH_3$          $C_4H_{10}O$

Trace quantities of aldehydes are emitted by diesel engines (see Chapter 15) and engines fueled with alcohols. Aldehydes contribute to photochemical smog. The general structure of an **aldehyde** is represented as follows:

$$
\begin{array}{c}
O \\
\| \\
C \\
\diagup \ \diagdown \\
R \qquad H
\end{array}
$$

The simplest aldehyde, formaldehyde, is a major intermediate species in the combustion of hydrocarbons, as shown in Figs. 5.4 and 5.5 in Chapter 5:

Formaldehyde                    $CH_2O$

Ketones are used as solvents and are formed as intermediate species in hydrocarbon combustion. The general structure of a ketone is

$$
\begin{array}{c}
O \\
\| \\
C \\
\diagup \ \diagdown \\
R \qquad R'
\end{array}
$$

The common solvent acetone is a ketone and has the following structure:

Acetone
(2-Propanone)                    $C_3H_6O$

**Esters** are important constituents of biodiesel and have the following general structure:

$$
\begin{array}{c}
O \\
\| \\
\diagup \ \diagdown \\
R \qquad OR'
\end{array}
$$

Esters have pleasant smells and are frequently used as fragrances and flavoring. One such ester is *n*-butyl acetate:

$$n\text{-Butyl acetate} \qquad \qquad \qquad \qquad C_6H_{12}O_2$$

---

# IMPORTANT PROPERTIES OF FUELS

## Ignition Characteristics

For reciprocating engines, the ignition quality of the fuel is extremely important. Because the combustion processes occurring in spark-ignition (premixed combustion) and diesel engines (nonpremixed combustion) result in different considerations of ignition quality, we consider them separately.

**Spark-Ignition Engines**   For spark-ignition engines, ignition quality relates to the prevention of engine **knock,** i.e., the uncontrolled spontaneous autoignition of the unburned mixture ahead of the flame. Knock in a spark-ignition engine is illustrated in Fig. 1.10 in Chapter 1. The **octane rating** or **octane number** is a measure of a spark-ignition engine fuel's ability to resist autoignition. You are probably familiar with this measure as different grades of gasoline are specified by their octane rating, which is prominently displayed at the pump where you purchase gasoline. Measuring the octane number of a fuel is a complex procedure, and various standardized tests [1, 2] have been developed, which are discussed in this chapter. In the United States and Canada, the octane rating specified at the pump is the **antiknock index (AKI)** and is an average of the laboratory-based **research octane number (RON)** [1] and **motor octane number (MON)** [2], i.e.,

$$AKI = (RON + MON)/2. \tag{17.1}$$

Regular grade gasoline typically has an AKI of 87; for automobiles requiring more knock-resistant fuels, higher-octane fuels are offered, e.g., those with ratings of 89, 91, or higher. Octane requirements decrease with altitude; thus, regular gasoline at high-altitude locations will have lower antiknock indices. For example, gasoline with an AKI of 85 is common in Denver, the "mile-high" city. Temperature and humidity also affect the antiknock performance of gasoline. Octane ratings based on driving a vehicle **(road octane numbers)** depend on a large number of variables; hence, laboratory methods have been adopted as standards. The octane number has no effect on the performance of an engine other than allowing the delivery of maximum power without knocking. If an engine operates without knocking, use of a higher-octane fuel will not result in more power or greater fuel economy.

ASTM International defines standard test procedures for determining research octane numbers [1] and motor octane numbers [2]. In these tests, a standardized, variable–compression ratio, single-cylinder engine is equipped with a knock sensor to measure knock intensity. In both procedures, the knock intensity of the fuel being rated is compared to that resulting from burning a blend of two primary reference fuels: isooctane (2,2,4-trimethyl pentane) and $n$-heptane. The octane number of isooctane is defined to be 100, and the octane number of $n$-heptane is defined to be zero. The percentage of the isooctane in the reference fuel blend, by volume, that matches the fuel being tested, defines the fuel's octane number. For example, a fuel that has the same knock index as a mixture of 80 percent isooctane and 20 percent $n$-heptane is designated to have an octane rating of 80. To rate fuels with octane numbers exceeding 100, e.g., aviation gasoline, a standard reference fuel is created by mixing tetraethyl lead with isooctane. For example, the octane rating of isooctane containing 2 ml tetraethyl lead per U.S. gallon has a research octane number of 112.8 [1]. The severity of the engine test conditions distinguishes the two test methods [3]: for the research method [1], conditions are less severe, with the engine operating at a moderate inlet mixture temperature and low engine speed ($600 \pm 6$ rpm); whereas, for the motor method [2], a higher mixture temperature and engine speed ($900 \pm 9$ rpm) are prescribed.

**Diesel (Compression-Ignition) Engines**    The combustion process in a diesel engine is initiated by the autoignition of the fuel–air mixture formed during the initial stages of the fuel-injection process. The time interval between the start of fuel injection and the onset of combustion, as distinguished by a pressure rise, is termed the **ignition delay.** Therefore, the ignition characteristics of the fuel determine the start and, hence, timing of the combustion process. (This is in contrast to spark-ignition engines, in which the combustion timing is established by the spark timing.) If a relatively large quantity of fuel is injected prior to ignition, the subsequent autoignition of the mixture results in a very rapid pressure rise and concomitant noise, i.e., knocking combustion; thus, short ignition delays are desired for diesel engine combustion. Thus, we see that the ignition characteristics of diesel fuels are in contradistinction to those of spark-ignition engine fuels: a good diesel fuel should be relatively easy to autoignite, whereas a good gasoline should be hard to autoignite.

The **cetane rating** or **cetane number** is a measure of a fuel's ability to autoignite. Low cetane numbers result in relatively long ignition delays and, conversely, higher cetane numbers imply relatively short ignition delays. ASTM International defines a standard test procedure for determining the cetane number of a fuel [4]. In this test, a standardized, variable–compression ratio, single-cylinder engine is equipped with sensors to measure ignition delay. In this procedure, the ignition delay of the fuel being rated is compared to that which results from operating with a blend of two primary reference fuels: $n$-cetane[1] and heptamethyl nonane. The cetane number of $n$-cetane ($C_{16}H_{34}$) is defined to be 100, and the

---

[1] $n$-Cetane is also known as $n$-hexadecane.

cetane number of heptamethyl nonane (also $C_{16}H_{34}$) is defined to be 15. Thus, the cetane number is determined by

$$\text{Cetane no.} \equiv \text{vol.\% } n\text{-cetane} + (0.15) \text{ vol.\% heptamethyl nonane.} \qquad (17.2)$$

The cetane scale ranges from 0 to 100, with typical fuels having cetane numbers in the range 30 to 65. The current use of heptamethyl nonane as a reference fuel replaces the original selection of $\alpha$-methylnaphthalene, which was defined to have a cetane number of zero. The poor storage characteristics of $\alpha$-methylnaphthalene required its replacement.

**Continuous Combustion Systems**    For continuous combustion systems, such as furnaces and jet engines, ignition quality is less important than in reciprocating engines; however, positive ignition must be assured for safety. In aircraft applications, jet engines must be capable of re-light in flight. Spray characteristics, which are affected by fuel density and viscosity, and fuel volatility control re-light ignition.

# Volatility

Different combustion applications require different fuel volatility characteristics. For example, spark-ignition engines use fuels that vaporize readily such that an essentially uniform fuel–air mixture is present at the time of spark; conversely, diesel engines use less volatile fuels, which prevent the formation of large quantities of combustible mixture during the ignition delay. Engine cold-starting characteristics, and other measures of driveability in automotive applications, depend strongly on fuel volatility. The tendency to develop **vapor lock,** that is, the formation of fuel vapor that can prevent flow within a fuel system, depends on a fuel's vapor pressure.[2] The volatility of a fuel also affects evaporative emissions and its safety and storage characteristics.

There are three primary measures of fuel volatility: (1) vapor pressure at a fixed temperature, (2) equilibrium air distillation profiles, and (3) liquid-vapor ratios at a fixed temperature. ASTM International provides standard test procedures for determining each of these measures.

For vapor pressure, the current accepted test method [5] is a modification of the Reid method.[3] In this test, the vapor pressure of the fuel, in the presence of a fixed quantity of air, is measured at a temperature of 38.8°C (100°F). Because of the presence of the air, the measured vapor pressure differs slightly from the true thermodynamic vapor pressure.

Because most fuels are blends of many components, there is no well-defined boiling point; hence, the distillation characteristics of the fuel are used to characterize a fuel's volatility. Such measures are usually expressed as the volume fraction

---

[2]Modern fuel-injected automobiles typically have their fuel pumps in the gas tank; hence, the fuel system is pressurized. This minimizes the problem of vapor lock that plagued vintage carbureted automobiles in hot weather.

[3]The term Reid vapor pressure (RVP) is frequently used to denote the result from this test method.

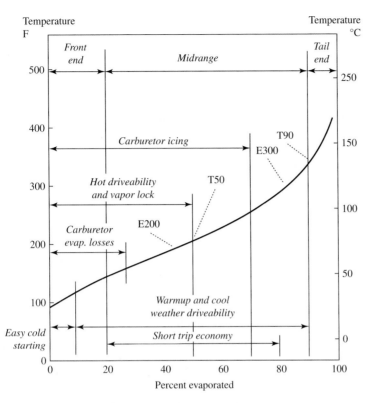

**Figure 17.1**     Correlation of distillation profile parameters [6] with gasoline performance. E200 designates the percentage (by volume) of the fuel evaporated at a temperature of 200 F; T50 designates the temperature at which 50 percent of the fuel has evaporated; E300 designates the percentage of the fuel evaporated at a temperature of 300 F; and T90 designates the temperature at which 90 percent of the fuel has evaporated.
| Adapted from Ref. [7] with permission.

evaporated at several selected temperatures within the range of 0–100 percent evaporated. ASTM D86 [6] prescribes standard test procedures for measuring distillation characteristics. Figure 17.1 illustrates the correlation of distillation parameters with the performance of automotive gasolines.

The temperature at which a fuel sample forms a vapor–liquid mixture in a ratio of 20:1 is a good measure of the vapor-lock tendency and other hot-fuel handling problems. Several standard procedures are available to make this measurement [3, 8].

In subsequent sections, note the importance of volatility in the standard specifications of various fuels.

## Energy Density

Energy density, expressed on either a mass or volumetric basis, is an important fuel property in many applications. For a fixed fuel tank volume, the range of a vehicle or aircraft depends on the fuel energy density. Among hydrocarbon fuels, variations in volumetric energy density are relatively small compared to those between a hydrocarbon and an alcohol. For example, the volumetric energy density of ethanol is about 30 percent less than that of gasoline. As we will see later, energy density is an important parameter in the specification of aircraft fuels.

Other fuel properties are of interest in particular applications. These and the properties above are noted in our subsequent discussion of specific fuels.

## CONVENTIONAL FUELS

### Gasoline

**Conventional Gasoline**    Gasoline is a blend of hydrocarbons, together with additives, formulated specifically for use in spark-ignition engines. Many requirements must be met by a gasoline: The fuel must resist autoignition (knocking); the fuel must be volatile enough to provide good cold-start characteristics and a homogeneous air-fuel mixture at the start of the combustion process; the fuel must not be too volatile such that it impairs the engine volumetric efficiency, or causes vapor lock, or has excessive evaporative emissions; the fuel must be compatible with the exhaust emissions control system, particularly the catalytic converter; the fuel must prevent the formation of deposits and gum; the fuel must be stable in storage; and the fuel must meet government regulations related to air quality concerns, among others. As a result of these, and other, requirements, the specifications presented in Table 17.2 control gasoline composition. This table is not intended to be exhaustive, and we refer the interested reader to ASTM Standard D4814–08b [3] and the other standards indicated in the table for more information.

Gasoline contains alkanes, alkenes, and aromatics. A general distribution of the various hydrocarbon types found in gasoline is shown in Table 17.3. Two hundred, or substantially more, individual hydrocarbon species make up gasoline [9]. The detailed specific composition of gasoline can vary widely; a gasoline produced from crude oil from the Arabian Gulf can differ significantly from those produced from Pennsylvania crude, or from a U.S. Gulf Coast crude oil. Various approximations are often used to represent gasoline. Thermodynamic properties for two such approximations, $C_{8.26}H_{15.5}$ and $C_{7.76}H_{13.1}$, are provided in Table B.2 in Appendix B.

To meet the volatility requirements defined in Table 17.4, gasoline contains relatively low–boiling point hydrocarbons, with molecules typically containing 4 to 12 carbon atoms [9]. The distillation characteristics for the least volatile distillation class (AA or summer) and the most volatile class (E or winter) gasoline are shown in Table 17.4. The greatest difference for these two specifications is at the lower end of the distillation curve; for example, the maximum temperature allowed to evaporate 10 percent of the mixture is $50°C$ for the winter (E class) gasoline, whereas the maximum temperature allowed to evaporate 10 percent of the mixture is $70°C$ for the summer (AA class) gasoline. The maximum end-point temperature ($225°C$) is the same for both specifications.

**Table 17.2**     Selected gasoline specifications for automotive spark-ignition engines

| Requirement | Parameter[a] | Typical (or Specified) Value(s) | U.S. Regulation, Standard or Test Method |
|---|---|---|---|
| Autoignition resistance | Antiknock index[b] = (RON + MON)/2 | | ASTM D2699 (RON) ASTM D2700 (MON) |
| Volatility[c] | Distillation temperatures for evaporation of 10%, 50%, 90%, and 100% by volume | | ASTM D4814–08b |
| Volatility[c] | Reid vapor pressure (RVP) = Vapor pressure (psi) at 100 F in presence of air | | ASTM 323 |
| Volatility[c] | Driveability Index[d] | | |
| Vapor Lock Protection[c] | Vapor lock protection class (1–6) relates the minimum temperature to produce a vapor-liquid ratio of 20 | | ASTM 2533 or ASTM 5188 |
| Exhaust emissions/ catalytic converter protection | Maximum lead content | 0.013 g/US gal | ASTM D4814–08b |
| Exhaust emissions/ catalytic converter protection | Maximum sulfur content | 80 ppm (by mass) | ASTM D4814–08b |
| Minimize fuel system corrosion | Copper strip corrosion number | No. 1 (max) | ASTM D130 |
| Minimize corrosion of fuel gauge in-tank sender units | Silver strip classification | 1 max (range 0–4) | ASTM D4814–08b |

[a]Detailed descriptions of these parameters are found in ASTM D4814–08b [3].

[b]The antiknock index is the average of the research octane number (RON), which is determined using a standardized single-cylinder test engine operating at a relatively low inlet temperature and engine speed (mild conditions), and the motor octane number (MON), which is determined using a standardized single-cylinder test engine operating at a relatively high inlet temperature and engine speed (more severe conditions).

[c]The various measures of volatility are used to establish various volatility classes. The U.S. EPA specifies which of these classes of gasoline are to be delivered to end-user distribution points depending on season and geographic location.

[d]ASTM D4814–08b [3]: "Driveability index (DI) is intended to provide control of distillation parameters and ethanol content that influence cold start and warm-up driveability. It is a function of the 10%, 50%, and 90% evaporated distillation temperatures…"

**Reformulated Gasoline (RFG)**     To improve air quality in selected regions of the United States, the 1990 Amendments to the Clean Air Act required gasoline producers to supply gasoline blends specifically tailored to reduce (1) volatile organic compound (VOC) emissions, (2) toxic air pollutant(s) (TAP) emissions (viz., benzene, 1,3-butadiene, polycyclic organic matter, formaldehyde, and acetaldehyde), and (3) oxides of nitrogen ($NO_x$) emissions [10]. Also, because of its carcinogenicity,

**Table 17.3**     General distribution of hydrocarbons in gasoline [9]

| Hydrocarbon Type | Proportion Range |
|---|---|
| Alkanes | 4–8% |
| Alkenes | 2–5% |
| Iso-alkanes | 25–40% |
| Cycloalkanes | 3–7% |
| Cycloalkenes | 1–4% |
| Aromatics | 20–50% |

benzene is restricted as a reformulated gasoline component to no more than 1 vol.%. The original provisions of the law specified a $\geq 2$ wt.% oxygen content, which was usually met with the addition of methyl tertiary-butyl ether (MTBE) or ethanol. The MTBE molecule, $C_5H_{12}O$, contains a single oxygen atom sandwiched between two carbon atoms. Because of concerns over MTBE contamination of groundwater, ethanol has become the oxygenate of choice for reformulated gasoline. The current regulations (§80.41 of [10]) do not require any specific oxygen content in reformulated gasoline, only that the fuel provides the specific reductions for VOCs, TAPs, and $NO_x$. The Energy Policy Act of 2005 [11], however, requires that gasoline contain a specified renewable fuel component, and this requirement is typically met with the ethanol.

Portions of 15 states require the use of reformulated gasoline, as shown in Fig. 17.2. The continental United States, including the District of Columbia, is divided into two VOC-control regions, with separate standards for the reduction of VOCs and $NO_x$ in each region. VOC-Control Region 1 comprises 23 mostly southern states, and VOC-Control Region 2 comprises 26 mostly northern states (§80.71 of [10]). The emission reduction standards and minimum benzene content for reformulated gasoline are shown in Table 17.5 (§80.41 of [10]). Table 17.6 presents the baseline

**Table 17.4**     Volatility requirements for summer and winter gasoline[a]

| Parameter | Summer Gasoline (Distillation Class AA) | Winter Gasoline (Distillation Class E) |
|---|---|---|
| Maximum distillation temperature (°C) for 10% (vol.) evaporated | 70 | 50 |
| Minimum distillation temperature (°C) for of 50% (vol.) evaporated | 77 | 66 |
| Maximum distillation temperature (°C) for of 50% (vol.) evaporated | 121 | 110 |
| Maximum distillation temperature (°C) for of 90% (vol.) evaporated | 190 | 185 |
| Maximum end-point temperature (°C) (100% evaporated) | 225 | 225 |
| Maximum Reid vapor pressure (kPa) | 54 | 103 |

[a]ASTM D4814 D–08b [3] defines six distillation classes. The two classes used here (AA and E) are the extremes of these six.

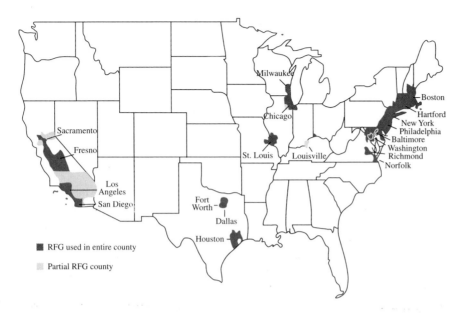

**Figure 17.2**    Map of areas requiring reformulated gasoline as of May 1, 2007.
Source: U.S. Environmental Protection Agency.

**Table 17.5**    Standards for Reformulated Gasoline[a]

| | |
|---|---|
| VOC emissions performance reduction (percent): | |
| Gasoline designated for VOC-Control Region 1 | |
|    Standard | $\geq 29.0$ |
|    Per-gallon minimum | $\geq 25.0$ |
| Adjusted VOC gasoline designated for VOC-Control Region 2 | |
|    Standard | $\geq 25.4$ |
|    Per-gallon minimum | $\geq 21.4$ |
| All other gasoline designated for VOC-Control Region 2 | |
|    Standard | $\geq 27.4$ |
|    Per-gallon minimum | $\geq 23.4$ |
| Toxic air pollutant emissions performance reduction (percent) | $\geq 21.5$ |
| $NO_x$ emissions performance reduction (percent): | |
|    Gasoline designated as VOC-controlled | $\geq 6.8$ |
|    Gasoline not designated as VOC-controlled | $\geq 1.5$ |
| Benzene (percent, by volume): | |
|    Standard | $\leq 0.95$ |
|    Per-gallon maximum | $\leq 1.30$ |

[a]Phase II Complex Model Averaged Standards from 40 CFR §80.41(f) [10].
Emissions reductions include both exhaust and non-exhaust components.

**Table 17.6**    Baseline exhaust emissions used to establish reformulated gasoline standards

| Exhaust Pollutant | Summer (mg/mi) | Winter (mg/mi) |
|---|---|---|
| Volatile organic compounds (VOC) | 907.0 | 1341.0 |
| Oxides of nitrogen ($NO_x$) | 1340.0 | 1540.0 |
| Benzene | 53.54 | 77.62 |
| Acetaldehyde | 4.44 | 7.25 |
| Formaldehyde | 9.70 | 15.34 |
| 1, 3-Butadiene | 9.38 | 15.84 |
| Polycyclic organic matter (POM) | 3.04 | 4.50 |

exhaust emissions (§80.45 of [10]) that serve as part of the basis for the reductions required by the standard. Non-exhaust emissions are also folded into the standards using complex prescribed formulas. The composition and properties of the baseline fuel, which approximate those of U.S. gasoline in 1990, are shown in Table 17.7.

To meet the required reductions for $NO_x$, fuel sulfur content was reduced to improve the performance of three-way catalyst systems (see Chapter 15). From 1997 to 2005, average annual sulfur content in all gasoline fell from about 300 ppm (by weight) to 90 ppm [12]. Additional legislative requirements [13] limited the average sulfur content to 30 ppm starting in 2006.

**Winter Oxygenated Fuel**    The 1990 amendments to the Clean Air Act require the use of oxygenated fuels to reduce vehicular carbon monoxide emissions during selected periods in the winter. The specific time periods and fuel oxygen contents are controlled by state governments as part of their State Implementation Plans (SIPs) to meet National Ambient Air Quality Standards (NAAQS). The situation is dynamic, as various areas in the United States attain or exceed air quality requirements. In 2008, eight metropolitan regions had winter oxygenated fuel requirements to attain or maintain NAAQS [14]: El Paso, Texas; Missoula, Montana; Las Vegas, Nevada; Phoenix, Arizona; Los Angeles, California; Reno, Nevada; Albuquerque, New Mexico; and Tucson, Arizona. Ethanol was the oxygenate of choice in all eight areas, with oxygen content ranging from 1.8 to 3.5 wt.%, depending on the area.

**Table 17.7**    Summer and winter baseline fuel properties [3]

| Fuel Property | Summer | Winter |
|---|---|---|
| Oxygen (wt.%) | 0.0 | 0.0 |
| Sulfur (ppm) | 339 | 338 |
| Reid vapor pressure (psi) | 8.7 | 11.5 |
| E200[a] (%) | 41.0 | 50.0 |
| E300[b] (%) | 83.0 | 83.0 |
| Aromatics (vol.%) | 32.0 | 26.4 |
| Olefins, i.e., alkenes (vol.%) | 9.2 | 11.9 |
| Benzene (vol.%) | 1.53 | 1.64 |

[a]Volume percentage evaporated at 200 F.
[b]Volume percentage evaporated at 300 F.

Determine the amount of ethanol that would have to be added to isooctane such that the ethanol–isooctane blend has a 2.7 wt.% oxygen content. Express the result as the ethanol mass fraction in the blend.

**Example 17.1**

**Solution**

We begin by defining the oxygen mass fraction in the blend in terms of the two components:

$$\frac{m_O}{m_{mix}} = \frac{m_O}{m_{EtOH} + m_{C_3H_8}} = 0.027.$$

The subscript *EtOH* designates ethanol. Dividing the above by the mass of the ethanol, we obtain an expression containing the ratio of the ethanol to the isooctane in the blend:

$$\frac{m_O/m_{EtOH}}{1 + m_{C_8H_{18}}/m_{EtOH}} = 0.027.$$

Recognizing that there is one mole of oxygen contained in one mole of ethanol, the mass of oxygen per mass of ethanol in the numerator is evaluated using the atomic weight of the oxygen and the molecular weight of the ethanol as follows:

$$\frac{m_O}{m_{EtOH}} = \frac{N_O}{N_{EtOH}} \frac{MW_O}{MW_{EtOH}} = \frac{(1)MW_O}{(1)MW_{EtOH}}.$$

Substituting this result into the prior equation and solving for $m_{EtOH}/m_{C_8H_{18}}$ yields

$$\frac{m_{EtOH}}{m_{C_8H_{18}}} = \frac{0.027}{\dfrac{MW_O}{MW_{EtOH}} - 0.027},$$

which we numerically evaluate as

$$\frac{m_{EtOH}}{m_{C_8H_{18}}} = \frac{0.027}{\dfrac{15.999}{46.069} - 0.027} = 0.0843.$$

From this result, we can easily obtain the required mass fraction through its definition:

$$Y_{EtOH} = \frac{m_{EtOH}}{m_{EtOH} + m_{C_8H_{18}}} = \frac{m_{EtOH}/m_{C_8H_{18}}}{1 + m_{EtOH}/m_{C_8H_{18}}}.$$

Substituting the numerical value for $m_{EtOH}/m_{C_8H_{18}}$, we obtain our final result:

$$\boxed{Y_{EtOH}} = \frac{0.0843}{1 + 0.0843} = \boxed{0.0777 \text{ or } 7.77 \text{ wt.\%}}$$

**Comment**

To obtain the 2.7 wt.% oxygen level typical of winterized gasoline requires a substantial ethanol addition, i.e., the ethanol addition is about 2.88 times the desired oxygen content.

## Diesel Fuels

Diesel fuel is a blend of hydrocarbons, sometimes containing additives, formulated specifically for use in compression-ignition (i.e., diesel) engines. In the United States, designations and properties of diesel fuels are specified by ASTM International Standard D975–08a [15]. There are three numerical grades of diesel fuel [15]: No. 1-D, a light middle distillate fuel; No. 2-D, a middle distillate fuel; and No. 4-D, a heavy distillate fuel, or a blend of distillate and residual oil. The mean molecular weight and viscosity of the fuels increase as the number designation increases. The numerical grades are also categorized with regard to their sulfur content to create a total of seven grades; the sulfur designations being S15, S500, and S5000, which refer to maximum sulfur contents of 15, 500, and 5000 ppm (by weight), respectively. For example, the three grades of No. 2-D fuel are denoted as No. 2-D S15, No. 2-D S500, and No. 2-D S5000, and similarly, for No. 1-D fuel. There is only one grade specification for No. 4-D, and thus no additional designation is applied. The maximum allowed sulfur content [15] for No. 4-D is 2 wt.%. Most on-road diesel-powered vehicles (automobiles, trucks, and buses) are fueled by No. 2-D S15 fuel. Low-sulfur fuels (S15) are used to minimize the emission of particulate matter and to prevent the degradation of exhaust treatment devices, as discussed in Chapter 15. The need for improved cold weather operability during the winter in certain regions results in modifying the manufacturing process at the refinery to avoid wax formation, or dilution of the No. 2 diesel with No. 1 diesel after its distribution from the refinery [16].

Many requirements must be met by diesel fuels: The fuel must provide good ignition and cold-start characteristics; the fuel must have sufficient lubricity and viscosity to provide proper functioning and wear characteristics of the fuel-injection system; the fuel must allow proper engine operation at low temperature; the fuel must be stable, noncorroding, and clean; the fuel must be compatible with the exhaust emissions control system; and the fuel must meet government regulations related to air quality concerns, among others. As a result of these, and other, requirements, the specifications presented in Table 17.8 control diesel fuel compositions. This table is not intended to be exhaustive, and we refer the interested reader to ASTM Standard D975–08a [15] and Ref. [16] for more information.

Diesel fuels consist of blends of primarily $C_9$–$C_{16}$ hydrocarbons (No. 1) and $C_{11}$–$C_{20}$ (No. 2) [17]. Approximate proportions of the various hydrocarbon families comprising diesel are shown in Table 17.9, and a more detailed analysis of a commercial diesel fuel is shown in Fig. 17.3. For simple combustion calculations, these fuels are frequently approximated as single molecules. For example, $C_{12}H_{22}$ and $C_{15}H_{25}$, respectively, are reasonable approximations of these fuels. Other values are also used.

## Heating Oils

Like diesel fuels, the lighter grade heating oils are designated as No. 1 and No. 2 fuel oils. The composition and properties of these heating oils are essentially the same as the corresponding grade of diesel fuel, with the differences primarily being the

**Table 17.8**    Selected specifications for diesel fuel oils[a]

| Property[b] | No. 1-D S15, S500, S5000 | No. 2-D S15, S500, S5000 | No. 4-D |
|---|---|---|---|
| Cetane number (minimum) | 40 | 40 | 30 |
| Distillation temperature, °C (evap. of 90% by volume) (minimum/maximum) | —/288 | 282/338 | —/— |
| Kinematic viscosity at 40 °C, mm²/s (minimum/maximum) | 1.3/2.4 | 1.9/4.1 | 5.5/24 |
| Water and sediment, % vol. (maximum) | 0.05 | 0.05 | 0.05 |
| Ash, % mass (maximum) | 0.01 | 0.01 | 0.10 |
| Copper strip corrosion number (maximum) | No. 3 | No. 3 | No. 3 |
| Aromatics,[c] % vol. (maximum) | 35 | 35 | — |
| Lubricity,[d] μm (maximum) | 520 | 520 | — |
| Conductivity,[e] pS/m | 25 | 25 | 25 |

[a]More complete specifications are contained in ASTM D975–08a [15].

[b]Detailed descriptions of these properties and required test methods are found in Ref. [15].

[c]This specification only applies to the S15 and S500 fuels. United State federal law (40 CFR part 80) limits the aromatics content to control NO$_x$ emissions. A cetane index (not to be confused with the cetane number) is also used as a surrogate specification of aromatics content.

[d]Wear scar diameter defined using high frequency reciprocating rig (HFFR) method.

[e]This specification only applies to the S15 and S500 fuels. Procedures used to reduce the sulfur content can result in reduced electrical conductivity of the fuel and allow dangerous charge buildup. Additives are often used to meet the conductivity specification [16].

additives and sulfur contents. Specifications for these grades, as well as the heavier No. 4, No. 5, and No. 6 oils, are provided in ASTM Standard D396 [19]. Domestic oil-burning heating systems typically use No. 2 fuel oil. The heavier oils are used in commercial and industrial applications; preheating may be needed for handling and proper atomization of these heavier fuel oils [19]. Specifications control flash point, water and sediment, distillation characteristics, kinematic viscosity, carbon residue, ash, sulfur, corrosion, density, and pour point [19].

## Aviation Fuels

Aviation fuels are differentiated by their applications. Aviation gasoline is employed in the spark-ignition engines typically used in general aviation. Turbo-shaft and

**Table 17.9**    General distribution of hydrocarbons in No. 1-D and No. 2-D fuels [17]

| Hydrocarbon Type | Approximate Proportion (Volume %) |
|---|---|
| Alkanes (*n*-, *iso*-, and *cyclo*-) | 64 |
| Alkenes | 1–2 |
| Aromatics | 35 |

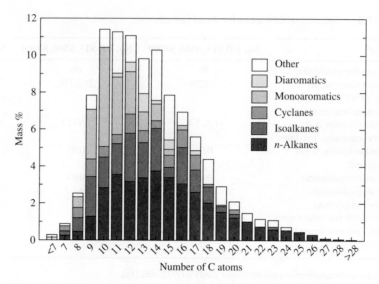

**Figure 17.3** Detailed composition of a commercial diesel fuel.
Adapted from [18] and used with permission.

jet engines, however, utilize kerosene-based fuels (Jet A and Jet A-1 for civil applications and JP-8 for military applications). In the United States, Jet A is used exclusively; whereas, Jet A-1, with a lower freezing point, is used in much of the rest of the world [20]. A wide-cut fuel (Jet B for civil applications and JP-4 for military applications) with distillation characteristics between gasoline and kerosene[4] is used in cold regions, such as some portions of Canada and Alaska [20]. Table 17.10 provides distillation and freezing point specifications for the three major classes of aviation fuels. Reference [20] provides a highly readable overview of aviation fuels.

**Table 17.10** Selected specifications for aviation gasoline, Jet B, and Jet A and Jet A-1 fuels

| Property | Aviation Gasoline [21] | Jet B [22] | Jet A and Jet A-1 [23] |
|---|---|---|---|
| Approximate range of component carbon numbers | $C_4-C_{10}$ | $C_5-C_{15}$ | $C_8-C_{16}$ |
| 10% evaporation distillation temperature, °C | 75 | 90 | 205 |
| 90% evaporation distillation temperature, °C | 135 | 245 | Final boiling point restricted to 340 °C. |
| Freezing point, °C (max.) | −58 | −50 | −40 Jet A; −47 Jet A-1 |
| Aromatics, % vol. (max.) | —[a] | 25 | 25 |

[a]Although there is no specification for aromatics, other specifications, in effect, limit the aromatic content to less than 25% vol. [21].

[4]Kerosene is a generic name for oil with distillation characteristics between gasoline and diesel oils [9]. Kerosine is an alternate spelling.

**Aviation Gasoline** Specifications for aviation gasoline [21] consider combustion characteristics and antiknock quality; fuel metering and aircraft range, as controlled by density and heat of combustion; carburetion and fuel vaporization, controlled by vapor pressure and distillation characteristics; corrosion; fluidity at low temperatures; and fuel cleanliness, handling, and storage stability. The basic composition of aviation gasoline differs from that of automotive gasoline; aviation gasoline consists of alkanes and *iso*-alkanes (50%–60%), cyclanes (20%–30%), small amounts of aromatics (<10%), and essentially no alkenes [9]. This composition contrasts with the somewhat smaller proportions of alkanes, *iso*-alkanes, and cycloalkanes; the presence of alkenes and cycloalkenes; and a greater proportion of aromatics (20%–50%) for automotive gasoline (see Table 17.3). The low aromatic content of aviation fuels results from the combined need to minimize the effects of the fuel on elastomers and to provide a high heating value and proper distillation characteristics [21]. The ASTM standard for aviation gasoline [21] indicates that an aromatics content of more than 25 percent is extremely unlikely. Tetraethyl lead is added to aviation gasoline to meet octane/performance number requirements. The decomposition products of tetraethyl lead scavenge radical species that lead to autoignition.

**Aviation Turbine Fuels** Specifications for Jet A and Jet A-1 turbine fuels [23] consider a large number of characteristics. Among them are the following: energy content, combustion, volatility, fluidity, corrosion, thermal stability, contaminants, and additives. The heat of combustion (see Chapter 2) is important as it controls the maximum range of an aircraft. The volumetric heating value (MJ/gallon) is the governing parameter for civil aviation [20], and a gravimetric specification (MJ/kg) is given in [23]. Minimizing the production of soot is important to meet emissions requirements and to minimize radiation to the combustor liner (see Chapters 10 and 15). To control soot formation, the aromatics content of the fuel is limited to 25 vol.%, and a minimum smoke point (see Tables 9.5 and 9.6 in Chapter 9) is specified. Aromatics are precursors to soot formation in the combustion of any fuel, and their presence as a fuel component promotes the production of soot. The freezing point is an important property as ambient temperatures at altitude can be quite low. Because the fuel is a mixture of many different hydrocarbons, the various components freeze (become wax) at various temperatures. The ASTM specification [23] of the freezing point is defined as the temperature at which the last wax crystals melt upon heating an initially completely solid fuel. Pumpability is the primary issue associated with fuel freezing, and most fuels will remain pumpable at temperatures slightly below the ASTM freezing point [20]. Jet A-1 fuel owes its existence to having a lower freezing point than Jet A. For more information on aviation fuels, we refer the reader to Refs. [20], [23], and [24].

## Natural Gas

Natural gas is typically found within or near oil fields. Natural gas is classified as *associated* or *nonassociated,* depending upon whether it is a product from an oil well (associated gas) or is the product of a gas well (nonassociated). Depending upon its composition, wellhead natural gas, particularly associated gas, must be processed before

**Table 17.11**   Typical values or ranges of specifications for pipeline-quality natural gas[a]

| Property or Specification | Typical Value or Range | Comment |
|---|---|---|
| Presence of solids | Commercially free | — |
| Oxygen ($O_2$) vol.% | <0.2%–1% | Two companies specified a significantly stricter requirement of <50 ppm. |
| Carbon dioxide ($CO_2$) and nitrogen ($N_2$) vol.% | <2% $CO_2$ and/or <4% $CO_2$ & $N_2$ combined | These represent typical specifications from the variety presented in [27].[b] |
| Liquid hydrocarbons | No liquid HCs at temperature and pressure of delivery point | — |
| Hydrogen sulfide ($H_2S$) | 5.7–23 mg/m$^3$ | — |
| Total sulfur | 17–460 mg/m$^3$ | In addition to $H_2S$, includes carbonyl sulfide, mercaptans, and mono-, di-, and poly sulfides. |
| Water ($H_2O$) | 65–110 mg/m$^3$ | — |
| Lower heating value | >36,000 kJ/m$^3$ typical | 34,500 to > 40,900 kJ/m$^3$ range |

[a]Information in this table was compiled from data presented for 18 pipeline companies in Ref. [27]. Values in U.S. customary units have been converted to SI units.

[b]$CO_2$ is removed both to prevent corrosion and to maintain an appropriately high heating value.

**Table 17.12**   Composition (mol%) and properties of natural gas from sources in the United States [28][a]

| Location | $CH_4$ | $C_2H_6$ | $C_3H_8$ | $C_4H_{10}$ | $CO_2$ | $N_2$ | Density[c] (kg/m$^3$) | HHV[d] (kJ/m$^3$) | HHV[d] (kJ/kg) |
|---|---|---|---|---|---|---|---|---|---|
| Alaska | 99.6 | — | — | — | — | 0.4 | 0.686 | 37,590 | 54,800 |
| Birmingham, AL | 90.0 | 5.0 | — | — | — | 5.0 | 0.735 | 37,260 | 50,690 |
| East Ohio[b] | 94.1 | 3.01 | 0.42 | 0.28 | 0.71 | 1.41 | 0.723 | 38,260 | 52,940 |
| Kansas City, MO | 84.1 | 6.7 | — | — | 0.8 | 8.4 | 0.772 | 36,140 | 46,830 |
| Pittsburgh, PA | 83.4 | 15.8 | — | — | — | 0.8 | 0.772 | 41,840 | 54,215 |

[a]Although not explicitly stated in Ref. [28], these gases appear to be pipeline gases.

[b]Also contains 0.01% $H_2$ and 0.01% $O_2$.

[c]At 1 atm and 15.6°C (60 F).

[d]Higher heating values for 1 atm and 15.6°C (60 F) [28].

**Table 17.13**   Composition (mol%) and properties of natural gas from worldwide sources [28][a]

| Location | $CH_4$ | $C_2H_6$ | $C_3H_8$ | $C_4H_{10}$ | $CO_2$ | $N_2$ | Density[b] (kg/m$^3$) | HHV[c] (kJ/m$^3$) | HHV[c] (kJ/kg) |
|---|---|---|---|---|---|---|---|---|---|
| Algeria LNG | 87.2 | 8.61 | 2.74 | 1.07 | — | 0.36 | 0.784 | 42,440 | 54,130 |
| Groningen, Netherlands | 81.2 | 2.9 | 0.36 | 0.14 | 0.87 | 14.4 | 0.784 | 33,050 | 42,150 |
| Kuwait, Bergen | 86.7 | 8.5 | 1.7 | 0.7 | 1.8 | 0.6 | 0.784 | 40,760 | 51,990 |
| Libya LNG | 70.0 | 15.0 | 10.0 | 3.5 | — | 0.90 | 0.956 | 49,890 | 52,210 |
| North Sea, Bacton | 93.63 | 3.25 | 0.69 | 0.27 | 0.13 | 1.78 | 0.723 | 38,450 | 53,200 |

[a]Although not explicitly stated in Ref. [28], these gases appear to be pipeline gases.

[b]At 1 atm and 15.6°C (60 F).

[c]Higher heating values for 1 atm and 15.6°C (60 F) [28].

it can enter distribution pipeline systems. Unprocessed natural gas is primarily methane, with smaller quantities of other light ($C_2$–$C_8$) hydrocarbons. Noncombustible gases, $N_2$, $CO_2$, and He, are also frequently present. Hydrogen sulfide, mercaptans, water, oxygen, and other trace contaminants may be present. Separation of dissolved associated gas from crude oil is frequently not economical [9]; nevertheless, the amount of gas flared or vented annually worldwide is huge—110 billion cubic meters, the equivalent to the combined annual natural gas consumption of France and Germany [25]. However, initiatives are in place to significantly reduce flaring of associated gas [25].

Although there are no industry or governmental standards for pipeline natural gas, contracts between producers and pipeline companies define general ranges of composition and other properties [26, 27]. Processing removes solid matter (e.g., sand), liquid hydrocarbons, sulfur compounds, water, nitrogen, carbon dioxide, helium, and any other undesirable compounds to meet contract specifications. The removal of sulfur compounds results in making an acidic, i.e., *sour,* gas *sweet.* Table 17.11 shows typical values, or ranges, of important properties of pipeline gas based on the *General Terms and Conditions* of a set of geographically dispersed pipeline companies in the United States and Canada.

The composition of natural gas varies widely depending upon the source. Examples for U.S. sources of natural gas are shown in Table 17.12. Compositions for natural gases from a variety of non-U.S. sources are provided in Table 17.13.

---

Using the 298.15 K reference state, calculate the higher heating value (HHV) for the natural gas from the Bergen field in Kuwait shown in Table 17.13. Compare the result with the value given in Table 17.13.

**Example 17.2**

**Solution**

Our solution follows that of Example 2.4. From Fig. 2.9, we see that the HHV can be expressed as

$$\text{HHV} = \Delta h_c = (H_{\text{reac}} - H_{\text{prod}})/MW_{\text{fuel}} \ (\text{kJ}/\text{kg}_{\text{fuel}})$$

where

$$H_{\text{reac}} = \sum_{\text{Reac}} N_i \bar{h}^o_{f,\,i} \quad \text{and} \quad H_{\text{prod}} = \sum_{\text{Prod}} N_i \bar{h}^o_{f,\,i}.$$

Using the given composition of the natural gas, we can calculate the apparent molecular weight of the fuel (natural gas) as

$$MW_{\text{fuel}} = \sum \chi_i MW_i = \chi_{\text{CH}_4} MW_{\text{CH}_4} + \chi_{\text{C}_2\text{H}_6} MW_{\text{C}_2\text{H}_6} + \chi_{\text{C}_3\text{H}_8} MW_{\text{C}_3\text{H}_8} + \chi_{\text{CO}_2} MW_{\text{CO}_2} + \chi_{\text{N}_2} MW_{\text{N}_2}.$$

Substituting numerical values, we obtain

$$MW_{\text{fuel}} = 0.867(16.043) + 0.085(30.069) + 0.017(44.096) + 0.018(44.011) + 0.006(28.013)$$
$$= 18.175 \ \text{kg}/\text{kmol}_{\text{fuel}}.$$

The reactant enthalpy $H_{\text{reac}}$ is evaluated using the given fuel composition and the enthalpies-of-formation:

| Constituent | $N_i$ or $\chi_i$ | $\bar{h}_{f,i}^o$ | Appendix Table |
|---|---|---|---|
| $CH_4$ | 0.867 | −74,831 | B.1 |
| $C_2H_6$ | 0.085 | −84,667 | B.1 |
| $C_3H_8$ | 0.017 | −103,847 | B.1 |
| $CO_2$ | 0.018 | −393,546 | A.2 |
| $N_2$ | 0.006 | 0 | A.7 |

Using the values from the table, we calculate

$$H_{reac} = 0.867(−74,831) + 0.085(−84,667) + 0.017(−103,847) + 0.018(−393,546) + 0.006(0)$$
$$= −80,924 \text{ kJ/kmol}_{fuel}.$$

To evaluate $H_{prod}$ we need to determine the composition of the products. For one mole of fuel, we write

$$0.867\,CH_4 + 0.085\,C_2H_6 + 0.017\,C_3H_8 + 0.018\,CO_2 + 0.006\,N_2 + a(O_2 + 3.76\,N_2)$$
$$\rightarrow b\,CO_2 + c\,H_2O + d\,N_2.$$

We evaluate $a$, $b$, $c$, and $d$ by applying element conservation for C, H, O, and N:

C-balance: $0.867(1) + 0.085(2) + 0.017(3) + 0.018(1) = b$  or  $b = 1.106$,
H-balance: $0.867(4) + 0.085(6) + 0.017(8) = 2c$  or  $c = 2.057$,
O-balance: $0.018(2) + 2a = 2b + c = 2(1.106) + 2.057$  or  $a = 2.1165$,
N-balance: $0.006(2) + 2(3.76a) = 2d$  or  $d = 7.964$.

Using the enthalpy-of-formation values from Table A.2 for $CO_2$ and Table A.6 for $H_2O$, we can evaluate the product mixture enthalpy at the reference state (298 K) as follows:

$$H_{prod} = 1.106(−393,546) + 2.057(−241,845 − 44,010) + 7.964(0) = −1,023,266 \text{ kJ/kmol}_{fuel},$$

where the 44,010 is the $H_2O$ enthalpy of vaporization (see Table A.6). This value is subtracted from the vapor-phase enthalpy of formation to obtain the liquid-phase value required to calculate the *higher* heating value of the fuel. Thus,

$$H_{reac} − H_{prod} = −80,924 − (−1,023,266) = 942,341 \text{ kJ/kmol}_{fuel}.$$

Using the previously calculated apparent molecular weight of the fuel, we obtain our final result:

$$HHV = \Delta h_c = (H_{reac} − H_{prod})/MW_{fuel} = 942,341/18.175 = 51,848 \text{ kJ/kg}_{fuel}$$

This value is quite close to the 51,990 kJ/kg$_{fuel}$ value given in Table 17.13.

**Comments**

(i) The tabulated HHV is about 0.27 percent higher than our calculation. This may be the result of the small difference in standard-state temperatures (298.15 K vs. 288.7 K) and slight differences in enthalpies of formation. (ii) Note how we used a *per-one-kmole-of-fuel* basis to write the combustion equation, which included the $CO_2$ and $N_2$ contained in the fuel. (iii) Note also how we needed to use the $H_2O$ enthalpy of vaporization to obtain the higher heating value. (iv) The higher heating value of pure methane is 55,528 kJ/kg (Table B.1). All of the other constituents in the Kuwait natural gas thus act to reduce the heating value from that of methane.

Calculate the constant-pressure adiabatic flame temperature for the Kuwait natural gas from Example 17.2. Use HPFLAME.

**Example 17.3**

### Solution

We begin by using the composition data to define a fuel molecule in the form $C_N H_M O_L N_K$. To find the number of carbon atoms in a "molecule" of the natural gas, we multiply the mole fraction of each C-containing constituent in the natural gas by the number of carbon atoms in the constituent and sum, i.e.,

$$\text{Number of C atoms} \equiv N = 0.867(1) + 0.085(2) + 0.017(3) + 0.018(1) = 1.106.$$

We proceed similarly for the H, O, and N atoms:

$$\text{Number of H atoms} \equiv M = 0.867(4) + 0.085(6) + 0.017(8) = 4.114,$$
$$\text{Number of O atoms} \equiv L = 0.018(2) = 0.036,$$
$$\text{Number of N atoms} \equiv K = 0.006(2) = 0.012.$$

Thus, our equivalent fuel molecule is $C_{1.106} H_{4.114} O_{0.036} N_{0.012}$. HPFLAME, and the other codes provided with this book, require the user to use integer values for $N$, $M$, $L$, and $K$. To use HPFLAME, we multiply $N$, $M$, $L$, and $K$ by 1000 to redefine the fuel as $C_{1106} H_{4114} O_{36} N_{12}$. To calculate the constant-pressure adiabatic flame temperature, HPFLAME requires an input value for the reactants enthalpy on a per-kmol-of-fuel basis. For the fuel expressed as $C_{1.106} H_{4.114} O_{0.036} N_{0.012}$, the reactants enthalpy was calculated in Example 17.2 to be $-80,924$ $kJ/kmol_{fuel}$. For reactants at 298 K, this enthalpy is just the enthalpy of formation of the fuel. For our redefined fuel, we obtain its enthalpy of formation by multiplying the $-80,924$ value by 1000; i.e.,

$$\bar{h}^o_{f, C_{1106}H_{4114}O_{36}N_{12}} = 1000 \bar{h}^o_{f, C_{1.106}H_{4.114}O_{0.0036}N_{0.0012}} = 1000(-80,924) = -80,924,000 \ kJ/kmol.$$

We now have all of the input information to use HPFLAME. For a stoichiometrc mixture at one atmosphere, we obtain

$$\boxed{T_{ad, C_{1106}H_{4114}O_{36}N_{12}} = 2228.1 \ K}$$

which is also the adiabatic flame temperature for $C_{1.106} H_{4.114} O_{0.036} N_{0.012}$.

### Comments

(i) Our calculated adiabatic flame temperature is very close to that of methane (2226 K). The nonmethane hydrocarbons in this natural gas all have adiabatic flame temperatures greater than $CH_4$; thus, we might expect a higher flame temperature for the Kuwait natural gas. The $CO_2$ and $N_2$ present in the natural gas, however, act as diluents and offset the effect of the higher hydrocarbons on flame temperature. (ii) Note the use of the multiplier employed to define the fuel molecule and its enthalpy of formation. This "trick" enabled us to use HPFLAME to solve this problem and showed how to add some versatility to a code that has limited applicability. Similarly, HPFLAME can be used to solve $H_2$ combustion problems by defining a fuel that is essentially all hydrogen, e.g., $CH_{10,000}$. This is necessary because the code requires that the fuel molecule contain carbon. The trace carbon-containing species can be ignored in the output, and the single C atom in the fuel has a negligible effect on the flame temperature. Likewise, carbon combustion can be approximated by defining the fuel to be $C_{10,000}H$, for example. Alternatively, one could employ one of the other equilibrium solvers discussed in Chapter 2 to solve such problems.

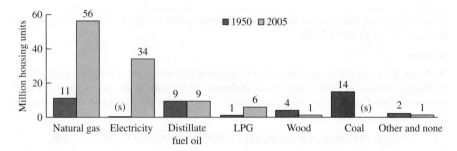

**Figure 17.4**        Type of heating in occupied housing units in the United States in 1950 and 2005.
From Ref. [29].

Natural gas is a relatively clean-burning fuel and is used in many applications. Natural gas is commonly used for space heating. In the United States in 2005, approximately 52 percent (or 56 million) of all housing units were heated by natural gas (Fig. 17.4). Natural gas ranks second to coal in the production of electricity in the United States. In 2007, 24.7 percent of the electricity generated from fossil fuels was from natural gas (Table 17.14). As can be seen in Table 17.14, natural gas plays an important role in all energy consumption sectors in the United States, with the exception of the transportation sector. In the transportation sector, the dominant use of natural gas is to power pipeline compressors [29].

## Coal

A primary factor complicating a description of coal and its combustion is the wide variability of its composition. For example, anthracite coal mined in Pennsylvania has a much different composition than a Montana bituminous coal. Coal is a sedimentary rock formed principally from organic matter, but it includes many minerals of

**Table 17.14**        Natural gas use in the United States for 2007

| Sector | Natural Gas[a] (trillion Btu) | Percentage of Total Fossil Fuel Energy from Natural Gas[b] | Percentage of Total Energy Use from Natural Gas[b] |
|---|---|---|---|
| Residential | 4842 | 79 % | 22.2 % |
| Commercial | 3083 | 81.6 % | 16.7 % |
| Industrial | 7999 | 41.2 % | 24.7 % |
| Transportation | 667 | 2.3 % | 2.3 % |
| Electric power | 7046 | 24.7 % | 17.4 % |
| Total | 23,637 | 27.4 % | 16.6 % |

[a]Data from Ref. [29].

[b]Percentages calculated from data presented in Ref. [29].

**Table 17.15**  Proximate analysis of coal

| Constituent Class | Mass Percentage Range (%)[a] |
|---|---|
| Moisture | 10–30 |
| Volatiles | 10–30 |
| Mineral matter | 10–30 |
| Char | Balance |

[a]These ranges are typical. A particular coal may fall outside these ranges.

different types and in different proportions. Table 17.15 shows the so-called **proximate analysis** of coal, where the constituents of the rock are grouped into four categories. From this table, we see that a large portion of the coal may be noncombustible (moisture and mineral matter). The volatiles and char contribute to the useful energy value of the coal, while the mineral matter produces an ash and/or slag, which contributes to operational difficulties (or complexities) and to environmental problems. The mineral matter typically occurs as inclusions in the carbonaceous rock matrix.

The elemental composition of the "useful" portion of the coal, the so-called dry, mineral-matter-free portion (volatiles + char), also has a widely variable composition, as can be seen in Table 17.16. Such a breakdown is referred to as an ultimate analysis. The sulfur and nitrogen contribute to air pollution problems and frequently require removal from the stack gases of utility boilers (see Chapter 15). Chlorine, phosphorous, mercury, and other elements are not included in a standard ultimate analysis [30].

Coals are classified according to the amount of geological metamorphism they have undergone, with hard coals (anthracites) having undergone the most metamorphism and lignites the least. Table 17.17 shows the standardized terminology used to describe coals and their formal definition based on composition and/or heating values [31]. Note the large range of heating values from high volatile A bituminous coal (32.6 kJ/kg) to lignite B (14.7 kJ/kg). Although the gross heating values shown in Table 17.17 are not strictly comparable to heating values for gaseous and liquid hydrocarbon fuels, the energy content of coals is substantially lower than that of

**Table 17.16**  Elemental composition of dry, mineral-matter-free coal

| Element | Typical Range of Composition (% by mass) |
|---|---|
| C | 65–95 |
| H | 2–6 |
| O | 2–25 |
| S | <10 |
| N | 1–2 |

**Table 17.17**  Classification of coals by rank (Adapted from Ref. [31])

| Class/Group | Fixed Carbon Limits (Dry, Mineral-Matter-Free Basis), % | | Volatile Matter Limits (Dry, Mineral-Matter-Free Basis), % | | Gross Calorific Value Limits (Moist, Mineral-Matter-Free Basis),[a] MJ/kg | | Agglomerating Character |
|---|---|---|---|---|---|---|---|
| | Equal or Greater Than | Less Than | Greater Than | Equal or Less Than | Equal or Greater Than | Less Than | |
| **Anthracitic:** | | | | | | | |
| Meta-anthracite | 98 | ... | ... | 2 | ... | ... | Nonagglomerating |
| Anthracite | 92 | 98 | 2 | 8 | ... | ... | " |
| Semianthracite[b] | 86 | 92 | 8 | 14 | ... | ... | " |
| **Bituminous:** | | | | | | | |
| Low volatile bituminous coal | 78 | 86 | 14 | 22 | ... | ... | Commonly agglomerating[c] |
| Medium volatile bituminous coal | 69 | 78 | 22 | 31 | ... | ... | " |
| High volatile A bituminous coal | ... | 69 | 31 | ... | 32.6[d] | ... | " |
| High volatile B bituminous coal | ... | ... | ... | ... | 30.2[d] | 32.6 | " |
| High volatile C bituminous coal | ... | ... | ... | ... | 26.7 | 30.2 | " |
| | | | | | 24.4 | 26.7 | Agglomerating |
| **Subbituminous:** | | | | | | | |
| Subbituminous A coal | ... | ... | ... | ... | 24.4 | 26.7 | Nonagglomerating |
| Subbituminous B coal | ... | ... | ... | ... | 22.1 | 24.4 | " |
| Subbituminous C coal | ... | ... | ... | ... | 19.3 | 22.1 | " |
| **Lignitic:** | | | | | | | |
| Lignite A | ... | ... | ... | ... | 14.7 | 19.3 | Nonagglomerating |
| Lignite B | ... | ... | ... | ... | ... | 14.7 | " |

[a] Moist refers to coal containing its natural inherent moisture but not including visible water on the surface of the coal.

[b] If agglomerating, classify in low volatile group of the bituminous class.

[c] It is recognized that there may be nonagglomerating varieties in these groups of the bituminous class and that there are notable exceptions in the high volatile C bituminous group.

[d] Coals having 69% or more fixed carbon on the dry, mineral-matter-free basis shall be classified according to fixed carbon, regardless of gross calorific value.

hydrocarbons.[5] For example, higher heating values for hydrocarbon fuels range from, say, 43 kJ/kg to 55 kJ/kg, compared to, say, 20 kJ/kg to 32 kJ/kg for coals. The following correlation [33] can be used to estimate the higher heating values of coals based on an ultimate analysis:

$$\text{HHV [MJ/kg]} = 0.3491\,C + 1.1783\,H$$
$$+ 0.1005\,S - 0.1034\,O - 0.015\,N - 0.0211\,A, \qquad (17.3)$$

where C, H, S, O, N, and A are the mass percentages (dry basis) of carbon, hydrogen, sulfur, oxygen, nitrogen, and ash in the fuel. The ranges of applicability of Eq. 17.3 are

$$0\% \leq C \leq 92.25\%,$$
$$0.43\% \leq H \leq 25.15\%,$$
$$0.00\% \leq O \leq 50.00\%,$$
$$0.00\% \leq N \leq 5.60\%,$$
$$0.00\% \leq S \leq 94.08\%,$$
$$0.00\% \leq A \leq 71.4\%,$$

and

$$4.745 \text{ MJ/kg} \leq \text{HHV} \leq 55.345 \text{ MJ/kg}.$$

This correlation can also be applied to conventional liquid and gaseous hydrocarbons, biofuels, and wastes, with accuracies generally within a few percent [33].

---

Estimate the higher heating value (HHV) of the #4 Alaska coal (Table 17.18) using the correlation given by Eqn. 17.3.

**Example 17.4**

**Solution**

From Table 17.18, the elemental composition (wt.%) of #4 Alaska coal on a dry, ash-free basis is

C: 69.94
H:  5.27
S:  0.24
O: 23.62
N:  0.93
A (ash): 0

The solution is a simple numerical substitution of these wt.% values into Eqn. 17.3:

$$\text{HHV (MJ/kg)} = 0.3491(C \text{ wt.\%}) + 1.1783(H \text{ wt.\%}) + 0.1005(S \text{ wt.\%}) - 0.1034(O \text{ wt.\%})$$
$$- 0.015(N \text{ wt.\%}) - 0.0211(A \text{ wt.\%})$$
$$= 0.3491(69.94) + 1.1783(5.27) + 0.1005(0.24) - 0.1034(23.62)$$
$$- 0.014(0.93) - 0.015(0)$$

$$\boxed{\text{HHV} = 28.19 \text{ MJ/kg}}$$

---

[5]The gross heating value of coal is a higher heating value defined on the basis of constant-volume combustion rather than constant-pressure combustion used to define the heating values of hydrocarbons. ASTM D5865 [32] provides a method to correct to constant-pressure heating values as well as making adjustment to account for moisture. Adjustments to account for differences in constant-volume values and constant-pressure values are small, say, a fractional percent.

**Table 17.18**  Compositions of selected U.S. coals[a]

| Coal Seam | State | ASTM Rank [b] | Proximate Analysis (% daf[c]) | | Ultimate Analysis (% daf[c]) | | | | | HGI[d] |
|---|---|---|---|---|---|---|---|---|---|---|
| | | | Volatiles | Fixed Carbon | C | H | N | S | O | |
| #4 | Alaska | subC | 59.21 | 40.79 | 69.94 | 5.27 | 0.93 | 0.24 | 23.62 | 31 |
| Upper Spadra | Arkansas | lvb | 17.88 | 82.12 | 88.55 | 4.19 | 1.48 | 2.95 | 2.83 | 85.1 |
| Colorado Q | Colorado | subA | 42.05 | 57.95 | 75.19 | 5.04 | 2.01 | 0.4 | 17.37 | — |
| Unnamed | Iowa | hvCb | 46.65 | 53.35 | 72.45 | 5.35 | 1.2 | 11.29 | 9.72 | 86.5 |
| Illinois #6 | Illinois | hvAb | 41.78 | 58.22 | 80.2 | 5.73 | 1.45 | 4.73 | 7.89 | 60.1 |
| Illinois #6 | Illinois | hvCb | 41.56 | 58.44 | 77.29 | 5.38 | 0.85 | 4.58 | 11.91 | — |
| Illinois #6 | Illinois | mvb | 32.74 | 67.26 | 77.54 | 5.41 | 1.58 | 5.29 | 10.17 | 56.7 |
| Elkhorn #3 | Kentucky | hvAb | 39.7 | 60.3 | 84.15 | 5.84 | 1.45 | 1.03 | 7.53 | 52.1 |
| Rosebud | Montana | subB | 42.23 | 57.77 | 73.68 | 5.27 | 0.55 | 0.61 | 19.88 | — |
| Fort Union Bed | Montana | lig | 46.86 | 53.14 | 68.64 | 4.72 | 1.26 | 0.64 | 24.74 | 103.1 |
| Ohio #4 | Ohio | hvCb | 47.84 | 52.16 | 77.13 | 5.4 | 1.33 | 6.68 | 9.47 | 52.5 |
| Pennsylvania #2 | Pennsylvania | sa | 10.49 | 89.51 | 92.58 | 3.39 | 0.87 | 0.72 | 2.44 | 79.5 |
| Lower Clarion | Pennsylvania | hvAb | 44.01 | 55.99 | 81.42 | 5.83 | 1.6 | 6.27 | 4.88 | 64.3 |
| Lykens Valley #2 | Pennsylvania | an | 5.08 | 94.92 | 90.33 | 4.01 | 0.8 | 0.56 | 4.3 | 31.6 |
| Redstone | West Virginia | hvAb | 43.03 | 56.97 | 82.54 | 5.77 | 1.25 | 2.41 | 8.04 | 57.4 |
| Dietz #3 | Wyoming | subB | 45.04 | 54.96 | 74.92 | 4.93 | 1.07 | 0.48 | 18.6 | 47.7 |

[a]Data on coals from the Penn State Coal Sample Bank and Database courtesy of Gareth Mitchell.

[b]ASTM abbreviations: an—anthracite; sa—semianthracite; lvb—low volatile bituminous; mvb—medium volatile bituminous; hvAb—high volatile A bituminous; hvCb—high volatile C bituminous; subA—subbituminous A; subB—subbituminous B; subC—subbituminous C; lig—lignite.

[c]Dry ash free

[d]Hardgrove Grindability Index [34]: Lower index values correspond to a higher resistance to grinding.

**Comments**

(i) We note that the estimated HHV is quite a bit larger than the upper limit for subbituminous C coal as given in Table 17.7 (22.1 MJ/kg). The values there, however, are for moist rather than dry coals; thus our estimate would tend to be high. (ii) Note that the presence of sulfur adds to the heating value, whereas oxygen, nitrogen, and ash subtract, as expected.

Properties of several selected U.S. coals are presented in Table 17.18. Coals were selected for this table to illustrate a wide range of coal ranks and for their geographical diversity within the United States. Indicated here are the name and location of the coal seam, the rank of the coal, and both the proximate and ultimate analyses. The last column of the table shows the Hardgrove Grindability Index (HGI), which is a measure of how easily a coal can be pulverized [34]. This index provides a measure of grinding power consumption and pulverizer capacities needed to prepare coal for burning. As expected, the general compositions of the various coals follow from their ASTM definition of rank (Table 17.17). We note that the Illinois #6 seam provides several different ranks. Although not shown in the table, this kind of variability occurs in other seams as well, for example, the Lower Kittanning seam in Pennsylvania. Of particular note is the wide variation in sulfur content in the samples shown: for this sample set, sulfur mass percentages range from 0.24 (Alaska) to 11.29 (Iowa). Low-sulfur coals are desirable to meet air quality standards for $SO_2$ and particulate matter. A substantial portion of the bound nitrogen in coal can be converted to $NO_x$, a controlled air pollutant. Coal also contains mercury and chlorine, which are air toxics. Coal combustion also results in substantially higher $CO_2$ emissions on an energy-weighted basis (i.e., $kg_{CO_2}/MJ$) than other fossil fuels. These issues are discussed in Chapter 15.

In addition to concerns for $SO_2$, $NO_x$, Hg, and Cl emissions, disposal of solid matter from coal-fired power plants is also a concern. Bottom ash, fly ash, slag, and flue gas desulfurization residues, i.e., coal combustion residuals, contain the mineral matter from the burned coal and can represent hazards. Metals such as arsenic, selenium, cadmium, lead, and mercury may be contained in these residuals. In 2009, the U.S. Environmental Protection Agency released a list of 44 locations at which the storage of these residuals represents a high hazard potential [35].

For readers interested in more detailed information on coal, the author recommends van Krevelen's authoritative reference book [36].

# ALTERNATIVE FUELS

For our purposes, we define an alternative fuel to be any fuel that has the potential for displacing or supplementing the traditionally used conventional fossil fuels discussed in the preceding section. Concerns over global warming, environmental degradation, and national energy independence have sparked interest in alternative fuels. Much research is in progress to discover and evaluate heretofore unknown fuels. Alternative fuels are not necessarily newly discovered fuels, but are often fuels that have been used under special circumstances, or that have been noneconomical to use, in the past. In this section we discuss several alternative fuels.

## Biofuels

Among alternative fuels, of particular current interest, are *renewable fuels* from biomass. Biomass is a wide-ranging term and includes crop materials specifically grown for fuel, waste from food crops, waste or by-products from forest products, animal wastes, algae, and many others. Biomass has been used directly as a fuel since the harnessing of fire by humans. A common example of direct combustion of a biofuel is a wood-fired stove used for space heating and/or cooking. Biomass can also be processed to create fuels. Here we consider three categories: biochemical processes, agrochemical processes, and thermochemical processes [37]. Production of ethanol from biomass is based on biochemical processes. Enzymes break down plant starch to sugars, which is followed by bacterial fermentation of the sugars to alcohol. An example of an agrochemical process is the production of biodiesel using oil extracted from seeds. Thermochemical processes include pyrolysis, direct liquefaction, gasification, and supercritical fluid extraction [37].

Current interest in biofuels is captured in the passage of the Energy Independence and Security Act of 2007 (EISA 2007) [38] by the United States Congress. This wide-sweeping legislation contains provisions (Subtitles A and B of Title II) mandating the production of renewable fuels for transportation. Two classes of renewable fuels are defined in the law: (1) *Conventional biofuel* is defined as a renewable fuel derived from corn starch. (2) *Advanced biofuels* are defined to be non-cornstarch-based renewable fuels that have life-cycle greenhouse gas emissions that are at least 50 percent less than the baseline emissions. Examples here include ethanol derived from cellulose or hemicellulose; ethanol from sugar or starch other than corn starch; ethanol derived from waste materials, e.g., crop residue; biomass-based diesel; biogas, e.g., landfill gas and sewage waste treatment gas; butanol or other alcohols produced from renewable biomass; and other fuels derived from cellulosic biomass [38]. Life-cycle greenhouse gas emissions include the greenhouse gas emissions from all stages of fuel production, distribution, and final use, where the contributions of the various gases are weighted according to their relative global warming potentials. EISA 2007 mandated the phasing in of renewable fuels for transportation. The law specifically requires the total production of renewable fuels by 2022 to be 36 billion gallons, of which 21 billion gallons are to be advanced biofuels. For perspective on this number, we note that gasoline consumption in the United States in 2008 was approximately 103.4 billion gallons [39]. A detailed timetable for achieving this goal is specified in the law [38]. The European Union [40] and other countries have also set targets for the use of biofuels for transportation.

**Ethanol**    Ethanol ($C_2H_5OH$) is frequently blended in gasoline to meet government-mandated reductions of tailpipe emissions (see Table 17.5) in specific regions of the United States (see Fig. 17.2), as well as to meet the requirements for winter carbon monoxide emissions in selected metropolitan areas. We discussed the use of ethanol in reformulated gasoline (RFG) and winter oxygenated fuels in a previous section. E10, or gasohol, is a gasoline blend containing 10 percent ethanol. Moreover, there has been interest in using ethanol as the primary constituent in automotive fuels, for example, E85, a fuel blend that contains 85 percent ethanol. ASTM specifications [41] for *fuel ethanol (Ed75-Ed85)* are presented in Table 17.19. Many vehicle manufacturers currently produce *flex-fuel* vehicles that are capable of operating with either conventional gasoline or high-ethanol-content fuels.

The case for ethanol as a transportation fuel is complicated. Small quantities of ethanol reduce tailpipe emissions of carbon monoxide (CO), volatile organic compounds (VOC), and particulate matter $PM_{10}$ [42], whereas emissions from E85-fueled vehicles show either increased emission or mixed results among the various pollutants [43, 44]. Life-cycle emissions for CO, VOC, $PM_{10}$, oxides of sulfur ($SO_x$), and oxides of nitrogen ($NO_x$) are all higher for a corn-based ethanol E85 blend [42, 44]. Primary arguments for ethanol use are that it has the potential to reduce dependence on petroleum and that it can reduce greenhouse gas emissions. Whether either of these benefits can be achieved depends on a complex analysis of all of the factors that are involved. Among these factors are growing the biomass, converting the biomass to fuel, and its distribution to fueling stations. Although $CO_2$ emissions may be reduced by using ethanol as a fuel, $N_2O$, a major greenhouse gas, is released when nitrogen fertilizers are needed to grow the feedstock [42]. A detailed analysis by Hill *et al.* [42] shows that greenhouse gas emissions from corn-based E85 show a relatively small reduction of 12 percent over gasoline on a fuel energy basis. Use of the cellulosic components of biomass to produce ethanol offers the potential for a

**Table 17.19** ASTM requirements for fuel ethanol (ED75-ED85) [41]

| Properties | Volatility Class 1[a] | Volatility Class 2 | Volatility Class 3 |
|---|---|---|---|
| Ethanol + higher alcohols, min., volume % | 79 | 74 | 70 |
| Hydrocarbon/aliphatic ether, volume % | 17–21 | 17–26 | 17–30 |
| Vapor pressure, kPa (psi) | 38–59 (5.5–8.5) | 48–65 (7.0–9.5) | 66–83 (9.5–12.0) |
| Sulfur, max., mg/kg | 80 | 80 | 80 |

| | All Classes |
|---|---|
| Methanol, volume %, max. | 0.5 |
| Higher alcohols ($C_3$–$C_8$), max., volume % | 2 |
| Acidity, (as acetic acid $CH_3COOH$), mass % (mg/L), max. | 0.005 |
| Solvent-washed gum content, max., mg/100 mL | 5 |
| $pH_e$ | 6.5 to 9.0 |
| Unwashed gum content, max., mg/100 mL | 20 |
| Inorganic chloride, max., mg/kg | 1 |
| Copper, max., mg/L | 0.07 |
| Water, max., mass % | 1.0 |
| Appearance | This product shall be visibly free of suspended or precipitated contaminants (clear and bright). This shall be determined at ambient temperature or 21°C (70 F), whichever is higher. |

[a]Volatility class is based on seasonal and geographical criteria specified in Ref. [41].

significantly increased net energy balance in its production and use compared with the production of ethanol from the starch of highly fertilized biomass (corn) grown on fertile land. EISA 2007 [38] recognizes the greater potential for cellulose-based fuels in its mandate for an increasing share of renewable fuels to be *advanced biofuels*. Trade-offs with food production complicate the issue, as discussed in Refs. [45–47]. The decade 2010–2020 should prove to be an interesting one for transportation fuels as society works through the many issues involved.

Selected properties of pure ethanol are shown in Table 17.20 along with properties for gasoline, No. 2 diesel, and biodiesel for comparison. Here we see that the boiling point of ethanol (78 °C) lies in the midrange of the distillation profile of gasoline (27 °C–225 °C) and that the Reid vapor pressure for ethanol (16 kPa) is substantially lower that that of gasoline (55 kPa–103 kPa). The energy content on both a mass and volumetric basis is substantially lower for ethanol in comparison with either gasoline or No. 2 diesel. For example, the volumetric LHV of ethanol is about 33 percent less than that of gasoline and about 40 percent less than diesel. The practical implication is that for a fixed fuel tank size, vehicle range would be diminished for operation with ethanol or with a high-ethanol-content fuel. Because

**Table 17.20**   Selected properties of ethanol, gasoline, No. 2 diesel, and biodiesel[a]

| Property | Ethanol | Gasoline[c] | No. 2 Diesel[d] | Biodiesel |
|---|---|---|---|---|
| Molecular weight (kg/kmol) | 46.0684 | 100–105 | ~200 | ~292 |
| Density at 1 atm and 15.6 °C (kg/m³) | 795 | 720–780 | 810–890 | 879 |
| Boiling temperature(s) (°C) | 78 | 27–225 | 190–340 | 315–350 |
| Enthalpy of vaporization at 1 atm and 25 °C (MJ/kg)[b] | 0.921 | 0.305 | 0.230 | — |
| Reid vapor pressure (kPa) | 16 | 55–103 | 1.4 | < 0.28 |
| Research octane number | 108 | 90–100 | — | — |
| Motor octane number | 92 | 81–90 | — | — |
| Cetane number | — | — | 40–55 | 48–65 |
| Higher heating values[f] | | | | |
| Mass-based (MJ/kg) | 29.8 | 43.7–47.5 | 44.6–46.5 | 40.2 |
| Volumetric (MJ/m³) | 23,400 | 34,800 | 38,700 | 35,700 |
| Lower heating values[f] | | | | |
| Mass-based (MJ/kg) | 26.7 | 41.9–44.2 | 41.9–44.2 | 37.5 |
| Volumetric (MJ/m³) | 21,200 | 32,000 | 35,800 | 33,300 |
| Stoichiometric air-fuel ratio (kg/kg) | 8.94 | ~14.7 | ~14.7 | ~13.8 |
| Viscosity at 1 atm and 15.6 °C (Pa-s) | $1.19 \cdot 10^{-3}$ | $(0.37–0.44) \cdot 10^{-3}$ | $(2.61–4.1) \cdot 10^{-3}$ | $6 \cdot 10^{-3}$ See note e |

[a]Properties adapted from Ref. [48] except as noted.
[b]From Ref. [49].
[c]See Table 17.2 for more information.
[d]See Table 17.8 for more information.
[e]Value at 5 °C (40 F) from EPA420-P-02-001 (2002) [58].
[f]See Table B.1 in Appendix B for mass-based HHVs and LHVs for pure hydrocarbons.

of the oxygen contained in ethanol, its stoichiometric air–fuel ratio is much less than those for gasoline and diesel fuel. Modern flex-fuel vehicles automatically adjust for such differences to provide good driveabilty and proper operation of the three-way catalyst emission control system (see Chapter 15).

**Example 17.5**    Calculate the stoichiometric air–fuel ratio for ethanol and compare this value to that for gasoline. Assume the gasoline can be represented as $C_{7.76}H_{13.1}$.

**Solution**

We begin by writing the stoichiometric combustion equation for ethanol ($C_2H_5OH$):

$$C_2H_5OH + a(O_2 + 3.76N_2) \rightarrow bCO_2 + cH_2O + dN_2 .$$

Knowing the value for $a$ allows us to calculate $(A/F)_{\text{stoic, EtOH}}$. To find $a$, we apply element conservation for C, H, and O, as follows:

$$C: 2 = b$$
$$H: 6 = 2c \text{ or } c = 3$$
$$O: 1 + 2a = 2b + c = 2(2) + 3 \text{ or } a = 3$$

We now apply the definition of the mass air–fuel ratio (Eqn. 2.32), i.e.,

$$(A/F)_{\text{stoic, EtOH}} = \frac{N_{\text{air}}MW_{\text{air}}}{N_{\text{EtOH}}MW_{\text{EtOH}}} = \frac{4.76aMW_{\text{air}}}{MW_{\text{EtOH}}} = \frac{4.76(3)28.85}{46.069}$$

$$(A/F)_{\text{stoic, EtOH}} = 8.94 \text{ kg}_{\text{air}}/\text{kg}_{\text{EtOH}}.$$

To calculate the stoichiometric air–fuel ratio for gasoline, we know from Chapter 2 that, for any hydrocarbon $C_xH_y$, $a = x + y/4$ (Eqn. 2.31); thus,

$$a = 7.76 + 13.1/4 = 11.035.$$

The apparent molecular weight of $C_{7.76}H_{13.1}$ is

$$MW_{\text{gas}} = 7.76(12.011) + 13.1(1.00794) = 106.5 \text{ kg}/\text{kmol}.$$

We again apply the definition of the air–fuel ratio:

$$(A/F)_{\text{stoic, gas}} = \frac{N_{\text{air}}MW_{\text{air}}}{N_{\text{gas}}MW_{\text{gas}}} = \frac{4.76aMW_{\text{air}}}{MW_{\text{gas}}} = \frac{4.76(11.035)28.85}{106.4}$$

$$(A/F)_{\text{stoic, gas}} = 14.24 \text{ kg}_{\text{air}}/\text{kg}_{\text{gas}}.$$

Comparing the two stoichiometric ratios, we see that the ethanol $A/F$ is about 63 percent smaller than that of gasoline ($8.94/14.24 = 0.628$).

### Comments

The fuel–air metering system on a flex-fuel vehicle must be able to deal with this large difference in the stoichiometric air–fuel ratios.

---

**Biodiesel**  The term biodiesel denotes a fuel produced from plant oils or animal fats suitable for use in diesel engines. Like ethanol, biodiesel offers an alternative to fossil fuels with a potential to reduce life-cycle greenhouse gas emissions. The Energy Independence and Security Act of 2007 classifies biodiesel as an "advanced biofuel."

Biodiesel fuels comprise mono-alkyl esters of long-chain fatty acids, as shown in Fig. 17.5. Sources of plant oils are soybeans, rapeseeds, sunflower seeds, palm fruit and kernels, and others, whereas animal fats (grease, tallow, and lard) used in cooking are frequently available as recyclables. Biodiesel can also be produced from oils from algae [51]. The production of biodiesel from a bio-feedstock involves the reaction of a triglyceride with an alcohol in the presence of a catalyst to form mono-alkyl esters and glycerol.[6] This reaction scheme can be represented as follows [52]:

$$
\begin{array}{llll}
\text{CH}_2-\text{OOC}-\text{R}_1 & & \text{R}_1-\text{COO}-\text{R}' & \text{CH}_2-\text{OH} \\
| & \overset{\text{Catalyst}}{\underset{}{\rightleftarrows}} & & | \\
\text{CH}_2-\text{OOC}-\text{R}_2 \;+\; 3\text{R}'\text{OH} & & \text{R}_2-\text{COO}-\text{R}' \;+\; \text{CH}_2-\text{OH} & \qquad (17.4)\\
| & & & | \\
\text{CH}_2-\text{OOC}-\text{R}_3 & & \text{R}_3-\text{COO}-\text{R}' & \text{CH}_2-\text{OH} \\
\text{Glyceride} & \text{Alcohol} & \text{Esters} & \text{Glycerol}
\end{array}
$$

---

[6]Glycerol is also known as glycerin or glycerine.

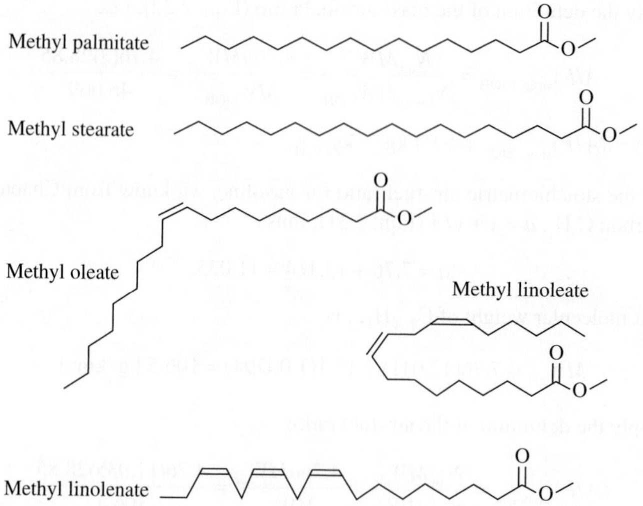

Methyl palmitate

Methyl stearate

Methyl oleate

Methyl linoleate

Methyl linolenate

**Figure 17.5** Five methyl esters. Each internal straight-line segment typically represents a C–H$_2$ group. The final segments are C–H$_3$ groups.
Adapted from Ref. [50].

where $R_1$, $R_2$, and $R_3$ are long-chain hydrocarbon radicals (see Fig. 17.5) and R′ is the hydrocarbon radical associated with the particular alcohol used. In plant oils, the long-chain radicals contain from 16 to 24 carbon atoms [53]. Typically, ethanol and methanol are used to produce biodiesel [52, 54]; hence, R′ = $C_2H_5$ and $CH_3$, respectively. Many types of catalysts (alkalis, acids, or enzymes) can be used; however, alkalis (e.g., sodium hydroxide, potassium hydroxide, and many others) are most often used in commercial production [52, 55].

Biodiesel blends are denoted by the letter B followed by a number denoting the volumetric percentage of biodiesel in the blend; for example, a 20 percent biodiesel blend is designated B20. Neat biodiesel is denoted as B100. Unmodified diesel engines can operate with diesel fuel blends containing up to 20 percent biodiesel. Concerns associated with the use of biodiesel are the potential for clogging fuel filters at low temperatures, formation of engine deposits, susceptibility to oxidative and biological instabilities, and corrosion or wear associated with impurities left from biodiesel processing (unreacted fatty acids, alcohol, glycerin, and catalyst) [50, 55]. Neat biodiesel has a greater viscosity than the maximum specified for No. 2 diesel [15], so blending must be done judiciously to provide acceptable viscosity. To address these concerns, and others, ASTM provides specifications for B100 used to create blends [56], as well as specifications for the blends themselves (B6–B20) [57].

Current estimates [44] indicate that the life-cycle greenhouse gas emissions for biodiesel are 50 percent to 80 percent less than those of conventional diesel fuel. In addition to this benefit, operation with biodiesel typically results in reductions in the exhaust emissions of carbon monoxide, particulate matter, and unburned

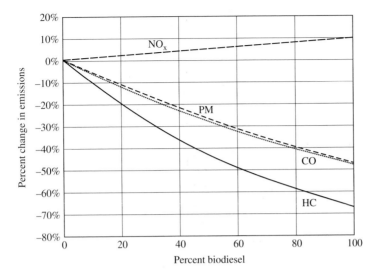

**Figure 17.6**    Heavy-duty highway diesel engine emission characteristics for operation with biodiesel. Curves represent regression analysis of numerous data sets. Significant reductions are seen for particulate matter (PM), carbon monoxide (CO), and unburned hydrocarbons (HC). A relatively smaller increase is observed for oxides of nitrogen ($NO_x$). The curves for particulate matter and carbon monoxide are nearly coincident.
From Ref. [58].

hydrocarbons, as shown in Fig. 17.6. Operation with B20 has the potential of 11 percent, 10 percent, and 21 percent reductions of these pollutants, respectively [58]. Unfortunately, oxides of nitrogen ($NO_x$) emissions increase slightly, and given the increasingly stringent $NO_x$ requirements in the United States and other countries, this is a source of concern. The average increase for operation with B20 is 2 percent [58]. Although many hypotheses have been put forward to explain this increase in $NO_x$ [59, 60], Lapuerta *et al.* [60] conclude that the most likely is the advanced (early) injection start for biodiesel compared to diesel. Emissions of aromatics and polycyclic hydrocarbons (see Chapter 15) are thought to be reduced by the use of biodiesel, whereas no conclusive trend has been observed for oxygenated hydrocarbons, e.g., aldehydes and ketones [60]. A considerable literature exists on the emissions from biodiesel-fueled engines, e.g., Refs. [58–62] and others; Lapuerta *et al.* [60] provide a review of this topic.

**Black Liquor**    In the production of paper, an energy-rich liquid, known as *black liquor*, is produced as a by-product of the pulping process. Black liquor contains dissolved organic solids, primarily lignin, and sodium and sulfur compounds from the paper-making process. Reference [63] reports a typical black liquor composition as $C_{10}H_{12.5}O_7Na_{2.4}S_{0.36}$. Paper mills typically burn black liquor to produce electricity and process steam to run the mill. Combustion occurs in a recovery boiler, a design unique to paper production, that recovers the sodium and sulfur compounds for reuse as well as producing heat for steam generation. The industry also has explored more

**Table 17.21**  Comparison of biomass pyrolysis and liquefaction processing of biomass (Adapted from Ref. [37])

| Process | Temperature (K) | Pressure (MPa) | Drying | Catalyst |
|---|---|---|---|---|
| Pyrolysis | 650–800 | 0.1–0.5 | Necessary | No |
| Liquefaction | 525–600 | 5–20 | Unnecessary | Yes |

economical and environmentally friendly ways to utilize black liquor [64, 65]. One such approach is to gasify black liquor to produce a synthesis gas, i.e., syngas, which consists of carbon monoxide and hydrogen. This syngas is burned in a gas turbine to produce electricity, and the exhaust of the turbine is then used to create process steam in a heat recovery steam generator. Supplemental fuels, such as biomass (bark, wood scraps, etc.) or coal, are part of the mix used to generate the syngas, as the energy from the black liquor is not sufficient to meet the entire energy needs of the paper mill. In 2007, 39.0 billion kilowatt-hours of electricity were generated from black liquor and wood and wood waste solids and liquids in the United States. For comparison, this represents about 15.8 percent of the 247.5 billion kilowatt-hours generated by conventional U.S. hydroelectric plants [66].

**Thermal-Conversion Liquids**   Biomass can be converted into liquids (bio-oil or bio-crude) by several so-called thermal conversion processes. A wide variety of biomass feedstocks can be used in thermal conversion processes. For example, He *et al.* [67] have demonstrated the feasibility of using swine manure for production of bio-oil. The most common thermal conversion processes are pyrolysis and direct liquefaction. These two processes are compared in Table 17.21. Pyrolysis yields the greatest amount of liquid for a given amount of biomass feedstock and, hence, is currently receiving the most attention for expanded commercial development [37, 68–70]. The liquids produced in these processes then can be upgraded to create fuel oils, similar to the refining of petroleum [70, 71]. Future developments in this area should be interesting in that these processes offer a means to create renewable transportation fuels with properties similar to those produced from petroleum.

**Biogas**   Gaseous biofuel, i.e., *biogas,* is produced by the bacterial decomposition of organic wastes. Two primary sources of biogas are *in situ* production in landfills (landfill gas) and specifically designed digester systems (digester gas). For landfills, the gas is mined by drilling wells or placing collection pipes as the landfill is built up. Digester systems typically use sewage, household waste, or manure as feedstocks, which are processed in continuously stirred reactors. Regardless of the production means, three sequential bacteria-controlled steps transform the organic waste into methane (and carbon dioxide) [72]: (1) A hydrolysis step, in which anaerobic bacteria and enzymes convert insoluble organic matter and higher-molecular-weight compounds to soluble organic compounds, (2) an acidogenesis step, in which the soluble organics are fermented into organic acids, hydrogen, and carbon dioxide, and (3) methane fermentation, in which methanogenic bacteria convert the products

**Table 17.22**  Compositions (mol%) of digester (sewage) and landfill biogases [28]

| Gas | CH$_4$ | Other HCs | H$_2$ | CO$_2$ | O$_2$ | N$_2$ | Density[a] (kg/m³) | HHV[b] (kJ/kg) |
|---|---|---|---|---|---|---|---|---|
| **Digester (NJ)** | 59.0 | 0.05 | – | 39.4 | 0.16 | 0.57 | 1.14 | 19,560 |
| **Digester (IL)** | 68.0 | – | 2.0 | 22.0 | – | 6.0 | 0.94 | 27,450 |
| **Landfill**[c] | 53.4 | 0.17 | 0.005 | 34.3 | 0.05 | 6.2 | 1.08 | 18,660 |

[a]At 1 atm and 15.6°C (60 F).

[b]Higher heating values for 1 atm and 15.6°C (60 F) [28].

[c]Also contains 0.005 mol% CO.

from step 2 to methane and carbon dioxide. Table 17.22 shows the composition of typical landfill and digester gases. Note that, in all cases, the biogases are dominated by methane (53.5–68%) and CO$_2$ (22–39.4%) and that the heating values are approximately half, or less, than those of natural gases (see Tables 17.12 and 17.13). Biogases also frequently contain a variety of trace contaminants, e.g., siloxanes from household and commercial waste [73, 74] and halonogenated and organosulfur compounds [75]. These contaminants, along with moisture and particulate matter, are often removed in the processing of the gas for downstream use. The combustion properties of landfill gases have been studied by Qin *et al.* [75].

The U.S. Environmental Protection Agency has a program to promote the recovery and use of landfill gas [76]. Currently, more than 400 landfill gas projects are producing gas for electricity generation or use in boilers, furnaces, kilns, and similar applications. Total electricity generation from landfill gas has increased by a factor of approximately nine from 1988 to 2008 [73]; however, compared to other means of producing electricity, the landfill gas contribution is small. In 2007, for example, 6.16 billion kilowatt-hours of electricity were generated from landfill gas compared to 247.5 billion kilowatt-hours from conventional hydroelectric generation [66]. Although principally known for its production of renewable energy from wind, Denmark has also developed a significant biogas production infrastructure [77]. In 2002, 20 commercial-scale and 35 farm-scale biogas digester plants were in operation. These plants processed approximately 3 percent of all the manure produced in Denmark and yielded an annual energy production of 2.6 petajoules [77], which, for comparison, is energetically equivalent to 0.7 billion kilowatt-hours or approximately 2 percent of Denmark's total electricity generation.

Figure 17.7 adds some perspective to our discussion of biofuels. In Fig. 17.7a, we see that the combined renewable energy sources are a relatively small fraction (6–7%) of the total energy consumption in the United States and range from about 75 percent to 86 percent of the nuclear energy contribution. Figure 17.7b shows the breakdown of the various renewable energy sources. The largest contribution to renewable energy is biomass, which shows steady growth from 2004 to 2008. Interestingly, energy from biomass exceeds that generated by water power (hydro). Wind energy, although significantly less than either biomass or hydro, also shows a strong growth trend.

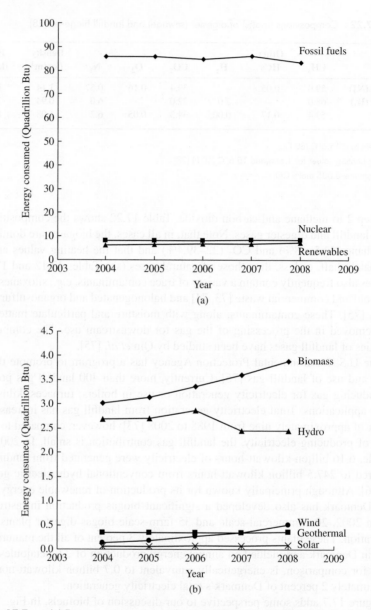

**Figure 17.7**    (a) Energy consumption in the United States is dominated by fossil fuels (top), with renewables providing approximately 6 to 7 percent of the total energy consumed. (b) The largest contribution to renewable energy is biomass, which shows steady growth from 2004 to 2008. Wind energy, although significantly less than either biomass or hydro, also shows a strong growth trend.

Data from Ref. [66].

## Fischer–Tropsch Liquid Fuels

Liquid transportation fuels can be made from a wide variety of feedstocks, such as natural gas, coal, coke, biomass, and residual oils in a sequence of processes as illustrated in Fig. 17.8. The key steps are (1) the creation from the feedstock material of a synthesis gas consisting of carbon monoxide and hydrogen, (2) the cleanup of the synthesis gas (i.e., syngas) to remove contaminants such as sulfur and particulate matter, (3) the synthesis of various hydrocarbons and associated products using the Fischer–Tropsch process, and (4) the upgrading of the Fischer–Tropsch products to either diesel fuel or gasoline using various combinations of hydrocracking, hydrotreating, and hydroisomerization [78]. Diesel fuel produced in this way has quite high cetane numbers (~70), and its production is less complicated than that of gasoline; thus, it is the more attractive product [79]. Other useful products, such as olefins for the manufacture of plastics, can be created in this process.

The Fischer–Tropsch process was invented in the 1920s [80] and has an interesting history [79, 81]. Absent of any governmental policies, the availability and price of crude petroleum determine whether or not transportation fuels can be produced economically using Fischer–Tropsch synthesis. Capital costs are generally quite high and are dominated by the equipment needed to produce and clean up the synthesis gas [78, 79]. Gasification costs can range from 60 to 70 percent of the total cost to produce a product [79]. The use of biomass as a feedstock is likely to result in an even higher gasification cost. Important to the process of converting synthesis gas to useful products are the use of catalysts—cobalt or iron at low temperatures (200–240 °C), or iron at

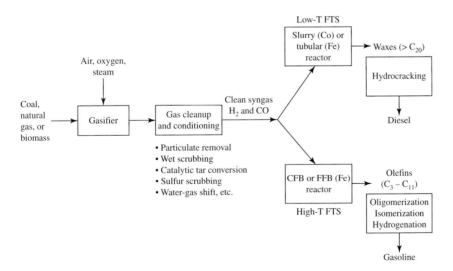

**Figure 17.8**     Liquid transportation fuels can be created from a variety of feedstocks (biomass, natural gas, etc.) by first creating a synthesis gas, i.e., syngas ($H_2$ and CO), which is then converted to either diesel fuel, via a low-temperature Fischer–Tropsch process using cobalt or iron catalysts followed by hydrocracking, or to gasoline, via a high-temperature Fischer–Tropsch process using an iron catalyst followed by oligomerization, isomerization, and hydrogenation. Figure adapted from Ref. [78].

high temperatures (300–350 °C)—and the use of high pressures (20–30 atm) [78, 79]. The low-temperature conversion process is used to produce higher-molecular-weight, straight-chain hydrocarbons that can be processed further to create high-cetane-rating diesel fuels. The high-temperature process is conducive to the production of lower-molecular-weight species more suitable for gasoline. Increasing the octane rating of the straight-run gasoline produced by the Fischer–Tropsch process is more complicated than increasing the cetane number of the straight-run diesel fuel; hence, the production of gasoline is less economical than the production of diesel fuel [79]. Because of the exothermicity of many of the synthesis reactions, energy management and temperature control are quite important [78]. Commercial Fischer–Tropsch synthesis plants have been in operation in Sasolberg, South Africa, for more than 50 years, where an abundance of coal and favorable government policies make them viable [79]. With current interest in biomass-based fuels to minimize greenhouse gas emissions and promote energy independence, Fischer–Tropsch synthesis of transportation fuels from biomass has seen renewed interest, as has liquid fuel synthesis from coal. Also, Fischer–Tropsch synthesis may be a partial solution to the problem of the wasteful and polluting flaring of associated natural gas. More detailed information on Fischer–Tropsch synthesis, as well as a review of other processes to convert biomass feedstocks to other fuels (methanol, ethanol, etc.) and chemicals, can be found in Ref. [78].

## Hydrogen

Hydrogen has frequently been proposed as a fuel for transportation systems and, as such, has a long history. Because hydrogen is not found plentifully in nature, it must be manufactured. The most straightforward approach is to electrolyze water, that is, to separate the $H_2$ and $O_2$ in water using electricity. Hydrogen can also be made from various hydrocarbons, or coal, with steam reforming. Currently, **steam reforming** of natural gas is the most common method of hydrogen production [82]. In this process, the natural gas reacts with steam to form carbon monoxide and hydrogen. This mixture is then reacted in the presence of a catalyst to force the water-gas shift reaction (Table. 2.3) to favor $CO_2$ and $H_2$. The $H_2$ is then separated from the $CO_2$. Large-scale production of hydrogen using steam reforming is not likely to be environmentally viable unless advances in $CO_2$ capture and storage technologies are made.

   Rather than thinking of hydrogen as an alternative fuel, it might better be viewed as an energy storage mode or energy carrier, like electricity. Hydrogen combustion ideally does not generate any pollutant emissions other than oxides of nitrogen, and if it is burned with pure oxygen rather than air, there are no oxides of nitrogen, either. Similarly, the combustion products ($H_2O$, $O_2$, and $N_2$) contain no greenhouse gases, noting that $H_2O$, although radiatively active, is equilibrated in the atmosphere.[7] However, pollutant and greenhouse gases are indeed associated with the manufacture of hydrogen, and these would need to be controlled. Because of hydrogen's low

---

[7]Because of the presence of the oceans and other bodies of water, the global average relative humidity is approximately constant and is not directly affected by the anthropogenic emission of $H_2O$ [83]. However, in a warming climate, the absolute amount of $H_2O$ in the atmosphere increases for a fixed relative humidity. This increased moisture thus produces a positive radiative forcing feedback.

density, its storage for transportation applications is problematic. For a review of the production and use of hydrogen, we refer the reader to Ref. [82].

## SUMMARY

In this chapter, you were introduced to the naming conventions employed for hydrocarbon and alcohols based on their bonding structures. We also examined important properties of transportation fuels: octane rating for spark-ignition engine fuels, cetane rating for diesel engine fuels, volatility characteristics, and energy density. Properties and specifications were presented for conventional fuels, i.e., gasoline, including reformulated gasoline and oxygenated winter gasoline; diesel fuel; heating oils; various aviation fuels; natural gas; and coal. Similarly, we highlighted interesting aspects of the production and/or use of several alternative fuels: ethanol and ethanol blends, biodiesel, thermal conversion liquids, biogas, black liquor, Fischer–Tropsch liquids, and hydrogen. Concerns over global warming, environmental degradation, and national energy independence, and others, will assure a continued interest in and development of alternate fuels.

## NOMENCLATURE AND ABBREVIATIONS

| | |
|---|---|
| $a$ | Molar oxygen-fuel ratio (kmol/kmol) |
| $A/F$ | Mass air-fuel ratio (kg/kg) |
| AKI | Antiknock index (dimensionless) |
| $\Delta h_c$ | Heat of combustion (heating value) (J/kg) |
| $H$ | Enthalpy (J) |
| $\bar{h}_f^o$ | Enthalpy of formation (J/kmol) |
| HGI | Hardgrove grindability index |
| HHV | Higher heating value (kJ/kg or MJ/kg) |
| LHV | Lower heating value (kJ/kg or MJ/kg) |
| $m$ | Mass (kg) |
| $MW$ | Molecular weight (kg/kmol) |
| MTBE | Methyl tertiary-butyl ether |
| MON | Motor octane number (dimensionless) |
| $N$ | Number of moles (kmol) |
| PM | Particulate matter |
| R | Arbitrary radical |
| RFG | Reformulated gasoline |
| RON | Research octane number (dimensionless) |
| RVP | Reid vapor pressure (kPa) |
| $T_{ad}$ | Adiabatic flame temperature (K) |
| TAP | Toxic air pollutant |
| VOC | Volatile organic compound |
| $x$ | Number of carbon atoms in fuel |

| $y$ | Number of hydrogen atoms in fuel |
|---|---|
| $Y$ | Mass fraction (kg/kg) |

**Greek Symbols**

| $\chi$ | Mole fraction (kmol/kmol) |
|---|---|

**Subscripts**

| $i$ | $i$th species |
|---|---|
| *prod* | Products |
| *reac* | Reactants |
| *s* | Stoichiometric |

# REFERENCES

1. Anon., "Standard Test Method for Research Octane Number of Spark-Ignition Engine Fuel, Designation: D2699–08," ASTM International, West Conshohocken, PA, 2008.

2. Anon., "Standard Test Method for Motor Octane Number of Spark-Ignition Engine Fuel, Designation: D2700–08," ASTM International, West Conshohocken, PA, 2008.

3. Anon., "Standard Specification for Automotive Spark-Ignition Engine Fuel, Designation: D4814–08b," ASTM International, West Conshohocken, PA, 2008.

4. Anon., "Standard Test Method for Cetane Number of Diesel Fuel Oil, Designation: D613–08," ASTM International, West Conshohocken, PA, 2008.

5. Anon., "Standard Test Method for Vapor Pressure of Gasoline and Gasoline-Oxygenate Blends (Dry Method), Designation: D4953–06," ASTM International, West Conshohocken, PA, 2006.

6. Anon., "Standard Test Method for Distillation of Petroleum Products at Atmospheric Pressure, Designation: D86–09," ASTM International, West Conshohocken, PA, 2009.

7. Chevron USA, Inc., "Motor Gasolines Technical Review," Chevron Products Company, San Ramon, CA, 1998, http://www.chevron.com/products/ourfuels/prodserv/fuels/documents/69083_MotorGas_Tech Review.pdf.

8. Anon., "Standard Test Method for Vapor-Liquid Ratio Temperature Determination of Fuels (Evacuated Chamber Method), Designation: D5188–04a," ASTM International, West Conshohocken, PA, 2004.

9. Speight, J. G., *The Chemistry and Technology of Petroleum,* 4th Ed., CRC Press, Boca Raton, FL, 2007.

10. Code of Federal Regulations, Title 40—Protection of the Environment, Chapter I—Environmental Protection Agency, Subchapter C—Air Programs, Part 80—Regulation of Fuels and Fuel Additives, Subpart D, Reformulated Gasoline, July 1, 2008 Edition.

11. Energy Policy Act of 2005, Public Law 109-58, 109th Congress. See Code of Federal Regulations, Title 40—Protection of the Environment, Chapter I—Environmental Protection Agency, Subchapter C—Air Programs, Part 80—Regulation of Fuels and Fuel Additives, Subpart K, Renewable Fuel Standard, July 1, 2008 Edition.

12. U.S. Environmental Protection Agency, "Fuel Trends Report: Gasoline 1995–2005," EPA420-S-08-001, January 2008.

13. Code of Federal Regulations, Title 40—Protection of the Environment, Chapter I—Environmental Protection Agency, Subchapter C—Air Programs, Part 80—Regulation of Fuels and Fuel Additives, Subpart H, Gasoline Sulfur, July 1, 2008 Edition.

14. U.S. Environmental Protection Agency, "State Winter Oxygenated Fuel Program Requirements for Attainment or Maintenance of CO NAAQS," EPA420-B-08-006, January 2008.

15. Anon., "Standard Specification for Diesel Fuel Oils, Designation: D975–08a," ASTM International, West Conshohocken, PA, 2008.

16. Bacha, J., *et al.,* "Diesel Fuels Technical Review," Chevron Products Company, San Ramon, CA, 1998, http://www.chevron.com/products/ourfuels/prodserv/fuels/documents/Diesel_Fuel_Tech_Review.pdf.

17. U.S. Department of Health and Human Services, Public Health Service, Agency for Toxic Substances and Disease Registry (ATSDR), "Toxicological Profile for Fuel Oils," Atlanta, GA, 1995.

18. Dagaut, P., "On the Kinetics of Hydrocarbons Oxidation from Natural Gas to Kerosene and Diesel Fuel," *Physical Chemistry and Chemical Physics,* 4: 2079–2094 (2002).

19. Anon., "Standard Specification for Fuel Oils, Designation: D396–09," ASTM International, West Conshohocken, PA, 2008.

20. Hemighaus, G. *et al.,* "Aviation Fuels Technical Review," Chevron Products Company, San Ramon, CA, 2006, http://www.chevron.com/products/ourfuels/prodserv/fuels/documents/aviation_fuels.pdf.

21. Anon., "Standard Specification for Aviation Gasolines, Designation: D910–07a," ASTM International, West Conshohocken, PA, 2008.

22. Anon., "Standard Specification for Jet B Wide-Cut Aviation Turbine Fuel, Designation: D6615–06," ASTM International, West Conshohocken, PA, 2008.

23. Anon., "Standard Specification for Aviation Turbine Fuels, Designation: D1655–08a," ASTM International, West Conshohocken, PA, 2008.

24. Coordinating Research Council, *Handbook of Aviation Fuel Properties,* SAE International, Warrendale, PA, 2004.

25. Gerner, F., and Svensson, B., "Regulation of Associated Gas Flaring and Venting: A Global Overview and Lessens," Report No. 3, Working Copy, Report No. 29554, World Bank Group, The World Bank, April 2004.

26. U.S. Energy Information Administration, Office of Oil and Gas, "Natural Gas Processing: The Crucial Link between Natural Gas Production and Its Transportation to Market," January 2006, http://www.eia.doe.gov/pub/oil_gas/natural_gas/feature_articles/2006/ngprocess/ngprocess.pdf.

27. Foss, M. M., "Interstate Natural Gas—Quality Specifications & Interchangeability," Center for Economics Research, The University of Texas at Austin, December 2004, www.beg.utexas.edu/energyecon/lng.

28. Reed, R. J., *North American Combustion Handbook,* 3rd Ed., Vol. I, North American Manufacturing Co., Cleveland, OH, 1986.

29. U. S. Energy Information Agency, "Annual Energy Review 2007," DOE/EIA-0383, 2008, http://www.eia.doe.gov/aer.

30. Anon., "Standard Terminology of Coal and Coke, Designation: D121–09," ASTM International, West Conshohocken, PA, 2009.

31. Anon., "Standard Classification of Coals by Rank, Designation: D388–05," ASTM International, West Conshohocken, PA, 2005.

32. Anon., "Standard Test Method for Gross Calorific Value of Coal and Coke, Designation: D5865–07a," ASTM International, West Conshohocken, PA, 2007.

33. Channiwala, S. A., and Parikh, P. P., "A Unified Correlation for Estimating HHV of Solid, Liquid and Gaseous Fuels," *Fuel,* 81: 1051–1063 (2002).

34. Hardgrove, R. M., "Grindability of Coal," *Transactions of the American Society of Mechanical Engineers,* 54: 37–46 (1932).

35. U.S. Environmental Protection Agency, "Fact Sheet: Coal Combustion Residues (CCR)—Surface Impoundments with High Hazard Potential Ratings," EPA530-F-09-006, June 2009.

36. Krevelen, D. W. van, *Coal—Typology, Physics, Chemistry, Constitution,* Elsevier, Amsterdam, 1993.

37. Demirbaş, A., "Biomass Resource Facilities and Biomass Conversion Processing for Fuels and Chemicals," *Energy Conversion and Management,* 42: 1357–2378 (2001).

38. U.S. Government Printing Office, Energy Independence and Security Act of 2007, Public Law 110-140, http://frwebgate.access.gpo.gov/cgi-bin/getdoc.cgi?dbname=110_cong_bills&docid=f:h6enr.txt.pdf.

39. U. S. Energy Information Agency, "Short-Term Energy Outlook," 7 July 2009, http://www.eia.doe.gov/steo.

40. European Commission, "Proposal for a Directive 2008/0016 (COD) of the European Parliament and of the Council on the Promotion of the Use of Energy from Renewable Sources," 2008. See also http://www.eea.europa.eu/themes/energy/bioenergy-and-biofuels-the-big-picture for related information.

41. Anon., "Standard Specification for Fuel Ethanol (Ed75-Ed85) for Automotive Spark-Ignition Engines, Designation: D5798–09b," ASTM International, West Conshohocken, PA, 2009.

42. Hill, J., Nelson, E., Tilman, D., Polasky, S., and Tiffany, D., "Environmental, Economic, and Energetic Costs and Benefits of Biodiesel and Ethanol Biofuels," *Proceedings of the National Academy of Science,* 103: 11206–11210 (2006).

43. Jacobson, M. Z., "Effects of Ethanol (E85) versus Gasoline Vehicles on Cancer and Mortality in the United States," *Environmental Science & Technology,* 41: 4150–4157 (2007).

44. Brinkman, N., Wang, M., Weber, T., and Darlington, T., "*Well-to-Wheels Analysis of Advanced Fuel / Vehicle Systems—A North American Study of Energy Use, Greenhouse Gas Emissions, and Criteria Pollutant Emissions,*" Argonne National Laboratory, May 2005.

45. Ohlrogge, J., Allen, D., Berguson, B., DellaPenna, D., Shachar-Hillo, Y., and Stymne, S., "Driving on Biomass," *Science,* 324: 1019–1020 (2009).

46. Campbell, J. E., Lobell, D. B., and Field, C. B., "Greater Transportation Energy and GHG Offsets from Bioelectricity than Ethanol," *Science,* 324: 1055–1057 (2009).

47. Wise, M., *et al.,* "Implications of Limiting $CO_2$ Concentrations for Land Use and Energy," *Science,* 324: 1183–1186 (2009).

48. Anon., Center for Transportation Analysis, Oak Ridge National Laboratory, http://cta .ornl.gov/bedb/biofuels/ethanol/Fuel_Property_Comparison_for_Ethanol-Gasoline-No2Diesel.xls.

49. Heywood, J. B., *Internal Combustion Engine Fundamentals,* McGraw-Hill, New York, 1988.

50. Meuller, C., "An Introduction to Biodiesel," Combustion Research Facility, Sandia National Laboratory, 2006, http://www.ca.sandia.gov/crf/viewArticle.php?cid=CRFV28N3-A.

51. Nagel, N., and Lemke, P., "Production of Methyl Fuel from Microalgae," *Applied Biochemistry and Biotechnology,* 24: 355–361 (1990).

52. Ma, F. and Hanna, M. A., "Biodiesel Production: A Review," *Bioresource Technology,* 70: 1–15 (1999).

53. Goering, C. E., Schwab, A. W., Daugherty, M. J., and Pryde, E. H., "Fuel Properties of Eleven Oils," *Transactions of the American Society of Agricultural Engineers,* 25: 1472–1483 (1982).

54. Encinar, J. M., González, J. F., Rodríguez, J. J., and Tejedor, A., "Biodiesel Fuels from Vegetable Oils: Transesterification of *Cynara Cardunculus* L. Oils with Ethanol," *Energy & Fuels,* 16: 443–450 (2002).

55. Waynick, J. A., "Characterization of Biodiesel Oxidation and Oxidation Products, CRC Project No. AVFL-2b," NREL/TP-540-39096, Coordinating Research Council and National Renewable Energy Laboratory, August 2005.

56. Anon., "Standard Specification for Biodiesel Fuel Blend Stock (B100) for Middle Distillate Fuels, Designation: D6751–09," ASTM International, West Conshohocken, PA, 2009.

57. Anon., "Standard Specification for Diesel Fuel Oil, Biodiesel Blend (B6 to B20), Designation: D7467–08," ASTM International, West Conshohocken, PA, 2009.

58. U.S. Environmental Protection Agency, "A Comprehensive Analysis of Biodiesel Impacts on Exhaust Emissions," Draft Technical Report, EPA420-P-02-001, October 2002.

59. Szybist, J. P., Song, J., Alam, M., and Boehman, A. L., "Biodiesel Combustion, Emissions and Emission Control," *Fuel Processing Technology,* 88: 679–691 (2007).

60. Lapuerta, M., Armas, O., and Rodríguez-Fernández, J., "Effect of Biodiesel Fuels on Diesel Engine Emissions," *Progress in Energy and Combustion Science,* 24: 198–223 (2008).

61. Pinto, A. C., *et al.,* "Biodiesel: An Overview," *Journal of the Brazilian Chemical Society,* 16: 1313–1330 (2005).

62. Demirbaş, A., "Biodiesel Impacts on Compression Ignition Engine (CIE): Analysis of Air Pollution Issues Relating to Exhaust Emissions," *Energy Sources,* 27: 549–558 (2005).

63. Demirbaş, A., "Pyrolysis and Steam Gasification Processes of Black Liquor," *Energy Conversion and Management,* 43: 877–884 (2002).

64. Larson, E. D., Kreutz, T. G., and Consonni, S., "Combined Biomass and Black Liquor Gasifier/Gas Turbine Cogeneration at Pulp and Paper Mills," *Journal of Engineering for Gas Turbines and Power,* 121: 394–400 (1999).

65. Larson, E. D., Consonni, S., and Kreutz, T. G., "Preliminary Economics of Black Liquor Gasifier/Gas Turbine Cogeneration at Pulp and Paper Mills," *Journal of Engineering for Gas Turbines and Power,* 122: 255–261 (2000).

66. U.S. Energy Information Administration, "Renewable Energy Trends in Consumption and Electricity 2007," April 2009. See also URL: www.eia.doe.gov/fuelrenewable.html.

67. He, B. J., Zhang, Y., Funk, T. L., Riskowski, G. L., and Yin, Y., "Thermochemical Conversion of Swine Manure: An Alternative Process for Waste Treatment and Renewable Energy Production," *Transactions of the ASAE,* 43: 1827–1833 (2000).

68. Bridgwater, A. V., Meier, D., Radlein, D., "An Overview of Fast Pyrolysis of Biomass," *Organic Geochemistry,* 30: 1479–1493 (1999).

69. Demirbaş, A., "Yields of Oil Products from Thermochemical Biomass Conversion Processes," *Energy Conversion and Management,* 39: 686–690 (1998).

70. Elliott, D. C., *et al.,* "Developments in Direct Thermochemical Liquefaction of Biomass: 1983–1990," *Energy & Fuels,* 5: 399–410 (1991).

71. Elliott, D. C., "Historical Developments in Hydroprocessing Bio-oils," *Energy & Fuels,* 21: 1792–1815 (2007).

72. Yadvika, Santosh, S., Sreekrishnan, T. R., Kohli, S., and Rana, V., "Enhancement of Biogas Production from Solid Substrates Using Different Techniques—A Review," *Bioresources Technology,* 95: 1–10 (2004).

73. U.S. Environmental Protection Agency, "LFG Energy Project Development Handbook," July 2009, http://www.epa.gov/lmop/res/handbook.htm.

74. Dewil, R., Appels, L., and Baeyens, J., "Energy Use of Biogas Hampered by the Presence of Siloxanes," *Energy Conversion and Management,* 47: 1711–1722 (2006).

75. Qin, W., Egolfopoulos, F. N., and Tsotsis, T. T., "Fundamental and Environmental Aspects of Landfill Gas Utilization for Power Generation," *Chemical Engineering Journal,* 82: 157–172 (2001).

76. U.S. Environmental Protection Agency, "Landfill Methane Outreach Program (LMOP)," URL: http://www.epa.gov/lmop/.

77. Raven, R. P. J. M., and Gregersen, K. H., "Biogas Plants in Denmark: Success and Setbacks," *Renewable and Sustainable Energy Reviews,* 11: 116–132 (2007).

78. Spath, R. L., and Dayton, D. C., "Preliminary Screening—Technical and Economic Assessment of Synthesis Gas to Fuels and Chemicals with Emphasis on the Potential for Biomass-Derived Syngas," NREL/TP-510-34929, National Renewable Energy Laboratory, December 2003.

79. Dry, M. E., "The Fischer–Tropsch Process: 1950–2000," *Catalysis Today,* 71: 227–241 (2002).

80. Fischer, F., and Tropsch, H., "Die Erdolsynthese bei Gewohnlichen Druck aus den Verasungsprodkten der Kohlen," *Brennstoff Chemie,* 7: 97–116 (1926).

81. Schulz, H., "Short History and Present Trends of Fischer–Tropsch Synthesis," *Applied Catalysis A: General,* 186: 3–12 (1999).

82. Koroneos, C., Dompros, A., Roumbas, G., and Moussiopoulos, N., "Life Cycle Assessment of Hydrogen Fuel Production Processes," *International Journal of Hydrogen Energy,* 29: 1443–1450 (2004).

83. Solomon, S., *et al.,* (eds.), *Climate Change 2007—The Physical Science Basis,* Contribution of Working Group I to the Fourth Assessment Report of the Intergovernmental Panel on Climate Change, Cambridge University Press, 2007.

## PROBLEMS

**17.1** Define RON, MON, and AKI and how they relate.

**17.2** In a test to determine the research octane number (RON) of a fuel, the mass flowrates of *n*-heptane and isooctane reference fuels were 0.062 g/s and 0.185 g/s, respectively, at the knock condition. The respective densities of the reference fuels are 697 and 684 kg/m$^3$. Determine the RON of this fuel. Is this a good fuel for use in a modern spark-ignition engine?

**17.3** Determine the cetane number for a test in which equal volumes of *n*-cetane and heptamethyl nonane reference fuels defined the knocking condition. Does this fuel meet the ASTM cetane number specification for No. 2 diesel fuel?

**17.4** Explain why pressurized fuel systems are less prone to vapor lock than those operating at, or slightly below, atmospheric pressure.

**17.5** Using the data from Table 17.4, estimate the percent evaporated at 100 °C for both summer and winter gasoline. Assume that the gasolines follow the maximum values in the distillation specifications.

**17.6** Consider 1 kmol of gasoline ($C_8H_{15}$). Determine the number of kmols of MTBE ($C_5H_{15}O$) needed to produce a fuel mixture that contains 2 wt.% oxygen.

**17.7** Determine the stoichiometric air–fuel ratio and the lower heating value for Kansas City natural gas (Table 17.12).

**17.8** Calculate the stoichiometric air–fuel ratio of a fuel blend containing 80 vol.% ethanol and 20 vol.% gasoline ($C_8H_{15}$). Assume the density of the gasoline to be 750 kg/m$^3$.

**17.9** Using Eq. 17.3, estimate the higher heating values for the coals with the lowest (Fort Union Bed, MT) and highest (Pennsylvania #2, PA) fixed carbon content based on the ultimate analysis shown in Table 17.18. How do your values compare with the values given in the table?

**17.10** Calculate the higher and lower heating values for gaseous hydrogen at 298 K and 1 atm. Express your results on both a mass and volume basis.

$HO_2 = \frac{1}{2} H_2 + O_2$

# Selected Thermodynamic Properties of Gases Comprising C–H–O–N System

## TABLES A.1 TO A.12

Ideal-gas values for standard reference state ($T = 298.15$ K, $P = 1$ atm) for

$$\bar{c}_p(T),\ \bar{h}^o(T) - \bar{h}^o_{f,\,\text{ref}},\ \bar{h}^o_f(T),\ \bar{s}^o(T),\ \bar{g}^o_f(T) \quad \text{for}$$

$$CO,\ CO_2,\ H_2,\ H,\ OH,\ H_2O,\ N_2,\ N,\ NO,\ NO_2,\ O_2,\ O.$$

Enthalpy of formation and Gibbs function of formation for compounds are calculated from the elements as

$$\bar{h}^o_{f,\,i}(T) = \bar{h}^o_i(T) - \sum_{j\ \text{elements}} v'_j \bar{h}^o_j(T)$$

$$\bar{g}^o_{f,\,i}(T) = \bar{g}^o_i(T) - \sum_{j\ \text{elements}} v'_j \bar{g}^o_j(T)$$

$$= \bar{h}^o_{f,\,i}(T) - T\bar{s}^o_i(T) - \sum_{j\ \text{elements}} v'_j[-T\bar{s}^o_j(T)]$$

SOURCE: Tables were generated from curvefit coefficients given in Kee, R. J., Rupley, F. M., and Miller, J. A., "The Chemkin Thermodynamic Data Base," Sandia Report, SAND87-8215B, March 1991.

## TABLE A.13

Curvefit coefficients for $\bar{c}_p(T)$ for the same gases as above.
SOURCE: ibid.

**Table A.1** Carbon monoxide (CO), MW = 28.010, enthalpy of formation @ 298 K (kJ/kmol) = −110,541

| $T$ (K) | $\bar{c}_p$ (kJ/kmol-K) | $(\bar{h}^o(T) - \bar{h}^o_f(298))$ (kJ/kmol) | $h^o_f(T)$ (kJ/kmol) | $\bar{s}^o(T)$ (kJ/kmol-K) | $\bar{g}^o_f(T)$ (kJ/kmol) |
|---|---|---|---|---|---|
| 200 | 28.687 | −2,835 | −111,308 | 186.018 | −128,532 |
| 298 | 29.072 | 0 | −110,541 | 197.548 | −137,163 |
| 300 | 29.078 | 54 | −110,530 | 197.728 | −137,328 |
| 400 | 29.433 | 2,979 | −110,121 | 206.141 | −146,332 |
| 500 | 29.857 | 5,943 | −110,017 | 212.752 | −155,403 |
| 600 | 30.407 | 8,955 | −110,156 | 218.242 | −164,470 |
| 700 | 31.089 | 12,029 | −110,477 | 222.979 | −173,499 |
| 800 | 31.860 | 15,176 | −110,924 | 227.180 | −182,473 |
| 900 | 32.629 | 18,401 | −111,450 | 230.978 | −191,386 |
| 1000 | 33.255 | 21,697 | −112,022 | 234.450 | −200,238 |
| 1100 | 33.725 | 25,046 | −112,619 | 237.642 | −209,030 |
| 1200 | 34.148 | 28,440 | −113,240 | 240.595 | −217,768 |
| 1300 | 34.530 | 31,874 | −113,881 | 243.344 | −226,453 |
| 1400 | 34.872 | 35,345 | −114,543 | 245.915 | −235,087 |
| 1500 | 35.178 | 38,847 | −115,225 | 248.332 | −243,674 |
| 1600 | 35.451 | 42,379 | −115,925 | 250.611 | −252,214 |
| 1700 | 35.694 | 45,937 | −116,644 | 252.768 | −260,711 |
| 1800 | 35.910 | 49,517 | −117,380 | 254.814 | −269,164 |
| 1900 | 36.101 | 53,118 | −118,132 | 256.761 | −277,576 |
| 2000 | 36.271 | 56,737 | −118,902 | 258.617 | −285,948 |
| 2100 | 36.421 | 60,371 | −119,687 | 260.391 | −294,281 |
| 2200 | 36.553 | 64,020 | −120,488 | 262.088 | −302,576 |
| 2300 | 36.670 | 67,682 | −121,305 | 263.715 | −310,835 |
| 2400 | 36.774 | 71,354 | −122,137 | 265.278 | −319,057 |
| 2500 | 36.867 | 75,036 | −122,984 | 266.781 | −327,245 |
| 2600 | 36.950 | 78,727 | −123,847 | 268.229 | −335,399 |
| 2700 | 37.025 | 82,426 | −124,724 | 269.625 | −343,519 |
| 2800 | 37.093 | 86,132 | −125,616 | 270.973 | −351,606 |
| 2900 | 37.155 | 89,844 | −126,523 | 272.275 | −359,661 |
| 3000 | 37.213 | 93,562 | −127,446 | 273.536 | −367,684 |
| 3100 | 37.268 | 97,287 | −128,383 | 274.757 | −375,677 |
| 3200 | 37.321 | 101,016 | −129,335 | 275.941 | −383,639 |
| 3300 | 37.372 | 104,751 | −130,303 | 277.090 | −391,571 |
| 3400 | 37.422 | 108,490 | −131,285 | 278.207 | −399,474 |
| 3500 | 37.471 | 112,235 | −132,283 | 279.292 | −407,347 |
| 3600 | 37.521 | 115,985 | −133,295 | 280.349 | −415,192 |
| 3700 | 37.570 | 119,739 | −134,323 | 281.377 | −423,008 |
| 3800 | 37.619 | 123,499 | −135,366 | 282.380 | −430,796 |
| 3900 | 37.667 | 127,263 | −136,424 | 283.358 | −438,557 |
| 4000 | 37.716 | 131,032 | −137,497 | 284.312 | −446,291 |
| 4100 | 37.764 | 134,806 | −138,585 | 285.244 | −453,997 |
| 4200 | 37.810 | 138,585 | −139,687 | 286.154 | −461,677 |
| 4300 | 37.855 | 142,368 | −140,804 | 287.045 | −469,330 |
| 4400 | 37.897 | 146,156 | −141,935 | 287.915 | −476,957 |
| 4500 | 37.936 | 149,948 | −143,079 | 288.768 | −484,558 |
| 4600 | 37.970 | 153,743 | −144,236 | 289.602 | −492,134 |
| 4700 | 37.998 | 157,541 | −145,407 | 290.419 | −499,684 |
| 4800 | 38.019 | 161,342 | −146,589 | 291.219 | −507,210 |
| 4900 | 38.031 | 165,145 | −147,783 | 292.003 | −514,710 |
| 5000 | 38.033 | 168,948 | −148,987 | 292.771 | −522,186 |

**Table A.2** Carbon dioxide ($CO_2$), MW = 44.011, enthalpy of formation @ 298 K (kJ/kmol) = −393,546

| $T$(K) | $\bar{c}_p$ (kJ/kmol-K) | $(\bar{h}^o(T)-\bar{h}_f^o(298))$ (kJ/kmol) | $\bar{h}_f^o(T)$ (kJ/kmol) | $\bar{s}^o(T)$ (kJ/kmol-K) | $\bar{g}_f^o(T)$ (kJ/kmol) |
|---|---|---|---|---|---|
| 200 | 32.387 | −3,423 | −393,483 | 199.876 | −394,126 |
| 298 | 37.198 | 0 | −393,546 | 213.736 | −394,428 |
| 300 | 37.280 | 69 | −393,547 | 213.966 | −394,433 |
| 400 | 41.276 | 4,003 | −393,617 | 225.257 | −394,718 |
| 500 | 44.569 | 8,301 | −393,712 | 234.833 | −394,983 |
| 600 | 47.313 | 12,899 | −393,844 | 243.209 | −395,226 |
| 700 | 49.617 | 17,749 | −394,013 | 250.680 | −395,443 |
| 800 | 51.550 | 22,810 | −394,213 | 257.436 | −395,635 |
| 900 | 53.136 | 28,047 | −394,433 | 263.603 | −395,799 |
| 1000 | 54.360 | 33,425 | −394,659 | 269.268 | −395,939 |
| 1100 | 55.333 | 38,911 | −394,875 | 274.495 | −396,056 |
| 1200 | 56.205 | 44,488 | −395,083 | 279.348 | −396,155 |
| 1300 | 56.984 | 50,149 | −395,287 | 283.878 | −396,236 |
| 1400 | 57.677 | 55,882 | −395,488 | 288.127 | −396,301 |
| 1500 | 58.292 | 61,681 | −395,691 | 292.128 | −396,352 |
| 1600 | 58.836 | 67,538 | −395,897 | 295.908 | −396,389 |
| 1700 | 59.316 | 73,446 | −396,110 | 299.489 | −396,414 |
| 1800 | 59.738 | 79,399 | −396,332 | 302.892 | −396,425 |
| 1900 | 60.108 | 85,392 | −396,564 | 306.132 | −396,424 |
| 2000 | 60.433 | 91,420 | −396,808 | 309.223 | −396,410 |
| 2100 | 60.717 | 97,477 | −397,065 | 312.179 | −396,384 |
| 2200 | 60.966 | 103,562 | −397,338 | 315.009 | −396,346 |
| 2300 | 61.185 | 109,670 | −397,626 | 317.724 | −396,294 |
| 2400 | 61.378 | 115,798 | −397,931 | 320.333 | −396,230 |
| 2500 | 61.548 | 121,944 | −398,253 | 322.842 | −396,152 |
| 2600 | 61.701 | 128,107 | −398,594 | 325.259 | −396,061 |
| 2700 | 61.839 | 134,284 | −398,952 | 327.590 | −395,957 |
| 2800 | 61.965 | 140,474 | −399,329 | 329.841 | −395,840 |
| 2900 | 62.083 | 146,677 | −399,725 | 332.018 | −395,708 |
| 3000 | 62.194 | 152,891 | −400,140 | 334.124 | −395,562 |
| 3100 | 62.301 | 159,116 | −400,573 | 336.165 | −395,403 |
| 3200 | 62.406 | 165,351 | −401,025 | 338.145 | −395,229 |
| 3300 | 62.510 | 171,597 | −401,495 | 340.067 | −395,041 |
| 3400 | 62.614 | 177,853 | −401,983 | 341.935 | −394,838 |
| 3500 | 62.718 | 184,120 | −402,489 | 343.751 | −394,620 |
| 3600 | 62.825 | 190,397 | −403,013 | 345.519 | −394,388 |
| 3700 | 62.932 | 196,685 | −403,553 | 347.242 | −394,141 |
| 3800 | 63.041 | 202,983 | −404,110 | 348.922 | −393,879 |
| 3900 | 63.151 | 209,293 | −404,684 | 350.561 | −393,602 |
| 4000 | 63.261 | 215,613 | −405,273 | 353.161 | −393,311 |
| 4100 | 63.369 | 221,945 | −405,878 | 353.725 | −393,004 |
| 4200 | 63.474 | 228,287 | −406,499 | 355.253 | −392,683 |
| 4300 | 63.575 | 234,640 | −407,135 | 356.748 | −392,346 |
| 4400 | 63.669 | 241,002 | −407,785 | 358.210 | −391,995 |
| 4500 | 63.753 | 247,373 | −408,451 | 359.642 | −391,629 |
| 4600 | 63.825 | 253,752 | −409,132 | 361.044 | −391,247 |
| 4700 | 63.881 | 260,138 | −409,828 | 362.417 | −390,851 |
| 4800 | 63.918 | 266,528 | −410,539 | 363.763 | −390,440 |
| 4900 | 63.932 | 272,920 | −411,267 | 365.081 | −390,014 |
| 5000 | 63.919 | 279,313 | −412,010 | 366.372 | −389,572 |

**Table A.3**    Hydrogen ($H_2$), MW = 2.016, enthalpy of formation @ 298 K (kJ/kmol) = 0

| $T(K)$ | $\bar{c}_p$ (kJ/kmol-K) | $(\bar{h}^o(T) - \bar{h}_f^o(298))$ (kJ/kmol) | $\bar{h}_f^o(T)$ (kJ/kmol) | $\bar{s}^o(T)$ (kJ/kmol-K) | $\bar{g}_f^o(T)$ (kJ/kmol) |
|---|---|---|---|---|---|
| 200 | 28.522 | −2,818 | 0 | 119.137 | 0 |
| 298 | 28.871 | 0 | 0 | 130.595 | 0 |
| 300 | 28.877 | 53 | 0 | 130.773 | 0 |
| 400 | 29.120 | 2,954 | 0 | 139.116 | 0 |
| 500 | 29.275 | 5,874 | 0 | 145.632 | 0 |
| 600 | 29.375 | 8,807 | 0 | 150.979 | 0 |
| 700 | 29.461 | 11,749 | 0 | 155.514 | 0 |
| 800 | 29.581 | 14,701 | 0 | 159.455 | 0 |
| 900 | 29.792 | 17,668 | 0 | 162.950 | 0 |
| 1000 | 30.160 | 20,664 | 0 | 166.106 | 0 |
| 1100 | 30.625 | 23,704 | 0 | 169.003 | 0 |
| 1200 | 31.077 | 26,789 | 0 | 171.687 | 0 |
| 1300 | 31.516 | 29,919 | 0 | 174.192 | 0 |
| 1400 | 31.943 | 33,092 | 0 | 176.543 | 0 |
| 1500 | 32.356 | 36,307 | 0 | 178.761 | 0 |
| 1600 | 32.758 | 39,562 | 0 | 180.862 | 0 |
| 1700 | 33.146 | 42,858 | 0 | 182.860 | 0 |
| 1800 | 33.522 | 46,191 | 0 | 184.765 | 0 |
| 1900 | 33.885 | 49,562 | 0 | 186.587 | 0 |
| 2000 | 34.236 | 52,968 | 0 | 188.334 | 0 |
| 2100 | 34.575 | 56,408 | 0 | 190.013 | 0 |
| 2200 | 34.901 | 59,882 | 0 | 191.629 | 0 |
| 2300 | 35.216 | 63,388 | 0 | 193.187 | 0 |
| 2400 | 35.519 | 66,925 | 0 | 194.692 | 0 |
| 2500 | 35.811 | 70,492 | 0 | 196.148 | 0 |
| 2600 | 36.091 | 74,087 | 0 | 197.558 | 0 |
| 2700 | 36.361 | 77,710 | 0 | 198.926 | 0 |
| 2800 | 36.621 | 81,359 | 0 | 200.253 | 0 |
| 2900 | 36.871 | 85,033 | 0 | 201.542 | 0 |
| 3000 | 37.112 | 88,733 | 0 | 202.796 | 0 |
| 3100 | 37.343 | 92,455 | 0 | 204.017 | 0 |
| 3200 | 37.566 | 96,201 | 0 | 205.206 | 0 |
| 3300 | 37.781 | 99,968 | 0 | 206.365 | 0 |
| 3400 | 37.989 | 103,757 | 0 | 207.496 | 0 |
| 3500 | 38.190 | 107,566 | 0 | 208.600 | 0 |
| 3600 | 38.385 | 111,395 | 0 | 209.679 | 0 |
| 3700 | 38.574 | 115,243 | 0 | 210.733 | 0 |
| 3800 | 38.759 | 119,109 | 0 | 211.764 | 0 |
| 3900 | 38.939 | 122,994 | 0 | 212.774 | 0 |
| 4000 | 39.116 | 126,897 | 0 | 213.762 | 0 |
| 4100 | 39.291 | 130,817 | 0 | 214.730 | 0 |
| 4200 | 39.464 | 134,755 | 0 | 215.679 | 0 |
| 4300 | 39.636 | 138,710 | 0 | 216.609 | 0 |
| 4400 | 39.808 | 142,682 | 0 | 217.522 | 0 |
| 4500 | 39.981 | 146,672 | 0 | 218.419 | 0 |
| 4600 | 40.156 | 150,679 | 0 | 219.300 | 0 |
| 4700 | 40.334 | 154,703 | 0 | 220.165 | 0 |
| 4800 | 40.516 | 158,746 | 0 | 221.016 | 0 |
| 4900 | 40.702 | 162,806 | 0 | 221.853 | 0 |
| 5000 | 40.895 | 166,886 | 0 | 222.678 | 0 |

**Table A.4** Hydrogen atom (H), MW = 1.008, enthalpy of formation @ 298 K (kJ/kmol) = 217,977

| $T(K)$ | $\bar{c}_p$ (kJ/kmol-K) | $(\bar{h}^o(T) - \bar{h}^o_f(298))$ (kJ/kmol) | $\bar{h}^o_f(T)$ (kJ/kmol) | $\bar{s}^o(T)$ (kJ/kmol-K) | $\bar{g}^o_f(T)$ (kJ/kmol) |
|---|---|---|---|---|---|
| 200 | 20.786 | −2,040 | 217,346 | 106.305 | 207,999 |
| 298 | 20.786 | 0 | 217,977 | 114.605 | 203,276 |
| 300 | 20.786 | 38 | 217,989 | 114.733 | 203,185 |
| 400 | 20.786 | 2,117 | 218,617 | 120.713 | 198,155 |
| 500 | 20.786 | 4,196 | 219,236 | 125.351 | 192,968 |
| 600 | 20.786 | 6,274 | 219,848 | 129.351 | 187,657 |
| 700 | 20.786 | 8,353 | 220,456 | 132.345 | 182,244 |
| 800 | 20.786 | 10,431 | 221,059 | 135.121 | 176,744 |
| 900 | 20.786 | 12,510 | 221,653 | 137.569 | 171,169 |
| 1000 | 20.786 | 14,589 | 222,234 | 139.759 | 165,528 |
| 1100 | 20.786 | 16,667 | 222,793 | 141.740 | 159,830 |
| 1200 | 20.786 | 18,746 | 223,329 | 143.549 | 154,082 |
| 1300 | 20.786 | 20,824 | 223,843 | 145.213 | 148,291 |
| 1400 | 20.786 | 22,903 | 224,335 | 146.753 | 142,461 |
| 1500 | 20.786 | 24,982 | 224,806 | 148.187 | 136,596 |
| 1600 | 20.786 | 27,060 | 225,256 | 149.528 | 130,700 |
| 1700 | 20.786 | 29,139 | 225,687 | 150.789 | 124,777 |
| 1800 | 20.786 | 31,217 | 226,099 | 151.977 | 118,830 |
| 1900 | 20.786 | 33,296 | 226,493 | 153.101 | 112,859 |
| 2000 | 20.786 | 35,375 | 226,868 | 154.167 | 106,869 |
| 2100 | 20.786 | 37,453 | 227,226 | 155.181 | 100,860 |
| 2200 | 20.786 | 39,532 | 227,568 | 156.148 | 94,834 |
| 2300 | 20.786 | 41,610 | 227,894 | 157.072 | 88,794 |
| 2400 | 20.786 | 43,689 | 228,204 | 157.956 | 82,739 |
| 2500 | 20.786 | 45,768 | 228,499 | 158.805 | 76,672 |
| 2600 | 20.786 | 47,846 | 228,780 | 159.620 | 70,593 |
| 2700 | 20.786 | 49,925 | 229,047 | 160.405 | 64,504 |
| 2800 | 20.786 | 52,003 | 229,301 | 161.161 | 58,405 |
| 2900 | 20.786 | 54,082 | 229,543 | 161.890 | 52,298 |
| 3000 | 20.786 | 56,161 | 229,772 | 162.595 | 46,182 |
| 3100 | 20.786 | 58,239 | 229,989 | 163.276 | 40,058 |
| 3200 | 20.786 | 60,318 | 230,195 | 163.936 | 33,928 |
| 3300 | 20.786 | 62,396 | 230,390 | 164.576 | 27,792 |
| 3400 | 20.786 | 64,475 | 230,574 | 165.196 | 21,650 |
| 3500 | 20.786 | 66,554 | 230,748 | 165.799 | 15,502 |
| 3600 | 20.786 | 68,632 | 230,912 | 166.384 | 9,350 |
| 3700 | 20.786 | 70,711 | 231,067 | 166.954 | 3,194 |
| 3800 | 20.786 | 72,789 | 231,212 | 167.508 | −2,967 |
| 3900 | 20.786 | 74,868 | 231,348 | 168.048 | −9,132 |
| 4000 | 20.786 | 76,947 | 231,475 | 168.575 | −15,299 |
| 4100 | 20.786 | 79,025 | 231,594 | 169.088 | −21,470 |
| 4200 | 20.786 | 81,104 | 231,704 | 169.589 | −27,644 |
| 4300 | 20.786 | 83,182 | 231,805 | 170.078 | −33,820 |
| 4400 | 20.786 | 85,261 | 231,897 | 170.556 | −39,998 |
| 4500 | 20.786 | 87,340 | 231,981 | 171.023 | −46,179 |
| 4600 | 20.786 | 89,418 | 232,056 | 171.480 | −52,361 |
| 4700 | 20.786 | 91,497 | 232,123 | 171.927 | −58,545 |
| 4800 | 20.786 | 93,575 | 232,180 | 172.364 | −64,730 |
| 4900 | 20.786 | 95,654 | 232,228 | 172.793 | −70,916 |
| 5000 | 20.786 | 97,733 | 232,267 | 173.213 | −77,103 |

**Table A.5**     Hydroxyl (OH), MW = 17.007, enthalpy of formation @ 298 K (kJ/kmol) = 38,985

| $T(K)$ | $\bar{c}_p$ (kJ/kmol-K) | $(\bar{h}^o(T) - \bar{h}_f^o(298))$ (kJ/kmol) | $\bar{h}_f^o(T)$ (kJ/kmol) | $\bar{s}^o(T)$ (kJ/kmol-K) | $\bar{g}_f^o(T)$ (kJ/kmol) |
|---|---|---|---|---|---|
| 200 | 30.140 | −2,948 | 38,864 | 171.607 | 35,808 |
| 298 | 29.932 | 0 | 38,985 | 183.604 | 34,279 |
| 300 | 29.928 | 55 | 38,987 | 183.789 | 34,250 |
| 400 | 29.718 | 3,037 | 39,030 | 192.369 | 32,662 |
| 500 | 29.570 | 6,001 | 39,000 | 198.983 | 31,072 |
| 600 | 29.527 | 8,955 | 38,909 | 204.369 | 29,494 |
| 700 | 29.615 | 11,911 | 38,770 | 208.925 | 27,935 |
| 800 | 29.844 | 14,883 | 38,599 | 212.893 | 26,399 |
| 900 | 30.208 | 17,884 | 38,410 | 216.428 | 24,885 |
| 1000 | 30.682 | 20,928 | 38,220 | 219.635 | 23,392 |
| 1100 | 31.186 | 24,022 | 38,039 | 222.583 | 21,918 |
| 1200 | 31.662 | 27,164 | 37,867 | 225.317 | 20,460 |
| 1300 | 32.114 | 30,353 | 37,704 | 227.869 | 19,017 |
| 1400 | 32.540 | 33,586 | 37,548 | 230.265 | 17,585 |
| 1500 | 32.943 | 36,860 | 37,397 | 232.524 | 16,164 |
| 1600 | 33.323 | 40,174 | 37,252 | 234.662 | 14,753 |
| 1700 | 33.682 | 43,524 | 37,109 | 236.693 | 13,352 |
| 1800 | 34.019 | 46,910 | 36,969 | 238.628 | 11,958 |
| 1900 | 34.337 | 50,328 | 36,831 | 240.476 | 10,573 |
| 2000 | 34.635 | 53,776 | 36,693 | 242.245 | 9,194 |
| 2100 | 34.915 | 57,254 | 36,555 | 243.942 | 7,823 |
| 2200 | 35.178 | 60,759 | 36,416 | 245.572 | 6,458 |
| 2300 | 35.425 | 64,289 | 36,276 | 247.141 | 5,099 |
| 2400 | 35.656 | 67,843 | 36,133 | 248.654 | 3,746 |
| 2500 | 35.872 | 71,420 | 35,986 | 250.114 | 2,400 |
| 2600 | 36.074 | 75,017 | 35,836 | 251.525 | 1,060 |
| 2700 | 36.263 | 78,634 | 35,682 | 252.890 | −275 |
| 2800 | 36.439 | 82,269 | 35,524 | 254.212 | −1,604 |
| 2900 | 36.604 | 85,922 | 35,360 | 255.493 | −2,927 |
| 3000 | 36.759 | 89,590 | 35,191 | 256.737 | −4,245 |
| 3100 | 36.903 | 93,273 | 35,016 | 257.945 | −5,556 |
| 3200 | 37.039 | 96,970 | 34,835 | 259.118 | −6,862 |
| 3300 | 37.166 | 100,681 | 34,648 | 260.260 | −8,162 |
| 3400 | 37.285 | 104,403 | 34,454 | 261.371 | −9,457 |
| 3500 | 37.398 | 108,137 | 34,253 | 262.454 | −10,745 |
| 3600 | 37.504 | 111,882 | 34,046 | 263.509 | −12,028 |
| 3700 | 37.605 | 115,638 | 33,831 | 264.538 | −13,305 |
| 3800 | 37.701 | 119,403 | 33,610 | 265.542 | −14,576 |
| 3900 | 37.793 | 123,178 | 33,381 | 266.522 | −15,841 |
| 4000 | 37.882 | 126,962 | 33,146 | 267.480 | −17,100 |
| 4100 | 37.968 | 130,754 | 32,903 | 268.417 | −18,353 |
| 4200 | 38.052 | 134,555 | 32,654 | 269.333 | −19,600 |
| 4300 | 38.135 | 138,365 | 32,397 | 270.229 | −20,841 |
| 4400 | 38.217 | 142,182 | 32,134 | 271.107 | −22,076 |
| 4500 | 38.300 | 146,008 | 31,864 | 271.967 | −23,306 |
| 4600 | 38.382 | 149,842 | 31,588 | 272.809 | −24,528 |
| 4700 | 38.466 | 153,685 | 31,305 | 273.636 | −25,745 |
| 4800 | 38.552 | 157,536 | 31,017 | 274.446 | −26,956 |
| 4900 | 38.640 | 161,395 | 30,722 | 275.242 | −28,161 |
| 5000 | 38.732 | 165,264 | 30,422 | 276.024 | −29,360 |

**Table A.6**    Water ($H_2O$), MW = 18.016, enthalpy of formation @ 298 K (kJ/kmol) = −241,845, enthalpy of vaporization (kJ/kmol) = 44,010

| $T(K)$ | $\bar{c}_p$ (kJ/kmol-K) | $(\bar{h}^o(T) - \bar{h}_f^o(298))$ (kJ/kmol) | $\bar{h}_f^o(T)$ (kJ/kmol) | $\bar{s}^o(T)$ (kJ/kmol-K) | $\bar{g}_f^o(T)$ (kJ/kmol) |
|---|---|---|---|---|---|
| 200 | 32.255 | −3,227 | −240,838 | 175.602 | −232,779 |
| 298 | 33.448 | 0 | −241,845 | 188.715 | −228,608 |
| 300 | 33.468 | 62 | −241,865 | 188.922 | −228,526 |
| 400 | 34.437 | 3,458 | −242,858 | 198.686 | −223,929 |
| 500 | 35.337 | 6,947 | −243,822 | 206.467 | −219,085 |
| 600 | 36.288 | 10,528 | −244,753 | 212.992 | −214,049 |
| 700 | 37.364 | 14,209 | −245,638 | 218.665 | −208,861 |
| 800 | 38.587 | 18,005 | −246,461 | 223.733 | −203,550 |
| 900 | 39.930 | 21,930 | −247,209 | 228.354 | −198,141 |
| 1000 | 41.315 | 25,993 | −247,879 | 232.633 | −192,652 |
| 1100 | 42.638 | 30,191 | −248,475 | 236.634 | −187,100 |
| 1200 | 43.874 | 34,518 | −249,005 | 240.397 | −181,497 |
| 1300 | 45.027 | 38,963 | −249,477 | 243.955 | −175,852 |
| 1400 | 46.102 | 43,520 | −249,895 | 247.332 | −170,172 |
| 1500 | 47.103 | 48,181 | −250,267 | 250.547 | −164,464 |
| 1600 | 48.035 | 52,939 | −250,597 | 253.617 | −158,733 |
| 1700 | 48.901 | 57,786 | −250,890 | 256.556 | −152,983 |
| 1800 | 49.705 | 62,717 | −251,151 | 259.374 | −147,216 |
| 1900 | 50.451 | 67,725 | −251,384 | 262.081 | −141,435 |
| 2000 | 51.143 | 72,805 | −251,594 | 264.687 | −135,643 |
| 2100 | 51.784 | 77,952 | −251,783 | 267.198 | −129,841 |
| 2200 | 52.378 | 83,160 | −251,955 | 269.621 | −124,030 |
| 2300 | 52.927 | 88,426 | −252,113 | 271.961 | −118,211 |
| 2400 | 53.435 | 93,744 | −252,261 | 274.225 | −112,386 |
| 2500 | 53.905 | 99,112 | −252,399 | 276.416 | −106,555 |
| 2600 | 54.340 | 104,524 | −252,532 | 278.539 | −100,719 |
| 2700 | 54.742 | 109,979 | −252,659 | 280.597 | −94,878 |
| 2800 | 55.115 | 115,472 | −252,785 | 282.595 | −89,031 |
| 2900 | 55.459 | 121,001 | −252,909 | 284.535 | −83,181 |
| 3000 | 55.779 | 126,563 | −253,034 | 286.420 | −77,326 |
| 3100 | 56.076 | 132,156 | −253,161 | 288.254 | −71,467 |
| 3200 | 56.353 | 137,777 | −253,290 | 290.039 | −65,604 |
| 3300 | 56.610 | 143,426 | −253,423 | 291.777 | −59,737 |
| 3400 | 56.851 | 149,099 | −253,561 | 293.471 | −53,865 |
| 3500 | 57.076 | 154,795 | −253,704 | 295.122 | −47,990 |
| 3600 | 57.288 | 160,514 | −253,852 | 296.733 | −42,110 |
| 3700 | 57.488 | 166,252 | −254,007 | 298.305 | −36,226 |
| 3800 | 57.676 | 172,011 | −254,169 | 299.841 | −30,338 |
| 3900 | 57.856 | 177,787 | −254,338 | 301.341 | −24,446 |
| 4000 | 58.026 | 183,582 | −254,515 | 302.808 | −18,549 |
| 4100 | 58.190 | 189,392 | −254,699 | 304.243 | −12,648 |
| 4200 | 58.346 | 195,219 | −254,892 | 305.647 | −6,742 |
| 4300 | 58.496 | 201,061 | −255,093 | 307.022 | −831 |
| 4400 | 58.641 | 206,918 | −255,303 | 308.368 | 5,085 |
| 4500 | 58.781 | 212,790 | −255,522 | 309.688 | 11,005 |
| 4600 | 58.916 | 218,674 | −255,751 | 310.981 | 16,930 |
| 4700 | 59.047 | 224,573 | −255,990 | 312.250 | 22,861 |
| 4800 | 59.173 | 230,484 | −256,239 | 313.494 | 28,796 |
| 4900 | 59.295 | 236,407 | −256,501 | 314.716 | 34,737 |
| 5000 | 59.412 | 242,343 | −256,774 | 315.915 | 40,684 |

**Table A.7**     Nitrogen ($N_2$), MW = 28.013, enthalpy of formation @ 298 K (kJ/kmol) = 0

| $T$(K) | $\bar{c}_p$ (kJ/kmol-K) | $(\bar{h}^o(T) - \bar{h}_f^o(298))$ (kJ/kmol) | $\bar{h}_f^o(T)$ (kJ/kmol) | $\bar{s}^o(T)$ (kJ/kmol-K) | $\bar{g}_f^o(T)$ (kJ/kmol) |
|---|---|---|---|---|---|
| 200 | 28.793 | −2,841 | 0 | 179.959 | 0 |
| 298 | 29.071 | 0 | 0 | 191.511 | 0 |
| 300 | 29.075 | 54 | 0 | 191.691 | 0 |
| 400 | 29.319 | 2,973 | 0 | 200.088 | 0 |
| 500 | 29.636 | 5,920 | 0 | 206.662 | 0 |
| 600 | 30.086 | 8,905 | 0 | 212.103 | 0 |
| 700 | 30.684 | 11,942 | 0 | 216.784 | 0 |
| 800 | 31.394 | 15,046 | 0 | 220.927 | 0 |
| 900 | 32.131 | 18,222 | 0 | 224.667 | 0 |
| 1000 | 32.762 | 21,468 | 0 | 228.087 | 0 |
| 1100 | 33.258 | 24,770 | 0 | 231.233 | 0 |
| 1200 | 33.707 | 28,118 | 0 | 234.146 | 0 |
| 1300 | 34.113 | 31,510 | 0 | 236.861 | 0 |
| 1400 | 34.477 | 34,939 | 0 | 239.402 | 0 |
| 1500 | 34.805 | 38,404 | 0 | 241.792 | 0 |
| 1600 | 35.099 | 41,899 | 0 | 244.048 | 0 |
| 1700 | 35.361 | 45,423 | 0 | 246.184 | 0 |
| 1800 | 35.595 | 48,971 | 0 | 248.212 | 0 |
| 1900 | 35.803 | 52,541 | 0 | 250.142 | 0 |
| 2000 | 35.988 | 56,130 | 0 | 251.983 | 0 |
| 2100 | 36.152 | 59,738 | 0 | 253.743 | 0 |
| 2200 | 36.298 | 63,360 | 0 | 255.429 | 0 |
| 2300 | 36.428 | 66,997 | 0 | 257.045 | 0 |
| 2400 | 36.543 | 70,645 | 0 | 258.598 | 0 |
| 2500 | 36.645 | 74,305 | 0 | 260.092 | 0 |
| 2600 | 36.737 | 77,974 | 0 | 261.531 | 0 |
| 2700 | 36.820 | 81,652 | 0 | 262.919 | 0 |
| 2800 | 36.895 | 85,338 | 0 | 264.259 | 0 |
| 2900 | 36.964 | 89,031 | 0 | 265.555 | 0 |
| 3000 | 37.028 | 92,730 | 0 | 266.810 | 0 |
| 3100 | 37.088 | 96,436 | 0 | 268.025 | 0 |
| 3200 | 37.144 | 100,148 | 0 | 269.203 | 0 |
| 3300 | 37.198 | 103,865 | 0 | 270.347 | 0 |
| 3400 | 37.251 | 107,587 | 0 | 271.458 | 0 |
| 3500 | 37.302 | 111,315 | 0 | 272.539 | 0 |
| 3600 | 37.352 | 115,048 | 0 | 273.590 | 0 |
| 3700 | 37.402 | 118,786 | 0 | 274.614 | 0 |
| 3800 | 37.452 | 122,528 | 0 | 275.612 | 0 |
| 3900 | 37.501 | 126,276 | 0 | 276.586 | 0 |
| 4000 | 37.549 | 130,028 | 0 | 277.536 | 0 |
| 4100 | 37.597 | 133,786 | 0 | 278.464 | 0 |
| 4200 | 37.643 | 137,548 | 0 | 279.370 | 0 |
| 4300 | 37.688 | 141,314 | 0 | 280.257 | 0 |
| 4400 | 37.730 | 145,085 | 0 | 281.123 | 0 |
| 4500 | 37.768 | 148,860 | 0 | 281.972 | 0 |
| 4600 | 37.803 | 152,639 | 0 | 282.802 | 0 |
| 4700 | 37.832 | 156,420 | 0 | 283.616 | 0 |
| 4800 | 37.854 | 160,205 | 0 | 284.412 | 0 |
| 4900 | 37.868 | 163,991 | 0 | 285.193 | 0 |
| 5000 | 37.873 | 167,778 | 0 | 285.958 | 0 |

**Table A.8**　　　Nitrogen atom (N), MW = 14.007, enthalpy of formation @ 298 K (kJ/kmol) = 472,629

| $T$(K) | $\bar{c}_p$ (kJ/kmol-K) | $(\bar{h}^o(T) - \bar{h}_f^o(298))$ (kJ/kmol) | $\bar{h}_f^o(T)$ (kJ/kmol) | $\bar{s}^o(T)$ (kJ/kmol-K) | $\bar{g}_f^o(T)$ (kJ/kmol) |
|---|---|---|---|---|---|
| 200 | 20.790 | −2,040 | 472,008 | 144.889 | 461,026 |
| 298 | 20.786 | 0 | 472,629 | 153.189 | 455,504 |
| 300 | 20.786 | 38 | 472,640 | 153.317 | 455,398 |
| 400 | 20.786 | 2,117 | 473,258 | 159.297 | 449,557 |
| 500 | 20.786 | 4,196 | 473,864 | 163.935 | 443,562 |
| 600 | 20.786 | 6,274 | 474,450 | 167.725 | 437,446 |
| 700 | 20.786 | 8,353 | 475,010 | 170.929 | 431,234 |
| 800 | 20.786 | 10,431 | 475,537 | 173.705 | 424,944 |
| 900 | 20.786 | 12,510 | 476,027 | 176.153 | 418,590 |
| 1000 | 20.786 | 14,589 | 476,483 | 178.343 | 412,183 |
| 1100 | 20.792 | 16,668 | 476,911 | 180.325 | 405,732 |
| 1200 | 20.795 | 18,747 | 477,316 | 182.134 | 399,243 |
| 1300 | 20.795 | 20,826 | 477,700 | 183.798 | 392,721 |
| 1400 | 20.793 | 22,906 | 478,064 | 185.339 | 386,171 |
| 1500 | 20.790 | 24,985 | 478,411 | 186.774 | 379,595 |
| 1600 | 20.786 | 27,064 | 478,742 | 188.115 | 372,996 |
| 1700 | 20.782 | 29,142 | 479,059 | 189.375 | 366,377 |
| 1800 | 20.779 | 31,220 | 479,363 | 190.563 | 359,740 |
| 1900 | 20.777 | 33,298 | 479,656 | 191.687 | 353,086 |
| 2000 | 20.776 | 35,376 | 479,939 | 192.752 | 346,417 |
| 2100 | 20.778 | 37,453 | 480,213 | 193.766 | 339,735 |
| 2200 | 20.783 | 39,531 | 480,479 | 194.733 | 333,039 |
| 2300 | 20.791 | 41,610 | 480,740 | 195.657 | 326,331 |
| 2400 | 20.802 | 43,690 | 480,995 | 196.542 | 319,612 |
| 2500 | 20.818 | 45,771 | 481,246 | 197.391 | 312,883 |
| 2600 | 20.838 | 47,853 | 481,494 | 198.208 | 306,143 |
| 2700 | 20.864 | 49,938 | 481,740 | 198.995 | 299,394 |
| 2800 | 20.895 | 52,026 | 481,985 | 199.754 | 292,636 |
| 2900 | 20.931 | 54,118 | 482,230 | 200.488 | 285,870 |
| 3000 | 20.974 | 56,213 | 482,476 | 201.199 | 279,094 |
| 3100 | 21.024 | 58,313 | 482,723 | 201.887 | 272,311 |
| 3200 | 21.080 | 60,418 | 482,972 | 202.555 | 265,519 |
| 3300 | 21.143 | 62,529 | 483,224 | 203.205 | 258,720 |
| 3400 | 21.214 | 64,647 | 483,481 | 203.837 | 251,913 |
| 3500 | 21.292 | 66,772 | 483,742 | 204.453 | 245,099 |
| 3600 | 21.378 | 68,905 | 484,009 | 205.054 | 238,276 |
| 3700 | 21.472 | 71,048 | 484,283 | 205.641 | 231,447 |
| 3800 | 21.575 | 73,200 | 484,564 | 206.215 | 224,610 |
| 3900 | 21.686 | 75,363 | 484,853 | 206.777 | 217,765 |
| 4000 | 21.805 | 77,537 | 485,151 | 207.328 | 210,913 |
| 4100 | 21.934 | 79,724 | 485,459 | 207.868 | 204,053 |
| 4200 | 22.071 | 81,924 | 485,779 | 208.398 | 197,186 |
| 4300 | 22.217 | 84,139 | 486,110 | 208.919 | 190,310 |
| 4400 | 22.372 | 86,368 | 486,453 | 209.431 | 183,427 |
| 4500 | 22.536 | 88,613 | 486,811 | 209.936 | 176,536 |
| 4600 | 22.709 | 90,875 | 487,184 | 210.433 | 169,637 |
| 4700 | 22.891 | 93,155 | 487,573 | 210.923 | 162,730 |
| 4800 | 23.082 | 95,454 | 487,979 | 211.407 | 155,814 |
| 4900 | 23.282 | 97,772 | 488,405 | 211.885 | 148,890 |
| 5000 | 23.491 | 100,111 | 488,850 | 212.358 | 141,956 |

**Table A.9**    Nitric oxide (NO), MW = 30.006, enthalpy of formation @ 298 K (kJ/kmol) = 90,297

| $T(K)$ | $\bar{c}_p$ (kJ/kmol-K) | $(\bar{h}^o(T) - \bar{h}_f^o(298))$ (kJ/kmol) | $\bar{h}_f^o(T)$ (kJ/kmol) | $\bar{s}^o(T)$ (kJ/kmol-K) | $\bar{g}_f^o(T)$ (kJ/kmol) |
|---|---|---|---|---|---|
| 200 | 29.374 | −2,901 | 90,234 | 198.856 | 87,811 |
| 298 | 29.728 | 0 | 90,297 | 210.652 | 86,607 |
| 300 | 29.735 | 55 | 90,298 | 210.836 | 86,584 |
| 400 | 30.103 | 3,046 | 90,341 | 219.439 | 85,340 |
| 500 | 30.570 | 6,079 | 90,367 | 226.204 | 84,086 |
| 600 | 31.174 | 9,165 | 90,382 | 231.829 | 82,828 |
| 700 | 31.908 | 12,318 | 90,393 | 236.688 | 81,568 |
| 800 | 32.715 | 15,549 | 90,405 | 241.001 | 80,307 |
| 900 | 33.489 | 18,860 | 90,421 | 244.900 | 79,043 |
| 1000 | 34.076 | 22,241 | 90,443 | 248.462 | 77,778 |
| 1100 | 34.483 | 25,669 | 90,465 | 251.729 | 76,510 |
| 1200 | 34.850 | 29,136 | 90,486 | 254.745 | 75,241 |
| 1300 | 35.180 | 32,638 | 90,505 | 257.548 | 73,970 |
| 1400 | 35.474 | 36,171 | 90,520 | 260.166 | 72,697 |
| 1500 | 35.737 | 39,732 | 90,532 | 262.623 | 71,423 |
| 1600 | 35.972 | 43,317 | 90,538 | 264.937 | 70,149 |
| 1700 | 36.180 | 46,925 | 90,539 | 267.124 | 68,875 |
| 1800 | 36.364 | 50,552 | 90,534 | 269.197 | 67,601 |
| 1900 | 36.527 | 54,197 | 90,523 | 271.168 | 66,327 |
| 2000 | 36.671 | 57,857 | 90,505 | 273.045 | 65,054 |
| 2100 | 36.797 | 61,531 | 90,479 | 274.838 | 63,782 |
| 2200 | 36.909 | 65,216 | 90,447 | 276.552 | 62,511 |
| 2300 | 37.008 | 68,912 | 90,406 | 278.195 | 61,243 |
| 2400 | 37.095 | 72,617 | 90,358 | 279.772 | 59,976 |
| 2500 | 37.173 | 76,331 | 90,303 | 281.288 | 58,711 |
| 2600 | 37.242 | 80,052 | 90,239 | 282.747 | 57,448 |
| 2700 | 37.305 | 83,779 | 90,168 | 284.154 | 56,188 |
| 2800 | 37.362 | 87,513 | 90,089 | 285.512 | 54,931 |
| 2900 | 37.415 | 91,251 | 90,003 | 286.824 | 53,677 |
| 3000 | 37.464 | 94,995 | 89,909 | 288.093 | 52,426 |
| 3100 | 37.511 | 98,744 | 89,809 | 289.322 | 51,178 |
| 3200 | 37.556 | 102,498 | 89,701 | 290.514 | 49,934 |
| 3300 | 37.600 | 106,255 | 89,586 | 291.670 | 48,693 |
| 3400 | 37.643 | 110,018 | 89,465 | 292.793 | 47,456 |
| 3500 | 37.686 | 113,784 | 89,337 | 293.885 | 46,222 |
| 3600 | 37.729 | 117,555 | 89,203 | 294.947 | 44,992 |
| 3700 | 37.771 | 121,330 | 89,063 | 295.981 | 43,766 |
| 3800 | 37.815 | 125,109 | 88,918 | 296.989 | 42,543 |
| 3900 | 37.858 | 128,893 | 88,767 | 297.972 | 41,325 |
| 4000 | 37.900 | 132,680 | 88,611 | 298.931 | 40,110 |
| 4100 | 37.943 | 136,473 | 88,449 | 299.867 | 38,900 |
| 4200 | 37.984 | 140,269 | 88,283 | 300.782 | 37,693 |
| 4300 | 38.023 | 144,069 | 88,112 | 301.677 | 36,491 |
| 4400 | 38.060 | 147,873 | 87,936 | 302.551 | 35,292 |
| 4500 | 38.093 | 151,681 | 87,755 | 303.407 | 34,098 |
| 4600 | 38.122 | 155,492 | 87,569 | 304.244 | 32,908 |
| 4700 | 38.146 | 159,305 | 87,379 | 305.064 | 31,721 |
| 4800 | 38.162 | 163,121 | 87,184 | 305.868 | 30,539 |
| 4900 | 38.171 | 166,938 | 86,984 | 306.655 | 29,361 |
| 5000 | 38.170 | 170,755 | 86,779 | 307.426 | 28,187 |

**Table A.10**    Nitrogen dioxide ($NO_2$), MW = 46.006, enthalpy of formation @ 298 K (kJ/kmol) = 33, 098

| $T(K)$ | $\bar{c}_p$ (kJ/kmol-K) | $(\bar{h}^o(T)-\bar{h}^o_f(298))$ (kJ/kmol) | $\bar{h}^o_f(T)$ (kJ/kmol) | $\bar{s}^o(T)$ (kJ/kmol-K) | $\bar{g}^o_f(T)$ (kJ/kmol) |
|---|---|---|---|---|---|
| 200 | 32.936 | −3,432 | 33,961 | 226.016 | 45,453 |
| 298 | 36.881 | 0 | 33,098 | 239.925 | 51,291 |
| 300 | 36.949 | 68 | 33,085 | 240.153 | 51,403 |
| 400 | 40.331 | 3,937 | 32,521 | 251.259 | 57,602 |
| 500 | 43.227 | 8,118 | 32,173 | 260.578 | 63,916 |
| 600 | 45.737 | 12,569 | 31,974 | 268.686 | 70,285 |
| 700 | 47.913 | 17,255 | 31,885 | 275.904 | 76,679 |
| 800 | 49.762 | 22,141 | 31,880 | 282.427 | 83,079 |
| 900 | 51.243 | 27,195 | 31,938 | 288.377 | 89,476 |
| 1000 | 52.271 | 32,375 | 32,035 | 293.834 | 95,864 |
| 1100 | 52.989 | 37,638 | 32,146 | 298.850 | 102,242 |
| 1200 | 53.625 | 42,970 | 32,267 | 303.489 | 108,609 |
| 1300 | 54.186 | 48,361 | 32,392 | 307.804 | 114,966 |
| 1400 | 54.679 | 53,805 | 32,519 | 311.838 | 121,313 |
| 1500 | 55.109 | 59,295 | 32,643 | 315.625 | 127,651 |
| 1600 | 55.483 | 64,825 | 32,762 | 319.194 | 133,981 |
| 1700 | 55.805 | 70,390 | 32,873 | 322.568 | 140,303 |
| 1800 | 56.082 | 75,984 | 32,973 | 325.765 | 146,620 |
| 1900 | 56.318 | 81,605 | 33,061 | 328.804 | 152,931 |
| 2000 | 56.517 | 87,247 | 33,134 | 331.698 | 159,238 |
| 2100 | 56.685 | 92,907 | 33,192 | 334.460 | 165,542 |
| 2200 | 56.826 | 98,583 | 33,233 | 337.100 | 171,843 |
| 2300 | 56.943 | 104,271 | 33,256 | 339.629 | 178,143 |
| 2400 | 57.040 | 109,971 | 33,262 | 342.054 | 184,442 |
| 2500 | 57.121 | 115,679 | 33,248 | 344.384 | 190,742 |
| 2600 | 57.188 | 121,394 | 33,216 | 346.626 | 197,042 |
| 2700 | 57.244 | 127,116 | 33,165 | 348.785 | 203,344 |
| 2800 | 57.291 | 132,843 | 33,095 | 350.868 | 209,648 |
| 2900 | 57.333 | 138,574 | 33,007 | 352.879 | 215,955 |
| 3000 | 57.371 | 144,309 | 32,900 | 354.824 | 222,265 |
| 3100 | 57.406 | 150,048 | 32,776 | 356.705 | 228,579 |
| 3200 | 57.440 | 155,791 | 32,634 | 358.529 | 234,898 |
| 3300 | 57.474 | 161,536 | 32,476 | 360.297 | 241,221 |
| 3400 | 57.509 | 167,285 | 32,302 | 362.013 | 247,549 |
| 3500 | 57.546 | 173,038 | 32,113 | 363.680 | 253,883 |
| 3600 | 57.584 | 178,795 | 31,908 | 365.302 | 260,222 |
| 3700 | 57.624 | 184,555 | 31,689 | 366.880 | 266,567 |
| 3800 | 57.665 | 190,319 | 31,456 | 368.418 | 272,918 |
| 3900 | 57.708 | 196,088 | 31,210 | 369.916 | 279,276 |
| 4000 | 57.750 | 201,861 | 30,951 | 371.378 | 285,639 |
| 4100 | 57.792 | 207,638 | 30,678 | 372.804 | 292,010 |
| 4200 | 57.831 | 213,419 | 30,393 | 374.197 | 298,387 |
| 4300 | 57.866 | 219,204 | 30,095 | 375.559 | 304,772 |
| 4400 | 57.895 | 224,992 | 29,783 | 376.889 | 311,163 |
| 4500 | 57.915 | 230,783 | 29,457 | 378.190 | 317,562 |
| 4600 | 57.925 | 236,575 | 29,117 | 379.464 | 323,968 |
| 4700 | 57.922 | 242,367 | 28,761 | 380.709 | 330,381 |
| 4800 | 57.902 | 248,159 | 28,389 | 381.929 | 336,803 |
| 4900 | 57.862 | 253,947 | 27,998 | 383.122 | 343,232 |
| 5000 | 57.798 | 259,730 | 27,586 | 384.290 | 349,670 |

**Table A.11**    Oxygen ($O_2$), MW = 31.999, enthalpy of formation @ 298 K (kJ/kmol) = 0

| $T(K)$ | $\bar{c}_p$ (kJ/kmol-K) | $(\bar{h}^o(T) - \bar{h}_f^o(298))$ (kJ/kmol) | $\bar{h}_f^o(T)$ (kJ/kmol) | $\bar{s}^o(T)$ (kJ/kmol-K) | $\bar{g}_f^o(T)$ (kJ/kmol) |
|---|---|---|---|---|---|
| 200 | 28.473 | −2,836 | 0 | 193.518 | 0 |
| 298 | 29.315 | 0 | 0 | 205.043 | 0 |
| 300 | 29.331 | 54 | 0 | 205.224 | 0 |
| 400 | 30.210 | 3,031 | 0 | 213.782 | 0 |
| 500 | 31.114 | 6,097 | 0 | 220.620 | 0 |
| 600 | 32.030 | 9,254 | 0 | 226.374 | 0 |
| 700 | 32.927 | 12,503 | 0 | 231.379 | 0 |
| 800 | 33.757 | 15,838 | 0 | 235.831 | 0 |
| 900 | 34.454 | 19,250 | 0 | 239.849 | 0 |
| 1000 | 34.936 | 22,721 | 0 | 243.507 | 0 |
| 1100 | 35.270 | 26,232 | 0 | 246.852 | 0 |
| 1200 | 35.593 | 29,775 | 0 | 249.935 | 0 |
| 1300 | 35.903 | 33,350 | 0 | 252.796 | 0 |
| 1400 | 36.202 | 36,955 | 0 | 255.468 | 0 |
| 1500 | 36.490 | 40,590 | 0 | 257.976 | 0 |
| 1600 | 36.768 | 44,253 | 0 | 260.339 | 0 |
| 1700 | 37.036 | 47,943 | 0 | 262.577 | 0 |
| 1800 | 37.296 | 51,660 | 0 | 264.701 | 0 |
| 1900 | 37.546 | 55,402 | 0 | 266.724 | 0 |
| 2000 | 37.788 | 59,169 | 0 | 268.656 | 0 |
| 2100 | 38.023 | 62,959 | 0 | 270.506 | 0 |
| 2200 | 38.250 | 66,773 | 0 | 272.280 | 0 |
| 2300 | 38.470 | 70,609 | 0 | 273.985 | 0 |
| 2400 | 38.684 | 74,467 | 0 | 275.627 | 0 |
| 2500 | 38.891 | 78,346 | 0 | 277.210 | 0 |
| 2600 | 39.093 | 82,245 | 0 | 278.739 | 0 |
| 2700 | 39.289 | 86,164 | 0 | 280.218 | 0 |
| 2800 | 39.480 | 90,103 | 0 | 281.651 | 0 |
| 2900 | 39.665 | 94,060 | 0 | 283.039 | 0 |
| 3000 | 39.846 | 98,036 | 0 | 284.387 | 0 |
| 3100 | 40.023 | 102,029 | 0 | 285.697 | 0 |
| 3200 | 40.195 | 106,040 | 0 | 286.970 | 0 |
| 3300 | 40.362 | 110,068 | 0 | 288.209 | 0 |
| 3400 | 40.526 | 114,112 | 0 | 289.417 | 0 |
| 3500 | 40.686 | 118,173 | 0 | 290.594 | 0 |
| 3600 | 40.842 | 122,249 | 0 | 291.742 | 0 |
| 3700 | 40.994 | 126,341 | 0 | 292.863 | 0 |
| 3800 | 41.143 | 130,448 | 0 | 293.959 | 0 |
| 3900 | 41.287 | 134,570 | 0 | 295.029 | 0 |
| 4000 | 41.429 | 138,705 | 0 | 296.076 | 0 |
| 4100 | 41.566 | 142,855 | 0 | 297.101 | 0 |
| 4200 | 41.700 | 147,019 | 0 | 298.104 | 0 |
| 4300 | 41.830 | 151,195 | 0 | 299.087 | 0 |
| 4400 | 41.957 | 155,384 | 0 | 300.050 | 0 |
| 4500 | 42.079 | 159,586 | 0 | 300.994 | 0 |
| 4600 | 42.197 | 163,800 | 0 | 301.921 | 0 |
| 4700 | 42.312 | 168,026 | 0 | 302.829 | 0 |
| 4800 | 42.421 | 172,262 | 0 | 303.721 | 0 |
| 4900 | 42.527 | 176,510 | 0 | 304.597 | 0 |
| 5000 | 42.627 | 180,767 | 0 | 305.457 | 0 |

**Table A.12**     Oxygen atom (O), MW = 16.000, enthalpy of formation @ 298 K (kJ/kmol) = 249,197

| $T(K)$ | $\bar{c}_p$ (kJ/kmol-K) | $(\bar{h}^o(T) - \bar{h}_f^o(298))$ (kJ/kmol) | $\bar{h}_f^o(T)$ (kJ/kmol) | $\bar{s}^o(T)$ (kJ/kmol-K) | $\bar{g}_f^o(T)$ (kJ/kmol) |
|---|---|---|---|---|---|
| 200 | 22.477 | −2,176 | 248,439 | 152.085 | 237,374 |
| 298 | 21.899 | 0 | 249,197 | 160.945 | 231,778 |
| 300 | 21.890 | 41 | 249,211 | 161.080 | 231,670 |
| 400 | 21.500 | 2,209 | 249,890 | 167.320 | 225,719 |
| 500 | 21.256 | 4,345 | 250,494 | 172.089 | 219,605 |
| 600 | 21.113 | 6,463 | 251,033 | 175.951 | 213,375 |
| 700 | 21.033 | 8,570 | 251,516 | 179.199 | 207,060 |
| 800 | 20.986 | 10,671 | 251,949 | 182.004 | 200,679 |
| 900 | 20.952 | 12,768 | 252,340 | 184.474 | 194,246 |
| 1000 | 20.915 | 14,861 | 252,698 | 186.679 | 187,772 |
| 1100 | 20.898 | 16,952 | 253,033 | 188.672 | 181,263 |
| 1200 | 20.882 | 19,041 | 253,350 | 190.490 | 174,724 |
| 1300 | 20.867 | 21,128 | 253,650 | 192.160 | 168,159 |
| 1400 | 20.854 | 23,214 | 253,934 | 193.706 | 161,572 |
| 1500 | 20.843 | 25,299 | 254,201 | 195.145 | 154,966 |
| 1600 | 20.834 | 27,383 | 254,454 | 196.490 | 148,342 |
| 1700 | 20.827 | 29,466 | 254,692 | 197.753 | 141,702 |
| 1800 | 20.822 | 31,548 | 254,916 | 198.943 | 135,049 |
| 1900 | 20.820 | 33,630 | 255,127 | 200.069 | 128,384 |
| 2000 | 20.819 | 35,712 | 255,325 | 201.136 | 121,709 |
| 2100 | 20.821 | 37,794 | 255,512 | 202.152 | 115,023 |
| 2200 | 20.825 | 39,877 | 255,687 | 203.121 | 108,329 |
| 2300 | 20.831 | 41,959 | 255,852 | 204.047 | 101,627 |
| 2400 | 20.840 | 44,043 | 256,007 | 204.933 | 94,918 |
| 2500 | 20.851 | 46,127 | 256,152 | 205.784 | 88,203 |
| 2600 | 20.865 | 48,213 | 256,288 | 206.602 | 81,483 |
| 2700 | 20.881 | 50,300 | 256,416 | 207.390 | 74,757 |
| 2800 | 20.899 | 52,389 | 256,535 | 208.150 | 68,027 |
| 2900 | 20.920 | 54,480 | 256,648 | 208.884 | 61,292 |
| 3000 | 20.944 | 56,574 | 256,753 | 209.593 | 54,554 |
| 3100 | 20.970 | 58,669 | 256,852 | 210.280 | 47,812 |
| 3200 | 20.998 | 60,768 | 256,945 | 210.947 | 41,068 |
| 3300 | 21.028 | 62,869 | 257,032 | 211.593 | 34,320 |
| 3400 | 21.061 | 64,973 | 257,114 | 212.221 | 27,570 |
| 3500 | 21.095 | 67,081 | 257,192 | 212.832 | 20,818 |
| 3600 | 21.132 | 69,192 | 257,265 | 213.427 | 14,063 |
| 3700 | 21.171 | 71,308 | 257,334 | 214.007 | 7,307 |
| 3800 | 21.212 | 73,427 | 257,400 | 214.572 | 548 |
| 3900 | 21.254 | 75,550 | 257,462 | 215.123 | −6,212 |
| 4000 | 21.299 | 77,678 | 257,522 | 215.662 | −12,974 |
| 4100 | 21.345 | 79,810 | 257,579 | 216.189 | −19,737 |
| 4200 | 21.392 | 81,947 | 257,635 | 216.703 | −26,501 |
| 4300 | 21.441 | 84,088 | 257,688 | 217.207 | −33,267 |
| 4400 | 21.490 | 86,235 | 257,740 | 217.701 | −40,034 |
| 4500 | 21.541 | 88,386 | 257,790 | 218.184 | −46,802 |
| 4600 | 21.593 | 90,543 | 257,840 | 218.658 | −53,571 |
| 4700 | 21.646 | 92,705 | 257,889 | 219.123 | −60,342 |
| 4800 | 21.699 | 94,872 | 257,938 | 219.580 | −67,113 |
| 4900 | 21.752 | 97,045 | 257,987 | 220.028 | −73,886 |
| 5000 | 21.805 | 99,223 | 258,036 | 220.468 | −80,659 |

**Table A.13** Curvefit coefficients for thermodynamic properties (C–H–O–N system)

$$\bar{c}_p/R_u = a_1 + a_2 T + a_3 T^2 + a_4 T^3 + a_5 T^4$$

$$\bar{h}^\circ/R_u T = a_1 + \frac{a_2}{2}T + \frac{a_3}{3}T^2 + \frac{a_4}{4}T^3 + \frac{a_5}{5}T^4 + \frac{a_6}{T}$$

$$\bar{s}^\circ/R_u = a_1 \ln T + a_2 T + \frac{a_3}{2}T^2 + \frac{a_4}{3}T^3 + \frac{a_5}{4}T^4 + a_7$$

| Species | T (K) | $a_1$ | $a_2$ | $a_3$ | $a_4$ | $a_5$ | $a_6$ | $a_7$ |
|---|---|---|---|---|---|---|---|---|
| CO | 1000–5000 | 0.03025078E+02 | 0.14426885E−02 | −0.05630827E−05 | 0.10185813E−09 | −0.06910951E−13 | −0.14268350E+05 | 0.06108217E+02 |
|  | 300–1000 | 0.03262451E+02 | 0.15119409E−02 | −0.03881755E−04 | 0.05581944E−07 | −0.02474951E−10 | −0.14310539E+05 | 0.04848897E+02 |
| CO$_2$ | 1000–5000 | 0.04453623E+02 | 0.03140168E−01 | −0.12784105E−05 | 0.02393996E−08 | −0.16690333E−13 | −0.04896696E+06 | −0.09553959E+01 |
|  | 300–1000 | 0.02275724E+02 | 0.09922072E−01 | −0.10409113E−04 | 0.06866686E−07 | −0.02117280E−10 | −0.04837314E+06 | 0.10188488E+02 |
| H$_2$ | 1000–5000 | 0.02991423E+02 | 0.07000644E−02 | −0.05633828E−06 | −0.09231578E−10 | 0.15827519E−14 | −0.08350340E+04 | −0.13551101E+01 |
|  | 300–1000 | 0.03298124E+02 | 0.08249441E−02 | −0.08143015E−05 | −0.09475434E−09 | 0.04134872E−11 | −0.10125209E+04 | −0.03294094E+02 |
| H | 1000–5000 | 0.02500000E+02 | 0.00000000E+00 | 0.00000000E+00 | 0.00000000E+00 | 0.00000000E+00 | 0.02547162E+06 | −0.04601176E+01 |
|  | 300–1000 | 0.02500000E+02 | 0.00000000E+00 | 0.00000000E+00 | 0.00000000E+00 | 0.00000000E+00 | 0.02547162E+06 | −0.04601176E+01 |
| OH | 1000–5000 | 0.02882730E+02 | 0.10139743E−02 | −0.02276877E−05 | 0.02174683E−09 | −0.05126305E−14 | 0.03886888E+05 | 0.05595712E+02 |
|  | 300–1000 | 0.03637266E+02 | 0.01850910E−02 | −0.16761646E−05 | 0.02387202E−07 | −0.08431442E−11 | 0.03606781E+05 | 0.13588605E+01 |
| H$_2$O | 1000–5000 | 0.02672145E+02 | 0.03056293E−01 | −0.08730260E−05 | 0.12009964E−09 | −0.06391618E−13 | −0.02989921E+06 | 0.06862817E+02 |
|  | 300–1000 | 0.03386842E+02 | 0.03474982E−01 | −0.06354696E−04 | 0.06968581E−07 | −0.02505888E−10 | −0.03020811E+06 | 0.02590232E+02 |
| N$_2$ | 1000–5000 | 0.02926640E+02 | 0.14879768E−02 | −0.05684760E−05 | 0.10097038E−09 | −0.06753351E−13 | −0.09227977E+04 | 0.05980528E+02 |
|  | 300–1000 | 0.03298677E+02 | 0.14082404E−02 | −0.03963222E−05 | 0.05641515E−07 | −0.02444854E−10 | −0.10208999E+04 | 0.03950372E+02 |
| N | 1000–5000 | 0.02450268E+02 | 0.10661458E−03 | −0.07465337E−06 | 0.01879652E−09 | −0.10259839E−14 | 0.05611604E+06 | 0.04448758E+02 |
|  | 300–1000 | 0.02503071E+02 | −0.02180018E−03 | 0.05420529E−06 | −0.05647560E−09 | 0.02099904E−12 | 0.05609890E+06 | 0.04167566E+02 |
| NO | 1000–5000 | 0.03245435E+02 | 0.12691383E−02 | −0.05015890E−05 | 0.09169283E−09 | −0.06275419E−13 | 0.09800840E+05 | 0.06417293E+02 |
|  | 300–1000 | 0.03376541E+02 | 0.12530634E−02 | −0.03302750E−05 | 0.05217810E−07 | −0.02446262E−10 | 0.09817961E+05 | 0.05829590E+02 |
| NO$_2$ | 1000–5000 | 0.04682859E+02 | 0.02462429E−01 | −0.10422585E−05 | 0.01976902E−08 | −0.13917168E−13 | 0.02261292E+05 | 0.09885985E+01 |
|  | 300–1000 | 0.02670600E+02 | 0.07838500E−01 | −0.08063864E−04 | 0.06161714E−07 | −0.02320150E−10 | 0.02896290E+05 | 0.11612071E+02 |
| O$_2$ | 1000–5000 | 0.03697578E+02 | 0.06135197E−02 | −0.12588420E−06 | 0.01775281E−09 | −0.11364354E−14 | −0.12339301E+04 | 0.03189165E+02 |
|  | 300–1000 | 0.03212936E+02 | 0.11274864E−02 | −0.05756150E−05 | 0.13138773E−08 | −0.08768554E−11 | −0.10052490E+04 | 0.06034737E+02 |
| O | 1000–5000 | 0.02542059E+02 | −0.02755061E−03 | −0.03102803E−07 | 0.04551067E−10 | −0.04368051E−14 | 0.02923080E+06 | 0.04920308E+02 |
|  | 300–1000 | 0.02946428E+02 | −0.16381665E−02 | 0.02421031E−04 | −0.16028431E−08 | 0.03890696E−11 | 0.02914764E+06 | 0.02963995E+02 |

SOURCE: Kee, R. J., Rupley, F. M., and Miller, J. A., "The Chemkin Thermodynamic Data Base," Sandia Report, SAND87-8215B, reprinted March 1991.

# Appendix
# B

# Fuel Properties

**Table B.1** Selected properties of hydrocarbon fuels: enthalpy of formation,[a] Gibbs function of formation,[a] entropy,[a] and higher and lower heating values all at 298.15 K and 1 atm; boiling points[b] and latent heat of vaporization[c] at 1 atm; constant–pressure adiabatic flame temperature at 1 atm;[d] liquid density[c]

| Formula | Fuel | MW (kg/kmol) | $\bar{h}_f^o$ (kJ/kmol) | $\bar{g}_f^o$ (kJ/kmol) | $\bar{s}^o$ (kJ/kmol-K) | HHV[†] (kJ/kg) | LHV[†] (kJ/kg) | Boiling pt. (°C) | $h_{fg}$ (kJ/kg) | $T_{ad}^{\ddagger}$ (K) | $\rho_{liq}^*$ (kg/m³) |
|---|---|---|---|---|---|---|---|---|---|---|---|
| $CH_4$ | Methane | 16.043 | −74,831 | −50,794 | 186.188 | 55,528 | 50,016 | −164 | 509 | 2226 | 300 |
| $C_2H_2$ | Acetylene | 26.038 | 226,748 | 209,200 | 200.819 | 49,923 | 48,225 | −84 | — | 2539 | — |
| $C_2H_4$ | Ethene | 28.054 | 52,283 | 68,124 | 219.827 | 50,313 | 47,161 | −103.7 | — | 2369 | — |
| $C_2H_6$ | Ethane | 30.069 | −84,667 | −32,886 | 229.492 | 51,901 | 47,489 | −88.6 | 488 | 2259 | 370 |
| $C_3H_6$ | Propene | 42.080 | 20,414 | 62,718 | 266.939 | 48,936 | 45,784 | −47.4 | 437 | 2334 | 514 |
| $C_3H_8$ | Propane | 44.096 | −103,847 | −23,489 | 269.910 | 50,368 | 46,357 | −42.1 | 425 | 2267 | 500 |
| $C_4H_8$ | 1-Butene | 56.107 | 1,172 | 72,036 | 307.440 | 48,471 | 45,319 | −63 | 391 | 2322 | 595 |
| $C_4H_{10}$ | n-Butane | 58.123 | −124,733 | −15,707 | 310.034 | 49,546 | 45,742 | −0.5 | 386 | 2270 | 579 |
| $C_5H_{10}$ | 1-Pentene | 70.134 | −20,920 | 78,605 | 347.607 | 48,152 | 45,000 | 30 | 358 | 2314 | 641 |
| $C_5H_{12}$ | n-Pentane | 72.150 | −146,440 | −8,201 | 348.402 | 49,032 | 45,355 | 36.1 | 358 | 2272 | 626 |
| $C_6H_6$ | Benzene | 78.113 | 82,927 | 129,658 | 269.199 | 42,277 | 40,579 | 80.1 | 393 | 2342 | 879 |
| $C_6H_{12}$ | 1-Hexene | 84.161 | −41,673 | 87,027 | 385.974 | 47,955 | 44,803 | 63.4 | 335 | 2308 | 673 |
| $C_6H_{14}$ | n-Hexane | 86.177 | −167,193 | 209 | 386.811 | 48,696 | 45,105 | 69 | 335 | 2273 | 659 |
| $C_7H_{14}$ | 1-Heptene | 98.188 | −62,132 | 95,563 | 424.383 | 47,817 | 44,665 | 93.6 | — | 2305 | — |
| $C_7H_{16}$ | n-Heptane | 100.203 | −187,820 | 8,745 | 425.262 | 48,456 | 44,926 | 98.4 | 316 | 2274 | 684 |
| $C_8H_{16}$ | 1-Octene | 112.214 | −82,927 | 104,140 | 462.792 | 47,712 | 44,560 | 121.3 | — | 2302 | — |
| $C_8H_{18}$ | n-Octane | 114.230 | −208,447 | 17,322 | 463.671 | 48,275 | 44,791 | 125.7 | 300 | 2275 | 703 |
| $C_9H_{18}$ | 1-Nonene | 126.241 | −103,512 | 112,717 | 501.243 | 47,631 | 44,478 | — | — | 2300 | — |
| $C_9H_{20}$ | n-Nonane | 128.257 | −229,032 | 25,857 | 502.080 | 48,134 | 44,686 | 150.8 | 295 | 2276 | 718 |
| $C_{10}H_{20}$ | 1-Decene | 140.268 | −124,139 | 121,294 | 539.652 | 47,565 | 44,413 | 170.6 | — | 2298 | — |
| $C_{10}H_{22}$ | n-Decane | 142.284 | −249,659 | 34,434 | 540.531 | 48,020 | 44,602 | 174.1 | 277 | 2277 | 730 |
| $C_{11}H_{22}$ | 1-Undecene | 154.295 | −144,766 | 129,830 | 578.061 | 47,512 | 44,360 | — | — | 2296 | — |
| $C_{11}H_{24}$ | n-Undecane | 156.311 | −270,286 | 43,012 | 578.940 | 47,926 | 44,532 | 195.9 | 265 | 2277 | 740 |
| $C_{12}H_{24}$ | 1-Dodecene | 168.322 | −165,352 | 138,407 | 616.471 | 47,468 | 44,316 | 213.4 | — | 2295 | — |
| $C_{12}H_{26}$ | n-Dodecane | 170.337 | −292,162 | — | — | 47,841 | 44,467 | 216.3 | 256 | 2277 | 749 |

[†]Based on gaseous fuel.

[‡]For stoichiometric combustion with air (79 percent $N_2$, 21 percent $O_2$).

[*]For liquids at 20°C or for gases at the boiling point of the liquefied gas.

SOURCES:

[a]Rossini, F. D., et al., *Selected Values of Physical and Thermodynamic Properties of Hydrocarbons and Related Compounds*, Carnegie Press, Pittsburgh, PA, 1953.

[b]Weast, R. C. (ed.), *Handbook of Chemistry and Physics*, 56th Ed., CRC Press, Cleveland, OH, 1976.

[c]Obert, E. F., *Internal Combustion Engines and Air Pollution*, Harper & Row, New York, 1973.

[d]Calculated using HPFLAME (Appendix F).

**Table B.2** Curvefit coefficients for fuel specific heat and standardized enthalpy[a] for reference state of zero enthalpy of the elements at 298.15 K, 1 atm

$$\bar{c}_p(kJ/kmol\text{-}K) = 4.184\,(a_1 + a_2\theta + a_3\theta^2 + a_4\theta^3 + a_5\theta^{-2}),$$

$$\bar{h}^\circ(kJ/kmol) = 4184\,(a_1\theta + a_2\theta^2/2 + a_3\theta^3/3 + a_4\theta^4/4 - a_5\theta^{-1} + a_6),$$

where $\theta \equiv T(K)/1000$

| Formula | Fuel | MW | $a_1$ | $a_2$ | $a_3$ | $a_4$ | $a_5$ | $a_6$ | $a_8^b$ |
|---|---|---|---|---|---|---|---|---|---|
| $CH_4$ | Methane | 16.043 | −0.29149 | 26.327 | −10.610 | 1.5656 | 0.16573 | −18.331 | 4.300 |
| $C_3H_8$ | Propane | 44.096 | −1.4867 | 74.339 | −39.065 | 8.0543 | 0.01219 | −27.313 | 8.852 |
| $C_6H_{14}$ | Hexane | 86.177 | −20.777 | 210.48 | −164.125 | 52.832 | 0.56635 | −39.836 | 15.611 |
| $C_8H_{18}$ | Isooctane | 114.230 | −0.55313 | 181.62 | −97.787 | 20.402 | −0.03095 | −60.751 | 20.232 |
| $CH_3OH$ | Methanol | 32.040 | −2.7059 | 44.168 | −27.501 | 7.2193 | 0.20299 | −48.288 | 5.3375 |
| $C_2H_5OH$ | Ethanol | 46.07 | 6.990 | 39.741 | −11.926 | 0 | 0 | −60.214 | 7.6135 |
| $C_{8.26}H_{15.5}$ | Gasoline | 114.8 | −24.078 | 256.63 | −201.68 | 64.750 | 0.5808 | −27.562 | 17.792 |
| $C_{7.76}H_{13.1}$ | | 106.4 | −22.501 | 227.99 | −177.26 | 56.048 | 0.4845 | −17.578 | 15.232 |
| $C_{10.8}H_{18.7}$ | Diesel | 148.6 | −9.1063 | 246.97 | −143.74 | 32.329 | 0.0518 | −50.128 | 23.514 |

[a]SOURCE: From Heywood, J. B., *Internal Combustion Engine Fundamentals*, McGraw-Hill, New York, 1988, by permission of McGraw-Hill, Inc.

[b]To obtain 0 K reference state for enthalpy, add $a_8$ to $a_6$.

## Table B.3  Curvefit coefficients for fuel vapor thermal conductivity, viscosity, and specific heat[a]

$$\left.\begin{array}{l} k\,(\text{W/m·K}) \\ \mu\,(\text{N·s/m}^2)\cdot 10^6 \\ c_p\,(\text{J/kg·K}) \end{array}\right\} = a_1 + a_2T + a_3T^2 + a_4T^3 + a_5T^4 + a_6T^5 + a_7T^6$$

| Formula | Fuel | T-range (K) | Property | $a_1$ | $a_2$ | $a_3$ | $a_4$ | $a_5$ | $a_6$ | $a_7$ |
|---|---|---|---|---|---|---|---|---|---|---|
| $CH_4$ | Methane | 100–1000 | $k$ | -1.34014990E-2 | 3.66307060E-4 | -1.82248608E-6 | 5.93987998E-9 | -9.14055050E-12 | 6.78968890E-15 | -1.95048736E-18 |
| | | 70–1000 | $\mu$ | 2.96826700E-1 | 3.71120100E-2 | 1.21829800E-5 | -7.02426000E-8 | 7.5432690E-11 | -2.72371660E-14 | 0 |
| | | | $c_p$ | See Table B.2 | | | | | | |
| $C_3H_8$ | Propane | 200–500 | $k$ | -1.07682209E-2 | 8.38590325E-5 | 4.22059864E-8 | 0 | 0 | 0 | 0 |
| | | 270–600 | $\mu$ | -3.54371100E-1 | 3.08009600E-2 | -6.99723000E-6 | 0 | 0 | 0 | 0 |
| | | | $c_p$ | See Table B.2 | | | | | | |
| $C_6H_{14}$ | n-Hexane | 150–1000 | $k$ | 1.28775700E-3 | -2.00499443E-5 | 2.37858831E-7 | -1.60944555E-10 | 7.71027290E-14 | 0 | 0 |
| | | 270–900 | $\mu$ | 1.54541200E+0 | 1.15080900E-2 | 2.72216500E-5 | -3.26900000E-8 | 1.24545900E-11 | 0 | 0 |
| | | | $c_p$ | See Table B.2 | | | | | | |
| $C_7H_{16}$ | n-Heptane | 250–1000 | $k$ | -4.60614700E-2 | 5.95652224E-4 | -2.98893153E-6 | 8.44612876E-9 | -1.22927E-11 | 9.0127E-15 | -2.62961E-18 |
| | | 270–580 | $\mu$ | 1.54009700E+0 | 1.09515700E-2 | 1.80006400E-5 | -1.3637900E-8 | 0 | 0 | 0 |
| | | 300–755 | $c_p$ | 9.46260000E+1 | 5.86099700E+0 | -1.98231320E-3 | -6.88699300E-8 | -1.93795260E-10 | 0 | 0 |
| | | 755–1365 | $c_p$ | -7.40308000E+2 | 1.08935370E+1 | 1.26512400E-2 | 9.84376300E-6 | -4.32282960E-9 | 7.86366300E-13 | 0 |
| $C_8H_{18}$ | n-Octane | 250–500 | $k$ | -4.01391940E-3 | 3.38796092E-5 | 8.19291819E-8 | 0 | 0 | 0 | 0 |
| | | 300–650 | $\mu$ | 8.32435400E-1 | 1.40045000E-2 | 8.79376500E-6 | -6.84030000E-9 | 0 | 0 | 0 |
| | | 275–755 | $c_p$ | 2.14419800E+2 | 5.35690500E+0 | -1.17497000E-3 | -6.99115500E-7 | 0 | 0 | 0 |
| | | 755–1365 | $c_p$ | 2.43596860E+3 | -4.46819470E+0 | 1.66843290E-2 | -1.78856050E-5 | 8.64282020E-9 | -1.61426500E-12 | 0 |
| $C_{10}H_{22}$ | n-Decane | 250–500 | $k$ | -5.88274000E-3 | 3.72449646E-5 | 7.55109624E-8 | 0 | 0 | 0 | 0 |
| | | | $\mu$ | Not available | | | | | | |
| | | 300–700 | $c_p$ | 2.40717800E+2 | 5.09965000E+0 | -6.29026000E-4 | -1.07155000E-6 | 0 | 0 | 0 |
| | | 700–1365 | $c_p$ | -1.35345890E+4 | 9.14879000E+1 | -2.20700000E-1 | 2.91406000E-4 | -2.15307400E-7 | 8.38600000E-11 | -1.34404000E-14 |
| $CH_3OH$ | Methanol | 300–550 | $k$ | -2.02986750E-2 | 1.21910927E-4 | -2.23748473E-8 | 0 | 0 | 0 | 0 |
| | | 250–650 | $\mu$ | 1.19790000E+0 | 2.45028000E-2 | 1.86162740E-5 | -1.30674820E-8 | 0 | 0 | 0 |
| | | | $c_p$ | See Table B.2 | | | | | | |
| $C_2H_5OH$ | Ethanol | 250–550 | $k$ | -2.46663000E-2 | 1.55892550E-4 | -8.22954822E-8 | 0 | 0 | 0 | 0 |
| | | 270–600 | $\mu$ | -6.33595000E-2 | 3.20713470E-2 | -6.25079576E-6 | 0 | 0 | 0 | 0 |
| | | | $c_p$ | See Table B.2 | | | | | | |

[a]SOURCE: Andrews, J. R., and Biblarz, O., "Temperature Dependence of Gas Properties in Polynomial Form," Naval Postgraduate School, NPS67-81-001, January 1981.

# Selected Properties of Air, Nitrogen, and Oxygen

**Table C.1**     Selected properties of air at 1 atm[a]

| T (K) | $\rho$ (kg/m³) | $c_p$ (kJ/kg-K) | $\mu \cdot 10^7$ (N-s/m²) | $\nu \cdot 10^6$ (m²/s) | $k \cdot 10^3$ (W/m-K) | $\alpha \cdot 10^6$ (m²/s) | Pr |
|---|---|---|---|---|---|---|---|
| 100 | 3.5562 | 1.032 | 71.1 | 2.00 | 9.34 | 2.54 | 0.786 |
| 150 | 2.3364 | 1.012 | 103.4 | 4.426 | 13.8 | 5.84 | 0.758 |
| 200 | 1.7458 | 1.007 | 132.5 | 7.590 | 18.1 | 10.3 | 0.737 |
| 250 | 1.3947 | 1.006 | 159.6 | 11.44 | 22.3 | 15.9 | 0.720 |
| 300 | 1.1614 | 1.007 | 184.6 | 15.89 | 26.3 | 22.5 | 0.707 |
| 350 | 0.9950 | 1.009 | 208.2 | 20.92 | 30.0 | 29.9 | 0.700 |
| 400 | 0.8711 | 1.014 | 230.1 | 26.41 | 33.8 | 38.3 | 0.690 |
| 450 | 0.7740 | 1.021 | 250.7 | 32.39 | 37.3 | 47.2 | 0.686 |
| 500 | 0.6964 | 1.030 | 270.1 | 38.79 | 40.7 | 56.7 | 0.684 |
| 550 | 0.6329 | 1.040 | 288.4 | 45.57 | 43.9 | 66.7 | 0.683 |
| 600 | 0.5804 | 1.051 | 305.8 | 52.69 | 46.9 | 76.9 | 0.685 |
| 650 | 0.5356 | 1.063 | 322.5 | 60.21 | 49.7 | 87.3 | 0.690 |
| 700 | 0.4975 | 1.075 | 338.8 | 68.10 | 52.4 | 98.0 | 0.695 |
| 750 | 0.4643 | 1.087 | 354.6 | 76.37 | 54.9 | 109 | 0.702 |
| 800 | 0.4354 | 1.099 | 369.8 | 84.93 | 57.3 | 120 | 0.709 |
| 850 | 0.4097 | 1.110 | 384.3 | 93.80 | 59.6 | 131 | 0.716 |
| 900 | 0.3868 | 1.121 | 398.1 | 102.9 | 62.0 | 143 | 0.720 |
| 950 | 0.3666 | 1.131 | 411.3 | 112.2 | 64.3 | 155 | 0.723 |
| 1000 | 0.3482 | 1.141 | 424.4 | 121.9 | 66.7 | 168 | 0.726 |
| 1100 | 0.3166 | 1.159 | 449.0 | 141.8 | 71.5 | 195 | 0.728 |
| 1200 | 0.2902 | 1.175 | 473.0 | 162.9 | 76.3 | 224 | 0.728 |
| 1300 | 0.2679 | 1.189 | 496.0 | 185.1 | 82 | 238 | 0.719 |
| 1400 | 0.2488 | 1.207 | 530 | 213 | 91 | 303 | 0.703 |
| 1500 | 0.2322 | 1.230 | 557 | 240 | 100 | 350 | 0.685 |
| 1600 | 0.2177 | 1.248 | 584 | 268 | 106 | 390 | 0.688 |
| 1700 | 0.2049 | 1.267 | 611 | 298 | 113 | 435 | 0.685 |
| 1800 | 0.1935 | 1.286 | 637 | 329 | 120 | 482 | 0.683 |
| 1900 | 0.1833 | 1.307 | 663 | 362 | 128 | 534 | 0.677 |
| 2000 | 0.1741 | 1.337 | 689 | 396 | 137 | 589 | 0.672 |
| 2100 | 0.1658 | 1.372 | 715 | 431 | 147 | 646 | 0.667 |
| 2200 | 0.1582 | 1.417 | 740 | 468 | 160 | 714 | 0.655 |
| 2300 | 0.1513 | 1.478 | 766 | 506 | 175 | 783 | 0.647 |
| 2400 | 0.1448 | 1.558 | 792 | 547 | 196 | 869 | 0.630 |
| 2500 | 0.1389 | 1.665 | 818 | 589 | 222 | 960 | 0.613 |
| 3000 | 0.1135 | 2.726 | 955 | 841 | 486 | 1,570 | 0.536 |

[a]SOURCE: Incropera, F. P., and DeWitt, D. P., *Fundamentals of Heat and Mass Transfer*, 3rd Ed. Reprinted by permission, © 1990, John Wiley & Sons, Inc.

**Table C.2** Selected properties of nitrogen and oxygen at 1 atm[a]

| T (K) | $\rho$ (kg/m³) | $c_p$ (kJ/kg-K) | $\mu \cdot 10^7$ (N-s/m²) | $\nu \cdot 10^6$ (m²/s) | $k \cdot 10^3$ (W/m-K) | $\alpha \cdot 10^6$ (m²/s) | Pr |
|---|---|---|---|---|---|---|---|
| *Nitrogen (N₂)* | | | | | | | |
| 100 | 3.4388 | 1.070 | 68.8 | 2.00 | 9.58 | 2.60 | 0.768 |
| 150 | 2.2594 | 1.050 | 100.6 | 4.45 | 13.9 | 5.86 | 0.759 |
| 200 | 1.6883 | 1.043 | 129.2 | 7.65 | 18.3 | 10.4 | 0.736 |
| 250 | 1.3488 | 1.042 | 154.9 | 11.48 | 22.2 | 15.8 | 0.727 |
| 300 | 1.1233 | 1.041 | 178.2 | 15.86 | 25.9 | 22.1 | 0.716 |
| 350 | 0.9625 | 1.042 | 200.0 | 20.78 | 29.3 | 29.2 | 0.711 |
| 400 | 0.8425 | 1.045 | 220.4 | 26.16 | 32.7 | 37.1 | 0.704 |
| 450 | 0.7485 | 1.050 | 239.6 | 32.01 | 35.8 | 45.6 | 0.703 |
| 500 | 0.6739 | 1.056 | 257.7 | 38.24 | 38.9 | 54.7 | 0.700 |
| 550 | 0.6124 | 1.065 | 274.7 | 44.86 | 41.7 | 63.9 | 0.702 |
| 600 | 0.5615 | 1.075 | 290.8 | 51.79 | 44.6 | 73.9 | 0.701 |
| 700 | 0.4812 | 1.098 | 321.0 | 66.71 | 49.9 | 94.4 | 0.706 |
| 800 | 0.4211 | 1.22 | 349.1 | 82.90 | 54.8 | 116 | 0.715 |
| 900 | 0.3743 | 1.146 | 375.3 | 100.3 | 59.7 | 139 | 0.721 |
| 1000 | 0.3368 | 1.167 | 399.9 | 118.7 | 64.7 | 165 | 0.721 |
| 1100 | 0.3062 | 1.187 | 423.2 | 138.2 | 70.0 | 193 | 0.718 |
| 1200 | 0.2807 | 1.204 | 445.3 | 158.6 | 75.8 | 224 | 0.707 |
| 1300 | 0.2591 | 1.219 | 466.2 | 179.9 | 81.0 | 256 | 0.701 |
| *Oxygen (O₂)* | | | | | | | |
| 100 | 3.945 | 0.962 | 76.4 | 1.94 | 9.25 | 2.44 | 0.796 |
| 150 | 2.585 | 0.921 | 114.8 | 4.44 | 13.8 | 5.80 | 0.766 |
| 200 | 1.930 | 0.915 | 147.5 | 7.64 | 18.3 | 10.4 | 0.737 |
| 250 | 1.542 | 0.915 | 178.6 | 11.58 | 22.6 | 16.0 | 0.723 |
| 300 | 1.284 | 0.920 | 207.2 | 16.14 | 26.8 | 22.7 | 0.711 |
| 350 | 1.100 | 0.929 | 233.5 | 21.23 | 29.6 | 29.0 | 0.733 |
| 400 | 0.9620 | 0.942 | 258.2 | 26.84 | 33.0 | 36.4 | 0.737 |
| 450 | 0.8554 | 0.956 | 281.4 | 32.90 | 36.3 | 44.4 | 0.741 |
| 500 | 0.7698 | 0.972 | 303.3 | 39.40 | 41.2 | 55.1 | 0.716 |
| 550 | 0.6998 | 0.988 | 324.0 | 46.30 | 44.1 | 63.8 | 0.726 |
| 600 | 0.6414 | 1.003 | 343.7 | 53.59 | 47.3 | 73.5 | 0.729 |
| 700 | 0.5498 | 1.031 | 380.8 | 69.26 | 52.8 | 93.1 | 0.744 |
| 800 | 0.4810 | 1.054 | 415.2 | 86.32 | 58.9 | 116 | 0.743 |
| 900 | 0.4275 | 1.074 | 447.2 | 104.6 | 64.9 | 141 | 0.740 |
| 1000 | 0.3848 | 1.090 | 477.0 | 124.0 | 71.0 | 169 | 0.733 |
| 1100 | 0.3498 | 1.103 | 505.5 | 144.5 | 75.8 | 196 | 0.736 |
| 1200 | 0.3206 | 1.115 | 532.5 | 166.1 | 81.9 | 229 | 0.725 |
| 1300 | 0.2960 | 1.125 | 588.4 | 188.6 | 87.1 | 262 | 0.721 |

[a]SOURCE: Incropera, F. P., and DeWitt, D. P., *Fundamentals of Heat and Mass Transfer,* 3rd Ed. Reprinted by permission, © 1990 John Wiley & Sons, Inc.

# D

# Binary Diffusion Coefficients and Methodology for their Estimation

**Table D.1**    Binary diffusion coefficients at 1 atm[a,b]

| Substance A | Substance B | $T$ (K) | $\mathcal{D}_{AB} \cdot 10^5$ (m$^2$/s) |
|---|---|---|---|
| Benzene | Air | 273 | 0.77 |
| Carbon dioxide | Air | 273 | 1.38 |
| Carbon dioxide | Nitrogen | 293 | 1.63 |
| Cyclohexane | Air | 318 | 0.86 |
| $n$-Decane | Nitrogen | 363 | 0.84 |
| $n$-Dodecane | Nitrogen | 399 | 0.81 |
| Ethanol | Air | 273 | 1.02 |
| $n$-Hexane | Nitrogen | 288 | 0.757 |
| Hydrogen | Air | 273 | 6.11 |
| Methanol | Air | 273 | 1.32 |
| $n$-Octane | Air | 273 | 0.505 |
| $n$-Octane | Nitrogen | 303 | 0.71 |
| Toluene | Air | 303 | 0.88 |
| 2,2,4-Trimethyl pentane (Isooctane) | Nitrogen | 303 | 0.705 |
| 2,2,3-Trimethyl heptane | Nitrogen | 363 | 0.684 |
| Water | Air | 273 | 2.2 |

[a]SOURCE: Perry, R. H., Green, D. W., and Maloney, J. O., *Perry's Chemical Engineers' Handbook*, 6th Ed., McGraw-Hill, New York, 1984.

[b]Assuming ideal-gas behavior, the pressure and temperature dependence of the binary diffusion coefficient can be estimated using $\mathcal{D}_{AB} \propto T^{3/2}/P$.

# PREDICTING BINARY DIFFUSION COEFFICIENTS FROM THEORY

The following approach for predicting binary diffusion coefficients is a brief summary of that presented by Reid *et al.* [1]. The methodology is based on the Chapman–Enskog theoretical description of binary mixtures of gases at low to moderate pressures. In this theory, the binary diffusion coefficient for the species pair A and B is

$$\mathcal{D}_{AB} = \frac{3}{16} \frac{(4\pi k_B T / MW_{AB})^{1/2}}{(P/R_u T)\pi\sigma_{AB}^2 \Omega_D} f_D, \tag{D.1}$$

where $k_B$ is the Boltzmann constant, $T$ (K) is the absolute temperature, $P$ (Pa) is the pressure, $R_u$ is the universal gas constant, and $f_D$ is a theoretical correction factor whose value is sufficiently close to unity to be assumed to be the same. The remaining terms are defined below:

$$MW_{AB} = 2[(1/MW_A) + (1/MW_B)]^{-1}, \tag{D.2}$$

where $MW_A$ and $MW_B$ are the molecular weights of species A and B, respectively;

$$\sigma_{AB} = (\sigma_A + \sigma_B)/2, \tag{D.3}$$

where $\sigma_A$ and $\sigma_B$ are the hard-sphere collision diameters of species A and B, respectively, values of which are shown in Table D.2 for several species of interest in combustion.

The collision integral, $\Omega_D$, is a dimensionless quantity calculated using the following expression:

$$\Omega_D = \frac{A}{(T^*)^B} + \frac{C}{\exp(DT^*)} + \frac{E}{\exp(FT^*)} + \frac{G}{\exp(HT^*)}, \tag{D.4}$$

where

$A = 1.06036,$    $B = 0.15610,$
$C = 0.19300,$    $D = 0.47635,$
$E = 1.03587,$    $F = 1.52996,$
$G = 1.76474,$    $H = 3.89411,$

and where the dimensionless temperature $T^*$ is defined by

$$T^* = k_B T / \varepsilon_{AB} = k_B T / (\varepsilon_A \varepsilon_B)^{1/2}. \tag{D.5}$$

Values of the characteristic Lennard–Jones energy, $\varepsilon_i$, are also tabulated in Table D.2 [1].

**Table D.2**     Lennard–Jones parameters for selected species [2]

| Species | $\sigma\,(\text{Å})$ | $\varepsilon/k_B\,(\text{K})$ | Species | $\sigma\,(\text{Å})$ | $\varepsilon/k_B\,(\text{K})$ |
|---|---|---|---|---|---|
| Air | 3.711 | 78.6 | $n\text{-}C_5H_{12}$ | 5.784 | 341.1 |
| Al | 2.655 | 2750 | $C_6H_6$ | 5.349 | 412.3 |
| Ar | 3.542 | 93.3 | $C_6H_{12}$ | 6.182 | 297.1 |
| B | 2.265 | 3331 | $n\text{-}C_6H_{14}$ | 5.949 | 399.3 |
| BO | 2.944 | 596 | H | 2.708 | 37.0 |
| $B_2O_3$ | 4.158 | 2092 | $H_2$ | 2.827 | 59.7 |
| C | 3.385 | 30.6 | $H_2O$ | 2.641 | 809.1 |
| CH | 3.370 | 68.7 | $H_2O_2$ | 4.196 | 389.3 |
| $CH_3OH$ | 3.626 | 481.8 | He | 2.551 | 10.22 |
| $CH_4$ | 3.758 | 148.6 | N | 3.298 | 71.4 |
| CN | 3.856 | 75 | $NH_3$ | 2.900 | 558.3 |
| CO | 3.690 | 91.7 | NO | 3.492 | 116.7 |
| $CO_2$ | 3.941 | 195.2 | $N_2$ | 3.798 | 71.4 |
| $C_2H_2$ | 4.033 | 231.8 | $N_2O$ | 3.828 | 232.4 |
| $C_2H_4$ | 4.163 | 224.7 | O | 3.050 | 106.7 |
| $C_2H_6$ | 4.443 | 215.7 | OH | 3.147 | 79.8 |
| $C_3H_8$ | 5.118 | 237.1 | $O_2$ | 3.467 | 106.7 |
| $n\text{-}C_3H_7OH$ | 4.549 | 576.7 | S | 3.839 | 847 |
| $n\text{-}C_4H_{10}$ | 4.687 | 531.4 | SO | 3.993 | 301 |
| $iso\text{-}C_4H_{10}$ | 5.278 | 330.1 | $SO_2$ | 4.112 | 335.4 |

Substituting numerical values for the constants in Eqn. D.1 results in

$$\mathcal{D}_{AB} = \frac{0.0266 T^{3/2}}{P M W_{AB}^{1/2} \sigma_{AB}^2 \Omega_D} \tag{D.6}$$

with the following associated units: $\mathcal{D}_{AB}[=]m^2/s$, $T[=]K$, $P[=]Pa$, and $\sigma_{AB}[=]\text{Å}$.

# REFERENCES

1. Reid, R. C., Prausnitz, J. M., and Poling, B. E., *The Properties of Gases and Liquids*, 4th Ed., McGraw-Hill, New York, 1987.

2. Svehla, R. A., "Estimated Viscosities and Thermal Conductivities of Gases at High Temperatures," NASA Technical Report R-132, 1962.

# Generalized Newton's Method for the Solution of Nonlinear Equations

The Newton–Raphson method, Eqn. E.1, can be extended and applied to a system of nonlinear equations, Eqn. E.2:

$$x_{k+1} = x_k - \frac{f(x_k)}{f'(x_k)} = x_k - \frac{f(x_k)}{\dfrac{df}{dx}(x_k)}; \qquad k \equiv \text{iteration}. \qquad (E.1)$$

System:

$$\begin{aligned} f_1(x_1, x_2, x_3, \ldots, x_n) &= 0, \\ f_2(x_1, x_2, x_3, \ldots, x_n) &= 0, \\ &\vdots \\ f_n(x_1, x_2, x_3, \ldots, x_n) &= 0. \end{aligned} \qquad (E.2)$$

Each of these may be expanded in Taylor's series form (truncating second-order and higher terms) as

$$f_i(\tilde{x} + \tilde{\delta}) = f_i(\tilde{x}) + \frac{\partial f_i}{\partial x_1}\delta_1 + \frac{\partial f_i}{\partial x_2}\delta_2 + \frac{\partial f_i}{\partial x_3}\delta_3 + \cdots + \frac{\partial f_i}{\partial x_n}\delta_n, \qquad (E.3)$$

for $i = 1, 2, 3, \ldots, n$, where

$$\tilde{x} \equiv \{x\}.$$

At the solution, $f(\tilde{x} + \tilde{\delta}) \to 0$; the above can be arranged as a set of *linear* equations in the matrix form,

$$\left[\frac{\partial f}{\partial x}\right]\{\delta\} = -\{f\};$$

that is,

$$
\begin{bmatrix}
\dfrac{\partial f_1}{\partial x_1} & \dfrac{\partial f_1}{\partial x_2} & \cdots & \dfrac{\partial f_1}{\partial x_n} \\[2mm]
\vdots & \vdots & & \vdots \\[2mm]
\dfrac{\partial f_n}{\partial x_1} & \dfrac{\partial f_n}{\partial x_2} & \cdots & \dfrac{\partial f_n}{\partial x_n}
\end{bmatrix}
\begin{Bmatrix} \delta_1 \\ \vdots \\ \delta_n \end{Bmatrix}
=
\begin{Bmatrix} -f_1 \\ \vdots \\ -f_n \end{Bmatrix}
\tag{E.4}
$$

where the coefficient matrix on the left-hand side is called the **Jacobian**.

Equation E.4 may be solved (for $\delta$) using Gauss elimination; once $\delta$ is known, the next (better) approximation is found from the recursion relation,

$$
\{x\}_{k+1} = \{x\}_k + \{\delta\}_k .
$$

The process of forming the Jacobian, solving Eqn. E.4, and calculating new values for $\{x\}$ is repeated until a stop criterion is met. The following is suggested by Suh and Radcliffe [1]:

| Stop criterion | Condition |
|---|---|
| $\left|\delta_j / x_j\right| \leq 10^{-7}$ | $\left|x_j\right| \geq 10^{-7}$ |
| or | |
| $\left|\delta_j\right| \leq 10^{-7}$ | $\left|x_j\right| \leq 10^{-7}$ |

for $j = 1, 2, 3, \ldots, n$.

Estimates to the partial derivatives may be formed numerically from

$$
\frac{\partial f_i}{\partial x_j} = \frac{f_i(x_1, x_2, \ldots, x_j + \varepsilon, \ldots, x_n) - f_i(x_1, x_2, \ x_3, \ldots, x_j, \ldots, x_n)}{\varepsilon}
$$

where

$$
\varepsilon = 10^{-5}\left|x_j\right| \ \text{ for } \ \left|x_j\right| > 1.0
$$

$$
\varepsilon = 10^{-5} \ \text{ for } \ \left|x_j\right| < 1.0 .
$$

**Instability** may (in many cases) be avoided as follows:

1. Compare the norm of the new function vector to the norm of the previous function vector, where

$$
\text{norm} = \sum_{i=1}^{n} \left|f_i(\tilde{x})\right| .
$$

2. If the norm of the new function vector is greater than that of the old, assume that the full step $\{\delta\}$ would not be productive and take a partial step $\{\delta\}/5$; otherwise, take a full step as usual.

The process of comparing norms and dividing $\{\delta\}$ by an arbitrary constant is termed "damping" and has proved successful in obtaining convergence even with very poor initial guesses.

A weakness of the Newton–Raphson method is that the Jacobian must be calculated at every step.

## REFERENCE

1. Suh, C. H., and Radcliffe, C. W., *Kinematics and Mechanisms Design*, John Wiley & Sons, New York, pp. 143–144, 1978.

# F

# Computer Codes for Equilibrium Products of Hydrocarbon–Air Combustion

Software for use with this book is available as a download from the publisher's website at www.mhhe.com/turns3e. This software contains the following files:

| File | Purpose |
| --- | --- |
| README | File containing instructions and other information concerning the use of the files listed below |
| Access to TPEQUIL, HPFLAME and UVFLAME Software | user interface |
| TPEQUIL | Executable module that calculates combustion products equilibrium composition and properties for specified fuel, equivalence ratio, temperature, and pressure |
| TPEQUIL.F | Fortran source listing for TPEQUIL |
| INPUT.TP | Input file read by TPEQUIL containing user specifications for fuel, equivalence ratio, temperature, and pressure |
| HPFLAME | Executable module that calculates the adiabatic flame temperature, equilibrium composition, and properties of the products of combustion for **adiabatic constant-pressure combustion** with specified fuel composition, reactant enthalpy, equivalence ratio, and pressure |

| HPFLAME.F | Fortran source listing for HPFLAME |
|---|---|
| INPUT.HP | Input file read by HPFLAME containing user specifications for fuel, reactant enthalpy (per kilomole of fuel), equivalence ratio, and pressure |
| UVFLAME | Executable module that calculates the adiabatic flame temperature, equilibrium composition, and properties of the products of combustion for **adiabatic constant-volume combustion** with specified fuel composition, reactant enthalpy, equivalence ratio, and initial temperature and pressure |
| UVFLAME.F | Fortran source listing for UVFLAME |
| INPUT.UV | Input file read by UVFLAME containing user specifications for fuel, reactant enthalpy (per kilomole of fuel), equivalence ratio, moles of reactants per mole of fuel, molecular weight of reactants, and initial temperature and pressure |
| GPROP.DAT | Thermodynamic property datafile |

The various codes above all incorporate the Olikara and Borman routines [1] for calculating equilibrium products of combustion for a fuel composed of C, H, O, and N atoms, given by $C_N H_M O_L N_K$ and air.[1] Thus, oxygenated fuels, such as alcohols, and fuels with bound nitrogen can be handled by the code. For simple hydrocarbons, the numbers of fuel oxygen and nitrogen atoms, $L$ and $K$, respectively, are set equal to zero in the user-modified input files. The oxidizer is assumed to be air with the simplified composition of 79 percent $N_2$ and 21 percent $O_2$, and is specified in the subroutine TABLES. A more complex oxidizer composition, including Ar, for example, can be obtained easily by modifying this subroutine and recompiling the source code. Eleven species are considered in the products of combustion: H, O, N, $H_2$, OH, CO, NO, $O_2$, $H_2O$, $CO_2$, and $N_2$. The code also considers Ar if it is included in the oxidizer. The Olikara and Borman routines [1] have been modified to deal in SI units. Other modifications to the original code include the way the JANAF thermodynamic data and equilibrium constants are input, as indicated in the source listings.

# REFERENCE

1. Olikara, C., and Borman, G. L., "A Computer Program for Calculating Properties of Equilibrium Combustion Products with Some Applications to I. C. Engines," SAE Paper 750468, 1975.

[1]The imbedded codes from Ref. [1], with modifications, are used with permission of the Society of Automotive Engineers, Inc., © 1975.

# INDEX

# NAME AND SUBJECT

## Atomic Weights of Selected Elements (1981)

| | | |
|---|---|---|
| Aluminum | Al | 26.9815 |
| Argon | Ar | 39.948 |
| Beryllium | Be | 9.01218 |
| Boron | B | 10.81 |
| Bromine | Br | 79.904 |
| Calcium | Ca | 40.08 |
| Carbon | C | 12.011 |
| Chlorine | Cl | 35.453 |
| Copper | Cu | 63.546 |
| Fluorine | F | 18.9984 |
| Helium | He | 4.00260 |
| Hydrogen | H | 1.00794 |
| Krypton | Kr | 83.80 |
| Magnesium | Mg | 24.305 |
| Nitrogen | N | 14.0067 |
| Oxygen | O | 15.9994 |
| Platinum | Pt | 195.08 |
| Silicon | Si | 28.0855 |
| Sodium | Na | 22.9898 |
| Sulfur | S | 32.06 |
| Xenon | Xe | 131.29 |

## Physical Constants[1]

| | | |
|---|---|---|
| Avogadro number | $N_{AV}$ | $6.022\ 136\ 7(36) \cdot 10^{23}$ molecules/mol |
| | | $6.022\ 136\ 7(36) \cdot 10^{26}$ molecules/kmol |
| Boltzmann constant | $k_B$ | $1.380\ 658(12) \cdot 10^{-23}$ J/K-molecule |
| Planck constant | $h$ | $6.626\ 075\ 5(40) \cdot 10^{-34}$ J-s/molecule |
| Speed of light in vacuum | $c$ | $299\ 792\ 458$ m/s |
| Standard acceleration of gravity | $g$ | $9.806\ 65$ m/s$^2$ |
| Standard atmosphere | atm | $101\ 325$ Pa |
| Stefan-Boltzmann constant | $\sigma$ | $5.670\ 51(19 \cdot 10^{-8}$ W/(m$^2$-K$^4$) |
| Universal gas constant | $R_u$ | $8\ 314.510(70)$ J/(kmol-K) |

[1]E. R. Cohen, and B. N. Taylor, "The Fundamental Physical Constants," *Physics Today*, August 1993, pp. 9–13.

**Conversion Factors**

| | | |
|---|---|---|
| Energy | 1 J | $= 9.478\ 17 \cdot 10^{-4}$ Btu |
| | | $= 2.388\ 5 \cdot 10^{-4}$ kcal |
| Energy rate | 1 W | $= 3.412\ 14$ Btu/hr |
| Force | 1 N | $= 0.224\ 809\ lb_f$ |
| Heat flux | 1 W/m$^2$ | $= 0.3171$ Btu/(hr-ft$^2$) |
| Kinematic viscosity and diffusivities | 1 m$^2$/s | $= 3.875 \cdot 10^4$ ft$^2$/hr |
| Length | 1 m | $= 39.370$ in |
| | | $= 3.280\ 8$ ft |
| Mass | 1 kg | $= 2.204\ 6\ lb_m$ |
| Mass density | 1 kg/m$^3$ | $= 0.062\ 428\ lb_m$/ft$^3$ |
| Mass flow rate | 1 kg/s | $= 7\ 936.6\ lb_m$/hr |
| Pressure | Pa | $= 1$ N/m$^2$ |
| | | $= 0.020\ 885\ 4\ lb_f$/ft$^2$ |
| | | $= 1.450\ 4 \cdot 10^{-4}\ lb_f$/in$^2$ |
| | | $= 4.015 \cdot 10^{-3}$ in water |
| | 0.1 MPa | $= 1$ bar |
| | 1 atm | $= 101.325$ kPa |
| | | $= 760$ mm Hg |
| | | $= 29.92$ in Hg |
| | | $= 14.70\ lb_f$/in$^2$ |
| Specific heat | 1 J/kg-K | $= 2.388\ 6 \cdot 10^{-4}$ Btu/(lb$_m$-°F) |
| Temperature | K | $= (5/9)$°R |
| | | $= (5/9)\ (°F + 459.67)$ |
| | | $= °C + 273.15$ |
| Time | 3600 s | $= 1$ hr |